PARALLELOGRAM

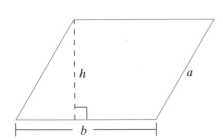

Perimeter: $P = 2a + 2b$
Area: $A = bh$

CIRCLE

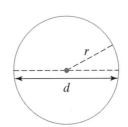

Circumference: $C = \pi d$
$C = 2\pi r$
Area: $A = \pi r^2$

RECTANGULAR SOLID

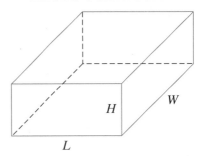

Volume: $V = LWH$
Surface Area: $A = 2HW + 2LW + 2LH$

CUBE

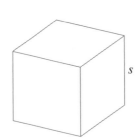

Volume: $V = s^3$
Surface Area: $A = 6s^2$

CONE

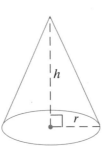

Volume: $V = \dfrac{1}{3}\pi r^2 h$
Lateral Surface Area: $A = \pi r \sqrt{r^2 + h^2}$

RIGHT CIRCULAR CYLINDER

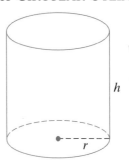

Volume: $V = \pi r^2 h$
Surface Area:
$A = 2\pi rh + 2\pi r^2$

OTHER FORMULAS

Distance: $d = rt$ (r = rate, t = time)

Temperature: $F = \dfrac{9}{5}C + 32 \qquad C = \dfrac{5}{9}(F - 32)$

Simple Interest: $I = Prt$
  (P = principal, r = annual interest rate, t = time in years)

Compound Interest: $A = P\left(1 + \dfrac{r}{n}\right)^{nt}$

  (P = principal, r = annual interest rate, t = time in years, n = number of compoundings per year)

ANNOTATED INSTRUCTOR'S EDITION

Beginning Algebra

FOURTH EDITION

K. Elayn Martin-Gay
University of New Orleans

This Annotated Instructor's Edition is exactly like your students' text, but it contains Teaching Tips, Classroom Examples, and all exercise answers including Vocabulary Checks. Where possible the answer is displayed on the same page as the exercise. Exercise answers that require graphical solutions, and answers to Spotlight on Decision Making and Chapter Projects, can be found in the Graphing Answer Section.

The Student Edition of the text contains answers to selected exercises. Located in the back of the Student Edition are answers to odd-numbered exercises, review, and cumulative review exercises, and all integrated review and chapter test exercises. It does not contain answers to even-numbered exercises, Spotlight on Decision Making, Chapter Projects, Vocabulary Checks, group activities, or instructor's notes.

PEARSON

Prentice
Hall

Upper Saddle River, New Jersey 07458

Editor in Chief: Christine Hoag
Senior Acquisitions Editor: Paul Murphy
Project Manager: Mary Beckwith
Assistant Editor: Christina Simoneau
Project Management: Elm Street Publishing Services, Inc.
Senior Managing Editor: Linda Mihatov Behrens
Executive Managing Editor: Kathleen Schiaparelli
Director/Vice President of Production and Manufacturing: David W. Riccardi
Assistant Manufacturing Manager/Buyer: Michael Bell
Manufacturing Manager: Trudy Pisciotti
Art Editor: Tom Benfatti
Executive Marketing Manager: Eilish Collins Main
Marketing Assistant: Annett Uebel
Marketing Project Manager: Barbara Herbst
Development Editor: Elka Block
Editor in Chief, Development: Carol Trueheart
Art Director/Cover Designer: Maureen Eide
Cover Art: Geoffrey Cassar
Assistant to Art Director: Dina Curro
Interior Designer: Donna Wickes
Creative Director: Carole Anson
Director of Creative Services: Paul Belfanti
Media Project Manager, Developmental Math: Audra J. Walsh

Assistant Managing Editor, Math Media Production: John Matthews
Composition: Progressive Information Technologies *and* Pearson formatting: Manager, Electronic Composition: Jim Sullivan; Assistant Formatting Manager: Allyson Graesser; Electronic Production Specialists: Karen Noferi, Karen Stephens
Director, Image Resource Center: Melinda Reo
Manager, Rights and Permissions: Zina Arabia
Interior Image Specialist: Beth Brenzel
Image Permission Coordinator: Lashonda Morris
Photo Researcher: Melinda Alexander
Art Studios: Artworks, Scientific Illustrators, Laserwords
  Artworks
  Managing Editor, AV Production & Management: Patricia Burns
  Production Manager: Ronda Whitson
  Production Technologies Manager: Matthew Haas
  Project Coordinator: Dan Missildine
  Art Supervisor: Kathryn Anderson
  Illustrators: Audrey Simonetti, Daniel Knopsnyder, Mark Landis, Nathan Storck, Ryan Currier, Stacy Smith, Scott Wieber
  Quality Supervisor: Pamela Taylor
  Quality Assurance: Cathy Shelly, Tim Nguyen, Ken Mooney

CONTENTS

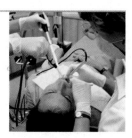

GRAPHING   163

SOLVING SYSTEMS OF LINEAR EQUATIONS
AND INEQUALITIES   243

EXPONENTS AND POLYNOMIALS   296

ABOUT THE BOOK

Beginning Algebra, Fourth Edition was written to provide a **solid foundation in algebra** for students who might have had no previous experience in algebra. Specific care has been taken to ensure that students have the most **up-to-date and relevant** text preparation for their next mathematics course, as well as to help students to succeed in nonmathematical courses that require a grasp of algebraic fundamentals. I have tried to achieve this by writing a user-friendly text that is keyed to objectives and contains many worked-out examples. The basic concepts of graphing are introduced early, and problem solving techniques, real-life and real-data applications, data interpretation, appropriate use of technology, mental mathematics, number sense, critical thinking, decision-making, and geometric concepts are emphasized and integrated throughout the book.

The new edition includes an increased emphasis on study and test preparation skills. In addition, the fourth edition now includes a new resource, the Chapter Test Prep Video CD. With this CD/Video, students have instant access to video solutions for each of the chapter test questions contained in the text. It is designed to help them study efficiently.

The many factors that contributed to the success of the previous editions have been retained. In preparing this edition, I considered the comments and suggestions of colleagues throughout the country, students, and many users of the prior editions. The AMATYC Crossroads in Mathematics: Standards for Introductory College Mathematics before Calculus and the MAA and NCTM standards (plus Addenda), together with advances in technology, also influenced the writing of this text.

Beginning Algebra, Fourth Edition is **part of a series of texts** that can include *Basic College Mathematics, Second Edition, Prealgebra, Fourth Edition, Intermediate Algebra, Fourth Edition,* or *Intermediate Algebra: A Graphing Approach, Third Edition,* and *Beginning and Intermediate Algebra, Third Edition,* a combined algebra text. Throughout the series, pedagogical features are designed to develop student proficiency in algebra and problem solving, and to prepare students for future courses.

NEW FEATURES AND CHANGES IN THE NEW EDITION

The following new features have been added to the fourth edition.

INCREASED EMPHASIS ON STUDY SKILLS

New! Study Skills Reminders integrated throughout the text to help students hone their study skills and serve as a point-of-use support resource, reinforcing the skills covered in Section 1.1, Tips for Success in Mathematics.

New! Integrated Reviews serve as mid-chapter reviews and help students assimilate new skills and concepts they have learned separately over several sections. The reviews provide students with another opportunity to practice with *mixed* exercises as they master the topics.

ENHANCED SECTION EXERCISE SETS

Mixed Practice Exercises. Exercise sets have been reorganized to include mixed practice exercises where appropriate. They give students the chance to assimilate the concepts and skills covered in separate objectives. Students have the opportunity to practice the kind of decision making they will encounter on tests.

New! Concept Extension Exercises. These have been added to the end of the section exercise sets. They extend the concepts and require students to combine several skills or concepts. They expose students to the way math ideas build upon each other and offer the opportunity for additional challenges.

MORE OPPORTUNITIES FOR STUDENTS TO CHECK THEIR UNDERSTANDING

New! Concept Checks are special exercises found in most sections following key examples. Working these will help students check their grasp of the concept being developed before moving to the next example.

INCREASED EMPHASIS ON IMPROVING TEST PREPARATION

New! Chapter Test Prep Video CD packaged with each text and presented by Elayn Martin-Gay provides students with a resource to take and correct sample tests as they prepare for exams. Step-by-Step solutions are presented for every Chapter Test exercise contained in the text. Easy video navigation allows students to instantly access the solutions to the exact exercises they need help with.

KEY CONTINUING FEATURES

The following key features have been retained from previous editions.

Readability and Connections   I have tried to make the writing style as clear as possible while still retaining the mathematical integrity of the content. When a new topic is presented, an effort has been made to **relate the new ideas to those that students may already know.** Constant reinforcement and connections within problem solving strategies, data interpretation, geometry, patterns, graphs, and situations from everyday life can help students gradually master both new and old information.

Problem Solving Process   This is formally introduced in Chapter 2 with a **four-step process that is integrated throughout the text.** The four steps are Understand, Translate, Solve, and Interpret. The repeated use of these steps throughout the text in a variety of examples shows their wide applicability. Reinforcing the steps can increase students' confidence in tackling problems.

Applications and Connections   Every effort was made to include as many accessible, interesting, and relevant real-life applications as possible throughout the text in both worked-out examples and exercise sets. The applications **strengthen students' understanding of mathematics in the real world** and help to motivate students. They show connections to a wide range of fields including agriculture, allied health, art, astronomy, automotive ownership, aviation, biology, business, chemistry, communication, computer technology, construction, consumer affairs, demographics, earth science, education, entertainment, environmental issues, finance and economics, food service, geography, government, history, hobbies, labor and career issues, life science, medicine, music, nutrition, physics, political science, population, recreation, sports, technology, transportation, travel, weather, and important related mathematical areas such as geometry and statistics. (See the Index of Applications on page xvi.) Many of the applications are based on **recent and interesting real-life data.** Sources for data include newspapers, magazines, government publications, publicly held companies, special interest groups, research organizations, and reference books. Opportunities for obtaining your own real data are also included.

Helpful Hints Helpful Hints contain practical advice on applying mathematical concepts. These are found throughout the text and **strategically placed** where students are most likely to need immediate reinforcement. They are highlighted in a box for quick reference and, as appropriate, an indicator line is used to precisely identify the particular part of a problem or concept being discussed. For instance, see pages 25 and 26.

Visual Reinforcement of Concepts The text contains numerous graphics, models, and illustrations to visually clarify and reinforce concepts. These include **new and updated** bar graphs, circle graphs in two and three dimensions, line graphs, calculator screens, application illustrations, photographs, and geometric figures.

Real World Chapter Openers The chapter openers focus on how math is used in a specific career, provide links to the World Wide Web, and reference a "Spotlight on Decision Making" feature within the chapter for further exploration of the **career and the relevance of algebra.**

Student Resource Icons At the beginning of each exercise set, videotape, tutorial software CD-Rom, Student Solutions Manual, Study Guide, and tutor center icons are displayed. These icons help reinforce that these learning aids are available should students wish to use them to review concepts and skills at their own pace. These items have **direct correlation to the text** and emphasize the text's methods of solution.

Chapter Highlights Found at the end of each chapter, the Chapter Highlights contain key definitions, concepts, *and* examples to **help students understand and retain** what they have learned.

Chapter Project This feature occurs at the end of each chapter, often serving as a chapter wrap-up. For **individual or group completion,** the multi-part Chapter Project, usually hands-on or data based, allows students to problem solve, make interpretations, and to think and write about algebra.

EXERCISE SETS

Each text section ends with an exercise set, usually divided into two parts. Both parts contain graded exercises. The **first part is carefully keyed** to at least one worked example in the text. Once a student has gained confidence in a skill, **the second part contains exercises not keyed to examples.** Exercises and examples marked with a video icon (▣) have been worked out step-by-step by the author in the lecture videos that accompany this text.

Throughout the text exercises there is an emphasis on data and graphical interpretation via tables, charts, and graphs. The ability to interpret data and read and create a variety of types of graphs is developed gradually so students become comfortable with it. Similarly, throughout the text there is integration of geometric concepts, such as perimeter and area. Exercises and examples marked with a geometry icon (△) have been identified for convenience.

Each exercise set contains one or more of the following features.

Mental Math These problems are found at the beginning of many exercise sets. They are mental warm-ups that **reinforce concepts** found in the accompanying section and increase students' confidence before they tackle an exercise set. By relying on their own mental skills, students increase not only their confidence in themselves, but also their number sense and estimation ability. This edition includes a greater number of Mental Math exercises.

Writing Exercises These exercises are found in almost every exercise set and are marked with the icon (＼). They require students to **assimilate information** and provide a written response to explain concepts or justify their thinking. Guidelines recommended by the American Mathematical Association of Two Year Colleges (AMATYC) and other professional groups recommend incorporating writing in mathematics courses to reinforce concepts.

Mixed Practice Exercises. Exercise sets have been reorganized to include mixed practice exercises where appropriate. They give students the chance to assimilate the concepts and skills covered in separate objectives. Students have the opportunity to practice the kind of decision making they will encounter on tests.

New! Concept Extension Exercises. These have been added to the end of the section exercise sets. They extend the concepts and require students to combine several skills or concepts. They expose students to the way math ideas build upon each other and offer the opportunity for additional challenge.

Data and Graphical Interpretation Throughout the text there is an emphasis on data interpretation in exercises via tables, bar charts, line graphs, or circle graphs. The ability to interpret data and read and create a variety of graphs is **developed gradually** so students become comfortable with it.

Calculator Explorations and Exercises These optional explorations offer guided instruction, through examples and exercises, on the proper use of **scientific and graphing calculators or computer graphing utilities as tools in the mathematical problem-solving process.** Placed appropriately throughout the text, these explorations reinforce concepts or motivate discovery learning.

Additional exercises building on the skills developed in the Explorations may be found in exercise sets throughout the text, and are marked with the icon 🖩 for scientific calculator use and with the icon 🖩 for graphing calculator use.

Review and Preview These exercises occur in each exercise set (except for those in Chapter 1). These problems are **keyed to earlier sections** and review concepts learned earlier in the text that are needed in the next section or in the next chapter. These exercises show the **links between earlier topics and later material.**

Vocabulary Checks Vocabulary checks provide an opportunity for students to become more familiar with the use of mathematical terms as they strengthen verbal skills. They are found at the end of each chapter.

Chapter Review and Chapter Test The end of each chapter contains a review of topics introduced in the chapter. The review problems are keyed to sections. The Chapter Test is not keyed to sections. The Chapter Test Prep Video CD provides solutions to every Chapter Test exercise in the text.

Cumulative Review Each chapter after the first contains a **cumulative review of all chapters beginning with the first** up through the chapter at hand. The odd problems contained in the cumulative reviews are actually an earlier worked example in the text. The even problems are not keyed.

KEY CONTENT CHANGES IN THE FOURTH EDITION

The following changes to content are included.

- **Chapter 3—Graphing**—now contains all of linear graphing, slope, equations of lines, and Section 3.7, Introduction to Functions. (Formerly sections covered in Chapters 3 and 7.)

- **Systems of Equations is now Chapter 4** (formerly Chapter 8). The reorganization allows instructors flexibility to introduce the topic as suits the needs of their course.

- **New Section 7.7: Variation and Problem Solving.** This is a new introductory section on types of direct and inverse variation containing all new applications.

- **Percents and Problem Solving** (formerly Section 2.7) has been streamlined and is now covered as an objective in Section 2.7, Further Problem Solving, versus a dedicated section. This change was made to improve the balance and progression of the problem solving sections.

Specific discussions have been enhanced in the fourth edition. Some of the changes include:

- **Increased coverage on factoring trinomials by grouping.** See Section 6.3.

- **Increased discussion about slope as a rate of change.** See Section 3.4.

- To help students prepare for further mathematics courses, **interval notation** is now introduced and used in this text.

- **Increased attention has been given to clarifying difficult areas for students** such as the difference between solving an equation with rational expressions and performing operations on rational expressions. See the Integrated Review in Chapter 7.

- More real-life data exercises are included: See Sections 2.2, 2.5, 4.4, 5.5, 7.7, 8.2, and 9.1.

- More allied health examples and exercises. For examples, see Section 2.3, exercises 73 and 74; Section 3.2, Example 7; Section 7.4, Exercise 85.

INSTRUCTOR AND STUDENT RESOURCES

The fourth edition is supported by comprehensive resources for instructors and students

INSTRUCTOR RESOURCES—PRINT

Annotated Instructor's Edition (ISBN 0-13-149893-2)

- Answers to exercises on the text page or in Graphing Answer Section

- Graphing Answer Section contains answers to exercises requiring graphical solutions, chapter projects, and Spotlight on Decision Making exercises

- Teaching Tips throughout the text placed at key points in the margin, found in places where students historically need extra help together with ideas on how to help students through these concepts, as well as placed appropriately to provide ideas for expanding upon a certain concept, other ways to present a concept, or ideas for classroom activities

- New Classroom Examples have been added. Each Classroom Example parallels the text example for an added resource during lecture.

Instructor's Solutions Manual (ISBN 0-13-144496-4)

- Detailed step-by-step solutions to even-numbered section exercises
- Solutions to every Spotlight on Decision Making exercise
- Solutions to every Calculator Exploration exercise
- Solutions to every Chapter Test and Chapter Review exercise
- Solution methods reflect those emphasized in the textbook

Instructor's Resource Manual with Tests (ISBN 0-13-144497-2)

- Notes to the Instructor that include new suggested assignments for each exercise set
- Eight Chapter Tests per chapter (5 free response, 3 multiple choice)
- Two Cumulative Review Tests (one free response, one multiple choice)
- Eight Final Exams (4 free response, 4 multiple choice)
- Twenty additional exercises per section for added test exercises or worksheets, if needed
- Group Activities (on average of two per chapter; providing short group activities in a convenient ready-to-use handout format)
- Answers to all items

INSTRUCTOR RESOURCES—MEDIA

TestGen with QuizMaster enables instructors to build, edit, print, and administer tests using a computerized bank of questions developed to cover all the objectives of the text. Instructors can modify test bank questions or add new questions by using the built-in question editor, which allows users to create graphs, import graphics, and insert math notation, variable numbers, or text. Tests can be printed or administered online via the Internet or another network. TestGen comes packaged with QuizMaster, which allows students to take tests on a local area network. The software is available on a dual-platform Windows/Macintosh CD-ROM.

New MyMathLab® (instructor) is a series of text-specific, easily customizable online courses for Prentice Hall mathematics textbooks. MyMathLab is powered by CourseCompass™—Pearson Education's online teaching and learning environment—and by MathXL®—our online homework, tutorial, and assessment system. MyMathLab gives you the tools you need to deliver all or a portion of your course online, whether your students are in a lab setting or working from home.

MyMathLab provides a rich and flexible set of course materials, featuring free-response exercises that are algorithmically generated for unlimited practice and mastery. Students can also use online tools such as video lectures, animations, and a multimedia textbook to independently improve their understanding and performance. Instructors can use MyMathLab's homework and test managers to select and assign online exercises correlated directly to the textbook, and they can also import TestGen tests into MyMathLab for added flexibility. MyMathLab's online gradebook—designed specifically for mathematics—automatically tracks students' homework and test results and gives the instructor control over how to calculate final grades.

MyMathLab is available to qualified adopters. For more information, visit our website at *www.mymathlab.com* or contact your Prentice Hall sales representative for a product demonstration.

MathXL® is a powerful online homework, tutorial, and assessment system that accompanies your Prentice Hall mathematics textbook. With MathXL, instructors can create, edit, and assign online homework and tests using algorithmically generated exercises correlated at the objective level to your textbook. All student work is tracked in MathXL's online gradebook. Students can take chapter tests in MathXL and receive personalized study plans based on their test results. The study plan diagnoses weaknesses and links students directly to tutorial exercises for the objectives they need to study and retest. Students can also access supplemental animations and video clips directly from selected exercises. MathXL is available to qualified adopters. For more information, visit our website at *www.mathxl.com* or contact your Prentice Hall sales representative for a product demonstration.

MathXL® Tutorials on CD This interactive tutorial CD-ROM provides algorithmically generated practice exercises that are correlated at the objective level to the exercises in the textbook. Every practice exercise is accompanied by an example and a guided solution designed to involve students in the solution process. Selected exercises may also include a video clip to help students visualize concepts. The software tracks student activity and scores and can generate printed summaries of students' progress.

"Instructor to Instructor" Videos Presented by Elayn Martin-Gay, these videos offer topical and teaching technique suggestions to new instructors and adjuncts to enhance effective classroom communication and provide seasoned faculty with additional teaching ideas and approaches. They also provide suggestions for presenting the topics in class, alternative approaches, time saving strategies, classroom activities, and much more.

STUDENT RESOURCES—PRINT

Student Solutions Manual (ISBN 0-13-144492-1)

- Detailed step-by-step solutions to odd-numbered section exercises
- Solutions to every (odd and even) Mental Math exercise
- Solutions to odd-numbered Calculator Exploration exercises
- Solutions to every (odd and even) exercise found in the Chapter Reviews and Chapter Tests
- Solution methods reflect those emphasized in the textbook

Study Guide (ISBN 0-13-144491-3)

- Additional step-by-step worked out examples and exercises
- Practice tests and final examination
- Includes Study Skills and note-taking suggestions
- Solutions to all exercises, tests, and final examination found in the Study Guide
- Solution methods reflect those emphasized in the text

STUDENT RESOURCES—MEDIA

New! Chapter Test Prep Video CD Provides a step-by-step video solution to each problem in the textbook's Chapter Tests, presented by Elayn Martin-Gay.

MyMathLab® (student) is a complete online course designed to help students succeed in learning and understanding mathematics. MyMathLab contains an online version of your textbook with links to multimedia resources—such as video clips, practice exercises, and animations—that are correlated to the examples and exercises in the text. MyMathLab also provides students with online homework and tests and generates a personalized study plan based on their test results. The study plan links directly to unlimited tutorial exercises for the areas students need to study and retest, so they can practice until they have mastered the skills and concepts in the textbook. All of the online homework, tests, and tutorial work students do is tracked in their MyMathLab gradebook.

MathXL® is a powerful online homework, tutorial, and assessment system that accompanies your Prentice Hall mathematics textbook. With MathXL, instructors can create, edit, and assign online homework and tests using algorithmically generated exercises correlated at the objective level to your textbook. All student work is tracked in MathXL's online gradebook. Students can take chapter tests in MathXL and receive personalized study plans based on their test results. The study plan diagnoses weaknesses and links students directly to tutorial exercises for the objectives they need to study and retest. Students can also access supplemental animations and video clips directly from selected exercises. MathXL is available to qualified adopters. For more information, visit our website at *www.mathxl.com* or contact your Prentice Hall sales representative for a product demonstration.

MathXL® Tutorials on CD This interactive tutorial CD-ROM provides algorithmically generated practice exercises that are correlated at the objective level to the exercises in the textbook. Every practice exercise is accompanied by an example and a guided solution designed to involve students in the solution process. Selected exercises may also include a video clip to help students visualize concepts. The software tracks student activity and scores and can generate printed summaries of students' progress.

Lecture Series Videos Digitized on CD-Rom (0-13-144489-1) and on VHS Tape (0-13-144495-6)

- Keyed to each section of the text
- Step-by-step solutions to exercises from each section of the text
- In-text exercises marked with a video icon appear on the videos
- Digitized videos offer convenient anytime access to video tutorial support when shrinkwrapped with the text.

PH Tutor Center (0-13-064604-0)

- Free tutorial support via phone, fax, or email
- Available Sunday—Thursday 5pm e.s.t. to midnight—5 days a week, 7 hours a day
- Accessed through a registration number that may be bundled with a new text or purchased separately with a used book.
- See *www.prenhall.com/tutorcenter* for FAQ

ACKNOWLEDGMENTS

First, as usual, I would like to thank my husband, Clayton, for his constant encouragement. I would also like to thank my children, Eric and Bryan, for continuing to eat my burnt meals. Thankfully, they have started to cook a little themselves.

I would also like to thank my extended family for their invaluable help and wonderful sense of humor. Their contributions are too numerous to list. They are Rod, Karen, and Adam Pasch; Peter, Michael, Christopher, Matthew, and Jessica Callac; Stuart, Earline, Melissa, Mandy, Bailey, and Ethan Martin; Mark, Sabrina, and Madison Martin; Leo and Barbara Miller; and Jewett Gay.

I would like to thank the following reviewers for their input and suggestions:

Cherie Bowers, *Santa Ana College*
Ann Hausecavriague, *Santiago College*
Alan Hayashi, *Oxnard College*
Kathryn Hodge, *Midland College*
Deanna Li, *North Seattle Community College*
Pam Lippert, *North Seattle Community College*
Donna Sperry, *University of Southern Maryland*
Gizelle Worley, *California State–Stanislaus*
Roy Tucker, *Palo Alto Community College*
Cynthia McGinnis, *Okaloosa Walton Community College*
Nancy Lang, *Invec Hills Community College*
Greg Daubenmire, *Las Positas Community College*
Carla Ainsworth, *Salt Lake Community College*

There were many people who helped me develop this text and I will attempt to thank some of them here. Cindy Trimble was invaluable for contributing to the overall accuracy of this text. Elka Block, Chris Callac, and Miriam Daunis were invaluable for their many suggestions and contributions during the development and writing of this fourth edition. Brandi Nelson provided guidance throughout the production process. I thank Richard Semmler and Carrie Green for all their work on the solutions, text, and accuracy.

Sadly, executive editor of this project, Karin Wagner, passed away. She will be dearly remembered by me and the rest of the staff at Prentice Hall for her integrity, wisdom, commitment, and especially her sense of humor and infectious laugh.

A very special thank you to my project manager, Mary Beckwith, for taking over during a difficult period for all of us.

Lastly, my thanks to the staff at Prentice Hall for all their support: Linda Behrens, Mike Bell, Patty Burns, Tom Benfatti, Paul Belfanti, Maureen Eide, Geoff Cassar, Jim Sullivan, Eilish Main, Patrice Jones, John Tweeddale, Chris Hoag, Paul Corey, and Tim Bozik.

K. Elayn Martin-Gay

ABOUT THE AUTHOR

K. Elayn Martin-Gay has taught mathematics at the University of New Orleans for more than 20 years. Her numerous teaching awards include the local University Alumni Association's Award for Excellence in Teaching, and Outstanding Developmental Educator at University of New Orleans, presented by the Louisiana Association of Developmental Educators.

Prior to writing textbooks, K. Elayn Martin-Gay developed an acclaimed series of lecture videos to support developmental mathematics students in their quest for success. These highly successful videos originally served as the foundation material for her texts. Today the tapes specifically support each book in the Martin-Gay series.

Elayn is the author of over nine published textbooks as well as multimedia interactive mathematics, all specializing in developmental mathematics courses such as basic mathematics, prealgebra, beginning and intermediate algebra. She has provided author participation across the broadest range of materials: textbook, videos, tutorial software, and Interactive Math courseware. All the components are designed to work together. This offers an opportunity of various combinations for an integrated teaching and learning package offering great consistency and comfort for the student.

APPLICATIONS INDEX

To my mother, Barbara M. Miller,
and her husband, Leo Miller,
and to the memory
of my father, Robert J. Martin

Highlights of *Beginning Algebra, Fourth Edition*

Every Student Can Succeed

Beginning Algebra, Fourth Edition has been written and designed to help you succeed in this course. Special care has been taken to ensure students have the most up-to-date and relevant text features, and as many real-world applications as possible to provide you with a solid foundation in algebra and prepare you for future courses.

Good study skills are essential for success in mathematics. This edition provides an increased emphasis on study and test preparation skills. Take a few minutes to examine the features and resources that have been incorporated into *Beginning Algebra, Fourth Edition* to help students excel.

CHAPTER

2

Equations, Inequalities, and Problem Solving

2.1 Simplifying Algebraic Expressions
2.2 The Addition Property of Equality
2.3 The Multiplication Property of Equality
2.4 Solving Linear Equations
    Integrated Review—Solving Linear Equations
2.5 An Introduction to Problem Solving
2.6 Formulas and Problem Solving
2.7 Further Problem Solving
2.8 Solving Linear Inequalities

CALLING ALL ACCOUNTANTS

There is an old saying: Nothing is certain, except death and taxes. And with ever-changing tax laws, is it any wonder that income tax preparation has become a multimillion-dollar industry? Companies such as H&R Block and Jackson-Hewitt employ accountants as both full-time and seasonal workers to prepare taxes and advise clients on financial strategies.

But accountants do so much more! Certified public accountants, or CPAs, also work for private companies or the federal government as auditors, consultants, and managers. Because they like to work with figures, their methodical approach to problem solving makes them valuable employees at all levels of business.

PAGE 80

◀ Real-World Chapter Openers

Real-world chapter openers focus on how algebraic concepts relate to the world around you. They also reference a *Spotlight on Decision Making* feature within the chapter for further exploration.

Spotlight on

DECISION MAKING

Suppose you are a personal income tax preparer. Your clients, Jose and Felicia Fernandez, are filing jointly Form 1040 as their individual income tax return. You know that medical expenses may be written off as an itemized deduction if the expenses exceed 7.5% of their adjusted gross income. Furthermore, only the portion of medical expenses that exceed 7.5% of their adjusted gross income can be deducted. Is the Fernandez family eligible to deduct their medical expenses? Explain.

Internal Revenue Service
Form 1040
Jose Fernandez
Felicia Fernandez
Adjusted Gross Income . . . $33,650

Fernandez Family Deductible Medical Expenses	
Medical bills	$1025
Dental bills	$ 325
Prescription drugs	$ 360
Medical Insurance premiums	$1200

PAGE 138

▲ Spotlight on Decision Making

These unique applications encourage students to develop decision making and problem solving abilities. Primarily workplace or career-related situations are highlighted in each feature.

Become a Confident Problem Solver!

A goal of this text is to help you develop problem solving abilities.

EXAMPLE 5

FINDING THE INVESTMENT AMOUNT

Rajiv Puri invested part of his $20,000 inheritance in a mutual funds account that pays 7% simple interest yearly and the rest in a certificate of deposit that pays 9% simple interest yearly. At the end of one year, Rajiv's investments earned $1550. Find the amount he invested at each rate.

Solution

1. UNDERSTAND. Read and reread the problem. Next, guess a solution. Suppose that Rajiv invested $8000 in the 7% fund and the rest, $12,000, in the fund paying 9%. To check, find his interest after one year. Recall the formula, $I = PRT$, so the interest from the 7% fund = $8000(0.07)(1) = $560. The interest from the 9% fund = $12,000(0.09)(1) = $1080. The sum of the interests is $560 + $1080 = $1640. Our guess is incorrect, since the sum of the interests is not $1550, but we now have a better understanding of the problem.

Let

$$x = \text{amount of money in the account paying 7\%.}$$

The rest of the money is $20,000 less x or

$$20,000 - x = \text{amount of money in the account paying 9\%.}$$

2. TRANSLATE. We apply the simple interest formula $I = PRT$ and organize our information in the following chart. Since there are two different rates of interest and two different amounts invested, we apply the formula twice.

	Principal ·	Rate ·	Time =	Interest
7% Fund	x	0.07	1	$x(0.07)(1)$ or $0.07x$
9% Fund	$20,000 - x$	0.09	1	$(20,000 - x)(0.09)(1)$ or $0.09(20,000 - x)$
Total	20,000			1550

The total interest earned, $1550, is the sum of the interest earned at 7% and the interest earned at 9%.

In words:	interest at 7%	+	interest at 9%	=	total interest
Translate:	$0.07x$	+	$0.09(20,000 - x)$	=	1550

3. SOLVE.

$$0.07x + 0.09(20,000 - x) = 1550$$
$$0.07x + 1800 - 0.09x = 1550 \quad \text{Apply the distributive property.}$$
$$1800 - 0.02x = 1550 \quad \text{Combine like terms.}$$
$$-0.02x = -250 \quad \text{Subtract 1800 from both sides.}$$
$$x = 12,500 \quad \text{Divide both sides by } -0.02.$$

4. INTERPRET.

Check: If $x = 12,500$, then $20,000 - x = 20,000 - 12,500$ or 7500. These solutions are reasonable, since their sum is $20,000 as required. The annual interest on $12,500 at 7% is $875; the annual interest on $7500 at 9% is $675, and $875 + $675 = $1550.
State: The amount invested at 7% is $12,500. The amount invested at 9% is $7500.

PAGE 137

◄ General Strategy for Problem Solving

Save time by having a plan. The organization of this text can help you. Note the outlined problem solving steps: *Understand, Translate, Solve,* and *Interpret.* Problem solving is introduced early, emphasized, and integrated throughout the chapters. The problem solving procedure is illustrated step-by-step in the in-text examples.

Geometry ►

Geometric concepts are integrated throughout the text in examples and exercises and are identified with a triangle icon. The inside front cover contains *Geometric Formulas* for convenient reference, and there are appendices on geometry in the back of the text.

57. The CART Fed Ex Championship Series is an open-wheeled race car competition based in the United States. A CART car has a maximum length of 199 inches, a maximum width of 78.5 inches, and a maximum height of 33 inches. When the CART series travels to another country for a grand prix, teams must ship their cars. Find the volume of the smallest shipping crate needed to ship a CART car of maximum dimensions. (*Source:* Championship Auto Racing Teams, Inc.)

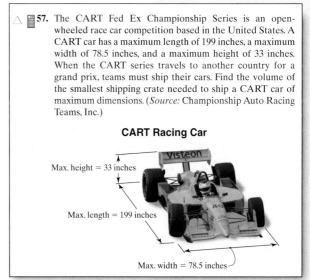

CART Racing Car

Max. height = 33 inches

Max. length = 199 inches

Max. width = 78.5 inches

PAGE 131

Learn Better—Master and Apply Skills and Concepts

K. Elayn Martin-Gay provides thorough explanations of concepts and enlivens the content by integrating successful and innovative instructional tools. These features have been included to enhance your understanding of mathematical concepts.

✔ **CONCEPT CHECK**

Suppose you have simplified several equations and obtain the following results. What can you conclude about the solutions to the original equation?

a. $7 = 7$ **b.** $x = 0$ **c.** $7 = -4$

5   We can apply our equation-solving skills to solving problems written in words. Many times, writing an equation that describes or models a problem involves a direct translation from a word sentence to an equation.

PAGE 107

◀ **New Concept Checks**

Concept Checks are special exercises found in most sections following key examples. Work these to help measure your grasp of the concept being explained before moving to the next example.

Mixed Practice Exercises ▶

The exercise sets have been reorganized and include *mixed practice* exercises. These exercises give students the chance to assimilate the concepts and skills that have been covered in separate objectives.

MIXED PRACTICE

Solve. See Examples 1 through 7.

31. $42 = 7x$ **32.** $81 = 3x$

33. $4.4 = -0.8x$ **34.** $6.3 = -0.6x$

35. $-\frac{3}{7}p = -2$ **36.** $-\frac{4}{5}r = -5$

37. $2x - 4 = 16$ **38.** $3x - 1 = 26$

39. $5 - 0.3k = 5$ **40.** $2 + 0.4p = 2$

41. $-\frac{4}{3}x = 12$ **42.** $-\frac{10}{3}x = 30$

43. $10 = 2x - 1$ **44.** $12 = 3j - 4$

45. $\frac{x}{3} + 2 = -5$ **46.** $\frac{b}{4} - 1 = -7$

47. $1 = 0.4x - 0.6x - 5$ **48.** $21 = 0.4x - 0.9x - 4$

49. $z - 5z = 7z - 9 - z$ **50.** $t - 6t = -13 + t - 3t$

51. $6 - 2x + 8 = 10$ **52.** $-5 - 6y + 6 = 19$

53. $-3a + 6 + 5a = 7a - 8a$

54. $4b - 8 - b = 10b - 3b$

55. The equation $3x + 6 = 2x + 10 + x - 4$ is true for all real numbers. Substitute a few real numbers for x to see that this is so and then try solving the equation.

PAGE 102

Concept Extensions

73. A licensed nurse practitioner is instructed to give a patient 2100 milligrams of an antibiotic over a period of 36 hours. If the antibiotic is to be given every 4 hours starting immediately, how much antibiotic should be given in each dose? To answer this question, solve the equation $9x = 2100$.

74. Suppose you are a pharmacist and a customer asks you the following question. His child is to receive 13.5 milliliters of a nausea medicine over a period of 54 hours. If the nausea medicine is to be administered every 6 hours starting immediately, how much medicine should be given in each dose?

◀ **New Concept Extensions**

Concept Extension exercises have also been added to the exercise sets. They extend the concepts and require students to combine several skills or concepts. These exercises expose students to the way math ideas build upon each other.

Study Better—Build Confidence and Develop Study Skills

Several features of this text can be helpful in building your confidence and mathematical competence. They will also help improve your study skills.

Tips for Success ▶

Coverage of study skills in Section 1.1 reinforces this important component to success in this course.

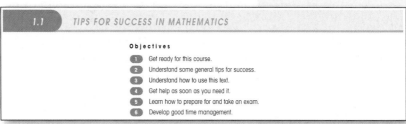

PAGE 2

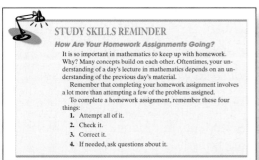

PAGE 117

◀ New Study Skills Reminders

Study Skills Reminders are integrated throughout the text to reinforce Section 1.1 and encourage the development of strong study skills.

Mental Math ▶

Mental Math warm-up exercises reinforce concepts found in the accompanying section and can increase your confidence before beginning an exercise set.

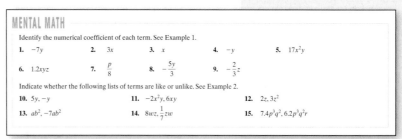

PAGE 86

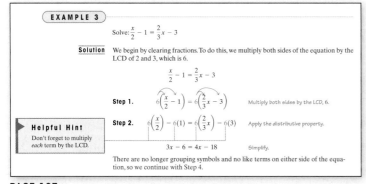

PAGE 105

◀ Helpful Hints

Found throughout the text, these features contain practical advice on applying mathematical concepts. They are strategically placed where students are most likely to need immediate reinforcement.

Vocabulary Checks ▶

Vocabulary Checks help students strengthen verbal skills and answer questions about a chapter's content by filling in the blank with the correct word from the vocabulary list.

2 CHAPTER VOCABULARY CHECK

Fill in each blank with one of the words or phrases listed below.

like terms   numerical coefficient   linear inequality in one variable
equivalent equations   formula   compound inequalities
linear equation in one variable

1. Terms with the same variables raised to exactly the same powers are called _____.
2. A _____ can be written in the form $ax + b = c$.
3. Equations that have the same solution are called _____.
4. Inequalities containing two inequality symbols are called _____.
5. An equation that describes a known relationship among quantities is called a _____.
6. A _____ can be written in the form $ax + b < c$, (or $>$, $\leq$, $\geq$).
7. The _____ of a term is its numerical factor.

PAGE 153

Test Better—Test Yourself and Check Your Understanding

Good exercise sets and an abundance of worked-out examples are essential for building confidence. The exercises in this text are intended to help you build skills and understand concepts as well as motivate and challenge you. In addition, features such as *Integrated Reviews, Chapter Highlights, Chapter Reviews, Chapter Tests,* and *Cumulative Reviews* are found in each chapter to help you study and organize your notes. A new resource—*Chapter Test Prep Video*—will help students study more effectively than ever before.

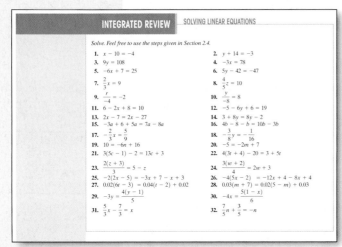

PAGE 111

◀ New Integrated Reviews

Integrated Reviews serve as mid-chapter reviews and help you learn the new skills you have been studying over several sections. This allows students to practice making decisions before taking a test.

Chapter Highlights ▶

Each chapter ends with *Chapter Highlights.* They contain key definitions, concepts, and examples to help you understand and retain what you have learned in the chapter.

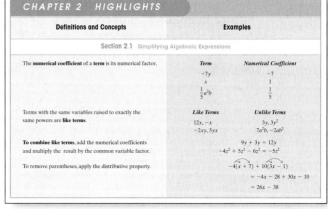

PAGE 153

◀ Chapter Tests

Take the Chapter Test at the end of each chapter to prepare for a class exam. With the new *Chapter Test Prep Video* (packaged with your text), you can instantly view the solution to the exact exercises you're working as part of your studying. See the next page for more details.

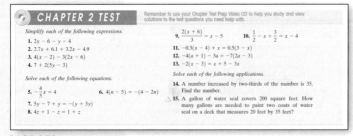

PAGE 159

Begin Studying More Effectively Today!

New Chapter Test Prep Video CD

Good study skills are essential for success in mathematics. Use the student resources that accompany *Beginning Algebra, Fourth Edition* to make the most of your valuable study time.

The *Chapter Test Prep Video* CD, presented by K. Elayn Martin-Gay, provides students with instant access to step-by-step video solutions for each of the Chapter Test questions in the text. To make the most of this resource when studying for a test, follow these three steps:

1. Take the Chapter Test at the end of each chapter.
2. Check your answers in the back of the text.
3. Use the *Chapter Test Prep Video* to review **every step** of the worked-out solution for those specific questions you answered incorrectly or didn't understand on the test.

The Video CD's easy navigation is designed to help students study efficiently. Students select the exact test questions they missed or don't understand and instantly view the solution worked out by the author.

Previous and Next buttons allow easy navigation within tests.

The Chapter Test menu allows students to select the test questions they wish to view.

Rewind, Pause, and Play buttons let students view the solutions at their own pace.

The Video CD includes an Introduction with instructions on how to use the *Chapter Test Prep Video CD* as well as test-taking study skills.

To begin studying and preparing for your test, see the Video CD and Note to Students in the back of your text. The Video CD is also available through your campus bookstore.

Get Motivated!

The fourth edition strongly emphasizes visualization. Graphing is introduced early and intuitively. Knowing how to read and use graphs is a valuable skill in the workplace as well as in this and other courses. In addition, this edition includes a wealth of real-world applications to show the relevance of math in everyday life.

Real-World Applications ▼

Many applications are included often based on real data drawn from current and familiar sources.

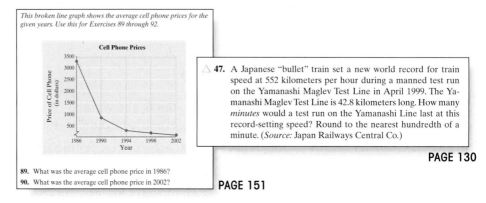

This broken line graph shows the average cell phone prices for the given years. Use this for Exercises 89 through 92.

Cell Phone Prices

89. What was the average cell phone price in 1986?
90. What was the average cell phone price in 2002?

PAGE 151

△ **47.** A Japanese "bullet" train set a new world record for train speed at 552 kilometers per hour during a manned test run on the Yamanashi Maglev Test Line in April 1999. The Yamanashi Maglev Test Line is 42.8 kilometers long. How many *minutes* would a test run on the Yamanashi Line last at this record-setting speed? Round to the nearest hundredth of a minute. (*Source:* Japan Railways Central Co.)

PAGE 130

Scientific and Graphing ▶ Calculator Explorations

These exploration features contain examples and exercises to reinforce concepts, help interpret graphs, and motivate discovery learning. Scientific and graphing calculator exercises can also be found in exercise sets.

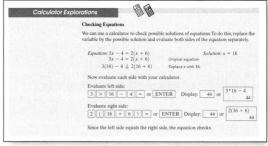

Calculator Explorations

Checking Equations

We can use a calculator to check possible solutions of equations. To do this, replace the variable by the possible solution and evaluate both sides of the equation separately.

$$\text{Equation: } 3x - 4 = 2(x + 6) \qquad \text{Solution: } x = 16$$
$$3x - 4 = 2(x + 6) \qquad \text{Original equation}$$
$$3(16) - 4 \stackrel{?}{=} 2(16 + 6) \qquad \text{Replace } x \text{ with 16.}$$

Now evaluate each side with your calculator.

Evaluate left side:
3 × 16 − 4 = or ENTER Display: 44 or $\dfrac{3*16 - 4}{44}$

Evaluate right side:
2 (16 + 6)) = or ENTER Display: 44 or $\dfrac{2(16 + 6)}{44}$

Since the left side equals the right side, the equation checks.

PAGE 108

△ **83.** A plot of land is in the shape of a triangle. If one side is x meters, a second side is $(2x - 3)$ meters and a third side is $(3x - 5)$ meters, express the perimeter of the lot as a simplified expression in x.

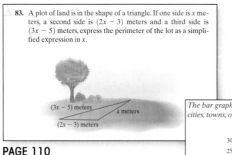

$(3x - 5)$ meters x meters
$(2x - 3)$ meters

PAGE 110

◀ Visualization of Topics

Many illustrations, models, photographs, tables, charts, and graphs provide visual reinforcement of concepts and opportunities for data interpretation.

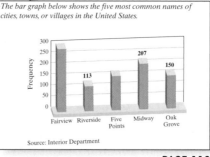

The bar graph below shows the five most common names of cities, towns, or villages in the United States.

Source: Interior Department

PAGE 110

Resources for Student Success

Beginning Algebra, Fourth Edition is supported by comprehensive resource packages for both instructors and students. Please refer to the Preface for detailed descriptions.

Highlights of the packages include the following:

Chapter Test Prep Video CD—included in every student text and annotated instructor's edition. It provides step-by-step worked out solutions to every question contained on the chapter tests in the text. Hosted by K. Elayn Martin-Gay.

MyMathLab®—a series of text-specific, easily customizable online courses for Prentice Hall mathematics texts.

MathXL®—a powerful online homework, tutorial, and assessment system.

MathXL® Tutorials on CD—provides algorithmically generated practice exercises that are correlated at the objective level to the exercises in the text.

CD Lecture Series—text-specific videos available on CD. They are hosted by K. Elayn Martin-Gay and cover each objective in every chapter section of the text.

Prentice Hall Tutor Center—provides text-specific tutoring via phone, fax, and email.

Student Solutions Manual and Student Study Guide

FOR INSTRUCTORS

Annotated Instructor's Edition—now includes one additional classroom example for each example in the text, Teaching Tips, and answers to all text exercises.

Instructor's Solutions Manual and Instructor's Resource Manual with Tests

TestGen—enables instructors to build, edit, print, and administer tests using a computerized bank of questions developed to cover all objectives of the text.

Instructor-to-Instructor Videos—provide suggestions for presenting topics, time-saving tips, alternative strategies, and much more.

Review of Real Numbers

The power of mathematics is its flexibility. We apply numbers to almost every aspect of our lives, from an ordinary trip to the grocery store to a rocket launched into space. The power of algebra is its generality. Using letters to represent numbers, we tie together the trip to the grocery store and the launched rocket.

In this chapter we review the basic symbols and words—the language—of arithmetic and introduce using variables in place of numbers. This is our starting place in the study of algebra.

HEALTH CARE PROFESSIONALS IN DEMAND

During a visit to the dentist's office for a regular checkup, patients usually spend most of their time with the dental hygienist. Hygienists are professional health care providers who examine and clean a patient's gums and teeth. They remove plaque from teeth, apply fluoride for cavity prevention, floss teeth, and take and develop dental X rays. Frequently, hygienists teach their patients, especially children, how to practice good oral hygiene.

According to the U.S. Department of Labor, the demand for dental hygienists in the workplace is expected to grow through the year 2010. This increase in demand is due to many factors, including population growth and greater retention of natural teeth.

In the Spotlight on Decision Making feature on page 44, you will have the opportunity to make a decision concerning the health of a patient's gums as a dental hygienist.
*Link: American Dental Hygienists' Association—*www.adha.org
Source of text: http://www.bls.gov/oco/ocos097.htm

1.1   *TIPS FOR SUCCESS IN MATHEMATICS*

Objectives

1   Get ready for this course.

2   Understand some general tips for success.

3   Understand how to use this text.

4   Get help as soon as you need it.

5   Learn how to prepare for and take an exam.

6   Develop good time management.

Before reading this section, remember that your instructor is your best source for information. Please see your instructor for any additional help or information.

1   **Getting Ready for This Course**

Now that you have decided to take this course, remember that a *positive attitude* will make all the difference in the world. Your belief that you can succeed is just as important as your commitment to this course. Make sure that you are ready for this course by having the time and positive attitude that it takes to succeed.

Next, make sure that you have scheduled your math course at a time that will give you the best chance for success. For example, if you are also working, you may want to check with your employer to make sure that your work hours will not conflict with your course schedule. Also, schedule your class during a time of day when you are more attentive and do your best work.

On the day of your first class period, double-check your schedule and allow yourself extra time to arrive in case of traffic problems or difficulty locating your classroom. Make sure that you bring at least your textbook, paper, and a writing instrument. Are you required to have a lab manual, graph paper, calculator, or some other supply besides this text? If so, also bring this material with you.

2   **General Tips for Success**

Below are some general tips that will increase your chance for success in a mathematics class. Many of these tips will also help you in other courses you may be taking.

Exchange names and phone numbers with at least one other person in class. This contact person can be a great help if you miss an assignment or want to discuss math concepts or exercises that you find difficult.

Choose to attend all class periods and be on time. If possible, sit near the front of the classroom. This way, you will see and hear the presentation better. It may also be easier for you to participate in classroom activities.

Do your homework. You've probably heard the phrase "practice makes perfect" in relation to music and sports. It also applies to mathematics. You will find that the more time you spend solving mathematics problems, the easier the process becomes. Be sure to schedule enough time to complete your assignments before the next class period.

Check your work. Review the steps you made while working a problem. Learn to check your answers to the original problems. You may also compare your answers with the answers to selected exercises section in the back of the

book. If you have made a mistake, try to figure out what went wrong. Then correct your mistake. If you can't find what went wrong, don't erase your work or throw it away. Bring your work to your instructor, a tutor in a math lab, or a classmate. It is easier for someone to find where you had trouble if they look at your original work.

Learn from your mistakes and be patient with yourself. Everyone, even your instructor, makes mistakes. (That definitely includes me—Elayn Martin-Gay.) Use your errors to learn and to become a better math student. The key is finding and understanding your errors.

Was your mistake a careless one, or did you make it because you can't read your own math writing? If so, try to work more slowly or write more neatly and make a conscious effort to carefully check your work.

Did you make a mistake because you don't understand a concept? Take the time to review the concept or ask questions to better understand it.

Did you skip too many steps? Skipping steps or trying to do too many steps mentally may lead to preventable mistakes.

Know how to get help if you need it. It's all right to ask for help. In fact, it's a good idea to ask for help whenever there is something that you don't understand. Make sure you know when your instructor has office hours and how to find his or her office. Find out whether math tutoring services are available on your campus. Check out the hours, location, and requirements of the tutoring service. Videotapes and software are available with this text. Learn how to access these resources.

Organize your class materials, including homework assignments, graded quizzes and tests, and notes from your class or lab. All of these items will be valuable references throughout your course especially when studying for upcoming tests and the final exam. Make sure that you can locate these materials when you need them.

Read your textbook before class. Reading a mathematics textbook is unlike leisure reading such as reading a novel or newspaper. Your pace will be much slower. It is helpful to have paper and a pencil with you when you read. Try to work out examples on your own as you encounter them in your text. You should also write down any questions that you want to ask in class. When you read a mathematics textbook, some of the information in a section may be unclear. But when you hear a lecture or watch a videotape on that section, you will understand it much more easily than if you had not read your text beforehand.

Don't be afraid to ask questions. Instructors are not mind readers. Many times we do not know a concept is unclear until a student asks a question. You are not the only person in class with questions. Other students are normally grateful that someone has spoken up.

Hand in assignments on time. This way you can be sure that you will not lose points for being late. Show every step of a problem and be neat and organized. Also be sure that you understand which problems are assigned for homework. You can always double-check this assignment with another student in your class.

3

Using This Text

There are many helpful resources that are available to you in this text. It is important that you become familiar with and use these resources. They should increase your chances for success in this course.

- The main section of exercises in each exercise set is referenced by an example(s). Use this referencing if you have trouble completing an assignment from the exercise set.

- If you need extra help in a particular section, look at the beginning of the section to see what videotapes and software are available.

- Make sure that you understand the meaning of the icons that are beside many exercises. The video icon 📷 tells you that the corresponding exercise may be viewed on the videotape that corresponds to that section. The pencil icon ✎ tells you that this exercise is a writing exercise in which you should answer in complete sentences. The △ icon tells you that the exercise involves geometry.

- Integrated Reviews in each chapter offer you a chance to practice—in one place—the many concepts that you have learned separately over several sections.

- There are many opportunities at the end of each chapter to help you understand the concepts of the chapter.

  Chapter Highlights contain chapter summaries and examples.

  Chapter Reviews contain review problems organized by section.

  Chapter Tests are sample tests to help you prepare for an exam.

  Cumulative Reviews are reviews consisting of material from the beginning of the book to the end of that particular chapter.

See the preface at the beginning of this text for a more thorough explanation of the features of this text.

4 **Getting Help**

If you have trouble completing assignments or understanding the mathematics, get help as soon as you need it! This tip is presented as an objective on its own because it is so important. In mathematics, usually the material presented in one section builds on your understanding of the previous section. This means that if you don't understand the concepts covered during a class period, there is a good chance that you will not understand the concepts covered during the next class period. If this happens to you, get help as soon as you can.

Where can you get help? Many suggestions have been made in the section on where to get help, and now it is up to you to do it. Try your instructor, a tutoring center, or a math lab, or you may want to form a study group with fellow classmates. If you do decide to see your instructor or go to a tutoring center, make sure that you have a neat notebook and are ready with your questions.

5 **Preparing for and Taking an Exam**

Make sure that you allow yourself plenty of time to prepare for a test. If you think that you are a little "math anxious," it may be that you are not preparing for a test in a way that will ensure success. The way that you prepare for a test in mathematics is important. To prepare for a test,

1. Review your previous homework assignments. You may also want to rework some of them.

2. Review any notes from class and section-level quizzes you have taken. (If this is a final exam, also review chapter tests you have taken.)

3. Review concepts and definitions by reading the Highlights at the end of each chapter.

4. Practice working out exercises by completing the Chapter Review found at the end of each chapter. (If this is a final exam, go through a Cumulative

Review. There is one found at the end of each chapter except Chapter 1. Choose the review found at the end of the latest chapter that you have covered in your course.) *Don't stop here!*

5. It is important that you place yourself in conditions similar to test conditions to find out how you will perform. In other words, as soon as you feel that you know the material, get a few blank sheets of paper and take a sample test. There is a Chapter Test available at the end of each chapter. During this sample test, do not use your notes or your textbook. Once you complete the Chapter Test, check your answers in the back of the book. If any answer is incorrect, there is a CD available with each exercise of each chapter test worked. Use this CD or your instructor to correct your sample test. Your instructor may also provide you with a review sheet. If you are not satisfied with the results, study the areas that you are weak in and try again.

6. Get a good night's sleep before the exam.

7. On the day of the actual test, allow yourself plenty of time to arrive at where you will be taking your exam.

When taking your test,

1. Read the directions on the test carefully.

2. Read each problem carefully as you take the test. Make sure that you answer the question asked.

3. Pace yourself by first completing the problems you are most confident with. Then work toward the problems you are least confident with. Watch your time so you do not spend too much time on one particular problem.

4. If you have time, check your work and answers.

5. Do not turn your test in early. If you have extra time, spend it double-checking your work.

6   **Managing Your Time**

As a college student, you know the demands that classes, homework, work, and family place on your time. Some days you probably wonder how you'll ever get everything done. One key to managing your time is developing a schedule. Here are some hints for making a schedule:

1. Make a list of all of your weekly commitments for the term. Include classes, work, regular meetings, extracurricular activities, etc. You may also find it helpful to list such things as laundry, regular workouts, grocery shopping, etc.

2. Next, estimate the time needed for each item on the list. Also make a note of how often you will need to do each item. Don't forget to include time estimates for reading, studying, and homework you do outside of your classes. You may want to ask your instructor for help estimating the time needed.

3. In the following exercise set, you are asked to block out a typical week on the schedule grid given. Start with items with fixed time slots like classes and work.

4. Next, include the items on your list with flexible time slots. Think carefully about how best to schedule some items such as study time.

5. Don't fill up every time slot on the schedule. Remember that you need to allow time for eating, sleeping, and relaxing! You should also allow a little extra time in case some items take longer than planned.

6. If you find that your weekly schedule is too full for you to handle, you may need to make some changes in your workload, classload, or in other areas of your life. You may want to talk to your advisor, manager or supervisor at work, or someone in your college's academic counseling center for help with such decisions.

Note: In this chapter, we begin a feature called Study Skills Reminder. The purpose of this feature is to remind you of some of the information given in this section and to further expand on some topics in this section.

EXERCISE SET 1.1

| STUDY GUIDE/SSM | CD/ VIDEO | PH MATH TUTOR CENTER | MathXL®Tutorials ON CD | MathXL® | MyMathLab® |

1. What is your instructor's name?

2. What are your instructor's office location and office hours?

3. What is the best way to contact your instructor?

4. What does the ✎ icon mean?

5. What does the 🔒 icon mean?

6. What does the △ icon mean?

7. Where are answers located in this text?

8. What Exercise Set answers are available to you in the answers section?

9. What Chapter Review, Chapter Test, and Cumulative Review answers are available to you in the answer section?

10. Are there worked-out solutions to exercises in this text?

11. If the answer to Exercise 10 is yes, what worked-out solutions are available to you in this text?

12. Go to the Highlights section at the end of this chapter. Describe how this section may be helpful to you when preparing for a test.

13. Do you have the name and contact information of at least one other student in class?

14. Will your instructor allow you to use a calculator in this class?

15. Are videotapes, CDs, and/or tutorial software available to you? If so, where?

16. Is there a tutoring service available? If so, what are its hours?

17. Have you attempted this course before? If so, write down ways that you might improve your chance of success during this second attempt.

18. List some steps that you can take if you begin having trouble understanding the material or completing an assignment.

19. Read or reread objective 6 and fill out the schedule grid below.

20. Study your filled-out grid from Exercise 19. Decide whether you have the time necessary to successfully complete this course and any other courses you may be registered for.

	Monday	Tuesday	Wednesday	Thursday	Friday	Saturday	Sunday
7:00 A.M.							
8:00 A.M.							
9:00 A.M.							
10:00 A.M.							
11:00 A.M.							
12:00 A.M.							
1:00 P.M.							
2:00 P.M.							
3:00 P.M.							
4:00 P.M.							
5:00 P.M.							
6:00 P.M.							
7:00 P.M.							
8:00 P.M.							
9:00 P.M.							

1.2 SYMBOLS AND SETS OF NUMBERS

Objectives

1 Use a number line to order numbers.

2 Translate sentences into mathematical statements.

3 Identify natural numbers, whole numbers, integers, rational numbers, irrational numbers, and real numbers.

4 Find the absolute value of a real number.

1   We begin with a review of the set of natural numbers and the set of whole numbers and how we use symbols to compare these numbers. A **set** is a collection of objects, each of which is called a **member** or **element** of the set. A pair of brace symbols { } encloses the list of elements and is translated as "the set of" or "the set containing."

Natural Numbers

The set of **natural numbers** is {1, 2, 3, 4, 5, 6, ... }.

Whole Numbers

The set of **whole numbers** is {0, 1, 2, 3, 4, ... }.

The three dots (an ellipsis) at the end of the list of elements of a set means that the list continues in the same manner indefinitely.

These numbers can be pictured on a **number line**. We will use number lines often to help us visualize distance and relationships between numbers. Visualizing mathematical concepts is an important skill and tool, and later we will develop and explore other visualizing tools.

To draw a number line, first draw a line. Choose a point on the line and label it 0. To the right of 0, label any other point 1. Being careful to use the same distance as from 0 to 1, mark off equally spaced distances. Label these points 2, 3, 4, 5, and so on. Since the whole numbers continue indefinitely, it is not possible to show every whole number on this number line. The arrow at the right end of the line indicates that the pattern continues indefinitely.

Picturing whole numbers on a number line helps us to see the order of the numbers. Symbols can be used to describe concisely in writing the order that we see.

The **equal symbol** = means "is equal to."

The symbol ≠ means "is not equal to."

These symbols may be used to form a **mathematical statement**. The statement might be true or it might be false. The two statements below are both true.

2 = 2   states that "two is equal to two"

2 ≠ 6   states that "two is not equal to six"

If two numbers are not equal, then one number is larger than the other. The symbol > means "is greater than." The symbol < means "is less than." For example,

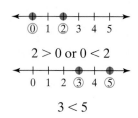

2 > 0 or 0 < 2

3 < 5

$2 > 0$ states that "two is greater than zero"

$3 < 5$ states that "three is less than five"

On a number line, we see that a number **to the right of** another number is **larger**. Similarly, a number **to the left of** another number is smaller. For example, 3 is to the left of 5 on a number line, which means that 3 is less than 5, or $3 < 5$. Similarly, 2 is to the right of 0 on a number line, which means 2 is greater than 0, or $2 > 0$. Since 0 is to the left of 2, we can also say that 0 is less than 2, or $0 < 2$.

The symbols $\neq$, $<$, and $>$ are called **inequality symbols**.

TEACHING TIP
If students are having trouble with inequality symbols, remind them to read the Helpful Hint. If the symbols "point" to the smaller number, the inequality statement will be correct. For example,

$$7 > 5$$
$$\uparrow$$
points to smaller number

> **Helpful Hint**
>
> Notice that $2 > 0$ has exactly the same meaning as $0 < 2$. Switching the order of the numbers and reversing the "direction of the inequality symbol" does not change the meaning of the statement.
>
> $$5 > 3 \text{ has the same meaning as } 3 < 5.$$
>
> Also notice that, when the statement is true, the inequality arrow points to the smaller number.

EXAMPLE 1

Insert $<$, $>$, or $=$ in the space between each pair of numbers to make each statement true.

a. 2   3   **b.** 7   4   **c.** 72   27

Solution
a. $2 < 3$ since 2 is to the left of 3 on the number line.

b. $7 > 4$ since 7 is to the right of 4 on the number line.

c. $72 > 27$ since 72 is to the right of 27 on the number line.

CLASSROOM EXAMPLE
Insert $<$, $>$, or $=$ between each pair of numbers.
a. 9   20   **b.** 100   99
answer: a. $<$ **b.** $>$

Two other symbols are used to compare numbers. The symbol $\leq$ means "is less than or equal to." The symbol $\geq$ means "is greater than or equal to." For example,

$$7 \leq 10 \text{ states that "seven is less than or equal to ten"}$$

This statement is true since $7 < 10$ is true. If either $7 < 10$ or $7 = 10$ is true, then $7 \leq 10$ is true.

$$3 \geq 3 \text{ states that "three is greater than or equal to three"}$$

This statement is true since $3 = 3$ is true. If either $3 > 3$ or $3 = 3$ is true, then $3 \geq 3$ is true.

The statement $6 \geq 10$ is false since neither $6 > 10$ nor $6 = 10$ is true. The symbols $\leq$ and $\geq$ are also called **inequality symbols**.

EXAMPLE 2

Tell whether each statement is true or false.

a. $8 \geq 8$ **b.** $8 \leq 8$ **c.** $23 \leq 0$ **d.** $23 \geq 0$

Solution
a. True, since $8 = 8$ is true.   **b.** True, since $8 = 8$ is true.

c. False, since neither $23 < 0$ nor $23 = 0$ is true.   **d.** True, since $23 > 0$ is true.

2   Now, let's use the symbols discussed above to translate sentences into mathematical statements.

EXAMPLE 3

Translate each sentence into a mathematical statement.

a. Nine is less than or equal to eleven.

b. Eight is greater than one.

c. Three is not equal to four.

Solution

a.	nine	is less than or equal to	eleven		**b.**	eight	is greater than	one
	9	$\leq$	11			8	$>$	1

c.	three	is not equal to	four
	3	$\neq$	4

3   Whole numbers are not sufficient to describe many situations in the real world. For example, quantities smaller than zero must sometimes be represented, such as temperatures less than 0 degrees.

We can picture numbers less than zero on a number line as follows:

Numbers less than 0 are to the left of 0 and are labeled $-1, -2, -3$, and so on. A $-$ sign, such as the one in -1, tells us that the number is to the left of 0 on a number line. In words, -1 is read "negative one." A $+$ sign or no sign tells us that a number lies to the right of 0 on the number line. For example, 3 and $+3$ both mean positive three.

The numbers we have pictured are called the set of **integers**. Integers to the left of 0 are called **negative integers**; integers to the right of 0 are called **positive integers**. The integer **0 is neither positive nor negative**.

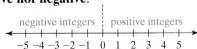

Integers

The set of **integers** is $\{ \ldots, -3, -2, -1, 0, 1, 2, 3, \ldots \}$.

Notice the ellipses (three dots) to the left and to the right of the list for the integers. This indicates that the positive integers and the negative integers continue indefinitely.

EXAMPLE 4

Use an integer to express the number in the following. "Pole of Inaccessibility, Antarctica, is the coldest location in the world, with an average annual temperature of 72 degrees below zero." (*Source: The Guinness Book of Records*)

Solution   The integer -72 represents 72 degrees below zero.

A problem with integers in real-life settings arises when quantities are smaller than some integer but greater than the next smallest integer. On a number line, these quantities may be visualized by points between integers. Some of these quantities between integers can be represented as a quotient of integers. For example,

The point on a number line halfway between 0 and 1 can be represented by $\frac{1}{2}$, a quotient of integers.

The point on a number line halfway between 0 and -1 can be represented by $-\frac{1}{2}$. Other quotients of integers and their graphs are shown.

The set numbers, each of which can be represented as a quotient of integers, is called the set of **rational numbers**. Notice that every integer is also a rational number since each integer can be expressed as a quotient of integers. For example, the integer 5 is also a rational number since $5 = \frac{5}{1}$.

> ### Rational Numbers
>
> The set of **rational numbers** is the set of all numbers that can be expressed as a quotient of integers with denominator not zero.

The number line also contains points that cannot be expressed as quotients of integers. These numbers are called **irrational numbers** because they cannot be represented by rational numbers. For example, $\sqrt{2}$ and π are irrational numbers.

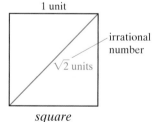

square

> ### Irrational Numbers
>
> The set of **irrational numbers** is the set of all numbers that correspond to points on the number line but that are not rational numbers. That is, an irrational number is a number that cannot be expressed as a quotient of integers.

Rational numbers and irrational numbers can be written as decimal numbers. The decimal equivalent of a rational number will either terminate or repeat in a pattern. For example, upon dividing we find that

$$\frac{3}{4} = 0.75 \text{ (decimal number terminates or ends) and}$$

$$\frac{2}{3} = 0.66666\ldots \text{ (decimal number repeats in a pattern)}$$

The decimal representation of an irrational number will neither terminate nor repeat. For example, the decimal representations of irrational numbers $\sqrt{2}$ and π are

$$\sqrt{2} = 1.414213562\ldots \text{ (decimal number does not terminate or repeat in a pattern)}$$

$$\pi = 3.141592653\ldots \text{ (decimal number does not terminate or repeat in a pattern)}$$

(For further review of decimals, see the Appendix.)

Combining the natural numbers with the irrational numbers gives the set of **real numbers**. One and only one point on a number line corresponds to each real number.

Real Numbers

The set of **real numbers** is the set of all numbers each of which corresponds to a point on a number line.

On the following number line, we see that real numbers can be positive, negative, or 0. Numbers to the left of 0 are called **negative numbers**; numbers to the right of 0 are called **positive numbers**. Positive and negative numbers are also called **signed numbers**.

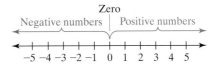

Several different sets of numbers have been discussed in this section. The following diagram shows the relationships among these sets of real numbers.

Common Sets of Numbers

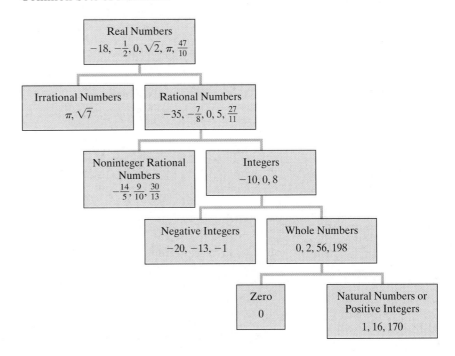

STUDY SKILLS REMINDER

Drawing a diagram, like the one shown on the previous page, for yourself in your study notes will help you understand and remember mathematical relationships.

EXAMPLE 5

Given the set $\left\{-2, 0, \frac{1}{4}, -1.5, 112, -3, 11, \sqrt{2}\right\}$, list the numbers in this set that belong to the set of:

a. Natural numbers   **b.** Whole numbers   **c.** Integers

d. Rational numbers   **e.** Irrational numbers   **f.** Real numbers

Solution

a. The natural numbers are 11 and 112.

b. The whole numbers are $0, 11,$ and $112.$

c. The integers are $-3, -2, 0, 11,$ and $112.$

d. Recall that integers are rational numbers also. The rational numbers are $-3, -2, -1.5, 0, \frac{1}{4}, 11,$ and $112.$

e. The irrational number is $\sqrt{2}.$

f. The real numbers are all numbers in the given set.

We can now extend the meaning and use of inequality symbols such as $<$ and $>$ to apply to all real numbers.

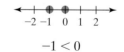

Order Property for Real Numbers

Given any two real numbers a and b, $a < b$ if a is to the left of b on a number line. Similarly, $a > b$ if a is to the right of b on a number line.

$$a < b \qquad\qquad a > b$$

$$\overset{a\quad\quad b}{\longleftrightarrow} \qquad\qquad \overset{b\quad\quad a}{\longleftrightarrow}$$

EXAMPLE 6

Insert $<, >,$ or $=$ in the appropriate space to make each statement true.

a. $-1 \quad\ 0$ **b.** $7 \quad \frac{14}{2}$ **c.** $-5 \quad\ -6$

Solution

a. $-1 < 0$ since -1 is to the left of 0 on a number line.

$$\overset{}{\underset{-2\;-1\;\;0\;\;1\;\;2}{\longleftrightarrow}}$$

$$-1 < 0$$

b. $7 = \frac{14}{2}$ since $\frac{14}{2}$ simplifies to 7.

c. $-5 > -6$ since -5 is to the right of -6 on the number line.

$$\overset{}{\underset{-7\;-6\;-5\;-4\;-3}{\longleftrightarrow}}$$

$$-5 > -6$$

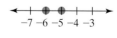

4   A number line not only gives us a picture of the real numbers, it also helps us visualize the distance between numbers. The distance between a real number a and 0 is given a special name called the **absolute value** of a. "The absolute value of a" is written in symbols as $|a|$.

> ## Absolute Value
>
> The absolute value of a real number a, denoted by $|a|$, is the distance between a and 0 on a number line.

For example, $|3| = 3$ and $|-3| = 3$ since both 3 and -3 are a distance of 3 units from 0 on a number line.

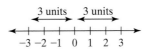

> ▶ **Helpful Hint**
> Since $|a|$ is a distance, $|a|$ is always either positive or 0, never negative. That is, **for any real number a, $|a| \geq 0$.**

 EXAMPLE 7

Find the absolute value of each number.

a. $|4|$ **b.** $|-5|$ **c.** $|0|$ **d.** $\left|-\frac{1}{2}\right|$ **e.** $|5.6|$

Solution   **a.** $|4| = 4$ since 4 is 4 units from 0 on a number line.

b. $|-5| = 5$ since -5 is 5 units from 0 on a number line.

c. $|0| = 0$ since 0 is 0 units from 0 on a number line.

d. $\left|-\frac{1}{2}\right| = \frac{1}{2}$ since $-\frac{1}{2}$ is $\frac{1}{2}$ units from 0 on a number line.

e. $|5.6| = 5.6$ since 5.6 is 5.6 units from 0 on a number line.

CLASSROOM EXAMPLE
Find the absolute value of each number.
a. $|7|$ b. $|-8|$
c. $\left|-\frac{2}{3}\right|$
answer: a. 7 b. 8 c. $\frac{2}{3}$

EXAMPLE 8

Insert $<$, $>$, or $=$ in the appropriate space to make each statement true.

a. $|0|$ 2   **b.** $|-5|$ 5   **c.** $|-3|$ $|-2|$ **d.** $|5|$ $|6|$ **e.** $|-7|$ $|6|$

Solution   **a.** $|0| < 2$ since $|0| = 0$ and $0 < 2$.

b. $|-5| = 5$ since $5 = 5$.

c. $|-3| > |-2|$ since $3 > 2$.

d. $|5| < |6|$ since $5 < 6$.

e. $|-7| > |6|$ since $7 > 6$.

CLASSROOM EXAMPLE
Insert $<$, $>$, or $=$ in the appropriate space to make each statement true.
a. $|-4|$ 4   b. -3 $|0|$
c. $|-2.7|$ $|-2|$
answer: a. $=$ b. $<$ c. $>$

Spotlight on

DECISION

of MAKING

Suppose you are a quality control engineering technician in a factory that makes machine screws. You have just helped to install programmable machinery on the production line that measures the length of each screw. If a screw's length is greater than 4.05 centimeters or less than or equal to 3.98 centimeters, the machinery is programmed to discard the screw. To check that the machinery works properly, you test six screws with known lengths. The results of the test are displayed. Is the new machinery working properly? Explain.

TEST RESULTS

Test Screw	Actual Length of Test Screw (cm)	Machine Action on Test Screw
A	4.03	Accept
B	3.96	Reject
C	4.05	Accept
D	4.08	Reject
E	3.98	Reject
F	4.01	Accept

EXERCISE SET 1.2

STUDY GUIDE/SSM · CD/ VIDEO · PH MATH TUTOR CENTER · MathXL®Tutorials ON CD · MathXL® · MyMathLab®

Insert <, >, *or* = *in the appropriate space to make the statement true. See Example 1.*

 1. 7 > 3

2. 9 < 15

3. 6.26 = 6.26

4. 2.13 > 1.13

5. 0 < 7

6. 20 > 0

7. −2 < 2

8. −4 > −6

9. The freezing point of water is 32° Fahrenheit. The boiling point of water is 212° Fahrenheit. Write an inequality statement using < or > comparing the numbers 32 and 212.   32 < 212

10. The freezing point of water is 0° Celsius. The boiling point of water is 100° Celsius. Write an inequality statement using < or > comparing the numbers 0 and 100.   0 < 100

11. The average salary in the United States for an experienced registered nurse is $44,300. The average salary for a drafter is $34,611. Write an inequality statement using < or > comparing the numbers 44,300 and 34,611. (*Source*: U.S. Department of Labor)   44,300 > 34,611

12. The state of New York is home to 312 institutions of higher learning. California claims a total of 384 colleges and universities. Write an inequality statement using < or > comparing the numbers 312 and 384. (*Source*: U.S. Department of Education)   312 < 384

Are the following statements true or false? See Example 2.

13. 11 ≤ 11   true

14. 4 ≥ 7   false

15. 10 > 11   false

16. 17 > 16   true

17. 3 + 8 ≥ 3(8)   false

18. 8·8 ≤ 8·7   false

19. 7 > 0   true

20. 4 < 7   true

△ **21.** An angle measuring 30° is shown and an angle measuring 45° is shown. Use the inequality symbol ≤ or ≥ to write a statement comparing the numbers 30 and 45.   30 ≤ 45

△ **22.** The sum of the measures of the angles of a triangle is 180°. The sum of the measures of the angles of a parallelogram is 360°. Use the inequality symbol ≤ or ≥ to write a statement comparing the numbers 360 and 180.   360 ≥ 180

Write each sentence as a mathematical statement. See Example 3.

23. Eight is less than twelve.   8 < 12

24. Fifteen is greater than five.   15 > 5

25. Five is greater than or equal to four.   5 ≥ 4

26. Negative ten is less than or equal to thirty-seven.   −10 ≤ 37

27. Fifteen is not equal to negative two.   15 ≠ −2

28. Negative seven is not equal to seven.   −7 ≠ 7

Use integers to represent the values in each statement. See Example 4.

29. Driskill Mountain, in Louisiana, has an altitude of 535 feet. New Orleans, Louisiana, lies 8 feet below sea level. (*Source:* U.S. Geological Survey)   535; −8

30. During a Green Bay Packers football game, the team gained 23 yards and then lost 12 yards on consecutive plays.   23; −12

31. From 1990 to 2000, the population of Washington, D.C. decreased by 34,841. (*Source: 2003 World Almanac*)   −34,841

32. From 1990 to 2000, the population of the state of Alaska rose by 406,513. (*Source: 2003 World Almanac*)   406,513

33. Aaron Miller deposited $350 in his savings account. He later withdrew $126.   350; −126

34. Aris Peña was deep-sea diving. During her dive, she ascended 30 feet and later descended 50 feet.   30; −50

The graph below is called a bar graph. This particular graph shows the average monthly home mortgage payment in the United States. Each bar represents a different year and the height of the bar represents the mortgage payment (principal and interest).

U.S. Average Monthly Mortgage Payments
(principal and interest)

Source: National Association of Realtors

35. In which year was the average mortgage payment the least? 1993

36. What is the greatest mortgage payment shown? $827

37. In what years was the average mortgage payment less than $600?   1993, 1994

38. In what years was the average mortgage payment greater than $800?   2000, 2002

39. Write an inequality statement using ≤ or ≥ comparing the average mortgage payment for 2002 and 2000.   827 ≥ 818

40. Do you notice any trends shown by this bar graph?
answers may vary

Tell which set or sets each number belongs to: natural numbers, whole numbers, integers, rational numbers, irrational numbers, and real numbers. See Example 5.

41. 0   **42.** $\frac{1}{4}$ **43.** −2   **44.** $-\frac{1}{2}$

45. 6   **46.** 5   **47.** $\frac{2}{3}$ **48.** $\sqrt{3}$

49. $-\sqrt{5}$ **50.** $-1\frac{5}{9}$ **48.–49.** irrational, real

41. whole, integers, rational, real   **42, 44, 47, 50.** rational, real   **43.** integers, rational, real   **45.–46.** natural, whole, integers, rational, real

Tell whether each statement is true or false.

51. Every rational number is also an integer.   false

52. Every negative number is also a rational number.   false

53. Every natural number is positive.   true

54. Every rational number is also a real number.   true

55. 0 is a real number.   true

56. Every real number is also a rational number.   false

57. Every whole number is an integer.   true

58. $\frac{1}{2}$ is an integer.   false

59. A number can be both rational and irrational.   false

60. Every whole number is positive.   false

Insert <, >, or = in the appropriate space to make a true statement. See Examples 6 through 8.

61. −10 > −100   **62.** −200 < −20   **63.** 32 > 5.2

64. 7 > −7   **65.** $\frac{18}{3} < \frac{24}{3}$ **66.** $\frac{8}{2} = \frac{12}{3}$

67. −51 < −50   **68.** |−20| > −200   **69.** |−5| > −4

70. 0 = |0| **71.** |−1| = |1| **72.** $\left|\frac{2}{5}\right| = \left|-\frac{2}{5}\right|$

73. |−2| < |−3| **74.** −500 < |−50| **75.** |0| < |−8|

76. $|−12| = \frac{24}{2}$

Concept Extensions

The apparent magnitude of a star is the measure of its brightness as seen by someone on Earth. The smaller the apparent magnitude, the brighter the star. Use the apparent magnitudes in the table to answer Exercises 77 through 82.

Star	Apparent Magnitude	Star	Apparent Magnitude
Arcturus	−0.04	Spica	0.98
Sirius	−1.46	Rigel	0.12
Vega	0.03	Regulus	1.35
Antares	0.96	Canopus	−0.72
Sun	−26.7	Hadar	0.61

(*Source: Norton's 2000: Star Atlas and Reference Handbook*, 18th ed., Longman Group, UK, 1989)

77. The apparent magnitude of the sun is −26.7. The apparent magnitude of the star Arcturus is −0.04. Write an inequality statement comparing the numbers −0.04 and −26.7.

78. The apparent magnitude of Antares is 0.96. The apparent magnitude of Spica is 0.98. Write an inequality statement comparing the numbers 0.96 and 0.98. $0.96 < 0.98$

79. Which is brighter, the sun or Arcturus?   sun

80. Which is dimmer, Antares or Spica?   Spica

81. Which star listed is the brightest?   sun

82. Which star listed is the dimmest?   Regulus

77. $-0.04 > -26.7$

Rewrite the following inequalities so that the inequality symbol points in the opposite direction and the resulting statement has the same meaning as the given one.

83. $25 \geq 20$ $20 \leq 25$

84. $-13 \leq 13$ $13 \geq -13$

85. $0 < 6$ $6 > 0$

86. $5 > 3$ $3 < 5$

87. $-10 > -12$ $-12 < -10$

88. $-4 < -2$ $-2 > -4$

89. In your own words, explain how to find the absolute value of a number.   answers may vary

90. Give an example of a real-life situation that can be described with integers but not with whole numbers.   answers may vary

---

1.3   FRACTIONS

Objectives

1 Write fractions in simplest form.

2 Multiply and divide fractions.

3 Add and subtract fractions.

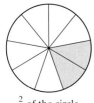

$\frac{2}{9}$ of the circle is shaded.

1   A quotient of two numbers such as $\frac{2}{9}$ is called a **fraction**. In the fraction $\frac{2}{9}$, the top number, 2, is called the **numerator** and the bottom number, 9, is called the **denominator**.

A fraction may be used to refer to part of a whole. For example, $\frac{2}{9}$ of the circle below is shaded. The denominator 9 tells us how many equal parts the whole circle is divided into and the numerator 2 tells us how many equal parts are shaded.

To simplify fractions, we can factor the numerator and the denominator. In the statement $3 \cdot 5 = 15$, 3 and 5 are called **factors** and 15 is the **product**. (The raised dot symbol indicates multiplication.)

$$
\begin{array}{ccccc}
3 & \cdot & 5 & = & 15 \\
\uparrow & & \uparrow & & \uparrow \\
\text{factor} & & \text{factor} & & \text{product}
\end{array}
$$

To **factor** 15 means to write it as a product. The number 15 can be factored as $3 \cdot 5$ or as $1 \cdot 15$.

A fraction is said to be **simplified** or in **lowest terms** when the numerator and the denominator have no factors in common other than 1. For example, the fraction $\frac{5}{11}$ is in lowest terms since 5 and 11 have no common factors other than 1.

To help us simplify fractions, we write the numerator and the denominator as a product of **prime numbers**.

Prime Number

A prime number is a whole number, other than 1, whose only factors are 1 and itself. The first few prime numbers are

$$2, 3, 5, 7, 11, 13, 17, 19, 23, 29, \text{ and so on.}$$

A natural number, other than 1, that is not a prime number is called a **composite number**. Every composite number can be written as a product of prime numbers. We call this product of prime numbers the prime factorization of the composite number.

EXAMPLE 1

Write each of the following numbers as a product of primes.

a. 40   **b.** 63

Solution   **a.** First, write 40 as the product of any two whole numbers, other than 1.

$$40 = 4 \cdot 10$$

Next, factor each of these numbers. Continue this process until all of the factors are prime numbers.

$$40 = \overset{\frown}{4} \cdot \overset{\frown}{10}$$
$$= 2 \cdot 2 \cdot 2 \cdot 5$$

All the factors are now prime numbers. Then 40 written as a product of primes is

$$40 = 2 \cdot 2 \cdot 2 \cdot 5$$

b. $63 = \overset{\frown}{9} \cdot 7$
$$= 3 \cdot 3 \cdot 7$$

CLASSROOM EXAMPLE
Write 60 as a product of primes.
answer: $2 \cdot 2 \cdot 3 \cdot 5$

TEACHING TIP

Help students understand that it makes no difference which two factors they start with. The resulting prime factorization is the same. For example:

$$40 = 5 \cdot 8$$
$$= 5 \cdot 2 \cdot 4$$
$$= 5 \cdot 2 \cdot 2 \cdot 2$$

To use prime factors to write a fraction in lowest terms, apply the fundamental principle of fractions.

Fundamental Principle of Fractions

If $\dfrac{a}{b}$ is a fraction and c is a nonzero real number, then

$$\frac{a \cdot c}{b \cdot c} = \frac{a}{b}$$

To understand why this is true, we use the fact that since c is not zero, then $\dfrac{c}{c} = 1$.

$$\frac{a \cdot c}{b \cdot c} = \frac{a}{b} \cdot \frac{c}{c} = \frac{a}{b} \cdot 1 = \frac{a}{b}$$

We will call this process dividing out the common factor of c.

EXAMPLE 2

Write each fraction in lowest terms.

a. $\dfrac{42}{49}$ **b.** $\dfrac{11}{27}$ **c.** $\dfrac{88}{20}$

Solution   **a.** Write the numerator and the denominator as products of primes; then apply the fundamental principle to the common factor 7.

$$\frac{42}{49} = \frac{2 \cdot 3 \cdot 7}{7 \cdot 7} = \frac{2 \cdot 3}{7} = \frac{6}{7}$$

b. $\dfrac{11}{27} = \dfrac{11}{3 \cdot 3 \cdot 3}$

There are no common factors other than 1, so $\dfrac{11}{27}$ is already in lowest terms.

c. $\dfrac{88}{20} = \dfrac{2 \cdot 2 \cdot 2 \cdot 11}{2 \cdot 2 \cdot 5} = \dfrac{22}{5}$

CLASSROOM EXAMPLE
Write $\dfrac{20}{35}$ in lowest terms.
answer: $\dfrac{4}{7}$

✔ **CONCEPT CHECK**

Explain the error in the following steps.

a. $\dfrac{15}{55} = \dfrac{1\,5}{5\,5} = \dfrac{1}{5}$ **b.** $\dfrac{6}{7} = \dfrac{5+1}{5+2} = \dfrac{1}{2}$

2   To multiply two fractions, multiply numerator times numerator to obtain the numerator of the product; multiply denominator times denominator to obtain the denominator of the product.

Multiplying Fractions

$$\frac{a}{b} \cdot \frac{c}{d} = \frac{a \cdot c}{b \cdot d}, \qquad \text{if } b \neq 0 \text{ and } d \neq 0$$

EXAMPLE 3

Multiply $\dfrac{2}{15}$ and $\dfrac{5}{13}$. Write the product in lowest terms.

Solution

$$\frac{2}{15} \cdot \frac{5}{13} = \frac{2 \cdot 5}{15 \cdot 13}$$ Multiply numerators.
Multiply denominators.

CLASSROOM EXAMPLE

Multiply $\dfrac{2}{7} \cdot \dfrac{3}{10}$. Write the product in low-

est terms.
answer: $\dfrac{3}{35}$

Next, simplify the product by dividing the numerator and the denominator by any common factors.

$$= \frac{2 \cdot 5}{3 \cdot 5 \cdot 13}$$

$$= \frac{2}{39}$$

Before dividing fractions, we first define **reciprocals**. Two fractions are reciprocals of each other if their product is 1. For example $\frac{2}{3}$ and $\frac{3}{2}$ are reciprocals since $\frac{2}{3} \cdot \frac{3}{2} = 1$. Also, the reciprocal of 5 is $\frac{1}{5}$ since $5 \cdot \frac{1}{5} = \frac{5}{1} \cdot \frac{1}{5} = 1$.

To divide fractions, multiply the first fraction by the reciprocal of the second fraction.

Dividing Fractions

$$\frac{a}{b} \div \frac{c}{d} = \frac{a}{b} \cdot \frac{d}{c}, \qquad \text{if } b \neq 0, d \neq 0, \text{ and } c \neq 0$$

EXAMPLE 4

Divide. Write all quotients in lowest terms.

Concept Check Answer:
answers may vary

a. $\dfrac{4}{5} \div \dfrac{5}{16}$ **b.** $\dfrac{7}{10} \div 14$ **c.** $\dfrac{3}{8} \div \dfrac{3}{10}$

Solution   **a.** $\dfrac{4}{5} \div \dfrac{5}{16} = \dfrac{4}{5} \cdot \dfrac{16}{5} = \dfrac{4 \cdot 16}{5 \cdot 5} = \dfrac{64}{25}$

b. $\dfrac{7}{10} \div 14 = \dfrac{7}{10} \div \dfrac{14}{1} = \dfrac{7}{10} \cdot \dfrac{1}{14} = \dfrac{7 \cdot 1}{2 \cdot 5 \cdot 2 \cdot 7} = \dfrac{1}{20}.$

c. $\dfrac{3}{8} \div \dfrac{3}{10} = \dfrac{3}{8} \cdot \dfrac{10}{3} = \dfrac{3 \cdot 2 \cdot 5}{2 \cdot 2 \cdot 2 \cdot 3} = \dfrac{5}{4}$

3   To add or subtract fractions with the same denominator, combine numerators and place the sum or difference over the common denominator.

> ## Adding and Subtracting Fractions with the Same Denominator
>
> $$\dfrac{a}{b} + \dfrac{c}{b} = \dfrac{a + c}{b}, \qquad \text{if } b \neq 0$$
>
> $$\dfrac{a}{b} - \dfrac{c}{b} = \dfrac{a - c}{b}, \qquad \text{if } b \neq 0$$

EXAMPLE 5

Add or subtract as indicated. Write each result in lowest terms.

a. $\dfrac{2}{7} + \dfrac{4}{7}$ **b.** $\dfrac{3}{10} + \dfrac{2}{10}$ **c.** $\dfrac{9}{7} - \dfrac{2}{7}$ **d.** $\dfrac{5}{3} - \dfrac{1}{3}$

Solution   **a.** $\dfrac{2}{7} + \dfrac{4}{7} = \dfrac{2 + 4}{7} = \dfrac{6}{7}$ **b.** $\dfrac{3}{10} + \dfrac{2}{10} = \dfrac{3 + 2}{10} = \dfrac{5}{10} = \dfrac{5}{2 \cdot 5} = \dfrac{1}{2}$

c. $\dfrac{9}{7} - \dfrac{2}{7} = \dfrac{9 - 2}{7} = \dfrac{7}{7} = 1$ **d.** $\dfrac{5}{3} - \dfrac{1}{3} = \dfrac{5 - 1}{3} = \dfrac{4}{3}$

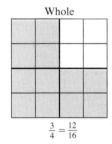

Whole

$\dfrac{3}{4} = \dfrac{12}{16}$

To add or subtract fractions without the same denominator, first write the fractions as **equivalent fractions** with a common denominator. Equivalent fractions are fractions that represent the same quantity. For example, $\frac{3}{4}$ and $\frac{12}{16}$ are equivalent fractions since they represent the same portion of a whole, as the diagram shows. Count the larger squares and the shaded portion is $\frac{3}{4}$. Count the smaller squares and the shaded portion is $\frac{12}{16}$. Thus, $\frac{3}{4} = \frac{12}{16}$.

We can write equivalent fractions by multiplying a given fraction by 1, as shown in the next example. Multiplying a fraction by 1 does not change the value of the fraction.

EXAMPLE 6

Write $\dfrac{2}{5}$ as an equivalent fraction with a denominator of 20.

Solution   Since $5 \cdot 4 = 20$, multiply the fraction by $\dfrac{4}{4}$. Multiplying by $\dfrac{4}{4} = 1$ does not change the value of the fraction.

Multiply by $\dfrac{4}{4}$ or **1**.

$$\dfrac{2}{5} = \dfrac{2}{5} \cdot \dfrac{4}{4} = \dfrac{2 \cdot 4}{5 \cdot 4} = \dfrac{8}{20}$$

EXAMPLE 7

Add or subtract as indicated. Write each answer in lowest terms.

a. $\dfrac{2}{5} + \dfrac{1}{4}$ **b.** $\dfrac{1}{2} + \dfrac{17}{22} - \dfrac{2}{11}$ **c.** $3\dfrac{1}{6} - 1\dfrac{11}{12}$

Solution

a. Fractions must have a common denominator before they can be added or subtracted. Since 20 is the smallest number that both 5 and 4 divide into evenly, 20 is the **least common denominator**. Write both fractions as equivalent fractions with denominators of 20. Since

$$\frac{2}{5} \cdot \frac{4}{4} = \frac{2 \cdot 4}{5 \cdot 4} = \frac{8}{20} \quad \text{and} \quad \frac{1}{4} \cdot \frac{5}{5} = \frac{1 \cdot 5}{4 \cdot 5} = \frac{5}{20}$$

then

$$\frac{2}{5} + \frac{1}{4} = \frac{8}{20} + \frac{5}{20} = \frac{13}{20}$$

b. The least common denominator for denominators 2, 22, and 11 is 22. First, write each fraction as an equivalent fraction with a denominator of 22. Then add or subtract from left to right.

$$\frac{1}{2} = \frac{1}{2} \cdot \frac{11}{11} = \frac{11}{22}, \quad \frac{17}{22} = \frac{17}{22}, \quad \text{and} \quad \frac{2}{11} = \frac{2}{11} \cdot \frac{2}{2} = \frac{4}{22}$$

Then

$$\frac{1}{2} + \frac{17}{22} - \frac{2}{11} = \frac{11}{22} + \frac{17}{22} - \frac{4}{22} = \frac{24}{22} = \frac{12}{11}$$

c. To find $3\dfrac{1}{6} - 1\dfrac{11}{12}$, lets use a vertical format.

$$
\begin{array}{ccccc}
 & & & \overbrace{\phantom{2+1\frac{2}{12}}}^{2+1\frac{2}{12}} & \\
3\dfrac{1}{6} & = & 3\dfrac{2}{12} & = & 2\dfrac{14}{12} \\[6pt]
-1\dfrac{11}{12} & = & -1\dfrac{11}{12} & = & -1\dfrac{11}{12} \\ \hline
 & & \underset{\text{Need to borrow}}{\uparrow} & & 1\dfrac{3}{12} \text{ or } 1\dfrac{1}{4}
\end{array}
$$

Spotlight on **DECISION MAKING**

Suppose you are fishing on a freshwater lake in Canada. You catch a whitefish weighing $14\frac{5}{32}$ pounds. According to the International Game Fish Association, the world's record for largest lake whitefish ever caught is $14\frac{3}{8}$ pounds. Did you set a new world's record? Explain. By how much did you beat or miss the existing world record?

CLASSROOM EXAMPLE
Add or subtract as indicated.

a. $\dfrac{3}{8} + \dfrac{1}{20}$ **b.** $18\dfrac{1}{4} - 6\dfrac{2}{3}$

answer: **a.** $\dfrac{17}{40}$ **b.** $11\dfrac{7}{12}$

TEACHING TIP

Once you have reviewed all four operations on fractions separately, ask students how they will perform each operation:

$\dfrac{1}{2} \cdot \dfrac{1}{3}$

$\dfrac{1}{2} \div \dfrac{1}{3}$

$\dfrac{1}{2} + \dfrac{1}{3}$

$\dfrac{1}{2} - \dfrac{1}{3}$

TEACHING TIP
A Group Activity for this section is available in the Instructor's Resource Manual.

MENTAL MATH

Represent the shaded part of each geometric figure by a fraction.

1. $\frac{3}{8}$ **2.** $\frac{1}{4}$ **3.** 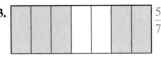 $\frac{5}{7}$ **4.** 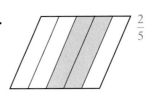 $\frac{2}{5}$

For Exercises 5 and 6, fill in the blank.

5. In the fraction $\frac{3}{5}$, 3 is called the <u>numerator</u> and 5 is called the <u>denominator</u>.

6. The reciprocal of $\frac{7}{11}$ is $\frac{11}{7}$.

EXERCISE SET 1.3

STUDY   CD/   PH MATH   MathXL®Tutorials   MathXL®   MyMathLab®
GUIDE/SSM   VIDEO   TUTOR CENTER   ON CD

1. $3 \cdot 11$ **2.** $2 \cdot 2 \cdot 3 \cdot 5$ **3.** $2 \cdot 7 \cdot 7$ **4.** $3 \cdot 3 \cdot 3$ **5.** $2 \cdot 2 \cdot 5$ **6.** $2 \cdot 2 \cdot 2 \cdot 7$ **7.** $3 \cdot 5 \cdot 5$ **8.** $2 \cdot 2 \cdot 2 \cdot 2 \cdot 2$ **9.** $3 \cdot 3 \cdot 5$ **10.** $2 \cdot 2 \cdot 2 \cdot 3$

Write each number as a product of primes. See Example 1.
1. 33   **2.** 60   **3.** 98   **4.** 27   **5.** 20
6. 56   **7.** 75   **8.** 32   **9.** 45   **10.** 24

Write the fraction in lowest terms. See Example 2.
11. $\frac{2}{4}$ $\frac{1}{2}$ **12.** $\frac{3}{6}$ $\frac{1}{2}$ **13.** $\frac{10}{15}$ $\frac{2}{3}$ **14.** $\frac{15}{20}$ $\frac{3}{4}$
15. $\frac{3}{7}$ $\frac{3}{7}$ **16.** $\frac{5}{9}$ $\frac{5}{9}$ **17.** $\frac{18}{30}$ $\frac{3}{5}$ **18.** $\frac{42}{45}$ $\frac{14}{15}$

Multiply or divide as indicated. Write the answer in lowest terms. See Examples 3 and 4.
19. $\frac{1}{2} \cdot \frac{3}{4}$ $\frac{3}{8}$ **20.** $\frac{10}{6} \cdot \frac{3}{5}$ 1 **21.** $\frac{2}{3} \cdot \frac{3}{4}$ $\frac{1}{2}$
22. $\frac{7}{8} \cdot \frac{3}{21}$ $\frac{1}{8}$ **23.** $\frac{1}{2} \div \frac{7}{12}$ $\frac{6}{7}$ **24.** $\frac{7}{12} \div \frac{1}{2}$ $\frac{7}{6}$
25. $\frac{3}{4} \div \frac{1}{20}$ 15 **26.** $\frac{3}{5} \div \frac{9}{10}$ $\frac{2}{3}$ **27.** $\frac{7}{10} \cdot \frac{5}{21}$ $\frac{1}{6}$
28. $\frac{3}{35} \cdot \frac{10}{63}$ $\frac{2}{147}$ **29.** $2\frac{7}{9} \cdot \frac{1}{3}$ $\frac{25}{27}$ **30.** $\frac{1}{4} \cdot 5\frac{5}{6}$ $1\frac{11}{24}$

The area of a plane figure is a measure of the amount of surface of the figure. Find the area of each figure below. (The area of a rectangle is the product of its length and width. The area of a triangle is $\frac{1}{2}$ the product of its base and height.) **31.** $\frac{11}{20}$ sq. mi   **32.** $\frac{5}{16}$ sq. m

△ **31.**   △ **32.**

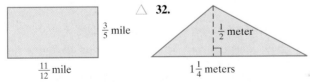

Add or subtract as indicated. Write the answer in lowest terms. See Example 5.
33. $\frac{4}{5} - \frac{1}{5}$ $\frac{3}{5}$ **34.** $\frac{6}{7} - \frac{1}{7}$ $\frac{5}{7}$ **35.** $\frac{4}{5} + \frac{1}{5}$ 1
36. $\frac{6}{7} + \frac{1}{7}$ 1 **37.** $\frac{17}{21} - \frac{10}{21}$ $\frac{1}{3}$ **38.** $\frac{18}{35} - \frac{11}{35}$ $\frac{1}{5}$

39. $\frac{23}{105} + \frac{4}{105}$ $\frac{9}{35}$ **40.** $\frac{13}{132} + \frac{35}{132}$ $\frac{4}{11}$

Write each fraction as an equivalent fraction with the given denominator. See Example 6.
41. $\frac{7}{10}$ with a denominator of 30 $\frac{21}{30}$
42. $\frac{2}{3}$ with a denominator of 9 $\frac{6}{9}$
43. $\frac{2}{9}$ with a denominator of 18 $\frac{4}{18}$
44. $\frac{8}{7}$ with a denominator of 56 $\frac{64}{56}$
45. $\frac{4}{5}$ with a denominator of 20 $\frac{16}{20}$
46. $\frac{4}{5}$ with a denominator of 25 $\frac{20}{25}$

Add or subtract as indicated. Write the answer in lowest terms. See Example 7.
47. $\frac{2}{3} + \frac{3}{7}$ $\frac{23}{21}$ **48.** $\frac{3}{4} + \frac{1}{6}$ $\frac{11}{12}$ **49.** $2\frac{13}{15} - 1\frac{1}{5}$ $1\frac{2}{3}$
50. $5\frac{2}{9} - 3\frac{1}{6}$ $2\frac{1}{18}$ **51.** $\frac{5}{22} - \frac{5}{33}$ $\frac{5}{66}$ **52.** $\frac{7}{10} - \frac{8}{15}$ $\frac{1}{6}$
53. $\frac{12}{5} - 1$ $\frac{7}{5}$ **54.** $2 - \frac{3}{8}$ $\frac{13}{8}$

Each circle below represents a whole, or 1. Use subtraction to determine the unknown part of the circle.

55. 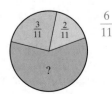 $\frac{1}{5}$ **56.** $\frac{6}{11}$

57. 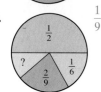 $\frac{3}{8}$

58. 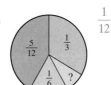 $\frac{1}{12}$

59. 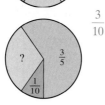 $\frac{1}{9}$

60. $\frac{3}{10}$

MIXED PRACTICE

Perform the following operations. Write answers in lowest terms.

61. $\frac{10}{21} + \frac{5}{21}$ $\frac{5}{7}$

62. $\frac{11}{35} + \frac{3}{35}$ $\frac{2}{5}$

 63. $\frac{10}{3} - \frac{5}{21}$ $\frac{65}{21}$

64. $\frac{11}{7} - \frac{3}{35}$ $\frac{52}{35}$

65. $\frac{2}{3} \cdot \frac{3}{5}$ $\frac{2}{5}$

66. $\frac{2}{3} \div \frac{3}{4}$ $\frac{8}{9}$

67. $\frac{3}{4} \div \frac{7}{12}$ $\frac{9}{7}$

68. $\frac{3}{5} + \frac{2}{3}$ $\frac{19}{15}$

69. $\frac{5}{12} + \frac{4}{12}$ $\frac{3}{4}$

70. $\frac{2}{7} + \frac{4}{7}$ $\frac{6}{7}$

71. $5 + \frac{2}{3}$ $\frac{17}{3}$

72. $7 + \frac{1}{10}$ $\frac{71}{10}$

73. $\frac{7}{8} \div 3\frac{1}{4}$ $\frac{7}{26}$

74. $3 \div \frac{3}{4}$ 4

75. $\frac{7}{18} \div \frac{14}{36}$ 1

76. $4\frac{3}{7} \div \frac{31}{7}$ 1

77. $\frac{23}{105} - \frac{2}{105}$ $\frac{1}{5}$

78. $\frac{57}{132} - \frac{13}{132}$ $\frac{1}{3}$

 79. $1\frac{1}{2} + 3\frac{2}{3}$ $5\frac{1}{6}$

80. $2\frac{3}{5} + 4\frac{7}{10}$ $7\frac{3}{10}$

81. $\frac{2}{3} - \frac{5}{9} + \frac{5}{6}$ $\frac{17}{18}$

82. $\frac{8}{11} - \frac{1}{4} + \frac{1}{2}$ $\frac{43}{44}$

The perimeter of a plane figure is the total distance around the figure. Find the perimeter of each figure in Exercises 83 and 84.

△ **83.** $55\frac{1}{4}$ ft

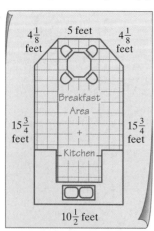

△ **84.** $57\frac{3}{4}$ ft

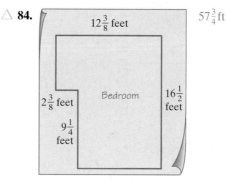

85. Svetlana Feofanova currently holds the Women's Indoor Pole Vault world record at $4\frac{3}{4}$ meters. The Men's Indoor Pole Vault world record is currently held by Sergei Bubka, at $1\frac{2}{5}$ meters higher than the Women's record. What is the current Men's Indoor Pole Vault Record? (*Source: 2003 World Almanac*) $6\frac{3}{20}$ m

86. In March 1999, a proposal to increase the size of rectangular escape vents in lobster traps was brought before the Maine state legislature. The proposed change would add $\frac{1}{16}$ of an inch to the current vent height of $1\frac{7}{8}$ inches. What would be the new vent height under the proposal? (*Source: The Boston Sunday Globe*, April 4, 1999) $1\frac{15}{16}$ in.

87. In your own words, explain how to add two fractions with different denominators.   answers may vary

88. In your own words, explain how to multiply two fractions.   answers may vary

The following trail chart is given to visitors at the Lakeview Forest Preserve.

Trail Name	Distance (miles)
Robin Path	$3\frac{1}{2}$
Red Falls	$5\frac{1}{2}$
Green Way	$2\frac{1}{8}$
Autumn Walk	$1\frac{3}{4}$

89. How much longer is Red Falls Trail than Green Way Trail?

90. Find the total distance traveled by someone who hiked along all four trails.   **89.** $3\frac{3}{8}$ mi   **90.** $12\frac{7}{8}$ mi

Concept Extensions

The breakdown of science and engineering doctorate degrees awarded in the United States is summarized in the graph on the next page, called a circle graph or a pie chart. Use the graph to answer the questions. (Source: National Science Foundation)

91. What fraction of science and engineering doctorates are awarded in the physical sciences? $\frac{7}{50}$

92. Engineering doctorates make up what fraction of all science and engineering doctorates awarded in the United States? $\frac{21}{100}$

Science and Engineering Doctorates Awarded, by Field of Study

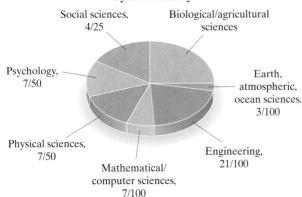

Social sciences, 4/25

Biological/agricultural sciences

Psychology, 7/50

Earth, atmospheric, ocean sciences, 3/100

Physical sciences, 7/50

Engineering, 21/100

Mathematical/ computer sciences, 7/100

93. What fraction of all science and engineering doctorates are awarded in the biological and agricultural sciences? $\frac{1}{4}$

94. Social sciences and psychology doctorates together make up what fraction of all science and engineering doctorates warded in the United States? $\frac{3}{10}$

As of February 2001, Gap Inc. operated a total of 3676 stores worldwide. The following chart shows the store breakdown by brand. (Source: Gap Inc.)

Brand	Number of Stores
Gap (Domestic)	2079
Gap (International)	529
Banana Republic	404
Old Navy	666
Total	3678

95. What fraction of Gap-brand stores were Old Navy stores? Simplify this fraction. $\frac{111}{613}$

96. What fraction of Gap-brand stores were either domestic or international Gap stores? Simplify this fraction. $\frac{1304}{1839}$

The area of a plane figure is a measure of the amount of surface of the figure. Find the area of each figure. (The area of a triangle is $\frac{1}{2}$ the product of its base and height. The area of a rectangle is the product of its length and width. Recall that area is measured in square units.)

△ **97.**

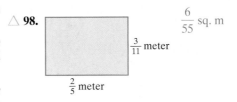

$\frac{4}{9}$ foot $\frac{7}{36}$ sq. ft

$\frac{7}{8}$ foot

△ **98.**

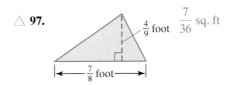

$\frac{6}{55}$ sq. m

$\frac{3}{11}$ meter

$\frac{2}{5}$ meter

1.4   INTRODUCTION TO VARIABLE EXPRESSIONS AND EQUATIONS

Objectives

1 Define and use exponents and the order of operations.

2 Evaluate algebraic expressions, given replacement values for variables.

3 Determine whether a number is a solution of a given equation.

4 Translate phrases into expressions and sentences into equations.

1   Frequently in algebra, products occur that contain repeated multiplication of the same factor. For example, the volume of a cube whose sides each measure 2 centimeters is $(2 \cdot 2 \cdot 2)$ cubic centimeters. We may use **exponential notation** to write such products in a more compact form. For example,

$$2 \cdot 2 \cdot 2 \quad \textit{may be written as} \quad 2^3.$$

2 cm

Volume is $(2 \cdot 2 \cdot 2)$
cubic centimeters.

The 2 in 2^3 is called the **base**; it is the repeated factor. The 3 in 2^3 is called the **exponent** and is the number of times the base is used as a factor. The expression 2^3 is called an **exponential expression**.

$$\underset{\text{base}}{\underbrace{\hspace{1cm}}} 2^3 = 2 \cdot 2 \cdot 2 = 8 \quad \overset{\text{exponent}}{}$$

2 is a factor 3 times

EXAMPLE 1

Evaluate the following:

a. 3^2 [read as "3 squared" or as "3 to second power"]

b. 5^3 [read as "5 cubed" or as "5 to the third power"]

c. 2^4 [read as "2 to the fourth power"]

d. 7^1 **e.** $\left(\dfrac{3}{7}\right)^2$

Solution

a. $3^2 = 3 \cdot 3 = 9$ **b.** $5^3 = 5 \cdot 5 \cdot 5 = 125$

c. $2^4 = 2 \cdot 2 \cdot 2 \cdot 2 = 16$ **d.** $7^1 = 7$

e. $\left(\dfrac{3}{7}\right)^2 = \left(\dfrac{3}{7}\right)\left(\dfrac{3}{7}\right) = \dfrac{9}{49}$

> **Helpful Hint**
> $2^3 \neq 2 \cdot 3$ since 2^3 indicates repeated **multiplication** of the same factor.
> $$2^3 = 2 \cdot 2 \cdot 2 = 8, \text{ whereas } 2 \cdot 3 = 6.$$

Using symbols for mathematical operations is a great convenience. However, the more operation symbols presented in an expression, the more careful we must be when performing the indicated operation. For example, in the expression $2 + 3 \cdot 7$, do we add first or multiply first? To eliminate confusion, **grouping symbols** are used. Examples of grouping symbols are parentheses ( ), brackets [ ], braces { }, and the fraction bar. If we wish $2 + 3 \cdot 7$ to be simplified by adding first, we enclose $2 + 3$ in parentheses.

$$(2 + 3) \cdot 7 = 5 \cdot 7 = 35$$

If we wish to multiply first, $3 \cdot 7$ may be enclosed in parentheses.

$$2 + (3 \cdot 7) = 2 + 21 = 23$$

To eliminate confusion when no grouping symbols are present, use the following agreed upon order of operations.

Order of Operations

Simplify expressions using the order below. If grouping symbols such as parentheses are present, simplify expressions within those first, starting with the innermost set. If fraction bars are present, simplify the numerator and the denominator separately.

1. Evaluate exponential expressions.
2. Perform multiplications or divisions in order from left to right.
3. Perform additions or subtractions in order from left to right.

Now simplify $2 + 3 \cdot 7$. There are no grouping symbols and no exponents, so we multiply and then add.

$$2 + 3 \cdot 7 = 2 + 21 \qquad \text{Multiply.}$$
$$ = 23 \qquad \text{Add.}$$

EXAMPLE 2

Simplify each expression.

a. $6 \div 3 + 5^2$ **b.** $\dfrac{2(12 + 3)}{|-15|}$ **c.** $3 \cdot 10 - 7 \div 7$ **d.** $3 \cdot 4^2$ **e.** $\dfrac{3}{2} \cdot \dfrac{1}{2} - \dfrac{1}{2}$

Solution

a. Evaluate 5^2 first.

$$6 \div 3 + 5^2 = 6 \div 3 + 25$$

Next divide, then add.

$$= 2 + 25 \qquad \text{Divide.}$$
$$= 27 \qquad \text{Add.}$$

b. First, simplify the numerator and the denominator separately.

$$\frac{2(12 + 3)}{|-15|} = \frac{2(15)}{15} \qquad \begin{array}{l}\text{Simplify numerator and}\\ \text{denominator separately.}\end{array}$$
$$= \frac{30}{15}$$
$$= 2 \qquad \text{Simplify.}$$

c. Multiply and divide from left to right. Then subtract.

$$3 \cdot 10 - 7 \div 7 = 30 - 1$$
$$= 29 \qquad \text{Subtract.}$$

d. In this example, only the 4 is squared. The factor of 3 is not part of the base because no grouping symbol includes it as part of the base.

$$3 \cdot 4^2 = 3 \cdot 16 \qquad \text{Evaluate the exponential expression.}$$
$$= 48 \qquad \text{Multiply.}$$

e. The order of operations applies to operations with fractions in exactly the same way as it applies to operations with whole numbers.

$$\frac{3}{2} \cdot \frac{1}{2} - \frac{1}{2} = \frac{3}{4} - \frac{1}{2} \qquad \text{Multiply.}$$
$$= \frac{3}{4} - \frac{2}{4} \qquad \text{The least common denominator is 4.}$$
$$= \frac{1}{4} \qquad \text{Subtract.}$$

> **Helpful Hint**
> Be careful when evaluating an exponential expression. In $3 \cdot 4^2$, the exponent 2 applies only to the base 4. In $(3 \cdot 4)^2$, we multiply first because of parentheses, so the exponent 2 applies to the product $3 \cdot 4$.
>
> $$3 \cdot 4^2 = 3 \cdot 16 = 48 \qquad (3 \cdot 4)^2 = (12)^2 = 144$$

Expressions that include many grouping symbols can be confusing. When simplifying these expressions, keep in mind that grouping symbols separate the expression into distinct parts. Each is then simplified separately.

EXAMPLE 3

Simplify $\dfrac{3 + |4 - 3| + 2^2}{6 - 3}$.

Solution   The fraction bar serves as a grouping symbol and separates the numerator and denominator. Simplify each separately. Also, the absolute value bars here serve as a grouping symbol. We begin in the numerator by simplifying within the absolute value bars.

$$\dfrac{3 + |4 - 3| + 2^2}{6 - 3} = \dfrac{3 + |1| + 2^2}{6 - 3} \qquad \text{Simplify the expression inside the absolute value bars.}$$

$$= \dfrac{3 + 1 + 2^2}{3} \qquad \text{Find the absolute value and simplify the denominator.}$$

$$= \dfrac{3 + 1 + 4}{3} \qquad \text{Evaluate the exponential expression.}$$

$$= \dfrac{8}{3} \qquad \text{Simplify the numerator.}$$

STUDY SKILLS REMINDER

Make a practice of neatly writing down enough steps so that you are comfortable with your computations.

EXAMPLE 4

Simplify $3[4 + 2(10 - 1)]$.

Solution   Notice that both parentheses and brackets are used as grouping symbols. Start with the innermost set of grouping symbols.

Helpful Hint

Be sure to follow order of operations and resist the temptation to incorrectly add 4 and 2 first.

$$3[4 + 2(10 - 1)] = 3[4 + 2(9)] \qquad \text{Simplify the expression in parentheses.}$$

$$= 3[4 + 18] \qquad \text{Multiply.}$$

$$= 3[22] \qquad \text{Add.}$$

$$= 66 \qquad \text{Multiply.}$$

EXAMPLE 5

Simplify $\dfrac{8 + 2 \cdot 3}{2^2 - 1}$.

Solution

$$\frac{8 + 2 \cdot 3}{2^2 - 1} = \frac{8 + 6}{4 - 1} = \frac{14}{3}$$

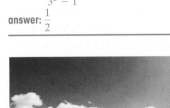

2   In algebra, we use symbols, usually letters such as x, y, or z, to represent unknown numbers. A symbol that is used to represent a number is called a **variable**. An **algebraic expression** is a collection of numbers, variables, operation symbols, and grouping symbols. For example,

$$2x, \qquad -3, \qquad 2x + 10, \qquad 5(p^2 + 1), \qquad \text{and} \qquad \frac{3y^2 - 6y + 1}{5}$$

are algebraic expressions. The expression $2x$ means $2 \cdot x$. Also, $5(p^2 + 1)$ means $5 \cdot (p^2 + 1)$ and $3y^2$ means $3 \cdot y^2$. If we give a specific value to a variable, we can **evaluate an algebraic expression**. To evaluate an algebraic expression means to find its numerical value once we know the values of the variables.

Algebraic expressions often occur during problem solving. For example, the expression

$$16t^2$$

gives the distance in feet (neglecting air resistance) that an object will fall in t seconds. (See Exercise 63 in this section.)

EXAMPLE 6

Evaluate each expression if $x = 3$ and $y = 2$.

   a. $2x - y$ **b.** $\dfrac{3x}{2y}$ **c.** $\dfrac{x}{y} + \dfrac{y}{2}$ **d.** $x^2 - y^2$

Solution   **a.** Replace x with 3 and y with 2.

$$
\begin{aligned}
2x - y &= 2(3) - 2 &&\text{Let } x = 3 \text{ and } y = 2. \\
&= 6 - 2 &&\text{Multiply.} \\
&= 4 &&\text{Subtract.}
\end{aligned}
$$

b. $\dfrac{3x}{2y} = \dfrac{3 \cdot 3}{2 \cdot 2} = \dfrac{9}{4}$ Let $x = 3$ and $y = 2$.

c. Replace x with 3 and y with 2. Then simplify.

$$\frac{x}{y} + \frac{y}{2} = \frac{3}{2} + \frac{2}{2} = \frac{5}{2}$$

d. Replace x with 3 and y with 2.

$$x^2 - y^2 = 3^2 - 2^2 = 9 - 4 = 5$$

3   Many times a problem-solving situation is modeled by an equation. An **equation** is a mathematical statement that two expressions have equal value. The equal symbol "$=$" is used to equate the two expressions. For example, $3 + 2 = 5$, $7x = 35$, $\dfrac{2(x - 1)}{3} = 0$, and $I = PRT$ are all equations.

> ## Helpful Hint
> An equation contains the equal symbol "=". An algebraic expression does not.

✔ **CONCEPT CHECK**

Which of the following are equations? Which are expressions?

a. $5x = 8$ **b.** $5x - 8$ **c.** $12y + 3x$ **d.** $12y = 3x$

When an equation contains a variable, deciding which values of the variable make an equation a true statement is called **solving** an equation for the variable. A **solution** of an equation is a value for the variable that makes the equation true. For example, 3 is a solution of the equation $x + 4 = 7$, because if x is replaced with 3 the statement is true.

$$x + 4 = 7$$
$$\downarrow$$
$$3 + 4 = 7 \qquad \text{Replace } x \text{ with 3.}$$
$$7 = 7 \qquad \text{True.}$$

Similarly, 1 is not a solution of the equation $x + 4 = 7$, because $1 + 4 = 7$ is **not** a true statement.

EXAMPLE 7

Decide whether 2 is a solution of $3x + 10 = 8x$.

Solution Replace x with 2 and see if a true statement results.

$$3x + 10 = 8x \qquad \text{Original equation}$$
$$3(2) + 10 \stackrel{?}{=} 8(2) \qquad \text{Replace } x \text{ with 2.}$$
$$6 + 10 \stackrel{?}{=} 16 \qquad \text{Simplify each side.}$$
$$16 = 16 \qquad \text{True.}$$

Since we arrived at a true statement after replacing x with 2 and simplifying both sides of the equation, 2 is a solution of the equation.

CLASSROOM EXAMPLE
Decide whether 3 is a solution of
$5x - 10 = x + 2$.
answer: yes

4 Now that we know how to represent an unknown number by a variable, let's practice translating phrases into algebraic expressions and sentences into equations. Oftentimes solving problems requires the ability to translate word phrases and sentences into symbols. Below is a list of some key words and phrases to help us translate.

TEACHING TIP
This is certainly an incomplete list of key words and phrases. Also, warn students that if a key word appears in a sentence, they must decide whether it translates to an operation. To see this, discuss the word "of" in these two examples. The sum of $\frac{2}{3}$ and 5. Find $\frac{2}{3}$ of 5.

ADDITION $(+)$	SUBTRACTION $(-)$	MULTIPLICATION $(\cdot)$	DIVISION $(\div)$	EQUALITY $(=)$
Sum	Difference of	Product	Quotient	Equals
Plus	Minus	Times	Divide	Gives
Added to	Subtracted from	Multiply	Into	Is/was/ should be
More than	Less than	Twice	Ratio	Yields
Increased by	Decreased by	Of	Divided by	Amounts to
Total	Less			Represents
				Is the same as

Concept Check Answer:
equations: a, d; expressions: b, c.

EXAMPLE 8

Write an algebraic expression that represents each phrase. Let the variable x represent the unknown number.

a. The sum of a number and 3

b. The product of 3 and a number

c. Twice a number

d. 10 decreased by a number

e. 5 times a number, increased by 7

TEACHING TIP
Tell students to be careful when translating phrases that include subtraction or division. With these operations, *order* makes a difference.

Solution

a. $x + 3$ since "sum" means to add

b. $3 \cdot x$ and $3x$ are both ways to denote the product of 3 and x

c. $2 \cdot x$ or $2x$

d. $10 - x$ because "decreased by" means to subtract

e. $\underbrace{5x}_{5 \text{ times a number}} + 7$

CLASSROOM EXAMPLE
Write an algebraic expression that represents each phrase. Let the variable x represent the unknown number.
a. The product of a number and 5
b. A number added to 7
c. Three times a number
d. A number subtracted from 8
e. Twice a number, plus 1
answer:
a. $5x$ **b.** $7 + x$ **c.** $3x$
d. $8 - x$ **e.** $2x + 1$

> **Helpful Hint**
> Make sure you understand the difference when translating phrases containing "decreased by," "subtracted from," and "less than."
>
> | Phrase | Translation |
> |---|---|
> | A number decreased by 10 | $x - 10$ |
> | A number subtracted from 10 | $10 - x$ |
> | 10 less than a number | $x - 10$ |
> | A number less 10 | $x - 10$ |
>
> Notice the order.

Now let's practice translating sentences into equations.

EXAMPLE 9

Write each sentence as an equation. Let x represent the unknown number.

a. The quotient of 15 and a number is 4.

b. Three subtracted from 12 is a number.

c. Four times a number, added to 17, is 21.

Solution

a. In words:

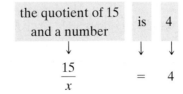

the quotient of 15 and a number	is	4
↓	↓	↓

Translate: $\dfrac{15}{x}$ $=$ 4

CLASSROOM EXAMPLE
Write each sentence as an equation. Let x represent the unknown number.
a. The difference of 10 and a number is 18.
b. Twice a number decreased by 1 is 99.
answer:
a. $10 - x = 18$ **b.** $2x - 1 = 99$

b. In words:

three subtracted **from** 12	is	a number
↓	↓	↓

Translate: $12 - 3$ $=$ x

Care must be taken when the operation is subtraction. The expression $3 - 12$ would be incorrect. Notice that $3 - 12 \neq 12 - 3$.

TEACHING TIP
Throughout this course, students need to know the difference between an expression and an equation. Start reminding them now that an equation contains an equal sign while an expression does not.

c. In words:

four times a number	added to	17	is	21
↓	↓	↓	↓	↓

Translate: $4x$ $+$ 17 $=$ 21

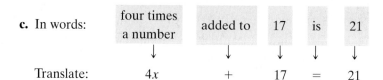

Calculator Explorations

Exponents

To evaluate exponential expressions on a scientific calculator, find the key marked $\boxed{y^x}$ or $\boxed{\wedge}$. To evaluate, for example, 3^5, press the following keys: $\boxed{3}\ \boxed{y^x}\ \boxed{5}\ \boxed{=}$ or $\boxed{3}\ \boxed{\wedge}\ \boxed{5}\ \boxed{=}$.

↕ or
$\boxed{\text{ENTER}}$

The display should read $\boxed{\qquad 243}$ or $\boxed{\begin{array}{l} 3 \wedge 5 \\ \qquad 243 \end{array}}$

Order of Operations

Some calculators follow the order of operations, and others do not. To see whether or not your calculator has the order of operations built in, use your calculator to find $2 + 3 \cdot 4$. To do this, press the following sequence of keys:

$\boxed{2}\ \boxed{+}\ \boxed{3}\ \boxed{\times}\ \boxed{4}\ \boxed{=}$.

↕ or
$\boxed{\text{ENTER}}$

The correct answer is 14 because the order of operations is to multiply before we add. If the calculator displays $\boxed{\qquad 14}$, then it has the order of operations built in.

Even if the order of operations is built in, parentheses must sometimes be inserted. For example, to simplify $\dfrac{5}{12-7}$, press the keys

$\boxed{5}\ \boxed{\div}\ \boxed{(}\ \boxed{1}\ \boxed{2}\ \boxed{-}\ \boxed{7}\ \boxed{)}\ \boxed{=}$.

↕ or
$\boxed{\text{ENTER}}$

The display should read $\boxed{\qquad 1}$ or $\boxed{\begin{array}{l} 5/(12-7) \\ \qquad\qquad 1 \end{array}}$

Use a calculator to evaluate each expression.

1. 5^4 625

2. 7^4 2401

3. 9^5 59,049

4. 8^6 262,144

5. $2(20 - 5)$ 30

6. $3(14 - 7) + 21$ 42

7. $24(862 - 455) + 89$ 9857

8. $99 + (401 + 962)$ 1462

9. $\dfrac{4623 + 129}{36 - 34}$ 2376

10. $\dfrac{956 - 452}{89 - 86}$ 168

Spotlight on

DECISION ✿ MAKING

Suppose you are a local area network (LAN) administrator for a small college and you are configuring a new LAN for the mathematics department. The department would like a network of 20 computers so that each user can transmit data over the network at a speed of 0.25 megabits per second. The collective speed for a LAN is given by the expression rn, where r is the data transmission speed needed by each of the n computers on the LAN. You know that the network will drastically lose its efficiency if the collective speed of the network exceeds 8 megabits per second. Decide whether the LAN requested by the math department will operate efficiently. Explain your reasoning.

MENTAL MATH

Fill in the blank with add, subtract, multiply, or divide.

1. To simplify the expression $1 + 3 \cdot 6$, first ___multiply___ .

2. To simplify the expression $(1 + 3) \cdot 6$, first ___add___ .

3. To simplify the expression $(20 - 4) \cdot 2$, first ___subtract___ .

4. To simplify the expression $20 - 4 \div 2$, first ___divide___ .

EXERCISE SET 1.4

STUDY GUIDE/SSM   CD/VIDEO   PH MATH TUTOR CENTER   MathXL®Tutorials ON CD   MathXL®   MyMathLab®

Evaluate. See Example 1.

1. 3^5 243

2. 5^3 125

3. 3^3 27

4. 4^4 256

5. 1^5 1

6. 1^8 1

7. 5^1 5

8. 8^1 8

9. $\left(\dfrac{1}{5}\right)^3$ $\dfrac{1}{125}$

10. $\left(\dfrac{6}{11}\right)^2$ $\dfrac{36}{121}$

11. $\left(\dfrac{2}{3}\right)^4$ $\dfrac{16}{81}$

12. $\left(\dfrac{1}{2}\right)^5$ $\dfrac{1}{32}$

13. 7^2 49

14. 9^2 81

15. 4^2 16

16. 2^4 16

17. $(1.2)^2$ 1.44

18. $(0.07)^2$ 0.0049

Simplify each expression. See Examples 2 through 5.

19. $5 + 6 \cdot 2$ 17

20. $8 + 5 \cdot 3$ 23

21. $4 \cdot 8 - 6 \cdot 2$ 20

22. $12 \cdot 5 - 3 \cdot 6$ 42

23. $2(8 - 3)$ 10

24. $5(6 - 2)$ 20

25. $2 + (5 - 2) + 4^2$ 21

26. $6 - 2 \cdot 2 + 2^5$ 34

27. $5 \cdot 3^2$ 45

28. $2 \cdot 5^2$ 50

29. $\dfrac{1}{4} \cdot \dfrac{2}{3} - \dfrac{1}{6}$ 0

30. $\dfrac{3}{4} \cdot \dfrac{1}{2} + \dfrac{2}{3}$ $\dfrac{25}{24}$

31. $\dfrac{6 - 4}{9 - 2}$ $\dfrac{2}{7}$

32. $\dfrac{8 - 5}{24 - 20}$ $\dfrac{3}{4}$

33. $2[5 + 2(8 - 3)]$ 30

34. $3[4 + 3(6 - 4)]$ 30

35. $\dfrac{19 - 3 \cdot 5}{6 - 4}$ 2

36. $\dfrac{4 \cdot 3 + 2}{4 + 3 \cdot 2}$ $\dfrac{7}{5}$

37. $\dfrac{|6 - 2| + 3}{8 + 2 \cdot 5}$ $\dfrac{7}{18}$

38. $\dfrac{15 - |3 - 1|}{12 - 3 \cdot 2}$ $\dfrac{13}{6}$

39. $\dfrac{3 + 3(5 + 3)}{3^2 + 1}$ $\dfrac{27}{10}$

40. $\dfrac{3 + 6(8 - 5)}{4^2 + 2}$ $\dfrac{7}{6}$

41. $\dfrac{6 + |8 - 2| + 3^2}{18 - 3}$ $\dfrac{7}{5}$

42. $\dfrac{16 + |13 - 5| + 4^2}{17 - 5}$ $\dfrac{10}{3}$

43. Are parentheses necessary in the expression $2 + (3 \cdot 5)$? Explain your answer. no

44. Are parentheses necessary in the expression $(2 + 3) \cdot 5$? Explain your answer. yes

For Exercises 45 and 46, match each expression in the first column with its value in the second column.

45.

a. $(6 + 2) \cdot (5 + 3)$ 64	19
b. $(6 + 2) \cdot 5 + 3$ 43	22
c. $6 + 2 \cdot 5 + 3$ 19	64
d. $6 + 2 \cdot (5 + 3)$ 22	43

46.

a. $(1 + 4) \cdot 6 - 3$ 27	15
b. $1 + 4 \cdot (6 - 3)$ 13	13
c. $1 + 4 \cdot 6 - 3$ 22	27
d. $(1 + 4) \cdot (6 - 3)$ 15	22

Evaluate each expression when $x = 1$, $y = 3$, and $z = 5$. See Example 6.

47. $3y$ 9

48. $4x$ 4

49. $\dfrac{z}{5x}$ 1

50. $\dfrac{y}{2z}$ $\dfrac{3}{10}$

51. $3x - 2$ 1

52. $6y - 8$ 10

53. $|2x + 3y|$ 11   **54.** $|5z - 2y|$ 19

55. $5y^2$ 45   **56.** $2z^2$ 50

Evaluate each expression if $x = 12, y = 8,$ *and* $z = 4.$ *See Example 6.*

57. $\dfrac{x}{z} + 3y$ 27   **58.** $\dfrac{y}{z} + 8x$ 98   **59.** $x^2 - 3y + x$ 132

60. $y^2 - 3x + y$ 36   **61.** $\dfrac{x^2 + z}{y^2 + 2z}$ $\dfrac{37}{18}$ **62.** $\dfrac{y^2 + x}{x^2 + 3y}$ $\dfrac{19}{42}$

Neglecting air resistance, the expression $16t^2$ *gives the distance in feet an object will fall in t seconds.*

63. Complete the chart below. To evaluate $16t^2$, remember to first find t^2, then multiply by 16.

Time t (in seconds)	Distance $16t^2$ (in feet)
1	16
2	64
3	144
4	256

64. Does an object fall the same distance *during* each second? Why or why not? (See Exercise 63.)   no

Decide whether the given number is a solution of the given equation. See Example 7.

65. Is 5 a solution of $3x - 6 = 9$?   yes
66. Is 6 a solution of $2x + 7 = 3x$?   no
67. Is 0 a solution of $2x + 6 = 5x - 1$?   no
68. Is 2 a solution of $4x + 2 = x + 8$?   yes
69. Is 8 a solution of $2x - 5 = 5$?   no
70. Is 6 a solution of $3x - 10 = 8$?   yes
71. Is 2 a solution of $x + 6 = x + 6$?   yes
72. Is 10 a solution of $x + 6 = x + 6$?   yes
73. Is 0 a solution of $x = 5x + 15$?   no
74. Is 1 a solution of $4 = 1 - x$?   no

Write each phrase as an algebraic expression. Let x represent the unknown number. See Example 8.

75. Fifteen more than a number $x + 15$
76. One-half times a number $\frac{1}{2}x$
77. Five subtracted from a number $x - 5$
78. The quotient of a number and 9 $\frac{x}{9}$
79. Three times a number, increased by 22 $3x + 22$
80. The product of 8 and a number $8x$

Write each sentence as an equation or inequality. Use x to represent any unknown number. See Example 9.

81. One increased by two equals the quotient of nine and three. $1 + 2 = 9 \div 3$
82. Four subtracted from eight is equal to two squared. $8 - 4 = 2^2$
83. Three is not equal to four divided by two. $3 \neq 4 \div 2$
84. The difference of sixteen and four is greater than ten. $16 - 4 > 10$
85. The sum of 5 and a number is 20. $5 + x = 20$
86. Twice a number is 17. $2x = 17$
87. Thirteen minus three times a number is 13. $13 - 3x = 13$
88. Seven subtracted from a number is 0. $x - 7 = 0$
89. The quotient of 12 and a number is $\frac{1}{2}$. $\dfrac{12}{x} = \dfrac{1}{2}$
90. The sum of 8 and twice a number is 42. $8 + 2x = 42$
91. In your own words, explain the difference between an expression and an equation.   answers may vary
92. Determine whether each is an expression or an equation.

    a. $3x^2 - 26$ **b.** $3x^2 - 26 = 1$
    c. $2x - 5 = 7x - 5$ **d.** $9y + x - 8$

92.a. expression
92.b. equation
92.c. equation
92.d. expression

Concept Extensions

93. Insert parentheses so that the following expression simplifies to 32. $(20 - 4) \cdot 4 \div 2$

$$20 - 4 \cdot 4 \div 2$$

94. Insert parentheses so that the following expression simplifies to 28. $2 \cdot (5 + 3^2)$

$$2 \cdot 5 + 3^2$$

Solve the following.

95. The perimeter of a figure is the distance around the figure. The expression $2l + 2w$ represents the perimeter of a rectangle when *l* is its length and *w* is its width. Find the perimeter of the following rectangle by substituting 8 for *l* and 6 for *w*.   28 m

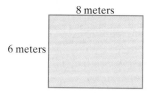

8 meters

6 meters

96. The expression $a + b + c$ represents the perimeter of a triangle when *a*, *b*, and *c* are the lengths of its sides. Find the perimeter of the following triangle. $\frac{11}{14}$ yd

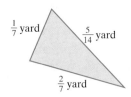

$\frac{1}{7}$ yard
$\frac{5}{14}$ yard
$\frac{2}{7}$ yard

97. The area of a figure is the total enclosed surface of the figure. Area is measured in square units. The expression *lw*

represents the area of a rectangle when l is its length and w is its width. Find the area of the following rectangular-shaped lot.   12,000 sq. ft

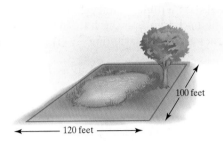

△ **98.** A trapezoid is a four-sided figure with exactly one pair of parallel sides. The expression $\frac{1}{2}h(B + b)$ represents its area, when B and b are the lengths of the two parallel sides and h is the height between these sides. Find the area if $B = 15$ inches, $b = 7$ inches, and $h = 5$ inches.   55 sq. in.

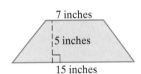

7 inches

5 inches

15 inches

99. The expression $\frac{I}{PT}$ represents the rate of interest being charged if a loan of P dollars for T years required I dollars in interest to be paid. Find the interest rate if a $650 loan for 3 years to buy a used IBM personal computer requires $126.75 in interest to be paid.   6.5%

100. The expression $\frac{d}{t}$ represents the average speed r in miles per hour if a distance of d miles is traveled in t hours. Find the rate to the nearest whole number if the distance between Dallas, Texas, and Kaw City, Oklahoma, is 432 miles, and it takes Peter Callac 8.5 hours to drive the distance.   51 mph

101. Sprint Communications Company offers a long-distance telephone plan called Sprint Simple 7 that charges $4.00 per month and $0.07 per minute of calling. The expression $4.00 + 0.07m$ represents the monthly long-distance bill for a customer who makes m minutes of long-distance calling on this plan. Find the monthly bill for a customer who makes 228 minutes of long-distance calls on the Sprint Simple 7 plan. $19.96

102. In forensics, the density of a substance is used to help identify it. The expression $\frac{M}{V}$ represents the density of an object with a mass of M grams and a volume of V milliliters. Find the density of an object having a mass of 29.76 grams and a volume of 12 milliliters.   2.48g/ml

---

1.5   ADDING REAL NUMBERS

Objectives

1 Add real numbers with the same sign.

2 Add real numbers with unlike signs.

3 Solve problems that involve addition of real numbers.

4 Find the opposite of a number.

1   Real numbers can be added, subtracted, multiplied, divided, and raised to powers, just as whole numbers can. We use a number line to help picture the addition of real numbers.

EXAMPLE 1

Add: $3 + 2$

Solution   Recall that 3 and 2 are called addends. We start at 0 on a number line, and draw an arrow representing the addend 3. This arrow is three units long and points to the right since 3 is positive. From the tip of this arrow, we draw another arrow representing the addend 2. The number below the tip of this arrow is the sum, 5.

CLASSROOM EXAMPLE
Add using a number line: $4 + 1$
answer:

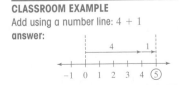

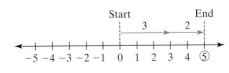

$$3 + 2 = 5$$

EXAMPLE 2

Add: $-1 + (-2)$

Solution Here, -1 and -2 are addends. We start at 0 on a number line, and draw an arrow representing -1. This arrow is one unit long and points to the left since -1 is negative. From the tip of this arrow, we draw another arrow representing -2. The number below the tip of this arrow is the sum, -3.

CLASSROOM EXAMPLE
Add using a number line: $-3 + (-4)$
answer:

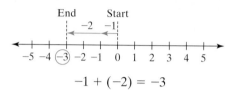

$$-1 + (-2) = -3$$

Thinking of signed numbers as money earned or lost might help make addition more meaningful. Earnings can be thought of as positive numbers. If $1 is earned and later another $3 is earned, the total amount earned is $4. In other words, $1 + 3 = 4$.

On the other hand, losses can be thought of as negative numbers. If $1 is lost and later another $3 is lost, a total of $4 is lost. In other words, $(-1) + (-3) = -4$.

Using a number line each time we add two numbers can be time consuming. Instead, we can notice patterns in the previous examples and write rules for adding signed numbers. When adding two numbers with the same sign, notice that the sign of the sum is the same as the sign of the addends.

Adding Two Numbers with the Same Sign

Add their absolute values. Use their common sign as the sign of the sum.

EXAMPLE 3

Add.

a. $-3 + (-7)$ **b.** $-1 + (-20)$ **c.** $-2 + (-10)$

Solution Notice that each time, we are adding numbers with the same sign.

CLASSROOM EXAMPLE
Add: $-5 + (-9)$
answer: -14

a. $-3 + (-7) = -10$ ← Add their absolute values: $3 + 7 = 10.$
                           Use their common sign.

b. $-1 + (-20) = -21$ ← Add their absolute values: $1 + 20 = 21.$
                            Common sign.

c. $-2 + (-10) = -12$ ← Add their absolute values.
                            Common sign.

2   Adding numbers whose signs are not the same can also be pictured on a number line.

EXAMPLE 4

Add: $-4 + 6$

Solution

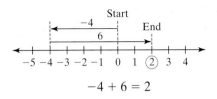

$$-4 + 6 = 2$$

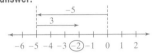

Using temperature as an example, if the thermometer registers 4 degrees below 0 degrees and then rises 6 degrees, the new temperature is 2 degrees above 0 degrees. Thus, it is reasonable that $-4 + 6 = 2$.

Once again, we can observe a pattern: when adding two numbers with different signs, the sign of the sum is the same as the sign of the addend whose absolute value is larger.

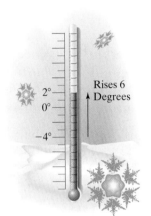

> ## Adding Two Numbers with Different Signs
>
> Subtract the smaller absolute value from the larger absolute value. Use the sign of the number whose absolute value is larger as the sign of the sum.

EXAMPLE 5

Add.

a. $3 + (-7)$ **b.** $-2 + 10$ **c.** $0.2 + (-0.5)$

Solution   Notice that each time, we are adding numbers with different signs.

a. $3 + (-7) = -4$ ← Subtract their absolute values: $7 - 3 = 4$.
           The negative number, -7, has the larger absolute value so the sum is negative.

b. $-2 + 10 = 8$ ← Subtract their absolute values: $10 - 2 = 8$.
           The positive number, 10, has the larger absolute value so the sum is positive.

c. $0.2 + (-0.5) = -0.3$ ← Subtract their absolute values: $0.5 - 0.2 = 0.3$.
           The negative number, -0.5, has the larger absolute value so the sum is negative.

EXAMPLE 6

Add.

a. $-8 + (-11)$ **b.** $-5 + 35$ **c.** $0.6 + (-1.1)$

d. $-\dfrac{7}{10} + \left(-\dfrac{1}{10}\right)$ **e.** $11.4 + (-4.7)$ **f.** $-\dfrac{3}{8} + \dfrac{2}{5}$

Solution **a.** $-8 + (-11) = -19$ *Same sign. Add absolute values and use the common sign.*

b. $-5 + 35 = 30$ *Different signs. Subtract absolute values and use the sign of the number with the larger absolute value.*

c. $0.6 + (-1.1) = -0.5$ *Different signs.*

d. $-\dfrac{7}{10} + \left(-\dfrac{1}{10}\right) = -\dfrac{8}{10} = -\dfrac{4}{5}$ *Same sign.*

e. $11.4 + (-4.7) = 6.7$

f. $-\dfrac{3}{8} + \dfrac{2}{5} = -\dfrac{15}{40} + \dfrac{16}{40} = \dfrac{1}{40}$

> **Helpful Hint**
> Don't forget that a common denominator is needed when adding or subtracting fractions. The common denominator here is 40.

EXAMPLE 7

CLASSROOM EXAMPLE
Add.
$[3 + (-13)] + [-4 + |-7|]$
answer: -7

Add.

a. $3 + (-7) + (-8)$ **b.** $[7 + (-10)] + [-2 + |-4|]$

Solution **a.** Perform the additions from left to right.

$$3 + (-7) + (-8) = -4 + (-8)$$ *Adding numbers with different signs.*
$$= -12$$ *Adding numbers with like signs.*

b. Simplify inside brackets first.

$$[7 + (-10)] + [-2 + |-4|] = [-3] + [-2 + 4]$$
$$= [-3] + [2]$$
$$= -1$$ Add.

> **Helpful Hint**
> Don't forget that brackets are grouping symbols. We simplify within them first.

3 Positive and negative numbers are often used in everyday life. Stock market returns show gains and losses as positive and negative numbers. Temperatures in cold climates often dip into the negative range, commonly referred to as "below zero" temperatures. Bank statements report deposits and withdrawals as positive and negative numbers.

EXAMPLE 8

CLASSROOM EXAMPLE
During a four-day period, a share of Walco stock recorded the following gains and losses:

Tuesday	Wednesday
a loss of $2	a loss of $1
Thursday	**Friday**
a gain of $3	a gain of $3

Find the overall gain or loss for the stock for the four days.
answer: A gain of $3.

FINDING THE GAIN OR LOSS OF A STOCK

During a three-day period, a share of Lamplighter's International stock recorded the following gains and losses:

| **Monday** | **Tuesday** | **Wednesday** |
| a gain of $2 | a loss of $1 | a loss of $3 |

Find the overall gain or loss for the stock for the three days.

Solution Gains can be represented by positive numbers. Losses can be represented by negative numbers. The overall gain or loss is the sum of the gains and losses.

In words: gain plus loss plus loss

Translate: $2 + (-1) + (-3) = -2$

The overall loss is $2.

ADDING REAL NUMBERS SECTION 1.5 **37**

TEACHING TIP
After covering adding numbers with the same sign and with different signs, take a moment to review this with students. Have them work these problems:

$$-5 + (-6)$$
$$7 + (-4)$$
$$-11 + 5$$

Give students the answers, and ask them if they knew how to apply the rules.

4 To help us subtract real numbers in the next section, we first review the concept of opposites. The graphs of 4 and −4 are shown on a number line below.

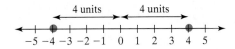

Notice that 4 and −4 lie on opposite sides of 0, and each is 4 units away from 0.

This relationship between −4 and +4 is an important one. Such numbers are known as **opposites** or **additive inverses** of each other.

Opposites or Additive Inverses

Two numbers that are the same distance from 0 but lie on opposite sides of 0 are called opposites or additive inverses of each other.

Let's discover another characteristic about opposites. Notice that the sum of a number and its opposite is 0.

$$10 + (-10) = 0$$
$$-3 + 3 = 0$$
$$\frac{1}{2} + \left(-\frac{1}{2}\right) = 0$$

In general, we can write the following:

The sum of a number a and its opposite $-a$ is 0.

$$a + (-a) = 0$$

This is why opposites are also called additive inverses. Notice that this also means that the opposite of 0 is then 0 since $0 + 0 = 0$.

EXAMPLE 9

Find the opposite or additive inverse of each number.

a. 5   **b.** −6   **c.** $\frac{1}{2}$ **d.** −4.5

Solution

a. The opposite of 5 is −5. Notice that 5 and −5 are on opposite sides of 0 when plotted on a number line and are equal distances away.

b. The opposite of −6 is 6.

c. The opposite of $\frac{1}{2}$ is $-\frac{1}{2}$.

d. The opposite of −4.5 is 4.5.

CLASSROOM EXAMPLE
Find the opposite of each number.
a. −35   **b.** 1.9   **c.** $-\frac{3}{11}$
answer:
a. 35   **b.** −1.9   **c.** $\frac{3}{11}$

We use the symbol "−" to represent the phrase "the opposite of" or "the additive inverse of." In general, if a is a number, we write the opposite or additive inverse of a as $-a$. We know that the opposite of -3 is 3. Notice that this translates as

the opposite of	-3	is	3
$\downarrow$	$\downarrow$	$\downarrow$	$\downarrow$
$-$	(-3)	$=$	3

This is true in general.

> If a is a number, then $-(-a) = a$.

EXAMPLE 10

Simplify each expression.

a. $-(-10)$ **b.** $-\left(-\dfrac{1}{2}\right)$ **c.** $-(-2x)$ **d.** $-|-6|$

Solution   **a.** $-(-10) = 10$ **b.** $-\left(-\dfrac{1}{2}\right) = \dfrac{1}{2}$ **c.** $-(-2x) = 2x$

d. Since $|-6| = 6$, then $-|-6| = -6$.

MENTAL MATH

Tell whether the sum is a positive number, a negative number, or 0. Do not actually find the sum.

1. $-80 + (-127)$ negative

2. $-162 + 164$ positive

3. $-162 + 162$ 0

4. $-1.26 + (-8.3)$ negative

5. $-3.68 + 0.27$ negative

6. $-\dfrac{2}{3} + \dfrac{2}{3}$ 0

EXERCISE SET 1.5

STUDY GUIDE/SSM   CD/ VIDEO   PH MATH TUTOR CENTER   MathXL®Tutorials ON CD   MathXL®   MyMathLab®

MIXED PRACTICE

Add. See Examples 1 through 7.

1. $6 + 3$ 9

2. $9 + (-12)$ -3

3. $-6 + (-8)$ -14

4. $-6 + (-14)$ -20

5. $8 + (-7)$ 1

6. $6 + (-4)$ 2

7. $-14 + 2$ -12

8. $-10 + 5$ -5

9. $-2 + (-3)$ -5

10. $-7 + (-4)$ -11

11. $-9 + (-3)$ -12

12. $7 + (-5)$ 2

13. $-7 + 3$ -4

14. $-5 + 9$ 4

15. $10 + (-3)$ 7

16. $8 + (-6)$ 2

17. $5 + (-7)$ -2

18. $3 + (-6)$ -3

19. $-16 + 16$ 0

20. $23 + (-23)$ 0

21. $27 + (-46)$ -19

22. $53 + (-37)$ 16

23. $-18 + 49$ 31

24. $-26 + 14$ -12

25. $-33 + (-14)$ -47

26. $-18 + (-26)$ -44

27. $6.3 + (-8.4)$ -2.1

28. $9.2 + (-11.4)$ -2.2

29. $|-8| + (-16)$ -8

30. $|-6| + (-61)$ -55

31. $117 + (-79)$ 38

32. $144 + (-88)$ 56

33. $-9.6 + (-3.5)$ -13.1 **34.** $-6.7 + (-7.6)$ -14.3

35. $-\dfrac{3}{8} + \dfrac{5}{8}$ $\dfrac{2}{8} = \dfrac{1}{4}$ **36.** $-\dfrac{5}{12} + \dfrac{7}{12}$ $\dfrac{2}{12} = \dfrac{1}{6}$

37. $-\dfrac{7}{16} + \dfrac{1}{4}$ $-\dfrac{3}{16}$ **38.** $-\dfrac{5}{9} + \dfrac{1}{3}$ $-\dfrac{2}{9}$

39. $-\dfrac{7}{10} + \left(-\dfrac{3}{5}\right)$ $-\dfrac{13}{10}$ **40.** $-\dfrac{5}{6} + \left(-\dfrac{2}{3}\right)$ $-\dfrac{9}{6} = -\dfrac{3}{2}$

41. $-15 + 9 + (-2)$ -8 **42.** $-9 + 15 + (-5)$ 1

43. $-21 + (-16) + (-22)$ -59 **44.** $-18 + (-6) + (-40)$ -64

45. $-23 + 16 + (-2)$ -9 **46.** $-14 + (-3) + 11$ -6

47. $|5 + (-10)|$ 5 **48.** $|7 + (-17)|$ 10

49. $6 + (-4) + 9$ 11 **50.** $8 + (-2) + 7$ 13

51. $[-17 + (-4)] + [-12 + 15]$ -18

52. $[-2 + (-7)] + [-11 + 22]$ 2

53. $|9 + (-12)| + |-16|$ 19

54. $|43 + (-73)| + |-20|$ 50

55. $-1.3 + [0.5 + (-0.3) + 0.4]$ -0.7

56. $-3.7 + [0.1 + (-0.6) + 8.1]$ 3.9

Solve. See Example 8.

57. The low temperature in Anoka, Minnesota, was $-15°$ last night. During the day it rose only $9°$. Find the high temperature for the day. $-6°$

58. On January 2, 1943, the temperature was $-4°$ at 7:30 A.M. in Spearfish, South Dakota. Incredibly, it got $49°$ warmer in the next 2 minutes. To what temperature did it rise by 7:32? $45°$

59. The lowest elevation on Earth is -1349 feet (that is, 1349 feet below sea level) at the Dead Sea. If you are standing 695 feet above the Dead Sea, what is your elevation? (*Source: National Geographic Society*) -654 ft

60. The lowest point in Africa is -512 feet at Lake Assal in Djibouti. If you are standing at a point 658 feet above Lake Assal, what is your elevation? (*Source: Microsoft Encarta*)   146 ft

A negative net income results when a company's expenses are more than the money it brings in.

61. The table below shows net incomes for Amazon.com for the years 2000, 2001, and 2002. Find the total net income for three years. $-\$2127$ million

Year	Net Income (in millions)
2000	$-\$1411$
2001	$-\$567$
2002	-149

(*Source*: amazon.com)

62. The table below shows net incomes for Continental Airlines for the years 2000, 2001, and 2002. Find the total net income for these years. $-\$204$ million

Year	Net Income (in millions)
2000	$\$342$
2001	$-\$95$
2002	$-\$451$

(*Source*: Continental Airlines)

In golf, scores that are under par for the entire round are shown as negative scores; positive scores are shown for scores that are over par, and 0 is par.

63. Candie Kung was the winner of the 2003 LPGA Takefuji Classic. Her scores were $5, -5, -2$. What was her overall score? (*Source*: Ladies Professional Golf Association) -2

64. During the 2003 LPGA Kraft Nabisco Championship, Patricia Meunier-Lebouc won with scores of $-2, -4, -2, 1$. What was her overall score? (*Source*: Ladies Professional Golf Association) -7

Find each additive inverse or opposite. See Example 9.

65. 6 -6 **66.** 4 -4 **67.** -2 2

68. -8 8 **69.** 0 0 **70.** $-\dfrac{1}{4}$ $\dfrac{1}{4}$

71. $|-6|$ -6 **72.** $|-11|$ -11

73. In your own words, explain how to find the opposite of a number.   answers may vary

74. In your own words, explain why 0 is the only number that is its own opposite.   answers may vary

Simplify each of the following. See Example 10.

75. $-|-2|$ -2 **76.** $-(-3)$ 3

77. $-|0|$ 0 **78.** $\left|-\dfrac{2}{3}\right|$ $\dfrac{2}{3}$

79. $-\left|-\dfrac{2}{3}\right|$ $-\dfrac{2}{3}$ **80.** $-(-7)$ 7

81. Explain why adding a negative number to another negative number always gives a negative sum. answers may vary

82. When a positive and a negative number are added, sometimes the sum is positive, sometimes it is zero, and sometimes it is negative. Explain why and when this happens. answers may vary

Decide whether the given number is a solution of the given equation.

83. Is −4 a solution of $x + 9 = 5$? yes

84. Is 10 a solution of $7 = -x + 3$? no

85. Is −1 a solution of $y + (-3) = -7$? no

86. Is −6 a solution of $1 = y + 7$? yes

Concept Extensions

The following bar graph shows each month's average daily low temperature in degrees Fahrenheit for Barrow, Alaska. Use this graph to answer Exercises 87 through 92.

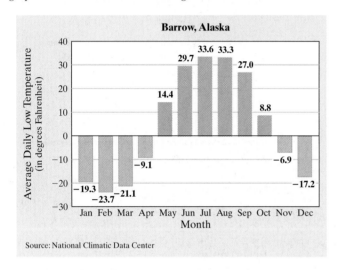

Barrow, Alaska

Source: National Climatic Data Center

88. February

87. For what month is the graphed temperature the highest? July

88. For what month is the graphed temperature the lowest?

89. For what month is the graphed temperature positive *and* closest to 0°? October

90. For what month is the graphed temperature negative *and* closest to 0°? November

91. Find the average of the temperatures shown for the months of April, May, and October. (To find the average of three temperatures, find their sum and divide by 3.) 4.7° F

92. Find the average of the temperatures shown for the months of January, September, and October. 5.5° F

If a is a positive number and b is a negative number, fill in the blanks with the words positive or negative.

93. $-a$ is a   negative .

94. $-b$ is a   positive .

95. $a + a$ is a   positive .

96. $b + b$ is a   negative .

1.6 SUBTRACTING REAL NUMBERS

Objectives

1. Subtract real numbers.
2. Add and subtract real numbers.
3. Evaluate algebraic expressions using real numbers.
4. Solve problems that involve subtraction of real numbers.

1   Now that addition of signed numbers has been discussed, we can explore subtraction. We know that $9 - 7 = 2$. Notice that $9 + (-7) = 2$, also. This means that

$$9 - 7 = 9 + (-7)$$

Notice that the difference of 9 and 7 is the same as the sum of 9 and the opposite of 7. In general, we have the following.

TEACHING TIP

Before Example 3, you may want to ask your students to mentally "subtract 8 from 10." Then ask them to write this subtraction problem in symbols and notice the order of the numbers.

> ### Subtracting Two Real Numbers
>
> If a and b are real numbers, then $a - b = a + (-b)$.

In other words, to find the difference of two numbers, add the first number to the opposite of the second number.

EXAMPLE 1

Subtract.

a. $-13 - 4$ **b.** $5 - (-6)$ **c.** $3 - 6$ **d.** $-1 - (-7)$

Solution

CLASSROOM EXAMPLE
Subtract.
a. $-7 - 1$ **b.** $9 - 19$
answer: a. -8 **b.** -10

a. $\overset{\text{add}}{-13 - 4} = -13 + \overset{\text{opposite}}{(-4)}$ Add -13 to the opposite of $+4$, which is -4.
$= -17$

b. $\overset{\text{add}}{5 - (-6)} = 5 + \overset{\text{opposite}}{(6)}$ Add 5 to the opposite of -6, which is 6.
$= 11$

c. $3 - 6 = 3 + (-6)$ Add 3 to the opposite of 6, which is -6.
$= -3$

d. $-1 - (-7) = -1 + (7) = 6$

TEACHING TIP

After covering addition and subtraction of numbers, ask students to work the following exercises.

$-7 + 2 \qquad -8 - (-4)$
$-7 - 3 \qquad -8 + (-4)$
$5 - (-1)$
$5 + (-1)$

Classify each problem as you review the results in class.

> ▶ **Helpful Hint**
>
> Study the patterns indicated.
>
> No change ———⟶ Change to addition.
> ——— Change to opposite.
>
> $5 - 11 = 5 + (-11) = -6$
> $-3 - 4 = -3 + (-4) = -7$
> $7 - (-1) = 7 + (1) = 8$

EXAMPLE 2

Subtract.

a. $5.3 - (-4.6)$ **b.** $-\dfrac{3}{10} - \dfrac{5}{10}$ **c.** $-\dfrac{2}{3} - \left(-\dfrac{4}{5}\right)$

Solution   **a.** $5.3 - (-4.6) = 5.3 + (4.6) = 9.9$

CLASSROOM EXAMPLE
Subtract.
a. $\dfrac{3}{8} - \left(-\dfrac{1}{8}\right)$ **b.** $-1.8 - (-2.7)$
answer: a. $\dfrac{1}{2}$ **b.** 0.9

b. $-\dfrac{3}{10} - \dfrac{5}{10} = -\dfrac{3}{10} + \left(-\dfrac{5}{10}\right) = -\dfrac{8}{10} = -\dfrac{4}{5}$

c. $-\dfrac{2}{3} - \left(-\dfrac{4}{5}\right) = -\dfrac{2}{3} + \left(\dfrac{4}{5}\right) = -\dfrac{10}{15} + \dfrac{12}{15} = \dfrac{2}{15}$ *The common denominator is* **15**.

EXAMPLE 3

Subtract 8 from -4.

Solution Be careful when interpreting this: The order of numbers in subtraction is important. 8 is to be subtracted **from** -4.

$$-4 - 8 = -4 + (-8) = -12$$

2 If an expression contains additions and subtractions, just write the subtractions as equivalent additions. Then simplify from left to right.

EXAMPLE 4

Simplify each expression.

a. $-14 - 8 + 10 - (-6)$ **b.** $1.6 - (-10.3) + (-5.6)$

Solution **a.** $-14 - 8 + 10 - (-6) = -14 + (-8) + 10 + 6$
$$= -6$$

b. $1.6 - (-10.3) + (-5.6) = 1.6 + 10.3 + (-5.6)$
$$= 6.3$$

When an expression contains parentheses and brackets, remember the order of operations. Start with the innermost set of parentheses or brackets and work your way outward.

EXAMPLE 5

Simplify each expression.

a. $-3 + [(-2 - 5) - 2]$ **b.** $2^3 - |10| + [-6 - (-5)]$

Solution **a.** Start with the innermost sets of parentheses. Rewrite $-2 - 5$ as a sum.

$$\begin{aligned} -3 + [(-2 - 5) - 2] &= -3 + [(-2 + (-5)) - 2] \\ &= -3 + [(-7) - 2] &&\text{Add: } -2 + (-5). \\ &= -3 + [-7 + (-2)] &&\text{Write } -7 - 2 \text{ as a sum.} \\ &= -3 + [-9] &&\text{Add.} \\ &= -12 &&\text{Add.} \end{aligned}$$

b. Start simplifying the expression inside the brackets by writing $-6 - (-5)$ as a sum.

$$\begin{aligned} 2^3 - |10| + [-6 - (-5)] &= 2^3 - |10| + [-6 + 5] \\ &= 2^3 - |10| + [-1] &&\text{Add.} \\ &= 8 - 10 + (-1) &&\text{Evaluate } 2^3 \text{ and } |10|. \\ &= 8 + (-10) + (-1) &&\text{Write } 8 - 10 \text{ as a sum.} \\ &= -2 + (-1) &&\text{Add.} \\ &= -3 &&\text{Add.} \end{aligned}$$

3 Knowing how to evaluate expressions for given replacement values is helpful when checking solutions of equations and when solving problems whose unknowns satisfy given expressions. The next example illustrates this.

EXAMPLE 6

Find the value of each expression when $x = 2$ and $y = -5$.

a. $\dfrac{x - y}{12 + x}$ **b.** $x^2 - y$

Solution

a. Replace x with 2 and y with -5. Be sure to put parentheses around -5 to separate signs. Then simplify the resulting expression.

$$\frac{x - y}{12 + x} = \frac{2 - (-5)}{12 + 2}$$
$$= \frac{2 + 5}{14}$$
$$= \frac{7}{14} = \frac{1}{2}$$

CLASSROOM EXAMPLE
Find the value of each expression when $x = 1$ and $y = -4$.
a. $\dfrac{x - y}{14 + x}$ **b.** $x^2 - y$
answer:
a. $\dfrac{1}{3}$ **b.** 5

b. Replace the x with 2 and y with -5 and simplify.

$$x^2 - y = 2^2 - (-5)$$
$$= 4 - (-5)$$
$$= 4 + 5$$
$$= 9$$

4   One use of positive and negative numbers is in recording altitudes above and below sea level, as shown in the next example.

EXAMPLE 7

FINDING THE DIFFERENCE IN ELEVATIONS

The lowest point on the surface of the Earth is the Dead Sea, at an elevation of 1349 feet below sea level. The highest point is Mt. Everest, at an elevation of 29,035 feet. How much of a variation in elevation is there between these two world extremes? (*Source*: National Geographic Society)

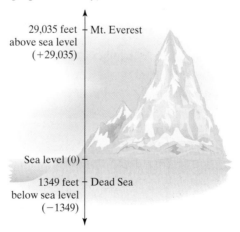

29,035 feet Mt. Everest
above sea level
(+29,035)

Sea level (0)

1349 feet Dead Sea
below sea level
(−1349)

Solution

To find the variation in elevation between the two heights, find the difference of the high point and the low point.

CLASSROOM EXAMPLE
At 6:00 P.M., the temperature at the Winter Olympics was 14°; by morning the temperature dropped to −23°. Find the overall change in temperature.
answer: −37°

In words:	high point	minus	low point	
	↓	↓	↓	
Translate:	29,035	−	(−1349)	= 29,035 + 1349
				= 30,384 feet

Thus, the variation in elevation is 30,384 feet.

Spotlight on

DECISION *of* MAKING

Suppose you are a dental hygienist. As part of a new patient assessment, you measure the depth of the gum tissue pocket around the patient's teeth with a dental probe and record the results. If these pockets deepen over time, this could indicate a problem with gum health or be an indication of gum disease. Now, a year later, you measure the patient's gum tissue pocket depth again to compare to the intial measurement. Based on these findings, would you alert the dentist to a problem with the health of the patien's gums? Explain.

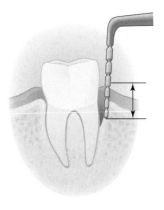

Dental Chart

Gum Tissue Pocket Depth (millimeters)

Tooth:	22	23	24	25	26	27
Initial	2	3	3	2	4	2
Current	2	2	4	5	6	5

A knowledge of geometric concepts is needed by many professionals, such as doctors, carpenters, electronic technicians, gardeners, machinists, and pilots, just to name a few. With this in mind, we review the geometric concepts of **complementary** and **supplementary angles**.

Complementary and Supplementary Angles

Two angles are **complementary** if their sum is 90°.

$x + y = 90°$

Two angles are **supplementary** if their sum is 180°.

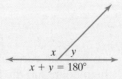

$x + y = 180°$

EXAMPLE 8

Find each unknown complementary or supplementary angle.

a.

a.

b.

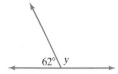

b.

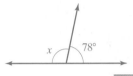

Solution

a. These angles are complementary, so their sum is $90°$. This means that x is $90° − 38°$.

$$x = 90° − 38° = 52°$$

b. These angles are supplementary, so their sum is $180°$. This means that y is $180° − 62°$.

$$y = 180° − 62° = 118°$$

answer: **a.** $102°$ **b.** $9°$

TEACHING TIP

A Group Activity for this section is available in the Instructor's Resource Manual.

EXERCISE SET 1.6

STUDY GUIDE/SSM   CD/ VIDEO   PH MATH TUTOR CENTER   MathXL®Tutorials ON CD   MathXL®   MyMathLab®

MIXED PRACTICE

Subtract. See Examples 1 through 5.

1. $−6 − 4$ $−10$
2. $−12 − 8$ $−20$
3. $4 − 9$ $−5$
4. $8 − 11$ $−3$
5. $16 − (−3)$ 19
6. $12 − (−5)$ 17
7. $\dfrac{1}{2} − \dfrac{1}{3}$ $\dfrac{1}{6}$
8. $\dfrac{3}{4} − \dfrac{7}{8}$ $−\dfrac{1}{8}$
9. $−16 − (−18)$ 2
10. $−20 − (−48)$ 28
11. $−6 − 5$ $−11$
12. $−8 − 4$ $−12$
13. $7 − (−4)$ 11
14. $3 − (−6)$ 9
15. $−6 − (−11)$ 5
16. $−4 − (−16)$ 12
17. $16 − (−21)$ 37
18. $15 − (−33)$ 48
19. $9.7 − 16.1$ $−6.4$
20. $8.3 − 11.2$ $−2.9$
21. $−44 − 27$ $−71$
22. $−36 − 51$ $−87$
23. $−21 − (−21)$ 0
24. $−17 − (−17)$ 0
25. $−2.6 − (−6.7)$ 4.1
26. $−6.1 − (−5.3)$ $−0.8$
27. $−\dfrac{3}{11} − \left(−\dfrac{5}{11}\right)$ $\dfrac{2}{11}$
28. $−\dfrac{4}{7} − \left(−\dfrac{1}{7}\right)$ $−\dfrac{3}{7}$
29. $−\dfrac{1}{6} − \dfrac{3}{4}$ $−\dfrac{11}{12}$
30. $−\dfrac{1}{10} − \dfrac{7}{8}$ $−\dfrac{39}{40}$
31. $8.3 − (−0.62)$ 8.92
32. $4.3 − (−0.87)$ 5.17

Perform the operation. See Example 3.

33. Subtract $−5$ from 8. 13
34. Subtract 3 from $−2$. $−5$
35. Subtract $−1$ from $−6$. $−5$
36. Subtract 17 from 1. $−16$
37. Subtract 8 from 7. $−1$
38. Subtract 9 from $−4$. $−13$
39. Decrease $−8$ by 15. $−23$

40. Decrease 11 by $−14$. 25
41. In your own words, explain why $5 − 8$ simplifies to a negative number.   answers may vary
42. Explain why $6 − 11$ is the same as $6 + (−11)$.   answers may vary

Simplify each expression. (Remember the order of operations.) See Examples 4 and 5.

43. $−10 − (−8) + (−4) − 20$ $−26$
44. $−16 − (−3) + (−11) − 14$ $−38$
45. $5 − 9 + (−4) − 8 − 8$ $−24$
46. $7 − 12 + (−5) − 2 + (−2)$ $−14$
47. $−6 − (2 − 11)$ 3
48. $−9 − (3 − 8)$ $−4$
49. $3^3 − 8·9$ $−45$
50. $2^3 − 6·3$ $−10$
51. $2 − 3(8 − 6)$ $−4$
52. $4 − 6(7 − 3)$ $−20$
53. $(3 − 6) + 4^2$ 13
54. $(2 − 3) + 5^2$ 24
55. $−2 + [(8 − 11) − (−2 − 9)]$ 6
56. $−5 + [(4 − 15) − (−6) − 8]$ $−18$
57. $|−3| + 2^2 + [−4 − (−6)]$ 9
58. $|−2| + 6^2 + (−3 − 8)$ 27

Evaluate each expression when $x = −5, y = 4,$ and $t = 10$. See Example 6.

59. $x − y$ $−9$
60. $y − x$ 9
61. $|x| + 2t − 8y$ $−7$
62. $|x + t − 7y|$ 23
63. $\dfrac{9 − x}{y + 6}$ $\dfrac{7}{5}$
64. $\dfrac{15 − x}{y + 2}$ $\dfrac{10}{3}$
65. $y^2 − x$ 21
66. $t^2 − x$ 105
67. $\dfrac{|x − (−10)|}{2t}$ $\dfrac{1}{4}$
68. $\dfrac{|5y − x|}{6t}$ $\dfrac{5}{12}$

Solve. See Example 7.

69. Within 24 hours in 1916, the temperature in Browning, Montana, fell from 44 degrees to -56 degrees. How large a drop in temperature was this? 100°

70. Much of New Orleans is below sea level. If George descends 12 feet from an elevation of 5 feet above sea level, what is his new elevation? -7 ft or 7 ft below sea level

71. In a series of plays, the San Francisco 49ers gain 2 yards, lose 5 yards, and then lose another 20 yards. What is their total gain or loss of yardage? -23 yd or 23 yd loss

72. In some card games, it is possible to have a negative score. Lavonne Schultz currently has a score of 15 points. She then loses 24 points. What is her new score? -9

73. Pythagoras died in the year -475 (or 475 B.C.). When was he born, if he was 94 years old when he died? -569 or 569 B.C.

74. The Greek astronomer and mathematician Geminus died in 60 A.D. at the age of 70. When was he born? -10 or 10 B.C.

75. A commercial jet liner hits an air pocket and drops 250 feet. After climbing 120 feet, it drops another 178 feet. What is its overall vertical change? -308 ft

76. Tyson Industries stock posted a loss of 1.625 points yesterday. If it drops another 0.75 point today, find its overall change for the two days. -2.375 points

77. The highest point in Africa is Mt. Kilimanjaro, Tanzania, at an elevation of 19,340 feet. The lowest point is Lake Assal, Djibouti, at 512 feet below sea level. How much higher is Mt. Kilimanjaro than Lake Assal? (*Source:* National Geographic Society) 19,852 ft

78. The airport in Bishop, California, is at an elevation of 4101 feet above sea level. The nearby Furnace Creek Airport in Death Valley, California, is at an elevation of 226 feet below sea level. How much higher in elevation is the Bishop Airport than the Furnace Creek Airport? (*Source:* National Climatic Data Center) 4327 ft

Find each unknown complementary or supplementary angle. See Example 8.

79.
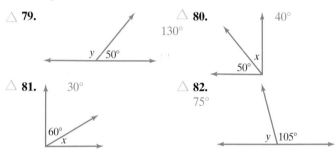
130° y 50°

80. 40° x 50°

81. 30° 60° x

82. 75° y 105°

Decide whether the given number is a solution of the given equation.

83. Is -4 a solution of $x - 9 = 5$? no

84. Is 3 a solution of $x - 10 = -7$? yes

85. Is -2 a solution of $-x + 6 = -x - 1$? no

86. Is -10 a solution of $-x - 6 = -x - 1$? no

87. Is 2 a solution of $-x - 13 = -15$? yes

88. Is 5 a solution of $4 = 1 - x$? no

Concept Extensions

Recall from the last section the bar graph below that shows each month's average daily low temperature in degrees Fahrenheit for Barrow, Alaska. Use this graph to answer Exercises 89 through 91.

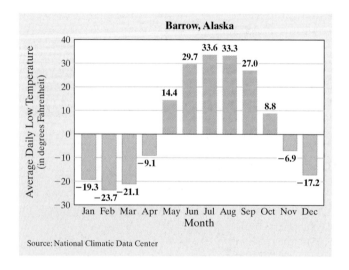

89. Record the monthly increases and decreases in the low temperature from the previous month.

Month	Monthly Increase or Decrease
February	−4.4°
March	2.6°
April	12°
May	23.5°
June	15.3°
July	3.9°
August	−0.3°
September	−6.3°
October	−18.2°
November	−15.7°
December	−10.3°

90. Which month had the greatest increase in temperature?   May
91. Which month had the greatest decrease in temperature?

If a is a positive number and b is a negative number, determine whether each statement is true or false.   **91.** October

92. $a - b$ is always a positive number.   true
93. $b - a$ is always a negative number.   true
94. $|b| - |a|$ is always a positive number.   false
95. $|b - a|$ is always a positive number.   true

Without calculating, determine whether each answer is positive or negative. Then use a calculator to find the exact difference.

96. $56,875 - 87,262$
negative, −30,387

97. $4.362 - 7.0086$
negative, −2.6466

1.7   MULTIPLYING AND DIVIDING REAL NUMBERS

Objectives

1 Multiply and divide real numbers.

2 Evaluate algebraic expressions using real numbers.

1   In this section, we discover patterns for multiplying and dividing real numbers. To discover sign rules for multiplication, recall that multiplication is repeated addition. Thus $3 \cdot 2$ means that 2 is an addend 3 times. That is,

$$2 + 2 + 2 = 3 \cdot 2$$

which equals 6. Similarly, $3 \cdot (-2)$ means -2 is an addend 3 times. That is,

$$(-2) + (-2) + (-2) = 3 \cdot (-2)$$

Since $(-2) + (-2) + (-2) = -6$, then $3 \cdot (-2) = -6$. This suggests that the product of a positive number and a negative number is a negative number.

What about the product of two negative numbers? To find out, consider the following pattern.

Factor decreases by **1** each time

$$
\left.
\begin{aligned}
-3 \cdot 2 &= -6 \\
-3 \cdot 1 &= -3 \\
-3 \cdot 0 &= 0
\end{aligned}
\right\} \text{Product increases by } \mathbf{3} \text{ each time.}
$$

This pattern continues as

Factor decreases by **1** each time

$$
\left.
\begin{aligned}
-3 \cdot -1 &= 3 \\
-3 \cdot -2 &= 6
\end{aligned}
\right\} \text{Product increases by } \mathbf{3} \text{ each time.}
$$

This suggests that the product of two negative numbers is a positive number.

> **Multiplying Real Numbers**
>
> 1. The product of two numbers with the *same* sign is a positive number.
> 2. The product of two numbers with *different* signs is a negative number.

EXAMPLE 1

Multiply.

a. $(-6)(4)$　　**b.** $2(-1)$　　**c.** $(-5)(-10)$

Solution　**a.** $(-6)(4) = -24$　　**b.** $2(-1) = -2$　　**c.** $(-5)(-10) = 50$

CLASSROOM EXAMPLE
Multiply.
a. $(-8)(-9)$ **b.** $(-4)(2)$
answer: **a.** 72 **b.** -8

We know that every whole number multiplied by zero equals zero. This remains true for real numbers.

> **Zero as a Factor**
>
> If b is a real number, then $b \cdot 0 = 0$. Also, $0 \cdot b = 0$.

EXAMPLE 2

Perform the indicated operations.

a. $(7)(0)(-6)$　　**b.** $(-2)(-3)(-4)$　　**c.** $(-1)(5)(-9)$　　**d.** $(-4)(-11) - (5)(-2)$

CLASSROOM EXAMPLE
Perform the indicated operations.
a. $(-2)(4)(-8)$
b. $-3(-9) - 4(-4)$
answer: **a.** 64 **b.** 43

Solution　**a.** By the order of operations, we multiply from left to right. Notice that, because one of the factors is 0, the product is 0.

$$(7)(0)(-6) = 0(-6) = 0$$

b. Multiply two factors at a time, from left to right.

$$(-2)(-3)(-4) = (6)(-4) \qquad \text{Multiply } (-2)(-3).$$
$$= -24$$

c. Multiply from left to right.

$$(-1)(5)(-9) = (-5)(-9) \qquad \text{Multiply } (-1)(5).$$
$$= 45$$

d. Follow the rules for order of operation.

$$(-4)(-11) - (5)(-2) = 44 - (-10) \qquad \text{Find each product.}$$
$$= 44 + 10 \qquad \text{Add } \mathbf{44} \text{ to the opposite of } -\mathbf{10}.$$
$$= 54 \qquad \text{Add.}$$

> **Helpful Hint**
>
> You may have noticed from the example that if we multiply:
>
> - an *even* number of negative numbers, the product is *positive*.
> - an *odd* number of negative numbers, the product is *negative*.

Multiplying signed decimals or fractions is carried out exactly the same way as multiplying by integers.

EXAMPLE 3

Multiply.

a. $(-1.2)(0.05)$ **b.** $\dfrac{2}{3} \cdot \left(-\dfrac{7}{10}\right)$ **c.** $\left(-\dfrac{4}{5}\right)(-20)$

Solution   **a.** The product of two numbers with different signs is negative.

$$(-1.2)(0.05) = -[(1.2)(0.05)]$$
$$= -0.06$$

b. $\dfrac{2}{3} \cdot \left(-\dfrac{7}{10}\right) = -\dfrac{2 \cdot 7}{3 \cdot 10} = -\dfrac{2 \cdot 7}{3 \cdot 2 \cdot 5} = -\dfrac{7}{15}$

c. $\left(-\dfrac{4}{5}\right)(-20) = \dfrac{4 \cdot 20}{5 \cdot 1} = \dfrac{4 \cdot 4 \cdot 5}{5 \cdot 1} = \dfrac{16}{1}$ or 16

Now that we know how to multiply positive and negative numbers, let's see how we find the values of $(-4)^2$ and -4^2, for example. Although these two expressions look similar, the difference between the two is the parentheses. In $(-4)^2$, the parentheses tell us that the base, or repeated factor, is -4. In -4^2, only 4 is the base. Thus,

$$(-4)^2 = (-4)(-4) = 16 \qquad \text{The base is } -4.$$
$$-4^2 = -(4 \cdot 4) = -16 \qquad \text{The base is } 4.$$

EXAMPLE 4

Evaluate.

a. $(-2)^3$ **b.** -2^3 **c.** $(-3)^2$ **d.** -3^2

Solution   **a.** $(-2)^3 = (-2)(-2)(-2) = -8$ The base is -2.
b. $-2^3 = -(2 \cdot 2 \cdot 2) = -8$ The base is 2.
c. $(-3)^2 = (-3)(-3) = 9$ The base is -3.
d. $-3^2 = -(3 \cdot 3) = -9$ The base is 3.

> **Helpful Hint**
> Be careful when identifying the base of an exponential expression.
>
> $(-3)^2$ -3^2
> Base is -3 Base is 3
> $(-3)^2 = (-3)(-3) = 9$ $-3^2 = -(3 \cdot 3) = -9$

Just as every difference of two numbers $a - b$ can be written as the sum $a + (-b)$, so too every quotient of two numbers can be written as a product. For example, the quotient $6 \div 3$ can be written as $6 \cdot \frac{1}{3}$. Recall that the pair of numbers 3 and $\frac{1}{3}$ has a special relationship. Their product is 1 and they are called reciprocals or **multiplicative inverses** of each other.

Reciprocals or Multiplicative Inverses

Two numbers whose product is 1 are called reciprocals or multiplicative inverses of each other.

Notice that **0 has no multiplicative inverse** since 0 multiplied by any number is never 1 but always 0.

EXAMPLE 5

Find the reciprocal of each number.

a. 22 **b.** $\dfrac{3}{16}$ **c.** -10 **d.** $-\dfrac{9}{13}$

Solution

a. The reciprocal of 22 is $\frac{1}{22}$ since $22 \cdot \frac{1}{22} = 1$.

b. The reciprocal of $\frac{3}{16}$ is $\frac{16}{3}$ since $\frac{3}{16} \cdot \frac{16}{3} = 1$.

c. The reciprocal of -10 is $-\frac{1}{10}$.

d. The reciprocal of $-\frac{9}{13}$ is $-\frac{13}{9}$.

We may now write a quotient as an equivalent product.

CLASSROOM EXAMPLE
Find the reciprocal.
a. $\dfrac{7}{15}$ **b.** -5
answer:
a. $\dfrac{15}{7}$ **b.** $-\dfrac{1}{5}$

Quotient of Two Real Numbers

If a and b are real numbers and b is not 0, then

$$a \div b = \frac{a}{b} = a \cdot \frac{1}{b}$$

In other words, the quotient of two real numbers is the product of the first number and the multiplicative inverse or reciprocal of the second number.

EXAMPLE 6

Use the definition of the quotient of two numbers to divide.

a. $-18 \div 3$ **b.** $\dfrac{-14}{-2}$ **c.** $\dfrac{20}{-4}$

Solution

a. $-18 \div 3 = -18 \cdot \dfrac{1}{3} = -6$ **b.** $\dfrac{-14}{-2} = -14 \cdot -\dfrac{1}{2} = 7$

c. $\dfrac{20}{-4} = 20 \cdot -\dfrac{1}{4} = -5$

CLASSROOM EXAMPLE
Divide.
a. $-12 \div 4$ **b.** $\dfrac{-80}{-10}$
answer:
a. -3 **b.** 8

Since the quotient $a \div b$ can be written as the product $a \cdot \frac{1}{b}$, it follows that sign patterns for dividing two real numbers are the same as sign patterns for multiplying two real numbers.

Multiplying and Dividing Real Numbers

1. The product or quotient of two numbers with the *same* sign is a positive number.

2. The product or quotient of two numbers with *different* signs is a negative number.

EXAMPLE 7

Divide.

a. $\dfrac{-24}{-4}$ **b.** $\dfrac{-36}{3}$ **c.** $\dfrac{2}{3} \div \left(-\dfrac{5}{4}\right)$ **d.** $-\dfrac{3}{2} \div 9$

Solution   **a.** $\dfrac{-24}{-4} = 6$ **b.** $\dfrac{-36}{3} = -12$ **c.** $\dfrac{2}{3} \div \left(-\dfrac{5}{4}\right) = \dfrac{2}{3} \cdot \left(-\dfrac{4}{5}\right) = -\dfrac{8}{15}$

d. $-\dfrac{3}{2} \div 9 = -\dfrac{3}{2} \cdot \dfrac{1}{9} = -\dfrac{3 \cdot 1}{2 \cdot 9} = -\dfrac{3 \cdot 1}{2 \cdot 3 \cdot 3} = -\dfrac{1}{6}$

The definition of the quotient of two real numbers does not allow for division by 0 because 0 does not have a multiplicative inverse. There is no number we can multiply 0 by to get 1. How then do we interpret $\dfrac{3}{0}$? We say that division by 0 is not allowed or not defined and that $\dfrac{3}{0}$ does not represent a real number. The denominator of a fraction can never be 0.

Can the numerator of a fraction be 0? Can we divide 0 by a number? Yes. For example,

$$\frac{0}{3} = 0 \cdot \frac{1}{3} = 0$$

In general, the quotient of 0 and any nonzero number is 0.

Zero as a Divisor or Dividend

1. The quotient of any nonzero real number and 0 is undefined. In symbols, if $a \neq 0$, $\dfrac{a}{0}$ is **undefined**.

2. The quotient of 0 and any real number except 0 is 0. In symbols, if $a \neq 0$, $\dfrac{0}{a} = 0$.

EXAMPLE 8

Perform the indicated operations.

a. $\dfrac{1}{0}$ **b.** $\dfrac{0}{-3}$ **c.** $\dfrac{0(-8)}{2}$

Solution   **a.** $\dfrac{1}{0}$ is undefined   **b.** $\dfrac{0}{-3} = 0$ **c.** $\dfrac{0(-8)}{2} = \dfrac{0}{2} = 0$

Notice that $\dfrac{12}{-2} = -6$, $-\dfrac{12}{2} = -6$, and $\dfrac{-12}{2} = -6$. This means that

$$\frac{12}{-2} = -\frac{12}{2} = \frac{-12}{2}$$

In words, a single negative sign in a fraction can be written in the denominator, in the numerator, or in front of the fraction without changing the value of the fraction. Thus,

$$\frac{1}{-7} = \frac{-1}{7} = -\frac{1}{7}$$

TEACHING TIP

Remind students that

$$\frac{39}{-5} = \frac{-39}{5} = -\frac{39}{5}.$$

In general, if a and b are real numbers, $b \neq 0$, $\dfrac{a}{-b} = \dfrac{-a}{b} = -\dfrac{a}{b}$.

Examples combining basic arithmetic operations along with the principles of order of operations help us to review these concepts.

EXAMPLE 9

Simplify each expression.

a. $\dfrac{(-12)(-3) + 3}{-7 - (-2)}$ b. $\dfrac{2(-3)^2 - 20}{-5 + 4}$

Solution   a. First, simplify the numerator and denominator separately, then divide.

CLASSROOM EXAMPLE

Simplify $\dfrac{5(-2)^3 + 52}{-4 + 1}$.

answer: -4

$$\frac{(-12)(-3) + 3}{-7 - (-2)} = \frac{36 + 3}{-7 + 2}$$

$$= \frac{39}{-5} \text{ or } -\frac{39}{5}$$

b. Simplify the numerator and denominator separately, then divide.

$$\frac{2(-3)^2 - 20}{-5 + 4} = \frac{2 \cdot 9 - 20}{-5 + 4} = \frac{18 - 20}{-5 + 4} = \frac{-2}{-1} = 2$$

2   Using what we have learned about multiplying and dividing real numbers, we continue to practice evaluating algebraic expressions.

EXAMPLE 10

If $x = -2$ and $y = -4$, evaluate each expression.

a. $5x - y$ b. $x^4 - y^2$ c. $\dfrac{3x}{2y}$

Solution   a. Replace x with -2 and y with -4 and simplify.

CLASSROOM EXAMPLE

If $x = -7$ and $y = -3$, evaluate $\dfrac{x^2 - y}{2}$.

answer: 26

$$5x - y = 5(-2) - (-4) = -10 - (-4) = -10 + 4 = -6$$

b. Replace x with -2 and y with -4.

$$x^4 - y^2 = (-2)^4 - (-4)^2 \qquad \text{Substitute the given values for the variables.}$$
$$= 16 - (16) \qquad \text{Evaluate exponential expressions.}$$
$$= 0 \qquad \text{Subtract.}$$

c. Replace x with -2 and y with -4 and simplify.

$$\frac{3x}{2y} = \frac{3(-2)}{2(-4)} = \frac{-6}{-8} = \frac{3}{4}$$

Calculator Explorations

Entering Negative Numbers on a Scientific Calculator

To enter a negative number on a scientific calculator, find a key marked $\boxed{+/-}$. (On some calculators, this key is marked $\boxed{\text{CHS}}$ for "change sign.") To enter -8, for example, press the keys $\boxed{8}$ $\boxed{+/-}$. The display will read $\boxed{-8}$.   *(continued)*

Entering Negative Numbers on a Graphing Calculator

To enter a negative number on a graphing calculator, find a key marked $\boxed{(-)}$. Do not confuse this key with the key $\boxed{-}$, which is used for subtraction. To enter -8, for example, press the keys $\boxed{(-)}$ $\boxed{8}$. The display will read $\boxed{-8}$.

Operations with Real Numbers

To evaluate $-2(7-9)-20$ on a calculator, press the keys

The display will read $\boxed{-16}$ or $\boxed{\begin{array}{l} -2(7-9)-20 \\ \qquad\qquad -16 \end{array}}$.

Use a calculator to simplify each expression.

1. $-38(26-27)$ 38

2. $-59(-8)+1726$ 2198

3. $134+25(68-91)$ -441

4. $45(32)-8(218)$ -304

5. $\dfrac{-50(294)}{175-265}$ $163.\overline{3}$

6. $\dfrac{-444-444.8}{-181-324}$ 1.76

7. 9^5-4550 $54,499$

8. 5^8-6259 $384,366$

9. $(-125)^2$ (Be careful.) $15,625$

10. -125^2 (Be careful.) $-15,625$

MENTAL MATH

Answer the following with positive or negative.

1. The product of two negative numbers is a <u>positive</u> number.
2. The quotient of two negative numbers is a <u>positive</u> number.
3. The quotient of a positive number and a negative number is a <u>negative</u> number.
4. The product of a positive number and a negative number is a <u>negative</u> number.
5. The reciprocal of a positive number is a <u>positive</u> number.
6. The opposite of a positive number is a <u>negative</u> number.

EXERCISE SET 1.7

STUDY GUIDE/SSM   CD/VIDEO   PH MATH TUTOR CENTER   MathXL®Tutorials ON CD   MathXL®   MyMathLab®

Multiply. See Examples 1 through 3.

 1. $-6(4)$ -24

2. $-8(5)$ -40

 3. $2(-1)$ -2

4. $7(-4)$ -28

 5. $-5(-10)$ 50

6. $-6(-11)$ 66

7. $-3 \cdot 4$ -12

8. $-2 \cdot 8$ -16

9. $-7 \cdot 0$ 0

10. $-6 \cdot 0$ 0

11. $2(-9)$ -18

12. $3(-5)$ -15

13. $-\dfrac{1}{2}\left(-\dfrac{3}{5}\right)$ $\dfrac{3}{10}$

14. $-\dfrac{1}{8}\left(-\dfrac{1}{3}\right)$ $\dfrac{1}{24}$

15. $-\dfrac{3}{4}\left(-\dfrac{8}{9}\right)$ $\dfrac{2}{3}$

16. $-\dfrac{5}{6}\left(-\dfrac{3}{10}\right)$ $\dfrac{1}{4}$

17. $5(-1.4)$ -7

18. $6(-2.5)$ -15

19. $-0.2(-0.7)$ 0.14

20. $-0.5(-0.3)$ 0.15

21. $-10(80)$ -800

22. $-20(60)$ -1200

23. $4(-7)$ -28

24. $5(-9)$ -45

25. $(-5)(-5)$ 25

26. $(-7)(-7)$ 49

 27. $\dfrac{2}{3}\left(-\dfrac{4}{9}\right)$ $-\dfrac{8}{27}$

28. $\dfrac{2}{7}\left(-\dfrac{2}{11}\right)$ $-\dfrac{4}{77}$

29. $-11(11)$ -121

30. $-12(12)$ -144

31. $-\dfrac{20}{25}\left(\dfrac{5}{16}\right)$ $-\dfrac{1}{4}$ **32.** $-\dfrac{25}{36}\left(\dfrac{6}{15}\right)$ $-\dfrac{5}{18}$

33. $(-1)(2)(-3)(-5)$ -30 **34.** $(-2)(-3)(-4)(-2)$ 48

Perform the indicated operation. See Example 2.

35. $(-2)(5) - (-11)(3)$ 23 **36.** $8(-3) - 4(-5)$ -4

37. $(-6)(-1)(-2) - (-5)$ -7

38. $20 - (-4)(3)(-2)$ -4

Decide whether each statement is true or false.

39. The product of three negative integers is negative. true
40. The product of three positive integers is positive. true
41. The product of four negative integers is negative. false
42. The product of four positive integers is positive. true

Evaluate. See Example 4.

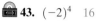 **43.** $(-2)^4$ 16 **44.** -2^4 -16
45. -1^5 -1 **46.** $(-1)^5$ -1
47. $(-5)^2$ 25 **48.** -5^2 -25
49. -7^2 -49 **50.** $(-7)^2$ 49

Find each reciprocal or multiplicative inverse. See Example 5.

51. 9 $\dfrac{1}{9}$ **52.** 100 $\dfrac{1}{100}$ **53.** $\dfrac{2}{3}$ $\dfrac{3}{2}$
54. $\dfrac{1}{7}$ 7 **55.** -14 $-\dfrac{1}{14}$ **56.** -8 $-\dfrac{1}{8}$
57. $-\dfrac{3}{11}$ $-\dfrac{11}{3}$ **58.** $-\dfrac{6}{13}$ $-\dfrac{13}{6}$ **59.** 0.2 $\dfrac{1}{0.2}$
60. 1.5 $\dfrac{1}{1.5}$ **61.** $\dfrac{1}{-6.3}$ -6.3 **62.** $\dfrac{1}{-8.9}$ -8.9

Divide. See Examples 6 through 8.

 63. $\dfrac{18}{-2}$ -9 **64.** $\dfrac{20}{-10}$ -2 **65.** $\dfrac{-16}{-4}$ 4
66. $\dfrac{-18}{-6}$ 3 **67.** $\dfrac{-48}{12}$ -4 **68.** $\dfrac{-60}{5}$ -12
69. $\dfrac{0}{-4}$ 0 **70.** $\dfrac{0}{-9}$ 0 **71.** $-\dfrac{15}{3}$ -5
72. $-\dfrac{24}{8}$ -3 **73.** $\dfrac{5}{0}$ undefined   **74.** $\dfrac{3}{0}$ undefined
75. $\dfrac{-12}{-4}$ 3 **76.** $\dfrac{-45}{-9}$ 5 **77.** $\dfrac{30}{-2}$ -15
78. $\dfrac{14}{-2}$ -7 **79.** $\dfrac{6}{7} \div \left(-\dfrac{1}{3}\right)$ **80.** $\dfrac{4}{5} \div \left(-\dfrac{1}{2}\right)$
81. $-\dfrac{5}{9} \div \left(-\dfrac{3}{4}\right)$ $\dfrac{20}{27}$ **82.** $-\dfrac{1}{10} \div \left(-\dfrac{8}{11}\right)$ $\dfrac{11}{80}$
83. $-\dfrac{4}{9} \div \dfrac{4}{9}$ -1 **84.** $-\dfrac{5}{12} \div \dfrac{5}{12}$ -1

MIXED PRACTICE

Simplify. See Example 9.   **79.** $-\dfrac{18}{7}$ **80.** $-\dfrac{8}{5}$

85. $\dfrac{-9(-3)}{-6}$ $-\dfrac{9}{2}$ **86.** $\dfrac{-6(-3)}{-4}$ $-\dfrac{9}{2}$

87. $\dfrac{12}{9 - 12}$ -4 **88.** $\dfrac{-15}{1 - 4}$ 5
89. $\dfrac{-6^2 + 4}{-2}$ 16 **90.** $\dfrac{3^2 + 4}{5}$ $\dfrac{13}{5}$
91. $\dfrac{8 + (-4)^2}{4 - 12}$ -3 **92.** $\dfrac{6 + (-2)^2}{4 - 9}$ -2
93. $\dfrac{22 + (3)(-2)}{-5 - 2}$ $-\dfrac{16}{7}$ **94.** $\dfrac{-20 + (-4)(3)}{1 - 5}$ 8
95. $\dfrac{-3 - 5^2}{2(-7)}$ 2 **96.** $\dfrac{-2 - 4^2}{3(-6)}$ 1
97. $\dfrac{6 - 2(-3)}{4 - 3(-2)}$ $\dfrac{6}{5}$ **98.** $\dfrac{8 - 3(-2)}{2 - 5(-4)}$ $\dfrac{7}{11}$
99. $\dfrac{-3 - 2(-9)}{-15 - 3(-4)}$ -5 **100.** $\dfrac{-4 - 8(-2)}{-9 - 2(-3)}$ -4
101. $\dfrac{|5 - 9| + |10 - 15|}{|2(-3)|}$ $\dfrac{3}{2}$ **102.** $\dfrac{|-3 + 6| + |-2 + 7|}{|-2 \cdot 2|}$ 2

If $x = -5$ and $y = -3$, evaluate each expression. See Example 10.

103. $3x + 2y$ -21 **104.** $4x + 5y$ -35
105. $2x^2 - y^2$ 41 **106.** $x^2 - 2y^2$ 7
107. $x^3 + 3y$ -134 **108.** $y^3 + 3x$ -42
109. $\dfrac{2x - 5}{y - 2}$ 3 **110.** $\dfrac{2y - 12}{x - 4}$ 2
111. $\dfrac{-3 - y}{x - 4}$ 0 **112.** $\dfrac{4 - 2x}{y + 3}$ undefined

113. At the end of 2002, Delta Airlines posted a net loss of $1272 million, which we will write as $-\$1272$ million. If this continued, what would Delta's income be after four years? (*Source*: Delta Airlines) $-\$5088$ million

114. Gateway sells personal computers through mail order and retail stores. In 2002, Gateway posted a net income of $-\$298$ million. If this continued, what would Gateway's income be after three years? (*Source*: Gateway)

Decide whether the given number is a solution of the given equation.   **114.** $-\$894$ million

115. Is 7 a solution of $-5x = -35$? yes
116. Is -4 a solution of $2x = x - 1$? no
117. Is -20 a solution of $\dfrac{x}{10} = 2$? no
118. Is -3 a solution of $\dfrac{45}{x} = -15$? yes
119. Is 5 a solution of $-3x - 5 = -20$? yes
120. Is -4 a solution of $2x + 4 = x + 8$? no

Concept Extensions

121. Explain why the product of an even number of negative numbers is a positive number. answers may vary
122. If a and b are any real numbers, is the statement $a \cdot b = b \cdot a$ always true? Why or why not? answers may vary
123. Find any real numbers that are their own reciprocal. $1, -1$
124. Explain why 0 has no reciprocal. answers may vary

If q is a negative number, r is a negative number, and t is a positive number, determine whether each expression simplifies to a positive or negative number. If it is not possible to determine, state so.

125. $\frac{q}{r \cdot t}$ positive

126. $q^2 \cdot r \cdot t$ negative

127. $q + t$ not possible

128. $t + r$ not possible

129. $t(q + r)$ negative

130. $r(q - t)$ positive

Write each of the following as an expression and evaluate.

131. The sum of -2 and the quotient of -15 and 3 $-2 + \frac{-15}{3}$; -7

132. The sum of 1 and the product of -8 and -5

133. Twice the sum of -5 and -3 $2[-5 + (-3)]$; -16

134. 7 subtracted from the quotient of 0 and 5 $\frac{0}{5} - 7$; -7

132. $1 + (-8)(-5)$; 41

OPERATIONS ON REAL NUMBERS   INTEGRATED REVIEW

Answer the following with positive or negative.

1. The product of two negative numbers is a __positive__ number.

2. The quotient of two negative numbers is a __positive__ number.

3. The quotient of a positive number and a negative number is a __negative__ number.

4. The product of a positive number and a negative number is a __negative__ number.

5. The reciprocal of a positive number is a __positive__ number.

6. The opposite of a positive number is a __negative__ number.

7. The sum of two negative numbers is a __negative__ number.

8. The absolute value of a negative number is a __positive__ number.

Perform each indicated operation and simplify.

9. $5(-7)$ -35

10. $-3(-10)$ 30

11. $\frac{-20}{-4}$ 5

12. $\frac{30}{-6}$ -5

13. $7 - (-3)$ 10

14. $-8 - 10$ -18

15. $-14 - (-12)$ -2

16. $-3 - (-1)$ -2

17. $-\frac{1}{2}\left(-\frac{3}{4}\right)$ $\frac{3}{8}$

18. $-\frac{2}{7}\left(\frac{11}{12}\right)$ $-\frac{11}{42}$

19. $\frac{-12}{0.2}$ -60

20. $\frac{-3.8}{-2}$ 1.9

21. $-19 + (-23)$ -42

22. $18 + (-25)$ -7

23. $-15 + 17$ 2

24. $-2 + (-37)$ -39

25. $(-8)^2$ 64

26. -9^2 -81

27. -3^3 -27

28. $(-2)^4$ 16

29. -1^{10} -1

30. $(-1)^{10}$ 1

31. $(-2)^5$ -32

32. -2^5 -32

33. $(2)(-8)(-3)$ 48

34. $3(-2)(5)$ -30

35. $-6(2) + 20 \div 2 - 4$ -6

36. $-4(-3) + 9 \div 3 - 6$ 9

37. $-3^2 - [6 + 5|-2 - 1|]$ -30 **38.** -44

38. $-5^2 - [4 + 3|-3 - 2|]$

39. $2(19 - 17)^3 - 3(7 - 9)^2$ 4

40. $3(10 - 9)^2 - 6(20 - 19)^3$ -3

41. $\frac{19 - 25}{3(-1)}$ 2

42. $\frac{8(-4)}{-2}$ 16

43. $\frac{-2(3 - 6) - 6(10 - 9)}{-6 - (-5)}$ 0

44. $\frac{5(7 - 9) - 3(100 - 97)}{4 - 5}$ 19

45. $\frac{-4(8 - 10)^3}{-2 - 1 - 12}$ $-\frac{32}{15}$

46. $\frac{6(7 - 10)^2}{6 - (-1) - 2}$ $\frac{54}{5}$

1.8   PROPERTIES OF REAL NUMBERS

Objectives

1   Use the commutative and associative properties.

2   Use the distributive property.

3   Use the identity and inverse properties.

1   In this section we give names to properties of real numbers with which we are already familiar. Throughout this section, the variables a, b, and c represent real numbers.

We know that order does not matter when adding numbers. For example, we know that $7 + 5$ is the same as $5 + 7$. This property is given a special name—the **commutative property of addition**. We also know that order does not matter when multiplying numbers. For example, we know that $-5(6) = 6(-5)$. This property means that multiplication is commutative also and is called the **commutative property of multiplication**.

Commutative Properties

Addition: $a + b = b + a$

Multiplication: $a \cdot b = b \cdot a$

These properties state that the *order* in which any two real numbers are added or multiplied does not change their sum or product. For example, if we let $a = 3$ and $b = 5$, then the commutative properties guarantee that

$$3 + 5 = 5 + 3 \qquad \text{and} \qquad 3 \cdot 5 = 5 \cdot 3$$

> **Helpful Hint**
>
> Is subtraction also commutative? Try an example. Is $3 - 2 = 2 - 3$? **No!** The left side of this statement equals 1; the right side equals -1. There is no commutative property of subtraction. Similarly, there is no commutative property for division. For example, $10 \div 2$ does not equal $2 \div 10$.

STUDY SKILLS REMINDER

You may want to start a list in your notes describing the situations in which you should take extra caution, such as

1. cannot divide by 0

2. subtraction is not commutative—order matters

EXAMPLE 1

Use a commutative property to complete each statement.

a. $x + 5 =$ _____   **b.** $3 \cdot x =$ _____

Solution   **a.** $x + 5 = 5 + x$ By the commutative property of addition

b. $3 \cdot x = x \cdot 3$ By the commutative property of multiplication

✔ **CONCEPT CHECK**

Which of the following pairs of actions are commutative?

a. "raking the leaves" and "bagging the leaves"
b. "putting on your left glove" and "putting on your right glove"
c. "putting on your coat" and "putting on your shirt"
d. "reading a novel" and "reading a newspaper"

Let's now discuss grouping numbers. We know that when we add three numbers, the way in which they are grouped or associated does not change their sum. For example, we know that $2 + (3 + 4) = 2 + 7 = 9$. This result is the same if we group the numbers differently. In other words, $(2 + 3) + 4 = 5 + 4 = 9$, also. Thus, $2 + (3 + 4) = (2 + 3) + 4$. This property is called the **associative property of addition**.

We also know that changing the grouping of numbers when multiplying does not change their product. For example, $2 \cdot (3 \cdot 4) = (2 \cdot 3) \cdot 4$ (check it). This is the **associative property of multiplication**.

Associative Properties

Addition: $(a + b) + c = a + (b + c)$
Multiplication: $(a \cdot b) \cdot c = a \cdot (b \cdot c)$

These properties state that the way in which three numbers are *grouped* does not change their sum or their product.

EXAMPLE 2

Use an associative property to complete each statement.

a. $5 + (4 + 6) =$ _____   **b.** $(-1 \cdot 2) \cdot 5 =$ _____

Solution   **a.** $5 + (4 + 6) = (5 + 4) + 6$ By the associative property of addition

b. $(-1 \cdot 2) \cdot 5 = -1 \cdot (2 \cdot 5)$ By the associative property of multiplication

▶ **Helpful Hint**

Remember the difference between the commutative properties and the associative properties. The commutative properties have to do with the *order* of numbers, and the associative properties have to do with the *grouping* of numbers.

Let's now illustrate how these properties can help us simplify expressions.

EXAMPLE 3

Simplify each expression.

a. $10 + (x + 12)$ **b.** $-3(7x)$

Solution

a. $10 + (x + 12) = 10 + (12 + x)$ By the commutative property of addition

$= (10 + 12) + x$ By the associative property of addition

$= 22 + x$ Add.

b. $-3(7x) = (-3 \cdot 7)x$ By the associative property of multiplication

$= -21x$ Multiply.

2 The **distributive property of multiplication over addition** is used repeatedly throughout algebra. It is useful because it allows us to write a product as a sum or a sum as a product.

We know that $7(2 + 4) = 7(6) = 42$. Compare that with $7(2) + 7(4) = 14 + 28 = 42$. Since both original expressions equal 42, they must equal each other, or

$$7(2 + 4) = 7(2) + 7(4)$$

This is an example of the distributive property. The product on the left side of the equal sign is equal to the sum on the right side. We can think of the 7 as being distributed to each number inside the parentheses.

Distributive Property of Multiplication Over Addition

$$a(b + c) = ab + ac$$

Since multiplication is commutative, this property can also be written as

$$(b + c)a = ba + ca$$

The distributive property can also be extended to more than two numbers inside the parentheses. For example,

$$3(x + y + z) = 3(x) + 3(y) + 3(z)$$
$$= 3x + 3y + 3z$$

Since we define subtraction in terms of addition, the distributive property is also true for subtraction. For example

$$2(x - y) = 2(x) - 2(y)$$
$$= 2x - 2y$$

EXAMPLE 4

Use the distributive property to write each expression without parentheses. Then simplify if possible.

a. $2(x + y)$ **b.** $-5(-3 + 2z)$ **c.** $5(x + 3y - z)$
d. $-1(2 - y)$ **e.** $-(3 + x - w)$ **f.** $4(3x + 7) + 10$

Solution   **a.** $2(x + y) = 2 \cdot x + 2 \cdot y$
$$= 2x + 2y$$

b. $-5(-3 + 2z) = -5(-3) + (-5)(2z)$
$$= 15 - 10z$$

c. $5(x + 3y - z) = 5(x) + 5(3y) - 5(z)$
$$= 5x + 15y - 5z$$

d. $-1(2 - y) = (-1)(2) - (-1)(y)$
$$= -2 + y$$

e. $-(3 + x - w) = -1(3 + x - w)$
$$= (-1)(3) + (-1)(x) - (-1)(w)$$
$$= -3 - x + w$$

> **Helpful Hint**
> Notice in part **e** that $-(3 + x - w)$ is first rewritten as $-1(3 + x - w)$.

f. $4(3x + 7) + 10 = 4(3x) + 4(7) + 10$ Apply the distributive property.
$$= 12x + 28 + 10$$ Multiply.
$$= 12x + 38$$ Add.

We can use the distributive property in reverse to write a sum as a product.

EXAMPLE 5

Use the distributive property to write each sum as a product.

a. $8 \cdot 2 + 8 \cdot x$ **b.** $7s + 7t$

Solution   **a.** $8 \cdot 2 + 8 \cdot x = 8(2 + x)$ **b.** $7s + 7t = 7(s + t)$

3   Next, we look at the **identity properties**.

The number 0 is called the identity for addition because when 0 is added to any real number, the result is the same real number. In other words, the *identity* of the real number is not changed.

The number 1 is called the identity for multiplication because when a real number is multiplied by 1, the result is the same real number. In other words, the *identity* of the real number is not changed.

> **Identities for Addition and Multiplication**
>
> 0 is the identity element for addition.
> $$a + 0 = a \quad \text{and} \quad 0 + a = a$$
> 1 is the identity element for multiplication.
> $$a \cdot 1 = a \quad \text{and} \quad 1 \cdot a = a$$

Notice that 0 is the *only* number that can be added to any real number with the result that the sum is the same real number. Also, 1 is the *only* number that can be multiplied by any real number with the result that the product is the same real number.

Additive inverses or **opposites** were introduced in Section 1.5. Two numbers are called additive inverses or opposites if their sum is 0. The additive inverse or opposite of 6 is -6 because $6 + (-6) = 0$. The additive inverse or opposite of -5 is 5 because $-5 + 5 = 0$.

Reciprocals or **multiplicative inverses** were introduced in Section 1.3. Two nonzero numbers are called reciprocals or multiplicative inverses if their product is 1. The reciprocal or multiplicative inverse of $\frac{2}{3}$ is $\frac{3}{2}$ because $\frac{2}{3} \cdot \frac{3}{2} = 1$. Likewise, the reciprocal of -5 is $-\frac{1}{5}$ because $-5\left(-\frac{1}{5}\right) = 1$.

✔ **CONCEPT CHECK**

Which of the following is the

a. opposite of $-\frac{3}{10}$, and which is the **b.** reciprocal of $-\frac{3}{10}$?

$$1, -\frac{10}{3}, \frac{3}{10}, 0, \frac{10}{3}, -\frac{3}{10}$$

Additive or Multiplicative Inverses

The numbers a and $-a$ are additive inverses or opposites of each other because their sum is 0; that is,

$$a + (-a) = 0$$

The numbers b and $\frac{1}{b}$ (for $b \neq 0$) are reciprocals or multiplicative inverses of each other because their product is 1; that is,

$$b \cdot \frac{1}{b} = 1$$

EXAMPLE 6

Name the property or properties illustrated by each true statement.

Solution
a. $3 \cdot y = y \cdot 3$ — *Commutative property of multiplication (order changed)*

b. $(x + 7) + 9 = x + (7 + 9)$ — *Associative property of addition (grouping changed)*

c. $(b + 0) + 3 = b + 3$ — *Identity element for addition*

d. $0.2 \cdot (z \cdot 5) = 0.2 \cdot (5 \cdot z)$ — *Commutative property of multiplication (order changed)*

e. $-2 \cdot \left(-\frac{1}{2}\right) = 1$ — *Multiplicative inverse property*

f. $-2 + 2 = 0$ — *Additive inverse property*

g. $-6 \cdot (y \cdot 2) = (-6 \cdot 2) \cdot y$ — *Commutative and associative properties of multiplication (order and grouping changed)*

Concept Check Answer:

a. $\frac{3}{10}$ **b.** $-\frac{10}{3}$

EXERCISE SET 1.8

STUDY GUIDE/SSM   CD/VIDEO   PH MATH TUTOR CENTER   MathXL®Tutorials ON CD   MathXL®   MyMathLab®

Use a commutative property to complete each statement. See Examples 1 and 3.

 1. $x + 16 = \underline{\quad 16 + x \quad}$

2. $4 + y = \underline{\quad y + 4 \quad}$

3. $-4 \cdot y = \underline{\quad y \cdot (-4) \quad}$

4. $-2 \cdot x = \underline{\quad x \cdot (-2) \quad}$

 5. $xy = \underline{\quad yx \quad}$

6. $ab = \underline{\quad ba \quad}$

7. $2x + 13 = \underline{\quad 13 + 2x \quad}$

8. $19 + 3y = \underline{\quad 3y + 19 \quad}$

Use an associative property to complete each statement. See Examples 2 and 3.   **11.** $(2 + a) + b$ **12.** $y + (4 + z)$

 9. $(xy) \cdot z = \underline{\quad x \cdot (yz) \quad}$

10. $3 \cdot (xy) = \underline{\quad (3x) \cdot y \quad}$

11. $2 + (a + b) = \underline{\qquad}$

12. $(y + 4) + z = \underline{\qquad}$

13. $4 \cdot (ab) = \underline{\quad (4a) \cdot b \quad}$

14. $(-3y) \cdot z = \underline{\quad -3 \cdot (yz) \quad}$

 15. $(a + b) + c = \underline{\quad a + (b + c) \quad}$

16. $6 + (r + s) = \underline{\quad (6 + r) + s \quad}$

Use the commutative and associative properties to simplify each expression. See Example 3.

17. $8 + (9 + b)$ $17 + b$

18. $(r + 3) + 11$ $r + 14$

19. $4(6y)$ $24y$

20. $2(42x)$ $84x$

21. $\frac{1}{5}(5y)$ y

22. $\frac{1}{8}(8z)$ z

23. $(13 + a) + 13$ $26 + a$

24. $7 + (x + 4)$ $11 + x$

25. $-9(8x)$ $-72x$

26. $-3(12y)$ $-36y$

27. $\frac{3}{4}\left(\frac{4}{3}s\right)$ s

28. $\frac{2}{7}\left(\frac{7}{2}r\right)$ r

29. Write an example that shows that division is not commutative.

30. Write an example that shows that subtraction is not commutative.   answers may vary   **29.** answers may vary

Use the distributive property to write each expression without parentheses. Then simplify the result. See Example 4.

31. $4(x + y)$ $4x + 4y$

32. $7(a + b)$ $7a + 7b$

33. $9(x - 6)$ $9x - 54$

34. $11(y - 4)$ $11y - 44$

35. $2(3x + 5)$ $6x + 10$

36. $5(7 + 8y)$ $35 + 40y$

37. $7(4x - 3)$ $28x - 21$

38. $3(8x - 1)$ $24x - 3$

39. $3(6 + x)$ $18 + 3x$

40. $2(x + 5)$ $2x + 10$

41. $-2(y - z)$ $-2y + 2z$

42. $-3(z - y)$ $-3z + 3y$

43. $-7(3y + 5)$ $-21y - 35$

44. $-5(2r + 11)$ $-10r - 55$

45. $5(x + 4m + 2)$

46. $8(3y + z - 6)$

47. $-4(1 - 2m + n)$

48. $-4(4 + 2p + 5)$

49. $-(5x + 2)$ $-5x - 2$

50. $-(9r + 5)$ $-9r - 5$

51. $-(r - 3 - 7p)$

52. $-(q - 2 + 6r)$

53. $\frac{1}{2}(6x + 8)$ $3x + 4$

54. $\frac{1}{4}(4x - 2)$ $x - \frac{1}{2}$

55. $-\frac{1}{3}(3x - 9y)$ $-x + 3y$

56. $-\frac{1}{5}(10a - 25b)$

57. $3(2r + 5) - 7$ $6r + 8$

58. $10(4s + 6) - 40$

59. $-9(4x + 8) + 2$ $-36x - 70$

60. $-11(5x + 3) + 10$

61. $-4(4x + 5) - 5$ $-16x - 25$

62. $-6(2x + 1) - 1$

Use the distributive property to write each sum as a product. See Example 5.   **67.** $-1(5 + x)$ **68.** $-3(a + b)$

63. $4 \cdot 1 + 4 \cdot y$ $4(1 + y)$

64. $14 \cdot z + 14 \cdot 5$ $14(z + 5)$

65. $11x + 11y$ $11(x + y)$

66. $9a + 9b$ $9(a + b)$

67. $(-1) \cdot 5 + (-1) \cdot x$

68. $(-3)a + (-3)b$

69. $30a + 30b$ $30(a + b)$

70. $25x + 25y$ $25(x + y)$

Name the properties illustrated by each true statement. See Example 6.

71. $3 \cdot 5 = 5 \cdot 3$ commutative property of multiplication

72. $4(3 + 8) = 4 \cdot 3 + 4 \cdot 8$ distributive property

73. $2 + (x + 5) = (2 + x) + 5$

74. $(x + 9) + 3 = (9 + x) + 3$

75. $9(3 + 7) = 9 \cdot 3 + 9 \cdot 7$ distributive property

76. $1 \cdot 9 = 9$ identity element of multiplication

77. $(4 \cdot y) \cdot 9 = 4 \cdot (y \cdot 9)$ associative property of multiplication

78. $6 \cdot \frac{1}{6} = 1$ multiplicative inverse property

79. $0 + 6 = 6$ identity element of addition

80. $(a + 9) + 6 = a + (9 + 6)$ associative property of addition

81. $-4(y + 7) = -4 \cdot y + (-4) \cdot 7$ distributive property

82. $(11 + r) + 8 = (r + 11) + 8$

83. $-4 \cdot (8 \cdot 3) = (8 \cdot -4) \cdot 3$

84. $r + 0 = r$ identity element of addition

82. commutative property of addition

83. commutative and associative properties of multiplication

Concept Extensions

Fill in the table with the opposite (additive inverse), and the reciprocal (multiplicative inverse). Assume that the value of each expression is not 0.

	Expression	Opposite	Reciprocal
85.	8	-8	$\frac{1}{8}$
86.	$-\frac{2}{3}$	$\frac{2}{3}$	$-\frac{3}{2}$
87.	x	$-x$	$\frac{1}{x}$
88.	$4y$	$-4y$	$\frac{1}{4y}$
89.	$2x$	$-2x$	$\frac{1}{2x}$
90.	$-7x$	$7x$	$-\frac{1}{7x}$

45. $5x + 20m + 10$ **46.** $24y + 8z - 48$ **47.** $-4 + 8m - 4n$ **48.** $-16 - 8p - 20$ **51.** $-r + 3 + 7p$ **52.** $-q + 2 - 6r$
56. $-2a + 5b$ **58.** $40s + 20$ **60.** $-55x - 23$ **62.** $-12x - 7$ **73.** associative property of addition   **74.** commutative property of addition

Determine which pairs of actions are commutative.

91. "taking a test" and "studying for the test"   no
92. "putting on your shoes" and "putting on your socks"   no
93. "putting on your left shoe" and "putting on your right shoe"

93. yes   **94.** yes   **95.–96.** answers may vary

94. "reading the sports section" and "reading the comics section"
95. Explain why 0 is called the identity element for addition.
96. Explain why 1 is called the identity element for multiplication.

1.9   READING GRAPHS

Objectives

1 Read bar graphs.
2 Read line graphs.

In today's world, where the exchange of information must be fast and entertaining, graphs are becoming increasingly popular. They provide a quick way of making comparisons, drawing conclusions, and approximating quantities. Bar and line graphs are common in previous sections, but in this section, we do not label the heights of the bars or points.

1   A **bar graph** consists of a series of bars arranged vertically or horizontally. The bar graph in Example 1 shows a comparison of the rates charged by selected electricity companies. The names of the companies are listed horizontally and a bar is shown for each company. Corresponding to the height of the bar for each company is a number along a vertical axis. These vertical numbers are cents charged for each kilowatt-hour of electricity used.

   EXAMPLE 1

The following bar graph shows the cents charged per kilowatt-hour for selected electricity companies.

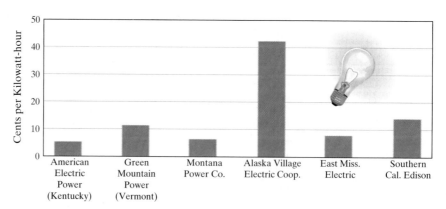

Source: Electric Company Listed

a. Which company charges the highest rate?
b. Which company charges the lowest rate?
c. Approximate the electricity rate charged by the first four companies listed.
d. Approximate the difference in the rates charged by the companies in parts (a) and (b).

Solution

a. The tallest bar corresponds to the company that charges the highest rate. Alaska Village Electric Cooperative charges the highest rate.

b. The shortest bar corresponds to the company that charges the lowest rate. American Electric Power in Kentucky charges the lowest rate.

c. To approximate the rate charged by American Electric Power, we go to the top of the bar that corresponds to this company. From the top of the bar, we move horizontally to the left until the vertical axis is reached.

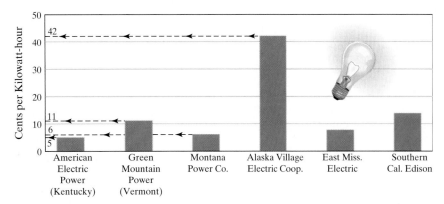

Source: Electric Company Listed

The height of the bar is approximately halfway between the 0 and 10 marks. We therefore conclude that

American Electric Power charges approximately 5¢ per kilowatt-hour.

Green Mountain Power charges approximately 11¢ per kilowatt-hour.

Montana Power Co. charges approximately 6¢ per kilowatt-hour.

Alaska Village Electric charges approximately 42¢ per kilowatt-hour.

d. The difference in rates for Alaska Village Electric Cooperative and American Electric Power is approximately 42¢ − 5¢ or 37¢.

EXAMPLE 2

The following bar graph shows the estimated worldwide number of Internet users by region as of 2002.

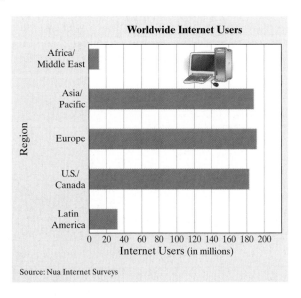

Worldwide Internet Users

Source: Nua Internet Surveys

a. Find the region that has the most Internet users and approximate the number of users.

b. How many more users are in Europe than Latin America?

Solution

a. Since these bars are arranged horizontally, we look for the longest bar, which is the bar representing Europe. To approximate the number associated with this region, we move from the right edge of this bar vertically downward to the Internet user axis. This region has approximately 190 million Internet users.

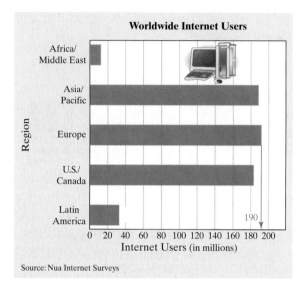

b. Europe has approximately 190 million Internet users. Latin America has approximately 33 million Internet users. To find how many more users are in Europe, we subtract 190 − 33 = 157 or 157 million more Internet users.

2 A **line graph** consists of a series of points connected by a line. The graph in Example 3 is a line graph.

EXAMPLE 3

The line graph below shows the relationship between the distance driven in a 14-foot U-Haul truck in one day and the total cost of renting this truck for that day. Notice that the horizontal axis is labeled Distance and the vertical axis is labeled Total Cost.

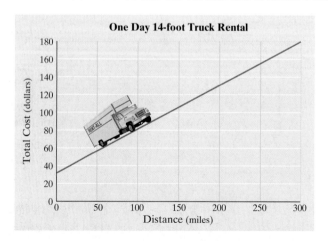

a. Find the total cost of renting the truck if 100 miles are driven.

b. Find the number of miles driven if the total cost of renting is $140.

Solution

a. Find the number 100 on the horizontal scale and move vertically upward until the line is reached. From this point on the line, we move horizontally to the left until the vertical scale is reached. We find that the total cost of renting the truck if 100 miles are driven is approximately $80.

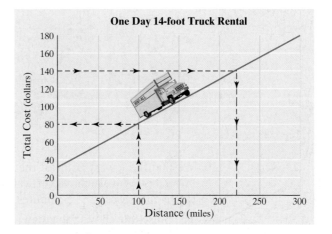

b. We find the number 140 on the vertical scale and move horizontally to the right until the line is reached. From this point on the line, we move vertically downward until the horizontal scale is reached. We find that the truck is driven approximately 225 miles.

From the previous example, we can see that graphing provides a quick way to approximate quantities. In Chapter 6 we show how we can use equations to find exact answers to the questions posed in Example 3. The next graph is another example of a line graph. It is also sometimes called a **broken line graph**.

 EXAMPLE 4

The line graph shows the relationship between time spent smoking a cigarette and pulse rate. Time is recorded along the horizontal axis in minutes, with 0 minutes being the moment a smoker lights a cigarette. Pulse is recorded along the vertical axis in heartbeats per minute.

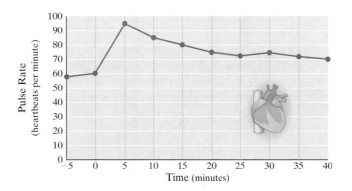

a. What is the pulse rate 15 minutes after lighting a cigarette?

b. When is the pulse rate the lowest?

c. When does the pulse rate show the greatest change?

Solution **a.** We locate the number 15 along the time axis and move vertically upward until the line is reached. From this point on the line, we move horizontally to the left until the pulse rate axis is reached. Reading the number of beats per minute, we find that the pulse rate is 80 beats per minute 15 minutes after lighting a cigarette.

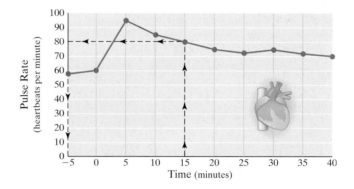

CLASSROOM EXAMPLE
Use the graph from Example 4 to answer the following.
a. What is the pulse rate 40 minutes after lighting a cigarette?
b. What is the pulse rate when the cigarette is being lit?
c. When is the pulse rate the highest?
answer: a. 70   **b.** 60
c. 5 minutes after lighting

b. We find the lowest point of the line graph, which represents the lowest pulse rate. From this point, we move vertically downward to the time axis. We find that the pulse rate is the lowest at −5 minutes, which means 5 minutes *before* lighting a cigarette.

TEACHING TIP
A Group Activity for this section is available in the Instructor's Resource Manual.

c. The pulse rate shows the greatest change during the 5 minutes between 0 and 5. Notice that the line graph is *steepest* between 0 and 5 minutes.

EXERCISE SET 1.9

STUDY GUIDE/SSM   CD/ VIDEO   PH MATH TUTOR CENTER   MathXL®Tutorials ON CD   MathXL®   MyMathLab®

The following bar graph shows the number of teenagers expected to use the Internet for the years shown. Use this graph to answer Exercises 1 through 4. See Example 1.

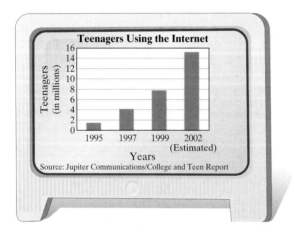

1. Approximate the number of teenagers who used the Internet in 1999.   approx. 7.8 million
2. Approximate the number of teenagers who used the Internet in 1995.   approx. 1.6 million
3. What year shows the greatest *increase* in number of teenagers using the Internet?   2002

4. How many more teenagers are expected to use the Internet in 2002 than in 1999?   approx. 7.4 million

As of August 2000, Crayola Crayons came in 120 colors. In addition to 2 blacks, 2 coppers, 2 grays, 1 gold, 1 silver, and 1 white, the following bar graph shows the number of shades of the seven other colors of Crayola Crayons. Use this graph to answer Exercises 5 through 8.

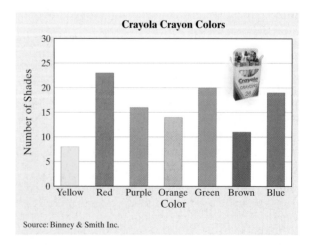

5. Find the Crayola color with the most shades and estimate the number of shades.   red: 23 shades

6. Find the Crayola color with the fewest shades and estimate the number of shades.   yellow: 8 shades   **7.** 9 shades
7. How many more shades of green are there than brown?
8. List the Crayola colors in order of number of shades, from least to greatest.   yellow, brown, orange, purple, blue, green, red

The following bar graph shows the top 10 tourist destinations and the number of tourists that visit each country per year. Use this graph to answer Exercises 9 through 14. See Examples 1 and 2.

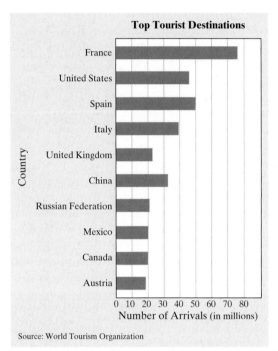

Top Tourist Destinations

Source: World Tourism Organization

9. France   **10.** Austria   **11.** France, U.S., Spain
9. Which country is the most popular tourist destination?
10. Which country shown is the least popular tourist destination?
11. Which countries have more than 40 million tourists per year?
12. Which countries shown have fewer than 20 million tourists per year?   Mexico, Canada, Austria
13. Estimate the number of tourists per year whose destination is Italy.   39 million
14. Estimate the number of tourists per year whose destination is France.   76 million
15. Africa/Middle East: 11 million users   **16.** 183 million users
Use the bar graph in Example 2 to answer Exercises 15 through 18.
15. Find the region that has the fewest Internet users and approximate the number of users.
16. Estimate the number of users in the U.S./Canada.
17. Estimate the number of users in Asia/Pacific. 187 million users
18. If the number of users in Africa is 6.3 million, estimate the number of users in the Middle East.   4.7 million users
22. 5 min before lighting
Use the line graph in Example 4 to answer Exercises 19 through 22.
19. Approximate the pulse rate 5 minutes before lighting a cigarette.   approx. 59 beats per min
20. Approximate the pulse rate 10 minutes after lighting a cigarette.   approx. 85 beats per min
21. Find the difference in pulse rate between 5 minutes before and 10 minutes after lighting a cigarette.   approx. 26 beats per min
22. When is the pulse rate fewer than 60 heartbeats per minute?

The following line graph shows the attendance at each Super Bowl game from 1995 through 2003. Use this graph to answer Exercises 23 through 30. See Examples 3 and 4.

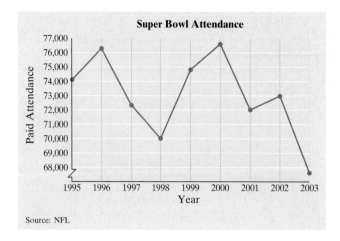

Super Bowl Attendance

Source: NFL

23. Estimate the Super Bowl attendance in 1999.   74,800
24. Estimate the Super Bowl attendance in 1997.   72,300
25. Find the year on the graph with the greatest Super Bowl attendance and approximate that attendance.   2000; 76,600
26. Find the year on the graph with the least Super Bowl attendance and approximate that attendance.   1998; 70,000
27. Name the year that Super Bowl attendance was less than 70,000.   2003
28. Name the years that Super Bowl attendance was greater than 75,000.   1996, 2000   **29.** answers may vary
29. If the interest in the game remains the same, name one reason why Super Bowl attendance might fluctuate.
30. Name the years that Super Bowl attendance was between 71,000 and 74,000.   1997, 2001, 2002

The line graph below shows the number of students per computer in U.S. public schools. Use this graph for Exercises 31 through 36. See Examples 3 and 4.

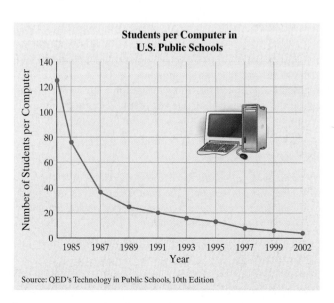

Students per Computer in U.S. Public Schools

Source: QED's Technology in Public Schools, 10th Edition

31. Approximate the number of students per computer in 1991. 20
32. Approximate the number of students per computer in 2002. 4
33. During what year was the greatest decrease in number of students per computer?   1985
34. What was the first year that the number of students per computer fell below 20?   1993
35. What was the first year shown that the number of students per computer fell below 10?   1997
36. Discuss any trends shown by this line graph. answers may vary

The special bar graph shown is called a double bar graph. This double bar graph is used to compare men and women in the U.S. labor force per year. Use this graph for Exercises 37 through 46.

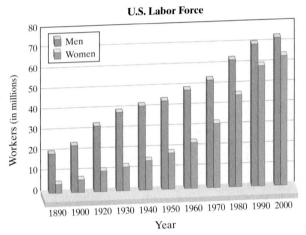

U.S. Labor Force

Source: U.S. Bureau of the Census

37. 18 million   **38.** 4 million   **39.** 63 million   **40.** 72 million

37. Estimate the number of men in the workforce in 1890.
38. Estimate the number of women in the workforce in 1890.
39. Estimate the number of women in the workforce in 2000.
40. Estimate the number of men in the workforce in 2000.
41. Give the first year that the number of men in the workforce rose above 20 million.   1900
42. Give the first year that the number of women in the workforce rose above 20 million.   1960
43. Estimate the difference in the number of men and women in the workforce in 1940.   27 million
44. Estimate the difference in the number of men and women in the workforce in 2000.   9 million
45. Discuss any trends shown by this graph.   answers may vary
46. Use the appearance of this bar graph to estimate when the number of women in the U.S. labor force may equal the number of men.   2010

Concept Extensions

The special line graph at the top of the next column is a double line graph. Here, we can compare the number of viewers who watch cable TV during the summer and the number of viewers who watch broadcast networks during the summer.

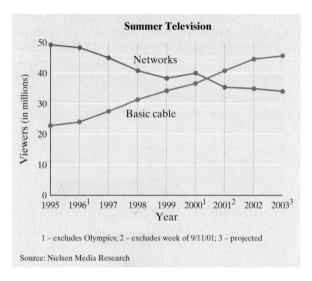

Summer Television

1 – excludes Olympics; 2 – excludes week of 9/11/01; 3 – projected

Source: Nielsen Media Research

47. During what summers did more viewers watch cable than the networks?   2001, 2002, 2003
48. During what summers did more viewers watch the networks than cable?   1995 through 2000
49. Estimate the number of cable viewers during the summer of 2003.   46 million
50. Estimate the number of broadcast viewers during the summer of 2003.   34.5 million

Geographic locations can be described by a gridwork of lines called latitudes and longitudes, as shown below. For example, the location of Houston, Texas, can be described by latitude 30° north and longitude 95° west. Use the map shown to answer Exercises 51 through 54.

52. answers may vary

51. Using latitude and longitude, describe the location of New Orleans, Louisiana.   30° north, 90° west
52. Use an atlas and describe the location of your hometown.
53. Using latitude and longitude, describe the location of Denver, Colorado.   40° north, 104° west
54. Give another name for 0° latitude.   equator

CHAPTER 1 PROJECT

Creating and Interpreting Graphs

Companies often rely on market research to investigate the types of consumers who buy their products and their competitors' products. One way of gathering this type of data is through the use of surveys. The raw data collected from surveys can be difficult to interpret without some organization. Graphs organize data visually. They also allow a user to interpret the data quickly.

In this project, you will conduct a brief survey and create tables and graphs to represent the results. This project may be completed by working in groups or individually.

1. Begin by conducting this survey with fellow students, either in this class or another class.

2. For each survey question, tally the results for each response category. Present the results in a table.

3. For each survey question, find the fraction of the total number of responses that fall in each response category.

4. For each survey question, decide which type of graph would best represent the data: bar graph, line graph, or circle graph. Then create an appropriate graph for each set of responses to the questions.

5. Use the data for the two Internet questions and the two newspaper questions to record the increase or decrease in the number of people who participated in these activities. Complete the table in the next column.

6. Study the tables and graphs. What may you conclude from them? What do they tell you about your survey respondents? Write a paragraph summarizing your findings.

Activity	Increase or Decrease from 3 Years Ago
Using the Internet more than 4 hours per week	
Reading a newspaper at least once per week	

Survey

What is your age?
Under 20   20s   30s   40s   50s   60 and older

What is your gender?
Female   Male

Do you currently use the Internet more than 4 hours per week?
Yes   No

Three years ago, did you use the Internet more than 4 hours per week?
Yes   No

Do you currently read a newspaper at least once per week?
Yes   No

Three years ago, did you read a newspaper at least once per week?
Yes   No

CHAPTER VOCABULARY CHECK

Fill in each blank with one of the words or phrases listed below.

set   inequality symbols   opposites   absolute value   numerator
denominator   grouping symbols   exponent   base   reciprocals
variable   equation   solution

1. The symbols ≠, <, and > are called <u>inequality symbols</u>.
2. A mathematical statement that two expressions are equal is called an <u>equation</u>.
3. The <u>absolute value</u> of a number is the distance between that number and 0 on the number line.
4. A symbol used to represent a number is called a <u>variable</u>.
5. Two numbers that are the same distance from 0 but lie on opposite sides of 0 are called <u>opposites</u>.
6. The number in a fraction above the fraction bar is called the <u>numerator</u>.
7. A <u>solution</u> of an equation is a value for the variable that makes the equation a true statement.

8. Two numbers whose product is 1 are called <u>reciprocals</u>.

9. In 2^3, the 2 is called the <u>base</u> and the 3 is called the <u>exponent</u>.

10. The number in a fraction below the fraction bar is called the <u>denominator</u>.

11. Parentheses and brackets are examples of <u>grouping symbols</u>.

12. A <u>set</u> is a collection of objects.

CHAPTER 1 HIGHLIGHTS

Definitions and Concepts	Examples

Section 1.2 Symbols and Sets of Numbers

A **set** is a collection of objects, called **elements**, enclosed in braces.

$\{a, c, e\}$

Natural Numbers: $\{1, 2, 3, 4, \ldots\}$
Whole Numbers: $\{0, 1, 2, 3, 4, \ldots\}$

Integers: $\{\ldots, -3, -2, -1, 0, 1, 2, 3, \ldots\}$

Rational Numbers: {real numbers that can be expressed as a quotient of integers}

Irrational Numbers: {real numbers that cannot be expressed as a quotient of integers}

Real Numbers: {all numbers that correspond to a point on the number line}

A line used to picture numbers is called a **number line**.

The **absolute value** of a real number a, denoted by $|a|$, is the distance between a and 0 on the number line.

Symbols: $=$ is equal to

$\neq$ is not equal to

$>$ is greater than

$<$ is less than

$\leq$ is less than or equal to

$\geq$ is greater than or equal to

Order Property for Real Numbers
For any two real numbers a and b, a is less than b if a is to the left of b on a number line.

Given the set $\{-3.4, \sqrt{3}, 0, \frac{2}{3}, 5, -4\}$, list the numbers that belong to the set of

Natural numbers: 5

Whole numbers: 0, 5

Integers: $-4, 0, 5$

Rational numbers: $-4, -3.4, 0, \frac{2}{3}, 5$

Irrational Numbers: $\sqrt{3}$

Real numbers: $-4, -3.4, 0, \frac{2}{3}, \sqrt{3}, 5$

$$-3\ -2\ -1\ \ 0\ \ 1\ \ 2\ \ 3$$

$|5| = 5 \qquad |0| = 0 \qquad |-2| = 2$

$-7 = -7$

$3 \neq -3$

$4 > 1$

$1 < 4$

$6 \leq 6$

$18 \geq -\dfrac{1}{3}$

$$-3\ -2\ -1\ \ 0\ \ 1\ \ 2\ \ 3$$

$-3 < 0 \qquad 0 > -3 \qquad 0 < 2.5 \qquad 2.5 > 0$

Section 1.3 Fractions

A quotient of two integers is called a **fraction**. The **numerator** of a fraction is the top number. The **denominator** of a fraction is the bottom number.

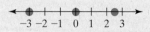

$\dfrac{13 \leftarrow \text{numerator}}{17 \leftarrow \text{denominator}}$

Definitions and Concepts	**Examples**

Section 1.3   Fractions

If $a \cdot b = c$, then a and b are **factors** and c is the **product**.

$$7 \quad \cdot \quad 9 \quad = \quad 63$$
$$\downarrow \qquad\quad \downarrow \qquad\qquad \downarrow$$
$$\text{factor} \qquad \text{factor} \qquad\quad \text{product}$$

A fraction is in **lowest terms** when the numerator and the denominator have no factors in common other than 1.

$\dfrac{13}{17}$ is in lowest terms.

To write a fraction in lowest terms, factor the numerator and the denominator; then apply the fundamental property.

Write in lowest terms.

$$\frac{6}{14} = \frac{2 \cdot 3}{2 \cdot 7} = \frac{3}{7}$$

Two fractions are **reciprocals** if their product is 1.
The reciprocal of $\frac{a}{b}$ is $\frac{b}{a}$.

The reciprocal of $\dfrac{6}{25}$ is $\dfrac{25}{6}$

To multiply fractions, numerator times numerator is the numerator of the product and denominator times denominator is the denominator of the product.

Perform the indicated operations.

$$\frac{2}{5} \cdot \frac{3}{7} = \frac{6}{35}$$

To divide fractions, multiply the first fraction by the reciprocal of the second fraction.

$$\frac{5}{9} \div \frac{2}{7} = \frac{5}{9} \cdot \frac{7}{2} = \frac{35}{18}$$

To add fractions with the same denominator, add the numerators and place the sum over the common denominator.

$$\frac{5}{11} + \frac{3}{11} = \frac{8}{11}$$

To subtract fractions with the same denominator, subtract the numerators and place the difference over the common denominator.
Fractions that represent the same quantity are called **equivalent fractions**.

$$\frac{13}{15} - \frac{3}{15} = \frac{10}{15} = \frac{2}{3}$$

$$\frac{1}{5} = \frac{1 \cdot 4}{5 \cdot 4} = \frac{4}{20}$$

$\dfrac{1}{5}$ and $\dfrac{4}{20}$ are equivalent fractions.

Section 1.4   Introduction to Variable Expressions and Equations

The expression a^n is an **exponential expression**. The number a is called the **base**; it is the repeated factor. The number n is called the **exponent**; it is the number of times that the base is a factor.

$$4^3 = 4 \cdot 4 \cdot 4 = 64$$
$$7^2 = 7 \cdot 7 = 49$$

Order of Operations
Simplify expressions in the following order. If grouping symbols are present, simplify expressions within those first, starting with the innermost set. Also, simplify the numerator and the denominator of a fraction separately.

1. Simplify exponential expressions.
2. Multiply or divide in order from left to right.
3. Add or subtract in order from left to right.

$$\frac{8^2 + 5(7 - 3)}{3 \cdot 7} = \frac{8^2 + 5(4)}{21}$$
$$= \frac{64 + 5(4)}{21}$$
$$= \frac{64 + 20}{21}$$
$$= \frac{84}{21}$$
$$= 4$$

Definitions and Concepts	**Examples**

Section 1.4 Introduction to Variable Expressions and Equations

A symbol used to represent a number is called a **variable**.	Examples of variables are: $$q, x, z$$
An **algebraic expression** is a collection of numbers, variables, operation symbols, and grouping symbols.	Examples of algebraic expressions are: $$5x, 2(y - 6), \frac{q^2 - 3q + 1}{6}$$
To evaluate an algebraic expression containing a variable, substitute a given number for the variable and simplify.	Evaluate $x^2 - y^2$ if $x = 5$ and $y = 3$. $$x^2 - y^2 = (5)^2 - 3^2$$ $$= 25 - 9$$ $$= 16$$
A mathematical statement that two expressions are equal is called an **equation**.	Equations: $$3x - 9 = 20$$ $$A = \pi r^2$$
A **solution** or **root** of an equation is a value for the variable that makes the equation a true statement.	Determine whether 4 is a solution of $5x + 7 = 27$. $$5x + 7 = 27$$ $$5(4) + 7 \stackrel{?}{=} 27$$ $$20 + 7 \stackrel{?}{=} 27$$ $$27 = 27 \quad \text{True}$$ 4 is a solution.

Section 1.5 Adding Real Numbers

To Add Two Numbers with the Same Sign 1. Add their absolute values. 2. Use their common sign as the sign of the sum.	Add. $$10 + 7 = 17$$ $$-3 + (-8) = -11$$
To Add Two Numbers with Different Signs 1. Subtract their absolute values. 2. Use the sign of the number whose absolute value is larger as the sign of the sum.	$$-25 + 5 = -20$$ $$14 + (-9) = 5$$
Two numbers that are the same distance from 0 but lie on opposite sides of 0 are called **opposites** or **additive inverses**. The opposite of a number a is denoted by $-a$.	The opposite of -7 is 7. The opposite of 123 is -123.
The sum of a number a and its opposite, $-a$, is 0. $$a + (-a) = 0$$ If a is a number, then $-(-a) = a$.	$$-4 + 4 = 0$$ $$12 + (-12) = 0$$ $$-(-8) = 8$$ $$-(-14) = 14$$

Definitions and Concepts	**Examples**

Section 1.6 Subtracting Real Numbers

To subtract two numbers a and b, add the first number a to the opposite of the second number b.

$$a - b = a + (-b)$$

Subtract.

$$3 - (-44) = 3 + 44 = 47$$
$$-5 - 22 = -5 + (-22) = -27$$
$$-30 - (-30) = -30 + 30 = 0$$

Section 1.7 Multiplying and Dividing Real Numbers

Quotient of two real numbers

$$\frac{a}{b} = a \cdot \frac{1}{b}$$

Multiplying and Dividing Real Numbers

The product or quotient of two numbers with the same sign is a positive number. The product or quotient of two numbers with different signs is a negative number.

Products and Quotients Involving Zero

The product of 0 and any number is 0.

$$b \cdot 0 = 0 \quad \text{and} \quad 0 \cdot b = 0$$

The quotient of a nonzero number and 0 is undefined.

$$\frac{b}{0} \text{ is undefined.}$$

The quotient of 0 and any nonzero number is 0.

$$\frac{0}{b} = 0$$

Multiply or divide.

$$\frac{42}{2} = 42 \cdot \frac{1}{2} = 21$$

$$7 \cdot 8 = 56 \quad -7 \cdot (-8) = 56$$
$$-2 \cdot 4 = -8 \quad 2 \cdot (-4) = -8$$

$$\frac{90}{10} = 9 \qquad \frac{-90}{-10} = 9$$
$$\frac{42}{-6} = -7 \qquad \frac{-42}{6} = -7$$

$$-4 \cdot 0 = 0 \qquad 0 \cdot \left(-\frac{3}{4}\right) = 0$$

$$\frac{-85}{0} \text{ is undefined.}$$

$$\frac{0}{18} = 0 \qquad \frac{0}{-47} = 0$$

Section 1.8 Properties of Real Numbers

Commutative Properties
Addition: $a + b = b + a$
Multiplication: $a \cdot b = b \cdot a$

Associative Properties
Addition: $(a + b) + c = a + (b + c)$
Multiplication: $(a \cdot b) \cdot c = a \cdot (b \cdot c)$

Two numbers whose product is 1 are called

multiplicative inverses or **reciprocals**. The reciprocal of a nonzero number a is $\dfrac{1}{a}$ because $a \cdot \dfrac{1}{a} = 1$.

$$3 + (-7) = -7 + 3$$
$$-8 \cdot 5 = 5 \cdot (-8)$$

$$(5 + 10) + 20 = 5 + (10 + 20)$$
$$(-3 \cdot 2) \cdot 11 = -3 \cdot (2 \cdot 11)$$

The reciprocal of 3 is $\frac{1}{3}$.

The reciprocal of $-\frac{2}{5}$ is $-\frac{5}{2}$.

Definitions and Concepts	**Examples**

Section 1.8 Properties of Real Numbers

Distributive Property $a(b + c) = a \cdot b + a \cdot c$

$$5(6 + 10) = 5 \cdot 6 + 5 \cdot 10$$
$$-2(3 + x) = -2 \cdot 3 + (-2)(x)$$

Identities $a + 0 = a$ $0 + a = a$

$\qquad\qquad a \cdot 1 = a$ $1 \cdot a = a$

$$5 + 0 = 5 \qquad 0 + (-2) = -2$$
$$-14 \cdot 1 = -14 \qquad 1 \cdot 27 = 27$$

Inverses

Addition or opposite: $a + (-a) = 0$

Multiplication or reciprocal: $b \cdot \dfrac{1}{b} = 1$

$$7 + (-7) = 0$$
$$3 \cdot \frac{1}{3} = 1$$

Section 1.9 Reading Graphs

To find the value on the vertical axis representing a location on a graph, move horizontally from the location on the graph until the vertical axis is reached. To find the value on the horizontal axis representing a location on a graph, move vertically from the location on the graph until the horizontal axis is reached.

The broken line graph to the right shows the average public classroom teachers' salaries for the school year ending in the years shown.

Estimate the average public teacher's salary for the school year ending in 1998.

Find the earliest year that the average salary rose above \$37,000.   The year was 1995.

The average salary is approximately \$39,400.

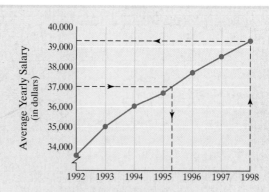

Source: U.S. Bureau of the Census, *Statistical Abstract of the United States: 1999* (114th edition) Washington, D.C., 1999

STUDY SKILLS REMINDER

Are You Preparing for a Test on Chapter 1?

Below I have listed some *common trouble areas* for topics covered in Chapter 1. After studying for your test—but before taking your test—read these.

▶ Do you know the difference between $|-3|$, $-|-3|$, and $-(-3)$?

$$|-3| = 3; \quad -|-3| = -3; \quad \text{and} \quad -(-3) = 3 \qquad \text{(Section 1.2)}$$

▶ Evaluate $x - y$ if $x = 7$ and $y = -3$.

$$x - y = 7 - (-3) = 10 \qquad \text{(Section 1.4)}$$

▶ Make sure you are familiar with order of operations. Sometimes the simplest-looking expressions can give you the most trouble.

$$1 + 2(3 + 6) = 1 + 2(9) = 1 + 18 = 19 \qquad \text{(Section 1.4)}$$

▶ Do you know the difference between $(-3)^2$ and -3^2?

$$(-3)^2 = 9 \quad \text{and} \quad -3^2 = -9 \qquad \text{(Section 1.7)}$$

▶ Do you know that these fractions are equivalent?

$$-\frac{1}{3} = \frac{-1}{3} = \frac{1}{-3} \qquad \text{(Section 1.8)}$$

▶ Do you know the difference between opposite and reciprocal? If not, study the table below.

Number	Opposite	Reciprocal	
5	-5	$\frac{1}{5}$	
$-\frac{4}{7}$	$\frac{4}{7}$	$-\frac{7}{4}$	(Sections 1.5 and 1.7)
$-\frac{1}{3}$	$\frac{1}{3}$	-3	

Remember: This is simply a checklist of selected topics given to check your understanding. For a review of Chapter 1 in the text, see the material at the end of Chapter 1.

CHAPTER REVIEW

(1.2) Insert $<$, $>$, or $=$ in the appropriate space to make the following statements true.

1. $8 \; < \; 10$

2. $7 \; > \; 2$

3. $-4 \; > \; -5$

4. $\dfrac{12}{2} \; > \; -8$

5. $|-7| \; < \; |-8|$

6. $|-9| \; > \; -9$

7. $-|-1| \; = \; -1$

8. $|-14| \; = \; -(-14)$

9. $1.2 \; > \; 1.02$

10. $-\dfrac{3}{2} \; < \; -\dfrac{3}{4}$

Translate each statement into symbols.

11. Four is greater than or equal to negative three. $4 \geq -3$

12. Six is not equal to five. $6 \neq 5$

13. 0.03 is less than 0.3. $0.03 < 0.3$

14. Lions and hyenas were featured in the Disney film *The Lion King*. For short distances, lions can run at a rate of 50 miles per hour whereas hyenas can run at a rate of 40 miles per hour. Write an inequality statement comparing the numbers 50 and 40. $50 > 40$

Given the following sets of numbers, list the numbers in each set that also belong to the set of:

a. Natural numbers   **b.** Whole numbers

c. Integers   **d.** Rational numbers

e. Irrational numbers   **f.** Real numbers

15. $\left\{ -6, 0, 1, 1\frac{1}{2}, 3, \pi, 9.62 \right\}$

16. $\left\{ -3, -1.6, 2, 5, \frac{11}{2}, 15.1, \sqrt{5}, 2\pi \right\}$

The following chart shows the gains and losses in dollars of Density Oil and Gas stock for a particular week.

Day	Gain or Loss in Dollars
Monday	$+1$
Tuesday	-2
Wednesday	$+5$
Thursday	$+1$
Friday	-4

15. a. $1, 3$ **b.** $0, 1, 3$ **c.** $-6, 0, 1, 3$ **d.** $-6, 0, 1, 1\frac{1}{2}, 3, 9.62$ **e.** π **f.** $-6, 0, 1, 1\frac{1}{2}, 3, \pi, 9.62$

16. a. $2, 5$ **b.** $2, 5$ **c.** $-3, 2, 5$ **d.** $-3, -1.6, 2, 5, \frac{11}{2}, 15.1$ **e.** $\sqrt{5}, 2\pi$ **f.** $-3, -1.6, 2, 5, \frac{11}{2}, 15.1, \sqrt{5}, 2\pi$

17. Which day showed the greatest loss?　Friday

18. Which day showed the greatest gain?　Wednesday

(1.3) Write the number as a product of prime factors.

19. 36　$2 \cdot 2 \cdot 3 \cdot 3$ **20.** 120　$2 \cdot 2 \cdot 2 \cdot 3 \cdot 5$

Perform the indicated operations. Write results in lowest terms.

21. $\frac{8}{15} \cdot \frac{27}{30}$　$\frac{12}{25}$ **22.** $\frac{7}{8} \div \frac{21}{32}$　$\frac{4}{3}$ **23.** $\frac{7}{15} + \frac{5}{6}$　$\frac{13}{10}$

24. $\frac{3}{4} - \frac{3}{20}$　$\frac{3}{5}$ **25.** $2\frac{3}{4} + 6\frac{5}{8}$　$9\frac{3}{8}$ **26.** $7\frac{1}{6} - 2\frac{2}{3}$　$4\frac{1}{2}$

27. $5 \div \frac{1}{3}$　15 **28.** $2 \cdot 8\frac{3}{4}$　$17\frac{1}{2}$

29. Determine the unknown part of the given circle.　$\frac{7}{12}$

30. $A = 1\frac{1}{6}$ sq. m
$P = 4\frac{5}{12}$ m

31. $A = \frac{34}{121}$ sq. in.
$P = 2\frac{4}{11}$ in.

Find the area and the perimeter of each figure.

△ **30.**

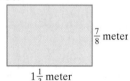

$\frac{7}{8}$ meter

$1\frac{1}{3}$ meter

△ **31.**

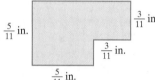

$\frac{3}{11}$ in.

$\frac{5}{11}$ in.

$\frac{3}{11}$ in.

$\frac{5}{11}$ in.

32. A trim carpenter needs a piece of quarter round molding $6\frac{1}{8}$ feet long for a bathroom. She finds a piece $7\frac{1}{2}$ feet long. How long a piece does she need to cut from the $7\frac{1}{2}$-foot-long molding in order to use it in the bathroom?　$1\frac{3}{8}$ ft

In December 1998, Nkem Chukwu gave birth to the world's first surviving octuplets in Houston, Texas. The following chart gives the octuplets' birthweights. The babies are listed in order of birth.

Baby's Name	Gender	Birthweight (pounds)
Ebuka	girl	$1\frac{1}{2}$
Chidi	girl	$1\frac{11}{16}$
Echerem	girl	$1\frac{3}{4}$
Chima	girl	$1\frac{5}{8}$
Odera	girl	$\frac{11}{16}$
Ikem	boy	$1\frac{1}{8}$
Jioke	boy	$1\frac{13}{16}$
Gorom	girl	$1\frac{1}{8}$

(*Source*: Texas Children's Hospital, Houston, Texas)

33. What was the total weight of the boy octuplets?　$2\frac{15}{16}$ lb

34. What was the total weight of the girl octuplets?　$8\frac{3}{8}$ lb

35. Find the combined weight of all eight octuplets.　$11\frac{5}{16}$ lb

36. Which baby weighed the most?　Jioke

37. Which baby weighed the least?　Odera

38. How much more did the heaviest baby weigh than the lightest baby?　$1\frac{1}{8}$ lb

39. By March 1999, Chima weighed $5\frac{1}{2}$ pounds. How much weight had she gained since birth?　$3\frac{7}{8}$ lb

40. By March 1999, Ikem weighed $4\frac{5}{32}$ pounds. How much weight had he gained since birth?　$3\frac{1}{32}$ lb

(1.4) Simplify each expression.

41. 2^4　16 **42.** 5^2　25

43. $\left(\frac{2}{7}\right)^2$　$\frac{4}{49}$ **44.** $\left(\frac{3}{4}\right)^3$　$\frac{27}{64}$

45. $6 \cdot 3^2 + 2 \cdot 8$　70 **46.** $68 - 5 \cdot 2^3$　28

47. $3(1 + 2 \cdot 5) + 4$　-37 **48.** $8 + 3(2 \cdot 6 - 1)$　41

49. $\dfrac{4 + |6 - 2| + 8^2}{4 + 6 \cdot 4}$　$\frac{18}{7}$ **50.** $5[3(2 + 5) - 5]$　80

Translate each word statement to symbols.　**52.** $\frac{9}{2} > -5$

51. The difference of twenty and twelve is equal to the product of two and four.　$20 - 12 = 2 \cdot 4$

52. The quotient of nine and two is greater than negative five.

Evaluate each expression if $x = 6$, $y = 2$, and $z = 8$.

53. $2x + 3y$　18 **54.** $x(y + 2z)$　108

55. $\dfrac{x}{y} + \dfrac{z}{2y}$　5 **56.** $x^2 - 3y^2$　24

△ **57.** The expression $180 - a - b$ represents the measure of the unknown angle of the given triangle. Replace a with 37 and b with 80 to find the measure of the unknown angle.　63°

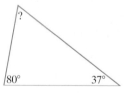

Decide whether the given number is a solution to the given equation.

58. Is $x = 3$ a solution of $7x - 3 = 18$?　yes

59. Is $x = 1$ a solution of $3x^2 + 4 = x - 1$?　no

(1.5) Find the additive inverse or the opposite.

60. -9　9 **61.** $\frac{2}{3}$　$-\frac{2}{3}$

62. $|-2|$　-2 **63.** $-|-7|$　7

Find the following sums.

64. $-15 + 4$ -11

65. $-6 + (-11)$ -17

66. $\dfrac{1}{16} + \left(-\dfrac{1}{4}\right)$ $-\dfrac{3}{16}$

67. $-8 + |-3|$ -5

68. $-4.6 + (-9.3)$ -13.9

69. $-2.8 + 6.7$ 3.9

70. The lowest elevation in North America is -282 feet at Death Valley in California. If you are standing at a point 728 feet above Death Valley, what is your elevation?(*Source*: National Geographic Society)   446 ft

(1.6) Perform the indicated operations.

71. $6 - 20$ -14

72. $-3.1 - 8.4$ -11.5

73. $-6 - (-11)$ 5

74. $4 - 15$ -11

75. $-21 - 16 + 3(8 - 2)$ -19

76. $\dfrac{11 - (-9) + 6(8 - 2)}{2 + 3 \cdot 4}$ 4

If $x = 3$, $y = -6$, and $z = -9$, evaluate each expression.

77. $2x^2 - y + z$ 15

78. $\dfrac{y - x + 5x}{2x}$ 1

(1.7) Find the multiplicative inverse or reciprocal.

79. -6 $-\dfrac{1}{6}$

80. $\dfrac{3}{5}$ $\dfrac{5}{3}$

Simplify each expression.

81. $6(-8)$ -48

82. $(-2)(-14)$ 28

83. $\dfrac{-18}{-6}$ 3

84. $\dfrac{42}{-3}$ -14

85. $\dfrac{4(-3) + (-8)}{2 + (-2)}$ undefined

86. $\dfrac{3(-2)^2 - 5}{-14}$ $-\dfrac{1}{2}$

87. $\dfrac{-6}{0}$ undefined

88. $\dfrac{0}{-2}$ 0

89. $-4^2 - (-3 + 5) \div (-1) \cdot 2$ -12

90. $-5^2 - (2 - 20) \div (-3) \cdot 3$ -43

If $x = -5$ and $y = -2$, evaluate each expression.

91. $x^2 - y^4$ 9

92. $x^2 - y^3$ 33

93. During the 1999 LPGA Sara Lee Classic, Michelle McGann had scores of -9, -7, and $+1$ in three rounds of golf. Find her average score per round. (*Source:* Ladies Professional Golf Association) -5

94. During the 1999 PGA Masters Tournament, Bob Estes had scores of $-1, 0, -3$, and 0 in four rounds of golf. Find his average score per round. (*Source*: Professional Golf Association) -1

(1.8) Name the property illustrated.

95. $-6 + 5 = 5 + (-6)$ commutative property of addition

96. $6 \cdot 1 = 6$ multiplicative identity property

97. $3(8 - 5) = 3 \cdot 8 + 3 \cdot (-5)$ distributive property

98. $4 + (-4) = 0$ additive inverse property

99. $2 + (3 + 9) = (2 + 3) + 9$

100. $2 \cdot 8 = 8 \cdot 2$ commutative property of multiplication

101. $6(8 + 5) = 6 \cdot 8 + 6 \cdot 5$ distributive property

102. $(3 \cdot 8) \cdot 4 = 3 \cdot (8 \cdot 4)$ associative property of multiplication

103. $4 \cdot \dfrac{1}{4} = 1$ multiplicative inverse

104. $8 + 0 = 8$ additive identity property

105. $4(8 + 3) = 4(3 + 8)$ commutative property of addition

99. associative property of addition

(1.9) Use the following graph to answer Exercises 106 through 109.

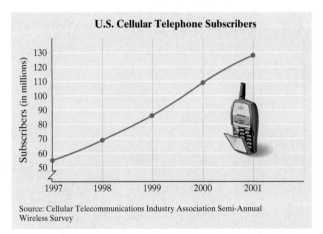

U.S. Cellular Telephone Subscribers

Source: Cellular Telecommunications Industry Association Semi-Annual Wireless Survey

106. Approximate the number of cellular phone subscribers in 2001.   128 million

107. Approximate the increase in cellular phone subscribers in 2001.   19 million

108. What year shows the greatest number of subscribers?   2000

109. What trend is shown by this graph?   number of subscribers is increasing

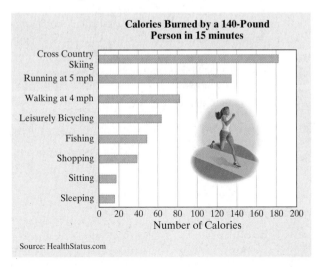

Calories Burned by a 140-Pound Person in 15 minutes

Source: HealthStatus.com

110. Which activity burns the most calories? Estimate the number of calories.   Cross country skiing; 181 calories

111. Which activity burns the fewest calories? Estimate the number of calories.   sleeping; 15 calories

112. Estimate how many more calories running at 5mph burns than walking at 4mph.   52 more calories

113. Estimate how many more calories fishing burns than shopping.   10 more calories

CHAPTER 1 TEST

Remember to use your Chapter Test Prep Video CD to help you study and view solutions to the test questions you need help with.

Translate the statement into symbols.

1. The absolute value of negative seven is greater than five. $|-7| > 5$

2. The sum of nine and five is greater than or equal to four. $9 + 5 \geq 4$

Simplify the expression.   **22.** $2221 < 10,993$

3. $-13 + 8$ -5

4. $-13 - (-2)$ -11

5. $12 \div 4 \cdot 3 - 6 \cdot 2$ -3

6. $(13)(-3)$ -39

7. $(-6)(-2)$ 12

8. $\dfrac{|-16|}{-8}$ -2

9. $\dfrac{-8}{0}$ undefined

10. $\dfrac{|-6| + 2}{5 - 6}$ -8

11. $\dfrac{1}{2} - \dfrac{5}{6}$ $-\dfrac{1}{3}$

12. $-1\dfrac{1}{8} + 5\dfrac{3}{4}$ $4\dfrac{5}{8}$

13. $(2 - 6) \div \dfrac{-2 - 6}{-3 - 1}$ $-\dfrac{1}{2}$

14. $3(-4)^2 - 80$ -32

15. $6[5 + 2(3 - 8) - 3]$

16. $\dfrac{-12 + 3 \cdot 8}{4}$ 3

17. $\dfrac{(-2)(0)(-3)}{-6}$ 0

Insert $<, >,$ or $=$ *in the appropriate space to make each of the following statements true.*

18. $-3 \; > \; -7$

19. $4 \; > \; -8$

20. $2 \; < \; |-3|$

21. $|-2| \; = \; -1 - (-3)$

22. In the state of Massachusetts, there are 2221 licensed child care centers and 10,993 licensed home-based child care providers. Write an inequality statement comparing the numbers 2221 and 10,993.(*Source:* Children's Foundation)

23. Given $\{-5, -1, 0, \frac{1}{4}, 1, 7, 11.6, \sqrt{7}, 3\pi\}$, list the numbers in this set that also belong to the set of:

 a. Natural numbers $1, 7$

 b. Whole numbers $0, 1, 7$

 c. Integers $-5, -1, 0, 1, 7$

 d. Rational numbers $-5, -1, 0, \frac{1}{4}, 1, 7, 11.6$

 e. Irrational numbers $\sqrt{7}, 3\pi$

 f. Real numbers $-5, -1, 0, \frac{1}{4}, 1, 7, 11.6, \sqrt{7}, 3\pi$

If $x = 6$, $y = -2$, and $z = -3$, evaluate each expression.

24. $x^2 + y^2$ 40

25. $x + yz$ 12

26. $2 + 3x - y$ 22

27. $\dfrac{y + z - 1}{x}$ -1

Identify the property illustrated by each expression.

28. $8 + (9 + 3) = (8 + 9) + 3$

29. $6 \cdot 8 = 8 \cdot 6$ commutative property of multiplication

30. $-6(2 + 4) = -6 \cdot 2 + (-6) \cdot 4$ distributive property

31. $\dfrac{1}{6}(6) = 1$ multiplicative inverse property

32. Find the opposite of -9. 9

33. Find the reciprocal of $-\dfrac{1}{3}$. -3

The New Orleans Saints were 22 yards from the goal when the following series of gains and losses occurred.

Gains and Losses in Yards	
First Down	5
Second Down	−10
Third Down	−2
Fourth Down	29

34. During which down did the greatest loss of yardage occur?   second down

35. Was a touchdown scored?   yes

13. $-\dfrac{5}{2}$ or $-2\dfrac{1}{2}$ **15.** -48 **28.** associative property of addition

36. The temperature at the Winter Olympics was a frigid 14 degrees below zero in the morning, but by noon it had risen 31 degrees. What was the temperature at noon?   17°

37. United HealthCare is a health insurance provider. In 3 consecutive recent years, it had net incomes of $356 million, $460 million, and −$166 million. What was United HealthCare's total net income for these three years? (*Source*: United HealthCare Corp.) $650 million

38. Jean Avarez decided to sell 280 shares of stock, which decreased in value by $1.50 per share yesterday. How much money did she lose? $420

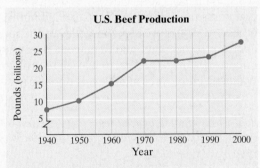

U.S. Beef Production

Source: U.S. Economic Research Service, Dept. of Agriculture

39. Which year shows the greatest production of beef? Estimate the production for that year.

40. Which year shows the least production of beef? Estimate the production for that year.

41. Which years had beef production greater than 20 billion pounds?   1970, 1980, 1990, 2000

42. Which year shows the greatest increase in beef production?   1970

The following bar graph shows the top steel-producing states ranked by total tons of raw steel produced in a recent year.

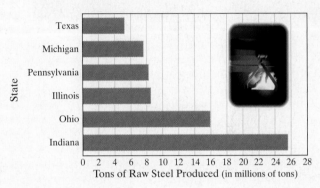

Source: American Iron & Steel Institute

43. Which state was the top steel producer? Approximate the amount of raw steel it produced.

44. Which of the top steel producing states produced the least steel? Approximate the amount of raw steel it produced.   Texas, 5 million tons

45. Approximate the amount of raw steel produced by Ohio.   16 million tons

46. Approximately how much more steel was produced in Pennsylvania than Texas?   3 million tons

39. 2000; 27 billion pounds   **40.** 1940; 7 billion pounds   **43.** Indiana, 25.2 million tons

Equations, Inequalities, and Problem Solving

Much of mathematics relates to deciding which statements are true and which are false. When a statement, such as an equation, contains variables, it is usually not possible to decide whether the equation is true or false until the variable has been replaced by a value. For example, the statement $x + 7 = 15$ is an equation stating that the sum $x + 7$ has the same value as 15. Is this statement true or false? It is false for some values of x and true for just one value of x, namely 8. Our purpose in this chapter is to learn ways of deciding which values make an equation or an inequality true.

CALLING ALL ACCOUNTANTS

There is an old saying: Nothing is certain, except death and taxes. And with ever-changing tax laws, is it any wonder that income tax preparation has become a multimillion-dollar industry? Companies such as H&R Block and Jackson-Hewitt employ accountants as both full-time and seasonal workers to prepare taxes and advise clients on financial strategies.

But accountants do so much more! Certified public accountants, or CPAs, also work for private companies or the federal government as auditors, consultants, and managers. Because they like to work with figures, their methodical approach to problem solving makes them valuable employees at all levels of business.

In the Spotlight on Decision Making feature on page 138, you will have the opportunity to make a decision involving itemized deductions as a personal income tax preparer.

Link: American Institute of Certified Public Accountants—
www.aicpa.org
Source of text: American Institute of Certified Public Accountants—
www.aicpa.org

SIMPLIFYING ALGEBRAIC EXPRESSIONS

Objectives

① Identify terms, like terms, and unlike terms.

② Combine like terms.

③ Use the distributive property to remove parentheses.

④ Write word phrases as algebraic expressions.

As we explore in this section, an expression such as $3x + 2x$ is not as simple as possible, because—even without replacing x by a value—we can perform the indicated addition.

① Before we practice simplifying expressions, some new language of algebra is presented. A **term** is a number or the product of a number and variables raised to powers.

Terms

$$-y, \quad 2x^3, \quad -5, \quad 3xz^2, \quad \frac{2}{y}, \quad 0.8z$$

The **numerical coefficient** (sometimes also simply called the **coefficient**) of a term is the numerical factor. The numerical coefficient of $3x$ is 3. Recall that $3x$ means $3 \cdot x$.

Term	*Numerical Coefficient*	
$3x$	3	
$\dfrac{y^3}{5}$	$\dfrac{1}{5}$	*since* $\dfrac{y^3}{5}$ *means* $\dfrac{1}{5} \cdot y^3$
$0.7ab^3c^5$	0.7	
z	1	
$-y$	-1	
-5	-5	

CLASSROOM EXAMPLE

Identify the numerical coefficient in each term.

a. $-5x$ **b.** $2x^2$

c. $-y$ **d.** $\dfrac{z}{3}$

answer:

a. -5 **b.** 2

c. -1 **d.** $\dfrac{1}{3}$

> **Helpful Hint**
>
> The term $-y$ means $-1y$ and thus has a numerical coefficient of -1. The term z means $1z$ and thus has a numerical coefficient of 1.

EXAMPLE 1

Identify the numerical coefficient in each term.

a. $-3y$ **b.** $22z^4$ **c.** y **d.** $-x$ **e.** $\dfrac{x}{7}$

Solution   **a.** The numerical coefficient of $-3y$ is -3.

b. The numerical coefficient of $22z^4$ is 22.

c. The numerical coefficient of y is 1, since y is $1y$.

d. The numerical coefficient of $-x$ is -1, since $-x$ is $-1x$.

e. The numerical coefficient of $\dfrac{x}{7}$ is $\dfrac{1}{7}$, since $\dfrac{x}{7}$ means $\dfrac{1}{7} \cdot x$.

TEACHING TIP

Remind students that a variable without a coefficient has an understood coefficient of 1. You may want to advise your students to write in the coefficient of 1.

Terms with the same variables raised to exactly the same powers are called **like terms**. Terms that aren't like terms are called **unlike terms**.

Like Terms	*Unlike Terms*	
$3x, 2x$	$5x, 5x^2$	Why? Same variable x, but different powers x and x^2
$-6x^2y, 2x^2y, 4x^2y$	$7y, 3z, 8x^2$	Why? Different variables
$2ab^2c^3, ac^3b^2$	$6abc^3, 6ab^2$	Why? Different variables and different powers

> ### Helpful Hint
>
> In like terms, each variable and its exponent must match exactly, but these factors don't need to be in the same order.
>
> $$2x^2y \text{ and } 3yx^2 \text{ are like terms.}$$

EXAMPLE 2

Determine whether the terms are like or unlike.

a. $2x, 3x^2$ **b.** $4x^2y, x^2y, -2x^2y$ **c.** $-2yz, -3zy$ **d.** $-x^4, x^4$

Solution   **a.** Unlike terms, since the exponents on x are not the same.

b. Like terms, since each variable and its exponent match.

c. Like terms, since $zy = yz$ by the commutative property.

d. Like terms.

2   An algebraic expression containing the sum or difference of like terms can be simplified by applying the distributive property. For example, by the distributive property, we rewrite the sum of the like terms $3x + 2x$ as

$$3x + 2x = (3 + 2)x = 5x$$

Also,

$$-y^2 + 5y^2 = (-1 + 5)y^2 = 4y^2$$

Simplifying the sum or difference of like terms is called **combining like terms**.

EXAMPLE 3

Simplify each expression by combining like terms.

a. $7x - 3x$ **b.** $10y^2 + y^2$ **c.** $8x^2 + 2x - 3x$

Solution   **a.** $7x - 3x = (7 - 3)x = 4x$

b. $10y^2 + y^2 = 10y^2 + 1y^2 = (10 + 1)y^2 = 11y^2$

c. $8x^2 + 2x - 3x = 8x^2 + (2 - 3)x = 8x^2 - x$

EXAMPLE 4

Simplify each expression by combining like terms.

a. $2x + 3x + 5 + 2$ **b.** $-5a - 3 + a + 2$ **c.** $4y - 3y^2$

d. $2.3x + 5x - 6$ **e.** $-\frac{1}{2}b + b$

Solution   Use the distributive property to combine the numerical coefficients of like terms.

a. $2x + 3x + 5 + 2 = (2 + 3)x + (5 + 2)$
$$= 5x + 7$$

b. $-5a - 3 + a + 2 = -5a + 1a + (-3 + 2)$
$$= (-5 + 1)a + (-3 + 2)$$
$$= -4a - 1$$

c. $4y - 3y^2$ These two terms cannot be combined because they are unlike terms.

d. $2.3x + 5x - 6 = (2.3 + 5)x - 6$
$$= 7.3x - 6$$

e. $-\dfrac{1}{2}b + b = \left(-\dfrac{1}{2} + 1\right)b = \dfrac{1}{2}b$

The examples above suggest the following:

CLASSROOM EXAMPLE

Simplify by combining like terms.

a. $-7a - 2 - a + 5$

b. $8.6x^2 - 4.3x$

c. $\dfrac{2}{3}x - 2x$

answer:

a. $-8a + 3$ **b.** $8.6x^2 - 4.3x$

c. $-\dfrac{4}{3}x$

> ## Combining Like Terms
>
> To **combine like terms**, add the numerical coefficients and multiply the result by the common variable factors.

3   Simplifying expressions makes frequent use of the distributive property to also remove parentheses.

EXAMPLE 5

Find each product by using the distributive property to remove parentheses.

a. $5(x + 2)$ **b.** $-2(y + 0.3z - 1)$ **c.** $-(x + y - 2z + 6)$

Solution   **a.** $5(x + 2) = 5 \cdot x + 5 \cdot 2$ Apply the distributive property.
$$= 5x + 10$$ Multiply.

b. $-2(y + 0.3z - 1) = -2(y) + (-2)(0.3z) + (-2)(-1)$ Apply the distributive property.
$$= -2y - 0.6z + 2$$ Multiply.

c. $-(x + y - 2z + 6) = -1(x + y - 2z + 6)$ Distribute -1 over each term.
$$= -1(x) - 1(y) - 1(-2z) - 1(6)$$
$$= -x - y + 2z - 6$$

CLASSROOM EXAMPLE

Find each product by using the distributive property to remove parentheses.

a. $-3(y + 6)$

b. $-(3x + 2y + z - 1)$

answer:

a. $-3y - 18$

b. $-3x - 2y - z + 1$

> ## ▶ Helpful Hint
>
> If a "$-$" sign precedes parentheses, the sign of each term inside the parentheses is changed when the distributive property is applied to remove parentheses.
>
> **Examples:**
>
> $$-(2x + 1) = -2x - 1 \qquad\qquad -(-5x + y - z) = 5x - y + z$$
> $$-(x - 2y) = -x + 2y \qquad\qquad -(-3x - 4y - 1) = 3x + 4y + 1$$

When simplifying an expression containing parentheses, we often use the distributive property in both directions—first to remove parentheses and then again to combine any like terms.

EXAMPLE 6

Simplify the following expressions.

a. $3(2x - 5) + 1$ **b.** $-2(4x + 7) - (3x - 1)$ **c.** $9 - 3(4x + 10)$

Solution   **a.** $3(2x - 5) + 1 = 6x - 15 + 1$ Apply the distributive property.
$= 6x - 14$ Combine like terms.

b. $-2(4x + 7) - (3x - 1) = -8x - 14 - 3x + 1$ Apply the distributive property.
$= -11x - 13$ Combine like terms.

c. $9 - 3(4x + 10) = 9 - 12x - 30$ Apply the distributive property.
$= -21 - 12x$ Combine like terms.

> **Helpful Hint**
> Don't forget to use the distributive property to multiply before adding or subtracting like terms.

EXAMPLE 7

Write the phrase below as an algebraic expression. Then simplify if possible. Subtract $4x - 2$ from $2x - 3$.

Solution   "Subtract $4x - 2$ **from** $2x - 3$" translates to $(2x - 3) - (4x - 2)$. Next, simplify the algebraic expression.

$(2x - 3) - (4x - 2) = 2x - 3 - 4x + 2$ Apply the distributive property.
$= -2x - 1$ Combine like terms.

4   Next, we practice writing word phrases as algebraic expressions.

EXAMPLE 8

Write the following phrases as algebraic expressions and simplify if possible. Let x represent the unknown number.

a. Twice a number, added to 6
b. The difference of a number and 4, divided by 7
c. Five added to 3 times the sum of a number and 1
d. The sum of twice a number, 3 times the number, and 5 times the number

Solution   **a.** In words:

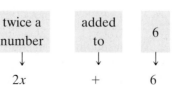

twice a number	added to	6
↓	↓	↓

Translate: $2x$ $+$ 6

CLASSROOM EXAMPLE

Write each phrase as an algebraic expression and simplify if possible.

a. Three times a number, *subtracted from* 10.

b. Five added to twice the sum of a number and seven.

answer:

a. $10 - 3x$ **b.** $2x + 19$

TEACHING TIP

A Group Activity for this section is available in the Instructor's Resource Manual.

b. In words:

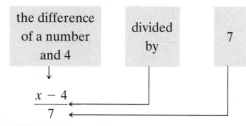

the difference of a number and 4   divided by   7

Translate: $\dfrac{x - 4}{7}$

c. In words:

five   added to   3 times   the sum of a number and 1

Translate: 5   +   3· $(x + 1)$

Next, we simplify this expression.

$$5 + 3(x + 1) = 5 + 3x + 3 \quad \text{Use the distributive property.}$$
$$= 8 + 3x \quad \text{Combine like terms.}$$

d. The phrase "the sum of" means that we add.

In words: twice a number   added to   3 times the number   added to   5 times the number

Translate: $2x$ + $3x$ + $5x$

Now let's simplify.

$$2x + 3x + 5x = 10x \quad \text{Combine like terms.}$$

STUDY SKILLS REMINDER

Are You Getting all the Mathematics Help that You Need?

Remember that in addition to your instructor, there are many places to get help with your mathematics course. For example, see the list below.

▶ There is an accompanying video lesson for every section in this text.

▶ The back of this book contains answers to odd-numbered exercises as well as answers to every exercise in the Integrated Reviews, Chapter Reviews, Chapter Tests, and Cumulative Reviews.

▶ MathPro is available with this text. It is a tutorial software program with lessons corresponding to each section in the text.

▶ There is a student solutions manual available that contains worked-out solutions to odd-numbered exercises as well as solutions to every exercise in the Integrated Reviews, Chapter Reviews, Chapter Tests, and Cumulative Reviews.

▶ Check with your instructor for other local resources available to you, such as the tutoring center.

MENTAL MATH

Identify the numerical coefficient of each term. See Example 1.

1. $-7y$ -7
2. $3x$ 3
3. x 1
4. $-y$ -1
5. $17x^2y$ 17

6. $1.2xyz$ 1.2
7. $\dfrac{p}{8}$ $\dfrac{1}{8}$
8. $-\dfrac{5y}{3}$ $-\dfrac{5}{3}$
9. $-\dfrac{2}{3}z$ $-\dfrac{2}{3}$

Indicate whether the following lists of terms are like or unlike. See Example 2.

10. $5y, -y$ like
11. $-2x^2y, 6xy$ unlike
12. $2z, 3z^2$ unlike

13. $ab^2, -7ab^2$ like
14. $8wz, \dfrac{1}{7}zw$ like
15. $7.4p^3q^2, 6.2p^3q^2r$ unlike

EXERCISE SET 2.1

STUDY GUIDE/SSM   CD/VIDEO   PH MATH TUTOR CENTER   MathXL®Tutorials ON CD   MathXL®   MyMathLab®

Simplify each expression by combining any like terms. See Examples 3 and 4.   **10.** $3g - 2$ **12.** $-3a - 2$

1. $7y + 8y$ $15y$
2. $3x + 2x$ $5x$
3. $-9n - 6n$ $-15n$
4. $-12p + 3p$ $-9p$
5. $3.5t - 4.5t$ $-1t$ or $-t$
6. $8.6y - 7.6y$ $1y$ or y
7. $8w - w + 6w$ $13w$
8. $c - 7c + 2c$ $-4c$
9. $3b - 5 - 10b - 4$ $-7b - 9$
10. $6g + 5 - 3g - 7$
11. $m - 4m + 2m - 6$ $-m - 6$
12. $a + 3a - 2 - 7a$
17. $-3x + 2y - 1$ **18.** $-y - 5z + 7$ **19.** $2x + 14$ **20.** $6x - 14$

Simplify each expression. First use the distributive property to remove any parentheses. See Examples 5 and 6.

13. $5(y - 4)$ $5y - 20$
14. $7(r - 3)$ $7r - 21$
15. $7(d - 3) + 10$ $7d - 11$
16. $9(z + 7) - 15$ $9z + 48$
17. $-(3x - 2y + 1)$
18. $-(y + 5z - 7)$
19. $5(x + 2) - (3x - 4)$
20. $4(2x - 3) - 2(x + 1)$

21. In your own words, explain how to combine like terms.
22. Do like terms contain the same numerical coefficients? Explain your answer.   **21.–22.** answers may vary

Write each of the following as an algebraic expression. Simplify if possible. See Example 7.   **23.** $10x - 3$ **24.** $4y + 11$

23. Add $6x + 7$ to $4x - 10$.
24. Add $3y - 5$ to $y + 16$.
25. Subtract $7x + 1$ from $3x - 8$. $-4x - 9$
26. Subtract $4x - 7$ from $12 + x$. $19 - 3x$

MIXED PRACTICE
Simplify each expression.   **29.** $4x - 3$ **30.** $27h - 6$

27. $7x^2 + 8x^2 - 10x^2$ $5x^2$
28. $8x + x - 11x$ $-2x$
29. $6x - 5x + x - 3 + 2x$
30. $8h + 13h - 6 + 7h - h$
31. $-5 + 8(x - 6)$ $8x - 53$
32. $-6 + 5(r - 10)$ $5r - 56$

33. $6.2x - 4 + x - 1.2$
34. $7.9y - 0.7 - y + 0.2$
35. $2k - k - 6$ $k - 6$
36. $7c - 8 - c$ $6c - 8$
37. $0.5(m + 2) + 0.4m$ $0.9m + 1$
38. $0.2(k + 8) - 0.1k$
39. $-4(3y - 4)$ $-12y + 16$
40. $-3(2x + 5)$ $-6x - 15$
41. $3(2x - 5) - 5(x - 4)$ $x + 5$
42. $2(6x - 1) - (x - 7)$
43. $6x + 0.5 - 4.3x - 0.4x + 3$ $1.3x + 3.5$
44. $0.4y - 6.7 + y - 0.3 - 2.6y$
45. $-2(3x - 4) + 7x - 6$
46. $8y - 2 - 3(y + 4)$ $5y - 14$
47. $-9x + 4x + 18 - 10x$
48. $5y - 14 + 7y - 20y$
49. $5k - (3k - 10)$ $2k + 10$
50. $-11c - (4 - 2c)$ $-9c - 4$
51. $(3x + 4) - (6x - 1)$
52. $(8 - 5y) - (4 + 3y)$

Write each of the following phrases as an algebraic expression and simplify if possible. Let x represent the unknown number. See Examples 7 and 8.

53. Twice a number, decreased by four $2x - 4$
54. The difference of a number and two, divided by five $\dfrac{x - 2}{5}$
55. Three-fourths of a number, increased by twelve $\dfrac{3}{4}x + 12$
56. Eight more than triple a number $3x + 8$
57. The sum of 5 times a number and -2, added to 7 times a number
58. The sum of 3 times a number and 10, **subtracted from** 9 times a number $6x - 10$
59. Subtract $5m - 6$ from $m - 9$. $-4m - 3$
60. Subtract $m - 3$ from $2m - 6$. $m - 3$
61. Eight times the sum of a number and six $8(x + 6)$
62. Five, subtracted from four times a number $4x - 5$
63. Double a number, minus the sum of the number and ten $x - 10$
64. Half a number, minus the product of the number and eight
65. Seven, multiplied by the quotient of a number and six $\dfrac{7x}{6}$
66. The product of a number and ten, less twenty $10x - 20$

33. $7.2x - 5.2$ **34.** $6.9y - 0.5$ **38.** $0.1k + 1.6$ **42.** $11x + 5$ **44.** $-1.2y - 7$ **45.** $x + 2$ **47.** $-15x + 18$ **48.** $-8y - 14$ **51.** $-3x + 5$
52. $4 - 8y$ **57.** $-2 + 12x$ **64.** $-7.5x$ **67.** $7x - 7$

67. The sum of 2, three times a number, -9, and four times a number

68. The sum of twice a number, -1, five times a number, and -12

△ **69.** Recall that the perimeter of a figure is the total distance around the figure. Given the following rectangle, express the perimeter as an algebraic expression containing the variable x.

68. $7x - 13$ **69.** $(18x - 2)$ ft

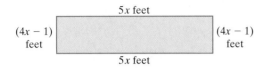
5x feet
$(4x - 1)$ feet $(4x - 1)$ feet
5x feet

△ **70.** Given the following triangle, express its perimeter as an algebraic expression containing the variable x. $(5x + 9)$ cm

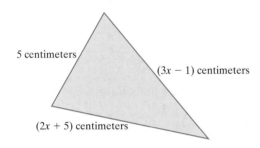
5 centimeters
$(3x - 1)$ centimeters
$(2x + 5)$ centimeters

REVIEW AND PREVIEW

Evaluate the following expressions for the given values. See Section 1.7.

71. If $x = -1$ and $y = 3$, find $y - x^2$.   2

72. If $g = 0$ and $h = -4$, find $gh - h^2$. -16

73. If $a = 2$ and $b = -5$, find $a - b^2$. -23

74. If $x = -3$, find $x^3 - x^2 + 4$. -32

75. If $y = -5$ and $z = 0$, find $yz - y^2$. -25

76. If $x = -2$, find $x^3 - x^2 - x$. -10

Concept Extensions

Given the following, determine whether each scale is balanced or not.

1 cone balances 1 cube

1 cylinder balances 2 cubes

77.

balanced

78.

not balanced

79.
balanced

80.
balanced

81. To convert from feet to inches, we multiply by 12. For example, the number of inches in 2 feet is $12 \cdot 2$ inches. If one board has a length of $(x + 2)$ *feet* and a second board has a length of $(3x - 1)$ *inches*, express their total length in inches as an algebraic expression. $(15x + 23)$ in.

82. The value of 7 nickels is $5 \cdot 7$ cents. Likewise, the value of x nickels is $5x$ cents. If the money box in a drink machine contains x *nickels*, $3x$ *dimes*, and $(30x - 1)$ *quarters*, express their total value in cents as an algebraic expression. $(785x - 25)$ ¢

For Exercises 83 through 88, see the example below.

Example

Simplify $-3xy + 2x^2y - (2xy - 1)$.

Solution

$-3xy + 2x^2y - (2xy - 1)$
$= -3xy + 2x^2y - 2xy + 1 = -5xy + 2x^2y + 1$

Simplify each expression.

83. $5b^2c^3 + 8b^3c^2 - 7b^3c^2$ **84.** $4m^4p^2 + m^4p^2 - 5m^2p^4$

85. $3x - (2x^2 - 6x) + 7x^2$ **86.** $9y^2 - (6xy^2 - 5y^2) - 8xy^2$

87. $-(2x^2y + 3z) + 3z - 5x^2y$ $-7x^2y$

88. $-(7c^3d - 8c) - 5c - 4c^3d$ $-11c^3d + 3c$

83. $5b^2c^3 + b^3c^2$ **84.** $5m^4p^2 - 5m^2p^4$ **85.** $5x^2 + 9x$
86. $14y^2 - 14xy^2$

2.2 THE ADDITION PROPERTY OF EQUALITY

Objectives

1. Define linear equation in one variable and equivalent equations.
2. Use the addition property of equality to solve linear equations.
3. Write word phrases as algebraic expressions.

1 Recall from Section 1.4 that an equation is a statement that two expressions have the same value. Also, a value of the variable that makes an equation a true statement is called a solution or root of the equation. The process of finding the solution of an equation is called **solving** the equation for the variable. In this section we concentrate on solving **linear equations** in one variable.

Linear Equation in One Variable

A linear equation in one variable can be written in the form

$$ax + b = c$$

where a, b, and c are real numbers and $a \neq 0$.

Evaluating a linear equation for a given value of the variable, as we did in Section 1.4, can tell us whether that value is a solution, but we can't rely on evaluating an equation as our method of solving it.

Instead, to solve a linear equation in x, we write a series of simpler equations, all *equivalent* to the original equation, so that the final equation has the form

$$x = \textbf{number} \quad \textbf{or} \quad \textbf{number} = x$$

Equivalent equations are equations that have the same solution. This means that the "number" above is the solution to the original equation.

2 The first property of equality that helps us write simpler equivalent equations is the **addition property of equality**.

Addition Property of Equality

If a, b, and c are real numbers, then

$$a = b \quad \text{and} \quad a + c = b + c$$

are equivalent equations.

This property guarantees that adding the same number to both sides of an equation does not change the solution of the equation. Since subtraction is defined in terms of addition, we may also **subtract the same number from both sides** without changing the solution.

A good way to picture a true equation is as a balanced scale. Since it is balanced, each side of the scale weighs the same amount.

If the same weight is added to or subtracted from each side, the scale remains balanced.

TEACHING TIP
Remind students that they may add or subtract *any* number to both sides of an equation, and the result is an equivalent equation. When solving equations, we try to add or subtract a number on both sides so that the equivalent equation is a simpler one to solve.

We use the addition property of equality to write equivalent equations until the variable is by itself on one side of the equation, and the equation looks like "x = number" or "number = x."

EXAMPLE 1

Solve $x - 7 = 10$ for x.

Solution   To solve for x, we want x alone on one side of the equation. To do this, we add 7 to both sides of the equation.

$$x - 7 = 10$$
$$x - 7 + 7 = 10 + 7 \qquad \text{Add 7 to both sides.}$$
$$x = 17 \qquad \text{Simplify.}$$

CLASSROOM EXAMPLE
Solve: $x - 5 = -2$

answer: 3

The solution of the equation $x = 17$ is obviously 17. Since we are writing equivalent equations, the solution of the equation $x - 7 = 10$ is also 17.

To check, replace x with 17 in the original equation.

$$x - 7 = 10$$
$$17 - 7 \stackrel{?}{=} 10 \qquad \text{Replace } x \text{ with 17 in the original equation.}$$
$$10 = 10 \qquad \text{True.}$$

Since the statement is true, 17 is the solution or we can say that the solution set is $\{17\}$.

✔ **CONCEPT CHECK**
Use the addition property to fill in the blank so that the middle equation simplifies to the last equation.

$$x - 5 = 3$$
$$x - 5 + \underline{} = 3 + \underline{}$$
$$x = 8$$

Concept Check Answer:
5

EXAMPLE 2

Solve $y + 0.6 = -1.0$ for y.

Solution To get y alone on one side of the equation, subtract 0.6 from both sides of the equation.

$$y + 0.6 = -1.0$$
$$y + 0.6 - 0.6 = -1.0 - 0.6 \quad \text{Subtract } \mathbf{0.6} \text{ from both sides.}$$
$$y = -1.6 \quad \text{Combine like terms.}$$

To check the proposed solution, -1.6, replace y with -1.6 in the original equation.

Check

$$y + 0.6 = -1.0$$
$$-1.6 + 0.6 \stackrel{?}{=} -1.0 \quad \text{Replace } y \text{ with } \mathbf{-1.6} \text{ in the original equation.}$$
$$-1.0 = -1.0 \quad \text{True.}$$

The solution is -1.6 or we can say that the solution set is $\{-1.6\}$.

EXAMPLE 3

Solve: $\dfrac{1}{2} = x - \dfrac{3}{4}$

Solution To get x alone, we add $\dfrac{3}{4}$ to both sides.

$$\frac{1}{2} = x - \frac{3}{4}$$
$$\frac{1}{2} + \frac{3}{4} = x - \frac{3}{4} + \frac{3}{4} \quad \text{Add } \frac{3}{4} \text{ to both sides.}$$
$$\frac{1}{2} \cdot \frac{2}{2} + \frac{3}{4} = x \quad \text{The LCD is } \mathbf{4}.$$
$$\frac{2}{4} + \frac{3}{4} = x \quad \text{Add the fractions.}$$
$$\frac{5}{4} = x$$

Check

$$\frac{1}{2} = x - \frac{3}{4} \quad \text{Original equation.}$$
$$\frac{1}{2} \stackrel{?}{=} \frac{5}{4} - \frac{3}{4} \quad \text{Replace } x \text{ with } \frac{5}{4}.$$
$$\frac{1}{2} \stackrel{?}{=} \frac{2}{4} \quad \text{Subtract.}$$
$$\frac{1}{2} = \frac{1}{2} \quad \text{True.}$$

The solution is $\dfrac{5}{4}$.

> **Helpful Hint**
> We may solve an equation so that the variable is alone on *either* side of the equation. For example, $\frac{5}{4} = x$ is equivalent to $x = \frac{5}{4}$.

EXAMPLE 4

Solve $5t - 5 = 6t + 2$ for t.

Solution

CLASSROOM EXAMPLE
Solve: $3a + 7 = 4a + 9$
answer: -2

To solve for t, we first want all terms containing t on one side of the equation and all other terms on the other side of the equation. To do this, first subtract $5t$ from both sides of the equation.

$$5t - 5 = 6t + 2$$
$$5t - 5 - 5t = 6t + 2 - 5t \qquad \text{Subtract } 5t \text{ from both sides.}$$
$$-5 = t + 2 \qquad \text{Combine like terms.}$$

Next, subtract 2 from both sides and the variable t will be isolated.

$$-5 = t + 2$$
$$-5 - 2 = t + 2 - 2 \qquad \text{Subtract 2 from both sides.}$$
$$-7 = t$$

Check the solution, -7, in the original equation. The solution is -7.

Many times, it is best to simplify one or both sides of an equation before applying the addition property of equality.

EXAMPLE 5

Solve: $2x + 3x - 5 + 7 = 10x + 3 - 6x - 4$

Solution

CLASSROOM EXAMPLE
Solve: $10w + 3 - 4w + 4$
$= -2w + 3 + 7w$
answer: -4

First we simplify both sides of the equation.

$$2x + 3x - 5 + 7 = 10x + 3 - 6x - 4$$
$$5x + 2 = 4x - 1 \qquad \begin{array}{l}\text{Combine like terms on each side} \\ \text{of the equation.}\end{array}$$

Next, we want all terms with a variable on one side of the equation and all numbers on the other side.

$$5x + 2 - 4x = 4x - 1 - 4x \qquad \text{Subtract } 4x \text{ from both sides.}$$
$$x + 2 = -1 \qquad \text{Combine like terms.}$$
$$x + 2 - 2 = -1 - 2 \qquad \begin{array}{l}\text{Subtract 2 from both sides} \\ \text{to get } x \text{ alone.}\end{array}$$
$$x = -3 \qquad \text{Combine like terms.}$$

Check

$$2x + 3x - 5 + 7 = 10x + 3 - 6x - 4 \qquad \text{Original equation.}$$
$$2(-3) + 3(-3) - 5 + 7 \stackrel{?}{=} 10(-3) + 3 - 6(-3) - 4 \qquad \text{Replace } x \text{ with } -3.$$
$$-6 - 9 - 5 + 7 \stackrel{?}{=} -30 + 3 + 18 - 4 \qquad \text{Multiply.}$$
$$-13 = -13 \qquad \text{True.}$$

The solution is -3.

If an equation contains parentheses, we use the distributive property to remove them, as before. Then we combine any like terms.

EXAMPLE 6

Solve: $6(2a - 1) - (11a + 6) = 7$

Solution

$6(2a - 1) - 1(11a + 6) = 7$

$6(2a) + 6(-1) - 1(11a) - 1(6) = 7$ Apply the distributive property.

$12a - 6 - 11a - 6 = 7$ Multiply.

$a - 12 = 7$ Combine like terms.

$a - 12 + 12 = 7 + 12$ Add **12** to both sides.

$a = 19$ Simplify.

CLASSROOM EXAMPLE
Solve: $3(2x - 5) - (5x + 1) = -3$
answer: 13

Check   Check by replacing a with 19 in the original equation.

EXAMPLE 7

Solve: $3 - x = 7$

Solution   First we subtract 3 from both sides.

$3 - x = 7$

$3 - x - 3 = 7 - 3$ Subtract 3 from both sides.

$-x = 4$ Simplify.

CLASSROOM EXAMPLE
Solve: $5 - x = 20$
answer: -15

We have not yet solved for x since x is not alone. However, this equation does say that the opposite of x is 4. If the opposite of x is 4, then x is the opposite of 4, or $x = -4$. If $-x = 4$, then $x = -4$.

Check

$3 - x = 7$ Original equation.

$3 - (-4) \stackrel{?}{=} 7$ Replace x with **-4**.

$3 + 4 \stackrel{?}{=} 7$ Add.

$7 = 7$ True.

The solution is -4.

TEACHING TIP

After solving Example 7, you may want to point out that there is more than one way to solve this problem.

$3 - x = 7$

$3 - x + x = 7 + x$

$3 = 7 + x$

$3 - 7 = 7 + x - 7$

$-4 = x$

3   Next, we practice writing word phrases as algebraic expressions.

EXAMPLE 8

a. The sum of two numbers is 8. If one number is 3, find the other number.

b. The sum of two numbers is 8. If one number is x, write an expression representing the other number.

c. An 8-foot board is cut into two pieces. If one piece is x feet, express the length of the other piece in terms of x.

Solution

a. If the sum of two numbers is 8 and one number is 3, we find the other number by subtracting 3 from 8. The other number is $8 - 3$ or 5.

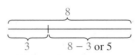

b. If the sum of two numbers is 8 and one number is x, we find the other number by subtracting x from 8. The other number is represented by $8 - x$.

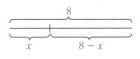

c. If an 8-foot board is cut into two pieces and one piece is x feet, we find the other length by subtracting x from 8. The other piece is $(8 - x)$ feet.

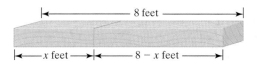

|←————— 8 feet —————→|

|←— x feet —→|←— $8 - x$ feet —→|

EXAMPLE 9

The Verrazano-Narrows Bridge in New York City is the longest suspension bridge in North America. The Golden Gate Bridge in San Francisco is 60 feet shorter than the Verrazano-Narrows Bridge. If the length of the Verrazano-Narrows Bridge is m feet, express the length of the Golden Gate Bridge as an algebraic expression in m. (*Source: Survey of State Highway Engineers*)

Solution   Since the Golden Gate is 60 feet shorter than the Verrazano-Narrows Bridge, we have that its length is

	Length of Verrazano-Narrows Bridge	minus	60
In words:			
Translate:	m	$-$	60

The Golden Gate Bridge is $(m - 60)$ feet long.

MENTAL MATH

Solve each equation mentally. See Examples 1 and 2.

1. $x + 4 = 6$ 2

2. $x + 7 = 10$ 3

3. $n + 18 = 30$ 12

4. $z + 22 = 40$ 18

5. $b - 11 = 6$ 17

6. $d - 16 = 5$ 21

EXERCISE SET 2.2

STUDY GUIDE/SSM   CD/VIDEO   PH MATH TUTOR CENTER   MathXL®Tutorials ON CD   MathXL®   MyMathLab®

Solve each equation. Check each solution. See Examples 1 through 3.

1. $x + 7 = 10$ 3

2. $x + 14 = 25$ 11

3. $x - 2 = -4$ -2

4. $y - 9 = 1$ 10

5. $-2 = t - 5$ 3

6. $-17 = x + 3$ -20

7. $r - 8.6 = -8.1$ 0.5

8. $t - 9.2 = -6.8$ 2.4

9. $\frac{3}{4} = \frac{1}{3} + f$ $\frac{5}{12}$

10. $\frac{3}{8} = c + \frac{1}{6}$ $\frac{5}{24}$

11. $5b - 0.7 = 6b$ -0.7

12. $9x + 5.5 = 10x$ 5.5

13. $7x - 3 = 6x$ 3

14. $18x - 9 = 19x$ -9

15. In your own words, explain what is meant by the solution of an equation.   answers may vary

16. In your own words, explain how to check a solution of an equation.   answers may vary

MIXED PRACTICE

Solve each equation. Don't forget to first simplify each side of the equation, if possible. Check each solution. See Examples 1 through 7.

17. $-8 = p - 4$ -4

18. $-10 = y + 2$ -12

19. $7x + 2x = 8x - 3$ -3

20. $3n + 2n = 7 + 4n$ 7

21. $2y + 10 = 5y - 4y$ -10

22. $4x - 4 = 10x - 7x$ 4

23. $3x - 6 = 2x + 5$ 11

24. $7y + 2 = 6y + 2$ 0

25. $5x - \frac{1}{6} = 6x - \frac{5}{6}$ $\frac{2}{3}$

26. $2x + \frac{1}{8} = x - \frac{3}{8}$ $-\frac{1}{2}$

27. $8y + 2 - 6y = 3 + y - 10$ -9

28. $4p - 11 - p = 2 + 2p - 20$ -7

29. $-6.5 - 4x - 1.6 - 3x = -6x + 9.8$ -17.9

30. $-1.4 - 7x - 3.6 - 2x = -8x + 4.4$ -9.4

31. $\frac{3}{8}x - \frac{1}{6} = -\frac{5}{8}x - \frac{2}{3}$ $-\frac{1}{2}$

32. $\frac{2}{5}x - \frac{1}{12} = -\frac{3}{5}x - \frac{3}{4}$ $-\frac{2}{3}$

33. $2(x - 4) = x + 3$ 11

34. $3(y + 7) = 2y - 5$ -26

35. $7(6 + w) = 6(2 + w)$ -30

36. $6(5 + c) = 5(c - 4)$ -50

37. $10 - (2x - 4) = 7 - 3x$

38. $15 - (6 - 7k) = 2 + 6k$

37. -7 **38.** -7

39. $-5(n - 2) = 8 - 4n$ 2

40. $-4(z - 3) = 2 - 3z$ 10

41. $-3\left(x - \frac{1}{4}\right) = -4x$ $-\frac{3}{4}$

42. $-2\left(x - \frac{1}{7}\right) = -3x$ $-\frac{2}{7}$

43. $3(n - 5) - (6 - 2n) = 4n$ 21

44. $5(3 + z) - (8z + 9) = -4z$ -6

45. $-2(x + 6) + 3(2x - 5) = 3(x - 4) + 10$ 25

46. $-5(x + 1) + 4(2x - 3) = 2(x + 2) - 8$ 13

47. $7(m - 2) - 6(m + 1) = -20$ 0

48. $-4(x - 1) - 5(2 - x) = -6$ 0

49. $0.8t + 0.2(t - 0.4) = 1.75$ 1.83

50. $0.6v + 0.4(0.3 + v) = 2.34$ 2.22

See Examples 8 and 9.

51. Two numbers have a sum of 20. If one number is p, express the other number in terms of p. $20 - p$

52. Two numbers have a sum of 13. If one number is y, express the other number in terms of y. $13 - y$

53. A 10-foot board is cut into two pieces. If one piece is x feet long, express the other length in terms of x. $(10 - x)$ ft

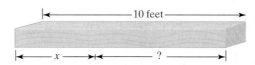

54. A 5-foot piece of string is cut into two pieces. If one piece is x feet long, express the other length in terms of x. $(5 - x)$ ft

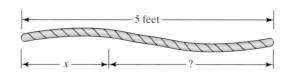

△ **55.** Two angles are *supplementary* if their sum is 180°. If one angle measures $x°$, express the measure of its supplement in terms of x. $(180 - x)°$

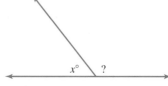

△ **56.** Two angles are *complementary* if their sum is 90°. If one angle measures $x°$, express the measure of its complement in terms of x. $(90 - x)°$

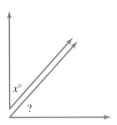

57. In a mayoral election, April Catarella received 284 more votes than Charles Pecot. If Charles received n votes, how many votes did April receive? $n + 284$

58. The length of the top of a computer desk is $1\frac{1}{2}$ feet longer than its width. If its width measures m feet, express its length as an algebraic expression in m. $(m + 1\frac{1}{2})$ ft

59. The longest interstate highway in the U.S. is I-90, which connects Seattle, Washington, and Boston, Massachusetts. The second longest interstate highway, I-80 (connecting San Francisco, California, and Teaneck, New Jersey), is 178.5 miles shorter than I-90. If the length of I-80 is m miles, express the length of I-90 as an algebraic expression in m. (*Source:* U.S. Department of Transportation–Federal Highway Administration) $m + 178.5$

60. In a recent election, Pat Ahumada ran against Solomon P. Ortiz for one of Texas's seats in the U.S. House of Representatives. Ahumada received 47,628 fewer votes than Ortiz. If Ahumada received n votes, how many did Ortiz receive? (*Source:* Voter News Service) $(n + 47,628)$ votes

61. The area of the Sahara Desert in Africa is 7 times the area of the Gobi Desert in Asia. If the area of the Gobi Desert is x square miles, express the area of the Sahara Desert as an algebraic expression in x. $7x$ sq. mi

62. The largest meteorite in the world is the Hoba West located in Namibia. Its weight is 3 times the weight of the Armanty meteorite located in Outer Mongolia. If the weight of the Armanty meteorite is y kilograms, express the weight of the Hoba West meteorite as an algebraic expression in y. $3y$ kg

REVIEW AND PREVIEW

Find the reciprocal or multiplicative inverse of each. See Section 1.7.

63. $\frac{5}{8}$ **64.** $\frac{7}{6}$ **65.** 2   **66.** 5   **67.** $-\frac{1}{9}$ **68.** $-\frac{3}{5}$

Perform each indicated operation and simplify. See Section 1.7.

69. $\frac{3x}{3}$ x **70.** $\frac{-2y}{-2}$ y **71.** $-5\left(-\frac{1}{5}y\right)$ y

72. $7\left(\frac{1}{7}r\right)$ r **73.** $\frac{3}{5}\left(\frac{5}{3}x\right)$ x **74.** $\frac{9}{2}\left(\frac{2}{9}x\right)$ x

Concept Extensions

△ **75.** The sum of the angles of a triangle is 180°. If one angle of a triangle measures $x°$ and a second angle measures $(2x + 7)°$, express the measure of the third angle in terms of x. Simplify the expression. $(173 - 3x)°$

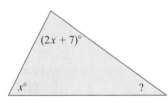

△ **76.** A quadrilateral is a four-sided figure like the one shown below whose angle sum is 360°. If one angle measures $x°$, a second angle measures $3x°$, and a third angle measures $5x°$, express the measure of the fourth angle in terms of x. Simplify the expression. $(360 - 9x)°$

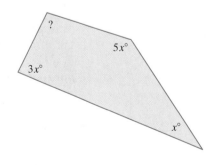

63. $\frac{8}{5}$ **64.** $\frac{6}{7}$ **65.** $\frac{1}{2}$ **66.** $\frac{1}{5}$ **67.** -9 **68.** $-\frac{5}{3}$

77. A nurse's aide recorded the following fluid intakes for a patient on her night shift: 200 ml, 150 ml, 400 ml. If the patient's doctor requested that a total of 1000 ml of fluid be taken by the patient overnight, how much more fluid must the nurse give the patient? To solve this problem, solve the equation $200 + 150 + 400 + x = 1000$.   250 ml

78. Let $x = 1$ and then $x = 2$ in the equation $x + 5 = x + 6$. Is either number a solution? How many solutions do you think this equation has? Explain your answer.   answers may vary

79. Let $x = 1$ and then $x = 2$ in the equation $x + 3 = x + 3$. Is either number a solution? How many solutions do you think this equation has? Explain your answer.   answers may vary

Use a calculator to determine whether the given value is a solution of the given equation.

80. $1.23x - 0.06 = 2.6x - 0.1285; x = 0.05$ solution

81. $8.13 + 5.85y = 20.05y - 8.91; y = 1.2$ solution

82. $3(a + 4.6) = 5a + 2.5; a = 6.3$ not a solution

83. $7(z - 1.7) + 9.5 = 5(z + 3.2) - 9.2; z = 4.8$ not a solution

---

2.3   THE MULTIPLICATION PROPERTY OF EQUALITY

Objectives

1. Use the multiplication property of equality to solve linear equations.
2. Use both the addition and multiplication properties of equality to solve linear equations.
3. Write word phrases as algebraic expressions.

TEACHING TIP

Remind students that a true equation is like a balanced scale. Then ask them: If you double the weight on one side, what must you do to the other side to keep it in balance?

1. As useful as the addition property of equality is, it cannot help us solve every type of linear equation in one variable. For example, adding or subtracting a value on both sides of the equation does not help solve

$$\frac{5}{2}x = 15.$$

Instead, we apply another important property of equality, the **multiplication property of equality**.

Multiplication Property of Equality

If a, b, and c are real numbers and $c \neq 0$, then

$$a = b \quad \text{and} \quad ac = bc$$

are equivalent equations.

This property guarantees that multiplying both sides of an equation by the same nonzero number does not change the solution of the equation. Since division is defined in terms of multiplication, we may also **divide both sides of the equation by the same nonzero number** without changing the solution.

EXAMPLE 1

Solve: $\dfrac{5}{2}x = 15$

Solution To get x alone, multiply both sides of the equation by the reciprocal of $\dfrac{5}{2}$, which is $\dfrac{2}{5}$.

$$\frac{5}{2}x = 15$$

$$\frac{2}{5} \cdot \frac{5}{2}x = \frac{2}{5} \cdot 15 \qquad \text{Multiply both sides by } \frac{2}{5}.$$

$$\left(\frac{2}{5} \cdot \frac{5}{2}\right)x = \frac{2}{5} \cdot 15 \qquad \text{Apply the associative property.}$$

$$1x = 6 \qquad \text{Simplify.}$$

or

$$x = 6$$

Check Replace x with 6 in the original equation.

$$\frac{5}{2}x = 15 \qquad \text{Original equation.}$$

$$\frac{5}{2}(6) \overset{?}{=} 15 \qquad \text{Replace } x \text{ with 6.}$$

$$15 = 15 \qquad \text{True.}$$

The solution is 6 or we say that the solution set is $\{6\}$

In the equation $\dfrac{5}{2}x = 15$, $\dfrac{5}{2}$ is the coefficient of x. When the coefficient of x is a *fraction*, we will get x alone by multiplying by the reciprocal. When the coefficient of x is an integer or a decimal, it is usually more convenient to divide both sides by the coefficient. (Dividing by a number is, of course, the same as multiplying by the reciprocal of the number.)

EXAMPLE 2

Solve: $-3x = 33$

Solution Recall that $-3x$ means $-3 \cdot x$. To get x alone, we divide both sides by the coefficient of x, that is, -3.

$$-3x = 33$$

$$\frac{-3x}{-3} = \frac{33}{-3} \qquad \text{Divide both sides by } -3.$$

$$1x = -11 \qquad \text{Simplify.}$$

$$x = -11$$

Check

$$-3x = 33 \qquad \text{Original equation.}$$

$$-3(-11) \overset{?}{=} 33 \qquad \text{Replace } x \text{ with } -11.$$

$$33 = 33 \qquad \text{True.}$$

The solution is -11, or the solution set is $\{-11\}$.

EXAMPLE 3

Solve: $\dfrac{y}{7} = 20$

Solution Recall that $\dfrac{y}{7} = \dfrac{1}{7}y$. To get y alone, we multiply both sides of the equation by 7, the reciprocal of $\dfrac{1}{7}$.

$$\dfrac{y}{7} = 20$$

$$\dfrac{1}{7}y = 20$$

$$7 \cdot \dfrac{1}{7}y = 7 \cdot 20 \qquad \text{Multiply both sides by 7.}$$

$$1y = 140 \qquad \text{Simplify.}$$

$$y = 140$$

Check

$$\dfrac{y}{7} = 20 \qquad \text{Original equation.}$$

$$\dfrac{140}{7} \stackrel{?}{=} 20 \qquad \text{Replace } y \text{ with } \mathbf{140}.$$

$$20 = 20 \qquad \text{True.}$$

The solution is 140.

EXAMPLE 4

Solve: $3.1x = 4.96$

Solution

$$3.1x = 4.96$$

$$\dfrac{3.1x}{3.1} = \dfrac{4.96}{3.1} \qquad \text{Divide both sides by 3.1.}$$

$$1x = 1.6 \qquad \text{Simplify.}$$

$$x = 1.6$$

Check Check by replacing x with 1.6 in the original equation. The solution is 1.6.

EXAMPLE 5

Solve: $-\dfrac{2}{3}x = -\dfrac{5}{2}$

Solution To get x alone, we multiply both sides of the equation by $-\dfrac{3}{2}$, the reciprocal of the co-efficient of x.

> **Helpful Hint**
> Don't forget to multiply both sides by $-\dfrac{3}{2}$.

$$-\dfrac{2}{3}x = -\dfrac{5}{2}$$

$$-\dfrac{3}{2} \cdot -\dfrac{2}{3}x = -\dfrac{3}{2} \cdot -\dfrac{5}{2} \qquad \text{Multiply both sides by } -\dfrac{3}{2}, \text{ the reciprocal of } -\dfrac{2}{3}.$$

$$x = \dfrac{15}{4} \qquad \text{Simplify.}$$

Check   Check by replacing x with $\dfrac{15}{4}$ in the original equation. The solution is $\dfrac{15}{4}$.   ⬭

2   We are now ready to combine the skills learned in the last section with the skills learned from this section to solve equations by applying more than one property.

EXAMPLE 6

Solve: $-z - 4 = 6$

Solution   First, get $-z$, the term containing the variable alone on one side. To do so, add 4 to both sides of the equation.

$$-z - 4 + 4 = 6 + 4 \qquad \text{Add 4 to both sides.}$$
$$-z = 10 \qquad \text{Simplify.}$$

Next, recall that $-z$ means $-1 \cdot z$. To get z alone, either multiply or divide both sides of the equation by -1. In this example, we divide.

$$-z = 10$$
$$\frac{-z}{-1} = \frac{10}{-1} \qquad \text{Divide both sides by the coefficient } -\mathbf{1}.$$
$$z = -10 \qquad \text{Simplify.}$$

Check   To check, replace z with -10 in the original equation. The solution is -10.   ⬭

EXAMPLE 7

Solve: $12a - 8a = 10 + 2a - 13 - 7$

Solution   First, simplify both sides of the equation by combining like terms.

$$12a - 8a = 10 + 2a - 13 - 7$$
$$4a = 2a - 10 \qquad \text{Combine like terms.}$$

To get all terms containing a variable on one side, subtract $2a$ from both sides.

$$4a - 2a = 2a - 10 - 2a \qquad \text{Subtract 2a from both sides.}$$
$$2a = -10 \qquad \text{Simplify.}$$
$$\frac{2a}{2} = \frac{-10}{2} \qquad \text{Divide both sides by 2.}$$
$$a = -5 \qquad \text{Simplify.}$$

Check   Check by replacing a with -5 in the original equation. The solution is -5.   ⬭

3   Next, we continue to sharpen our problem-solving skills by writing word phrases as algebraic expressions.

EXAMPLE 8

If x is the first of three consecutive integers, express the sum of the three integers in terms of x. Simplify if possible.

Solution An example of three consecutive integers is

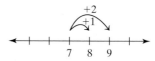

The second consecutive integer is always 1 more than the first, and the third consecutive integer is 2 more than the first. If x is the first of three consecutive integers, the three consecutive integers are

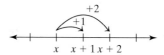

Their sum is

In words:

first integer	+	second integer	+	third integer
x	+	$(x + 1)$	+	$(x + 2)$

Translate:

which simplifies to $3x + 3$.

Below are examples of consecutive even and odd integers.

Even integers:

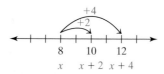

Odd integers:

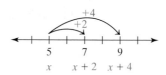

> ## Helpful Hint
>
> If x is an odd integer, then $x + 2$ is the next odd integer. This 2 simply means that odd integers are always 2 units from each other. (The same is true for even integers. They are always 2 units from each other.)
>
>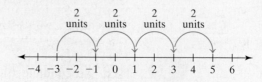

STUDY SKILLS REMINDER
Are You Organized?

Have you ever had trouble finding a completed assignment? When it's time to study for a test, are your notes neat and organized? Have your ever had trouble reading your own mathematics handwriting? (Be honest—I have had trouble reading my own handwriting before.)

When any of these things happen, it's time to get organized. Here are a few suggestions:

Write your notes and complete your homework assignment in a notebook with pockets (spiral or ring binder). Take class notes in this notebook, and then follow the notes with your completed homework assignment. When you receive graded papers or handouts, place them in the notebook pocket so that you will not lose them.

Place a mark (possibly an exclamation point) beside any note(s) that seem especially important to you. Also place a mark (possibly a question mark) beside any note(s) or homework that you are having trouble with. Don't forget to see your instructor, a tutor, or your fellow classmates to help you understand the concepts or exercises you have marked.

Also, if you are having trouble reading your own handwriting, *slow down* and write your mathematics work clearly!

MENTAL MATH

Solve each equation mentally. See Examples 2 and 3.

1. $3a = 27$ 9

2. $9c = 54$ 6

3. $5b = 10$ 2

4. $7t = 14$ 2

5. $6x = -30$ −5

6. $8r = -64$ −8

1. −4   **2.** 7   **3.** 0   **4.** 0   **5.** 12   **6.** −8   **7.** −12   **8.** −20   **9.** 3   **10.** 2   **11.** 2   **12.** 30

EXERCISE SET 2.3

STUDY GUIDE/SSM   CD/VIDEO   PH MATH TUTOR CENTER   MathXL®Tutorials ON CD   MathXL®   MyMathLab®

Solve each equation. Check each solution. See Examples 1 through 5.

 1. $-5x = 20$

2. $7x = 49$

3. $3x = 0$

4. $2x = 0$

5. $-x = -12$

6. $-y = 8$

 7. $\frac{2}{3}x = -8$

8. $\frac{3}{4}n = -15$

9. $\frac{1}{6}d = \frac{1}{2}$

10. $\frac{1}{8}v = \frac{1}{4}$

11. $\frac{a}{2} = 1$

12. $\frac{d}{15} = 2$

13. $\frac{k}{-7} = 0$ 0

14. $\frac{f}{-5} = 0$ 0

15. $1.7x = 10.71$ 6.3

16. $8.5y = 18.7$ 2.2

Solve each equation. Check each solution. See Examples 6 and 7.

17. $-x + 2 = 22$ −20

18. $-x + 4 = -24$ 28

 19. $6a + 3 = 3$ 0

20. $8t + 5 = 5$ 0

21. $6x + 10 = -20$ −5

22. $-10y + 15 = 5$ 1

23. $-2x + \dfrac{1}{2} = \dfrac{7}{2}$ $-\frac{3}{2}$

24. $-3n - \dfrac{1}{3} = \dfrac{8}{3}$ -1

25. $6z - 8 - z + 3 = 0$ 1

26. $4a + 1 + a - 11 = 0$ 2

27. $10 - 3x - 6 - 9x = 7$ $-\frac{1}{4}$

28. $12x + 30 + 8x - 6 = 10$ $-\frac{7}{10}$ **29.** $0.4x - 0.6x - 5 = 1$ -30

30. $0.4x - 0.9x - 6 = 19$ -50

MIXED PRACTICE

Solve. See Examples 1 through 7.

31. $42 = 7x$ 6

32. $81 = 3x$ 27

33. $4.4 = -0.8x$ -5.5

34. $6.3 = -0.6x$ -10.5

35. $-\dfrac{3}{7}p = -2$ $\frac{14}{3}$

36. $-\dfrac{4}{5}r = -5$ $\frac{25}{4}$

37. $2x - 4 = 16$ 10

38. $3x - 1 = 26$ 9

39. $5 - 0.3k = 5$ 0

40. $2 + 0.4p = 2$ 0

41. $-\dfrac{4}{3}x = 12$ -9

42. $-\dfrac{10}{3}x = 30$ -9

43. $10 = 2x - 1$ $\frac{11}{2}$

44. $12 = 3j - 4$ $\frac{16}{3}$

45. $\dfrac{x}{3} + 2 = -5$ -21

46. $\dfrac{b}{4} - 1 = -7$ -24

47. $1 = 0.4x - 0.6x - 5$ -30

48. $21 = 0.4x - 0.9x - 4$

49. $z - 5z = 7z - 9 - z$ $\frac{9}{10}$

50. $t - 6t = -13 + t - 3t$

51. $6 - 2x + 8 = 10$ 2

52. $-5 - 6y + 6 = 19$ -3

53. $-3a + 6 + 5a = 7a - 8a$ -2 **48.** -50 **50.** $\frac{13}{3}$

54. $4b - 8 - b = 10b - 3b$ -2

55. The equation $3x + 6 = 2x + 10 + x - 4$ is true for all real numbers. Substitute a few real numbers for x to see that this is so and then try solving the equation.

56. The equation $6x + 2 - 2x = 4x + 1$ has no solution. Try solving this equation for x and see what happens.

57. From the results of Exercises 55 and 56, when do you think an equation has all real numbers as its solution set?

58. From the results of Exercises 55 and 56, when do you think an equation has no solution?

55.–58. answers may vary

Write each algebraic expression described. Simplify if possible. See Example 8.

59. If x represents the first of two consecutive odd integers, express the sum of the two integers in terms of x. $2x + 2$

60. If x is the first of four consecutive even integers, write their sum as an algebraic expression in x. $4x + 12$

61. If x is the first of three consecutive integers, express the sum of the first integer and the third integer as an algebraic expression containing the variable x. $2x + 2$

62. If x is the first of two consecutive integers, express the sum of 20 and the second consecutive integer as an algebraic expression containing the variable x. $x + 21$

REVIEW AND PREVIEW **64.** $-8y - 3$ **66.** $-a + 3$

Simplify each expression. See Section 2.1.

63. $5x + 2(x - 6)$ $7x - 12$

64. $-7y + 2y - 3(y + 1)$

65. $-(x - 1) + x$ 1

66. $-(3a - 3) + 2a - 6$

Insert $<$, $>$, or $=$ in the appropriate space to make each statement true. See Sections 1.2 and 1.7.

67. $(-3)^2 \; > \; -3^2$

68. $(-2)^4 \; > \; -2^4$

69. $(-2)^3 \; = \; -2^3$

70. $(-4)^3 \; = \; -4^3$

71. $-|-6| \; < \; 6$

72. $-|-0.7| \; = \; -0.7$

Concept Extensions

73. A licensed nurse practitioner is instructed to give a patient 2100 milligrams of an antibiotic over a period of 36 hours. If the antibiotic is to be given every 4 hours starting immediately, how much antibiotic should be given in each dose? To answer this question, solve the equation $9x = 2100$. $\frac{700}{3}$ mg

74. Suppose you are a pharmacist and a customer asks you the following question. His child is to receive 13.5 milliliters of a nausea medicine over a period of 54 hours. If the nausea medicine is to be administered every 6 hours starting immediately, how much medicine should be given in each dose? 1.5 ml

Solve each equation.

75. $-3.6x = 10.62$ -2.95

76. $4.95y = -31.185$ -6.3

77. $7x - 5.06 = -4.92$

78. $0.06y + 2.63 = 2.5562$

77. 0.02

78. -1.23

2.4 | *SOLVING LINEAR EQUATIONS*

Objectives

1 Apply the general strategy for solving a linear equation.

2 Solve equations containing fractions.

3 Solve equations containing decimals.

4 Recognize identities and equations with no solution.

5 Write sentences as equations and solve.

1   We now present a general strategy for solving linear equations. One new piece of strategy is a suggestion to "clear an equation of fractions" as a first step. Doing so makes the equation more manageable, since operating on integers is more convenient than operating on fractions.

TEACHING TIP

Before students are shown the steps for solving a linear equation, let them come up with their own steps for solving a linear equation with your guidance. Then compare their steps with these. They will remember a set of steps better if they are involved in writing them.

Solving Linear Equations in One Variable

Step 1. Multiply on both sides by the LCD to clear the equation of fractions if they occur.

Step 2. Use the distributive property to remove parentheses if they occur.

Step 3. Simplify each side of the equation by combining like terms.

Step 4. Get all variable terms on one side and all numbers on the other side by using the addition property of equality.

Step 5. Get the variable alone by using the multiplication property of equality.

Step 6. Check the solution by substituting it into the original equation.

EXAMPLE 1

Solve: $4(2x - 3) + 7 = 3x + 5$

Solution   There are no fractions, so we begin with Step 2.

CLASSROOM EXAMPLE

Solve: $5(3x - 1) + 2 = 12x + 6$

answer: 3

$$4(2x - 3) + 7 = 3x + 5$$

Step 2. $8x - 12 + 7 = 3x + 5$ *Apply the distributive property.*

Step 3. $8x - 5 = 3x + 5$ *Combine like terms.*

Step 4. Get all variable terms on the same side of the equation by subtracting $3x$ from both sides, then adding 5 to both sides.

$$8x - 5 - 3x = 3x + 5 - 3x$$ *Subtract 3x from both sides.*
$$5x - 5 = 5$$ *Simplify.*
$$5x - 5 + 5 = 5 + 5$$ *Add 5 to both sides.*
$$5x = 10$$ *Simplify.*

Step 5. Use the multiplication property of equality to get x alone.

$$\frac{5x}{5} = \frac{10}{5}$$ *Divide both sides by 5.*
$$x = 2$$ *Simplify.*

Step 6. Check.

$$4(2x - 3) + 7 = 3x + 5 \qquad \text{Original equation}$$

$$4[2(2) - 3] + 7 \stackrel{?}{=} 3(2) + 5 \qquad \text{Replace } x \text{ with } 2.$$

$$4(4 - 3) + 7 \stackrel{?}{=} 6 + 5$$

$$4(1) + 7 \stackrel{?}{=} 11$$

$$4 + 7 \stackrel{?}{=} 11$$

$$11 = 11 \qquad \text{True.}$$

The solution is 2 or the solution set is {2}.

> **Helpful Hint**
> When checking solutions, use the original written equation.

EXAMPLE 2

Solve: $8(2 - t) = -5t$

Solution First, we apply the distributive property.

CLASSROOM EXAMPLE

Solve: $9(5 - x) = -3x$

answer: $\dfrac{15}{2}$

$$8(2 - t) = -5t$$

Step 2. $16 - 8t = -5t$ Use the distributive property.

Step 4. $16 - 8t + 8t = -5t + 8t$ To get variable terms on one side, add **8t** to both sides.

$$16 = 3t \qquad \text{Combine like terms.}$$

Step 5. $\dfrac{16}{3} = \dfrac{3t}{3}$ Divide both sides by **3**.

$$\dfrac{16}{3} = t \qquad \text{Simplify.}$$

Step 6. Check.

$$8(2 - t) = -5t \qquad \text{Original equation}$$

$$8\left(2 - \dfrac{16}{3}\right) \stackrel{?}{=} -5\left(\dfrac{16}{3}\right) \qquad \text{Replace } t \text{ with } \dfrac{16}{3}.$$

$$8\left(\dfrac{6}{3} - \dfrac{16}{3}\right) \stackrel{?}{=} -\dfrac{80}{3} \qquad \text{The LCD is 3.}$$

$$8\left(-\dfrac{10}{3}\right) \stackrel{?}{=} -\dfrac{80}{3} \qquad \text{Subtract fractions.}$$

$$-\dfrac{80}{3} = -\dfrac{80}{3} \qquad \text{True.}$$

The solution is $\dfrac{16}{3}$.

---

2   If an equation contains fractions, we can clear the equation of fractions by multiplying both sides by the LCD of all denominators. By doing this, we avoid working with time-consuming fractions.

EXAMPLE 3

Solve: $\dfrac{x}{2} - 1 = \dfrac{2}{3}x - 3$

Solution   We begin by clearing fractions. To do this, we multiply both sides of the equation by the LCD of 2 and 3, which is 6.

$$\frac{x}{2} - 1 = \frac{2}{3}x - 3$$

CLASSROOM EXAMPLE

Solve: $\dfrac{5}{2}x - 1 = \dfrac{3}{2}x - 4$

answer: -3

Step 1. $6\left(\dfrac{x}{2} - 1\right) = 6\left(\dfrac{2}{3}x - 3\right)$ Multiply both sides by the LCD, 6.

> **Helpful Hint**
>
> Don't forget to multiply *each* term by the LCD.

Step 2. $6\left(\dfrac{x}{2}\right) - 6(1) = 6\left(\dfrac{2}{3}x\right) - 6(3)$ Apply the distributive property.

$$3x - 6 = 4x - 18 \qquad \text{Simplify.}$$

There are no longer grouping symbols and no like terms on either side of the equation, so we continue with Step 4.

$$3x - 6 = 4x - 18$$

Step 4. $3x - 6 - 3x = 4x - 18 - 3x$ To get variable terms on one side, subtract 3x from both sides.

$$-6 = x - 18 \qquad \text{Simplify.}$$

$$-6 + 18 = x - 18 + 18 \qquad \text{Add 18 to both sides.}$$

$$12 = x \qquad \text{Simplify.}$$

Step 5.   The variable is now alone, so there is no need to apply the multiplication property of equality.

Step 6.   Check.

$$\frac{x}{2} - 1 = \frac{2}{3}x - 3 \qquad \text{Original equation}$$

$$\frac{12}{2} - 1 \stackrel{?}{=} \frac{2}{3} \cdot 12 - 3 \qquad \text{Replace } x \text{ with 12.}$$

$$6 - 1 \stackrel{?}{=} 8 - 3 \qquad \text{Simplify.}$$

$$5 = 5 \qquad \text{True.}$$

The solution is 12.

EXAMPLE 4

Solve: $\dfrac{2(a + 3)}{3} = 6a + 2$

Solution   We clear the equation of fractions first.

$$\frac{2(a + 3)}{3} = 6a + 2$$

CLASSROOM EXAMPLE

Solve: $\dfrac{3(x - 2)}{5} = 3x + 6$

answer: -3

Step 1. $3 \cdot \dfrac{2(a + 3)}{3} = 3(6a + 2)$ Clear the fraction by multiplying both sides by the LCD, 3.

$$2(a + 3) = 3(6a + 2)$$

Step 2.   Next, we use the distributive property and remove parentheses.

$$2a + 6 = 18a + 6$$ Apply the distributive property.

Step 4. $2a + 6 - 6 = 18a + 6 - 6$ Subtract 6 from both sides.

$$2a = 18a$$

$$2a - 18a = 18a - 18a$$ Subtract **18a** from both sides.

$$-16a = 0$$

Step 5. $\dfrac{-16a}{-16} = \dfrac{0}{-16}$ Divide both sides by **−16.**

$$a = 0$$ Write the fraction in simplest form.

Step 6. To check, replace a with 0 in the original equation. The solution is 0.

3   When solving a problem about money, you may need to solve an equation containing decimals. If you choose, you may multiply to clear the equation of decimals.

 EXAMPLE 5

Solve: $0.25x + 0.10(x - 3) = 0.05(22)$

Solution First we clear this equation of decimals by multiplying both sides of the equation by 100. Recall that multiplying a decimal number by 100 has the effect of moving the decimal point 2 places to the right.

$$0.25x + 0.10(x - 3) = 0.05(22)$$

Helpful Hint

By the distributive property, 0.10 is multiplied by x and −3. Thus to multiply each term here by 100, we only need to multiply 0.10 by 100.

Step 1. $0.25x + 0.10(x - 3) = 0.05(22)$ Multiply both sides by **100.**
$$25x + 10(x - 3) = 5(22)$$

Step 2. $25x + 10x - 30 = 110$ Apply the distributive property.

Step 3. $35x - 30 = 110$ Combine like terms.

Step 4. $35x - 30 + 30 = 110 + 30$ Add **30** to both sides.

$$35x = 140$$ Combine like terms.

Step 5. $\dfrac{35x}{35} = \dfrac{140}{35}$ Divide both sides by **35.**

$$x = 4$$

Step 6. To check, replace x with 4 in the original equation. The solution is 4.

CLASSROOM EXAMPLE

Solve: $0.06x - 0.10(x - 2)$
$= -0.02(8)$

answer: 9

TEACHING TIP

Help your students understand that each term must be multiplied by 100. In the case of $0.10(x - 3)$, this is accomplished by multiplying 0.10 by 100. Tell your students why.

4   So far, each equation that we have solved has had a single solution. However, not every equation in one variable has a single solution. Some equations have no solution, while others have an infinite number of solutions. For example,

$$x + 5 = x + 7$$

has no solution since no matter which **real number** we replace x with, the equation is false.

real number + 5 = same **real number** + 7   **FALSE**

On the other hand,

$$x + 6 = x + 6$$

has infinitely many solutions since x can be replaced by any real number and the equation is always true.

real number + 6 = same **real number** + 6   **TRUE**

The equation $x + 6 = x + 6$ is called an **identity**. The next few examples illustrate special equations like these.

> ### EXAMPLE 6
>
> Solve: $-2(x - 5) + 10 = -3(x + 2) + x$

Solution

$$-2(x - 5) + 10 = -3(x + 2) + x$$
$$-2x + 10 + 10 = -3x - 6 + x \qquad \text{Apply the distributive property on both sides.}$$
$$-2x + 20 = -2x - 6 \qquad \text{Combine like terms.}$$
$$-2x + 20 + 2x = -2x - 6 + 2x \qquad \text{Add } \mathbf{2x} \text{ to both sides.}$$
$$20 = -6 \qquad \text{Combine like terms.}$$

CLASSROOM EXAMPLE
Solve: $5(2 - x) + 8x = 3(x - 6)$
answer: no solution

The final equation contains no variable terms, and there is no value for x that makes $20 = -6$ a true equation. We conclude that there is **no solution** to this equation. In set notation, we can indicate that there is no solution with the empty set, $\{\ \}$, or use the empty set or null set symbol, $\varnothing$. In this chapter, we will simply write *no solution*.

> ### EXAMPLE 7
>
> Solve: $3(x - 4) = 3x - 12$

Solution

$$3(x - 4) = 3x - 12$$
$$3x - 12 = 3x - 12 \qquad \text{Apply the distributive property.}$$

CLASSROOM EXAMPLE
Solve: $-6(2x + 1) - 14$
$\qquad = -10(x + 2) - 2x$
answer: all real numbers

The left side of the equation is now identical to the right side. Every real number may be substituted for x and a true statement will result. We arrive at the same conclusion if we continue.

$$3x - 12 = 3x - 12$$
$$3x - 12 + 12 = 3x - 12 + 12 \qquad \text{Add } \mathbf{12} \text{ to both sides.}$$
$$3x = 3x \qquad \text{Combine like terms.}$$
$$3x - 3x = 3x - 3x \qquad \text{Subtract } \mathbf{3x} \text{ from both sides.}$$
$$0 = 0$$

Again, one side of the equation is identical to the other side. Thus, $3(x - 4) = 3x - 12$ is an **identity** and **all real numbers** are solutions. In set notation, this is $\{$all real numbers$\}$.

✓ **CONCEPT CHECK**

Suppose you have simplified several equations and obtain the following results. What can you conclude about the solutions to the original equation?

    a. $7 = 7$ **b.** $x = 0$ **c.** $7 = -4$

5   We can apply our equation-solving skills to solving problems written in words. Many times, writing an equation that describes or models a problem involves a direct translation from a word sentence to an equation.

> ### EXAMPLE 8

FINDING AN UNKNOWN NUMBER

Twice a number, added to seven, is the same as three subtracted from the number. Find the number.

Concept Check Answer:
a. Every real number is a solution.
b. The solution is 0.
c. There is no solution.

Solution

Translate the sentence into an equation and solve.

In words:

twice a number	added to	seven	is the same as	three subtracted from the number

Translate: $2x$ $+$ 7 $=$ $x - 3$

To solve, begin by subtracting x from both sides to isolate the variable term.

$$2x + 7 = x - 3$$
$$2x + 7 - x = x - 3 - x \qquad \text{Subtract } x \text{ from both sides.}$$
$$x + 7 = -3 \qquad \text{Combine like terms.}$$
$$x + 7 - 7 = -3 - 7 \qquad \text{Subtract } 7 \text{ from both sides.}$$
$$x = -10 \qquad \text{Combine like terms.}$$

Check the solution in the problem as it was originally stated. To do so, replace "number" in the sentence with -10. Twice "-10" added to 7 is the same as 3 subtracted from "-10."

$$2(-10) + 7 = -10 - 3$$
$$-13 = -13$$

The unknown number is -10.

> **Helpful Hint**
>
> When checking solutions, go back to the original stated problem, rather than to your equation in case errors have been made in translating to an equation.

Calculator Explorations

Checking Equations

We can use a calculator to check possible solutions of equations. To do this, replace the variable by the possible solution and evaluate both sides of the equation separately.

Equation: $3x - 4 = 2(x + 6)$ *Solution:* $x = 16$

$$3x - 4 = 2(x + 6) \qquad \text{Original equation}$$
$$3(16) - 4 \overset{?}{=} 2(16 + 6) \qquad \text{Replace } x \text{ with 16.}$$

Now evaluate each side with your calculator.

Evaluate left side:

$\boxed{3}\ \boxed{\times}\ \boxed{16}\ \boxed{-}\ \boxed{4}\ \boxed{=}$ or $\boxed{\text{ENTER}}$ Display: $\boxed{44}$ or $\begin{array}{l} 3*16 - 4 \\ \hfill 44 \end{array}$

Evaluate right side:

$\boxed{2}\ \boxed{(}\ \boxed{16}\ \boxed{+}\ \boxed{6}\ \boxed{)}\ \boxed{=}$ or $\boxed{\text{ENTER}}$ Display: $\boxed{44}$ or $\begin{array}{l} 2(16 + 6) \\ \hfill 44 \end{array}$

Since the left side equals the right side, the equation checks.

(continued)

Use a calculator to check the possible solutions to each equation.

1. $2x = 48 + 6x$; $x = -12$ solution

2. $-3x - 7 = 3x - 1$; $x = -1$ solution

3. $5x - 2.6 = 2(x + 0.8)$; $x = 4.4$

4. $-1.6x - 3.9 = -6.9x - 25.6$; $x = 5$

5. $\dfrac{564x}{4} = 200x - 11(649)$; $x = 121$

6. $20(x - 39) = 5x - 432$; $x = 23.2$

3. not a solution
5. solution

4. not a solution
6. solution

TEACHING TIP
A Group Activity for this section is available in the Instructor's Resource Manual.

EXERCISE SET 2.4

STUDY
GUIDE/SSM

CD/
VIDEO

PH MATH
TUTOR CENTER

MathXL®Tutorials
ON CD

MathXL®

MyMathLab®

Solve each equation. See Examples 1 and 2.

1. $-2(3x - 4) = 2x$ 1

2. $-(5x - 1) = 9$ $-\frac{8}{5}$

3. $4(2n - 1) = (6n + 4) + 1$ $\frac{9}{2}$

4. $3(4y + 2) = 2(1 + 5y) + 8$ 2

5. $5(2x - 1) - 2(3x) = 1$ $\frac{3}{2}$

6. $3(2 - 5x) + 4(6x) = 12$ $\frac{2}{3}$

7. $6(x - 3) + 10 = -8$ 0

8. $-4(2 + n) + 9 = 1$ 0

Solve each equation. See Examples 3 through 5.

9. $\dfrac{3}{4}x - \dfrac{1}{2} = 1$ 2

10. $\dfrac{2}{3}x + \dfrac{5}{3} = \dfrac{5}{3}$ 0

11. $x + \dfrac{5}{4} = \dfrac{3}{4}x$ -5

12. $\dfrac{7}{8}x + \dfrac{1}{4} = \dfrac{3}{4}x$ -2

13. $\dfrac{x}{2} - 1 = \dfrac{x}{5} + 2$ 10

14. $\dfrac{x}{5} - 2 = \dfrac{x}{3}$ -15

15. $\dfrac{6(3 - z)}{5} = -z$ 18

16. $\dfrac{4(5 - w)}{3} = -w$ 20

17. $\dfrac{2(x + 1)}{4} = 3x - 2$ 1

18. $\dfrac{3(y + 3)}{5} = 2y + 6$ -3

19. $0.50x + 0.15(70) = 0.25(142)$ 50

20. $0.40x + 0.06(30) = 0.20(49)$ 20

21. $0.12(y - 6) + 0.06y = 0.08y - 0.07(10)$ 0.2

22. $0.60(z - 300) + 0.05z = 0.70z - 0.41(500)$ 500

Solve each equation. See Examples 6 and 7.   **25.–28.** no solution

23. $5x - 5 = 2(x + 1) + 3x - 7$ all real numbers

24. $3(2x - 1) + 5 = 6x + 2$ all real numbers

25. $\dfrac{x}{4} + 1 = \dfrac{x}{4}$

26. $\dfrac{x}{3} - 2 = \dfrac{x}{3}$

27. $3x - 7 = 3(x + 1)$

28. $2(x - 5) = 2x + 10$

29. Explain the difference between simplifying an expression and solving an equation.   **29.–32.** answers may vary

30. When solving an equation, if the final equivalent equation is $0 = 5$, what can we conclude? If the final equivalent equation is $-2 = -2$, what can we conclude?

31. On your own, construct an equation for which every real number is a solution.

32. On your own, construct an equation that has no solution.

MIXED PRACTICE

Solve each equation.

33. $4x + 3 = 2x + 11$ 4

34. $6y - 8 = 3y + 7$ 5

35. $-2y - 10 = 5y + 18$ -4

36. $7n + 5 = 10n - 10$ 5

37. $0.6x - 0.1 = 0.5x + 0.2$ 3

38. $0.2x - 0.1 = 0.6x - 2.1$ 5

39. $2y + 2 = y$ -2

40. $7y + 4 = -3$ -1

41. $3(5c - 1) - 2 = 13c + 3$ 4

42. $4(3t + 4) - 20 = 3 + 5t$ 1

43. $x + \dfrac{7}{6} = 2x - \dfrac{7}{6}$ $\frac{7}{3}$

44. $\dfrac{5}{2}x - 1 = x + \dfrac{1}{4}$ $\frac{5}{6}$

45. $2(x - 5) = 7 + 2x$ no solution

46. $-3(1 - 3x) = 9x - 3$ all real numbers

47. $\dfrac{2(z + 3)}{3} = 5 - z$ $\frac{9}{5}$

48. $\dfrac{3(w + 2)}{4} = 2w + 3$ $-\frac{6}{5}$

49. $\dfrac{4(y - 1)}{5} = -3y$ $\frac{4}{19}$

50. $\dfrac{5(1 - x)}{6} = -4x$ $-\frac{5}{19}$

51. $8 - 2(a - 1) = 7 + a$ 1

52. $5 - 6(2 + b) = b - 14$ 1

53. $2(x + 3) - 5 = 5x - 3(1 + x)$ no solution

54. $4(2 + x) + 1 = 7x - 3(x - 2)$ no solution

55. $\dfrac{5x - 7}{3} = x$ $\frac{7}{2}$

56. $\dfrac{7n + 3}{5} = -n$ $-\frac{1}{4}$

57. $\dfrac{9 + 5v}{2} = 2v - 4$ -17

58. $\dfrac{6 - c}{2} = 5c - 8$ 2

59. $-3(t - 5) + 2t = 5t - 4$ $\frac{19}{6}$

60. $-(4a - 7) - 5a = 10 + a$ $-\frac{3}{10}$

61. $0.02(6t - 3) = 0.12(t - 2) + 0.18$ all real numbers

62. $0.03(2m + 7) = 0.06(5 + m) - 0.09$ all real numbers

63. $0.06 - 0.01(x + 1) = -0.02(2 - x)$ 3

64. $-0.01(5x + 4) = 0.04 - 0.01(x + 4)$ -1

65. $\dfrac{3(x - 5)}{2} = \dfrac{2(x + 5)}{3}$ 13

66. $\dfrac{5(x - 1)}{4} = \dfrac{3(x + 1)}{2}$ -11

Write each of the following as equations. Then solve. See Example 8.

67. The sum of twice a number and 7 is equal to the sum of a number and 6. Find the number. $2x + 7 = x + 6; -1$

68. The difference of three times a number and 1 is the same as twice a number. Find the number. $3x - 1 = 2x; 1$

69. Three times a number, minus 6, is equal to two times a number, plus 8. Find the number. $3x - 6 = 2x + 8; 14$

70. The sum of 4 times a number and −2 is equal to the sum of 5 times a number and −2. Find the number.

71. One-third of a number is five-sixths. Find the number.

72. Seven-eighths of a number is one-half. Find the number.

73. If the quotient of a number and 4 is added to $\frac{1}{2}$, the result is $\frac{3}{4}$. Find the number. $\frac{x}{4} + \frac{1}{2} = \frac{3}{4}; 1$

74. If $\frac{3}{4}$ is added to three times a number, the result is $\frac{1}{2}$ subtracted from twice the number. Find the number.

75. Five times a number, subtracted from ten, is triple the number. Find the number. $10 - 5x = 3x; \frac{5}{4}$

76. Nine is equal to ten subtracted from double a number. Find the number. $9 = 2x - 10; \frac{19}{2}$

△ **77.** The perimeter of a geometric figure is the sum of the lengths of its sides. If the perimeter of the following pentagon (five-sided figure) is 28 centimeters, find the length of each side.

$x = 4$ cm, $2x = 8$ cm

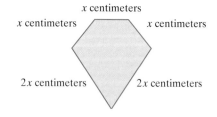

△ **78.** The perimeter of the following triangle is 35 meters. Find the length of each side.

$x = 6$ m; $2x + 1 = 13$ m; $3x - 2 = 16$ m

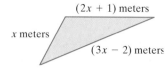

70. $4x - 2 = 5x - 2; 0$ **71.** $\frac{1}{3}x = \frac{5}{6}; \frac{5}{2}$ **72.** $\frac{7}{8}x = \frac{1}{2}; \frac{4}{7}$ **74.** $\frac{3}{4} + 3x = 2x - \frac{1}{2}; -\frac{5}{4}$

REVIEW AND PREVIEW

Evaluate. See Section 1.7.

79. $|2^3 - 3^2| - |5 - 7|$ −1   **80.** $|5^2 - 2^2| + |9 + (-3)|$ 27

81. $\dfrac{5}{4 + 3 \cdot 7}$ $\frac{1}{5}$ **82.** $\dfrac{8}{24 - 8 \cdot 2}$ 1

See Section 2.1.

△ **83.** A plot of land is in the shape of a triangle. If one side is x meters, a second side is $(2x - 3)$ meters and a third side is $(3x - 5)$ meters, express the perimeter of the lot as a simplified expression in x. $(6x - 8)$ m

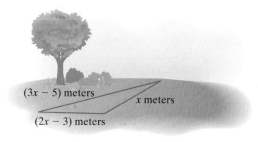

84. A portion of a board has length x feet. The other part has length $(7x - 9)$ feet. Express the total length of the board as a simplified expression in x. $(8x - 9)$ ft

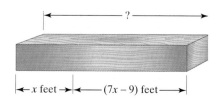

Concept Extensions

The bar graph below shows the five most common names of cities, towns, or villages in the United States.

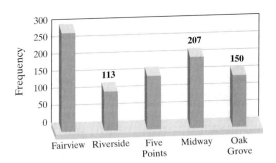

Source: Interior Department

85. What is the most popular name of a city, town, or village in the United States? Fairview

86. How many more cities, towns, or villages are named Oak Grove than named Riverside? 37

87. Let x represent "the number of cities, towns, or villages named Five Points" and use the information given to determine the unknown number. "The number of cities, towns, or villages named Five Points" added to 65 is equal to twice "the number of cities, towns, or villages named Five Points" minus 90. Check your answer by noticing the height of the bar representing Five Points. Is your answer reasonable?   155

88. Let x represent "the number of cities, towns, or villages named Fairview" and use the information given to determine the unknown number. Three times "the number of cities, towns, or villages named Fairview" added to 119 is equal to 168 subtracted from 4 times "the number of cities, towns, or villages named Fairview." Check your answer by noticing the height of the bar representing Fairview. Is your answer reasonable?   287

89. $1000(7x - 10) = 50(412 + 100x)$ 15.3

90. $10,000(x + 4) = 100(16 + 75x)$ -15.36

91. $0.035x + 5.112 = 0.010x + 5.107$ -0.2

92. $0.127x - 2.685 = 0.027x - 2.38$ 3.05

For Exercises 93 through 96, see the example below.

Example
Solve: $t(t + 4) = t^2 - 2t + 6$.

Solution
$$t(t + 4) = t^2 - 2t + 6$$
$$t^2 + 4t = t^2 - 2t + 6$$
$$t^2 + 4t - t^2 = t^2 - 2t + 6 - t^2$$
$$4t = -2t + 6$$
$$4t + 2t = -2t + 6 + 2t$$
$$6t = 6$$
$$t = 1$$

Solve each equation.

93. $x(x - 3) = x^2 + 5x + 7$ $-\frac{7}{8}$

94. $t^2 - 6t = t(8 + t)$ 0

95. $2z(z + 6) = 2z^2 + 12z - 8$ no solution

96. $y^2 - 4y + 10 = y(y - 5)$ -10

INTEGRATED REVIEW   SOLVING LINEAR EQUATIONS

Solve. Feel free to use the steps given in Section 2.4.

1. $x - 10 = -4$ 6

2. $y + 14 = -3$ -17

3. $9y = 108$ 12

4. $-3x = 78$ -26

5. $-6x + 7 = 25$ -3

6. $5y - 42 = -47$ -1

7. $\frac{2}{3}x = 9$ $\frac{27}{2}$

8. $\frac{4}{5}z = 10$ $\frac{25}{2}$

9. $\frac{r}{-4} = -2$ 8

10. $\frac{y}{-8} = 8$ -64

11. $6 - 2x + 8 = 10$ 2

12. $-5 - 6y + 6 = 19$ -3

13. $2x - 7 = 2x - 27$ no solution

14. $3 + 8y = 8y - 2$ no solution

15. $-3a + 6 + 5a = 7a - 8a$ -2

16. $4b - 8 - b = 10b - 3b$ -2

17. $-\frac{2}{3}x = \frac{5}{9}$ $-\frac{5}{6}$

18. $-\frac{3}{8}y = -\frac{1}{16}$ $\frac{1}{6}$

19. $10 = -6n + 16$ 1

20. $-5 = -2m + 7$ 6

21. $3(5c - 1) - 2 = 13c + 3$ 4

22. $4(3t + 4) - 20 = 3 + 5t$ 1

23. $\frac{2(z + 3)}{3} = 5 - z$ $\frac{9}{5}$

24. $\frac{3(w + 2)}{4} = 2w + 3$ $-\frac{6}{5}$

25. $-2(2x - 5) = -3x + 7 - x + 3$

26. $-4(5x - 2) = -12x + 4 - 8x + 4$

27. $0.02(6t - 3) = 0.04(t - 2) + 0.02$ 0*

28. $0.03(m + 7) = 0.02(5 - m) + 0.03$ -1.6

29. $-3y = \frac{4(y - 1)}{5}$ $\frac{4}{19}$

30. $-4x = \frac{5(1 - x)}{6}$ $-\frac{5}{19}$

31. $\frac{5}{3}x - \frac{7}{3} = x$ $\frac{7}{2}$

32. $\frac{7}{5}n + \frac{3}{5} = -n$ $-\frac{1}{4}$

25.–26.˙ all real numbers

2.5 AN INTRODUCTION TO PROBLEM SOLVING

Objective

1. Apply the steps for problem solving.

1. In previous sections, you practiced writing word phrases and sentences as algebraic expressions and equations to help prepare for problem solving. We now use these translations to help write equations that model a problem. The problem-solving steps given next may be helpful.

TEACHING TIP

Spend some time helping students with step 1. We often jump to step 2 before students really have a chance to understand the problem.

General Strategy for Problem Solving

1. UNDERSTAND the problem. During this step, become comfortable with the problem. Some ways of doing this are:

   Read and reread the problem.
   Choose a variable to represent the unknown.
   Construct a drawing, whenever possible.
   Propose a solution and check. Pay careful attention to how you check your proposed solution. This will help when writing an equation to model the problem.

2. TRANSLATE the problem into an equation.
3. SOLVE the equation.
4. INTERPRET the results: *Check* the proposed solution in the stated problem and state your conclusion.

Much of problem solving involves a direct translation from a sentence to an equation. Although we have been practicing these translations in previous sections, this section will often have translations that require a placement of parentheses.

EXAMPLE 1

FINDING AN UNKNOWN NUMBER

Twice the sum of a number and 4 is the same as four times the number decreased by 12. Find the number.

Solution

1. UNDERSTAND. Read and reread the problem. If we let

$$x = \text{the unknown number, then}$$

"the sum of a number and 4" translates to "$x + 4$" and "four times the number" translates to "$4x$."

2. TRANSLATE.

CLASSROOM EXAMPLE

Three times the difference of a number and 5 is the same as twice the number, decreased by 3. Find the number.
answer: 12

twice	sum of a number and 4	is the same as	four times the number	decreased by	12
↓	↓	↓	↓	↓	↓
2	$(x + 4)$	=	$4x$	−	12

3. SOLVE.

$$2(x + 4) = 4x - 12$$
$$2x + 8 = 4x - 12 \qquad \text{Apply the distributive property.}$$
$$2x + 8 - 4x = 4x - 12 - 4x \qquad \text{Subtract } 4x \text{ from both sides.}$$
$$-2x + 8 = -12$$
$$-2x + 8 - 8 = -12 - 8 \qquad \text{Subtract } 8 \text{ from both sides.}$$
$$-2x = -20$$
$$\frac{-2x}{-2} = \frac{-20}{-2} \qquad \text{Divide both sides by } -2.$$
$$x = 10$$

4. INTERPRET.

Check: Check this solution in the problem as it was originally stated. To do so, re-place "number" with 10. Twice the sum of "10" and 4 is 28, which is the same as 4 times "10" decreased by 12.

State: The number is 10.

Next, we continue to review consecutive integers.

(**EXAMPLE 2**)

Some states have a single area code for the entire state. Two such states have area codes that are consecutive odd integers. If the sum of these integers is 1208, find the two area codes. (*Source: World Almanac*, 2003)

Solution   **1.** UNDERSTAND. Read and reread the problem. If we let

$$x = \text{the first odd integer, then}$$
$$x + 2 = \text{the next odd integer}$$

> **Helpful Hint**
> Remember, the 2 here means that odd integers are 2 units apart, for ex-ample, the odd integers 13 and $13 + 2 = 15$.

2. TRANSLATE.

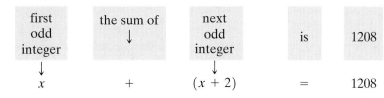

first odd integer	the sum of ↓	next odd integer	is	1208
↓		↓		
x	$+$	$(x + 2)$	$=$	1208

3. SOLVE.

$$x + x + 2 = 1208$$
$$2x + 2 = 1208$$
$$2x + 2 - 2 = 1208 - 2$$
$$2x = 1206$$
$$\frac{2x}{2} = \frac{1206}{2}$$
$$x = 603$$

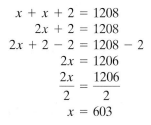

4. INTERPRET.

Check: If $x = 603$, then the next odd integer $x + 2 = 603 + 2 = 605$. Notice their sum, $603 + 605 = 1208$, as needed.

State: The area codes are 603 and 605.

Note: New Hampshire's area code is 603 and South Dakota's area code is 605.

EXAMPLE 3

FINDING THE LENGTH OF A BOARD

A 10-foot board is to be cut into two pieces so that the longer piece is 4 times the shorter. Find the length of each piece.

Solution

1. UNDERSTAND the problem. To do so, read and reread the problem. You may also want to propose a solution. For example, if 3 feet represents the length of the shorter piece, then $4(3) = 12$ feet is the length of the longer piece, since it is 4 times the length of the shorter piece. This guess gives a total board length of 3 feet + 12 feet = 15 feet, too long. However, the purpose of proposing a solution is not to guess correctly, but to help better understand the problem and how to model it.

Since the length of the longer piece is given in terms of the length of the shorter piece, let's let

$$x = \text{length of shorter piece, then}$$
$$4x = \text{length of longer piece}$$

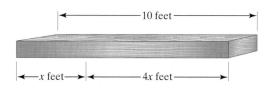

2. TRANSLATE the problem. First, we write the equation in words.

length of shorter piece	added to	length of longer piece	equals	total length of board
↓	↓	↓	↓	↓
x	$+$	$4x$	$=$	10

3. SOLVE.

$$x + 4x = 10$$
$$5x = 10 \quad \text{Combine like terms.}$$
$$\frac{5x}{5} = \frac{10}{5} \quad \text{Divide both sides by 5.}$$
$$x = 2$$

4. INTERPRET.

Check: Check the solution in the stated problem. If the shorter piece of board is 2 feet, the longer piece is $4 \cdot (2 \text{ feet}) = 8 \text{ feet}$ and the sum of the two pieces is 2 feet + 8 feet = 10 feet.

State: The shorter piece of board is 2 feet and the longer piece of board is 8 feet.

> **Helpful Hint**
> Make sure that units are included in your answer, if appropriate.

EXAMPLE 4

FINDING THE NUMBER OF REPUBLICAN AND DEMOCRATIC SENATORS

In a recent year, the U.S. House of Representatives had a total of 431 Democrats and Republicans. There were 15 more Republican representatives than

Democratic representatives. Find the number of representatives from each party. (*Source:* Office of the Clerk of the U.S. House of Representatives)

Solution   1. UNDERSTAND the problem. Read and reread the problem. Let's suppose that there were 200 Democratic representatives. Since there were 15 more Republicans than Democrats, there must have been 200 + 15 = 215 Republicans. The total number of Democrats and Republicans was then 200 + 215 = 415. This is incorrect since the total should be 431, but we now have a better understanding of the problem.

In general, if we let

$$x = \text{number of Democrats, then}$$
$$x + 15 = \text{number of Republicans}$$

2. TRANSLATE the problem. First, we write the equation in words.

number of Democrats	added to	number of Republicans	equals	431
↓	↓	↓	↓	↓
x	$+$	$(x + 15)$	$=$	431

3. SOLVE.

$$x + (x + 15) = 431$$
$$2x + 15 = 431 \qquad \text{Combine like terms.}$$
$$2x + 15 - 15 = 431 - 15 \qquad \text{Subtract 15 from both sides.}$$
$$2x = 416$$
$$\frac{2x}{2} = \frac{416}{2} \qquad \text{Divide both sides by 2.}$$
$$x = 208$$

4. INTERPRET.

Check: If there were 208 Democratic representatives, then there were 208 + 15 = 223 Republican representatives. The total number of representatives is then 208 + 223 = 431. The results check.

State: There were 208 Democratic and 223 Republican representatives in Congress.

EXAMPLE 5

CALCULATING HOURS ON JOB

A computer science major at a local university has a part-time job working on computers for his clients. He charges $20 to come to your home or office and then $25 per hour. During one month he visited 10 homes or offices and his total income was $575. How many hours did he spend working on computers?

Solution   1. UNDERSTAND. Read and reread the problem. Let's propose that the student spent 20 hours working on computers. Pay careful attention as to how his income is calculated. For 20 hours and 10 visits, his income is 20($25) + 10($20) = $700, more than $575. We now have a better understanding of the problem and know that the time working on computers is less than 20 hours.

Let's let

$$x = \text{hours working on computers. Then}$$
$$25x = \text{amount of money made while working on computers}$$

2. TRANSLATE.

money made while working on computers	plus	money made for visits	is equal to	575
↓	↓	↓	↓	↓
$25x$	$+$	$10(20)$	$=$	575

3. SOLVE.

$$25x + 200 = 575$$
$$25x + 200 - 200 = 575 - 200 \qquad \text{Subtract } \textbf{200} \text{ from both sides.}$$
$$25x = 375 \qquad \text{Simplify.}$$
$$\frac{25x}{25} = \frac{375}{25} \qquad \text{Divide both sides by } \textbf{25.}$$
$$x = 15 \qquad \text{Simplify.}$$

4. INTERPRET.

Check: If the student works 15 hours and makes 10 visits, his income is 15($25) + 10($20) = $575.

State: The student spent 15 hours working on computers.

 EXAMPLE 6

FINDING ANGLE MEASURES

If the two walls of the Vietnam Veterans Memorial in Washington D.C. were connected, an isosceles triangle would be formed. The measure of the third angle is 97.5° more than the measure of either of the other two equal angles. Find the measure of the third angle. (*Source:* National Park Service)

Solution

1. UNDERSTAND. Read and reread the problem. We then draw a diagram (recall that an isosceles triangle has two angles with the same measure) and let

$$x = \text{degree measure of one angle}$$
$$x = \text{degree measure of the second equal angle}$$
$$x + 97.5 = \text{degree measure of the third angle}$$

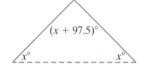

2. TRANSLATE. Recall that the sum of the measures of the angles of a triangle equals 180.

measure of first angle		measure of second angle		measure of third angle	equals	180
↓		↓		↓	↓	↓
x	$+$	x	$+$	$(x + 97.5)$	$=$	180

3. SOLVE.

$$x + x + (x + 97.5) = 180$$
$$3x + 97.5 = 180 \qquad \text{Combine like terms.}$$
$$3x + 97.5 - 97.5 = 180 - 97.5 \qquad \text{Subtract } \textbf{97.5} \text{ from both sides.}$$
$$3x = 82.5$$
$$\frac{3x}{3} = \frac{82.5}{3} \qquad \text{Divide both sides by } \textbf{3.}$$
$$x = 27.5$$

4. INTERPRET.

Check: If $x = 27.5$, then the measure of the third angle is $x + 97.5 = 125$. The sum of the angles is then $27.5 + 27.5 + 125 = 180$, the correct sum.
State: The third angle measures 125°.*
(* The two walls actually meet at an angle of 125 degrees 12 minutes. The measurement of 97.5° given in the problem is an approximation.)

STUDY SKILLS REMINDER

How Are Your Homework Assignments Going?

It is so important in mathematics to keep up with homework. Why? Many concepts build on each other. Oftentimes, your understanding of a day's lecture in mathematics depends on an understanding of the previous day's material.

Remember that completing your homework assignment involves a lot more than attempting a few of the problems assigned.

To complete a homework assignment, remember these four things:

1. Attempt all of it.

2. Check it.

3. Correct it.

4. If needed, ask questions about it.

EXERCISE SET 2.5

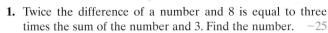

| STUDY GUIDE/SSM | CD/ VIDEO | PH MATH TUTOR CENTER | MathXL®Tutorials ON CD | MathXL® | MyMathLab® |

Solve. See Example 1.   **3.** $-\frac{3}{4}$

1. Twice the difference of a number and 8 is equal to three times the sum of the number and 3. Find the number. -25

2. Five times the sum of a number and -1 is the same as 6 times the number. Find the number. -5

3. The product of twice a number and three is the same as the difference of five times the number and $\frac{3}{4}$. Find the number.

4. If the difference of a number and four is doubled, the result is $\frac{1}{4}$ less than the number. Find the number. $\frac{31}{4}$

Solve. See Example 2.

5. The left and right page numbers of an open book are two consecutive integers whose sum is 469. Find these page numbers. $234, 235$

6. The room numbers of two adjacent classrooms are two consecutive even numbers. If their sum is 654, find the classroom numbers. $326, 328$

7. To make an international telephone call, you need the code for the country you are calling. The codes for Belgium, France, and Spain are three consecutive integers whose sum is 99. Find the code for each country. (*Source: The World Almanac and Book of Facts*, 2003)   Belgium: 32; France: 33; Spain: 34

8. To make an international telephone call, you need the code for the country you are calling. The codes for Mali Republic, Côte d'Ivoire, and Niger are three consecutive odd integers whose sum is 675. Find the code for each country.
Mali Republic: 223; Côte d'Ivoire: 225; Niger: 227
Solve. See Example 3.

9. A 40-inch board is to be cut into three pieces so that the second piece is twice as long as the first piece and the third piece is 5 times as long as the first piece. If *x* represents the length of the first piece, find the lengths of all three pieces.
1st piece: 5 in.; 2nd piece: 10 in.; 3rd piece: 25 in.

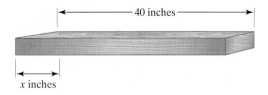

10. A 21-foot beam is to be divided so that the longer piece is 1 foot more than 3 times the shorter piece. If *x* represents the length of the shorter piece, find the lengths of both pieces.
shorter: 5 ft; longer: 16 ft

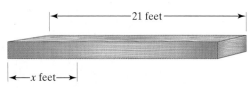

11. governor of Nebraska: $65,000; governor of Washington: $130,000

Solve. See Example 4.

11. The governor of Washington state makes twice as much money as the governor of Nebraska. If the total of their salaries is $195,000, find the salary of each.

12. In the 2002 Winter Olympics, Germany won 2 more gold medals than the United States. If the total number of gold medals for both is 22, find the number of gold medals that each team won. (*Source: World Almanac,* 2003)
U.S.: 10 medals; Germany: 12 medals
Solve. See Example 5. **13.** 172 mi

13. A car rental agency advertised renting a Buick Century for $24.95 per day and $0.29 per mile. If you rent this car for 2 days, how many whole miles can you drive on a $100 budget?

14. A plumber gave an estimate for the renovation of a kitchen. Her hourly pay is $27 per hour and the plumber's parts will cost $80. If her total estimate is $404, how many hours does she expect this job to take? 12 hr

15. In one U.S. city, the taxi cost is $3 plus $0.80 per mile. If you are traveling from the airport, there is an additional charge of $4.50 for tolls. How far can you travel from the airport by taxi for $27.50? 25 mi

16. A professional carpet cleaning service charges $30 plus $25.50 per hour to come to your home. If your total bill from this company is $119.25 before taxes, how many hours were you charged? 3.5 hr

Solve. See Example 6.

△ 17. The flag of Equatorial Guinea contains an isosceles triangle. (Recall that an isosceles triangle contains two angles with the same measure.) If the measure of the third angle of the triangle is 30° more than twice the measure of either of the other two angles, find the measure of each angle of the triangle. (*Hint:* Recall that the sum of the measures of the angles of a triangle is 180°.) 1st angle: 37.5°; 2nd angle: 37.5°; 3rd angle: 105°

△ 18. The flag of Brazil contains a parallelogram. One angle of the parallelogram is 15° less than twice the measure of the angle next to it. Find the measure of each angle of the parallelogram. (*Hint:* Recall that opposite angles of a parallelogram have the same measure and that the sum of the measures of the angles is 360°.) 1st angle: 65°; 2nd angle: 115°

19. The sum of the measures of the angles of a parallelogram is 360°. In the parallelogram below, angles *A* and *D* have the same measure as well as angles *C* and *B*. If the measure of angle *C* is twice the measure of angle *A*, find the measure of each angle. *A:* 60°; *B:* 120°; *C:* 120°; *D:* 60°

20. Recall that the sum of the measures of the angles of a triangle is 180°. In the triangle below, angle *C* has the same measure as angle *B*, and angle *A* measures 42° less than angle *B*. Find the measure of each angle. *B:* 74°; *C:* 74°; *A:* 32°

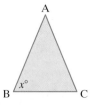

MIXED PRACTICE   **21.** 5 ft, 12 ft   **22.** 3 ft, 15 ft

21. A 17-foot piece of string is cut into two pieces so that one piece is 2 feet longer than twice the shorter piece. If the shorter piece is x feet long, find the lengths of both pieces.

22. An 18-foot wire is to be cut so that the longer piece is 5 times longer than the shorter piece. Find the length of each piece.

23. From 1997 to 2001, the number of prescriptions written for ADHD drugs increased by 5.5 million. If the sum of the number of prescriptions for these two years is 35.7 million, find the number of prescriptions for each year. Check to see that your results agree with the heights of the bars in the graph.   1997: 15.1 million prescriptions; 2001: 20.6 million prescriptions

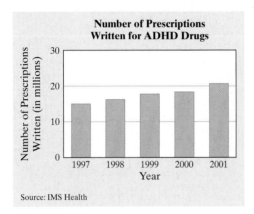

Source: IMS Health

24. The Pentagon Building in Washington, D.C., is the headquarters for the U.S. Department of Defense. The Pentagon is also the world's largest office building in terms of ground space with a floor area of over 6.5 million square feet. This is three times the floor area of the Empire State Building. About how much floor space does the Empire State Building have? Round to the nearest tenth.   2.2 million sq. ft

 25. Two angles are supplementary if their sum is 180°. One angle measures three times the measure of a smaller angle. If x represents the measure of the smaller angle and these two angles are supplementary, find the measure of each angle.   45°, 135°

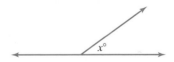

△ **26.** Two angles are complementary if their sum is 90°. Given the measures of the complementary angles shown, find the measure of each angle.   31°, 59°

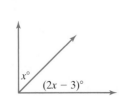

△ **27.** The measures of the angles of a triangle are 3 consecutive even integers. Find the measure of each angle.   58°, 60°, 62°

△ **28.** A quadrilateral is a polygon with 4 sides. The sum of the measures of the 4 angles in a quadrilateral is 360°. If the measures of the angles of a quadrilateral are consecutive odd integers, find the measures.   87°, 89°, 91°, 93°

29. Hertz Car Rental charges a daily rate of $39 plus $0.20 per mile for a certain car. Suppose that you rent that car for a day and your bill (before taxes) is $95. How many miles did you drive?   280 mi

30. A woman's $15,000 estate is to be divided so that her husband receives twice as much as her son. If x represents the amount of money that her son receives, find the amount of money that her husband receives and the amount of money that her son receives.   son: $5000; husband: $10,000

31. In SuperBowl XXXVII in San Diego, California, the Tampa Bay Buccaneers won over the Oakland Raiders with a 27 point lead. If the total of the two scores was 69, find the individual team scores. (*Source: World Almanac*, 2003)

32. During the 2002 Houston Bowl, Oklahoma State beat Southern Miss by 10 points. If their combined scores total 56, find the individual team scores. (*Source*: ESPN)

33. In the 2000 Summer Olympics, China won more medals than Australia, who won more medals than Germany. If the number of medals won by each country is three consecutive integers whose sum is 174, how many medals did each country win? (*Source: The World Almanac and Book of Facts 2001*)

34. The number of counties in California and the number of counties in Montana are consecutive even integers whose sum is 114. If California has more counties than Montana, how many counties does each state have? (*Source: The World Almanac and Book of Facts 2001*)   56, 58

31. Tampa Bay: 48; Oakland: 21
32. Oklahoma State: 33; Southern Miss: 23
33. China: 59 medals; Australia: 58 medals; Germany: 57 medals

35. Starting in 2004, the state of California will have 21 more electoral votes for president than the state of Texas. If the total electoral votes for these two states is 89, find the number of electoral votes for each state.   California: 55; Texas: 34

36. After a recent election, there were 2 more Republican governors than Democratic governors in the United States. How many Democrats and how many Republicans held governor's offices after this election? (*Source: The World Almanac and Book of Facts, 2003*)

37. Over the past few years the satellite Voyager II has passed by the planets Saturn, Uranus, and Neptune, continually updating information about these planets, including the number of moons for each. Uranus is now believed to have 13 more moons than Neptune. Also, Saturn is now believed to have 2 more than twice the number of moons of Neptune. If the total number of moons for these planets is 47, find the number of moons for each planet. (*Source*: National Space Science Data Center)   Neptune: 8 moons; Uranus: 21 moons; Saturn: 18 moons

36. 24 Democratic governors; 26 Republican governors
41. Brown: 66.362; Randall: 53.074

38. On April 7, 2001, the Mars Odyssey spacecraft was launched, beginning a multi-year mission to observe and map the planet Mars. Mars Odyssey was launched on Boeing's Delta II 7925 launch vehicle using nine strap-on solid rocket motors. Each solid rocket motor has a height that is 8 meters more than 5 times its diameter. If the sum of the height and the diameter for a single solid rocket motor is 14 meters, find each dimension. (*Source*: NASA)   diameter: 1 m; height: 13 m

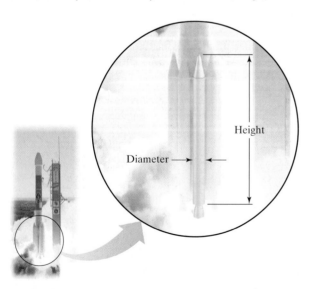

Height

Diameter → ←

39. If the sum of a number and five is tripled, the result is one less than twice the number. Find the number.   −16

40. Twice the sum of a number and six equals three times the sum of the number and four. Find the number.   0

41. In a recent election in Florida for a seat in the United States House of Representatives, Corrine Brown received 13,288 more votes than Bill Randall. If the total number of votes was 119,436, find the number of votes for each candidate.

42. In a recent election in Texas for a seat in the United States House of Representatives, Max Sandlin received 25,557 more votes than opponent Dennis Boerner. If the total number of votes was 135,821, find the number of votes for each candidate.(*Source*: Voter News Service)   Sandlin: 80,689; Boerner: 55,132

The graph below shows the states with the highest tourism budgets. Use the graph for Exercises 43 through 47.

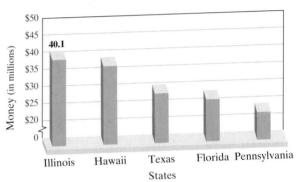

Source: Travel Industry Association of America

44. Texas and Florida

43. Which state spends the most money on tourism?   Illinois

44. Which states spend between $25 and $30 million on tourism?

45. The states of Texas and Florida spend a total of $56.6 million for tourism. The state of Texas spends $2.2 million more than the state of Florida. Find the amount that each state spends on tourism.   Texas: $29.4 million; Florida: $27.2 million

46. The states of Hawaii and Pennsylvania spend a total of $60.9 million for tourism. The state of Hawaii spends $8.1 million less than twice the amount of money that the state of Pennsylvania spends. Find the amount that each state spends on tourism.   Hawaii: $37.9 million; Pennsylvania: $23 million

47. Compare the heights of the bars in the graph with your results of Exercises 45 and 46. Are your answers reasonable?   answers may vary

REVIEW AND PREVIEW

Perform the indicated operations. See Sections 1.5 and 1.6.

48. $3 + (-7)$ −4

49. $-2 + (-8)$ −10

50. $4 - 10$ −6

51. $-11 + 2$ −9

52. $-5 - (-1)$ −4

53. $-12 - 3$ −15

Concept Extensions

△ **54.** The golden rectangle is a rectangle whose length is approximately 1.6 times its width. The early Greeks thought that a rectangle with these dimensions was the most pleasing to the eye and examples of the golden rectangle are found in many early works of art. For example, the Parthenon in Athens contains many examples of golden rectangles. Mike Hallahan would like to plant a rectangular garden in the shape of a golden rectangle. If he has 78 feet of fencing available, find the dimensions of the garden.   15 ft by 24 ft

The length-width rectangle approximates the golden rectangle as well as the width-height rectangle.

55. Dr. Dorothy Smith gave the students in her geometry class at the University of New Orleans the following question. Is it possible to construct a triangle such that the second angle of the triangle has a measure that is twice the measure of the first angle and the measure of the third angle is 3 times the measure of the first? If so, find the measure of each angle. (*Hint:* Recall that the sum of the measures of the angles of a triangle is 180°.)   yes: 30°, 60°, 90°

56. The human eye blinks once every 5 seconds on average. How many times does the average eye blink in one hour? In one 16-hour day while awake? In one year?

57. Give an example of how you recently solved a problem using mathematics.   **57.–60.** answers may vary

58. In your own words, explain why a solution of a word problem should be checked using the original wording of the problem and not the equation written from the wording.

Recall from Exercise 54 that the golden rectangle is a rectangle whose length is approximately 1.6 times its width.

△ **59.** It is thought that for about 75% of adults, a rectangle in the shape of the golden rectangle is the most pleasing to the eye.

56. 720 blinks per hour; 11,520 blinks per day; 4,204,800 blinks per year

Draw 3 rectangles, one in the shape of the golden rectangle, and poll your class. Do the results agree with the percentage given above?

△ **60.** Examples of golden rectangles can be found today in architecture and manufacturing packaging. Find an example of a golden rectangle in your home. A few suggestions: the front face of a book, the floor of a room, the front of a box of food.

△ **61.** Measure the dimensions of each rectangle and decide which one best approximates the shape of the golden rectangle.   c

a.

b.

c.

2.6   FORMULAS AND PROBLEM SOLVING

Objectives

1 Use formulas to solve problems.

2 Solve a formula or equation for one of its variables.

1   An equation that describes a known relationship among quantities, such as distance, time, volume, weight, and money is called a **formula**. These quantities are represented by letters and are thus variables of the formula. Here are some common formulas and their meanings.

$A = lw$
Area of a rectangle = length $\cdot$ width

$I = PRT$
Simple Interest = Principal $\cdot$ Rate $\cdot$ Time

$P = a + b + c$
Perimeter of a triangle = side a + side b + side c

$d = rt$
distance = rate $\cdot$ time

$V = lwh$
Volume of a rectangular solid = length $\cdot$ width $\cdot$ height

$$F = \left(\frac{9}{5}\right)C + 32$$

degrees Fahrenheit = $\left(\frac{9}{5}\right) \cdot$ degrees Celsius + 32

TEACHING TIP
Remind students that the front and back covers of this text contain formulas that they may need and have forgotten.

Formulas are valuable tools because they allow us to calculate measurements as long as we know certain other measurements. For example, if we know we traveled a distance of 100 miles at a rate of 40 miles per hour, we can replace the variables d and r in the formula $d = rt$ and find our time, t.

$$d = rt \qquad \text{Formula.}$$
$$100 = 40t \qquad \text{Replace } d \text{ with } 100 \text{ and } r \text{ with } 40.$$

This is a linear equation in one variable, t. To solve for t, divide both sides of the equation by 40.

$$\frac{100}{40} = \frac{40t}{40} \qquad \text{Divide both sides by 40.}$$

$$\frac{5}{2} = t \qquad \text{Simplify.}$$

The time traveled is $\frac{5}{2}$ hours or $2\frac{1}{2}$ hours.

In this section we solve problems that can be modeled by known formulas. We use the same problem-solving steps that were introduced in the previous section. These steps have been slightly revised to include formulas.

EXAMPLE 1

FINDING TIME GIVEN RATE AND DISTANCE

A glacier is a giant mass of rocks and ice that flows downhill like a river. Portage Glacier in Alaska is about 6 miles, or 31,680 *feet*, long and moves 400 *feet* per year. Icebergs are created when the front end of the glacier flows into Portage Lake. How long does it take for ice at the head (beginning) of the glacier to reach the lake?

Solution

1. UNDERSTAND.   Read and reread the problem. The appropriate formula needed to solve this problem is the distance formula, $d = rt$. To become familiar with this formula, let's find the distance that ice traveling at a rate of 400 feet per year travels in 100 years. To do so, we let time t be 100 years and rate r be the given 400 feet per year, and substitute these values into the formula $d = rt$. We then have that distance $d = 400(100) = 40,000$ feet. Since we are interested in finding how long it takes ice to travel 31,680 feet, we now know that it is less than 100 years.

Since we are using the formula $d = rt$, we let

t = the time in years for ice to reach the lake

r = rate or speed of ice

d = distance from beginning of glacier to lake

2. TRANSLATE.   To translate to an equation, we use the formula $d = rt$ and let distance $d = 31,680$ feet and rate $r = 400$ feet per year.

$$d = r \cdot t$$
$$31{,}680 = 400 \cdot t \qquad \text{Let } d = \textbf{31,680} \text{ and } r = \textbf{400.}$$

3. SOLVE. Solve the equation for t. To solve for t, divide both sides by 400.

$$\frac{31{,}680}{400} = \frac{400 \cdot t}{400} \qquad \text{Divide both sides by } \textbf{400.}$$
$$79.2 = t \qquad \text{Simplify.}$$

> **Helpful Hint**
> Don't forget to include units, if appropriate.

4. INTERPRET.

Check: To check, substitute 79.2 for t and 400 for r in the distance formula and check to see that the distance is 31,680 feet.

State: It takes 79.2 years for the ice at the head of Portage Glacier to reach the lake.

EXAMPLE 2

CALCULATING THE LENGTH OF A GARDEN

Charles Pecot can afford enough fencing to enclose a rectangular garden with a perimeter of 140 feet. If the width of his garden must be 30 feet, find the length.

TEACHING TIP
You may want to take a moment to remind students that perimeter is measured in units, area in square units, and volume in cubic units.

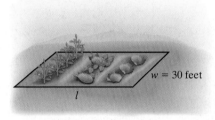

$w = 30$ feet

l

Solution

CLASSROOM EXAMPLE
A wood deck is being built behind a house. The width of the deck is 14 feet. If there is 168 square feet of decking material, find the length of the deck.
answer: 12 ft

1. UNDERSTAND. Read and reread the problem. The formula needed to solve this problem is the formula for the perimeter of a rectangle, $P = 2l + 2w$. Before continuing, let's become familiar with this formula.

   l = the length of the rectangular garden
   w = the width of the rectangular garden
   P = perimeter of the garden

2. TRANSLATE. To translate to an equation, we use the formula $P = 2l + 2w$ and let perimeter $P = 140$ feet and width $w = 30$ feet.

$$P = 2l + 2w$$
$$\downarrow \qquad\qquad \searrow$$
$$140 = 2l + 2(30) \qquad \text{Let } P = 140 \text{ and } w = 30.$$

3. SOLVE.

$$
\begin{aligned}
140 &= 2l + 2(30) & \\
140 &= 2l + 60 & \text{Multiply } 2(30). \\
140 - 60 &= 2l + 60 - 60 & \text{Subtract } 60 \text{ from both sides.} \\
80 &= 2l & \text{Combine like terms.} \\
40 &= l & \text{Divide both sides by } 2.
\end{aligned}
$$

4. INTERPRET.

   Check: Substitute 40 for l and 30 for w in the perimeter formula and check to see that the perimeter is 140 feet.
   State: The length of the rectangular garden is 40 feet.

EXAMPLE 3

FINDING AN EQUIVALENT TEMPERATURE

The average maximum temperature for January in Algerias, Algeria, is 59° Fahrenheit. Find the equivalent temperature in degrees Celsius.

Solution

1. UNDERSTAND. Read and reread the problem. A formula that can be used to solve this problem is the formula for converting degrees Celsius to degrees Fahrenheit, $F = \frac{9}{5}C + 32$. Before continuing, become familiar with this formula. Using

this formula, we let

$$C = \text{temperature in degrees Celsius, and}$$
$$F = \text{temperature in degrees Fahrenheit.}$$

2. TRANSLATE. To translate to an equation, we use the formula $F = \frac{9}{5}C + 32$ and let degrees Fahrenheit $F = 59$.

Formula: $\qquad\qquad\qquad\qquad F = \frac{9}{5}C + 32$

Substitute: $\qquad\qquad\qquad 59 = \frac{9}{5}C + 32 \qquad\qquad$ Let $F = 59$.

3. SOLVE.

$$59 = \frac{9}{5}C + 32$$

$$59 - 32 = \frac{9}{5}C + 32 - 32 \qquad \text{Subtract \textbf{32} from both sides.}$$

$$27 = \frac{9}{5}C \qquad\qquad\qquad \text{Combine like terms.}$$

$$\frac{5}{9} \cdot 27 = \frac{5}{9} \cdot \frac{9}{5}C \qquad\qquad \text{Multiply both sides by } \tfrac{5}{9}.$$

$$15 = C \qquad\qquad\qquad\quad \text{Simplify.}$$

4. INTERPRET.

Check: To check, replace C with 15 and F with 59 in the formula and see that a true statement results.

State: Thus, 59° Fahrenheit is equivalent to 15° Celsius.

In the next example, we again use the formula for perimeter of a rectangle as in Example 2. In Example 2, we knew the width of the rectangle. In this example, both the length and width are unknown.

EXAMPLE 4

FINDING ROAD SIGN DIMENSIONS

The length of a rectangular road sign is 2 feet less than three times its width. Find the dimensions if the perimeter is 28 feet.

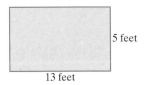

5 feet

13 feet

WESTVIEW	92
RIVERSIDE	162
SUNDALE	205

Solution

1. UNDERSTAND. Read and reread the problem. Recall that the formula for the perimeter of a rectangle is $P = 2l + 2w$. Draw a rectangle and guess the solution. If the width of the rectangular sign is 5 feet, its length is 2 feet less than 3 times the width or $3(5 \text{ feet}) - 2 \text{ feet} = 13 \text{ feet}$. The perimeter P of the rectangle is then $2(13 \text{ feet}) + 2(5 \text{ feet}) = 36 \text{ feet}$, too much. We now know that the width is less than 5 feet.

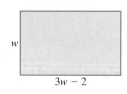

Let
$$w = \text{the width of the rectangular sign; then}$$
$$3w - 2 = \text{the length of the sign.}$$

Draw a rectangle and label it with the assigned variables.

2. TRANSLATE.

Formula: $P = 2l + 2w$ or
Substitute: $28 = 2(3w - 2) + 2w.$

3. SOLVE.

$$28 = 2(3w - 2) + 2w$$
$$28 = 6w - 4 + 2w \qquad \text{Apply the distributive property.}$$
$$28 = 8w - 4$$
$$28 + 4 = 8w - 4 + 4 \qquad \text{Add 4 to both sides.}$$
$$32 = 8w$$
$$\frac{32}{8} = \frac{8w}{8} \qquad \text{Divide both sides by 8.}$$
$$4 = w$$

4. INTERPRET.

Check: If the width of the sign is 4 feet, the length of the sign is $3(4 \text{ feet})$ $- 2$ feet $= 10$ feet. This gives a perimeter of $P = 2(4 \text{ feet}) + 2(10 \text{ feet}) = 28$ feet, the correct perimeter.

State: The width of the sign is 4 feet and the length of the sign is 10 feet.

2   We say that the formula $F = \frac{9}{5}C + 32$ is solved for F because F is alone on one side of the equation and the other side of the equation contains no F's. Suppose that we need to convert many Fahrenheit temperatures to equivalent degrees Celsius. In this case, it is easier to perform this task by solving the formula $F = \frac{9}{5}C + 32$ for C. (See Example 8.) For this reason, it is important to be able to solve an equation for any one of its specified variables. For example, the formula $d = rt$ is solved for d in terms of r and t. We can also solve $d = rt$ for t in terms of d and r. To solve for t, divide both sides of the equation by r.

$$d = rt$$
$$\frac{d}{r} = \frac{rt}{r} \qquad \text{Divide both sides by } r.$$
$$\frac{d}{r} = t \qquad \text{Simplify.}$$

To solve a formula or an equation for a specified variable, we use the same steps as for solving a linear equation. These steps are listed next.

Solving Equations for a Specified Variable

Step 1. Multiply on both sides to clear the equation of fractions if they occur.

Step 2. Use the distributive property to remove parentheses if they occur.

Step 3. Simplify each side of the equation by combining like terms.

Step 4. Get all terms containing the specified variable on one side and all other terms on the other side by using the addition property of equality.

Step 5. Get the specified variable alone by using the multiplication property of equality.

EXAMPLE 5

Solve $V = lwh$ for l.

Solution   This formula is used to find the volume of a box. To solve for l, divide both sides by wh.

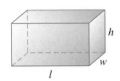

$$V = lwh$$
$$\frac{V}{wh} = \frac{lwh}{wh} \qquad \text{Divide both sides by } \boldsymbol{wh}.$$
$$\frac{V}{wh} = l \qquad \text{Simplify.}$$

CLASSROOM EXAMPLE
Solve: $C = 2\pi r$ for r.
answer: $r = \dfrac{C}{2\pi}$

Since we have l alone on one side of the equation, we have solved for l in terms of V, w, and h. Remember that it does not matter on which side of the equation we isolate the variable.

EXAMPLE 6

Solve $y = mx + b$ for x.

Solution   The term containing the variable we are solving for, mx, is on the right side of the equation. Get mx alone by subtracting b from both sides.

CLASSROOM EXAMPLE
Solve: $y = 3x - 7$ for x.
answer: $x = \dfrac{y + 7}{3}$

$$y = mx + b$$
$$y - b = mx + b - b \qquad \text{Subtract } \boldsymbol{b} \text{ from both sides.}$$
$$y - b = mx \qquad \text{Combine like terms.}$$

Next, solve for x by dividing both sides by m.

$$\frac{y - b}{m} = \frac{mx}{m}$$
$$\frac{y - b}{m} = x \qquad \text{Simplify.}$$

 EXAMPLE 7

Solve $P = 2l + 2w$ for w.

Solution   This formula relates the perimeter of a rectangle to its length and width. Find the term containing the variable w. To get this term, $2w$, alone subtract $2l$ from both sides.

> **Helpful Hint**
> The 2's may *not* be divided out here. Although 2 is a factor of the denominator, 2 is *not* a factor of the numerator since it is not a factor of both terms in the numerator.

$$P = 2l + 2w$$
$$P - 2l = 2l + 2w - 2l \qquad \text{Subtract } \boldsymbol{2l} \text{ from both sides.}$$
$$P - 2l = 2w \qquad \text{Combine like terms.}$$
$$\frac{P - 2l}{2} = \frac{2w}{2} \qquad \text{Divide both sides by } \boldsymbol{2}.$$
$$\frac{P - 2l}{2} = w \qquad \text{Simplify.}$$

CLASSROOM EXAMPLE
Solve: $P = 2a + b - c$ for a.
answer: $a = \dfrac{P - b + c}{2}$

The next example has an equation containing a fraction. We will first clear the equation of fractions and then solve for the specified variable.

EXAMPLE 8

Solve $F = \frac{9}{5}C + 32$ for C.

Solution

$$F = \frac{9}{5}C + 32$$

$$5(F) = 5\left(\frac{9}{5}C + 32\right)$$ Clear the fraction by multiplying both sides by the LCD.

$$5F = 9C + 160$$ Distribute the 5.

$$5F - 160 = 9C + 160 - 160$$ To get the term containing the variable C alone, subtract 160 from both sides.

$$5F - 160 = 9C$$ Combine like terms.

$$\frac{5F - 160}{9} = \frac{9C}{9}$$ Divide both sides by 9.

$$\frac{5F - 160}{9} = C$$ Simplify.

Spotlight on

DECISION
𝒳 MAKING

Suppose you have just purchased a house. You are required by your mortgage company to buy home-owner's insurance. In choosing a policy, you must also choose a deductible level. A deductible is how much you, the home-owner, pay to repair or replace a loss before the insurance company starts paying. Typically, deductibles start at $250. However, other levels such as $500 or $1000 are also available. One way to save money when purchasing your homeowner's policy is to raise the deductible amount. However, doing so means you will have to pay more out of pocket in case of theft or fire. Study the accompanying chart. Which amount of deductible would you choose? Why?

Increase Your Deductible to:	
$500	save up to 12%
$1000	save up to 24%
$2500	save up to 30%
$5000	save up to 37%

on the cost of a homeowner's policy, depending on your insurance company.

Source: Insurance Information Institute

EXERCISE SET 2.6

STUDY GUIDE/SSM	CD/ VIDEO	PH MATH TUTOR CENTER	MathXL®Tutorials ON CD	MathXL®	MyMathLab®

Substitute the given values into each given formula and solve for the unknown variable. If necessary, round to one decimal place. See Examples 1 through 3.

△ **1.** $A = bh$; $A = 45, b = 15$ (Area of a parallelogram) $h = 3$

2. $d = rt$; $d = 195, t = 3$ (Distance formula) $r = 65$

△ **3.** $S = 4lw + 2wh$; $S = 102, l = 7, w = 3$ (Surface area of a special rectangular box) $h = 3$

△ **4.** $V = lwh$; $l = 14, w = 8, h = 3$ (Volume of a rectangular box) $V = 336$

△ **5.** $A = \frac{1}{2}h(B + b)$; $A = 180, B = 11, b = 7$ (Area of a trapezoid) $h = 20$

△ **6.** $A = \frac{1}{2}h(B + b)$; $A = 60, B = 7, b = 3$ (Area of a trapezoid) $h = 12$

7. $P = a + b + c$; $P = 30, a = 8, b = 10$ (Perimeter of a triangle) $c = 12$

8. $V = \frac{1}{3}Ah$; $V = 45, h = 5$ (Volume of a pyramid) $A = 27$

 9. $C = 2\pi r$; $C = 15.7$ (use the approximation 3.14 or a calculator approximation for π) (Circumference of a circle) $r = 2.5$

10. $A = \pi r^2$; $r = 4.5$ (use the approximation 3.14 or a calculator approximation for π)(Area of a circle) $A = 63.6$

11. $I = PRT$; $I = 3750, P = 25,000, R = 0.05$ (Simple interest formula) $T = 3$

12. $I = PRT$; $I = 1,056,000, R = 0.055, T = 6$ (Simple interest formula) $P = 3,200,000$

13. $V = \frac{1}{3}\pi r^2 h$; $V = 565.2, r = 6$ (use a calculator approximation for π)(Volume of a cone) $h = 15$

14. $V = \frac{4}{3}\pi r^3$; $r = 3$ (use a calculator approximation for π) (Volume of a sphere) $V = 113.1$

Solve each formula for the specified variable. See Examples 5 through 8.   **20.** $y = x + 13$ **21.** $R = \frac{A - P}{PT}$ **22.** $T = \frac{A - P}{PR}$

15. $f = 5gh$ for h $h = \frac{f}{5g}$ **16.** $C = 2\pi r$ for r $r = \frac{C}{2\pi}$

 17. $V = lwh$ for w $w = \frac{V}{lh}$ **18.** $T = mnr$ for n $n = \frac{T}{mr}$

19. $3x + y = 7$ for y $y = 7 - 3x$ **20.** $-x + y = 13$ for y

21. $A = P + PRT$ for R **22.** $A = P + PRT$ for T

23. $V = \frac{1}{3}Ah$ for A $A = \frac{3V}{h}$ **24.** $D = \frac{1}{4}fk$ for k $k = \frac{4D}{f}$

25. $P = a + b + c$ for a $a = P - b - c$

26. $PR = s_1 + s_2 + s_3 + s_4$ for s_3 $s_3 = PR - s_1 - s_2 - s_4$

 27. $S = 2\pi rh + 2\pi r^2$ for h **28.** $S = 4lw + 2wh$ for h

Solve. See Examples 1 through 4.   **27.** $\frac{S - 2\pi r^2}{2\pi r}$ **28.** $\frac{S - 4lw}{2w}$

29. The world's largest sign for Coca-Cola is located in Arica, Chile. The rectangular sign has a length of 400 feet and has an area of 52,400 square feet. Find the width of the sign. (*Source:* Fabulous Facts about Coca-Cola, Atlanta, GA)   131 ft

30. The length of a rectangular garden is 6 meters. If 21 meters of fencing are required to fence the garden, find its width.   4.5 m

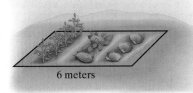

6 meters

31. −10°C   **32.** 23°F

31. Convert Nome, Alaska's 14°F high temperature to Celsius.

32. Convert Paris, France's low temperature of −5°C to Fahrenheit.

33. The X-30 is a "space plane" that skims the edge of space at 4000 miles per hour. Neglecting altitude, if the circumference of the Earth is approximately 25,000 miles, how long will it take for the X-30 to travel around the Earth?   6.25 hr

34. In the United States, a notable hang glider flight was a 303-mile, $8\frac{1}{2}$ hour flight from New Mexico to Kansas. What was the average rate during this flight? $35\frac{11}{17}$ mph

35. An architect designs a rectangular flower garden such that the width is exactly two-thirds of the length. If 260 feet of antique picket fencing are to be used, find the dimensions of the garden. length: 78 ft; width: 52 ft

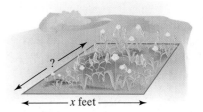

x feet

36. If the length of a rectangular parking lot is 10 meters less than twice its width, and the perimeter is 400 meters, find the length of the parking lot.   130 m

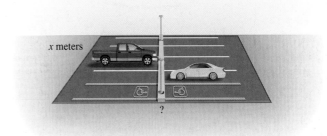

x meters

37. A flower bed is in the shape of a triangle with one side twice the length of the shortest side, and the third side is 30 feet more than the length of the shortest side. Find the dimensions if the perimeter is 102 feet.   18 ft, 36 ft, 48 ft

38. The perimeter of a yield sign in the shape of an isosceles triangle is 22 feet. If the shortest side is 2 feet less than the other two sides, find the length of the shortest side. (*Hint:* An isosceles triangle has two sides the same length.)   6 ft

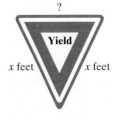

?

Yield

x feet   *x* feet

39. The Cat is a high-speed catamaran auto ferry that operates between Bar Harbor, Maine, and Yarmouth, Nova Scotia. The Cat can make the trip in about $2\frac{1}{2}$ hours at a speed of 55 mph. About how far apart are Bar Harbor and Yarmouth? (*Source:* Bay Ferries)   137.5 mi

40. A family is planning their vacation to Disney World. They will drive from a small town outside New Orleans, Louisiana, to Orlando, Florida, a distance of 700 miles. They plan to average a rate of 55 mph. How long will this trip take? $12\frac{8}{11}$ hr

41. Piranha fish require 1.5 cubic feet of water per fish to maintain a healthy environment. Find the maximum number of piranhas you could put in a tank measuring 8 feet by 3 feet by 6 feet.   96 piranhas

42. Find how many goldfish you can put in a cylindrical tank whose diameter is 8 meters and whose height is 3 meters, if each goldfish needs 2 cubic meters of water.   75 goldfish

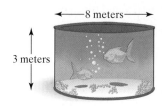

43. A lawn is in the shape of a trapezoid with a height of 60 feet and bases of 70 feet and 130 feet. How many bags of fertilizer must be purchased to cover the lawn if each bag covers 4000 square feet?   2 bags

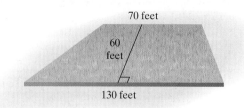

44. If the area of a right-triangularly shaped sail is 20 square feet and its base is 5 feet, find the height of the sail.   8 ft

45. Maria's Pizza sells one 16-inch cheese pizza or two 10-inch cheese pizzas for $9.99. Determine which size gives more pizza.   one 16-in. pizza

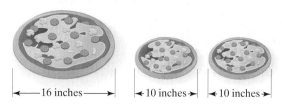

46. Find how much rope is needed to wrap around the Earth at the equator, if the radius of the Earth is 4000 miles. (*Hint:* Use 3.14 for π and the formula for circumference.)   25,120 mi

47. A Japanese "bullet" train set a new world record for train speed at 552 kilometers per hour during a manned test run on the Yamanashi Maglev Test Line in April 1999. The Yamanashi Maglev Test Line is 42.8 kilometers long. How many *minutes* would a test run on the Yamanashi Line last at this record-setting speed? Round to the nearest hundredth of a minute. (*Source:* Japan Railways Central Co.)   4.65 min

48. In 1983, the Hawaiian volcano Kilauea began erupting in a series of episodes still occurring at the time of this writing. At times, the lava flows advanced at speeds of up to 0.5 kilometer per hour. In 1983 and 1984 lava flows destroyed 16 homes in the Royal Gardens subdivision, about 6 km away from the eruption site. Roughly how long did it take the lava to reach Royal Gardens? (*Source:* U.S. Geological Survey Hawaiian Volcano Observatory)   12 hr

49. The perimeter of an equilateral triangle is 7 inches more than the perimeter of a square, and the side of the triangle is 5 inches longer than the side of the square. Find the side of the triangle. (*Hint:* An equilateral triangle has three sides the same length.)   13 in.

50. A square animal pen and a pen shaped like an equilateral triangle have equal perimeters. Find the length of the sides of each pen if the sides of the triangular pen are fifteen less than twice a side of the square pen.

51. Find how long it takes Tran Nguyen to drive 135 miles on I-10 if he merges onto I-10 at 10 A.M. and drives nonstop with his cruise control set on 60 mph.   2.25 hr

52. Beaumont, Texas, is about 150 miles from Toledo Bend. If Leo Miller leaves Beaumont at 4 A.M. and averages 45 mph, when should he arrive at Toledo Bend?   7:20 A.M.

53. The longest runway at Los Angeles International Airport has the shape of a rectangle and an area of 1,813,500 square

50. square's side length: 22.5 units; triangle's side length: 30 units

feet. This runway is 150 feet wide. How long is the runway? (*Source:* Los Angeles World Airports)   12,090 ft

54. Bolts of lightning can travel at the speed of 270,000 miles per second. How many times can a lightning bolt travel around the world in one second? (See Exercise 46. Round to the nearest tenth.)   10.7

55. The highest temperature ever recorded in Europe was 122°F in Seville, Spain, in August of 1881. Convert this record high temperature to Celsius. (*Source:* National Climatic Data Center)   50°C

56. The lowest temperature ever recorded in Oceania was −10°C at the Haleakala Summit in Maui, Hawaii, in January 1961. Convert this record low temperature to Fahrenheit. (*Source:* National Climatic Data Center)   14°F

△ 📠 57. The CART Fed Ex Championship Series is an open-wheeled race car competition based in the United States. A CART car has a maximum length of 199 inches, a maximum width of 78.5 inches, and a maximum height of 33 inches. When the CART series travels to another country for a grand prix, teams must ship their cars. Find the volume of the smallest shipping crate needed to ship a CART car of maximum dimensions. (*Source:* Championship Auto Racing Teams, Inc.)   515,509.5 cu. in.

CART Racing Car

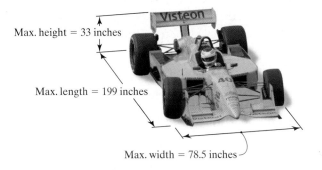

Max. height = 33 inches

Max. length = 199 inches

Max. width = 78.5 inches

58. On a road course, a CART car's speed can average up to around 105 mph. Based on this speed, how long would it take a CART driver to travel from Los Angeles to New York City, a distance of about 2810 miles by road, without stopping? Round to the nearest tenth of an hour.   26.8 hr

△ 📠 59. The Hoberman Sphere is a toy ball that expands and contracts. When it is completely closed, it has a diameter of 9.5 inches. Find the volume of the Hoberman Sphere when it is completely closed. Use 3.14 for π. Round to the nearest whole cubic inch. (*Source:* Hoberman Designs, Inc.)   449 cu. in.

△ 📠 60. When the Hoberman Sphere (see Exercise 59) is completely expanded, its diameter is 30 inches. Find the volume of the Hoberman Sphere when it is completely expanded. Use 3.14 for π. Round to the nearest whole cubic inch. (*Source:* Hoberman Designs, Inc.)   14,130 cu. in.

61. The average temperature on the planet Mercury is 167°C. Convert this temperature to degrees Fahrenheit. (*Source:* National Space Science Data Center)   332.6°F

62. The average temperature on the planet Jupiter is −227°F. Convert this temperature to degrees Celsius. Round to the nearest degree. (*Source:* National Space Science Data Center)

REVIEW AND PREVIEW   62. −144°C

Write the following phrases as algebraic expressions. See Section 2.1.

63. Nine divided by the sum of a number and 5 $\frac{9}{x+5}$

64. Half the product of a number and five $\frac{1}{2}(5x)$

65. Three times the sum of a number and four $3(x+4)$

66. Double the sum of ten and four times a number $2(10+4x)$

67. Triple the difference of a number and twelve $3(x-12)$

68. A number minus the sum of the number and six $x-(x+6)$

Concept Extensions   71. 500 sec or $8\frac{1}{3}$ min

📠 69. Dry ice is a name given to solidified carbon dioxide. At −78.5° Celsius it changes directly from a solid to a gas. Convert this temperature to Fahrenheit.   −109.3°F

📠 70. Lightning bolts can reach a temperature of 50,000° Fahrenheit. Convert this temperature to Celsius.   27,760°C

📠 71. The distance from the sun to the Earth is approximately 93,000,000 miles. If light travels at a rate of 186,000 miles per second, how long does it take light from the sun to reach us?

📠 72. Light travels at a rate of 186,000 miles per second. If our moon is 238,860 miles from the Earth, how long does it take light from the moon to reach us? (Round to the nearest tenth of a second.)   1.3 sec

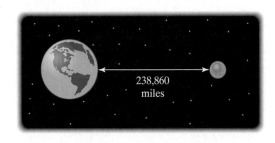

238,860 miles

73. A glacier is a giant mass of rocks and ice that flows downhill like a river. Exit Glacier, near Seward, Alaska, moves at a rate of 20 inches a day. Find the distance in feet the glacier moves in a year. (Assume 365 days a year. Round to 2 decimal places.)

73. 608.33 ft

74. Flying fish do not *actually* fly, but glide. They have been known to travel a distance of 1300 feet at a rate of 20 miles per hour. How many seconds did it take to travel this distance? (*Hint:* First convert miles per hour to feet per second. Recall that 1 mile = 5280 feet. Round to the nearest tenth of a second.)
44.3 sec

75. Stalactites join stalagmites to form columns. A column found at Natural Bridge Caverns near San Antonio, Texas, rises 15 feet and has a *diameter* of only 2 inches. Find the volume of this column in cubic inches. (*Hint:* Use the formula for volume of a cylinder and use a calculator approximation for π.)
565.5 cu. in.

76. Find the temperature at which the Celsius measurement and Fahrenheit measurement are the same number. $-40°$

77. Normal room temperature is about 78°F. Convert this temperature to Celsius. $25\frac{5}{9}°C$

78. The formula $V = LWH$ is used to find the volume of a box. If the length of a box is doubled, the width is doubled, and the height is doubled, how does this affect the volume?

79. The formula $A = bh$ is used to find the area of a parallelogram. If the base of a parallelogram is doubled and its height is doubled, how does this affect the area?

78. It multiplies the volume by 8.   **79.** It multiplies the area by 4.

2.7   *FURTHER PROBLEM SOLVING*

Objectives

1   Solve problems involving percents.

2   Solve problems involving distance.

3   Solve problems involving mixtures.

4   Solve problems involving interest.

This section is devoted to solving problems in the categories listed. The same problem-solving steps used in previous sections are also followed in this section. They are listed below for review.

> **General Strategy for Problem Solving**
>
> 1. UNDERSTAND the problem. During this step, become comfortable with the problem. Some ways of doing this are:
>   Read and reread the problem.
>   Choose a variable to represent the unknown.
>   Construct a drawing, whenever possible.
>   Propose a solution and check. Pay careful attention to how you check your proposed solution. This will help writing an equation to model the problem.
> 2. TRANSLATE the problem into an equation.
> 3. SOLVE the equation.
> 4. INTERPRET the results: *Check* the proposed solution in the stated problem and *state* your conclusion.

1   The first two examples involve percents.

Percent increase or percent decrease is a common way to describe how some measurement has increased or decreased. For example, crime increased by 8%, teachers received a 5.5% increase in salary, or a company decreased its employees by 10%. The next example is a review of percent increase.

EXAMPLE 1

CALCULATING THE COST OF ATTENDING COLLEGE

The cost of attending a public college rose from $5324 in 1990 to $8086 in 2000. Find the percent increase. (*Source:* U.S. Department of Education. *Note:* These costs include room and board.)

Solution

1. UNDERSTAND. Read and reread the problem. Let's guess that the percent increase is 20%. To see if this is the case, we find 20% of $5324 to find the *increase* in cost. Then we add this increase to $5324 to find the *new cost*. In other words, 20% ($5324) = 0.20($5324) = $1064.80, the *increase* in cost. The new cost then would be $5324 + $1064.80 = $6388.80, less than the actual new cost of $8086. We now know that the increase is greater than 20% and we know how to check our proposed solution.

   Let x = the percent increase.

2. TRANSLATE. First, find the **increase**, and then the **percent increase**. The increase in cost is found by:

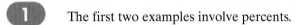

In words: increase   =   new cost   −   old cost   or

Translate: increase   = $8086   − $5324

                        = $2762

Next, find the percent increase. The percent increase or percent decrease is always a percent of the original number or in this case, the old cost.

In words: increase   is   what percent increase   of   old cost

Translate: $2762   = x · $5324

3. SOLVE.

$$2762 = 5324x$$ Divide both sides by **5324** and
$$0.519 \approx x$$ round to **3** decimal places.
$$51.9\% \approx x$$ Write as a percent.

CLASSROOM EXAMPLE

If a number decreases from 200 to 120, find the percent decrease.
answer: 40%

4. INTERPRET.

Check: Check the proposed solution.

State: The percent increase in cost is approximately 51.9%.

EXAMPLE 2

Most of the movie screens in the United States project analog films, but the number of cinemas using digital are increasing. Find the number of digital screens last year if after a 175% increase, the number this year is 124. Round to the nearest whole.

Solution

CLASSROOM EXAMPLE
Find the original price of a suit if the sale price is $46 after a 20% discount.
answer: $57.50

1. UNDERSTAND. Read and reread the problem. Let's guess a solution and see how we would check our guess. If the number of digital screens last year was 50, we would see if 50 plus the increase is 124; that is,

$$50 + 175\%(50) = 50 + 1.75(50) = 2.75(50) = 137.50$$

Since 137.5 is too large, we know that our guess of 50 is too large. We also have a better understanding of the problem. Let

$$x = \text{number of digital screens last year}$$

2. TRANSLATE. To translate to an equation, we remember that

In words:

number of digital screen last year	plus	increase	equals	number of digital screens this year
x	$+$	$1.75x$	$=$	124

Translate:

3. SOLVE.

$$2.75x = 124$$
$$x = \frac{124}{2.75}$$
$$x \approx 45$$

4. INTERPRET.

Check: Recall that x represents the number of digital screens last year. If this number is approximately 45, let's see if 45 plus the increase is close to 124. (We use the word "close" since 45 is rounded.)

$$45 + 175\%(45) = 45 + 1.75(45) = 2.75(45) = 123.75,$$

which is close to 124.

State: There were approximately 45 digital screens last year.

2 The next example involves distance.

EXAMPLE 3

FINDING TIME GIVEN RATE AND DISTANCE

Marie Antonio, a bicycling enthusiast, rode her 21-speed at an average speed of 18 miles per hour on level roads and then slowed down to an average of 10 mph on the hilly roads of the trip. If she covered a distance of 98 miles, how long did the entire trip take if traveling the level roads took the same time as traveling the hilly roads?

Solution

1. UNDERSTAND the problem. To do so, read and reread the problem. The formula $d = r \cdot t$ is needed. At this time, let's guess a solution. Suppose that she spent 2 hours traveling on the level roads. This means that she also spent 2 hours traveling on the hilly roads, since the times spent were the same. What is her total distance? Her distance on the level road is rate·time = 18(2) = 36 miles. Her distance on the hilly roads is rate·time = 10(2) = 20 miles. This gives a total distance of 36 miles + 20 miles = 56 miles, not the correct distance of 98 miles. Remember that the purpose of guessing a solution is not to guess correctly (although this may happen) but to help better understand the problem and how to model it with an equation.

We are looking for the length of the entire trip, so we begin by letting

$$x = \text{the time spent on level roads.}$$

Because the same amount of time is spent on hilly roads, then also

$$x = \text{the time spent on hilly roads.}$$

2. TRANSLATE. To help us translate to an equation, we now summarize the information from the problem on the following chart. Fill in the rates given, the variables used to represent the times, and use the formula $d = r \cdot t$ to fill in the distance column.

	Rate · Time = Distance		
Level	18	x	$18x$
Hilly	10	x	$10x$

Since the entire trip covered 98 miles, we have that

In words:	total distance	=	level distance	+	hilly distance
Translate:	98	=	$18x$	+	$10x$

3. SOLVE.

$$98 = 28x \qquad \text{Add like terms.}$$
$$\frac{98}{28} = \frac{28x}{28} \qquad \text{Divide both sides by 28.}$$
$$3.5 = x$$

4. INTERPRET the results.

Check: Recall that x represents the time spent on the level portion of the trip and also the time spent on the hilly portion. If Marie rides for 3.5 hours at 18 mph, her distance is 18(3.5) = 63 miles. If Marie rides for 3.5 hours at 10 mph, her distance is 10(3.5) = 35 miles. The total distance is 63 miles + 35 miles = 98 miles, the required distance.

State: The time of the entire trip is then 3.5 hours + 3.5 hours or 7 hours.

3   Mixture problems involve two or more different quantities being combined to form a new mixture. These applications range from Dow Chemical's need to form a chemical mixture of a required strength to Planter's Peanut Company's need to find the correct mixture of peanuts and cashews, given taste and price constraints.

EXAMPLE 4

CALCULATING PERCENT FOR A LAB EXPERIMENT

A chemist working on his doctoral degree at Massachusetts Institute of Technology needs 12 liters of a 50% acid solution for a lab experiment. The stockroom has only

40% and 70% solutions. How much of each solution should be mixed together to form 12 liters of a 50% solution?

Solution

1. UNDERSTAND. First, read and reread the problem a few times. Next, guess a solution. Suppose that we need 7 liters of the 40% solution. Then we need $12 - 7 = 5$ liters of the 70% solution. To see if this is indeed the solution, find the amount of pure acid in 7 liters of the 40% solution, in 5 liters of the 70% solution, and in 12 liters of a 50% solution, the required amount and strength.

number of liters	×	acid strength	=	amount of pure acid
7 liters	×	40%	=	7(0.40) or 2.8 liters
5 liters	×	70%	=	5(0.70) or 3.5 liters
12 liters	×	50%	=	12(0.50) or 6 liters

Since 2.8 liters + 3.5 liters = 6.3 liters and not 6, our guess is incorrect, but we have gained some valuable insight into how to model and check this problem.

Let

$$x = \text{number of liters of 40\% solution; then}$$
$$12 - x = \text{number of liters of 70\% solution.}$$

x liters (12−x) liters (12−x) liters x liters
40% solution + 70% solution = 50% solution
12 liters

2. TRANSLATE. To help us translate to an equation, the following table summarizes the information given. Recall that the amount of acid in each solution is found by multiplying the acid strength of each solution by the number of liters.

	No. of Liters	·	Acid Strength	=	Amount of Acid
40% Solution	x		40%		$0.40x$
70% Solution	$12 - x$		70%		$0.70(12 - x)$
50% Solution Needed	12		50%		$0.50(12)$

The amount of acid in the final solution is the sum of the amounts of acid in the two beginning solutions.

In words:	acid in 40% solution	+	acid in 70% solution	=	acid in 50% mixture
Translate:	$0.40x$	+	$0.70(12 - x)$	=	$0.50(12)$

3. SOLVE.

$$0.40x + 0.70(12 - x) = 0.50(12)$$

$$0.4x + 8.4 - 0.7x = 6 \quad \text{Apply the distributive property.}$$
$$-0.3x + 8.4 = 6 \quad \text{Combine like terms.}$$
$$-0.3x = -2.4 \quad \text{Subtract } \mathbf{8.4} \text{ from both sides.}$$
$$x = 8 \quad \text{Divide both sides by } -\mathbf{0.3}.$$

4. INTERPRET.

Check: To check, recall how we checked our guess.
State: If 8 liters of the 40% solution are mixed with $12 - 8$ or 4 liters of the 70% solution, the result is 12 liters of a 50% solution.

4   The next example is an investment problem.

EXAMPLE 5

FINDING THE INVESTMENT AMOUNT

Rajiv Puri invested part of his $20,000 inheritance in a mutual funds account that pays 7% simple interest yearly and the rest in a certificate of deposit that pays 9% simple interest yearly. At the end of one year, Rajiv's investments earned $1550. Find the amount he invested at each rate.

Solution

1. UNDERSTAND. Read and reread the problem. Next, guess a solution. Suppose that Rajiv invested $8000 in the 7% fund and the rest, $12,000, in the fund paying 9%. To check, find his interest after one year. Recall the formula, $I = PRT$, so the interest from the 7% fund = $8000(0.07)(1) = $560. The interest from the 9% fund = $12,000(0.09)(1) = $1080. The sum of the interests is $560 + $1080 = $1640. Our guess is incorrect, since the sum of the interests is not $1550, but we now have a better understanding of the problem.

Let

$$x = \text{amount of money in the account paying 7\%.}$$

The rest of the money is $20,000 less x or

$$20,000 - x = \text{amount of money in the account paying 9\%.}$$

2. TRANSLATE. We apply the simple interest formula $I = PRT$ and organize our information in the following chart. Since there are two different rates of interest and two different amounts invested, we apply the formula twice.

	Principal ·	*Rate* ·	*Time* =	*Interest*
7% Fund	x	0.07	1	$x(0.07)(1)$ or $0.07x$
9% Fund	$20,000 - x$	0.09	1	$(20,000 - x)(0.09)(1)$ or $0.09(20,000 - x)$
Total	20,000			1550

The total interest earned, $1550, is the sum of the interest earned at 7% and the interest earned at 9%.

In words:	interest at 7%	+	interest at 9%	=	total interest
Translate:	$0.07x$	+	$0.09(20,000 - x)$	=	1550

3. SOLVE.

$$0.07x + 0.09(20,000 - x) = 1550$$
$$0.07x + 1800 - 0.09x = 1550 \qquad \text{Apply the distributive property.}$$
$$1800 - 0.02x = 1550 \qquad \text{Combine like terms.}$$
$$-0.02x = -250 \qquad \text{Subtract } \mathbf{1800} \text{ from both sides.}$$
$$x = 12,500 \qquad \text{Divide both sides by } -\mathbf{0.02}.$$

4. INTERPRET.

Check: If $x = 12,500$, then $20,000 - x = 20,000 - 12,500$ or 7500. These solutions are reasonable, since their sum is $20,000 as required. The annual interest on $12,500 at 7% is $875; the annual interest on $7500 at 9% is $675, and $875 + $675 = $1550.
State: The amount invested at 7% is $12,500. The amount invested at 9% is $7500.

Spotlight on

DECISION
of MAKING

Suppose you are a personal income tax preparer. Your clients, Jose and Felicia Fernandez, are filing jointly Form 1040 as their individual income tax return. You know that medical expenses may be written off as an itemized deduction if the expenses exceed 7.5% of their adjusted gross income. Furthermore, only the portion of medical expenses that exceed 7.5% of their adjusted gross income can be deducted. Is the Fernandez family eligible to deduct their medical expenses? Explain.

Internal Revenue Service
Form 1040

Jose Fernandez
Felicia Fernandez

Adjusted Gross Income . . . $33,650

Fernandez Family **Deductible Medical Expenses**	
Medical bills	$1025
Dental bills	$ 325
Prescription drugs	$ 360
Medical Insurance premiums	$1200

EXERCISE SET 2.7

STUDY GUIDE/SSM · CD/VIDEO · PH MATH TUTOR CENTER · MathXL®Tutorials ON CD · MathXL® · MyMathLab®

Solve. See Examples 1 and 2.   **2.** $0.14 increase; $1.09 new price

1. Nordstrom's advertised a 25% off sale. If a London Fog coat originally sold for $256, find the decrease in price and the sale price. $64 decrease; $192 sale price

2. Time Saver increased the price of a $0.95 cola by 15%. Find the increase in price and the new price.

3. Find the original price of a pair of shoes if the sale price is $78 after a 25% discount. $104

4. Find last year's salary if after a 4% pay raise, this year's salary is $33,800. $32,500

5. The number of fraud complaints (usually ID theft) rose from 220,000 in 2001 to 380,000 in 2002. Find the percent increase. Round to the nearest whole percent.   73%

6. The number of text messages rose from 996 million in June to 1100 million in December. Find the percent increase. Round to the nearest whole percent.   10%

Solve. See Example 3.

7. A jet plane traveling at 500 mph overtakes a propeller plane traveling at 200 mph that had a 2-hour head start. How far from the starting point are the planes? $666\frac{2}{3}$ mi

8. How long will it take a bus traveling at 60 miles per hour to overtake a car traveling at 40 mph if the car had a 1.5-hour head start?   3 hr

9. The Jones family drove to Disneyland at 50 miles per hour and returned on the same route at 40 mph. Find the distance to Disneyland if the total driving time was 7.2 hours.   160 mi

10. A bus traveled on a level road for 3 hours at an average speed 20 miles per hour faster than it traveled on a winding road. The time spent on the winding road was 4 hours. Find the average speed on the level road if the entire trip was 305 miles.   55 mph

Solve. See Example 4.

11. How much pure acid should be mixed with 2 gallons of a 40% acid solution in order to get a 70% acid solution?   2 gal

12. How many cubic centimeters of a 25% antibiotic solution should be added to 10 cubic centimeters of a 60% antibiotic solution in order to get a 30% antibiotic solution?   60 cc

13. Planter's Peanut Company wants to mix 20 pounds of peanuts worth $3 a pound with cashews worth $5 a pound in order to make an experimental mix worth $3.50 a pound. How many pounds of cashews should be added to the peanuts? $6\frac{2}{3}$ lb

14. Community Coffee Company wants a new flavor of Cajun coffee. How many pounds of coffee worth $7 a pound should be added to 14 pounds of coffee worth $4 a pound to get a mixture worth $5 a pound?   7 lb

15. Is it possible to mix a 30% antifreeze solution with a 50% antifreeze solution and obtain a 70% antifreeze solution? Why or why not?   no; answers may vary

16. A trail mix is made by combining peanuts worth $3 a pound, raisins worth $2 a pound, and M & M's worth $4 a pound. Would it make good business sense to sell the trail mix for $1.98 a pound? Why or why not?   no; answers may vary

Solve. See Example 5.   **20.** $8500 @ 10%; $17,000 @ 12%

17. Zoya Lon invested part of her $25,000 advance at 8% annual simple interest and the rest at 9% annual simple interest. If her total yearly interest from both accounts was $2135, find the amount invested at each rate. $11,500 @ 8%; $13,500 @ 9%

18. Karen Waugtal invested some money at 9% annual simple interest and $250 more than that amount at 10% annual simple interest. If her total yearly interest was $101, how much was invested at each rate? $400 @ 9%; $650 @ 10%

19. Sam Mathius invested part of his $10,000 bonus in a fund that paid an 11% profit and invested the rest in stock that suffered a 4% loss. Find the amount of each investment if his overall net profit was $650. $7000 @ 11% profit; $3000 @ 4% loss

20. Bruce Blossum invested a sum of money at 10% annual simple interest and invested twice that amount at 12% annual simple interest. If his total yearly income from both investments was $2890, how much was invested at each rate?

MIXED PRACTICE

Solve. **23.** $30,000 @ 8%; $24,000 @ 10%

21. A student at the University of New Orleans makes money by buying and selling used cars. Charles bought a used car and later sold it for a 20% profit. If he sold it for $4680, how much did Charles pay for the car? $3900

22. After a downturn in the economy, a company downsized by 35%. If there are still 78 employees, how many employees were there prior to the layoffs?   120 employees

23. How can $54,000 be invested, part at 8% annual simple interest and the remainder at 10% annual simple interest, so that the interest earned by the two accounts will be equal?

24. Ms. Mills invested her $20,000 bonus in two accounts. She took a 4% loss on one investment and made a 12% profit on another investment, but ended up breaking even. How much was invested in each account? $5000 @ 12%; $15,000 @ 4%

25. Kathleen and Cade Williams leave simultaneously from the same point hiking in opposite directions, Kathleen walking at 4 miles per hour and Cade at 5 mph. How long can they talk on their walkie-talkies if the walkie-talkies have a 20-mile radius? $2\frac{2}{9}$ hrs

26. Alan and Dave Schaferkötter leave from the same point driving in opposite directions, Alan driving at 55 miles per hour and Dave at 65 mph. Alan has a one-hour head start. How long will they be able to talk on their car phones if the phones have a 250-mile range?   2 hr and $37\frac{1}{2}$ min

27. How much of an alloy that is 20% copper should be mixed with 200 ounces of an alloy that is 50% copper in order to get an alloy that is 30% copper?   400 oz

28. How much water should be added to 30 gallons of a solution that is 70% antifreeze in order to get a mixture that is 60% antifreeze?   5 gal

29. Iceberg lettuce is grown and shipped to stores for about 40 cents a head, and consumers purchase it for about 70 cents a head. Find the percent increase.   75% increase

30. The lettuce consumption per capita in 1968 was about 21.5 pounds, and in 2000 the consumption rose to 24.3 pounds. Find the percent increase. (Round to the nearest tenth of a percent.)   13.0%

31. If $3000 is invested at 6% annual simple interest, how much should be invested at 9% annual simple interest so that the total yearly income from both investments is $585? $4500

32. Trudy Waterbury, a financial planner, invested a certain amount of money at 9% annual simple interest, twice that amount at 10% annual simple interest, and three times that amount at 11% annual simple interest. Find the amount invested at each rate if her total yearly income from the investments was $2790. $4500 @ 9%; $9000 @ 10%; $13,500 @ 11%

33. Smart Cards (cards with an embedded computer chip) have been growing in popularity in recent years. In 2006, 500 million Smart Cards are expected to be issued. This represents a 117% increase from the number of cards that were issued in 2001. How many Smart Cards were issued in 2001? Round to the nearest million. (*Source:* The Freedonia Group)   230 million

34. April Thrower spent $32.25 to take her daughter's birthday party guests to the movies. Adult tickets cost $5.75 and children tickets cost $3.00. If 8 persons were at the party, how many adult tickets were bought?   3 adult tickets

35. Two hikers are 11 miles apart and walking toward each other. They meet in 2 hours. Find the rate of each hiker if one hiker walks 1.1 mph faster than the other.   2.2 mph; 3.3 mph

36. On a 255-mile trip, Gary Alessandrini traveled at an average speed of 70 mph, got a speeding ticket, and then traveled at 60 mph for the remainder of the trip. If the entire trip took 4.5 hours and the speeding ticket stop took 30 minutes, how long did Gary speed before getting stopped? $1\frac{1}{2}$ hr

37. The ACT Assessment is a college entrance exam taken by about 60% of college-bound students. The national average score was 20.7 in 1993 and rose to 21.0 in 2001. Find the percent increase. (Round to the nearest hundredth of a percent.)   1.45% increase   **38.** 166,567%

38. The first Barbie doll was introduced in March 1959 and cost $3. This same 1959 Barbie doll now costs up to $5000. Find the percent increase rounded to the nearest whole percent.

39. Nedra and Latonya Dominguez are 12 miles apart hiking toward each other. How long will it take them to meet if Nedra walks at 3 mph and Latonya walks 1 mph faster? $1\frac{5}{7}$ hr

40. Mark Martin can row upstream at 5 mph and downstream at 11 mph. If Mark starts rowing upstream until he gets tired and then rows downstream to his starting point, how far did Mark row if the entire trip took 4 hours?   27.5 mi

REVIEW AND PREVIEW

Perform the indicated operations. See Section 1.6.

41. $3 + (-7)$ -4 **42.** $(-2) + (-8)$ -10

43. $\frac{3}{4} - \frac{3}{16}$ $\frac{9}{16}$ **44.** $-11 + 2.9$ -8.1

45. $-5 - (-1)$ -4 **46.** $-12 - 3$ -15

Place $<$, $>$, or $=$ in the appropriate space to make each a true statement. See Sections 1.1 and 1.7.

47. $-5 > -7$ **48.** $\frac{12}{3} = 2^2$

49. $|-5| = -(-5)$ **50.** $-3^3 = (-3)^3$

Concept Extensions   54. 25 monitors

To "break even" in a manufacturing business, revenue R (income) must equal the cost C of production, or R = C.

51. The cost C to produce x number of skateboards is given by $C = 100 + 20x$. The skateboards are sold wholesale for $24 each, so revenue R is given by $R = 24x$. Find how many skateboards the manufacturer needs to produce and sell to break even. (*Hint*: Set the expression for R equal to the expression for C, then solve for x.)   25 skateboards

52. The revenue R from selling x number of computer boards is given by $R = 60x$, and the cost C of producing them is given by $C = 50x + 5000$. Find how many boards must be sold to break even. Find how much money is needed to produce the break-even number of boards.   500; $30,000

53. The cost C of producing x number of paperback books is given by $C = 4.50x + 2400$. Income R from these books is given by $R = 7.50x$. Find how many books should be produced and sold to break even.   800 books

54. Find the break-even quantity for a company that makes x number of computer monitors at a cost C given by $C = 870 + 70x$ and receives revenue R given by $R = 105x$.

55. Problems 51 through 54 involve finding the break-even point for manufacturing. Discuss what happens if a company makes and sells fewer products than the break-even point. Discuss what happens if more products than the break-even point are made and sold.   answers may vary

Standardized nutrition labels like the ones below have been displayed on food items since 1994. The percent column on the right shows the percent of daily values based on a 2000-calorie diet shown at the bottom of the label. For example, a serving of this food contains 4 grams of total fat, where the recommended daily fat based on a 2000-calorie diet is 65 grams of fat. This means that $\frac{4}{65}$ or approximately 6% (as shown) of your daily recommended fat is taken in by eating a serving of this food.

56. Based on a 2000-calorie diet, what percent of the daily value of sodium is contained in a serving of this food? In other words, find x. (Round to the nearest tenth of a percent.)   9.6%

57. Based on a 2000-calorie diet, what percent of the daily value of total carbohydrate is contained in a serving of this food? In other words, find y. (Round to the nearest tenth of a percent.)   7.7%

58. Notice on the nutrition label that one serving of this food contains 130 calories and 35 of these calories are from fat. Find the percent of calories from fat. (Round to the nearest tenth of a percent.) It is recommended that no more than 30% of calorie intake come from fat. Does this food satisfy this recommendation?   26.9%; yes

Below is a nutrition label for a particular food.

NUTRITIONAL INFORMATION PER SERVING

Serving Size: 9.8 oz.   **Servings Per Container: 1**

Calories.................280	Polyunsaturated Fat.........1g
Protein..................12g	Saturated Fat..............3g
Carbohydrate.............45g	Cholesterol.............20mg
Fat......................6g	Sodium...............520mg
Percent of Calories from Fat...... ?	Potassium.............220mg

59. If fat contains approximately 9 calories per gram, find the percent of calories from fat in one serving of this food. (Round to the nearest tenth of a percent.)   19.3%

60. If protein contains approximately 4 calories per gram, find the percent of calories from protein from one serving of this food. (Round to the nearest tenth of a percent.)   17.1%

61. Find a food that contains more than 30% of its calories per serving from fat. Analyze the nutrition label and verify that the percents shown are correct.   answers may vary

Nutrition Facts

Serving Size 18 crackers (31g)
Servings Per Container About 9

Amount Per Serving

Calories 130 Calories from Fat 35

	% Daily Value*
Total Fat 4g	**6%**
Saturated Fat 0.5g	**3%**
Polyunsaturated Fat 0g	
Monounsaturated Fat 1.5g	
Cholesterol 0mg	**0%**
Sodium 230mg	*x*
Total Carbohydrate 23g	*y*
Dietary Fiber 2g	**8%**
Sugars 3g	
Protein 2g	

Vitamin A 0%	•	Vitamin C 0%
Calcium 2%	•	Iron 6%

*Percent Daily Values are based on a 2,000 calorie diet. Your daily values may be higher or lower depending on your calorie needs.

	Calories:	2,000	2,500
Total Fat	Less than	65g	80g
Sat Fat	Less than	20g	25g
Cholesterol	Less than	300mg	300mg
Sodium	Less than	2400mg	2400mg
Total Carbohydrate		300g	375g
Dietary Fiber		25g	30g

2.8   SOLVING LINEAR INEQUALITIES

Objectives

1. Define linear inequality in one variable.
2. Graph solution sets on a number line and use interval notation.
3. Solve linear inequalities.
4. Solve compound inequalities.
5. Solve inequality applications.

1 In Chapter 1, we reviewed these inequality symbols and their meanings:

$<$ means "is less than" $\leq$ means "is less than or equal to"
$>$ means "is greater than" $\geq$ means "is greater than or equal to"

Equations	Inequalities
$x = 3$	$x \leq 3$
$5n - 6 = 14$	$5n - 6 > 14$
$12 = 7 - 3y$	$12 \leq 7 - 3y$
$\dfrac{x}{4} - 6 = 1$	$\dfrac{x}{4} - 6 > 1$

A linear inequality is similar to a linear equation except that the equality symbol is replaced with an inequality symbol.

Linear Inequality in One Variable

A **linear inequality in one variable** is an inequality that can be written in the form

$$ax + b < c$$

where a, b, and c are real numbers and a is not 0.

This definition and all other definitions, properties, and steps in this section also hold true for the inequality symbols, $>$, $\geq$, and $\leq$.

2   A **solution of an inequality** is a value of the variable that makes the inequality a true statement. The solution set is the set of all solutions. For the inequality $x < 3$, replacing x with any number less than 3, that is, to the left of 3 on a number line, makes the resulting inequality true. This means that any number less than 3 is a solution of the inequality $x < 3$.

Since there are infinitely many such numbers, we cannot list all the solutions of the inequality. We *can* use set notation and write

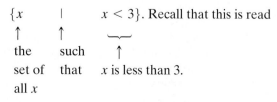

$\{x \quad | \quad x < 3\}$. Recall that this is read

the set of all x such that x is less than 3.

We can also picture the solutions on a number line. If we use open/closed-circle notation, the graph of $\{x | x < 3\}$ looks like the following.

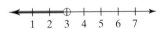

In this text, a convenient notation, called **interval notation**, will be used to write solution sets of inequalities. To help us understand this notation, a different graphing notation will be used. Instead of an open circle, we use a parenthesis; instead of a closed circle, we use a bracket. With this new notation, the graph of $\{x | x < 3\}$ now looks like

and can be represented in interval notation as $(-\infty, 3)$. The symbol $-\infty$, read as "negative infinity," does not indicate a number, but does indicate that the shaded arrow to the left never ends. In other words, the interval $(-\infty, 3)$ includes *all* numbers less than 3.

Picturing the solutions of an inequality on a number line is called **graphing** the solutions or graphing the inequality, and the picture is called the **graph** of the inequality.

To graph $\{x | x \leq 3\}$ or simply $x \leq 3$, shade the numbers to the left of 3 and place a bracket at 3 on the number line. The bracket indicates that 3 **is** a solution: 3 **is** less than or equal to 3. In interval notation, we write $(-\infty, 3]$.

> **Helpful Hint**
>
> When writing an inequality in interval notation, it may be easier to first graph the inequality, then write it in interval notation. To help, think of the number line as approaching $-\infty$ to the left and $+\infty$ or ∞ to the right. Then simply write the interval notation by following your shading from left to right.
>
> $$x > 5 \qquad \infty \qquad\qquad -\infty \qquad x \leq -7$$
>
> ```
>   +---+---(---+---+---+->      <-+---+---]---+---+-
>   3   4   5   6   7            -9 -8 -7 -6 -5
>        (5, ∞)                      (-∞, -7]
> ```

$\boxed{\text{EXAMPLE 1}}$

Graph $x \geq -1$. Then write the solutions in interval notation.

Solution We place a bracket at -1 since the inequality symbol is $\geq$ and -1 is greater than or equal to -1. Then we shade to the right of -1.

```
<-+---+---+---[---+---+---+---+->
 -4  -3  -2  -1   0   1   2   3
```

In interval notation, this is $[-1, \infty)$.

3   When solutions of a linear inequality are not immediately obvious, they are found through a process similar to the one used to solve a linear equation. Our goal is to get the variable alone, and we use properties of inequality similar to properties of equality.

> ### Addition Property of Inequality
>
> If a, b, and c are real numbers, then
>
> $$a < b \quad \text{and} \quad a + c < b + c$$
>
> are equivalent inequalities.

This property also holds true for subtracting values, since subtraction is defined in terms of addition. In other words, adding or subtracting the same quantity from both sides of an inequality does not change the solution of the inequality.

EXAMPLE 2

Solve $x + 4 \leq -6$ for x. Graph the solution set and write it in interval notation.

Solution   To solve for x, subtract 4 from both sides of the inequality.

$$\begin{array}{ll} x + 4 \leq -6 & \text{Original inequality.} \\ x + 4 - 4 \leq -6 - 4 & \text{Subtract 4 from both sides.} \\ x \leq -10 & \text{Simplify.} \end{array}$$

The solution set is $(-\infty, -10]$.

-12 -11 -10 -9 -8 -7 -6

CLASSROOM EXAMPLE
Solve $x - 6 \geq -11$. Graph the solution set and write it in interval notation.
answer: $[-5, \infty)$

-8 -7 -6 -5 -4 -3

TEACHING TIP
Ask students to give a few specific solutions to Example 2. Stress that *all* numbers less than or equal to −10 are solutions.

> ▶ ### Helpful Hint
>
> Notice that any number less than or equal to −10 is a solution to $x \leq -10$. For example, solutions include
>
> $$-10, \quad -200, \quad -11\frac{1}{2}, \quad -7\pi, \quad -\sqrt{130}, \quad -50.3$$

An important difference between linear equations and linear inequalities is shown when we multiply or divide both sides of an inequality by a nonzero real number. For example, start with the true statement $6 < 8$ and multiply both sides by 2. As we see below, the resulting inequality is also true.

$$\begin{array}{ll} 6 < 8 & \text{True.} \\ 2(6) < 2(8) & \text{Multiply both sides by 2.} \\ 12 < 16 & \text{True.} \end{array}$$

But if we start with the same true statement $6 < 8$ and multiply both sides by −2, the resulting inequality is not a true statement.

$$\begin{array}{ll} 6 < 8 & \text{True.} \\ -2(6) < -2(8) & \text{Multiply both sides by } -2. \\ -12 < -16 & \text{False.} \end{array}$$

Notice, however, that if we reverse the direction of the inequality symbol, the resulting inequality is true.

$$\begin{array}{ll} -12 < -16 & \text{False.} \\ -12 > -16 & \text{True.} \end{array}$$

This demonstrates the multiplication property of inequality.

> ### Multiplication Property of Inequality
>
> **1.** If a, b, and c are real numbers, and c is **positive**, then
>
> $$a < b \quad \text{and} \quad ac < bc$$
>
> are equivalent inequalities.
>
> **2.** If a, b, and c are real numbers, and c is **negative**, then
>
> $$a < b \quad \text{and} \quad ac > bc$$
>
> are equivalent inequalities.

Because division is defined in terms of multiplication, this property also holds true when dividing both sides of an inequality by a nonzero number. If we multiply or divide both sides of an inequality by a negative number, **the direction of the inequality sign must be reversed for the inequalities to remain equivalent**.

CLASSROOM EXAMPLE
Solve $-3x \le 12$.
answer: $[-4, \infty)$

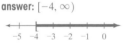

> ### Helpful Hint
> Whenever both sides of an inequality are multiplied or divided by a negative number, the direction of the inequality symbol **must be** reversed to form an equivalent inequality.

EXAMPLE 3

Solve $-2x \le -4$. Graph the solution set and write it in interval notation.

Solution Remember to reverse the direction of the inequality symbol when dividing by a negative number.

> ### Helpful Hint
> Don't forget to reverse the direction of the inequality sign.

$$-2x \le -4$$
$$\frac{-2x}{-2} \ge \frac{-4}{-2} \qquad \text{Divide both sides by } -2 \text{ and reverse the direction of the inequality sign.}$$
$$x \ge 2 \qquad \text{Simplify.}$$

The solution set $[2, \infty)$ is graphed as shown.

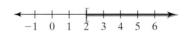

EXAMPLE 4

Solve $2x < -4$. Graph the solution set and write it in interval notation.

Solution

CLASSROOM EXAMPLE
Solve $5x < -15$.
answer: $(-\infty, -3)$

> ### Helpful Hint
> Do not reverse the inequality sign.

$$2x < -4$$
$$\frac{2x}{2} < \frac{-4}{2} \qquad \begin{array}{l}\text{Divide both sides by 2.}\\ \text{Do not reverse the direction of the inequality sign.}\end{array}$$
$$x < -2 \qquad \text{Simplify.}$$

The solution set $(-\infty, -2)$ is graphed as shown.

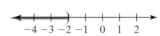

✔ **CONCEPT CHECK**

Fill in the blank with $<$, $>$, $\le$, or $\ge$.

a. Since $-8 < -4$, then $3(-8)$____$3(-4)$.

b. Since $5 \ge -2$, then $\dfrac{5}{-7}$ ____ $\dfrac{-2}{-7}$.

c. If $a < b$, then $2a$ ____ $2b$.

d. If $a \ge b$, then $\dfrac{a}{-3}$ ____ $\dfrac{b}{-3}$.

The following steps may be helpful when solving inequalities. Notice that these steps are similar to the ones given in Section 2.4 for solving equations.

Solving Linear Inequalities in One Variable

Step 1. Clear the inequality of fractions by multiplying both sides of the inequality by the lowest common denominator (LCD) of all fractions in the inequality.

Step 2. Remove grouping symbols such as parentheses by using the distributive property.

Step 3. Simplify each side of the inequality by combining like terms.

Step 4. Write the inequality with variable terms on one side and numbers on the other side by using the addition property of inequality.

Step 5. Get the variable alone by using the multiplication property of inequality.

▶ **Helpful Hint**

Don't forget that if both sides of an inequality are multiplied or divided by a negative number, the direction of the inequality sign must be reversed.

EXAMPLE 5

Solve $-4x + 7 \ge -9$. Graph the solution set and write it in interval notation.

Solution

$$-4x + 7 \ge -9$$
$$-4x + 7 - 7 \ge -9 - 7 \qquad \text{Subtract 7 from both sides.}$$
$$-4x \ge -16 \qquad \text{Simplify.}$$
$$\frac{-4x}{-4} \le \frac{-16}{-4} \qquad \text{Divide both sides by } -4 \text{ and reverse the direction of the inequality sign.}$$
$$x \le 4 \qquad \text{Simplify.}$$

The solution set $(-\infty, 4]$ is graphed as shown.

CLASSROOM EXAMPLE

Solve $-3x + 11 \le -13$.

answer: $[8, \infty)$

EXAMPLE 6

Solve $2x + 7 \le x - 11$. Graph the solution set and write it in interval notation.

Solution

$$2x + 7 \le x - 11$$
$$2x + 7 - x \le x - 11 - x \qquad \text{Subtract } x \text{ from both sides.}$$
$$x + 7 \le -11 \qquad \text{Combine like terms.}$$
$$x + 7 - 7 \le -11 - 7 \qquad \text{Subtract } 7 \text{ from both sides.}$$
$$x \le -18 \qquad \text{Combine like terms.}$$

The graph of the solution set $(-\infty, -18]$ is shown.

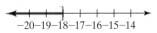

CLASSROOM EXAMPLE
Solve $5x + 3 \ge 4x + 3$.
answer: $[0, \infty)$

EXAMPLE 7

Solve $-5x + 7 < 2(x - 3)$. Graph the solution set and write it in interval notation.

Solution

$$-5x + 7 < 2(x - 3)$$
$$-5x + 7 < 2x - 6 \qquad \text{Apply the distributive property.}$$
$$-5x + 7 - 2x < 2x - 6 - 2x \qquad \text{Subtract } 2x \text{ from both sides.}$$
$$-7x + 7 < -6 \qquad \text{Combine like terms.}$$
$$-7x + 7 - 7 < -6 - 7 \qquad \text{Subtract } 7 \text{ from both sides.}$$
$$-7x < -13 \qquad \text{Combine like terms.}$$
$$\frac{-7x}{-7} > \frac{-13}{-7} \qquad \begin{array}{l}\text{Divide both sides by } -7 \text{ and reverse} \\ \text{the direction of the inequality sign.}\end{array}$$
$$x > \frac{13}{7} \qquad \text{Simplify.}$$

The graph of the solution set $\left(\frac{13}{7}, \infty\right)$ is shown.

CLASSROOM EXAMPLE
Solve $-6x - 3 > -4(x + 1)$.
answer: $\left(-\infty, \frac{1}{2}\right)$

EXAMPLE 8

Solve $2(x - 3) - 5 \le 3(x + 2) - 18$. Graph the solution set and write it in interval notation.

Solution

$$2(x - 3) - 5 \le 3(x + 2) - 18$$
$$2x - 6 - 5 \le 3x + 6 - 18 \qquad \text{Apply the distributive property.}$$
$$2x - 11 \le 3x - 12 \qquad \text{Combine like terms.}$$
$$-x - 11 \le -12 \qquad \text{Subtract } 3x \text{ from both sides.}$$
$$-x \le -1 \qquad \text{Add } 11 \text{ to both sides.}$$

CLASSROOM EXAMPLE

Solve

$3(x + 5) - 1 \geq 5(x - 1) + 7.$

answer: $(-\infty, 6]$

$$\frac{-x}{-1} \geq \frac{-1}{-1}$$

Divide both sides by -1 and reverse the direction of the inequality sign.

$$x \geq 1$$

Simplify.

The graph of the solution set $[1, \infty)$ is shown.

4 Inequalities containing one inequality symbol are called **simple inequalities**, while inequalities containing two inequality symbols are called **compound inequalities**. A compound inequality is really two simple inequalities in one. The compound inequality

$$3 < x < 5 \quad \text{means} \quad 3 < x \textbf{ and } x < 5$$

This can be read "x is greater than 3 and less than 5."

A solution of a compound inequality is a value that is a solution of both of the simple inequalities that make up the compound inequality. For example,

$$4\frac{1}{2} \text{ is a solution of } 3 < x < 5 \text{ since } 3 < 4\frac{1}{2} \textbf{ and } 4\frac{1}{2} < 5.$$

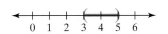

To graph $3 < x < 5$, place parentheses at both 3 and 5 and shade between.

EXAMPLE 9

Graph $2 < x \leq 4$. Write the solutions in interval notation.

Solution Graph all numbers greater than 2 and less than or equal to 4. Place a parenthesis at 2, a bracket at 4, and shade between.

CLASSROOM EXAMPLE

Graph $-2 \leq x < 0$. Write the solutions in interval notation.

answer: $[-2, 0)$

When we solve a simple inequality, we isolate the variable on one side of the inequality. When we solve a compound inequality, we isolate the variable in the middle part of the inequality. Also, when solving a compound inequality, we must perform the same operation to all **three** parts of the inequality: left, middle, and right.

EXAMPLE 10

Solve $-1 \leq 2x - 3 < 5$. Graph the solution set and write it in interval notation.

Solution

CLASSROOM EXAMPLE

Solve $-24 < 5x + 1 \leq 6$.

answer: $(-5, 1]$

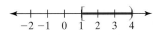

$$-1 \leq 2x - 3 < 5$$
$$-1 + 3 \leq 2x - 3 + 3 < 5 + 3 \quad \text{Add } \textbf{3} \text{ to all three parts.}$$
$$2 \leq 2x < 8 \quad \text{Combine like terms.}$$
$$\frac{2}{2} \leq \frac{2x}{2} < \frac{8}{2} \quad \text{Divide all three parts by } \textbf{2.}$$
$$1 \leq x < 4 \quad \text{Simplify.}$$

The graph of the solution set $[1, 4)$ is shown.

EXAMPLE 11

Solve $3 \leq \dfrac{3x}{2} + 4 \leq 5$. Graph the solution set and write it in interval notation.

Solution

$$3 \leq \frac{3x}{2} + 4 \leq 5$$

$$2(3) \leq 2\left(\frac{3x}{2} + 4\right) \leq 2(5) \qquad \text{Multiply all three parts by 2 to clear the fraction.}$$

$$6 \leq 3x + 8 \leq 10 \qquad \text{Distribute.}$$

$$-2 \leq 3x \leq 2 \qquad \text{Subtract 8 from all three parts.}$$

$$\frac{-2}{3} \leq \frac{3x}{3} \leq \frac{2}{3} \qquad \text{Divide all three parts by 3.}$$

$$-\frac{2}{3} \leq x \leq \frac{2}{3} \qquad \text{Simplify.}$$

The graph of the solution set $\left[-\frac{2}{3}, \frac{2}{3}\right]$ is shown.

CLASSROOM EXAMPLE

Solve $-4 \leq \dfrac{2x}{5} - 1 < 0$.

answer: $\left[-\frac{15}{2}, \frac{5}{2}\right)$

5 Problems containing words such as "at least," "at most," "between," "no more than," and "no less than" usually indicate that an inequality should be solved instead of an equation. In solving applications involving linear inequalities, use the same procedure you use to solve applications involving linear equations.

EXAMPLE 12

STAYING WITHIN BUDGET

Marie Chase and Jonathan Edwards are having their wedding reception at the Gallery Reception Hall. They may spend at most $2000 for the reception. If the reception hall charges a $100 cleanup fee plus $36 per person, find the greatest number of people that they can invite and still stay within their budget.

Solution

1. UNDERSTAND. Read and reread the problem. Next, guess a solution. If 40 people attend the reception, the cost is $100 + $36(40) = $100 + $1440 = $1540. Let x = the number of people who attend the reception.

2. TRANSLATE.

In words:	cleanup fee	+	cost per person	must be less than or equal to	$2000
Translate:	100	+	36x	≤	2000

3. SOLVE.

$$100 + 36x \leq 2000$$

$$36x \leq 1900 \qquad \text{Subtract 100 from both sides.}$$

$$x \leq 52\frac{7}{9} \qquad \text{Divide both sides by 36.}$$

4. INTERPRET.

Check: Since x represents the number of people, we round down to the nearest whole, or 52. Notice that if 52 people attend, the cost is

$100 + $36(52) = $1972. If 53 people attend, the cost is
$100 + $36(53) = $2008, which is more than the given $2000.

State: Marie Chase and Jonathan Edwards can invite at most 52 people to the reception.

MENTAL MATH

Solve each of the following inequalities.

1. $5x > 10$ $x > 2$ **2.** $4x < 20$ $x < 5$ **3.** $2x \geq 16$ $x \geq 8$ **4.** $9x \leq 63$ $x \leq 7$

EXERCISE SET 2.8

STUDY GUIDE/SSM | CD/VIDEO | PH MATH TUTOR CENTER | MathXL®Tutorials ON CD | MathXL® | MyMathLab®

Graph each set of numbers given in interval notation. Then write an inequality statement in x describing the numbers graphed.

1. $[2, \infty)$ $,x \geq 2$

2. $(-3, \infty)$ $,x > -3$

3. $(-\infty, -5)$ $,x < -5$

4. $(-\infty, 4]$ $,x \leq 4$

Graph each inequality on a number line. Then write the solutions in interval notation. See Example 1.

5. $x \leq -1$ $,(-\infty, -1]$

6. $y < 0$ $,(-\infty, 0)$

7. $x < \dfrac{1}{2}$ $,\left(-\infty, \dfrac{1}{2}\right)$

8. $z < -\dfrac{2}{3}$ $,\left[-\infty, -\dfrac{2}{3}\right)$

9. $y \geq 5$ $,[5, \infty)$

10. $x > 3$ $,(3, \infty)$

Solve each inequality. Graph the solution set and write it in interval notation. See Examples 2 through 4.

11. $2x < -6$ $,(-\infty, -3)$

12. $3x > -9$ $,(-3, \infty)$

13. $x - 2 \geq -7$ $,[-5, \infty)$

14. $x + 4 \leq 1$ $,(-\infty, -3]$

15. $-8x \leq 16$ $,[-2, \infty)$

16. $-5x < 20$ $,(-4, \infty)$

Solve each inequality. Graph the solution set and write it in interval notation. See Examples 5 and 6.

17. $3x - 5 > 2x - 8$ $,(-3, \infty)$

18. $3 - 7x \geq 10 - 8x$ $,[7, \infty)$

19. $4x - 1 \leq 5x - 2x$ $,(-\infty, 1]$

20. $7x + 3 < 9x - 3x$ $,(-\infty, -3)$

Solve each inequality. Graph the solution set and write it in interval notation. See Examples 7 and 8.

21. $x - 7 < 3(x + 1)$ $,(-5, \infty)$

22. $3x + 9 \leq 5(x - 1)$ $,[7, \infty)$

23. $-6x + 2 \geq 2(5 - x)$ $,(-\infty, -2]$

24. $-7x + 4 > 3(4 - x)$ $,(-\infty, -2)$

25. $4(3x - 1) \leq 5(2x - 4)$ $,(-\infty, -8]$

26. $3(5x - 4) \leq 4(3x - 2)$ $,\left(-\infty, \dfrac{4}{3}\right]$

27. $3(x + 2) - 6 > -2(x - 3) + 14$ $,(4, \infty)$

28. $7(x - 2) + x \leq -4(5 - x) - 12$ $,\left(-\infty, -\dfrac{9}{2}\right]$

MIXED PRACTICE

Solve the following inequalities. Graph each solution set and write it in interval notation.

29. $-2x \leq -40$ $,[20, \infty)$

30. $-7x > 21$ $,(-\infty, -3)$

31. $-9 + x > 7$ $\longleftarrow$ ($\longrightarrow$, $(16, \infty)$
              16

32. $y - 4 \leq 1$ $\longleftarrow$] $\longrightarrow$, $(-\infty, 5]$
             5

33. $3x - 7 < 6x + 2$ $\longleftarrow$ ($\longrightarrow$, $(-3, \infty)$
            -3

34. $2x - 1 \geq 4x - 5$ $\longleftarrow$] $\longrightarrow$, $(-\infty, 2]$
            2

35. $5x - 7x \geq x + 2$ $\longleftarrow$] $+$ $\longrightarrow$, $\left(-\infty, -\frac{2}{3}\right]$
        $-\frac{2}{3}$
       -1 0

36. $4 - x < 8x + 2x$ $\longleftarrow$ ($+$ $\longrightarrow$, $\left(\frac{4}{11}, \infty\right)$
        $\frac{4}{11}$
       0 1

37. $\frac{3}{4}x > 2$ $\longleftarrow$ + ($\longrightarrow$, $\left(\frac{8}{3}, \infty\right)$
       $\frac{8}{3}$
     2 3

38. $\frac{5}{6}x \geq -8$ $\longleftarrow$ [+ $\longrightarrow$, $\left[-\frac{48}{5}, \infty\right)$
       $-\frac{48}{5}$
    -10 -9

39. $3(x - 5) < 2(2x - 1)$ $\longleftarrow$ ($\longrightarrow$, $(-13, \infty)$
             -13

40. $5(x + 4) < 4(2x + 3)$ $\longleftarrow$ + ($\longrightarrow$, $\left(\frac{8}{3}, \infty\right)$
        $\frac{8}{3}$
     2 3

41. $4(2x + 1) < 4$ $\longleftarrow$) $\longrightarrow$, $(-\infty, 0)$
          0

42. $6(2 - x) \geq 12$ $\longleftarrow$] $\longrightarrow$, $(-\infty, 0]$
          0

43. $-5x + 4 \geq -4(x - 1)$ $\longleftarrow$] $\longrightarrow$, $(-\infty, 0]$
            0

44. $-6x + 2 < -3(x + 4)$ $\longleftarrow$ + ($\longrightarrow$, $\left(\frac{14}{3}, \infty\right)$
        $\frac{14}{3}$
     4 5

45. $-2(x - 4) - 3x < -(4x + 1) + 2x$ $\longleftarrow$ ($\longrightarrow$, $(3, \infty)$
                            3

46. $-5(1 - x) + x \leq -(6 - 2x) + 6$ $\longleftarrow$ +] $\longrightarrow$, $\left(-\infty, \frac{5}{4}\right]$
                     $\frac{5}{4}$
                   1 2

47. $-3x + 6 \geq 2x + 6$ $\longleftarrow$] $\longrightarrow$, $(-\infty, 0]$
            0

48. $-(x - 4) < 4$ $\longleftarrow$ ($\longrightarrow$, $(0, \infty)$
         0

49. Explain how solving a linear inequality is similar to solving a linear equation.   answers may vary

50. Explain how solving a linear inequality is different from solving a linear equation.   answers may vary

Graph each inequality. Then write the solutions in interval notation. See Example 9.

51. $-1 < x < 3$ $\longleftarrow$ ( ) $\longrightarrow$, $(-1, 3)$ **52.** $2 \leq y \leq 3$
       -1 3
                           $\longleftarrow$ [ ] $\longrightarrow$, $[2, 3]$
                             2 3

53. $0 \leq y < 2$ $\longleftarrow$ [ ) $\longrightarrow$, $[0, 2)$ **54.** $-1 \leq x \leq 4$
       0 2
                           $\longleftarrow$ [ ] $\longrightarrow$, $[-1, 4]$
                           -1 4

Solve each inequality. Graph the solution set and write it in interval notation. See Examples 10 and 11.

55. $-3 < 3x < 6$ $\longleftarrow$ ( ) $\longrightarrow$, $(-1, 2)$
         -1 2

56. $-5 < 2x < -2$ $\longleftarrow$ + ( ) $\longrightarrow$, $\left(-\frac{5}{2}, -1\right)$
           $-\frac{5}{2}$
      -3 -2 -1

57. $2 \leq 3x - 10 \leq 5$ $\longleftarrow$ [ ] $\longrightarrow$, $[4, 5]$
            4 5

58. $4 \leq 5x - 6 \leq 19$ $\longleftarrow$ [ ] $\longrightarrow$, $[2, 5]$
            2 5

59. $-4 < 2(x - 3) \leq 4$ $\longleftarrow$ ( ] $\longrightarrow$, $(1, 5]$
            1 5

60. $0 < 4(x + 5) \leq 8$ $\longleftarrow$ ( ] $\longrightarrow$, $(-5, -3]$
           -5 -3

61. $-2 < 3x - 5 < 7$ $\longleftarrow$ ( ) $\longrightarrow$, $(1, 4)$
           1 4

62. $1 < 4 + 2x \leq 7$ $\longleftarrow$ ( ] $\longrightarrow$, $\left(-\frac{3}{2}, \frac{3}{2}\right]$
       -2 -1 1 2

63. $-6 < 3(x - 2) \leq 8$ $\frac{14}{3}$
                           $\longleftarrow$ ( ] $\longrightarrow$, $\left(0, \frac{14}{3}\right]$
                          0 4 5

64. $-5 \leq 2(x + 4) < 8$ $-\frac{13}{2}$
                     $\longleftarrow$ [ ) $\longrightarrow$, $\left[-\frac{13}{2}, 0\right)$
                -7 -6 0

65. Explain how solving a linear inequality is different from solving a compound inequality.   answers may vary

66. Explain how solving a linear inequality is similar to solving a compound inequality.   answers may vary

67. Six more than twice a number is greater than negative fourteen. Find all numbers that make this statement true. $x > -10$

68. Five times a number, increased by one, is less than or equal to ten. Find all such numbers. $x \leq \frac{9}{5}$

69. Dennis and Nancy Wood are celebrating their 30th wedding anniversary by having a reception at Tiffany Oaks reception hall. They have budgeted $3000 for their reception. If the reception hall charges a $50.00 cleanup fee plus $34 per person, find the greatest number of people that they may invite and still stay within their budget.   86 people

70. A surprise retirement party is being planned for Pratep Puri. A total of $860 has been collected for the event, which is to be held at a local reception hall. This reception hall charges a cleanup fee of $40 and $15 per person for drinks and light snacks. Find the greatest number of people that may be invited and still stay within $860.   54 people

71. Find the values for x so that the perimeter of this rectangle is no greater than 100 centimeters. $x \leq 35$

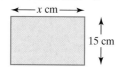

72. Find the values for x so that the perimeter of this triangle is no longer than 87 inches. $x \leq 15$

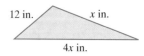

73. A financial planner has a client with $15,000 to invest. If he invests $10,000 in a certificate of deposit paying 11% annual simple interest, at what rate does the remainder of the money need to be invested so that the two investments together yield at least $1600 in yearly interest?   at least 10%

74. Alex earns $600 per month plus 4% of all his sales over $1000. Find the minimum sales that will allow Alex to earn at least $3000 per month. $61,000

75. Ben Holladay bowled 146 and 201 in his first two games. What must he bowl in his third game to have an average of at least 180?   at least 193

76. On an NBA team the two forwards measure 6′8″ and 6′6″ and the two guards measure 6′0″ and 5′9″ tall. How tall a center should they hire if they wish to have a starting team average height of at least 6′5″?   at least 7′2″

Solve the following. See Example 12.

77. High blood cholesterol levels increase the risk of heart disease in adults. Doctors recommend that total blood cholesterol be less than 200 milligrams per deciliter. Total cholesterol levels from 200 up to 240 milligrams per deciliter are considered borderline. Any total cholesterol reading above 240 milligrams per deciliter is considered high.

Letting x represent a patient's total blood cholesterol level, write a series of three inequalities that describe the ranges corresponding to recommended, borderline, and high levels of total blood cholesterol.

78. T. Theodore Fujita created the Fujita Scale (or F-Scale), which uses ratings from 0 to 5 to classify tornadoes. An F-0 tornado has wind speeds between 40 and 72 mph, inclusive. The winds in an F-1 tornado range from 73 to 112 mph. In an F-2 tornado, winds are from 113 to 157 mph. An F-3 tornado has wind speeds ranging from 158 to 206 mph. Wind speeds in an F-4 tornado are clocked at 207 to 260 mph. The most violent tornadoes are ranked at F-5, with wind speeds of at least 261 mph. (*Source:* National Weather Service)

Letting y represent a tornado's wind speed, write a series of six inequalities that describe the wind speed ranges corresponding to each Fujita Scale rank.

$40 \leq y \leq 72$ F-0
$73 \leq y \leq 112$ F-1
$113 \leq y \leq 157$ F-2
$158 \leq y \leq 206$ F-3
$207 \leq y \leq 260$ F-4
$y \geq 261$ F-5

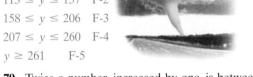

79. Twice a number, increased by one, is between negative five and seven. Find all such numbers. $-3 < x < 3$

80. Half a number, decreased by four, is between two and three. Find all such numbers. $12 < x < 14$

81. The temperatures in Ohio range from −39°C to 45°C. Use a compound inequality to convert these temperatures to Fahrenheit temperatures. (*Hint:* Use $C = \frac{5}{9}(F - 32)$.)

82. Mario Lipco has scores of 85, 95, and 92 on his algebra tests. Use a compound inequality to find the range of scores he can make on his final exam in order to receive an A in the course. The final exam counts as three tests, and an A is received if the final course average is from 90 to 100. (*Hint:* The average of a list of numbers is their sum divided by the number of numbers in the list.) $89.3 \leq x \leq 100$

REVIEW AND PREVIEW   **81.** $-38.2° \leq F \leq 113°$

Evaluate the following. See Section 1.3.

83. $(2)^3$ 8   **84.** $(3)^3$ 27   **85.** $(1)^{12}$ 1

77. $x < 200$ recommended; $200 \leq x \leq 240$ borderline; $x > 240$ high

86. 0^5 0   **87.** $\left(\frac{4}{7}\right)^2$ $\frac{16}{49}$ **88.** $\left(\frac{2}{3}\right)^3$ $\frac{8}{27}$

This broken line graph shows the average cell phone prices for the given years. Use this for Exercises 89 through 92.   **91.** 1986–1990

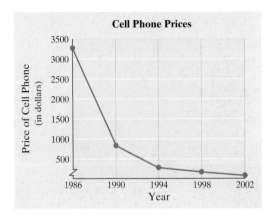

89. What was the average cell phone price in 1986? $3225

90. What was the average cell phone price in 2002? $150

91. Which period had the greatest drop in cell phone prices?

92. What years had cell phone prices over $500?   1986, 1990

Concept Extensions   **93.** $0.924 \leq d \leq 0.987$

93. The formula $C = 3.14d$ can be used to approximate the circumference of a circle given its diameter. Waldo Manufacturing manufactures and sells a certain washer with an outside circumference of 3 centimeters. The company has decided that a washer whose actual circumference is in the interval $2.9 \leq C \leq 3.1$ centimeters is acceptable. Use a compound inequality and find the corresponding interval for diameters of these washers. (Round to 3 decimal places.)

94. Bunnie Supplies manufactures plastic Easter eggs that open. The company has determined that if the circumference of the opening of each part of the egg is in the interval $118 \leq C \leq 122$ millimeters, the eggs will open and close comfortably. Use a compound inequality and find the corresponding interval for diameters of these openings. (Round to 2 decimal places.) $37.58 \leq d \leq 38.85$

Concept Extensions

For Exercises 95 through 98, see the Example below.

Solve $x(x - 6) > x^2 - 5x + 6$. Graph the solution set and write it in interval notation.

Solution

$$x(x - 6) > x^2 - 5x + 6$$
$$x^2 - 6x > x^2 - 5x + 6$$
$$x^2 - 6x - x^2 > x^2 - 5x + 6 - x^2$$
$$-6x > -5x + 6$$
$$-x > 6$$
$$\frac{-x}{-1} < \frac{6}{-1}$$
$$x < -6$$

The solution set $(-\infty, -6)$ is graphed as shown.

-7 -6 -5 -4 -3 -2 -1

Solve each inequality. Graph the solution set and write it in interval notation.

95. $x(x + 4) > x^2 - 2x + 6$ $, (1, \infty)$

96. $x(x - 3) \geq x^2 - 5x - 8$ $, [-4, \infty)$

97. $x^2 + 6x - 10 < x(x - 10)$ $, \left(-\infty, \frac{5}{8}\right)$

98. $x^2 - 4x + 8 < x(x + 8)$ $, \left(\frac{2}{3}, \infty\right)$

CHAPTER 2 PROJECT

Developing a Budget

Whether you are rolling in dough or pinching every penny, a budget can help put you in control of your personal finances. Listing your income lets you see where your money comes from, and by tracking your expenses you can analyze your spending practices. Putting these together in a budget helps you make informed decisions about what you spend.

In this project, you will have the opportunity to develop a personal budget. This project may be completed by working in groups or individually.

1. Keep track of your spending for a week or two. Be sure to write down *everything* that you spend during that time—no matter how small—from a daily newspaper or a can of soda to your tuition bill for the term.

2. Decide on a timeframe for your budget: a month, a term, a school year, or the entire year. How many weeks are in your timeframe?

3. Make a list of all of your income for your timeframe of choice. Be sure to include sources like grants or scholarships, portion of savings earmarked for college expenses for this timeframe, take-home pay from a job, gifts, financial support from family, and interest. Total your income from all sources.

4. Make a list of all of your expected expenses for your timeframe of choice. Be sure to include major expenses such as car

payments, auto and health insurance, room and board or rent/mortgage payment, and tuition. Try to estimate more variable expenses like telephone and utilities, books and supplies, groceries and/or restaurant meals, entertainment, transportation or parking, personal care, clothing, dues and/or lab fees, etc. Don't forget to take into account small or irregular purchases. Use the expense record you made in Question 1 to help you gauge your levels of spending and identify spending categories. Total your expected expenses in all categories.

5. In the equation $I = E + wx$, I represents your income from Question 3, E represents your expected expenses from Question 4, w represents the number of weeks in your timeframe of choice (from Question 2), and x represents the weekly shortage or surplus in your budget. If x is positive, this is the extra amount you have in your budget each week to save or spend as "pocket money." If x is negative, this is the amount by which you fall short each week. To keep your budget balanced, you should try to reduce nonessential expenses by this amount each week.

Substitute for I, E, and w in the equation and solve for x. Interpret your result.

6. If you have a weekly shortage, which expenses could you try to reduce to balance your budget? If you have a weekly surplus, what would you do with this extra money?

2 CHAPTER VOCABULARY CHECK

Fill in each blank with one of the words or phrases listed below.

like terms	numerical coefficient	linear inequality in one variable
equivalent equations	formula	compound inequalities
linear equation in one variable		

1. Terms with the same variables raised to exactly the same powers are called like terms.
2. A linear equation in one variable can be written in the form $ax + b = c$.
3. Equations that have the same solution are called equivalent equations.
4. Inequalities containing two inequality symbols are called compound inequalities.
5. An equation that describes a known relationship among quantities is called a formula .
6. A linear inequality in one variable can be written in the form $ax + b < c$, (or $>$, $\leq$, $\geq$).
7. The numerical coefficient of a term is its numerical factor.

CHAPTER 2   HIGHLIGHTS

Definitions and Concepts	Examples

Section 2.1 Simplifying Algebraic Expressions

Definitions and Concepts	Examples
The **numerical coefficient** of a **term** is its numerical factor.	**Term** **Numerical Coefficient** $-7y$ -7 x 1 $\frac{1}{5}a^2b$ $\frac{1}{5}$
Terms with the same variables raised to exactly the same powers are **like terms**.	**Like Terms** **Unlike Terms** $12x, -x$ $3y, 3y^2$ $-2xy, 5yx$ $7a^2b, -2ab^2$
To combine like terms, add the numerical coefficients and multiply the result by the common variable factor. To remove parentheses, apply the distributive property.	$9y + 3y = 12y$ $-4z^2 + 5z^2 - 6z^2 = -5z^2$ $-4(x + 7) + 10(3x - 1)$ $= -4x - 28 + 30x - 10$ $= 26x - 38$

Definitions and Concepts	**Examples**

Section 2.2 The Addition Property of Equality

A **linear equation in one variable** can be written in the form $ax + b = c$ where a, b, and c are real numbers and $a \neq 0$.

Linear Equations

$$-3x + 7 = 2$$

$$3(x - 1) = -8(x + 5) + 4$$

Equivalent equations are equations that have the same solution.

$x - 7 = 10$ and $x = 17$
are equivalent equations.

Addition Property of Equality

Adding the same number to or subtracting the same number from both sides of an equation does not change its solution.

$$y + 9 = 3$$
$$y + 9 - 9 = 3 - 9$$
$$y = -6$$

Section 2.3 The Multiplication Property of Equality

Multiplication Property of Equality

Multiplying both sides or dividing both sides of an equation by the same nonzero number does not change its solution.

$$\frac{2}{3}a = 18$$
$$\frac{3}{2}\left(\frac{2}{3}a\right) = \frac{3}{2}(18)$$
$$a = 27$$

Section 2.4 Solving Linear Equations

To Solve Linear Equations

Solve: $\dfrac{5(-2x + 9)}{6} + 3 = \dfrac{1}{2}$

1. Clear the equation of fractions.

  1. $6 \cdot \dfrac{5(-2x + 9)}{6} + 6 \cdot 3 = 6 \cdot \dfrac{1}{2}$

    $5(-2x + 9) + 18 = 3$

2. Remove any grouping symbols such as parentheses.
3. Simplify each side by combining like terms.
4. Write variable terms on one side and numbers on the other side using the addition property of equality.

  2. $-10x + 45 + 18 = 3$ Distributive property.
  3. $-10x + 63 = 3$ Combine like terms.
  4. $-10x + 63 - 63 = 3 - 63$ Subtract 63.
    $-10x = -60$

5. Get the variable alone using the multiplication property of equality.

  5. $\dfrac{-10x}{-10} = \dfrac{-60}{-10}$ Divide by -10.
    $x = 6$

6. Check by substituting in the original equation.

  6. $\dfrac{5(-2x + 9)}{6} + 3 = \dfrac{1}{2}$

    $\dfrac{5(-2 \cdot 6 + 9)}{6} + 3 \stackrel{?}{=} \dfrac{1}{2}$

    $\dfrac{5(-3)}{6} + 3 \stackrel{?}{=} \dfrac{1}{2}$

    $-\dfrac{5}{2} + \dfrac{6}{2} \stackrel{?}{=} \dfrac{1}{2}$

    $\dfrac{1}{2} = \dfrac{1}{2}$ True.

Definitions and Concepts	**Examples**

Section 2.5 An Introduction to Problem Solving

Problem-Solving Steps

1. UNDERSTAND the problem.

The height of the Hudson volcano in Chili is twice the height of the Kiska volcano in the Aleutian Islands. If the sum of their heights is 12,870 feet, find the height of each.

1. Read and reread the problem. Guess a solution and check your guess.

  Let x be the height of the Kiska volcano. Then $2x$ is the height of the Hudson volcano.

  $$x\rfloor \qquad 2x\rfloor$$
  Kiska   Hudson

2. TRANSLATE the problem.

2. In words:

height of Kiska	added to	height of Hudson	is	12,870
Translate: x	$+$	$2x$	$=$	12,870

3. SOLVE.

3.

$$\begin{aligned} x + 2x &= 12{,}870 \\ 3x &= 12{,}870 \\ x &= 4290 \end{aligned}$$

4. INTERPRET the results.

4. *Check*: If x is 4290 then $2x$ is 2(4290) or 8580. Their sum is 4290 + 8580 or 12,870, the required amount.

  State: Kiska volcano is 4290 feet high and Hudson volcano is 8580 feet high.

Section 2.6 Formulas and Problem Solving

An equation that describes a known relationship among quantities is called a **formula**.

To solve a formula for a specified variable, use the same steps as for solving a linear equation. Treat the specified variable as the only variable of the equation.

Formulas

$A = lw$ (area of a rectangle)
$I = PRT$ (simple interest)

Solve $P = 2l + 2w$ for l.

$$P = 2l + 2w$$
$$P - 2w = 2l + 2w - 2w \qquad \text{Subtract } 2w.$$
$$P - 2w = 2l$$
$$\frac{P - 2w}{2} = \frac{2l}{2} \qquad \text{Divide by 2.}$$
$$\frac{P - 2w}{2} = l \qquad \text{Simplify.}$$

If all values for the variables in a formula are known except for one, this unknown value may be found by substituting in the known values and solving.

If $d = 182$ miles and $r = 52$ miles per hour in the formula $d = r \cdot t$, find t.

$$d = r \cdot t$$
$$182 = 52 \cdot t \qquad \text{Let } d = 182 \text{ and } r = 52.$$
$$3.5 = t$$

The time is 3.5 hours.

Definitions and Concepts	**Examples**

How many liters of a 20% acid solution must be mixed with a 50% acid solution in order to obtain 12 liters of a 30% solution?

1. UNDERSTAND.

1. Read and reread. Guess a solution and check.
Let x = number of liters of 20% solution.
Then $12 - x$ = number of liters of 50% solution.

2. TRANSLATE.

2.

	No. of Liters ·	Strength =	*Acid* Amount of Acid
20% Solution	x	20%	$0.20x$
50% Solution	$12 - x$	50%	$0.50(12 - x)$
30% Solution Needed	12	30%	$0.30(12)$

In words:

acid in 20% solution	+	acid in 50% solution	=	acid in 30% solution

Translate: $0.20x$ + $0.50(12 - x)$ = $0.30(12)$

3. SOLVE.

3. Solve $0.20x + 0.50(12 - x) = 0.30(12)$

$$0.20x + 6 - 0.50x = 3.6 \quad \text{Apply the distributive property.}$$
$$-0.30x + 6 = 3.6$$
$$-0.30x = -2.4 \quad \text{Subtract 6.}$$
$$x = 8 \quad \text{Divide by } -0.30.$$

4. INTERPRET.

4. *Check, then state.*
If 8 liters of a 20% acid solution are mixed with $12 - 8$ or 4 liters of a 50% acid solution, the result is 12 liters of a 30% solution.

A **linear inequality in one variable** is an inequality that can be written in one of the forms:

$$ax + b < c \qquad ax + b \leq c$$
$$ax + b > c \qquad ax + b \geq c$$

where a, b, and c are real numbers and a is not 0.

Linear Inequalities

$$2x + 3 < 6 \qquad 5(x - 6) \geq 10$$

$$\frac{x - 2}{5} > \frac{5x + 7}{2} \qquad \frac{-(x + 8)}{9} \leq \frac{-2x}{11}$$

Addition Property of Inequality

Adding the same number to or subtracting the same number from both sides of an inequality does not change the solutions.

$$y + 4 \leq -1$$
$$y + 4 - 4 \leq -1 - 4 \quad \text{Subtract 4.}$$
$$y \leq -5$$

$$\begin{array}{ccccccccc} & & & & & & & & \\ \hline -6 & -5 & -4 & -3 & -2 & -1 & 0 & 1 & 2 \end{array}$$

Definitions and Concepts	**Examples**

Multiplication Property of Inequality

Multiplying or dividing both sides of an inequality by the same positive number does not change its solutions.

$$\frac{1}{3}x > -2$$

$$3\left(\frac{1}{3}x\right) > 3 \cdot -2 \quad \text{Multiply by 3.}$$

$$x > -6$$

Multiplying or dividing both sides of an inequality by the same **negative number and reversing the direction of the inequality sign** does not change its solutions.

$$-2x \le 4$$

$$\frac{-2x}{-2} \ge \frac{4}{-2} \quad \text{Divide by } -2, \text{ reverse inequality sign.}$$

$$x \ge -2$$

To Solve Linear Inequalities

1. Clear the equation of fractions.
2. Remove grouping symbols.
3. Simplify each side by combining like terms.
4. Write variable terms on one side and numbers on the other side using the addition property of inequality.
5. Get the variable alone using the multiplication property of inequality.

Solve: $3(x + 2) \le -2 + 8$

1. No fractions to clear. $3(x + 2) \le -2 + 8$
2. $3x + 6 \le -2 + 8 \quad$ Distributive property.
3. $3x + 6 \le 6 \quad$ Combine like terms.
4. $3x + 6 - 6 \le 6 - 6 \quad$ Subtract 6.
   $3x \le 0$
5. $\dfrac{3x}{3} \le \dfrac{0}{3} \quad$ Divide by 3.
   $x \le 0$

Inequalities containing two inequality symbols are called **compound inequalities**.

To solve a compound inequality, isolate the variable in the middle part of the inequality. Perform the same operation to all three parts of the inequality: left, middle, right.

Compound Inequalities

$-2 < x < 6$

$5 \le 3(x - 6) < \dfrac{20}{3}$

Solve: $-2 < 3x + 1 < 7$

$-2 - 1 < 3x + 1 - 1 < 7 - 1 \quad$ Subtract 1.

$-3 < 3x < 6$

$\dfrac{-3}{3} < \dfrac{3x}{3} < \dfrac{6}{3} \quad$ Divide by 3.

$-1 < x < 2$

2 CHAPTER REVIEW

(2.1) Simplify the following expressions. 3. $4x - 2$ 5. $3n - 18$

1. $5x - x + 2x$ $6x$
2. $0.2z - 4.6x - 7.4z$ $-4.6x - 7.2z$
3. $\frac{1}{2}x + 3 + \frac{7}{2}x - 5$
4. $\frac{4}{5}y + 1 + \frac{6}{5}y + 2$ $2y + 3$
5. $2(n - 4) + n - 10$
6. $3(w + 2) - (12 - w)$ $4w - 6$
7. Subtract $7x - 2$ from $x + 5$. $-6x + 7$
8. Subtract $1.4y - 3$ from $y - 0.7$. $-0.4y + 2.3$

Write each of the following as algebraic expressions.

9. Three times a number decreased by 7 $3x - 7$
10. Twice the sum of a number and 2.8 added to 3 times a number $2(x + 2.8) + 3x$

(2.2) Solve the following.

11. $8x + 4 = 9x$ 4
12. $5y - 3 = 6y$ -3
13. $3x - 5 = 4x + 1$ -6
14. $2x - 6 = x - 6$ 0
15. $4(x + 3) = 3(1 + x)$ -9
16. $6(3 + n) = 5(n - 1)$
16. -23

Use the addition property to fill in the blank so that the middle equation simplifies to the last equation.

17. $ x - 5 = 3$
$x - 5 + \underline{\ 5\ } = 3 + \underline{\ 5\ }$
$ x = 8$

18. $ x + 9 = -2$
$x + 9 - \underline{\ 9\ } = -2 - \underline{\ 9\ }$
$ x = -11$

Write each as an algebraic expression. 20. $(x - 5)$ in.

19. The sum of two numbers is 10. If one number is x, express the other number in terms of x. $10 - x$
20. Mandy is 5 inches taller than Melissa. If x inches represents the height of Mandy, express Melissa's height in terms of x.
△ 21. If one angle measures $(x + 5)°$, express the measure of its supplement in terms of x. $(175 - x)°$

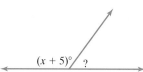

$(x + 5)°$?

(2.3) Solve each equation.

22. $\frac{3}{4}x = -9$ -12
23. $\frac{x}{6} = \frac{2}{3}$ 4
24. $-3x + 1 = 19$ -6
25. $5x + 25 = 20$ -1
26. $5x + x = 9 + 4x - 1 + 6$ 7
27. $-y + 4y = 7 - y - 3 - 8$ -1
28. Express the sum of three even consecutive integers as an expression in x. Let x be the first even integer. $3x + 6$

(2.4) Solve the following.

29. $\frac{2}{7}x - \frac{5}{7} = 1$ 6
30. $\frac{5}{3}x + 4 = \frac{2}{3}x$ -4

31. $-(5x + 1) = -7x + 3$ 2
32. $-4(2x + 1) = -5x + 5$
33. $-6(2x - 5) = -3(9 + 4x)$
34. $3(8y - 1) = 6(5 + 4y)$
35. $\frac{3(2 - z)}{5} = z$ $\frac{3}{4}$
36. $\frac{4(n + 2)}{5} = -n$ $-\frac{8}{9}$
37. $5(2n - 3) - 1 = 4(6 + 2n)$ 20 32. -3
38. $-2(4y - 3) + 4 = 3(5 - y)$ -1 33.–34. no solution
39. $9z - z + 1 = 6(z - 1) + 7$ 0 42. $-\frac{6}{23}$
40. $5t - 3 - t = 3(t + 4) - 15$ 0
41. $-n + 10 = 2(3n - 5)$ $\frac{20}{7}$ 42. $-9 - 5a = 3(6a - 1)$
43. $\frac{5(c + 1)}{6} = 2c - 3$ $\frac{23}{7}$ 44. $\frac{2(8 - a)}{3} = 4 - 4a$ $-\frac{2}{5}$
45. $200(70x - 3560) = -179(150x - 19{,}300)$ 102
46. $1.72y - 0.04y = 0.42$ 0.25

Solve.

47. The quotient of a number and 3 is the same as the difference of the number and two. Find the number. 3
48. Double the sum of a number and six is the opposite of the number. Find the number. -4

(2.5) Solve each of the following.

49. The height of the Eiffel Tower is 68 feet more than three times a side of its square base. If the sum of these two dimensions is 1380 feet, find the height of the Eiffel Tower. 1052 ft

50. A 12-foot board is to be divided into two pieces so that one piece is twice as long as the other. If x represents the length of the shorter piece, find the length of each piece. 4 ft; 8 ft

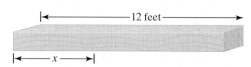

51. One area code in Ohio is 34 more than three times another area code used in Ohio. If the sum of these area codes is 1262, find the two area codes.   307; 955

52. Find three consecutive even integers whose sum is -114. $-40, -38, -36$

(2.6) Substitute the given values into the given formulas and solve for the unknown variable.

△ **53.** $P = 2l + 2w$; $P = 46, l = 14$ $w = 9$

△ **54.** $V = lwh$; $V = 192, l = 8, w = 6$ $h = 4$

Solve each of the following for the indicated variable.

55. $y = mx + b$ for m **56.** $r = vst - 9$ for s

57. $2y - 5x = 7$ for x **58.** $3x - 6y = -2$ for y

△ **59.** $C = \pi D$ for π $\pi = \frac{C}{D}$ △ **60.** $C = 2\pi r$ for π $\pi = \frac{C}{2r}$

△ **61.** A swimming pool holds 900 cubic meters of water. If its length is 20 meters and its height is 3 meters, find its width.   15 m

55. $m = \frac{y - b}{x}$

56. $s = \frac{r + 9}{vt}$

57. $x = \frac{2y - 7}{5}$

58. $y = \frac{2 + 3x}{6}$

62. The highest temperature on record in Rome, Italy, is 104° Fahrenheit. Convert this temperature to Celsius.   40°C

63. A charity 10K race is given annually to benefit a local hospice organization. How long will it take to run/walk a 10K race (10 kilometers or 10,000 meters) if your average pace is 125 **meters** per minute?   1 hr and 20 min

(2.7) Solve each of the following.

64. A $50,000 retirement pension is to be invested into two accounts: a money market fund that pays 8.5% and a certificate of deposit that pays 10.5%. How much should be invested at each rate in order to provide a yearly interest income of $4550? $35,000 @ 8.5%; $15,000 @ 10.5%

65. A pay phone is holding its maximum number of 500 coins consisting of nickels, dimes, and quarters. The number of quarters is twice the number of dimes. If the value of all the coins is $88.00, how many nickels were in the pay phone?   80 nickels

66. How long will it take an Amtrak passenger train to catch up to a freight train if their speeds are 60 and 45 mph and the freight train had an hour and a half head start?   4.5 hrs

67. In 2003, Lance Armstrong became the 4th cyclist in history to win the Tour de France 5 years in a row. Suppose he rides a bicycle up a mountain trail at 8 mph and down the same trail at 12 mph. Find the round-trip distance traveled if the total travel time was 5 hours.   48 mi

(2.8) Solve and graph the solution of each of the following inequalities.   68.–79. See graphing answer section.

68. $x \leq -2$ $(-\infty, -2]$ **69.** $x > 0$ $(0, \infty)$

70. $-1 < x < 1$ $(-1, 1)$ **71.** $0.5 \leq y < 1.5$ $[0.5, 1.5)$

72. $-2x \geq -20$ $(-\infty, 10]$ **73.** $-3x > 12$ $(-\infty, -4)$

74. $5x - 7 > 8x + 5$ **75.** $x + 4 \geq 6x - 16$ $(-\infty, 4]$

76. $2 \leq 3x - 4 < 6$ $[2, \frac{10}{3})$ **77.** $-3 < 4x - 1 < 2$ $(-\frac{1}{2}, \frac{3}{4})$

78. $-2(x - 5) > 2(3x - 2)$ **79.** $4(2x - 5) \leq 5x - 1$ $(-\infty, \frac{19}{3}]$

80. Tina earns $175 per week plus a 5% commission on all her sales. Find the minimum amount of sales to ensure that she earns at least $300 per week. $2500

81. Ellen Catarella shot rounds of 76, 82, and 79 golfing. What must she shoot on her next round so that her average will be below 80?   score must be less than 83

74. $(-\infty, -4)$ **78.** $(-\infty, \frac{7}{4})$

CHAPTER 2 TEST

Remember to use your Chapter Test Prep Video CD to help you study and view solutions to the test questions you need help with.

Simplify each of the following expressions.

1. $2y - 6 - y - 4$ $y - 10$

2. $2.7x + 6.1 + 3.2x - 4.9$ $5.9x + 1.2$

3. $4(x - 2) - 3(2x - 6)$ $-2x + 10$

4. $7 + 2(5y - 3)$ $10y + 1$

Solve each of the following equations.

5. $-\frac{4}{5}x = 4$ -5 **6.** $4(n - 5) = -(4 - 2n)$ 8

7. $5y - 7 + y = -(y + 3y)$ $\frac{7}{10}$

8. $4z + 1 - z = 1 + z$ 0

9. $\frac{2(x + 6)}{3} = x - 5$ 27   **10.** $\frac{1}{2} - x + \frac{3}{2} = x - 4$ 3

11. $-0.3(x - 4) + x = 0.5(3 - x)$ 0.25

12. $-4(a + 1) - 3a = -7(2a - 3)$ $\frac{25}{7}$

13. $-2(x - 3) = x + 5 - 3x$ no solution

Solve each of the following applications.

14. A number increased by two-thirds of the number is 35. Find the number.   21

△ **15.** A gallon of water seal covers 200 square feet. How many gallons are needed to paint two coats of water seal on a deck that measures 20 feet by 35 feet?   7 gal

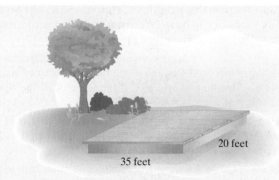

20 feet

35 feet

16. Sedric Angell invested an amount of money in Amoxil stock that earned an annual 10% return, and then he invested twice the original amount in IBM stock that earned an annual 12% return. If his total return from both investments was $2890, find how much he invested in each stock. $8500 @ 10%; $17,000 @ 12%

17. Two trains leave Los Angeles simultaneously traveling on the same track in opposite directions at speeds of 50 and 64 mph. How long will it take before they are 285 miles apart? $2\frac{1}{2}$ hrs

18. Find the value of x if $y = -14$, $m = -2$, and $b = -2$ in the formula $y = mx + b$. $x = 6$

Solve each of the following equations for the indicated variable.

△ **19.** $V = \pi r^2 h$ for h $h = \dfrac{V}{\pi r^2}$

20. $3x - 4y = 10$ for y $y = \frac{3x - 10}{4}$

Solve and graph each of the following inequalities.

21. $3x - 5 \geq 7x + 3$ $(-\infty, -2]$ **22.** $x + 6 > 4x - 6$ $(-\infty, 4)$

23. $-2 < 3x + 1 < 8$ $(-1, \frac{7}{3})$ **24.** $\dfrac{2(5x + 1)}{3} > 2$ $(\frac{2}{5}, \infty)$

21.–24. See graphing answer section.

CHAPTER CUMULATIVE REVIEW

7. $\frac{8}{20}$ (Sec. 1.3, Ex. 6)   **8.** $\frac{16}{24}$ (Sec. 1.3)   **11.** 2 is a solution (Sec. 1.4, Ex. 7)   **12.** 3 is not a solution (Sec. 1.4)

1. Given the set $\{-2, 0, \frac{1}{4}, 112, -3, 11, \sqrt{2}\}$, list the numbers in this set that belong to the set of:

  a. Natural numbers   11, 112

  b. Whole numbers   0, 11, 112

  c. Integers   −3, −2, 0, 11, 112

  d. Rational numbers $-3, -2, 0, \frac{1}{4}, 11, 112$

  e. Irrational numbers $\sqrt{2}$

  f. Real numbers   all numbers in the given set (Sec. 1.2, Ex. 5)

2. Given the set $\{7, 2, -\frac{1}{5}, 0, \sqrt{3}, -185, 8\}$, list the numbers in this set that belong to the set of:

  a. Natural numbers   2, 7, 8

  b. Whole Numbers   0, 2, 7, 8

  c. Integers   −185, 0, 2, 7, 8

  d. Rational Numbers $-185, -\frac{1}{5}, 0, 2, 7, 8$

  e. Irrational numbers $\sqrt{3}$

  f. Real numbers   all numbers in the given set (Sec. 1.2)

3. Find the absolute value of each number.

  a. $|4|$ 4   **b.** $|-5|$ 5   **c.** $|0|$ 0 (Sec. 1.2, Ex. 7)

4. Find the absolute value of each number.

  a. $|5|$ 5   **b.** $|-8|$ 8   **c.** $\left|-\frac{2}{3}\right|$ $\frac{2}{3}$ (Sec. 1.2)

5. Write each of the following numbers as a product of primes.

  a. 40 $2 \cdot 2 \cdot 2 \cdot 5$ **b.** 63 $3 \cdot 3 \cdot 7$ (Sec. 1.3, Ex. 1)

6. Write each number as a product of primes.

  a. 44 $2 \cdot 2 \cdot 11$ **b.** 90 $2 \cdot 3 \cdot 3 \cdot 5$ (Sec. 1.3)

7. Write $\frac{2}{5}$ as an equivalent fraction with a denominator of 20.

8. Write $\frac{2}{3}$ as an equivalent fraction with a denominator of 24.

9. Simplify $3[4(5 + 2) - 10]$.   54 (Sec. 1.4, Ex. 4)

10. Simplify $5[16 - 4(2 + 1)]$.   20 (Sec. 1.4)

11. Decide whether 2 is a solution of $3x + 10 = 8x$.

12. Decide whether 3 is a solution of $5x - 2 = 4x$.

Add.

13. $-1 + (-2)$ −3 (Sec. 1.5, Ex. 2)   **14.** $(-2) + (-8)$ −10

15. $-4 + 6$ 2 (Sec. 1.5, Ex. 4)   **16.** $-3 + 10$ 7

17. Simplify each expression.

  a. $-(-10)$ 10

  b. $-\left(-\frac{1}{2}\right)$ $\frac{1}{2}$

  c. $-(-2x)$ $2x$

  d. $-|-6|$ −6 (Sec. 1.5, Ex. 10)

18. Simplify each expression.

  a. $-(-5)$ 5

  b. $-\left(-\frac{2}{3}\right)$ $\frac{2}{3}$

  c. $-(-a)$ a

  d. $-|-3|$ −3 (Sec. 1.5)

19. Subtract.

  a. $5.3 - (-4.6)$ 9.9

  b. $-\frac{3}{10} - \frac{5}{10}$ $-\frac{4}{5}$

  c. $-\frac{2}{3} - \left(-\frac{4}{5}\right)$ $\frac{2}{15}$ (Sec. 1.6, Ex. 2)

20. Subtract

  a. $-2.7 - 8.4$ −11.1

  b. $-\frac{4}{5} - \left(-\frac{3}{5}\right)$ $-\frac{1}{5}$

  c. $\frac{1}{4} - \left(-\frac{1}{2}\right)$ $\frac{3}{4}$ (Sec. 1.6)

21. Find each unknown complementary or supplementary angle.

a.

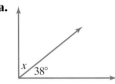

52°

b.

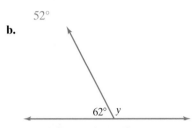

118° (Sec. 1.6, Ex. 8)

22. Find each unknown complementary or supplementary angle.

a.   18° (Sec. 1.6)

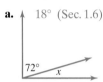

b.   133°

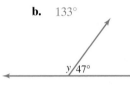

23. Find each product.

a. $(-1.2)(0.05)$ -0.06

b. $\dfrac{2}{3} \cdot -\dfrac{7}{10}$ $-\frac{7}{15}$ (Sec. 1.7, Ex. 3)

24. Find each product.

a. $(4.5)(-0.08)$ -0.36 **b.** $-\dfrac{3}{4} \cdot -\dfrac{8}{17}$ $\frac{6}{17}$ (Sec. 1.7)

25. Find each quotient.

a. $\dfrac{-24}{-4}$ 6   **b.** $\dfrac{-36}{3}$ -12

c. $\dfrac{2}{3} \div \left(-\dfrac{5}{4}\right)$ $-\frac{8}{15}$ (Sec. 1.7, Ex. 7)

26. Find each quotient.

a. $\dfrac{-32}{8}$ -4 **b.** $\dfrac{-108}{-12}$ 9

c. $\dfrac{-5}{7} \div \left(\dfrac{-9}{2}\right)$ $\frac{10}{63}$ (Sec. 1.7)

27. Use a commutative property to complete each statement.

a. $x + 5 =$ ___ $5 + x$ **b.** $3 \cdot x =$ ___

28. Use a commutative property to complete each statement.

a. $y + 1 =$ ___ $1 + y$ **b.** $y \cdot 4 =$ ___ $4 \cdot y$

29. Use the distributive property to write each sum as a product.

a. $8 \cdot 2 + 8 \cdot x$ $8(2 + x)$ **b.** $7s + 7t$

30. Use the distributive property to write each sum as a product.

a. $4 \cdot y + 4 \cdot \dfrac{1}{3}$ $4\left(y + \frac{1}{3}\right)$ **b.** $0.10x + 0.10y$

31. Subtract $4x - 2$ from $2x - 3$. $-2x - 1$ (Sec. 2.1, Ex. 7)

32. Subtract $10x + 3$ from $-5x + 1$. $-15x - 2$ (Sec. 2.1)

Solve.

33. $\dfrac{1}{2} = x - \dfrac{3}{4}$ $\frac{5}{4}$ (Sec. 2.2, Ex. 3)   **34.** $\dfrac{5}{6} + x = \dfrac{2}{3}$ $-\frac{1}{6}$

35. $6(2a - 1) - (11a + 6) = 7$ 19 (Sec. 2.2, Ex. 6)

36. $-3x + 1 - (-4x - 6) = 10$ 3 (Sec. 2.2)

37. $\dfrac{y}{7} = 20$ 140 (Sec. 2.3, Ex. 3)   **38.** $\dfrac{x}{4} = 18$ 72

39. $4(2x - 3) + 7 = 3x + 5$ 2 (Sec. 2.4, Ex. 1)

40. $6x + 5 = 4(x + 4) - 1$ 5 (Sec. 2.4)

41. Twice the sum of a number and 4 is the same as four times the number decreased by 12. Find the number.

42. A number increased by 4 is the same as 3 times the number decreased by 8. Find the number.   6 (Sec. 2.5)

43. Solve $V = lwh$ for l. $\dfrac{V}{wh} = l$ (Sec. 2.6, Ex. 5)

44. Solve $C = 2\pi r$ for r. $\dfrac{C}{2\pi} = r$ (Sec. 2.6)

45. Solve $x + 4 \leq -6$ for x. Graph the solution set and write it in interval notation. $(-\infty, -10]$ (Sec. 2.8, Ex. 2)

46. Solve $x - 3 > 2$ for x. Graph the solution set and write it in interval notation. $(5, \infty)$ $\left\{ x | x > 5 \right\}$ (Sec. 2.8)

27. b. $x \cdot 3$ (Sec. 1.8, Ex. 1)   **29. b.** $7(s + t)$ (Sec. 1.8, Ex. 5)   **30. b.** $0.10(x + y)$ (Sec. 1.8)   **41.** 10 (Sec. 2.5, Ex. 1)

Graphing

In the previous chapter we learned to solve and graph the solutions of linear equations and inequalities in one variable. Now we define and present techniques for solving and graphing linear equations and inequalities in two variables.

FOR SALE: RESIDENCE

The purchase and sale of a house are likely the largest monetary transactions people make in their lifetimes. Although some people attempt to handle the sale on their own, most seek the help of a professional real estate agent. Realtors give their clients advice on all aspects of purchase and sale, including price negotiation, home warranties, market trends, and interest rates. An agent for the seller could provide information on how a remodeling project might affect the marketability of a home, whereas the buyer's agent can provide tips on the availability of different types of financing. In exchange for their services, real estate agents usually collect a commission based on a percent of the selling price of the home.

In the Spotlight on Decision Making feature on page 200, you will have the opportunity to choose a real estate agency as the seller of a home.
*Link: National Association of Realtors—*www.realtor.org

3.1 THE RECTANGULAR COORDINATE SYSTEM

Objectives

1 Define the rectangular coordinate system and plot ordered pairs of numbers.

2 Graph paired data to create a scatter diagram.

3 Determine whether an ordered pair is a solution of an equation in two variables.

4 Find the missing coordinate of an ordered pair solution, given one coordinate of the pair.

1   In Section 1.9, we learned how to read graphs. Example 4 in Section 1.9 presented the graph below showing the relationship between time since smoking a cigarette and pulse rate. Notice in this graph that there are two numbers associated with each point of the graph. For example, we discussed earlier that 15 minutes after "lighting up," the pulse rate is 80 beats per minute. If we agree to write the time first and the pulse rate second, we can say there is a point on the graph corresponding to the **ordered pair** of numbers (15, 80). A few more ordered pairs are listed alongside their corresponding points.

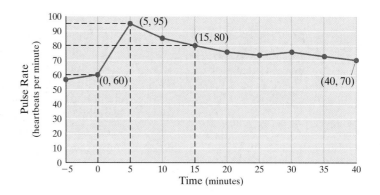

In general, we use this same ordered pair idea to describe the location of a point in a plane (such as a piece of paper). We start with a horizontal and a vertical axis. Each axis is a number line, and for the sake of consistency we construct our axes to intersect at the 0 coordinate of both. This point of intersection is called the **origin**. Notice that these two number lines or axes divide the plane into four regions called **quadrants**. The quadrants are usually numbered with Roman numerals as shown. The axes are not considered to be in any quadrant.

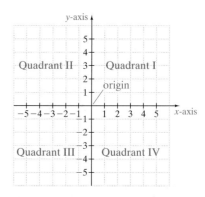

It is helpful to label axes, so we label the horizontal axis the **x-axis** and the vertical axis the **y-axis**. We call the system described above the **rectangular coordinate system**.

Just as with the pulse rate graph, we can then describe the locations of points by ordered pairs of numbers. We list the horizontal **x-axis** measurement first and the vertical **y-axis** measurement second.

To plot or graph the point corresponding to the ordered pair

$$(a, b)$$

we start at the origin. We then move a units left or right (right if a is positive, left if a is negative). From there, we move b units up or down (up if b is positive, down if b is negative). For example, to plot the point corresponding to the ordered pair $(3, 2)$, we start at the origin, move 3 units right, and from there move 2 units up. (See the figure below.) The x-value, 3, is called the **x-coordinate** and the y-value, 2, is called the **y-coordinate**. From now on, we will call the point with coordinates $(3, 2)$ simply the point $(3, 2)$. The point $(-2, 5)$ is graphed below also.

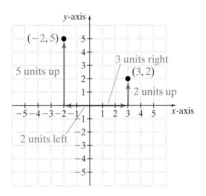

Does the order in which the coordinates are listed matter? Yes! Notice that the point corresponding to the ordered pair $(2, 3)$ is in a different location than the point corresponding to $(3, 2)$. These two ordered pairs of numbers describe two different points of the plane.

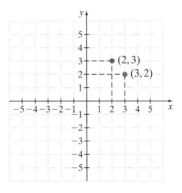

TEACHING TIP

Remind students that ordered pair values are given in alphabetical order. (x, y).

TEACHING TIP

Throughout this section, remind students of the contents of this Helpful Hint.

Concept Check Answer:

The graph of point $(-5, 1)$ lies in quadrant II and the graph of point $(1, -5)$ lies in quadrant IV. They are *not* in the same location.

✔ **CONCEPT CHECK**

Is the graph of the point $(-5, 1)$ in the same location as the graph of the point $(1, -5)$? Explain.

> **Helpful Hint**
>
> Don't forget that **each ordered pair corresponds to exactly one point in the plane and that each point in the plane corresponds to exactly one ordered pair.**

EXAMPLE 1

On a single coordinate system, plot each ordered pair. State in which quadrant, if any, each point lies.

a. $(5, 3)$ **b.** $(-5, 3)$ **c.** $(-2, -4)$ **d.** $(1, -2)$

e. $(0, 0)$ **f.** $(0, 2)$ **g.** $(-5, 0)$ **h.** $\left(0, -5\frac{1}{2}\right)$

Solution

Point $(5, 3)$ lies in quadrant I.

Point $(-5, 3)$ lies in quadrant II.

Point $(-2, -4)$ lies in quadrant III.

Point $(1, -2)$ lies in quadrant IV.

Points $(0, 0)$, $(0, 2)$, $(-5, 0)$, and $\left(0, -5\frac{1}{2}\right)$ lie on axes, so they are not in any quadrant.

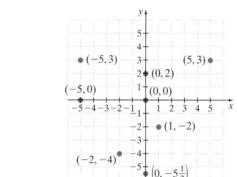

CLASSROOM EXAMPLE

On a single coordinate system, plot each ordered pair.

a. $(4, 2)$ **b.** $(-1, -3)$

c. $(2, -2)$ **d.** $(-5, 1)$

e. $(0, 3)$ **f.** $(3, 0)$

g. $(0, -4)$ **h.** $\left(-2\frac{1}{2}, 0\right)$

answer:

From Example 1, notice that the y-coordinate of any point on the x-axis is 0. For example, the point $(-5, 0)$ lies on the x-axis. Also, the x-coordinate of any point on the y-axis is 0. For example, the point $(0, 2)$ lies on the y-axis.

✔ **CONCEPT CHECK**

For each description of a point in the rectangular coordinate system, write an ordered pair that represents it.

a. Point A is located three units to the left of the y-axis and five units above the x-axis.

b. Point B is located six units below the origin.

2   Data that can be represented as an ordered pair is called **paired data**. Many types of data collected from the real world are paired data. For instance, the annual measurement of a child's height can be written as an ordered pair of the form (year, height in inches) and is paired data. The graph of paired data as points in the rectangular coordinate system is called a **scatter diagram**. Scatter diagrams can be used to look for patterns and trends in paired data.

EXAMPLE 2

Concept Check Answer:
a. $(-3, 5)$ **b.** $(0, -6)$

The table gives the annual net sales for Wal-Mart Stores for the years shown. (*Source:* Wal-Mart Stores, Inc.)

Year	Wal-Mart Net Sales (in billions of dollars)
1997	105
1998	118
1999	138
2000	165
2001	191
2002	218
2003	245

a. Write this paired data as a set of ordered pairs of the form (year, sales in billions of dollars).

b. Create a scatter diagram of the paired data.

c. What trend in the paired data does the scatter diagram show?

Solution   **a.** The ordered pairs are (1997, 105), (1998, 118), (1999, 138), (2000, 165), (2001, 191), (2002, 218), and (2003, 245).

b. We begin by plotting the ordered pairs. Because the x-coordinate in each ordered pair is a year, we label the x-axis "Year" and mark the horizontal axis with the years given. Then we label the y-axis or vertical axis "Net Sales (in billions of dollars)." In this case it is convenient to mark the vertical axis in multiples of 20. Since no net sale is less than 100, we use the notation $\frac{l}{Z}$ to skip to 100, then proceed by multiples of 20.

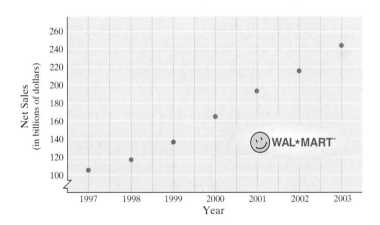

c. The scatter diagram shows that Wal-Mart net sales steadily increased over the years 1997–2003.

CLASSROOM EXAMPLE

The table gives the number of tornadoes that have occurred in the United States for the years shown. (*Source:* Storm Prediction Center, National Weather Service)

Year	Tornadoes
1995	1234
1996	1173
1997	1148
1998	1424
1999	1343
2000	997

Create a scatter diagram of the paired data.

answer:

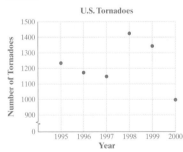

3   Let's see how we can use ordered pairs to record solutions of equations containing two variables. An equation in one variable such as $x + 1 = 5$ has one solution, which is 4: the number 4 is the value of the variable x that makes the equation true.

An equation in two variables, such as $2x + y = 8$, has solutions consisting of two values, one for x and one for y. For example, $x = 3$ and $y = 2$ is a solution of $2x + y = 8$ because, if x is replaced with 3 and y with 2, we get a true statement.

$$2x + y = 8$$
$$2(3) + 2 = 8$$
$$8 = 8 \quad \text{True.}$$

The solution $x = 3$ and $y = 2$ can be written as $(3, 2)$, an **ordered pair** of numbers. The first number, 3, is the x-value and the second number, 2, is the y-value.

In general, an ordered pair is a **solution** of an equation in two variables if replacing the variables by the values of the ordered pair results in a true statement.

EXAMPLE 3

Determine whether each ordered pair is a solution of the equation $x - 2y = 6$.

a. $(6, 0)$ **b.** $(0, 3)$ **c.** $\left(1, -\dfrac{5}{2}\right)$

Solution   **a.** Let $x = 6$ and $y = 0$ in the equation $x - 2y = 6$.

$$x - 2y = 6$$
$$6 - 2(0) = 6 \qquad \text{Replace } x \text{ with 6 and } y \text{ with 0.}$$
$$6 - 0 = 6 \qquad \text{Simplify.}$$
$$6 = 6 \qquad \text{True.}$$

$(6, 0)$ is a solution, since $6 = 6$ is a true statement.

b. Let $x = 0$ and $y = 3$.

$$x - 2y = 6$$
$$0 - 2(3) = 6 \qquad \text{Replace } x \text{ with 0 and } y \text{ with 3.}$$
$$0 - 6 = 6$$
$$-6 = 6 \qquad \text{False.}$$

$(0, 3)$ is *not* a solution, since $-6 = 6$ is a false statement.

c. Let $x = 1$ and $y = -\dfrac{5}{2}$ in the equation.

$$x - 2y = 6$$
$$1 - 2\left(-\dfrac{5}{2}\right) = 6 \qquad \text{Replace } x \text{ with 1 and } y \text{ with } -\dfrac{5}{2}.$$
$$1 + 5 = 6$$
$$6 = 6 \qquad \text{True.}$$

$\left(1, -\dfrac{5}{2}\right)$ is a solution, since $6 = 6$ is a true statement.

4   If one value of an ordered pair solution of an equation is known, the other value can be determined. To find the unknown value, replace one variable in the equation by its known value. Doing so results in an equation with just one variable that can be solved for the variable using the methods of Chapter 2.

EXAMPLE 4

Complete the following ordered pair solutions for the equation $3x + y = 12$.

a. $(0, \ \ )$ **b.** $(\ \ , 6)$ **c.** $(-1, \ \ )$

Solution   **a.** In the ordered pair $(0, \quad)$, the x-value is 0. Let $x = 0$ in the equation and solve for y.

$$3x + y = 12$$
$$3(0) + y = 12 \qquad \text{Replace } x \text{ with } 0.$$
$$0 + y = 12$$
$$y = 12$$

The completed ordered pair is $(0, 12)$.

b. In the ordered pair $(\quad, 6)$, the y-value is 6. Let $y = 6$ in the equation and solve for x.

$$3x + y = 12$$
$$3x + 6 = 12 \qquad \text{Replace } y \text{ with } 6.$$
$$3x = 6 \qquad \text{Subtract 6 from both sides.}$$
$$x = 2 \qquad \text{Divide both sides by 3.}$$

The ordered pair is $(2, 6)$.

c. In the ordered pair $(-1, \quad)$, the x-value is -1. Let $x = -1$ in the equation and solve for y.

$$3x + y = 12$$
$$3(-1) + y = 12 \qquad \text{Replace } x \text{ with } -1.$$
$$-3 + y = 12$$
$$y = 15 \qquad \text{Add 3 to both sides.}$$

The ordered pair is $(-1, 15)$.

Solutions of equations in two variables can also be recorded in a **table of values**, as shown in the next example.

EXAMPLE 5

Complete the table for the equation $y = 3x$.

	x	y
a.	-1	
b.		0
c.		-9

Solution   **a.** Replace x with -1 in the equation and solve for y.

$$y = 3x$$
$$y = 3(-1) \qquad \text{Let } x = -1.$$
$$y = -3$$

The ordered pair is $(-1, -3)$.

b. Replace y with 0 in the equation and solve for x.

$$y = 3x$$
$$0 = 3x \qquad \text{Let } y = 0.$$
$$0 = x \qquad \text{Divide both sides by 3.}$$

The ordered pair is $(0, 0)$.

x	y
−1	−3
0	0
−3	−9

c. Replace y with -9 in the equation and solve for x.

$$y = 3x$$
$$-9 = 3x \qquad \text{Let } y = -9.$$
$$-3 = x \qquad \text{Divide both sides by } 3.$$

The ordered pair is $(-3, -9)$. The completed table is shown to the left.

EXAMPLE 6

Complete the table for the equation $y = 3$.

x	y
−2	
0	
−5	

x	y
−2	3
0	3
−5	3

Solution The equation $y = 3$ is the same as $0x + y = 3$. Replace x with -2 and we have $0(-2) + y = 3$ or $y = 3$. Notice that no matter what value we replace x by, y always equals 3. The completed table is shown on the right.

By now, you have noticed that equations in two variables often have more than one solution. We discuss this more in the next section.

A table showing ordered pair solutions may be written vertically or horizontally as shown in the next example.

EXAMPLE 7

FINDING THE VALUE OF A COMPUTER

A computer was recently purchased for a small business for $2000. The business manager predicts that the computer will be used for 5 years and the value in dollars y of the computer in x years is $y = -300x + 2000$. Complete the table.

x	0	1	2	3	4	5
y						

Computers for the Home

Solution To find the value of y when x is 0, replace x with 0 in the equation. We use this same procedure to find y when x is 1 and when x is 2.

When $x = 0$,	**When $x = 1$,**	**When $x = 2$,**
$y = -300x + 2000$	$y = -300x + 2000$	$y = -300x + 2000$
$y = -300 \cdot 0 + 2000$	$y = -300 \cdot 1 + 2000$	$y = -300 \cdot 2 + 2000$
$y = 0 + 2000$	$y = -300 + 2000$	$y = -600 + 2000$
$y = 2000$	$y = 1700$	$y = 1400$

We have the ordered pairs $(0, 2000)$, $(1, 1700)$, and $(2, 1400)$. This means that in 0 years the value of the computer is $2000, in 1 year the value of the computer is $1700, and in 2 years the value is $1400. To complete the table of values, we continue the procedure for $x = 3$, $x = 4$, and $x = 5$.

When $x = 3$,

$y = -300x + 2000$
$y = -300 \cdot 3 + 2000$
$y = -900 + 2000$
$y = 1100$

When $x = 4$,

$y = -300x + 2000$
$y = -300 \cdot 4 + 2000$
$y = -1200 + 2000$
$y = 800$

When $x = 5$,

$y = -300x + 2000$
$y = -300 \cdot 5 + 2000$
$y = -1500 + 2000$
$y = 500$

The completed table is

x	0	1	2	3	4	5
y	2000	1700	1400	1100	800	500

The ordered pair solutions recorded in the completed table for the example above are graphed below. Notice that the graph gives a visual picture of the decrease in value of the computer.

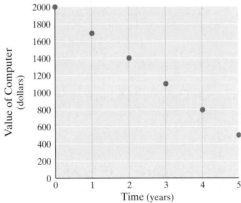

Computer Value

x	y
0	2000
1	1700
2	1400
3	1100
4	800
5	500

Spotlight on

DECISION ✕ MAKING

Suppose you own a small mail-order business. One of your employees brings you this graph showing the expected retail revenue from on-line shopping sales on the Internet. Should you consider expanding your business to include taking orders over the Internet? Explain your reasoning. What other factors would you want to consider?

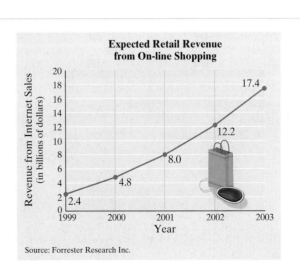

Expected Retail Revenue from On-line Shopping

Source: Forrester Research Inc.

STUDY SKILLS REMINDER

Are You Satisfied with Your Performance on a Particular Quiz or Exam?

If not, analyze your quiz or exam like you would a good mystery novel. Look for common themes in your errors.

Were most of your errors a result of

- *Carelessness?* If your errors were careless, did you turn in your work before the allotted time expired? If so, resolve to use the entire time allotted next time. Any extra time can be spent checking your work.
- *Running out of time?* If so, make a point to better manage your time on your next exam. A few suggestions are to work any questions that you are unsure of last and to check your work after all questions have been answered.
- *Not understanding a concept?* If so, review that concept and correct your work. Remember next time to make sure that all concepts on a quiz or exam are understood before the exam.

1. answers may vary: Ex. $(5,5), (7,3)$ **2.** answers may vary: Ex. $(0,6), (6,0)$ **3.** answers may vary: Ex. $(3,5), (3,0)$ **4.** answers may vary: Ex. $(0,-2), (1,-2)$

MENTAL MATH

Give two ordered pair solutions for each of the following linear equations.

1. $x + y = 10$ **2.** $x + y = 6$ **3.** $x = 3$ **4.** $y = -2$

EXERCISE SET 3.1

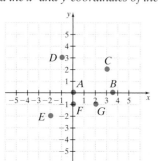

STUDY GUIDE/SSM   CD/VIDEO   PH MATH TUTOR CENTER   MathXL®Tutorials ON CD   MathXL®   MyMathLab®

Plot the ordered pairs. State in which quadrant, if any, each point lies. See Example 1.   **1.–12.** See graphing answer section.

1. $(1,5)$ **2.** $(-5,-2)$ **3.** $(-3,0)$

4. $(0,-1)$ **5.** $(2,-4)$ **6.** $\left(-1, 4\frac{1}{2}\right)$

7. $\left(4\frac{3}{4}, 0\right)$ **8.** $\left(0, \frac{7}{8}\right)$ **9.** $(0,0)$

10. $(5,0)$ **11.** $(0,4)$ **12.** $(-3,-3)$

Find the x- and y-coordinates of the following labeled points.

13.

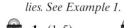

$A\ (0,0)$;
$B\ (3\frac{1}{2}, 0)$;
$C\ (3,2)$;
$D\ (-1,3)$;
$E\ (-2,-2)$;
$F\ (0,-1)$;
$G\ (2,-1)$

14.

$A\ (2,0)$;
$B\ (0,-3)$;
$C\ (-2,3)$;
$D\ (1,3)$;
$E\ (1,-1)$;
$F\ (-3,-1)$;
$G\ (-2,0)$

15. Plot the points $(-1,5), (3,5), (3,-4)$, and $(-1,-4)$ on a rectangular coordinate system. Find the perimeter of the rectangle whose vertices are these points.   26 units

16. Plot the points $(5,2), (5,-6), (0,-6)$, and $(0,2)$ on a rectangular coordinate system. Find the area of the rectangle whose vertices are these points.   40 sq. units

Solve. See Example 2.

17. The table shows the average price of a gallon of regular unleaded gasoline (in dollars) for the years shown. (*Source*: Energy Information Administration)

17. a.	Year	Price per Gallon of Unleaded Gasoline (in dollars)
(1994, 1.11)	1994	1.11
(1995, 1.15)	1995	1.15
(1996, 1.23)	1996	1.23
(1997, 1.23)	1997	1.23
(1998, 1.06)	1998	1.06
(1999, 1.17)	1999	1.17
(2000, 1.51)	2000	1.51
(2001, 1.46)	2001	1.46
(2002, 1.29)0	2002	1.29

17. b. See graphing answer section.

   a. Write each paired data as an ordered pair of the form (year, gasoline price).

   b. Draw a grid such as the one in Example 2 and create a scatter diagram of the paired data.

18. The table shows the number of regular-season NFL football games won by the winner of the Super Bowl for the years shown. (*Source*: National Football League)

18. a.	Year	Regular-Season Games Won by Super Bowl Winner
(1996, 12)	1996	12
(1997, 13)	1997	13
(1998, 12)	1998	12
(1999, 14)	1999	14
(2000, 13)	2000	13
(2001, 11)	2001	11
(2002, 9)	2002	9

18. b. See graphing answer section.

   a. Write each paired data as an ordered pair of the form (year, games won).

   b. Draw a grid such as the one in Example 2 and create a scatter diagram of the paired data.

19. The table shows the average monthly mortgage payment for median priced homes made by Americans during the years shown. (*Source*: National Association of REALTORS®)

19. a.	Year	Average Monthly Mortgage Payment for Median-Priced Homes (in dollars)
(1998, 690)	1998	690
(1999, 733)	1999	733
(2000, 818)	2000	818
(2001, 789)	2001	789
(2002, 827)	2002	827

   a. Write each paired data as an ordered pair of the form (year, mortgage payment).

   b. Draw a grid such as the one in Example 2 and create a scatter diagram of the paired data.

   c. What trend in the paired data does the scatter diagram show?   **19. b.** See graphing answer section.

19. c. Average monthly mortgage payment increases each year.

20. The table shows the enrollment in public schools in the United States for the years shown. (*Source*: U.S. Department of Education)

20. a.	Year	Enrollment in Public Schools (in millions)
(1950, 25)	1950	25
(1960, 35)	1960	35
(1970, 46)	1970	46
(1980, 42)	1980	42
(1990, 41)	1990	41
(2000, 47)	2000	47
(2010, 47)	2010*	47

*projected

20. b. See graphing answer section.

   a. Write each paired data as an ordered pair of the form (year, number of institutions).

   b. Draw a grid such as the one in Example 2 and create a scatter diagram of the paired data.

   c. What trend in the paired data does the scatter diagram show?   answers may vary

21. The table shows the distance from the equator (in miles) and the average annual snowfall (in inches) for each of eight selected U.S. cities. (*Sources*: National Climatic Data Center, Wake Forest University Albatross Project)

21. a. (2313, 2), (2085, 1), (2711, 21), (2869, 39), (2920, 42), (4038, 99), (1783, 0), (2493, 9)

City	Distance from Equator (in miles)	Average Annual Snowfall (in inches)
1. Atlanta, GA	2313	2
2. Austin, TX	2085	1
3. Baltimore, MD	2711	21
4. Chicago, IL	2869	39
5. Detroit, MI	2920	42
6. Juneau, AK	4038	99
7. Miami, FL	1783	0
8. Winston-Salem, NC	2493	9

   a. Write this paired data as a set of ordered pairs of the form (distance from equator, average annual snowfall).

b. Create a scatter diagram of the paired data. Be sure to label the axes appropriately.

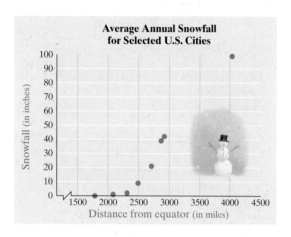

Average Annual Snowfall for Selected U.S. Cities

c. What trend in the paired data does the scatter diagram show?   The farther from the equator, the more snowfall.

22. The table shows the average farm size (in acres) in the United States during the years shown. (*Source*: National Agricultural Statistics Service)

	Year	*Average Farm Size (in acres)*
22. a.	1995	438
(1995, 438),	1996	438
(1996, 438),	1997	436
(1997, 436),	1998	435
(1998, 435),	1999	432
(1999, 432),	2000	434
(2000, 434),	2001	436
(2001, 436)		

a. Write this paired data as a set of ordered pairs of the form (year, average farm size).

b. Create a scatter diagram of the paired data. Be sure to label the axes appropriately.

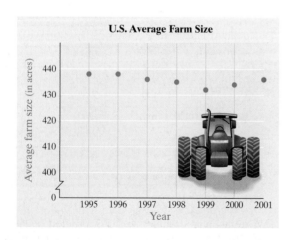

U.S. Average Farm Size

Determine whether each ordered pair is a solution of the given linear equation. See Example 3.

23. $2x + y = 7$; $(3, 1), (7, 0), (0, 7)$ yes; no; yes

24. $x - y = 6$; $(5, -1), (7, 1), (0, -6)$ yes; yes; yes

25. $y = -5x$; $(-1, -5), (0, 0), (2, -10)$ no; yes; yes

26. $x = 2y$; $(0, 0), (2, 1), (-2, -1)$ yes; yes; yes

27. $x = 5$; $(4, 5), (5, 4), (5, 0)$ no; yes; yes

28. $y = -2$; $(-2, 2), (2, -2), (0, -2)$ no; yes; yes

29. $x + 2y = 9$; $(5, 2), (0, 9)$ yes; no

30. $3x + y = 8$; $(2, 3), (0, 8)$ no; yes

31. $2x - y = 11$; $(3, -4), (9, 8)$ no; no

32. $x - 4y = 14$; $(2, -3), (14, 6)$ yes; no

33. $x = \frac{1}{3}y$; $(0, 0), (3, 9)$ yes; yes

34. $y = -\frac{1}{2}x$; $(0, 0), (4, 2)$ yes; no

39. $(11, -7)(11, -7)$; answers may vary, Ex. $(2, -7)$

Complete each ordered pair so that it is a solution of the given linear equation. See Examples 4 through 6.

35. $x - 4y = 4$; $(\;\;, -2), (4, \;\;)$ $(-4, -2), (4, 0)$

36. $x - 5y = -1$; $(\;\;, -2), (4, \;\;)$ $(-11, -2), (4, 1)$

37. $3x + y = 9$; $(0, \;\;), (\;\;, 0)$ $(0, 9), (3, 0)$

38. $x + 5y = 15$; $(0, \;\;), (\;\;, 0)$ $(0, 3), (15, 0)$

39. $y = -7$; $(11, \;\;), (\;\;, -7)$

40. $x = \frac{1}{2}$; $(\;\;, 0), \left(\frac{1}{2}, \;\;\right)$ $(\frac{1}{2}, 0)$; answers may vary, Ex. $(\frac{1}{2}, 4)$

41. $-2x + 7y = -3$ $(\;\;, 1) (1, \;\;)$ $(5, 1)(1, -\frac{1}{7})$

42. $-3x + 4y = -1$ $(\;\;, 2) (-2, \;\;)$ $(3, 2)(-2, -\frac{7}{4})$

Complete the table of values for each given equation; then plot each solution. Use a single coordinate system for each equation. See Examples 4 through 6. **43.–48.** See graphing answer section.

43. $x + 3y = 6$

x	y
0	2
6	0

44. $2x + y = 4$

x	y
0	4
2	0

45. $2x - y = 12$

x	y
0	-12
5	-2
-3	-18

46. $-5x + y = 10$

x	y
-2	0
-1	5
2	20

47. $2x + 7y = 5$

x	y
0	$\frac{5}{7}$
$\frac{5}{2}$	0
-1	1

48. $x - 6y = 3$

x	y
0	$-\frac{1}{2}$
1	$-\frac{1}{3}$
-3	-1

49.–52. See graphing answer section.

49. $x = 3$

x	y
3	0
3	−0.5
3	$\frac{1}{4}$

50. $y = -1$

x	y
−2	−1
0	−1
−1	−1

51. $x = -5y$

x	y
0	0
−5	1
10	−2

52. $y = -3x$

x	y
0	0
−2	6

53. Discuss any similarities in the graphs of the ordered pair solutions for Exercises 43–52.   answers may vary

54. Explain why equations in two variables have more than one solution.   answers may vary

Solve. See Example 7.

55. The cost in dollars y of producing x computer desks is given by $y = 80x + 5000$.

a. Complete the following table and graph the results.
See graphing answer section.

x	100	200	300
y	13,000	21,000	29,000

b. Find the number of computer desks that can be produced for $8600. (*Hint:* Find x when $y = 8600$.)   45 desks

56. The hourly wage y of an employee at a certain production company is given by $y = 0.25x + 9$ where x is the number of units produced in an hour.

a. Complete the table and graph the results.
See graphing answer section.

x	0	1	5	10
y	9	9.25	10.25	11.50

b. Find the number of units that must be produced each hour to earn an hourly wage of $12.25. (*Hint:* Find x when $y = 12.25$.)   13 units

57. The percent y of recorded music sales that were in cassette format from 1995 through 2000 is given by $y = -3.95x + 24.93$. In the equation, x represents the number of years after 1995. (*Source*: Recording Industry Association of America)

a. Complete the table.

x	1	3	5
y	20.98	13.08	5.18

b. Find the year in which approximately 17% of recorded music sales were cassettes. (*Hint:* Find x when $y = 17$ and round to the nearest whole number.)   1997

58. The amount y of land operated by farms in the United States (in million acres) from 1990 through 2000 is given by $y = -4.22x + 985.02$. In the equation, x represents the number of years after 1990. (*Source*: National Agricultural Statistics Service)

a. Complete the table.

x	4	7	10
y	968.14	955.48	942.82

b. Find the year in which there were approximately 947 million acres of land operated by farms. (*Hint:* Find x when $y = 947$ and round to the nearest whole number.)   1999

The graph below shows the number of Target stores for each year. Use this graph to answer Exercises 59–62.

59. In 1995, there were 670 Target stores.

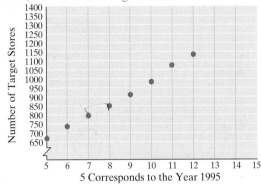

Source: *USA Today* 1/14/99

60. In 2000, there are 984 Target stores.

59. The ordered pair $(5, 670)$ is a point of the graph. Write a sentence describing the meaning of this ordered pair.

60. The ordered pair $(10, 984)$ is a point of the graph. Write a sentence describing the meaning of this ordered pair.

61. Estimate the increase in Target stores for years 6, 7, and 8.

62. Use a straightedge or ruler and this graph to predict the number of Target stores in the year 2009.   approx. 1310 stores

63. When is the graph of the ordered pair (a, b) the same as the graph of the ordered pair (b, a)? $a = b$

64. In your own words, describe how to plot an ordered pair.
answers may vary

REVIEW AND PREVIEW

61. year 6: 66 stores; year 7: 60 stores; year 8: 55 stores
Solve each equation for y. See Section 2.6.   **68.** $y = -\frac{5}{2}x + \frac{7}{2}$

65. $x + y = 5$ $y = 5 - x$

66. $x - y = 3$ $y = x - 3$

67. $2x + 4y = 5$ $y = -\frac{1}{2}x + \frac{5}{4}$

68. $5x + 2y = 7$

69. $10x = -5y$ $y = -2x$

70. $4y = -8x$ $y = -2x$

71. $x - 3y = 6$ $y = \frac{1}{3}x - 2$

72. $2x - 9y = -20$

Concept Extensions   **72.** $y = \frac{2}{9}x + \frac{20}{9}$

Determine the quadrant or quadrants in which the points described below lie.

73. The first coordinate is positive and the second coordinate is negative.   quadrant IV

74. Both coordinates are negative.   quadrant III

75. The first coordinate is negative.   quadrants II or III

76. The second coordinate is positive.   quadrants I or II

77. Three vertices of a rectangle are $(-2, -3)$, $(-7, -3)$, and $(-7, 6)$.   **77. a.** $(-2, 6)$

    a. Find the coordinates of the fourth vertex of a rectangle.

    b. Find the perimeter of the rectangle.   28 units

    c. Find the area of the rectangle.   45 sq. units

78. Three vertices of a square are $(-4, -1)$, $(-4, 8)$, and $(5, 8)$.

    a. Find the coordinates of the fourth vertex of the square.

    b. Find the perimeter of the square.   36 units

    c. Find the area of the square.   81 sq. units

78. a. $(5, -1)$

3.2   GRAPHING LINEAR EQUATIONS

Objectives

1   Identify linear equations.

2   Graph a linear equation by finding and plotting ordered pair solutions.

1   In the previous section, we found that equations in two variables may have more than one solution. For example, both $(6, 0)$ and $(2, -2)$ are solutions of the equation $x - 2y = 6$. In fact, this equation has an infinite number of solutions. Other solutions include $(0, -3)$, $(4, -1)$, and $(-2, -4)$. If we graph these solutions, notice that a pattern appears.

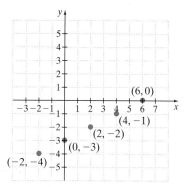

These solutions all appear to lie on the same line, which has been filled in below. It can be shown that every ordered pair solution of the equation corresponds to a point on this line, and every point on this line corresponds to an ordered pair solution. Thus, we say that this line is the graph of the equation $x - 2y = 6$.

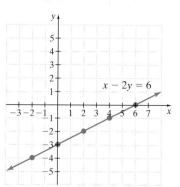

The equation $x - 2y = 6$ is called a **linear equation in two variables** and **the graph of every linear equation in two variables is a line.**

Linear Equation in Two Variables

A linear equation in two variables is an equation that can be written in the form

$$Ax + By = C$$

where A, B, and C are real numbers and A and B are not both 0. **The graph of a linear equation in two variables is a straight line.**

TEACHING TIP

For contrast, you may want to give some examples of equations in two variables that are not linear. For instance,

$$x^2 = 6y + 4$$
$$y = 3^x$$
$$y + 9 = \sqrt{x}$$
$$3x - \frac{1}{y} = 5$$

The form $Ax + By = C$ is called **standard form**.

> **Helpful Hint**
> Notice in the form $Ax + By = C$, the understood exponent on both x and y is 1.

Examples of Linear Equations in Two Variables

$$2x + y = 8 \qquad -2x = 7y \qquad y = \frac{1}{3}x + 2 \qquad y = 7$$

Before we graph linear equations in two variables, let's practice identifying these equations.

(EXAMPLE 1)

Identify the linear equations in two variables.

a. $x - 1.5y = -1.6$ **b.** $y = -2x$ **c.** $x + y^2 = 9$ **d.** $x = 5$

Solution

a. This is a linear equation in two variables because it is written in the form $Ax + By = C$ with $A = 1$, $B = -1.5$, and $C = -1.6$.

b. This is a linear equation in two variables because it can be written in the form $Ax + By = C$.

$$y = -2x$$
$$2x + y = 0 \qquad \text{Add } 2x \text{ to both sides.}$$

c. This is *not* a linear equation in two variables because y is squared.

d. This is a linear equation in two variables because it can be written in the form $Ax + By = C$.

$$x = 5$$
$$x + 0y = 5 \qquad \text{Add } 0 \cdot y.$$

---

2   From geometry, we know that a straight line is determined by just two points. Graphing a linear equation in two variables, then, requires that we find just two of its infinitely many solutions. Once we do so, we plot the solution points and draw the line connecting the points. Usually, we find a third solution as well, as a check.

EXAMPLE 2

Graph the linear equation $2x + y = 5$.

Solution Find three ordered pair solutions of $2x + y = 5$. To do this, choose a value for one variable, x or y, and solve for the other variable. For example, let $x = 1$. Then $2x + y = 5$ becomes

$$2x + y = 5$$
$$2(1) + y = 5 \qquad \text{Replace } x \text{ with } \mathbf{1}.$$
$$2 + y = 5 \qquad \text{Multiply.}$$
$$y = 3 \qquad \text{Subtract } \mathbf{2} \text{ from both sides.}$$

Since $y = 3$ when $x = 1$, the ordered pair $(1, 3)$ is a solution of $2x + y = 5$. Next, let $x = 0$.

$$2x + y = 5$$
$$2(0) + y = 5 \qquad \text{Replace } x \text{ with } \mathbf{0}.$$
$$0 + y = 5$$
$$y = 5$$

The ordered pair $(0, 5)$ is a second solution.

The two solutions found so far allow us to draw the straight line that is the graph of all solutions of $2x + y = 5$. However, we find a third ordered pair as a check. Let $y = -1$.

$$2x + y = 5$$
$$2x + (-1) = 5 \qquad \text{Replace } y \text{ with } -\mathbf{1}.$$
$$2x - 1 = 5$$
$$2x = 6 \qquad \text{Add } \mathbf{1} \text{ to both sides.}$$
$$x = 3 \qquad \text{Divide both sides by } \mathbf{2}.$$

The third solution is $(3, -1)$. These three ordered pair solutions are listed in table form as shown. The graph of $2x + y = 5$ is the line through the three points.

x	y
1	3
0	5
3	−1

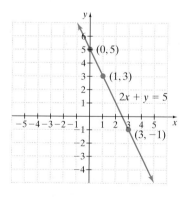

TEACHING TIP

Point out the arrowheads on the graph in Example 2. See if the students understand their meaning. Remind them that when they are graphing an equation, they are illustrating all the solutions of the equation.

> ▶ **Helpful Hint**
> All three points should fall on the same straight line. If not, check your ordered pair solutions for a mistake.

EXAMPLE 3

Graph the linear equation $-5x + 3y = 15$.

Solution   Find three ordered pair solutions of $-5x + 3y = 15$.

Let $x = 0$.	**Let $y = 0$.**	**Let $x = -2$.**
$-5x + 3y = 15$	$-5x + 3y = 15$	$-5x + 3y = 15$
$-5 \cdot 0 + 3y = 15$	$-5x + 3 \cdot 0 = 15$	$-5(-2) + 3y = 15$
$0 + 3y = 15$	$-5x + 0 = 15$	$10 + 3y = 15$
$3y = 15$	$-5x = 15$	$3y = 5$
$y = 5$	$x = -3$	$y = \frac{5}{3}$

The ordered pairs are $(0, 5)$, $(-3, 0)$, and $(-2, \frac{5}{3})$. The graph of $-5x + 3y = 15$ is the line through the three points.

x	y
0	5
−3	0
−2	$\frac{5}{3} = 1\frac{2}{3}$

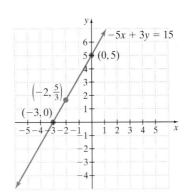

EXAMPLE 4

Graph the linear equation $y = 3x$.

Solution   To graph this linear equation, we find three ordered pair solutions. Since this equation is solved for y, choose three x values.

If $x = 2$, $y = 3 \cdot 2 = 6$.
If $x = 0$, $y = 3 \cdot 0 = 0$.
If $x = -1$, $y = 3 \cdot -1 = -3$.

x	y
2	6
0	0
−1	−3

Next, graph the ordered pair solutions listed in the table above and draw a line through the plotted points as shown on the next page. The line is the graph of $y = 3x$. Every point on the graph represents an ordered pair solution of the equation and every ordered pair solution is a point on this line.

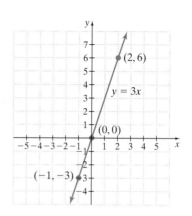

EXAMPLE 5

Graph the linear equation $y = -\frac{1}{3}x$.

Solution

CLASSROOM EXAMPLE
Graph: $y = -\frac{1}{2}x$
answer:

Find three ordered pair solutions, graph the solutions, and draw a line through the plotted solutions. To avoid fractions, choose x values that are multiples of 3 to substitute in the equation. When a multiple of 3 is multiplied by $-\frac{1}{3}$, the result is an integer. See the calculations shown to the right of the table below.

If $x = 6$, then $y = -\frac{1}{3} \cdot 6 = -2$.

If $x = 0$, then $y = -\frac{1}{3} \cdot 0 = 0$.

If $x = -3$, then $y = -\frac{1}{3} \cdot -3 = 1$.

x	y
6	-2
0	0
-3	1

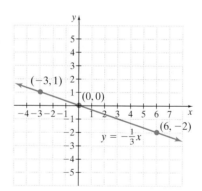

Let's compare the graphs in Examples 4 and 5. The graph of $y = 3x$ tilts upward (as we follow the line from left to right) and the graph of $y = -\frac{1}{3}x$ tilts downward (as we follow the line from left to right). Also notice that both lines go through the origin or that $(0, 0)$ is an ordered pair solution of both equations. In general, the graphs of $y = 3x$ and $y = -\frac{1}{3}x$ are of the form $y = mx$ where m is a constant. The graph of an equation in this form always goes through the origin $(0, 0)$ because when x is 0, $y = mx$ becomes $y = m \cdot 0 = 0$.

EXAMPLE 6

Graph the linear equation $y = 3x + 6$ and compare this graph with the graph of $y = 3x$ in Example 4.

Solution   Find ordered pair solutions, graph the solutions, and draw a line through the plotted solutions. We choose x values and substitute in the equation $y = 3x + 6$.

If $x = -3$, then $y = 3(-3) + 6 = -3$.
If $x = 0$, then $y = 3(0) + 6 = 6$.
If $x = 1$, then $y = 3(1) + 6 = 9$.

x	y
-3	-3
0	6
1	9

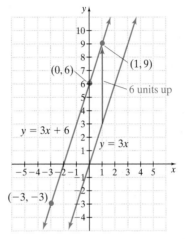

CLASSROOM EXAMPLE

Graph the linear equation $y = 2x + 3$ and compare this graph with the graph of $y = 2x$.

answer:

Same as the graph of $y = 2x$ except that the graph of $y = 2x + 3$ is moved 3 units upward.

The most startling similarity is that both graphs appear to have the same upward tilt as we move from left to right. Also, the graph of $y = 3x$ crosses the y-axis at the origin, while the graph of $y = 3x + 6$ crosses the y-axis at 6. In fact, the graph of $y = 3x + 6$ is the same as the graph of $y = 3x$ moved vertically upward 6 units.

Notice that the graph of $y = 3x + 6$ crosses the y-axis at 6. This happens because when $x = 0$, $y = 3x + 6$ becomes $y = 3 \cdot 0 + 6 = 6$. The graph contains the point $(0, 6)$, which is on the y-axis.

In general, if a linear equation in two variables is solved for y, we say that it is written in the form $y = mx + b$. The graph of this equation contains the point $(0, b)$ because when $x = 0$, $y = mx + b$ is $y = m \cdot 0 + b = b$.

> The graph of $y = mx + b$ crosses the y-axis at $(0, b)$.

We will review this again in Section 3.5.

Linear equations are often used to model real data as seen in the next example.

EXAMPLE 7

ESTIMATING THE NUMBER OF MEDICAL ASSISTANTS

One of the occupations expected to have the most growth in the next few years is medical assistant. The number of people y (in thousands) employed as medical assistants in the United States can be estimated by the linear equation $y = 31.8x + 180$, where x is the number of years after the year 1995. (*Source:* based on data from the Bureau of Labor Statistics)

Graph the equation and use the graph to predict the number of medical assistants in the year 2010.

Solution To graph $y = 31.8x + 180$, choose x-values and substitute in the equation.

If $x = 0$, then $y = 31.8(0) + 180 = 180$.

If $x = 2$, then $y = 31.8(2) + 180 = 243.6$.

If $x = 7$, then $y = 31.8(7) + 180 = 402.6$.

x	y
0	180
2	243.6
7	402.6

CLASSROOM EXAMPLE

Use the graph in Example 7 to predict the number of medical assistants in 2004.

answer: 465 thousand

(Graph: *Number of Medical Assistants (in thousands)* on the vertical axis vs. *Years after 1995* on the horizontal axis.)

To use the graph to *predict* the number of medical assistants in the year 2010, we need to find the y-coordinate that corresponds to $x = 15$. (15 years after 1995 is the year 2010.) To do so, find 15 on the x-axis. Move vertically upward to the graphed line and then horizontally to the left. We approximate the number on the y-axis to be 655. Thus in the year 2010, we predict that there will be 655 thousand medical assistants. (The actual value, using 15 for x, is 657.)

Graphing Calculator Explorations

TEACHING TIP

Point out that for some graphing calculators, coefficients that are fractions must be entered with parentheses. For instance, $y = \frac{1}{4}x - 2$ may need to be entered as $y = (1/4)x - 2$.

In this section, we begin an optional study of graphing calculators and graphing software packages for computers. These graphers use the same point plotting technique that was introduced in this section. The advantage of this graphing technology is, of course, that graphing calculators and computers can find and plot ordered pair solutions much faster than we can. Note, however, that the features described in these boxes may not be available on all graphing calculators.

The rectangular screen where a portion of the rectangular coordinate system is displayed is called a **window**. We call it a **standard window** for graphing when both the x- and y-axes show coordinates between -10 and 10. This information is often displayed in the window menu on a graphing calculator as

Xmin $= -10$
Xmax $= 10$
Xscl $= 1$ The scale on the x-axis is one unit per tick mark.
Ymin $= -10$
Ymax $= 10$
Yscl $= 1$ The scale on the y-axis is one unit per tick mark.

To use a graphing calculator to graph the equation $y = 2x + 3$, press the $\boxed{Y=}$ key and enter the keystrokes $\boxed{2}$ $\boxed{x}$ $\boxed{+}$ $\boxed{3}$. The top row should now read

$Y_1 = 2x + 3$. Next press the [**GRAPH**] key, and the display should look like this:

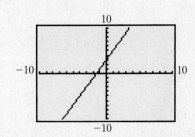

Use a standard window and graph the following linear equations. (Unless otherwise stated, use a standard window when graphing.)   See graphing answer section.

1. $y = -3x + 7$ **2.** $y = -x + 5$

3. $y = 2.5x - 7.9$ **4.** $y = -1.3x + 5.2$

5. $y = -\dfrac{3}{10}x + \dfrac{32}{5}$ **6.** $y = \dfrac{2}{9}x - \dfrac{22}{3}$

EXERCISE SET 3.2

STUDY GUIDE/SSM   CD/VIDEO   PH MATH TUTOR CENTER   MathXL®Tutorials ON CD   MathXL®   MyMathLab®

Determine whether each equation is a linear equation in two variables. See Example 1.

1. $-x = 3y + 10$ yes   **2.** $y = x - 15$ yes

3. $x = y$ yes   **4.** $x = y^3$ no

5. $x^2 + 2y = 0$ no   **6.** $0.01x - 0.2y = 8.8$ yes

7. $y = -1$ yes   **8.** $x = 25$ yes

MIXED PRACTICE   9.–32. See graphing answer section.

Graph each linear equation. See Examples 2 through 5.

9. $x + y = 4$ **10.** $x + y = 7$

11. $x - y = -2$ **12.** $-x + y = 6$

13. $x - 2y = 6$ **14.** $-x + 5y = 5$

15. $y = 6x + 3$ **16.** $y = -2x + 7$

17. $x - 2y = -6$ **18.** $-x + 2y = 5$

19. $y = 6x$ **20.** $x = -2y$

21. $3y - 10 = 5x$ **22.** $-2x + 7 = 2y$

23. $x + 3y = 9$ **24.** $2x + y = -2$

25. $y - x = -1$ **26.** $x - y = 5$

27. $x = -3y$ **28.** $y = -x$

29. $5x - y = 10$ **30.** $7x - y = 2$

31. $y = \dfrac{1}{2}x + 2$ **32.** $y = -\dfrac{1}{5}x - 1$

Graph each pair of linear equations on the same set of axes. Discuss how the graphs are similar and how they are different. See Example 6.   33.–36. See graphing answer section.

33. $y = 5x; y = 5x + 4$ **34.** $y = 2x; y = 2x + 5$

35. $y = -2x; y = -2x - 3$ **36.** $y = x; y = x - 7$

37. $y = \dfrac{1}{2}x; y = \dfrac{1}{2}x + 2$ **38.** $y = -\dfrac{1}{4}x; y = -\dfrac{1}{4}x + 3$

The graph of $y = 5x$ is given below as well as Figures a–d. For Exercises 39 through 42, match each equation with its graph. Hint: Recall that if an equation is written in the form $y = mx + b$, its graph crosses the y-axis at b.

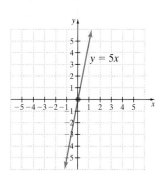

a.

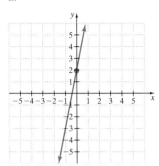

b.

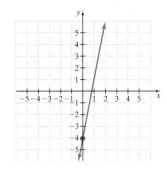

c.

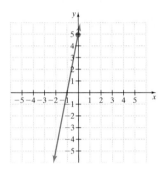

d.

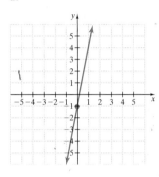

39. $y = 5x + 5$ c

40. $y = 5x - 4$ b

41. $y = 5x - 1$ d

42. $y = 5x + 2$ a

Solve. See Example 7.

43. Skateboarding has been growing in popularity in recent years, and the sales of skateboarding apparel and hardware have been increasing at a steady rate. The sales of skateboard apparel and hardware (in millions of dollars) from the years 1996 to 2001 are given by the equation $y = 180x + 450$ where x is the number of years after 1996. If this trend continues, what will the total sales be in 2009? (*Source*: Transworld SKATEboarding Magazine) $2790 million

44. The revenue y (in billions of dollars) for Home Depot stores during the years 1998 through 2001 is given by the equation $y = 8x + 30$, where x is the number of years after 1998. Graph this equation and use the graph to predict the revenue

for Home Depot Stores in the year 2010. (*Source*: Based on data from The Home Depot Inc.)   126 billion dollars

45. The average weekly earnings (in dollars) by U.S. production workers can be approximated by the equation $y = 16x + 144$ from 1998 to 2001 where x is the number of years after 1998. Graph this equation and use it to predict the average weekly earnings in 2007. (*Source*: U.S. Bureau of Labor Statistics) $288 per week

46. The U.S. silver production (in metric tons) from 1998 to 2001 has been steadily dropping, and can be approximated by the equation $y = -105x + 2060$ where x is the number of years after 1998. If this current trend continues, use the equation to estimate the U.S. silver production in 2008. (*Source*: U.S. Geological Survey)   1010 metric tons

REVIEW AND PREVIEW 48. $(-3, 4), (2, 4); (-3, -6), (2, -6)$

△ **47.** The coordinates of three vertices of a rectangle are $(-2, 5), (4, 5),$ and $(-2, -1)$. Find the coordinates of the fourth vertex. See Section 3.1. $(4, -1)$

△ **48.** The coordinates of two vertices of a square are $(-3, -1)$ and $(2, -1)$. Find the coordinates of two pairs of points possible for the third and fourth vertices. See Section 3.1.

Solve the following equations. See Section 2.4.   **49.** -5 **50.** 0

49. $3(x - 2) + 5x = 6x - 16$ **50.** $5 + 7(x + 1) = 12 + 10x$

51. $3x + \dfrac{2}{5} = \dfrac{1}{10} - \dfrac{1}{10}$ **52.** $\dfrac{1}{6} + 2x = \dfrac{2}{3} \quad \dfrac{1}{4}$

Complete each table. See Section 3.1.

53. $x - y = -3$

x	y
0	3
−3	0

54. $y - x = 5$

x	y
0	5
−5	0

55. $y = 2x$

x	y
0	0
0	0

56. $x = -3y$

x	y
0	0
0	0

Concept Extensions

Write each statement as an equation in two variables. Then graph the equation.   **57.–58.** See graphing answer section.

57. The y-value is 5 more than the x-value. $y = x + 5$

58. The y-value is twice the x-value. $y = 2x$

59.–60. See graphing answer section.

59. Two times the *x*-value, added to three times the *y*-value is 6.

60. Five times the *x*-value, added to twice the *y*-value is −10.

61. Explain how to find ordered pair solutions of linear equations in two variables.   answers may vary

△ **62.** The perimeter of the trapezoid below is 22 centimeters. Write a linear equation in two variables for the perimeter. Find *y* if *x* is 3 cm. $x + y = 12; y = 9$ cm

59. $2x + 3y = 6$ **60.** $5x + 2y = -10$

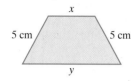

5 cm   5 cm

63. If (a, b) is an ordered pair solution of $x + y = 5$, is (b, a) also a solution? Explain why or why not.   answers may vary

△ **64.** The perimeter of the rectangle below is 50 miles. Write a linear equation in two variables for this perimeter. Use this equation to find *x* when *y* is 20. $x + y = 25; x = 5$

65. Graph the nonlinear equation $y = x^2$ by completing the table shown. Plot the ordered pairs and connect them with a smooth curve.   See graphing answer section.

x	*y*
0	0
1	1
−1	1
2	4
−2	4

66. Graph the nonlinear equation $y = |x|$ by completing the table shown. Plot the ordered pairs and connect them. This curve is "V" shaped.   See graphing answer section.

$y = |x|$

x	*y*
0	0
1	1
−1	1
2	2
−2	2

3.3   INTERCEPTS

Objectives

1 Identify intercepts of a graph.

2 Graph a linear equation by finding and plotting intercepts.

3 Identify and graph vertical and horizontal lines.

1   In this section, we graph linear equations in two variables by identifying intercepts. For example, the graph of $y = 4x - 8$ is shown below. Notice that this graph crosses the *y*-axis at the point $(0, -8)$. This point is called the **y-intercept**. Likewise, the graph crosses the *x*-axis at $(2, 0)$, and this point is called the **x-intercept**.

TEACHING TIP

Remind students that all points on the *x*-axis have a *y*-value of 0, and all points on the *y*-axis have an *x*-value of 0. Also, the point $(0, 0)$ lies on both axes.

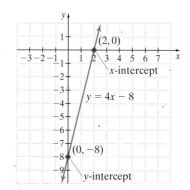

> **Helpful Hint**
>
> If a graph crosses the x-axis at $(-3, 0)$ and the y-axis at $(0, 7)$, then
>
> $$\underset{\underset{\text{x-intercept}}{\uparrow}}{(-3, 0)} \qquad \underset{\underset{\text{y-intercept}}{\uparrow}}{(0, 7)}$$
>
> Notice that for the y-intercept, the x-value is 0 and for the x-intercept, the y-value is 0.
>
> Note: Sometimes in mathematics, you may see just the number 7 stated as the y-intercept, and -3 stated as the x-intercept.

EXAMPLE 1

Identify the x- and y-intercepts.

a.

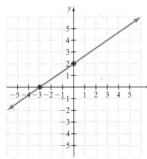

b.

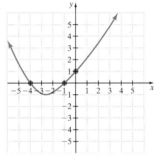

TEACHING TIP

Some questions you may want to ask your students:

▲ Can the same point ever be the x-intercept point and the y-intercept point of a graph?

▲ Excluding the line $y = 0$, what is the maximum number of x-intercepts a line will have?

▲ Will a line always have at least one y-intercept?

c.

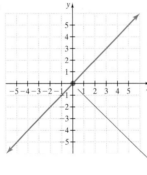

d.

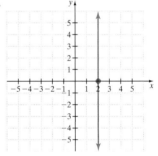

e.

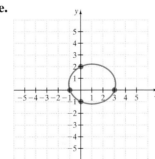

> **Helpful Hint**
>
> Notice that any time $(0, 0)$ is a point of a graph, then it is an x-intercept and a y-intercept.

Solution   **a.** The graph crosses the x-axis at -3, so the x-intercept is $(-3, 0)$. The graph crosses the y-axis at 2, so the y-intercept is $(0, 2)$.

b. The graph crosses the x-axis at -4 and -1, so the x-intercepts are $(-4, 0)$ and $(-1, 0)$. The graph crosses the y-axis at 1, so the y-intercept is $(0, 1)$.

c. The x-intercept and the y-intercept are both $(0, 0)$.

d. The x-intercept is $(2, 0)$. There is no y-intercept.

e. The x-intercepts are $(-1, 0)$ and $(3, 0)$. The y-intercepts are $(0, -1)$, and $(0, 2)$.

2   Given the equation of a line, intercepts are usually easy to find since one coordinate is 0.

One way to find the y-intercept of a line, given its equation, is to let $x = 0$, since a point on the y-axis has an x-coordinate of 0. To find the x-intercept of a line, let $y = 0$, since a point on the x-axis has a y-coordinate of 0.

Finding x- and y-intercepts

To find the x-intercept, let $y = 0$ and solve for x.

To find the y-intercept, let $x = 0$ and solve for y.

EXAMPLE 2

Graph $x - 3y = 6$ by finding and plotting intercepts.

Solution   Let $y = 0$ to find the x-intercept and let $x = 0$ to find the y-intercept.

$$
\begin{array}{ll}
\text{Let } y = 0 & \text{Let } x = 0 \\
x - 3y = 6 & x - 3y = 6 \\
x - 3(0) = 6 & 0 - 3y = 6 \\
x - 0 = 6 & -3y = 6 \\
x = 6 & y = -2
\end{array}
$$

The x-intercept is $(6, 0)$ and the y-intercept is $(0, -2)$. We find a third ordered pair solution to check our work. If we let $y = -1$, then $x = 3$. Plot the points $(6, 0)$, $(0, -2)$, and $(3, -1)$. The graph of $x - 3y = 6$ is the line drawn through these points, as shown.

x	y
6	0
0	-2
3	-1

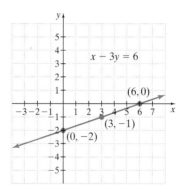

EXAMPLE 3

Graph $x = -2y$ by plotting intercepts.

Solution   Let $y = 0$ to find the x-intercept and $x = 0$ to find the y-intercept.

$$
\begin{array}{ll}
\text{Let } y = 0 & \text{Let } x = 0 \\
x = -2y & x = -2y \\
x = -2(0) & 0 = -2y \\
x = 0 & 0 = y
\end{array}
$$

Both the x-intercept and y-intercept are $(0, 0)$. In other words, when $x = 0$, then $y = 0$, which gives the ordered pair $(0, 0)$. Also, when $y = 0$, then $x = 0$, which gives the same ordered pair $(0, 0)$. This happens when the graph passes through the origin. Since two points are needed to determine a line, we must find at least one more ordered pair that satisfies $x = -2y$. Let $y = -1$ to find a second ordered pair solution and let $y = 1$ as a checkpoint.

Let $y = -1$	Let $y = 1$
$x = -2(-1)$	$x = -2(1)$
$x = 2$	$x = -2$

The ordered pairs are $(0, 0)$, $(2, -1)$, and $(-2, 1)$. Plot these points to graph $x = -2y$.

x	y
0	0
2	-1
-2	1

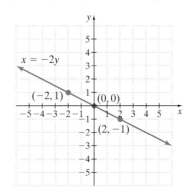

EXAMPLE 4

Graph $4x = 3y - 9$.

Solution Find the x- and y-intercepts, and then choose $x = 2$ to find a third checkpoint.

Let $y = 0$	Let $x = 0$	Let $x = 2$
$4x = 3(0) - 9$	$4 \cdot 0 = 3y - 9$	$4(2) = 3y - 9$
$4x = -9$	$9 = 3y$	$8 = 3y - 9$
Solve for x.	Solve for y.	Solve for y.
$x = -\dfrac{9}{4}$ or $-2\dfrac{1}{4}$	$3 = y$	$17 = 3y$
		$\dfrac{17}{3} = y$ or $y = 5\dfrac{2}{3}$

The ordered pairs are $\left(-2\frac{1}{4}, 0\right)$, $(0, 3)$, and $\left(2, 5\frac{2}{3}\right)$. The equation $4x = 3y - 9$ is graphed as follows.

x	y
$-2\frac{1}{4}$	0
0	3
2	$5\frac{2}{3}$

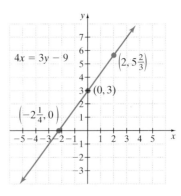

3 The equation $x = c$, where c is a real number constant, is a linear equation in two variables because it can be written in the form $x + 0y = c$. The graph of this equation is a vertical line as shown in the next example.

EXAMPLE 5

Graph $x = 2$.

Solution   The equation $x = 2$ can be written as $x + 0y = 2$. For any y-value chosen, notice that x is 2. No other value for x satisfies $x + 0y = 2$. Any ordered pair whose x-coordinate is 2 is a solution of $x + 0y = 2$. We will use the ordered pair solutions $(2, 3)$, $(2, 0)$, and $(2, -3)$ to graph $x = 2$.

CLASSROOM EXAMPLE
Graph $x = -3$.
answer:

x	y
2	3
2	0
2	−3

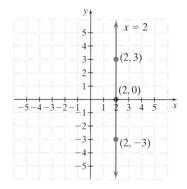

The graph is a vertical line with x-intercept $(2, 0)$. Note that this graph has no y-intercept because x is never 0.

> **Vertical Lines**
>
> The graph of $x = c$, where c is a real number, is a vertical line with x-intercept $(c, 0)$.
>
>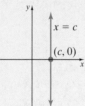

EXAMPLE 6

Graph $y = -3$.

Solution   The equation $y = -3$ can be written as $0x + y = -3$. For any x-value chosen, y is -3. If we choose 4, 1, and -2 as x-values, the ordered pair solutions are $(4, -3)$, $(1, -3)$, and $(-2, -3)$. Use these ordered pairs to graph $y = -3$. The graph is a horizontal line with y-intercept $(0, -3)$ and no x-intercept.

CLASSROOM EXAMPLE
Graph $y = 4$.
answer:

x	y
4	−3
1	−3
−2	−3

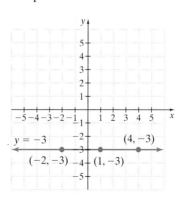

Horizontal Lines

The graph of $y = c$, where c is a real number, is a horizontal line with y-intercept $(0, c)$.

Graphing Calculator Explorations

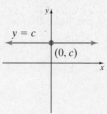

You may have noticed that to use the $\boxed{Y=}$ key on a grapher to graph an equation, the equation must be solved for y. For example, to graph $2x + 3y = 7$, we solve this equation for y.

$$2x + 3y = 7$$

$$3y = -2x + 7 \qquad \text{Subtract } 2x \text{ from both sides.}$$

$$\frac{3y}{3} = -\frac{2x}{3} + \frac{7}{3} \qquad \text{Divide both sides by 3.}$$

$$y = -\frac{2}{3}x + \frac{7}{3} \qquad \text{Simplify.}$$

To graph $2x + 3y = 7$ or $y = -\dfrac{2}{3}x + \dfrac{7}{3}$, press the $\boxed{Y=}$ key and enter

$$Y_1 = -\frac{2}{3}x + \frac{7}{3}$$

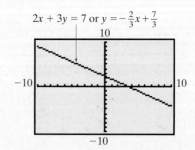

Graph each linear equation. See graphing answer section.

1. $x = 3.78y$ **2.** $-2.61y = x$

3. $3x + 7y = 21$ **4.** $-4x + 6y = 12$

5. $-2.2x + 6.8y = 15.5$ **6.** $5.9x - 0.8y = -10.4$

MENTAL MATH

Answer the following true or false.

1. The graph of $x = 2$ is a horizontal line.   false
2. All lines have an *x*-intercept *and* a *y*-intercept.   false
3. The graph of $y = 4x$ contains the point $(0, 0)$.   true
4. The graph of $x + y = 5$ has an *x*-intercept of $(5, 0)$ and a *y*-intercept of $(0, 5)$.   true
5. The graph of $y = 5x$ contains the point $(5, 1)$.   false
6. The graph of $y = 5$ is a horizontal line.   true

EXERCISE SET 3.3

STUDY GUIDE/SSM   CD/VIDEO   PH MATH TUTOR CENTER   MathXL®Tutorials ON CD   MathXL®   MyMathLab®

Identify the intercepts. See Example 1.

 1. $(-1, 0)$; $(0, 1)$

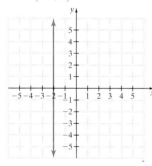

2. $(-2, 0)$; $(2, 0)$; $(0, -2)$

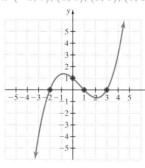

3. $(-2, 0)$

4. $(-2, 0)$; $(1, 0)$; $(3, 0)$; $(0, 1)$

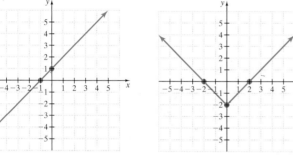

5. $(-1, 0)$; $(1, 0)$; $(0, 1)$; $(0, -2)$ **6.** $(0, 0)$; $(0, 2)$

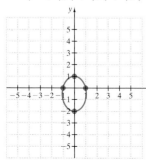 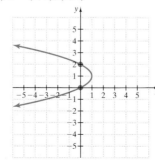

7. What is the greatest number of intercepts for a line?   infinite
8. What is the least number of intercepts for a line?   1
9. What is the least number of intercepts for a circle?   0
10. What is the greatest number of intercepts for a circle?   4

Graph each linear equation by finding x- and y-intercepts. See Examples 2 through 4.   **11.–50.**   See graphing answer section.

11. $x - y = 3$

12. $x - y = -4$

13. $x = 5y$

14. $2x = y$

15. $-x + 2y = 6$

16. $x - 2y = -8$

17. $2x - 4y = 8$

18. $2x + 3y = 6$

Graph each linear equation. See Examples 5 and 6.

19. $x = -1$

20. $y = 5$

21. $y = 0$

22. $x = 0$

23. $y + 7 = 0$

24. $x - 2 = 0$

MIXED PRACTICE

Graph each linear equation.

25. $x + 2y = 8$ **26.** $x - 3y = 3$

27. $x - 7 = 3y$ **28.** $y - 3x = 2$

29. $x = -3$ **30.** $y = 3$ **31.** $3x + 5y = 7$

32. $3x - 2y = 5$ **33.** $x = y$ **34.** $x = -y$

35. $x + 8y = 8$ **36.** $x - 3y = 9$ **37.** $5 = 6x - y$

38. $4 = x - 3y$ **39.** $-x + 10y = 11$

40. $-x + 9 = -y$ **41.** $y = 4.5$

42. $x = -\dfrac{1}{2}$ **43.** $y = \dfrac{1}{2}x$ **44.** $y = -2x$

45. $x + 4 = 0$ **46.** $y - 5.2 = 0$

47. $3x - 4y = -12$ **48.** $-2x + 5y = 10$

49. $2x + 3y = 6$ **50.** $4x + y = 5$

For Exercises 51 through 56, match each equation with its graph.

A.

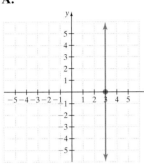

B.

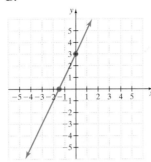

C.

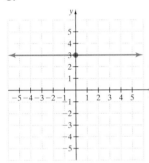

D.

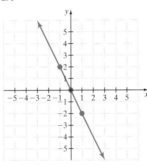

E.

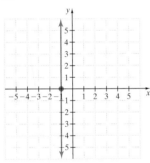

F.

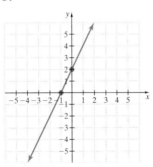

51. $y = 3$ C

52. $y = 2x + 2$ F

53. $x = -1$ E

54. $x = 3$ A

55. $y = 2x + 3$ B

56. $y = -2x$ D

REVIEW AND PREVIEW

Simplify.

57. $\dfrac{-6 - 3}{2 - 8}$ $\dfrac{3}{2}$

58. $\dfrac{4 - 5}{-1 - 0}$ 1

59. $\dfrac{-8 - (-2)}{-3 - (-2)}$ 6

60. $\dfrac{12 - 3}{10 - 9}$ 9

61. $\dfrac{0 - 6}{5 - 0}$ $-\dfrac{6}{5}$

62. $\dfrac{2 - 2}{3 - 5}$ 0

Concept Extensions

63. The average price of a digital camera (y) can be modeled by the linear equation $y = -78.1x + 569.9$ where x represents the number of years after 1999. (*Source*: NPD Techworld)

a. Find the y-intercept of this equation. $(0, 569.9)$

b. What does this y-intercept mean?

63. b. In 1999, the average price of a digital camera was $569.90.

64. Since the summer of 1999, the number of G- and PG-rated movies has been steadily increasing. The number of G and PG movies released each year can be estimated by the equation $y = 4x + 49$ where x represents the number of years after 1999. (*Source*: Nielsen EDI)

a. Find the x-intercept of this equation. $(-12.25, 0)$

b. What does this x-intercept mean?

c. Use part b to comment on the limitations of using equations modeling real data.   answers may vary

64. b. 12.25 years before 1999, there were no G or PG movies.

65. b. 7.1 years after 1995, no music cassettes will be shipped.

65. The number of music cassettes y (in millions) shipped to retailers in the United States from 1995 through 2000 can be modeled by the equation $y = -37.2x + 264.4$, where x represents the number of years after 1995. (*Source*: Recording Industry Association of America)

a. Find the x-intercept of this equation. (Round to the nearest tenth.) $(7.1, 0)$

b. What does this x-intercept mean?

c. Use part b to comment on the limitations of using equations to model real data.   answers may vary

66. The number of Disney Stores y for the years 1996–2000 can be modeled by the equation $y = 51.6x + 560.2$, where x represents the number of years after 1996. (*Source*: The Walt Disney Company Fact Book 2000*)

a. Find the y-intercept of this equation. $(0, 560.2)$

b. What does this y-intercept mean?

66. b. In 1996, the number of Disney Stores was about 560.2.

67. The production supervisor at Alexandra's Office Products finds that it takes 3 hours to manufacture a particular office chair and 6 hours to manufacture an office desk. A total of 1200 hours is available to produce office chairs and desks of this style. The linear equation that models this situation is $3x + 6y = 1200$, where x represents the number of chairs produced and y the number of desks manufactured.

   a. Complete the ordered pair solution $(0, \quad)$ of this equation. Describe the manufacturing situation that corresponds to this solution. $(0, 200)$; answers may vary

   b. Complete the ordered pair solution $(\quad, 0)$ of this equation. Describe the manufacturing situation that corresponds to this solution. $(400, 0)$; answers may vary

   c. Use the ordered pairs found above and graph the equation $3x + 6y = 1200$. See graphing answer section.

   d. If 50 desks are manufactured, find the greatest number of chairs that they can make. 300 chairs

Two lines in the same plane that do not intersect are called **parallel lines.**

△ **68.** Draw a line parallel to the line $x = 5$ that intersects the x-axis at $(1, 0)$. What is the equation of this line? $x = 1$

△ **69.** Draw a line parallel to the line $y = -1$ that intersects the y-axis at $(0, -4)$. What is the equation of this line? $y = -4$

╲ **70.** Discuss whether a vertical line ever has a y-intercept.

╲ **71.** Explain why it is a good idea to use three points to graph a linear equation.

╲ **72.** Discuss whether a horizontal line ever has an x-intercept.

╲ **73.** Explain how to find intercepts.

   70.–73. answers may vary

3.4   SLOPE AND RATE OF CHANGE

Objectives

1 Find the slope of a line given two points of the line.

2 Find the slopes of horizontal and vertical lines.

3 Compare the slopes of parallel and perpendicular lines.

4 Solve applications of slope.

1   Thus far, much of this chapter has been devoted to graphing lines. You have probably noticed by now that a key feature of a line is its slant or steepness. In mathematics, the slant or steepness of a line is formally known as its **slope**. We measure the slope of a line by the ratio of vertical change to the corresponding horizontal change as we move along the line.

   On the line below, for example, suppose that we begin at the point $(1, 2)$ and move to the point $(4, 6)$. The vertical change is the change in y-coordinates: $6 - 2$ or 4 units. The corresponding horizontal change is the change in x-coordinates: $4 - 1 = 3$ units. The ratio of these changes is

$$\text{slope} = \frac{\text{change in } y \text{ (vertical change)}}{\text{change in } x \text{ (horizontal change)}} = \frac{4}{3}$$

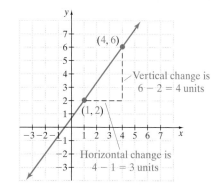

The slope of this line, then, is $\frac{4}{3}$: for every 4 units of change in y-coordinates, there is a corresponding change of 3 units in x-coordinates.

> ### Helpful Hint
>
> It makes no difference what two points of a line are chosen to find its slope. The slope of a line is the same everywhere on the line.

To find the slope of a line, then, choose two points of the line. Label the two x-coordinates of two points, x_1 and x_2 (read "x sub one" and "x sub two"), and label the corresponding y-coordinates y_1 and y_2.

The vertical change or **rise** between these points is the difference in the y-coordinates: $y_2 - y_1$. The horizontal change or **run** between the points is the difference of the x-coordinates: $x_2 - x_1$. The slope of the line is the ratio of $y_2 - y_1$ to $x_2 - x_1$, and we traditionally use the letter m to denote slope $m = \dfrac{y_2 - y_1}{x_2 - x_1}$.

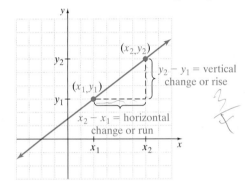

Slope of a Line

The slope m of the line containing the points (x_1, y_1) and (x_2, y_2) is given by

$$m = \frac{\text{rise}}{\text{run}} = \frac{\text{change in } y}{\text{change in } x} = \frac{y_2 - y_1}{x_2 - x_1}, \qquad \text{as long as } x_2 \neq x_1$$

EXAMPLE 1

Find the slope of the line through $(-1, 5)$ and $(2, -3)$. Graph the line.

Solution If we let (x_1, y_1) be $(-1, 5)$, then $x_1 = -1$ and $y_1 = 5$. Also, let (x_2, y_2) be $(2, -3)$ so that $x_2 = 2$ and $y_2 = -3$. Then, by the definition of slope,

$$m = \frac{y_2 - y_1}{x_2 - x_1}$$

$$= \frac{-3 - 5}{2 - (-1)}$$

$$= \frac{-8}{3} = -\frac{8}{3}$$

The slope of the line is $-\dfrac{8}{3}$.

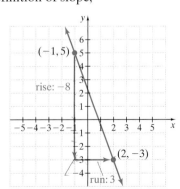

TEACHING TIP

For Example 1, have students place their finger on $(-1, 5)$ and make a vertical and horizontal move to $(2, -3)$. Then have them write their move as a ratio of the vertical move to the horizontal move using the following convention: upward moves are positive, downward moves are negative, rightward moves are positive, leftward moves are negative. Now have them put their finger on $(2, -3)$ and move to $(-1, 5)$ using a vertical and horizontal move. Have them write this move as a ratio. How did the ratios change? Are the ratios equal?

✔ **CONCEPT CHECK**

The points $(-2, -5)$, $(0, -2)$, $(4, 4)$, and $(10, 13)$ all lie on the same line. Work with a partner and verify that the slope is the same no matter which points are used to find slope.

> **Helpful Hint**
>
> When finding slope, it makes no difference which point is identified as (x_1, y_1) and which is identified as (x_2, y_2). Just remember that whatever y-value is first in the numerator, its corresponding x-value is first in the denominator. Another way to calculate the slope in Example 1 is:
>
> $$m = \frac{y_2 - y_1}{x_2 - x_1} = \frac{5 - (-3)}{-1 - 2} = \frac{8}{-3} \quad \text{or} \quad -\frac{8}{3} \leftarrow \text{Same slope as found in Example 1.}$$

EXAMPLE 2

Find the slope of the line through $(-1, -2)$ and $(2, 4)$. Graph the line.

Solution

CLASSROOM EXAMPLE

Find the slope of the line through $(-2, 1)$ and $(3, 5)$. Graph the line.

answer: $m = \frac{4}{5}$

y-value

$$m = \frac{-2 - 4}{-1 - 2} = \frac{-6}{-3} = 2$$

corresponding x-value

The slope is the same if we begin with the other y-value.

y-value

$$m = \frac{4 - (-2)}{2 - (-1)} = \frac{6}{3} = 2$$

corresponding x-value

The slope is 2.

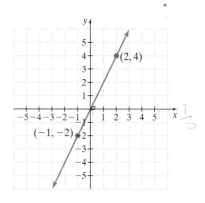

✔ **CONCEPT CHECK**

What is wrong with the following slope calculation for the points $(3, 5)$ and $(-2, 6)$?

$$m = \frac{5 - 6}{-2 - 3} = \frac{-1}{-5} = \frac{1}{5}$$

Concept Check Answer:

$$m = \frac{3}{2}$$

The order in which the x- and y-values are used must be the same.

$$m = \frac{5 - 6}{3 - (-2)} = \frac{-1}{5} = -\frac{1}{5}$$

Notice that the slope of the line in Example 1 is negative, whereas the slope of the line in Example 2 is positive. Let your eye follow the line with negative slope from left to right and notice that the line "goes down." Following the line with positive slope from left to right, notice that the line "goes up." This is true in general.

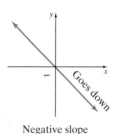

Negative slope

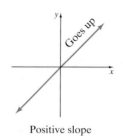

Positive slope

> **Helpful Hint**
>
> To decide whether a line "goes up" or "goes down", always follow the line from left to right.

2   If a line tilts upward from left to right, its slope is positive. If a line tilts downward from left to right, its slope is negative. Let's now find the slopes of two special lines, horizontal and vertical lines.

EXAMPLE 3

Find the slope of the line $y = -1$.

Solution   Recall that $y = -1$ is a horizontal line with y-intercept $(0, -1)$. To find the slope, find two ordered pair solutions of $y = -1$. Solutions of $y = -1$ must have a y-value of -1. Let's use points $(2, -1)$ and $(-3, -1)$, which are on the line.

$$m = \frac{y_2 - y_1}{x_2 - x_1} = \frac{-1 - (-1)}{-3 - 2} = \frac{0}{-5} = 0$$

The slope of the line $y = -1$ is 0. The graph of $y = -1$ is given below.

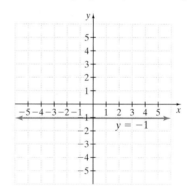

Any two points of a horizontal line will have the same y-values. This means that the y-values will always have a difference of 0 for all horizontal lines. Thus, **all horizontal lines have a slope 0**.

EXAMPLE 4

Find the slope of the line $x = 5$.

Solution   Recall that the graph of $x = 5$ is a vertical line with x-intercept $(5, 0)$.

To find the slope, find two ordered pair solutions of $x = 5$. Solutions of $x = 5$ must have an x-value of 5. Let's use points $(5, 0)$ and $(5, 4)$, which are on the line.

$$m = \frac{y_2 - y_1}{x_2 - x_1} = \frac{4 - 0}{5 - 5} = \frac{4}{0}$$

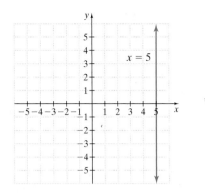

Since $\frac{4}{0}$ is undefined, we say the slope of the vertical line $x = 5$ is undefined.

Any two points of a vertical line will have the same x-values. This means that the x-values will always have a difference of 0 for all vertical lines. Thus **all vertical lines have undefined slope**.

> ### Helpful Hint
> Slope of 0 and undefined slope are not the same. Vertical lines have undefined slope or no slope, while horizontal lines have a slope of 0.

Here is a general review of slope.

Summary of Slope

Slope m of the line through (x_1, y_1) and (x_2, y_2) is given by the equation $m = \dfrac{y_2 - y_1}{x_2 - x_1}$.

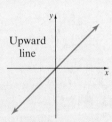

Upward line

Positive slope: $m > 0$

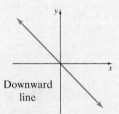

Downward line

Negative slope: $m < 0$

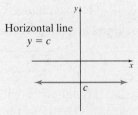

Horizontal line $y = c$

Zero slope: $m = 0$

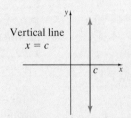

Vertical line $x = c$

Undefined slope or no slope

---

3 Two lines in the same plane are **parallel** if they do not intersect. Slopes of lines can help us determine whether lines are parallel. Parallel lines have the same steepness, so it follows that they have the same slope.

Parallel Lines

Nonvertical parallel lines have the same slope.

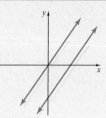

How do the slopes of perpendicular lines compare? Two lines that intersect at right angles are said to be **perpendicular**. The product of the slopes of two perpendicular lines is -1.

Perpendicular Lines

If the product of the slopes of two lines is -1, the lines are perpendicular.

(Two nonvertical lines are perpendicular if the slope of one is the negative reciprocal of the slope of the other.)

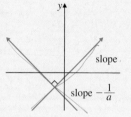

TEACHING TIP

Some questions you might want to ask your students:

▲ Can parallel lines have the same y-intercept?

▲ Can perpendicular lines have the same y-intercept?

▲ Do perpendicular lines always have the same y-intercept?

Helpful Hint

Here are examples of numbers that are negative (opposite) reciprocals.

Number	Negative Reciprocal	Their product is -1.
$\dfrac{2}{3}$	$-\dfrac{3}{2}$	$\dfrac{2}{3} \cdot -\dfrac{3}{2} = -\dfrac{6}{6} = -1$
-5 or $-\dfrac{5}{1}$	$\dfrac{1}{5}$	$-5 \cdot \dfrac{1}{5} = -\dfrac{5}{5} = -1$

Helpful Hint

Here are a few important facts about vertical and horizontal lines.

• Two distinct vertical lines are parallel.

• Two distinct horizontal lines are parallel.

• A horizontal line and a vertical line are always perpendicular.

EXAMPLE 5

Is the line passing through the points $(-6, 0)$ and $(-2, 3)$ parallel to the line passing through the points $(5, 4)$ and $(9, 7)$?

Solution To see if these lines are parallel, we find and compare slopes. The line passing through the points $(-6, 0)$ and $(-2, 3)$ has slope

$$m = \frac{3 - 0}{-2 - (-6)} = \frac{3}{4}$$

The line passing through the points $(5, 4)$ and $(9, 7)$ has slope

$$m = \frac{7-4}{9-5} = \frac{3}{4}$$

Since the slopes are the same, these lines are parallel.

EXAMPLE 6

Find the slope of a line perpendicular to the line passing through the points $(-1, 7)$ and $(2, 2)$.

Solution   First, let's find the slope of the line through $(-1, 7)$ and $(2, 2)$. This line has slope

$$m = \frac{2-7}{2-(-1)} = \frac{-5}{3}$$

The slope of every line perpendicular to the given line has a slope equal to the negative reciprocal of

$$-\frac{5}{3} \quad \text{or} \quad -\left(-\frac{3}{5}\right) = \frac{3}{5}$$

negative   reciprocal

The slope of a line perpendicular to the given line has slope $\frac{3}{5}$.

4   Slope can also be interpreted as a rate of change. In other words, slope tells us how fast y is changing with respect to x. To see this, lets look at a few of the many real-world applications of slope. For example, the pitch of a roof, used by builders and architects, is its slope. The pitch of the roof on the left is $\frac{7}{10}\left(\frac{\text{rise}}{\text{run}}\right)$. This means that the roof rises vertically 7 feet for every horizontal 10 feet. The rate of change for the roof is 7 vertical feet (y) per 10 horizontal feet (x).

The grade of a road is its slope written as a percent. A 7% grade, as shown below, means that the road rises (or falls) 7 feet for every horizontal 100 feet. (Recall that $7\% = \frac{7}{100}$.) Here, the slope of $\frac{7}{100}$ gives us the rate of change. The road rises (in our diagram) 7 vertical feet (y) for every 100 horizontal feet (x).

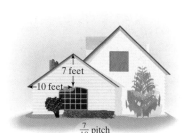

$\frac{7}{10}$ pitch

$\frac{7}{100}$ = 7% grade   7 feet   100 feet

EXAMPLE 7

FINDING THE GRADE OF A ROAD

At one part of the road to the summit of Pikes Peak, the road rises at a rate of 15 vertical feet for a horizontal distance of 250 feet. Find the grade of the road.

Solution   Recall that the grade of a road is its slope written as a percent.

$$\text{grade} = \frac{\text{rise}}{\text{run}} = \frac{15}{250} = 0.06 = 6\%$$

250 feet   15 feet

The grade is 6%.

EXAMPLE 8

FINDING THE SLOPE OF A LINE

The following graph shows the cost y (in cents) of an in-state long-distance telephone call in Massachusetts where x is the length of the call in minutes. Find the slope of the line and attach the proper units for the rate of change.

Solution Use $(2, 48)$ and $(5, 81)$ to calculate slope.

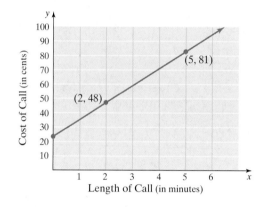

$$m = \frac{81 - 48}{5 - 2} = \frac{33}{3} = \frac{11 \text{ cents}}{1 \text{ minute}}$$

This means that the rate of change of a phone call is 11 cents per 1 minute or the cost of the phone call increases 11 cents per minute.

Spotlight on
DECISION MAKING

Suppose you own a house that you are trying to sell. You want a real estate agency to handle the sale of your house. You begin by scanning your local newspaper to find likely candidates, and spot the two ads shown.

Which real estate agency would you want to check into first? Explain. What other factors would you want to consider?

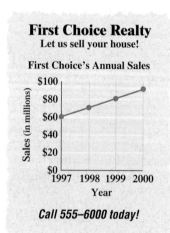

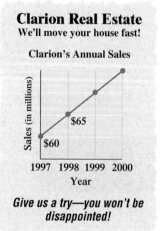

Graphing Calculator Explorations

It is possible to use a grapher to sketch the graph of more than one equation on the same set of axes. This feature can be used to confirm our findings from Section 3.2 when we learned that the graph of an equation written in the form $y = mx + b$ has a y-intercept of b. For example, graph the equations $y = \frac{2}{5}x$, $y = \frac{2}{5}x + 7$, and $y = \frac{2}{5}x - 4$ on the same set of axes. To do so, press the $\boxed{\text{Y=}}$ key and enter the equations on the first three lines.

$$Y_1 = \left(\frac{2}{5}\right)x$$

$$Y_2 = \left(\frac{2}{5}\right)x + 7$$

$$Y_3 = \left(\frac{2}{5}\right)x - 4$$

The screen should look like:

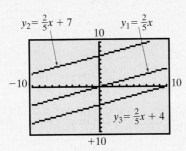

TEACHING TIP

Consider exploring slopes with a graphing calculator. Have students graph the following equations on a graphing calculator and sketch all the results of each set on the same axes.

First Set	Second Set	First Set	Second Set
$y = 10x$	$y = -\frac{1}{10}x$	$y = \frac{1}{2}x$	$y = -2x$
$y = 5x$	$y = -\frac{1}{5}x$	$y = \frac{1}{5}x$	$y = -5x$
$y = 2x$	$y = -\frac{1}{2}x$	$y = \frac{1}{10}x$	$y = -10x$
$y = x$	$y = -x$		

Then ask how a line of slope 0 would be positioned.

Notice that all three graphs appear to have the same positive slope. The graph of $y = \frac{2}{5}x + 7$ is the graph of $y = \frac{2}{5}x$ moved 7 units upward with a y-intercept of 7. Also, the graph of $y = \frac{2}{5}x - 4$ is the graph of $y = \frac{2}{5}x$ moved 4 units downward with a y-intercept of -4.

Graph the equations on the same set of axes. Describe the similarities and differences in their graphs.   See graphing answer section.

1. $y = 3.8x$, $y = 3.8x - 3$, $y = 3.8x + 9$

2. $y = -4.9x$, $y = -4.9x + 1$, $y = -4.9x + 8$

3. $y = \frac{1}{4}x$; $y = \frac{1}{4}x + 5$, $y = \frac{1}{4}x - 8$

4. $y = -\frac{3}{4}x$, $y = -\frac{3}{4}x - 5$, $y = -\frac{3}{4}x + 6$

TEACHING TIP

A Group Activity for this section is available in the Instructor's Resource Manual.

MENTAL MATH

Decide whether a line with the given slope is upward, downward, horizontal, or vertical.

1. $m = \frac{7}{6}$ upward   **2.** $m = -3$ downward   **3.** $m = 0$ horizontal   **4.** m is undefined.   vertical

EXERCISE SET 3.4

STUDY
GUIDE/SSM
CD/
VIDEO
PH MATH
TUTOR CENTER
MathXL®Tutorials
ON CD
MathXL®
MyMathLab®

Find the slope of each line if it exists. See Example 1.

1. $m = -\frac{4}{3}$

2. $m = \frac{5}{2}$

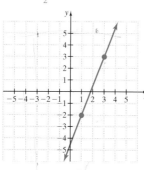

3. undefined slope

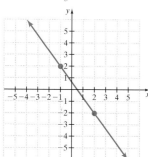

4. $m = 0$

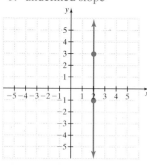

5. $m = \frac{5}{2}$

6. $m = -\frac{7}{4}$

Find the slope of the line that goes through the given points. See Examples 1 and 2. **13.–14.** undefined

7. $(0,0)$ and $(7,8)$ $\quad \frac{8}{7}$

8. $(-1,5)$ and $(0,0)$ $\quad -5$

9. $(-1,5)$ and $(6,-2)$ $\quad -1$

10. $(-1,9)$ and $(-3,4)$ $\quad \frac{5}{2}$

11. $(1,4)$ and $(5,3)$ $\quad -\frac{1}{4}$

12. $(3,1)$ and $(2,6)$ $\quad -5$

13. $(-4,3)$ and $(-4,5)$

14. $(6,-6)$ and $(6,2)$

15. $(-2,8)$ and $(1,6)$ $\quad -\frac{2}{3}$

16. $(4,-3)$ and $(2,2)$ $\quad -\frac{5}{2}$

17. $(1,0)$ and $(1,1)$ $\quad$ undefined

18. $(0,13)$ and $(-4,13)$ $\quad 0$

19. $(5,1)$ and $(-2,1)$ $\quad 0$

20. $(5,4)$ and $(0,5)$ $\quad -\frac{1}{5}$

For each graph, determine which line has the greater slope.

21.

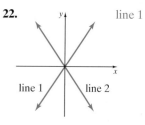

line 1

line 1

line 2

22.

line 1

line 1   line 2

23.

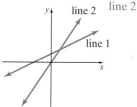

line 2   line 2

line 1

24.

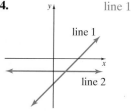

line 1

line 1

line 2

In Exercises 25–30, match each line with its slope.

A. $m = 0$ **B.** undefined slope   **C.** $m = 3$

D. $m = 1$ **E.** $m = -\frac{1}{2}$ **F.** $m = -\frac{3}{4}$

25. D

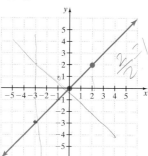

26. A

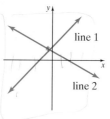

27. B

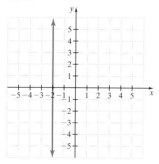

28. C

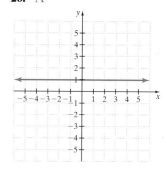

29. E

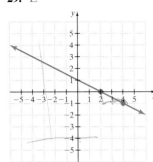

30. F

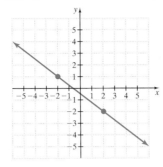

Find the slope of each line. See Examples 3 and 4.

31. $x = 6$ undefined slope   **32.** $y = 4$ $m = 0$

33. $y = -4$ $m = 0$ **34.** $x = 2$ undefined slope

35. $x = -3$ undefined slope   **36.** $x = -11$ undefined slope

37. $y = 0$ $m = 0$ **38.** $x = 0$ undefined slope

△ *Find the slope of the line* ***a.*** *parallel and* ***b.*** *perpendicular to the line through each pair of points. See Examples 5 and 6.*

39. $(-3, -3)$ and $(0, 0)$ **a.** 1 **b.** -1 **45.** perpendicular

40. $(6, -2)$ and $(1, 4)$ **a.** $-\frac{6}{5}$ **b.** $\frac{5}{6}$

41. $(-8, -4)$ and $(3, 5)$ **a.** $\frac{9}{11}$ **b.** $-\frac{11}{9}$

42. $(6, -1)$ and $(-4, -10)$ **a.** $\frac{9}{10}$ **b.** $-\frac{10}{9}$

△ *Determine whether the lines through each pair of points are parallel, perpendicular, or neither. See Examples 5 and 6.*   **49.** neither

43. $(0, 6)$ and $(-2, 0)$
    $(0, 5)$ and $(1, 8)$ parallel

44. $(1, -1)$ and $(-1, -11)$
    $(-2, -8)$ and $(0, 2)$ parallel

45. $(2, 6)$ and $(-2, 8)$
    $(0, 3)$ and $(1, 5)$

46. $(-1, 7)$ and $(1, 10)$
    $(0, 3)$ and $(1, 5)$ neither

47. $(3, 6)$ and $(7, 8)$
    $(0, 6)$ and $(2, 7)$ parallel

48. $(4, 2)$ and $(6, 6)$
    $(0, -2)$ and $(1, 0)$ parallel

49. $(2, -3)$ and $(6, -5)$
    $(5, -2)$ and $(-3, -4)$

50. $(-1, -5)$ and $(4, 4)$
    $(10, 8)$ and $(-7, 4)$ neither

51. $(-4, -3)$ and $(-1, 0)$
    $(4, -4)$ and $(0, 0)$

52. $(2, -2)$ and $(4, -8)$
    $(0, 7)$ and $(3, 8)$

△ *Solve. See Examples 5 and 6.*   **51.–52.** perpendicular

53. Find the slope of a line parallel to the line passing through $(-7, -5)$ and $(-2, -6)$. $-\frac{1}{5}$

54. Find the slope of a line parallel to the line passing through the origin and $(-2, 10)$. -5

55. Find the slope of a line perpendicular to the line passing through the origin and $(1, -3)$. $\frac{1}{3}$

56. Find the slope of the line perpendicular to the line passing through $(-1, 3)$ and $(2, -8)$. $\frac{3}{11}$

57. Find the slope of a line parallel to the line passing through $(3, 3)$ and $(-3, -3)$.   1

58. Find the slope of a line parallel to the line passing through $(0, 3)$ and $(6, 0)$. $-\frac{1}{2}$

The pitch of a roof is its slope. Find the pitch of each roof shown. See Example 7.

59. $\frac{3}{5}$

6 feet

←10 feet→

60. $\frac{1}{2}$

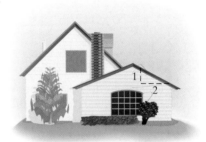

1
2

The grade of a road is its slope written as a percent. Find the grade of each road shown. See Example 7.

61.   12.5%

2 meters

16 meters

62.   16%

16 feet

100 feet

63. One of Japan's superconducting "bullet" trains is researched and tested at the Yamanashi Maglev Test Line near Otsuki City. The steepest section of the track has a rise of 2580 meters for a horizontal distance of 6450 meters. What is the grade for this section of track? (*Source:* Japan Railways Central Co.)   40%

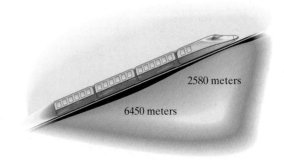

2580 meters

6450 meters

64. The steepest street is Baldwin Street in Dunedin, New Zealand. It has a maximum rise of 10 meters for a horizontal distance of 12.66 meters. Find the grade for this section of road. Round to the nearest whole percent. (*Source: The Guinness Book of Records*)   79%

65. Professional plumbers suggest that a sewer pipe should rise 0.25 inch for every horizontal foot. Find the recommended slope for a sewer pipe. Round to the nearest hundredth.   0.02

0.25 inch

12 inches

66. According to federal regulations, a wheel chair ramp should rise no more than 1 foot for a horizontal distance of 12 feet. Write the slope as a grade. Round to the nearest tenth of a percent.   8.3%

Find the slope of each line and write the slope as a rate of change. Don't forget to attach the proper units. See Example 8.

67. This graph shows a projection of the increase of cell phone users *y* (in millions) for year *x*.

Every 1 year, there are 5 million more cell phone users

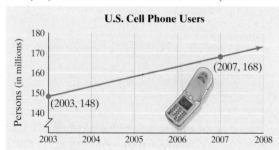

U.S. Cell Phone Users

(2007, 168)

(2003, 148)

Persons (in millions)

2003 2004 2005 2006 2007 2008

68. This graph approximates the total number of cosmetic surgeons for year *x*.

Every 1 year, there are 150 more cosmetic surgeons.

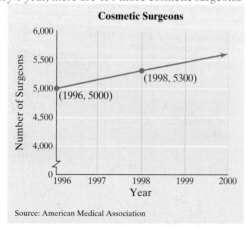

Cosmetic Surgeons

(1998, 5300)

(1996, 5000)

Number of Surgeons

1996 1997 1998 1999 2000
Year

Source: American Medical Association

69. The graph below shows the total cost *y* (in dollars) of owning and operating a compact car where *x* is the number of miles driven.

It costs $0.36 per 1 mile to own and operate a compact car.

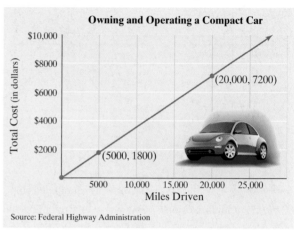

Owning and Operating a Compact Car

Total Cost (in dollars)

(20,000, 7200)

(5000, 1800)

5000 10,000 15,000 20,000 25,000
Miles Driven

Source: Federal Highway Administration

70. The graph below shows the total cost *y* (in dollars) of owning and operating a full-size pickup truck, where *x* is the number of miles driven.   It costs $0.40 per 1 mile to own and operate a full-size pickup truck.

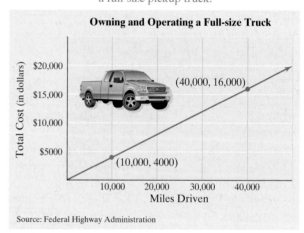

Owning and Operating a Full-size Truck

Total Cost (in dollars)

(40,000, 16,000)

(10,000, 4000)

10,000 20,000 30,000 40,000
Miles Driven

Source: Federal Highway Administration

REVIEW AND PREVIEW

Solve each equation for y. See Section 2.6.   **72.** $y = x - 7$

71. $x + y = 10$ $y = -x + 10$ **72.** $-x + y = -7$

73. $x + 2y = -12$ $y = -\frac{1}{2}x - 6$ **74.** $x - 3y = 15$

75. $5x - y = 17$ $y = 5x - 17$ **76.** $4x - y = 20$

74. $y = \frac{1}{3}x - 5$ **76.** $y = 4x - 20$

Concept Extensions

The line graph on the next page shows the average fuel economy (in miles per gallon) by passenger automobiles produced during each of the model years shown. Use this graph to answer Exercises 77 through 82.

77. What was the average fuel economy (in miles per gallon) for automobiles produced during 1995?   28.3

78. Find the decrease in average fuel economy for automobiles for the years 1991 to 1992.   0.4 miles per gal

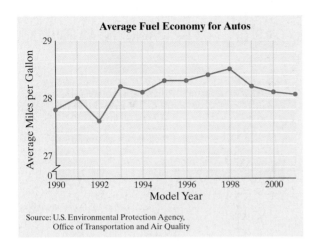

Average Fuel Economy for Autos

Source: U.S. Environmental Protection Agency,
Office of Transportation and Air Quality

79. During which of the model years shown was average fuel economy the lowest?   1992

What was the average fuel economy for that year?   27.6

80. During which of the model years shown was average fuel economy the highest?   1998

What was the average fuel economy for that year?   28.5

81. What line segment has the greatest slope?   from 1992 to 1993

82. What line segment has the smallest slope?   from 1991 to 1992

83. Find x so that the pitch of the roof is $\frac{1}{3}$. $x = 6$

84. Find x so that the pitch of the roof is $\frac{2}{5}$. $x = 20$

85. The average price of an acre of U.S. farmland was \$782 in 1994. In 2001, the price of an acre rose to approximately \$1132. (*Source:* National Agricultural Statistics Service)

85. a. $(1994, 782), (2001, 1132)$

a. Write two ordered pairs of the form (year, price of acre).

b. Find the slope of the line through the two points.   50

c. Write a sentence explaining the meaning of the slope as a rate of change.

86. There were approximately 13,290 kidney transplants performed in the United States in 2000. In 2002, the number of kidney transplants performed in the United States rose to 14,774. (*Source:* Organ Procurement and Transplantation Network)

a. Write two ordered pairs of the form (year, number of kidney transplants). $(2000, 13{,}290), (2002, 14{,}774)$

b. Find the slope of the line between the two points.   742

c. Write a sentence explaining the meaning of the slope as a rate of change.

87. Show that a triangle with vertices at the points $(1, 1), (-4, 4)$, and $(-3, 0)$ is a right triangle.

88. Show that the quadrilateral with vertices $(1, 3), (2, 1), (-4, 0)$, and $(-3, -2)$ is a parallelogram.

Find the slope of the line through the given points.

89. $(2.1, 6.7)$ and $(-8.3, 9.3)$ -0.25

90. $(-3.8, 1.2)$ and $(-2.2, 4.5)$ 2.0625

91. $(2.3, 0.2)$ and $(7.9, 5.1)$ 0.875

92. $(14.3, -10.1)$ and $(9.8, -2.9)$ -1.6

93. The graph of $y = -\frac{1}{3}x + 2$ has a slope of $-\frac{1}{3}$. The graph of $y = -2x + 2$ has a slope of -2. The graph of $y = -4x + 2$ has a slope of -4. Graph all three equations on a single coordinate system. As the absolute value of the slope becomes larger, how does the steepness of the line change?

94. The graph of $y = \frac{1}{2}x$ has a slope of $\frac{1}{2}$. The graph of $y = 3x$ has a slope of 3. The graph of $y = 5x$ has a slope of 5. Graph all three equations on a single coordinate system. As slope becomes larger, how does the steepness of the line change?

93. the line becomes steeper **94.** the line becomes steeper

85. c. For the years 1994 through 2001, the price per acre of U.S. farmland rose \$50 every year.
86. c. From 2000 to 2002, the number of kidney transplants increased at a rate of 742 per year.
87. The slope through $(-3, 0)$ and $(1, 1)$ is $\frac{1}{4}$. The slope through $(-3, 0)$ and $(-4, 4)$ is -4. The product of the slope is -1. So the sides are perpendicular.
88. Opposite sides are parallel lines since their slopes are equal.

INTEGRATED REVIEW — SUMMARY ON SLOPE AND GRAPHING LINEAR EQUATIONS

Find the slope of each line.

1.

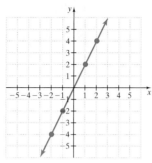

$m = 2$

2.

$m = 0$

3.

$m = -\dfrac{2}{3}$

4.

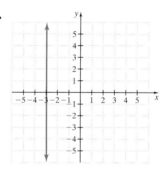

slope is undefined

Graph each linear equation.   **5.–12.** See graphing answer section.

5. $y = -2x$

6. $x + y = 3$

7. $x = -1$

8. $y = 4$

9. $x - 2y = 6$

10. $y = 3x + 2$

11. $5x + 3y = 15$

12. $2x - 4y = 8$

Determine whether the lines through the points are parallel, perpendicular, or neither.

13. $(-1, 3)$ and $(1, -3)$
$(2, -1)$ and $(4, -7)$ parallel

14. $(-6, -6)$ and $(-1, -2)$
$(-4, 3)$ and $(3, -3)$ neither

15. In the years 2000 through 2002 (the number of admissions to movie theaters in the U.S. can be modeled by the linear equation $y = 110x + 1407$ where x is years after 2000 and y is admissions in millions. (*Source:* Motion Picture Assn. of America)

  a. Find the y-intercept of this line.   (0,1407)

  b. Write a sentence explaining the meaning of this intercept.

  c. Find the slope of this line.   110

  d. Write a sentence explaining the meaning of the slope as a rate of change.

15. b. In 2000, there were 1407 million admissions to movie theaters in the U.S.

15. d. For the years 2000 through 2002, the number of movie theater admissions increased at a rate of 110 million per year.

3.5 THE SLOPE-INTERCEPT FORM

Objectives

△ 1 Use the slope-intercept form to find the slope and the y-intercept of a line.

△ 2 Use the slope-intercept form to determine whether two lines are parallel, perpendicular, or neither.

3 Use the slope-intercept form to write an equation of a line.

4 Use the slope-intercept form to graph a linear equation.

1 In Section 3.4, we learned that the slant of a line is called its slope. Recall that the slope of a line m is the ratio of vertical change (change in y) to the corresponding horizontal change (change in x) as we move along the line.

> **Helpful Hint**
> Don't forget that the slope of a line is the same no matter which ordered pairs of the line are used to calculate slope.

EXAMPLE 1

Find the slope of the line whose equation is $y = \frac{3}{4}x + 6$.

Solution To find the slope of this line, find any two ordered pair solutions of the equation. Let's find and use intercept points as our two ordered pair solutions. Recall from Section 3.2 that the graph of an equation of the form $y = mx + b$ has y-intercept (0, b).

CLASSROOM EXAMPLE

Find the slope of the line whose equation is $y = \frac{5}{9}x + 3$.

answer: $m = \frac{5}{9}$

This means that the graph of $y = \frac{3}{4}x + 6$ has a y-intercept of (0, 6) as shown below.

To find the y-intercept, let $x = 0$.

Then $y = \frac{3}{4}x + 6$ becomes

$$y = \frac{3}{4}\cdot 0 + 6 \text{ or}$$

$$y = 6$$

The y-intercept is (0, 6), as expected.

To find the x-intercept, let $y = 0$.

Then $y = \frac{3}{4}x + 6$ becomes

$$0 = \frac{3}{4}x + 6 \text{ or}$$

$$0 = 3x + 24 \quad \text{Multiply both sides by 4.}$$

$$-24 = 3x \quad \text{Subtract 24 from both sides.}$$

$$-8 = x \quad \text{Divide both sides by 3.}$$

The x-intercept is $(-8, 0)$.

Use the points $(0, 6)$ and $(-8, 0)$ to find the slope. Then

$$m = \frac{\text{change in } y}{\text{change in } x} = \frac{0 - 6}{-8 - 0} = \frac{-6}{-8} = \frac{3}{4}$$

The slope of the line is $\frac{3}{4}$.

Analyzing the results of Example 1, you may notice a striking pattern:

The slope of $y = \frac{3}{4}x + 6$ is $\frac{3}{4}$, the same as the coefficient of x.

Also, as mentioned earlier, the y-intercept is $(0, 6)$. Notice that the y-value 6 is the same as the constant term.

When a linear equation is written in the form $y = mx + b$, not only is $(0, b)$ the y-intercep°t of the line, but m is its slope. The form $y = mx + b$ is appropriately called the **slope-intercept form**.

$$\underset{\underset{(\mathbf{0},\, \mathbf{b})}{\text{slope } \text{y-intercept}}}{\uparrow \qquad \uparrow}$$

TEACHING TIP
Stress this concept to students in words: If a linear equation in two variables is solved for y, the coefficient of x is the slope, and the constant term is the y-intercept.

Slope-Intercept Form

When a linear equation in two variables is written in slope-intercept form,

$$y = mx + b$$

m is the slope of the line and $(0, b)$ is the y-intercept of the line.

EXAMPLE 2

Find the slope and the y-intercept of the line whose equation is $5x + y = 2$.

Solution Write the equation in slope-intercept form by solving the equation for y.

$$5x + y = 2$$
$$y = -5x + 2 \qquad \text{Subtract } 5x \text{ from both sides.}$$

CLASSROOM EXAMPLE
Find the slope and the y-intercept of $-4x + y = 7$,
answer: $m = 4$, y-intercept $(0, 7)$

The coefficient of x, -5, is the slope and the constant term, 2, is the y-value of the y-intercept, $(0, 2)$.

EXAMPLE 3

Find the slope and the y-intercept of the line whose equation is $3x - 4y = 4$.

Solution Write the equation in slope-intercept form by solving for y.

$$3x - 4y = 4$$
$$-4y = -3x + 4 \qquad \text{Subtract } 3x \text{ from both sides.}$$
$$\frac{-4y}{-4} = \frac{-3x}{-4} + \frac{4}{-4} \qquad \text{Divide both sides by } -4.$$
$$y = \frac{3}{4}x - 1 \qquad \text{Simplify.}$$

CLASSROOM EXAMPLE
Find the slope and the y-intercept of $7x - 2y = 8$.
answer: $m = \frac{7}{2}$, y-intercept $(0, -4)$

The coefficient of x, $\frac{3}{4}$, is the slope, and the y-intercept is $(0, -1)$.

2   The slope-intercept form can be used to determine whether two lines are parallel or perpendicular. Recall that nonvertical parallel lines have the same slope and different y-intercepts. Also, nonvertical perpendicular lines have slopes whose product is -1.

EXAMPLE 4

Determine whether the graphs of $y = -\dfrac{1}{5}x + 1$ and $2x + 10y = 30$ are parallel lines, perpendicular lines, or neither.

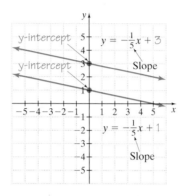

Solution The graph of $y = -\dfrac{1}{5}x + 1$ is a line with slope $-\dfrac{1}{5}$ and with y-intercept $(0, 1)$. To find the slope and the y-intercept of the graph of $2x + 10y = 30$, write this equation in slope-intercept form. To do this, solve the equation for y.

$$2x + 10y = 30$$

$$10y = -2x + 30 \qquad \text{Subtract } 2x \text{ from both sides.}$$

$$y = -\frac{1}{5}x + 3 \qquad \text{Divide both sides by } 10.$$

The graph of this equation is a line with slope $-\dfrac{1}{5}$ and y-intercept $(0, 3)$. Because both lines have the *same* slope but different y-intercepts, these lines are parallel.

✔ **CONCEPT CHECK**

Write the equations of any three parallel lines.

3   The slope-intercept form can also be used to write the equation of a line given its slope and y-intercept.

EXAMPLE 5

Find an equation of the line with y-intercept $(0, -3)$ and slope of $\dfrac{1}{4}$.

Solution We are given the slope and the y-intercept. Let $m = \dfrac{1}{4}$ and $b = -3$, and write the equation in slope-intercept form, $y = mx + b$.

Concept Check Answer:
For example,
$y = 2x - 3$, $y = 2x - 1$, $y = 2x$

$$y = mx + b$$

$$y = \frac{1}{4}x + (-3) \qquad \text{Let } m = \frac{1}{4} \text{ and } b = -3.$$

$$y = \frac{1}{4}x - 3 \qquad \text{Simplify.}$$

4   We now use the slope-intercept form of the equation of a line to graph a linear equation.

EXAMPLE 6

Graph: $y = \frac{3}{5}x - 2$.

Solution   Since the equation $y = \frac{3}{5}x - 2$ is written in slope-intercept form $y = mx + b$, the slope of its graph is $\frac{3}{5}$ and the y-intercept is $(0, -2)$. To graph, begin by plotting the intercept $(0, -2)$. From this point, find another point of the graph by using the slope $\frac{3}{5}$ and recalling that slope is $\frac{\text{rise}}{\text{run}}$. Start at the intercept point and move 3 units up since the numerator of the slope is 3; then move 5 units to the right since the denominator of the slope is 5. Stop at the point $(5, 1)$. The line through $(0, -2)$ and $(5, 1)$ is the graph of $y = \frac{3}{5}x - 2$.

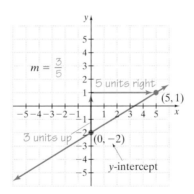

EXAMPLE 7

Use the slope-intercept form to graph the equation $4x + y = 1$.

Solution   First, write the given equation in slope-intercept form.

$$4x + y = 1$$
$$y = -4x + 1$$

The graph of this equation will have slope -4 and y-intercept $(0, 1)$. To graph this line, first plot the intercept point $(0, 1)$. To find another point of the graph, use the slope -4, which can be written as $\frac{-4}{1}$. Start at the point $(0, 1)$ and move 4 units down (since the numerator of the slope is -4); then move 1 unit to the right (since the denominator of the slope is 1). We arrive at the point $(1, -3)$. The line through $(0, 1)$ and $(1, -3)$ is the graph of $4x + y = 1$.

$$m = \frac{-4}{1}$$

4 units down

(0, 1)

(1, −3)

1 unit right

> ### Helpful Hint
>
> In Example 7, if we interpret the slope of -4 as $\dfrac{4}{-1}$, we arrive at $(-1, 5)$ for a second point. Notice that this point is also on the line.

Graphing Calculator Explorations

A grapher is a very useful tool for discovering patterns. To discover the change in the graph of a linear equation caused by a change in slope, try the following. Use a standard window and graph a linear equation in the form $y = mx + b$. Recall that the graph of such an equation will have slope m and y-intercept b.

First graph $y = x + 3$. To do so, press the $\boxed{Y=}$ key and enter $Y_1 = x + 3$. Notice that this graph has slope 1 and that the y-intercept is 3. Next, on the same set of axes, graph $y = 2x + 3$ and $y = 3x + 3$ by pressing $\boxed{Y=}$ and entering $Y_2 = 2x + 3$ and $Y_3 = 3x + 3$.

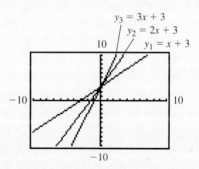

Notice the difference in the graph of each equation as the slope changes from 1 to 2 to 3. How would the graph of $y = 5x + 3$ appear? To see the change in the graph caused by a change in negative slope, try graphing $y = -x + 3$, $y = -2x + 3$, and $y = -3x + 3$ on the same set of axes.

Use a grapher to graph the following equations. For each exercise, graph the first equation and use its graph to predict the appearance of the other equations. Then graph the other equations on the same set of axes and check your prediction.

See graphing answer section.

1. $y = x; y = 6x, y = -6x$ **2.** $y = -x; y = -5x, y = -10x$

3. $y = \dfrac{1}{2}x + 2; y = \dfrac{3}{4}x + 2, y = x + 2$ **4.** $y = x + 1; y = \dfrac{5}{4}x + 1, y = \dfrac{5}{2}x - 1$

5. $y = -7x + 5; y = 7x + 5$ **6.** $y = 3x - 1; y = -3x - 1$

STUDY SKILLS REMINDER

Tips for Studying for an Exam

To prepare for an exam, try the following study techniques.

▶ Start the study process days before your exam.

▶ Make sure that you are current and up-to-date on your assignments.

▶ If there is a topic that you are unsure of, use one of the many resources that are available to you. For example,

See your instructor.

Visit a learning resource center on campus where math tutors are available.

Read the textbook material and examples on the topic.

View a videotape on the topic.

▶ Reread your notes and carefully review the Chapter Highlights at the end of the chapter.

▶ Work the review exercises at the end of the chapter and check your answers. Make sure that you correct any missed exercises. If you have trouble on a topic, use a resource listed above.

▶ Find a quiet place to take the Chapter Test found at the end of the chapter. Do not use any resources when taking this sample test. This way you will have a clear indication of how prepared you are for your exam. Check your answers and make sure that you correct any missed exercises.

▶ Get lots of rest the night before the exam. It's hard to show how well you know the material if your brain is foggy from lack of sleep.

Good luck and keep a positive attitude.

MENTAL MATH

Identify the slope and the *y*-intercept of the graph of each equation.

1. $y = 2x - 1$ $m = 2; (0, -1)$

2. $y = -7x + 3$ $m = -7; (0, 3)$

3. $y = x + \dfrac{1}{3}$ $m = 1; \left(0, \dfrac{1}{3}\right)$

4. $y = -x - \dfrac{2}{9}$ $m = -1, \left(0, -\dfrac{2}{9}\right)$

5. $y = \dfrac{5}{7}x - 4$ $m = \dfrac{5}{7}; (0, -4)$

6. $y = -\dfrac{1}{4}x + \dfrac{3}{5}$ $m = -\dfrac{1}{4}; \left(0, \dfrac{3}{5}\right)$

EXERCISE SET 3.5

STUDY GUIDE/SSM   CD/ VIDEO   PH MATH TUTOR CENTER   MathXL®Tutorials ON CD   MathXL®   MyMathLab®

2. $m = 3; (0, -5)$ **4.** $m = -\dfrac{1}{7}; \left(0, \dfrac{6}{7}\right)$ **6.** $m = \dfrac{1}{3}; (0, -2)$ **8.** $m = 1; (0, 0)$ **10.** $m = 0; (0, -2)$ **12.** $m = \dfrac{8}{3}; \left(0, \dfrac{11}{3}\right)$

Determine the slope and the y-intercept of the graph of each equation. See Examples 1 through 3.

1. $2x + y = 4$ $m = -2; (0, 4)$

2. $-3x + y = -5$

🔒 **3.** $x + 9y = 1$ $m = -\dfrac{1}{9}; \left(0, \dfrac{1}{9}\right)$

4. $x + 7y = 6$

5. $4x - 3y = 12$ $m = \dfrac{4}{3}; (0, -4)$

6. $2x - 6y = 12$

7. $x + y = 0$ $m = -1; (0, 0)$

8. $x - y = 0$

9. $y = -3$ $m = 0; (0, -3)$

10. $y + 2 = 0$

11. $-x + 5y = 20$ $m = \dfrac{1}{5}; (0, 4)$

12. $-8x + 3y = 11$

Match each linear equation with its graph.

13. $y = 2x + 1$ B

14. $y = -x + 1$ C

15. $y = -3x - 2$ D

16. $y = \frac{5}{3}x - 2$ A

A.

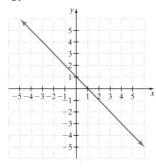

B.

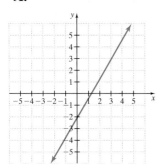

C.

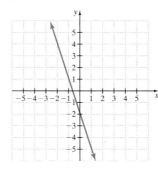

D.

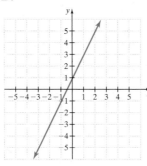

Determine whether the lines are parallel, perpendicular, or neither. See Example 4.

18. perpendicular   **21.** perpendicular   **24.** perpendicular

△ **17.** $x - 3y = -6$
$3x - y = 0$ neither

△ **18.** $-5x + y = -6$
$x + 5y = 5$

△ **19.** $2x - 7y = 1$
$2y = 7x - 2$ neither

△ **20.** $y = 4x - 2$
$4x + y = 5$ neither

🔒 **21.** $10 + 3x = 5y$
$5x + 3y = 1$

△ **22.** $-x + 2y = -2$
$2x = 4y + 3$ parallel

△ **23.** $6x = 5y + 1$
$-12x + 10y = 1$ parallel

△ **24.** $11x + 10y = 3$
$11y = 10x + 5$

25. Explain how to write the equation of a line if you are given the slope of the line and its y-intercept.   answers may vary

26. Explain why the graphs of $y = x$ and $y = -x$ are perpendicular lines.   answers may vary
△

Use the slope-intercept form of the linear equation to write an equation of each line with given slope and y-intercept. See Example 5.

27. Slope -1; y-intercept $(0, 1)$ $y = -x + 1$

28. Slope $\frac{1}{2}$; y-intercept $(0, -6)$ $y = \frac{1}{2}x - 6$

🔒 **29.** Slope 2; y-intercept $\left(0, \frac{3}{4}\right)$ $y = 2x + \frac{3}{4}$

30. Slope -3; y-intercept $\left(0, -\frac{1}{5}\right)$ $y = -3x - \frac{1}{5}$

31. Slope $\frac{2}{7}$; y-intercept $(0, 0)$ $y = \frac{2}{7}x$

32. Slope $-\frac{4}{5}$; y-intercept $(0, 0)$ $y = -\frac{4}{5}x$

Use the slope-intercept form to graph each equation. See Examples 6 and 7.   **33.–48.** See graphing answer section.

33. $y = \frac{2}{3}x + 5$

34. $y = \frac{1}{4}x - 3$

35. $y = -\frac{3}{5}x - 2$

36. $y = -\frac{5}{7}x + 5$

37. $y = 2x + 1$

38. $y = -4x - 1$

🔒 **39.** $y = -5x$

40. $y = 6x$

41. $4x + y = 6$

42. $-3x + y = 2$

43. $x - y = -2$

44. $x + y = 3$

45. $3x + 5y = 10$

46. $2x + 3y = 5$

🔒 **47.** $4x - 7y = -14$

48. $3x - 4y = 4$

REVIEW AND PREVIEW

49. $y = 2x - 14$ **50.** $y = -9x + 61$ **51.** $y = -6x - 11$
Solve each equation for y. See Section 2.6.

49. $y - (-6) = 2(x - 4)$

50. $y - 7 = -9(x - 6)$

51. $y - 1 = -6(x - (-2))$

52. $y - (-3) = 4(x - (-5))$

52. $y = 4x + 17$

In Exercises 53–58, match each line with its slope.

A. $m = 0$ **B.** undefined slope   **C.** $m = 3$

D. $m = 1$ **E.** $m = -\frac{1}{2}$ **F.** $m = -\frac{3}{4}$

53. D

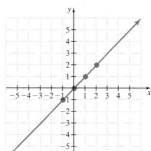

54. A

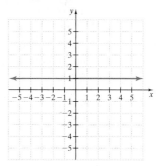

55. B

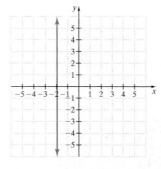

56. C

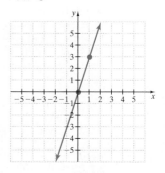

57. E

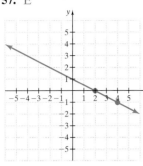

58. F

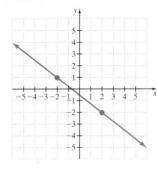

Concept Extensions

Solve.

59. American Automobile Association (AAA) membership was 21 million in 1980. In 2002, their membership was 45 million. (*Source*: AAA)

  a. Write two ordered pairs of the form (years after 1980, membership in millions).   (0,21), (22,45)

  b. The relationship between the years after 1980 and membership is approximately linear over this period. Use the ordered pairs from part (a) to write an equation of the line relating year to AAA membership. $y = \frac{12}{11}x + 21$

60. In 1998, 1.7 million electronic bill statements were delivered and payments occurred. In 2002, that number rose to 9.9 million. (*Source*: USA Today)

  a. Write two ordered pairs of the form (years after 1998, millions of electronic bills).   (0, 1.7), (4, 9.9)

  b. The relationship between years after 1998 and 2002 is approximately linear over this period. Use the ordered pairs from part a to write an equation of the line relating to electronic bills. $y = 2.05x + 1.7$

  c. Use the linear equation from part (b) to estimate the electronic bills to be delivered and paid in 2006.   18.1 million

61. Write an equation of the line parallel to $2y + 4x = 12$ that has a y-intercept of $(0, 5)$. $y = -2x + 5$

62. Write an equation of the line perpendicular to $3y - 9x = 15$ that has a y-intercept of $(0, -2)$. $y = \frac{-1}{3}x - 2$

63. The equation relating Celsius temperature and Fahrenheit temperature is linear and its graph is shown below. This line contains the point $(0, 32)$, which means that $0°$ Celsius is equivalent to $32°$ Fahrenheit. This line also passes through the point $(100, 212)$.

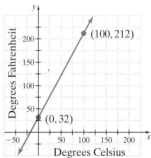

63. a. The temperature $100°$ Celsius is equivalent to $212°$ Fahrenheit.

  a. Write the meaning of the ordered pair $(100, 212)$ in words.

  b. Use the graph and approximate the Fahrenheit temperature equivalent to $20°$ Celsius. $68°F$

  c. Use the graph and approximate the Celsius temperature equivalent to $80°$ Fahrenheit. $27°C$

  d. Use the ordered pairs $(0, 32)$ and $(100, 212)$ and the slope-intercept form and write an equation of the line relating Celsius temperatures and Fahrenheit temperatures. (*Hint:* Use both ordered pairs to find slope. Then substitute the slope and y-intercept in slope-intercept form.) $F = \frac{9}{5}C - 32$

  e. Use the linear equation found in part (d) to check your approximations from (b) and (c).

64. The equation relating temperatures on the Kelvin scale and the Celsius scale is linear. If $0°$ Celsius is equivalent to 273 Kelvin and $100°$ Celsius is equivalent to 373 Kelvin, find the linear equation that models the relationship between these two temperature scales. (*Hint:* See Exercise 63 and use the ordered pairs $(0, 273)$ and $(100, 373)$.) $K = C + 273$

65. Show that the slope of a line whose equation is $y = mx + b$ is m. To do so, find two ordered pair solutions of this equation and use the slope formula. For example, if $x = 0$, then $y = 0 \cdot x + b$ or $y = b$. Thus, one ordered pair solution is $(0, b)$. Find one more ordered pair solution and find the slope.   answers may vary

3.6   THE POINT-SLOPE FORM

Objectives

1. Use the point-slope form to find an equation of a line given its slope and a point of the line.
2. Use the point-slope form to find an equation of a line given two points of the line.
3. Find equations of vertical and horizontal lines.
4. Use the point-slope form to solve problems.

1   From the last section, we know that if the y-intercept of a line and its slope are given, then the line can be graphed and an equation of this line can be found. Our goal in this section is to answer the following: If any two points of a line are given, can an equation of the line be found? To answer this question, first let's see if we can find an equation of a line given one point (not necessarily the y-intercept) and the slope of the line.

Suppose that the slope of a line is -3 and the line contains the point $(2, 4)$. For any other point with ordered pair (x, y) to be on the line, its coordinates must satisfy the slope equation

$$\frac{y - 4}{x - 2} = -3$$

Now multiply both sides of this equation by $x - 2$.

$$y - 4 = -3(x - 2)$$

This equation is a linear equation whose graph is a line that contains the point $(2, 4)$ and has slope -3. This form of a linear equation is called the **point-slope form.**

In general, when the slope of a line and any point on the line are known, the equation of the line can be found. To do this, use the slope formula to write the slope of a line that passes through points (x, y) and (x_1, y_1). We have

$$\frac{y - y_1}{x - x_1} = m$$

Multiply both sides of this equation by $x - x_1$ to obtain

$$y - y_1 = m(x - x_1)$$
$$\uparrow$$
$$slope$$

Point-Slope Form of the Equation of a Line

The point-slope form of the equation of a line is $y - y_1 = m(x - x_1)$, where m is the slope of the line and (x_1, y_1) is a point on the line.

EXAMPLE 1

Find an equation of the line passing through $(-1, 5)$ with slope -2. Write the equation in standard form: $Ax + By = C$.

Solution   Since the slope and a point on the line are given, use point-slope form $y - y_1 = m(x - x_1)$ to write the equation. Let $m = -2$ and $(-1, 5) = (x_1, y_1)$.

$$y - y_1 = m(x - x_1)$$
$$y - 5 = -2[x - (-1)] \quad \text{Let } m = -2 \text{ and } (x_1, y_1) = (-1, 5).$$
$$y - 5 = -2(x + 1) \quad \text{Simplify.}$$
$$y - 5 = -2x - 2 \quad \text{Use the distributive property.}$$
$$y = -2x + 3 \quad \text{Add 5 to both sides.}$$
$$2x + y = 3 \quad \text{Add } 2x \text{ to both sides.}$$

In standard form, the equation is $2x + y = 3$.

2   We may also find the equation of a line given any two points of the line, as seen in the next example.

EXAMPLE 2

Find an equation of the line through $(2, 5)$ and $(-3, 4)$. Write the equation in standard form.

Solution   First, use the two given points to find the slope of the line.

$$m = \frac{4 - 5}{-3 - 2} = \frac{-1}{-5} = \frac{1}{5}$$

Next, use the slope $\frac{1}{5}$ and either one of the given points to write the equation in point-slope form. We use $(2, 5)$. Let $x_1 = 2$, $y_1 = 5$, and $m = \frac{1}{5}$.

$$y - y_1 = m(x - x_1) \quad \text{Use point-slope form.}$$
$$y - 5 = \frac{1}{5}(x - 2) \quad \text{Let } x_1 = 2, y_1 = 5, \text{ and } m = \frac{1}{5}.$$
$$5(y - 5) = 5 \cdot \frac{1}{5}(x - 2) \quad \text{Multiply both sides by 5 to clear fractions.}$$
$$5y - 25 = x - 2 \quad \text{Use the distributive property and simplify.}$$
$$-x + 5y - 25 = -2 \quad \text{Subtract } x \text{ from both sides.}$$
$$-x + 5y = 23 \quad \text{Add 25 to both sides.}$$

> **Helpful Hint**
>
> Multiply both sides of the equation $-x + 5y = 23$ by -1, and it becomes $x - 5y = -23$. Both $-x + 5y = 23$ and $x - 5y = -23$ are in standard form, and they are equations of the same line.

3   Recall from Section 3.3 that:

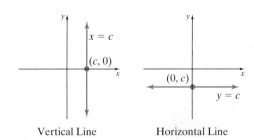

Vertical Line   Horizontal Line

EXAMPLE 3

Find an equation of the vertical line through $(-1, 5)$.

Solution   The equation of a vertical line can be written in the form $x = c$, so an equation for a vertical line passing through $(-1, 5)$ is $x = -1$.

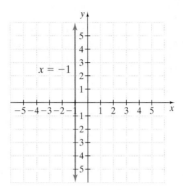

EXAMPLE 4

Find an equation of the line parallel to the line $y = 5$ and passing through $(-2, -3)$.

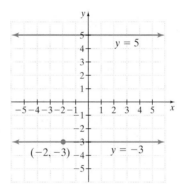

Solution   Since the graph of $y = 5$ is a horizontal line, any line parallel to it is also horizontal. The equation of a horizontal line can be written in the form $y = c$. An equation for the horizontal line passing through $(-2, -3)$ is $y = -3$.

4   Problems occurring in many fields can be modeled by linear equations in two variables. The next example is from the field of marketing and shows how consumer demand of a product depends on the price of the product.

EXAMPLE 5

PREDICTING THE SALES OF FRISBEES

The Whammo Company has learned that by pricing a newly released Frisbee at $6, sales will reach 2000 Frisbees per day. Raising the price to $8 will cause the sales to fall to 1500 Frisbees per day.

a. Assume that the relationship between sales price and number of Frisbees sold is linear and write an equation describing this relationship. Write the equation in slope-intercept form.

b. Predict the daily sales of Frisbees if the price is $7.50.

Solution

a. First, use the given information and write two ordered pairs. Ordered pairs will be in the form (sales price, number sold) so that our ordered pairs are (6, 2000) and (8, 1500). Use the point-slope form to write an equation. To do so, we find the slope of the line that contains these points.

$$m = \frac{2000 - 1500}{6 - 8} = \frac{500}{-2} = -250$$

Next, use the slope and either one of the points to write the equation in point-slope form. We use (6, 2000).

$$y - y_1 = m(x - x_1) \qquad \text{Use point-slope form.}$$
$$y - 2000 = -250(x - 6) \qquad \text{Let } x_1 = 6, y_1 = 2000, \text{ and } m = -250.$$
$$y - 2000 = -250x + 1500 \qquad \text{Use the distributive property.}$$
$$y = -250x + 3500 \qquad \text{Write in slope-intercept form.}$$

b. To predict the sales if the price is \$7.50, we find y when $x = 7.50$.

$$y = -250x + 3500$$
$$y = -250(7.50) + 3500 \qquad \text{Let } x = 7.50.$$
$$y = -1875 + 3500$$
$$y = 1625$$

If the price is \$7.50, sales will reach 1625 Frisbees per day.

The preceding example may also be solved by using ordered pairs of the form (number sold, sales price).

Forms of Linear Equations

$Ax + By = C$	**Standard form** of a linear equation. A and B are not both 0.
$y = mx + b$	**Slope-intercept form** of a linear equation. The slope is m and the y-intercept is $(0, b)$.
$y - y_1 = m(x - x_1)$	**Point-slope form** of a linear equation. The slope is m and (x_1, y_1) is a point on the line.
$y = c$	**Horizontal line** The slope is 0 and the y-intercept is $(0, c)$.
$x = c$	**Vertical line** The slope is undefined and the x-intercept is $(c, 0)$.

Parallel and Perpendicular Lines

Nonvertical parallel lines have the same slope.
The product of the slopes of two nonvertical perpendicular lines is -1.

Spotlight on DECISION & MAKING

Suppose you are a physical therapist. At weekly sessions, you are administering treadmill treatment to a patient. In the first phase of the treatment plan, the patient is to walk on a treadmill at a speed of 2 mph until fatigue sets in. Once the patient is able to walk 10 minutes at this speed, he will be ready for the next phase of treatment. However, you may change the treatment plan if it looks like it will take longer than 10 weeks to build up to 10 minutes of walking at 2 mph.

You record and plot the patient's progress on his chart for 5 weeks. Do you think you may need to change the first phase of treatment? Explain your reasoning. If not, when do you think the patient will be ready for the next phase of treatment?

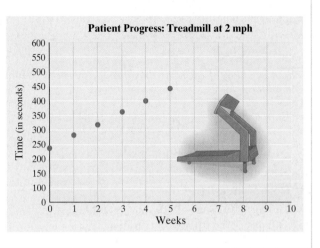

Patient Progress: Treadmill at 2 mph

2. $m = 5$; answers may vary, Ex. $(2, 1)$ **4.** $m = -7$; answers may vary, Ex. $(2, -6)$ **6.** $m = \frac{3}{7}$; answers may vary, Ex. $(-4, 0)$

MENTAL MATH

The graph of each equation below is a line. Use the equation to identify the slope and a point of the line.

1. $y - 8 = 3(x - 4)$ $m = 3$; answers may vary, Ex. $(4, 8)$

2. $y - 1 = 5(x - 2)$

3. $y + 3 = -2(x - 10)$ $m = -2$; answers may vary, Ex. $(10, -3)$

4. $y + 6 = -7(x - 2)$

5. $y = \frac{2}{5}(x + 1)$ $m = \frac{2}{5}$; answers may vary, Ex. $(-1, 0)$

6. $y = \frac{3}{7}(x + 4)$

TEACHING TIP

A Group Activity for this section is available in the Instructor's Resource Manual.

EXERCISE SET 3.6

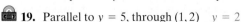

STUDY GUIDE/SSM   CD/ VIDEO   PH MATH TUTOR CENTER   MathXL®Tutorials ON CD   MathXL®   MyMathLab®

Use the point-slope form of the linear equation to find an equation of each line with the given slope and passing through the given point. Then write the equation in standard form. See Example 1.

1. Slope 6; through $(2, 2)$ $6x - y = 10$

2. Slope 4; through $(1, 3)$ $4x - y = 1$

3. Slope -8; through $(-1, -5)$ $8x + y = -13$

4. Slope -2; through $(-11, -12)$ $2x + y = -34$

5. Slope $\frac{1}{2}$; through $(5, -6)$ $x - 2y = 17$

6. Slope $\frac{2}{3}$; through $(-8, 9)$ $2x - 3y = -43$

Find an equation of the line through the given points. Write the equation in standard form. See Example 2.

7. Through $(3, 2)$ and $(5, 6)$ $2x - y = 4$

8. Through $(6, 2)$ and $(8, 8)$ $3x - y = 16$

9. Through $(-1, 3)$ and $(-2, -5)$ $8x - y = -11$

10. Through $(-4, 0)$ and $(6, -1)$ $x + 10y = -4$

11. Through $(2, 3)$ and $(-1, -1)$ $4x - 3y = -1$

12. Through $(0, 0)$ and $\left(\frac{1}{2}, \frac{1}{3}\right)$ $2x - 3y = 0$

Find an equation of each line. See Example 3.

13. Vertical line through $(0, 2)$ $x = 0$

14. Horizontal line through $(1, 4)$ $y = 4$

15. Horizontal line through $(-1, 3)$ $y = 3$

16. Vertical line through $(-1, 3)$ $x = -1$

17. Vertical line through $\left(-\frac{7}{3}, -\frac{2}{5}\right)$ $x = -\frac{7}{3}$

18. Horizontal line through $\left(\frac{2}{7}, 0\right)$ $y = 0$

Find an equation of each line. See Example 4.

19. Parallel to $y = 5$, through $(1, 2)$ $y = 2$

20. Perpendicular to $y = 5$, through $(1, 2)$ $x = 1$

△ **21.** Perpendicular to $x = -3$, through $(-2, 5)$ $y = 5$

△ **22.** Parallel to $y = -4$, through $(0, -3)$ $y = -3$

△ **23.** Parallel to $x = 0$, through $(6, -8)$ $x = 6$

△ **24.** Perpendicular to $x = 7$, through $(-5, 0)$ $y = 0$

MIXED PRACTICE

Find an equation of each line described. Write each equation in standard form.

25. With slope $-\dfrac{1}{2}$, through $\left(0, \dfrac{5}{3}\right)$ $3x + 6y = 10$

26. With slope $\dfrac{5}{7}$, through $(0, -3)$ $5x - 7y = 21$

27. Slope 1, through $(-7, 9)$ $x - y = -16$

28. Slope 5, through $(6, -8)$ $5x - y = 38$

29. Through $(10, 7)$ and $(7, 10)$ $x + y = 17$

30. Through $(5, -6)$ and $(-6, 5)$ $x + y = -1$

△ **31.** Through $(6, 7)$, parallel to the *x*-axis $y = 7$

△ **32.** Through $(0, -5)$, parallel to the *y*-axis $x = 0$

33. Slope $-\dfrac{4}{7}$, through $(-1, -2)$ $4x + 7y = -18$

34. Slope $-\dfrac{3}{5}$, through $(4, 4)$ $3x + 5y = 32$

35. Through $(-8, 1)$ and $(0, 0)$ $x + 8y = 0$

36. Through $(2, 3)$ and $(0, 0)$ $3x - 2y = 0$

37. Through $(0, 0)$ with slope 3 $3x - y = 0$

38. Through $(0, -2)$ with slope -1 $x + y = -2$

39. Through $(-6, -6)$ and $(0, 0)$ $x - y = 0$

40. Through $(0, 0)$ and $(4, 4)$ $x - y = 0$

41. Slope -5, *y*-intercept 7 $5x + y = 7$

42. Slope -2; *y*-intercept -4 $2x + y = -4$

43. Through $(-1, 5)$ and $(0, -6)$ $11x + y = -6$

44. Through $(4, 0)$ and $(0, -5)$ $5x - 4y = 20$

🔒 **45.** With undefined slope, through $\left(-\dfrac{3}{4}, 1\right)$ $x = -\dfrac{3}{4}$

46. With slope 0, through $(6.7, 12.1)$ $y = 12.1$

△ **47.** Through $(-2, -3)$, perpendicular to the *y*-axis $y = -3$

△ **48.** Through $(0, 12)$, perpendicular to the *x*-axis $x = 0$

49. With slope 7, through $(1, 3)$ $7x - y = 4$

50. With slope -10, through $(5, -1)$ $10x + y = 49$

Solve. Assume that each exercise describes a linear relationship. See Example 5.   **51. a.** $y = 640x + 4760$

51. In 2001 there were 6680 electric-powered vehicles in use in the United States. In 1998 only 4760 electric vehicles were being used. (*Source*: U.S. Energy Information Administration)

   a. Write an equation describing the relationship between time and number of electric-powered vehicles. Use ordered pairs of the form (years past 1998, number of vehicles).

   b. Use this equation to predict the number of electric-powered vehicles in use in 2008.   11,160 vehicles

52. In 1996 there were 457 thousand eating establishments in the United States. In 2000, there were 483 thousand eating establishments.

   a. Write an equation describing the relationship between time and number of eating establishments. Use ordered pairs of the form (years past 1996, number of eating establishments in thousands). $y = 6.5x + 457$

   b. Use this equation to predict the number of eating establishments in 2009.   541,500 or approx. 542 thousand eating establishments

53. A rock is dropped from the top of a 400-foot cliff. After 1 second, the rock is traveling 32 feet per second. After 3 seconds, the rock is traveling 96 feet per second.

53. a. $s = 32t$ or $y = 32x$

   a. Assume that the relationship between time and speed is linear and write an equation describing this relationship. Use ordered pairs of the form (time, speed).

   b. Use this equation to determine the speed of the rock 4 seconds after it was dropped.   128 ft/sec

54. A Hawaiian fruit company is studying the sales of a pineapple sauce to see if this product is to be continued. At the end of its first year, profits on this product amounted to $30,000. At the end of the fourth year, profits are $66,000.

   a. Assume that the relationship between years on the market and profit is linear and write an equation describing this relationship. Use ordered pairs of the form (years on the market, profit). $p = 12{,}000t + 18{,}000$

b. Use this equation to predict the profit at the end of 7 years. $102,000   **55. a.** $y = 0.93x + 70.3$

55. In 2000, the U.S. population per square mile of land area was 79.6. In 1990, the person per square mile population was 70.3.

a. Write an equation describing the relationship between year and persons per square mile. Use ordered pairs of the form (years past 1990, persons per square mile).

b. Use this equation to predict the person per square mile population in 2007.   86.11 persons per sq. mi

56. In 1996, there were 135 thousand apparel and accessory stores in the United States. In 2001, there were a total of 152 thousand apparel and accessory stores. (*Source*: U.S. Bureau of the Census. *County Business Patterns*, annual)

a. Write an equation describing this relationship. Use ordered pairs of the form (years past 1996, number of stores in thousands). $y = 3.4x + 135$

b. Use this equation to predict the number of apparel and accessory stores in 2009.

57. In 1995, the sales of battery-operated ride-on toys in the United States were $191 million. In 2000, the sales of these ride-on toys had risen to $260 million. (*Source*: Toy Manufacturers of America, Inc.)   **57. a.** $(0, 191), (5, 260)$

a. Write two ordered pairs of the form (years after 1995, sales of battery-operated ride-on toys) for this situation.

b. The relationship between years after 1995 and sales of battery-operated ride-on toys is linear over this period. Use the ordered pairs from part (a) to write an equation of the line relating year to sales of battery-operated ride-on toys.

c. Use the linear equation from part (b) to estimate the sales of ride-on toys in 1999. $246.2 million

58. In 2000, crude oil production by OPEC countries was 29.2 million barrels per day. In 2002, OPEC crude oil production had risen to about 28.7 million barrels per day. (*Source*: Energy Information Administration)

a. Write two ordered pairs of the form (years after 2000, crude oil production) for this situation. $(0, 29.2), (2, 28.7)$

b. Assume that the relationship between years after 1997 and crude oil production is linear over this period. Use the ordered pairs from part (a) to write an equation of the line relating year to crude oil production. $y = -0.25x + 29.2$

c. Use the linear equation from part (b) to estimate the crude oil production by OPEC countries in 2001.

△ **59.** A linear equation can be written that relates the radius of a circle to its circumference. A circle with a radius of 10 centimeters has a circumference of approximately 63 centimeters. A circle with a radius of 15 centimeters has a circumference of approximately 94 centimeters. Use the ordered pairs (10, 63) and (15, 94) to write a linear equation in standard form that approximates the relationship between the radius of the circle and its circumference. $31x - 5y = -5$

60. The value of a computer bought in 1998 depreciates or decreases as time passes. Two years after the computer was bought, it was worth $2600 and 5 years after it was bought, it was worth $2000.

a. If the relationship between number of years past 1998 and value of the computer is linear, write an equation describing

this relationship. Use ordered pairs of the form (years past 1998, value of computer). $y = -200x + 3000$

b. Use this equation to estimate the value of the computer in the year 2008. $1000

61. The Pool Fun Company has learned that, by pricing a newly released Fun Noodle at $3, sales will reach 10,000 Fun Noodles per day during the summer. Raising the price to $5 will cause the sales to fall to 8000 Fun Noodles per day.

a. Assume that the relationship between sales price and number of Fun Noodles sold is linear and write an equation describing this relationship. Use ordered pairs of the form (sales price, number sold). $S = -1000p + 13,000$

b. Predict the daily sales of Fun Noodles if the price is $3.50.
61. b. 9500 Fun Noodles

62. a. $V = -5000t + 200,000$

62. The value of a building bought in 1985 may be depreciated (or decreased) as time passes for income tax purposes. Seven years after the building was bought, this value was $165,000 and 12 years after it was bought, this value was $140,000.

a. If the relationship between number of years past 1985 and the depreciated value of the building is linear, write an equation describing this relationship. Use ordered pairs of the form (years past 1985, value of building).

b. Use this equation to estimate the depreciated value of the building in 2005. $100,000

REVIEW AND PREVIEW

Find the value of $x^2 - 3x + 1$ for each given value of x. See Section 1.7.

63. 2 -1 **64.** 5   11   **65.** -1 5   **66.** -3 19

For each graph, determine whether any x-values correspond to two or more y-values. See Section 6.1.

67. no   **68.** no

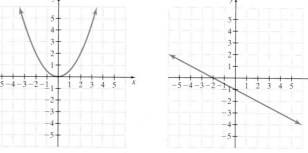

56. b. approx. 179 thousand apparel and accessory stores   **57. b.** $y = 13.8x + 191$ **58. c.** 28.95 million barrels per day

69. yes

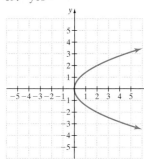

70. yes

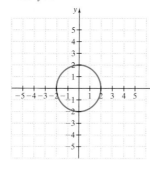

△ **73.** Write an equation in standard form of the line that contains the point $(-1, 2)$ and is
    a. parallel to the line $y = 3x - 1$. $3x - y = -5$
    b. perpendicular to the line $y = 3x - 1$. $x + 3y = 5$

△ **74.** Write an equation in standard form of the line that contains the point $(4, 0)$ and is
    a. parallel to the line $y = -2x + 3$. $2x + y = 8$
    b. perpendicular to the line $y = -2x + 3$. $x - 2y = 4$

△ **75.** Write an equation in standard form of the line that contains the point $(3, -5)$ and is
    a. parallel to the line $3x + 2y = 7$. $3x + 2y = -1$
    b. perpendicular to the line $3x + 2y = 7$. $2x - 3y = 21$

Concept Extensions   **71.–72.** answers may vary.

71. Given the equation of a nonvertical line, explain how to find the slope without finding two points on the line.

72. Given two points on a nonvertical line, explain how to use the point-slope form to find the equation of the line.

△ **76.** Write an equation in standard form of the line that contains the point $(-2, 4)$ and is
    a. parallel to the line $x + 3y = 6$. $x + 3y = 10$
    b. perpendicular to the line $x + 3y = 6$. $3x - y = -10$

3.7   FUNCTIONS

Objectives

1   Identify relations, domains, and ranges.

2   Identify functions.

3   Use the vertical line test.

4   Use function notation.

1   In previous sections, we have discussed the relationships between two quantities. For example, the relationship between the length of the side of a square x and its area y is described by the equation $y = x^2$. These variables x and y are related in the following way: for any given value of x, we can find the corresponding value of y by squaring the x-value. Ordered pairs can be used to write down solutions of this equation. For example, $(2, 4)$ is a solution of $y = x^2$, and this notation tells us that the x-value 2 is related to the y-value 4 for this equation. In other words, when the length of the side of a square is 2 units, its area is 4 square units.

A set of ordered pairs is called a **relation**. The set of all x-coordinates is called the **domain** of a relation, and the set of all y-coordinates is called the **range** of a relation. Equations such as $y = x^2$ are also called relations since equations in two variables define a set of ordered pair solutions.

EXAMPLE 1

Find the domain and the range of the relation $\{(0, 2), (3, 3), (-1, 0), (3, -2)\}$.

Solution   The domain is the set of all x-values or $\{-1, 0, 3\}$, and the range is the set of all y-values, or $\{-2, 0, 2, 3\}$.

2   Some relations are also functions.

> ### Function
> A function is a set of ordered pairs that assigns to each *x*-value exactly one *y*-value.

EXAMPLE 2

TEACHING TIP

Remind students that all functions are relations, but not all relations are functions.

CLASSROOM EXAMPLE

Which of the following relations are also function?
a. $\{(2, 5), (-3, 7), (4, 5), (0, -1)\}$
b. $\{(1, 4), (6, 6), (1, -3), (7, 5)\}$
answer:
a. function   **b.** not a function

Which of the following relations are also functions?

a. $\{(-1, 1), (2, 3), (7, 3), (8, 6)\}$ **b.** $\{(0, -2), (1, 5), (0, 3), (7, 7)\}$

Solution   **a.** Although the ordered pairs $(2, 3)$ and $(7, 3)$ have the same *y*-value, each *x*-value is assigned to only one *y*-value so this set of ordered pairs is a function.

b. The *x*-value 0 is assigned to two *y*-values, -2 and 3, so this set of ordered pairs is not a function.

Relations and functions can be described by a graph of their ordered pairs.

EXAMPLE 3

CLASSROOM EXAMPLE

Which graph is the graph of a function?
a.

b.

answer:
a. function   **b.** not a function

Which graph is the graph of a function?

a. 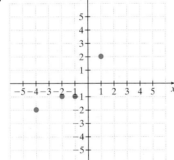   **b.**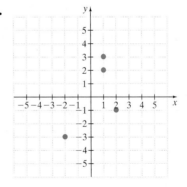

Solution   **a.** This is the graph of the relation $\{(-4, -2), (-2, -1)(-1, -1), (1, 2)\}$. Each *x*-coordinate has exactly one *y*-coordinate, so this is the graph of a function.

b. This is the graph of the relation $\{(-2, -3), (1, 2), (1, 3), (2, -1)\}$. The *x*-coordinate 1 is paired with two *y*-coordinates, 2 and 3, so this is not the graph of a function.

3   The graph in Example 3(b) was not the graph of a function because the *x*-coordinate 1 was paired with two *y*-coordinates, 2 and 3. Notice that when an *x*-coordinate is paired with more than one *y*-coordinate, a vertical line can be drawn that will intersect the graph at more than one point. We can use this fact to determine whether a relation is also a function. We call this the **vertical line test**.

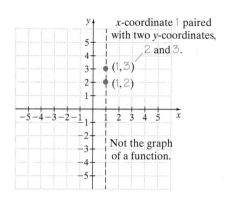

TEACHING TIP
Show students many visual examples of this to help them remember.

Vertical Line Test

If a vertical line can be drawn so that it intersects a graph more than once, the graph is not the graph of a function.

This vertical line test works for all types of graphs on the rectangular coordinate system.

EXAMPLE 4

Use the vertical line test to determine whether each graph is the graph of a function.

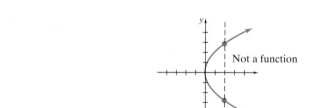

a. b. c. d.

Solution **a.** This graph is the graph of a function since no vertical line will intersect this graph more than once.

CLASSROOM EXAMPLE
Determine whether each graph is a function.
a.

b.

answer: **a.** yes **b.** no

b. This graph is also the graph of a function; no vertical line will intersect it more than once.

c. This graph is not the graph of a function. Vertical lines can be drawn that intersect the graph in two points. An example of one is shown.

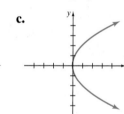

Not a function

d. This graph is not the graph of a function. A vertical line can be drawn that intersects this line at every point.

Recall that the graph of a linear equation is a line, and a line that is not vertical will pass the vertical line test. **Thus, all linear equations are functions except those of the form $x = c$, which are vertical lines.**

EXAMPLE 5

Which of the following linear equations are functions?

a. $y = x$ **b.** $y = 2x + 1$ **c.** $y = 5$ **d.** $x = -1$

Solution   **a, b**, and **c** are functions because their graphs are nonvertical lines. **d** is not a function because its graph is a vertical line.

Examples of functions can often be found in magazines, newspapers, books, and other printed material in the form of tables or graphs such as that in Example 6.

EXAMPLE 6

The graph shows the sunrise time for Indianapolis, Indiana, for the year. Use this graph to answer the questions.

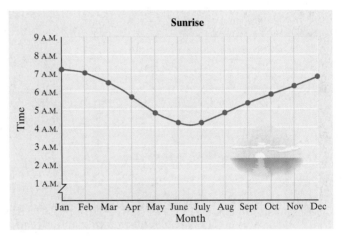

a. Approximate the time of sunrise on February 1.

b. Approximately when does the sun rise at 5 A.M.?

c. Is this the graph of a function?

Solution   **a.** To approximate the time of sunrise on February 1, we find the mark on the horizontal axis that corresponds to February 1. From this mark, we move vertically upward until the graph is reached. From that point on the graph, we move horizontally to the left until the vertical axis is reached. The vertical axis there reads 7 A.M.

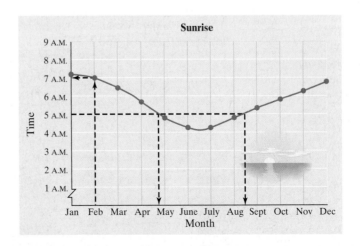

b. To approximate when the sun rises at 5 A.M., we find 5 A.M. on the time axis and move horizontally to the right. Notice that we will reach the graph twice, corresponding to two dates for which the sun rises at 5 A.M. We follow both points on the graph vertically downward until the horizontal axis is reached. The sun rises at 5 A.M. at approximately the end of the month of April and the middle of the month of August.

c. The graph is the graph of a function since it passes the vertical line test. In other words, for every day of the year in Indianapolis, there is exactly one sunrise time.

4   The graph of the linear equation $y = 2x + 1$ passes the vertical line test, so we say that $y = 2x + 1$ is a function. In other words, $y = 2x + 1$ gives us a rule for writing ordered pairs where every x-coordinate is paired with one y-coordinate.

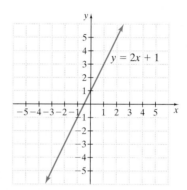

We often use letters such as f, g, and h to name functions. For example, the symbol $f(x)$ means *function of x* and is read "f of x." This notation is called **function notation**. The equation $y = 2x + 1$ can be written as $f(x) = 2x + 1$ using function notation, and these equations mean the same thing. In other words, $y = f(x)$.

The notation $f(1)$ means to replace x with 1 and find the resulting y or function value. Since

$$f(x) = 2x + 1$$

then

$$f(1) = 2(1) + 1 = 3$$

This means that, when $x = 1$, y or $f(x) = 3$, and we have the ordered pair $(1, 3)$. Now let's find $f(2), f(0)$, and $f(-1)$.

$$f(x) = 2x + 1 \qquad\qquad f(x) = 2x + 1 \qquad\qquad f(x) = 2x + 1$$
$$f(2) = 2(2) + 1 \qquad\quad f(0) = 2(0) + 1 \qquad\quad f(-1) = 2(-1) + 1$$
$$= 4 + 1 \qquad\qquad\qquad = 0 + 1 \qquad\qquad\qquad = -2 + 1$$
$$= 5 \qquad\qquad\qquad\quad = 1 \qquad\qquad\qquad\quad = -1$$

Ordered
Pair: $(2, 5)$ $(0, 1)$ $(-1, -1)$

> ### Helpful Hint
> Note that $f(x)$ is a special symbol in mathematics used to denote a function. The symbol $f(x)$ is read "f of x." It does **not** mean $f \cdot x$ (f times x).

EXAMPLE 7

Given $g(x) = x^2 - 3$, find the following. Then write down the corresponding ordered pairs generated.

a. $g(2)$ **b.** $g(-2)$ **c.** $g(0)$

Solution

a. $g(x) = x^2 - 3$
$g(2) = 2^2 - 3$
$= 4 - 3$
$= 1$

b. $g(x) = x^2 - 3$
$g(-2) = (-2)^2 - 3$
$= 4 - 3$
$= 1$

c. $g(x) = x^2 - 3$
$g(0) = 0^2 - 3$
$= 0 - 3$
$= -3$

Ordered Pair:

$g(2) = 1$ gives $(2, 1)$

$g(-2) = 1$ gives $(-2, 1)$

$g(0) = -3$ gives $(0, -3)$

We now practice finding the domain and the range of a function. The domain of our functions will be the set of all possible real numbers that x can be replaced by. The range is the set of corresponding y-values.

EXAMPLE 8

Find the domain of each function.

a. $g(x) = \dfrac{1}{x}$ **b.** $f(x) = 2x + 1$

Solution

a. Recall that we cannot divide by 0 so that the domain of $g(x)$ is the set of all real numbers except 0. In interval notation, we can write $(-\infty, 0) \cup (0, \infty)$.

b. In this function, x can be any real number. The domain of $f(x)$ is the set of all real numbers, or $(-\infty, \infty)$ in interval notation.

✔ **CONCEPT CHECK**

Suppose that the value of f is -7 when the function is evaluated at 2. Write this situation in function notation.

EXAMPLE 9

Find the domain and the range of each function graphed. Use interval notation.

a.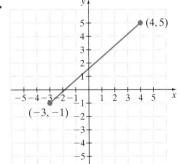

b.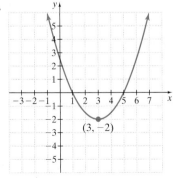

Solution **a.**

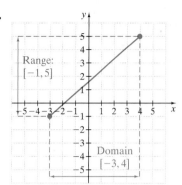

Range: [−1, 5]

Domain [−3, 4]

b.

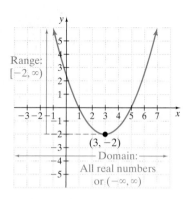

Range: [−2, ∞)

(3, −2)

Domain: All real numbers or (−∞, ∞)

Spotlight on DECISION MAKING

Suppose you are the property and grounds manager for a small company in Portland, Oregon. A recent heat wave has caused employees to ask for air-conditioning in the building. As you begin to research air-conditioning installation, you discover this graph of average high temperatures for Portland. Because the building is surrounded by trees, the current ventilation system (without air-conditioning) is able to keep the temperature 10°F cooler than the outside temperature on a warm day. A comfortable temperature range for working is 68°F to 74°F. Would you recommend installing air-conditioning in the building? Explain your reasoning. What other factors would you want to consider?

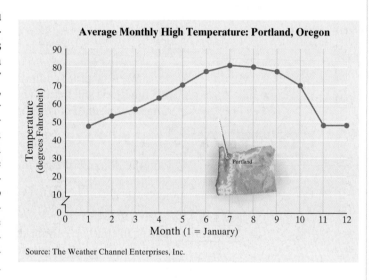

Average Monthly High Temperature: Portland, Oregon

Portland

Month (1 = January)

Source: The Weather Channel Enterprises, Inc.

TEACHING TIP

A Group Activity for this section is available in the Instructor's Resource Manual.

EXERCISE SET 3.7

STUDY GUIDE/SSM   CD/VIDEO   PH MATH TUTOR CENTER   MathXL®Tutorials ON CD   MathXL®   MyMathLab®

1. domain: $\{-7, 0, 2, 10\}$; range: $\{-7, 0, 4, 10\}$ **2.** domain: $\{-2, 1, 3\}$; range: $\{-6, -2, 4\}$ **4.** domain: $\{5\}$; range: $\{-3, 0, 3, 4\}$

Find the domain and the range of each relation. See Example 1.

1. $\{(2, 4), (0, 0), (-7, 10), (10, -7)\}$

2. $\{(3, -6), (1, 4), (-2, -2)\}$

3. $\{(0, -2), (1, -2)(5, -2)\}$ domain: $\{0, 1, 5\}$; range: $\{-2\}$

4. $\{(5, 0), (5, -3), (5, 4), (5, 3)\}$

Determine which relations are also functions. See Example 2.

5. $\{(1, 1), (2, 2), (-3, -3), (0, 0)\}$ yes

6. $\{(1, 2), (3, 2), (4, 2)\}$ yes

7. $\{(-1, 0), (-1, 6), (-1, 8)\}$ no

8. $\{(11, 6), (-1, -2), (0, 0), (3, -2)\}$ yes

Use the vertical line test to determine whether each graph is the graph of a function. See Example 4.

9. no

10. yes

11. yes

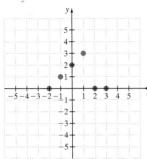

12. no

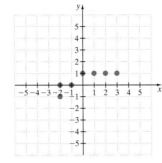

13. yes

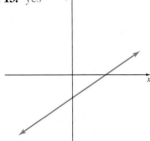

14. yes

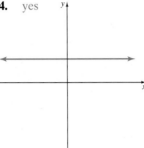

15. no

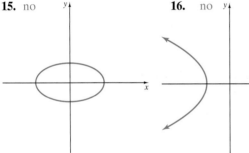

17. no

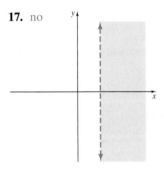

18. no

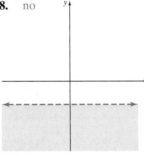

19. yes

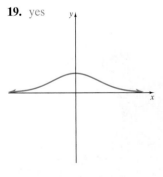

20. no

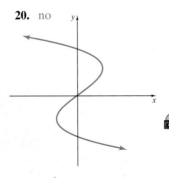

Decide whether the equation describes a function. See Example 5.

21. $y = x + 1$ yes

22. $y = x - 1$ yes

23. $y - x = 7$ yes

24. $2x - 3y = 9$ yes

25. $y = 6$ yes

26. $x = 3$ no

27. $x = -2$ no

28. $y = -9$ yes

29. $x = y^2$ no

30. $y = x^2 - 3$ yes

Use the graph in Example 6 to answer Exercises 31 through 34.
33. answers may vary

31. Approximate the time of sunrise on September 1 in Indianapolis.   5:20 A.M.

32. Approximate the date(s) when the sun rises in Indianapolis at 7 A.M.   Feb. 1

33. Describe the change in sunrise over the year for Indianapolis.

34. When, in Indianapolis, is the earliest sunrise? What point on the graph does this correspond to?   mid-June; lowest point

This graph shows the U.S. hourly minimum wage at the beginning of each year shown. Use this graph to answer Exercises 35 through 40.

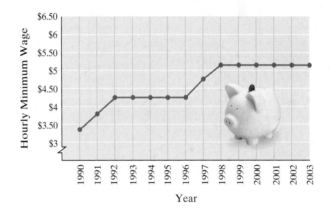

35. $4.75 per hour **36.** $5.15 per hour **39.–40.** answers may vary

35. Approximate the minimum wage at the beginning of 1997.

36. Approximate the minimum wage at the beginning of 1999.

37. Approximate the year when the minimum wage increased to over $4.50 per hour.   1996

38. Approximate the year when the minimum wage increased to over $5.00 per hour.   1997

39. Is this graph the graph of a function? Why or why not?

40. Do you think that a similar graph of your hourly wage on January 1 of every year (whether you are working or not) will be the graph of a function? Why or why not?

Given the following functions, find $f(-2)$, $f(0)$, and $f(3)$. See Example 7.

41. $f(x) = 2x - 5$ $-9, -5, 1$ **42.** $f(x) = 3 - 7x$ $17, 3, -18$

43. $f(x) = x^2 + 2$ $6, 2, 11$ **44.** $f(x) = x^2 - 4$ $0, -4, 5$

45. $f(x) = x^3$ $-8, 0, 27$ **46.** $f(x) = -x^3$ $8, 0, -27$

47. $f(x) = |x|$ $2, 0, 3$ **48.** $f(x) = |2 - x|$ $4, 2, 1$

49. $-5, 0, 20; (-1, -5), (0, 0), (4, 20)$ **50.** $3, 0, -12; (-1, 3), (0, 0), (4, -12)$
51. $5, 3, 35; (-1, 5), (0, 3), (4, 35)$ **52.** $-1, 0, -16; (-1, -1), (0, 0), (4, -16)$ **53.** $4, 3, -21; (-1, 4), (0, 3), (4, -21)$

Given the following functions, find $h(-1), h(0),$ and $h(4)$. Then write the corresponding ordered pair. See Example 7.

54. $-8, -3, -3; (-1, -8), (0, -3), (4, -3)$

49. $h(x) = 5x$ **50.** $h(x) = -3x$

51. $h(x) = 2x^2 + 3$ **52.** $h(x) = -x^2$

53. $h(x) = -x^2 - 2x + 3$ **54.** $h(x) = -x^2 + 4x - 3$

55. $h(x) = 6$ **56.** $h(x) = -12$

55. $6, 6, 6: (-1, 6), (0, 6), (4, 6)$

Find the domain of each function. See Example 8.

56. $-12, -12, -12 (-1, -12), (0, -12), (4, -12)$

57. $f(x) = 3x - 7$ $(-\infty, \infty)$ **58.** $g(x) = 5 - 2x$ $(-\infty, \infty)$

59. $h(x) = \dfrac{1}{x + 5}$ **60.** $f(x) = \dfrac{1}{x - 6}$

61. $g(x) = |x + 1|$ $(-\infty, \infty)$ **62.** $h(x) = |2x|$ $(-\infty, \infty)$

59. all real numbers except -5 or $(-\infty, -5) \cup (-5, \infty)$

Find the domain and the range of each function graphed. See Example 9.   **60.** all real numbers except 6 or $(-\infty, 6) \cup (6, \infty)$

63. domain: $(-\infty, \infty)$ range: $[-4, \infty)$

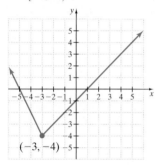

$(-3, -4)$

64. domain: $(-\infty, \infty)$; range: $(-\infty, 5]$

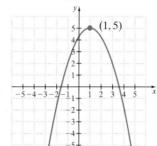

$(1, 5)$

65. domain: $(-\infty, \infty)$; range: $(-\infty, \infty)$

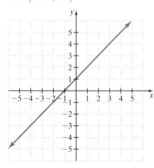

66. domain: $(-\infty, \infty)$; range: $(-\infty, \infty)$

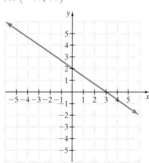

67. domain: $(-\infty, \infty)$; range: $\{2\}$

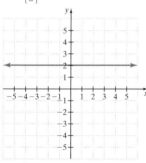

68. domain: $\{3\}$; range: $(-\infty, \infty)$

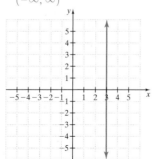

REVIEW AND PREVIEW

Find the coordinates of the point of intersection. See Section 3.1.

69. $(-2, 1)$

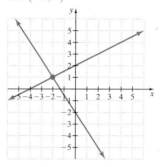

70. $(3, 0)$

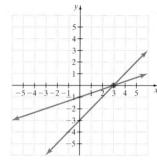

71. $(-3, -1)$

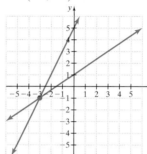

72. $(-3, -3)$

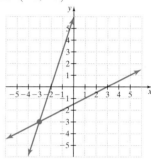

Concept Extensions

73. Forensic scientists use the function

$$H(x) = 2.59x + 47.24$$

to estimate the height of a woman in centimeters given the length x of her femur bone.

a. Estimate the height of a woman whose femur measures 46 centimeters.   166.38 cm

b. Estimate the height of a woman whose femur measures 39 centimeters.   148.25 cm

74. The dosage in milligrams D of Ivermectin, a heartworm preventive for a dog who weighs x pounds, is given by the function

$$D(x) = \dfrac{136}{25}x$$

a. Find the proper dosage for a dog that weighs 35 pounds.

b. Find the proper dosage for a dog that weighs 70 pounds.

75. In your own words define **(a)** function; **(b)** domain; **(c)** range.

76. Explain the vertical line test and how it is used.

77. Since $y = x + 7$ is a function, rewrite the equation using function notation. $f(x) = x + 7$

74.a. 190.4 mg   **74.b.** 380.8 mg   **75–76.** answers may vary

See the example below for Exercises 78 through 81.

Example

If $f(x) = x^2 + 2x + 1$, find $f(\pi)$.

Solution:

$f(x) = x^2 + 2x + 1$
$f(\pi) = \pi^2 + 2\pi + 1$

Given the following functions, find the indicated values.

78. $f(x) = 2x + 7$
   a. $f(2)$ 11   **b.** $f(a)$ $2a + 7$

79. $g(x) = -3x + 12$
   a. $g(s)$ $-3s + 12$ **b.** $g(r)$ $-3r + 12$

80. $h(x) = x^2 + 7$
   a. $h(3)$ 16   **b.** $h(a)$ $a^2 + 7$

81. $f(x) = x^2 - 12$
   a. $f(12)$ 132   **b.** $f(a)$ $a^2 - 12$

CHAPTER 3 PROJECT

Financial Analysis

Investment analysts investigate a company's sales, net profit, debt, and assets to decide whether investing in it is a wise choice. One way to analyze this data is to graph it and look for trends over time. Another way is to find algebraically the rate at which the data change over time.

The table below gives the net profits in millions of dollars for the leading U.S. businesses in the pharmaceutical industry for the years 1997 and 1998. In this project, you will analyze the performances of these companies and, based on this information alone, make an investment recommendation. This project may be completed by working in groups or individually.

PHARMACEUTICAL INDUSTRY NET PROFITS (IN MILLIONS OF DOLLARS)

Company	2000	2001
Merck	$40,363	$47,716
Pfizer	$29,574	$32,259
Johnson & Johnson	$29,139	$33,004
Bristol-Myers Squibb	$21,331	$21,717
Pharmacia	$18,150	$19,299
Abbott Laboratories	$13,746	$16,285
Eli Lilly	$10,862	$11,543
Schering-Plough	$ 9815	$ 9802

(*Source: Fortune Magazine*)

1. Scan the table. Did any of the companies have a loss during the years shown? If so, which company and when? What does this mean?

2. Write the data for each company as two ordered pairs of the form (year, net profit). Assuming that the trends in net profit are linear, use graph paper to graph the line represented by the ordered pairs for each company. Describe the trend shown by each graph.

3. Find the slope of the line for each company.

4. Which of the lines, if any, have positive slopes? What does that mean in this context? Which of the lines, if any, have negative slopes? What does that mean in this context?

Of these pharmaceutical companies, which one(s) would you recommend as an investment choice? Why?

Do you think it is wise to make a decision after looking at only 2 years of net profits? What other factors do you think should be taken into consideration when making an investment choice?

(Optional) Use financial magazines, company annual reports, or online investing information to find net profit information for two different years for two to four companies in the same industry. Analyze the net profits and make an investment recommendation.

CHAPTER VOCABULARY CHECK

Fill in each blank with one of the words listed below.

relation	function	domain	range	standard	slope-intercept
y-axis	*x*-axis	solution	linear	slope	point-slope
x-intercept	*y*-intercept	*y*	*x*		

1. An ordered pair is a solution of an equation in two variables if replacing the variables by the coordinates of the ordered pair results in a true statement.

2. The vertical number line in the rectangular coordinate system is called the *y*-axis .

3. A linear equation can be written in the form $Ax + By = C$.

4. A(n) *x*-intercept is a point of the graph where the graph crosses the *x*-axis.

5. The form $Ax + By = C$ is called standard form.

6. A(n) *y*-intercept is a point of the graph where the graph crosses the *y*-axis.

7. The equation $y = 7x - 5$ is written in slope-intercept form.

8. The equation $y + 1 = 7(x - 2)$ is written in point-slope form.

9. To find an *x*-intercept of a graph, let *y* = 0.

10. The horizontal number line in the rectangular coordinate system is called the *x*-axis .

11. To find a *y*-intercept of a graph, let *x* = 0.

12. The slope of a line measures the steepness or tilt of a line.

13. A set of ordered pairs that assigns to each *x*-value exactly one *y*-value is called a function .

14. The set of all *x*-coordinates of a relation is called the domain of the relation.

15. The set of all *y*-coordinates of a relation is called the range of the relation.

16. A set of ordered pairs is called a relation .

CHAPTER 3   HIGHLIGHTS

Definitions and Concepts	**Examples**

Section 3.1   The Rectangular Coordinate System

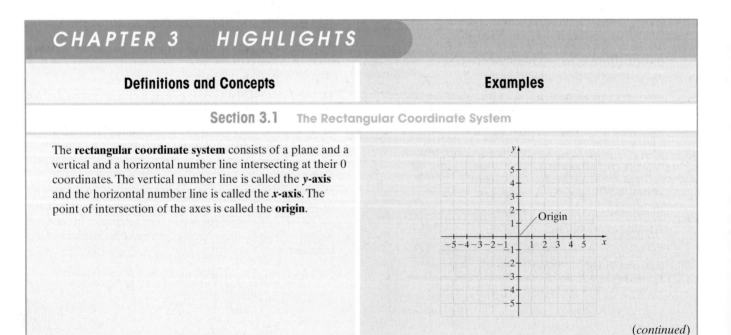

The **rectangular coordinate system** consists of a plane and a vertical and a horizontal number line intersecting at their 0 coordinates. The vertical number line is called the **y-axis** and the horizontal number line is called the **x-axis**. The point of intersection of the axes is called the **origin**.

(continued)

Definitions and Concepts	**Examples**

Section 3.1 The Rectangular Coordinate System

To **plot** or **graph** an ordered pair means to find its corresponding point on a rectangular coordinate system.

To plot or graph an ordered pair such as $(3, -2)$, start at the origin. Move 3 units to the right and from there, 2 units down.

To plot or graph $(-3, 4)$ start at the origin. Move 3 units to the left and from there, 4 units up.

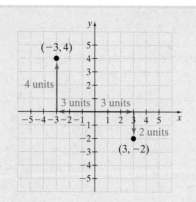

An ordered pair is a **solution** of an equation in two variables if replacing the variables by the coordinates of the ordered pair results in a true statement.

Determine whether $(-1, 5)$ is a solution of $2x + 3y = 13$.
$$2x + 3y = 13$$
$$2(-1) + 3 \cdot 5 = 13 \quad \text{Let } x = -1, y = 5$$
$$-2 + 15 = 13$$
$$13 = 13 \quad \text{True.}$$

If one coordinate of an ordered pair solution is known, the other value can be determined by substitution.

Complete the ordered pair solution $(0, \quad)$ for the equation $x - 6y = 12$.

$$x - 6y = 12$$
$$0 - 6y = 12 \quad \text{Let } x = 0.$$
$$\frac{-6y}{-6} = \frac{12}{-6} \quad \text{Divide by } -6.$$
$$y = -2$$

The ordered pair solution is $(0, -2)$.

Section 3.2 Graphing Linear Equations

A **linear equation in two variables** is an equation that can be written in the form $Ax + By = C$ where A and B are not both 0. The form $Ax + By = C$ is called **standard form**.

To graph a linear equation in two variables, find three ordered pair solutions. Plot the solution points and draw the line connecting the points.

Linear Equations

$$3x + 2y = -6 \qquad x = -5$$
$$y = 3 \qquad y = -x + 10$$

$x + y = 10$ is in standard form.

Graph $x - 2y = 5$.

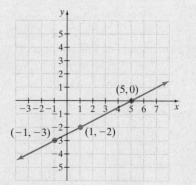

x	y
5	0
1	-2
-1	-3

(continued)

Definitions and Concepts	**Examples**

Section 3.3   Intercepts

An **intercept** of a graph is a point where the graph intersects an axis. If a graph intersects the x-axis at a, then $(a, 0)$ is the **x-intercept**. If a graph intersects the y-axis at b, then $(0, b)$ is the **y-intercept**.

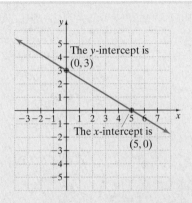

To find the x-intercept, let $y = 0$ and solve for x.

To find the y-intercept, let $x = 0$ and solve for y.

Graph $2x - 5y = -10$ by finding intercepts.

$$
\begin{array}{cc}
\text{If } y = 0, \text{ then} & \text{If } x = 0, \text{ then} \\
2x - 5 \cdot 0 = -10 & 2 \cdot 0 - 5y = -10 \\
2x = -10 & -5y = -10 \\
\dfrac{2x}{2} = \dfrac{-10}{2} & \dfrac{-5y}{-5} = \dfrac{-10}{-5} \\
x = -5 & y = 2
\end{array}
$$

The x-intercept is $(-5, 0)$. The y-intercept is $(0, 2)$.

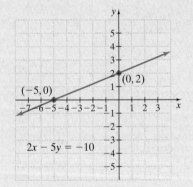

The graph of $x = c$ is a vertical line with x-intercept $(c, 0)$.

The graph of $y = c$ is a horizontal line with y-intercept $(0, c)$.

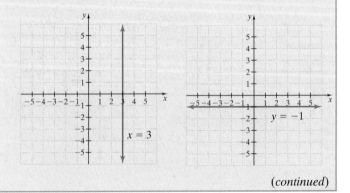

(continued)

Definitions and Concepts	**Examples**

Section 3.4   Slope and Rate of Change

The **slope** m of the line through points (x_1, y_1) and (x_2, y_2) is given by

$$m = \frac{y_2 - y_1}{x_2 - x_1} \qquad \text{as long as } x_2 \neq x_1$$

A horizontal line has slope 0.

The slope of a vertical line is undefined.

Nonvertical parallel lines have the same slope.

Two nonvertical lines are perpendicular if the slope of one is the negative reciprocal of the slope of the other.

The slope of the line through points $(-1, 6)$ and $(-5, 8)$ is

$$m = \frac{y_2 - y_1}{x_2 - x_1} = \frac{8 - 6}{-5 - (-1)} = \frac{2}{-4} = -\frac{1}{2}$$

The slope of the line $y = -5$ is 0.

The line $x = 3$ has undefined slope.

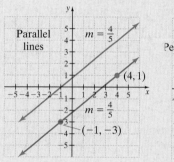 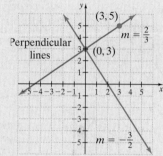

Section 3.5   The Slope-Intercept Form

Slope-Intercept Form

$$y = mx + b$$

m is the slope of the line.
$(0, b)$ is the y-intercept.

Find the slope and the y-intercept of the line whose equation is $2x + 3y = 6$.

Solve for y:

$$2x + 3y = 6$$
$$3y = -2x + 6 \qquad \text{Subtract } 2x.$$
$$y = -\frac{2}{3}x + 2 \qquad \text{Divide by 3.}$$

The slope of the line is $-\frac{2}{3}$ and the y-intercept is $(0, 2)$.

Find an equation of the line with slope 3 and y-intercept $(0, -1)$.

The equation is $y = 3x - 1$.

Section 3.6   The Point-Slope Form

Point-Slope Form

$$y - y_1 = m(x - x_1)$$

m is the slope.
(x_1, y_1) is a point on the line.

Find an equation of the line with slope $\frac{3}{4}$ that contains the point $(-1, 5)$.

$$y - 5 = \frac{3}{4}[x - (-1)]$$

$$4(y - 5) = 3(x + 1) \qquad \text{Multiply by 4.}$$
$$4y - 20 = 3x + 3 \qquad \text{Distribute.}$$
$$-3x + 4y = 23 \qquad \text{Subtract } 3x \text{ and add 20.}$$

(continued)

Definitions and Concepts	Examples

Section 3.7   Functions

A set of ordered pairs is a **relation**. The set of all *x*-coordinates is called the **domain** of the relation and the set of all *y*-coordinates is called the **range** of the relation.

The domain of the relation $\{(0, 5), (2, 5), (4, 5), (5, -2)\}$ is $\{0, 2, 4, 5\}$. The range is $\{-2, 5\}$.

A **function** is a set of ordered pairs that assigns to each *x*-value exactly one *y*-value.

Which are graphs of functions?

Vertical Line Test

If a vertical line can be drawn so that it intersects a graph more than once, the graph is not the graph of a function.

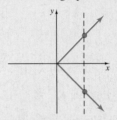   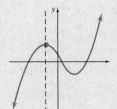

This graph is not the graph of a function.   This graph is the graph of a function.

The symbol $f(x)$ means **function of x**. This notation is called **function** notation.

If $f(x) = 2x^2 + 6x - 1$, find $f(3)$.

$$\begin{aligned} f(3) &= 2(3)^2 + 6 \cdot 3 - 1 \\ &= 2 \cdot 9 + 18 - 1 \\ &= 18 + 18 - 1 \\ &= 35 \end{aligned}$$

3 CHAPTER REVIEW

8. a. (1996, 10.5) (1997, 10) (1998, 9.8) (1999, 9.9) (2000, 9.6) (2001, 9.8)

(3.1) *Plot the following ordered pairs on a Cartesian coordinate system.*   **1.–6.** See graphing answer section.

1. $(-7, 0)$

2. $\left(0, 4\frac{4}{5}\right)$

3. $(-2, -5)$

4. $(1, -3)$

5. $(0.7, 0.7)$

6. $(-6, 4)$

7. A local lumberyard uses quantity pricing. The table shows the price per board for different amounts of lumber purchased.

7. a. (8.00, 1), (7.50, 10), (6.50, 25), (5.00, 50), (2.00, 100)

Price per Board (in dollars)	Number of Boards Purchased
8.00	1
7.50	10
6.50	25
5.00	50
2.00	100

a. Write each paired data as an ordered pair of the form (price per board, number of boards purchased).

b. Create a scatter diagram of the paired data. Be sure to label the axes appropriately.   See graphing answer section.

8. The table shows the annual overnight stays in national parks (*Source*: National Park Service)

Year	Overnight Stays in National Parks (in millions)
1996	10.5
1997	10
1998	9.8
1999	9.9
2000	9.6
2001	9.8

a. Write each paired data as an ordered pair of the form (year, number of overnight stays).

b. Create a scatter diagram of the paired data. Be sure to label the axes properly.   See graphing answer section.

Determine whether each ordered pair is a solution of the given equation.

9. $7x - 8y = 56; (0, 56), (8, 0)$ no; yes

10. $-2x + 5y = 10; (-5, 0), (1, 1)$ yes; no

11. $x = 13; (13, 5), (13, 13)$ yes; yes

12. $y = 2; (7, 2), (2, 7)$ yes; no

Complete the ordered pairs so that each is a solution of the given equation.

13. $-2 + y = 6x; (7, \ )$ (7, 44)

14. $y = 3x + 5; (\ , -8)$ $\left(-\frac{13}{3}, -8\right)$

Complete the table of values for each given equation; then plot the ordered pairs. Use a single coordinate system for each exercise.
15.–17. See graphing answer section.

15. $9 = -3x + 4y$

x	y
-3	0
1	3
9	9

16. $y = 5$

x	y
7	5
-7	5
0	5

17. $x = 2y$

x	y
0	0
10	5
-10	-5

18. The cost in dollars of producing x compact disk holders is given by $y = 5x + 2000$.

a. Complete the following table.

x	y
1	2005
100	2500
1000	7000

b. Find the number of compact disk holders that can be produced for $6430. 886 compact disk holders

(3.2) Graph each linear equation.

19. $x - y = 1$ **20.** $x + y = 6$

21. $x - 3y = 12$ **22.** $5x - y = -8$

23. $x = 3y$ **24.** $y = -2x$

25. $2x - 3y = 6$ **26.** $4x - 3y = 12$

27. The projected U.S. long-distance revenue (in billions of dollars) from 1999 to 2004 is given by the equation, $y = 3x + 111$

19.–26. See graphing answer section.

where x is the number of years after 1999. Graph this equation and use it to estimate the amount of long-distance revenue in 2007. (*Source*: Giga Information Group)
See graphing answer section; $135 billion

(3.3) Identify the intercepts.

28. $(4, 0); (0, -2)$

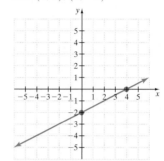

29. $(0, -3)$

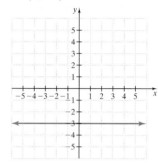

30.

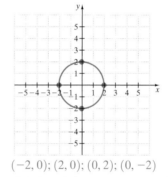

31.

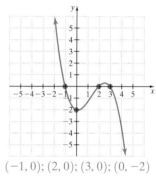

$(-2, 0); (2, 0); (0, 2); (0, -2)$ $(-1, 0); (2, 0); (3, 0); (0, -2)$

Graph each linear equation by finding its intercepts.

32. $x - 3y = 12$ **33.** $-4x + y = 8$

34. $y = -3$ **35.** $x = 5$

36. $y = -3x$ **37.** $x = 5y$

38. $x - 2 = 0$ **39.** $y + 6 = 0$

32.–39. See graphing answer section.

(3.4) Find the slope of each line.

40. $m = -\frac{3}{4}$

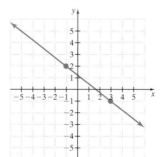

41. $m = \frac{1}{5}$

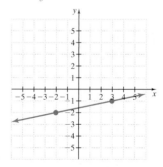

In Exercises 42–45, match each line with its slope.

a.

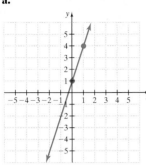

b.

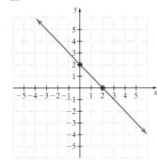

c.

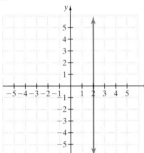

d.

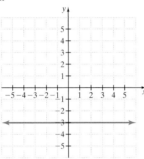

e.

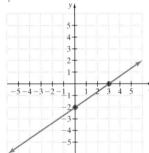

42. $m = 0$ d
43. $m = -1$ b
44. undefined slope   c
45. $m = 3$ a

46. $m = \dfrac{2}{3}$ e

Find the slope of the line that goes through the given points.
47. $(2, 5)$ and $(6, 8)$ $\frac{3}{4}$
48. $(4, 7)$ and $(1, 2)$ $\frac{5}{3}$
49. $(1, 3)$ and $(-2, -9)$ 4
50. $(-4, 1)$ and $(3, -6)$ -1

Find the slope of each line.

51. $x = 5$ undefined
52. $y = -1$ 0
53. $y = -2$ 0
54. $x = 0$ undefined

Determine whether the lines through the pairs of points are parallel, perpendicular, or neither.

△ **55.** $(-3, 1)$ and $(1, -2)$, $(2, 4)$ and $(6, 1)$ parallel
△ **56.** $(-7, 6)$ and $(0, 4)$, $(-9, -3)$ and $(1, 5)$ neither
△ **57.** $(9, 10)$ and $(8, -7)$, $(-1, -3)$ and $(2, -8)$ neither
△ **58.** $(-1, 3)$ and $(3, -2)$, $(-2, -2)$ and $(3, 2)$ perpendicular

Find the slope of each line and write the slope as a rate of change. Don't forget to attach the proper units.

59. The graph below shows the average monthly day care cost for a 3-year-old attending 8 hours a day, 5 days a week. every 1 year, monthly day care cost increases by $17.75

Monthly Day Care Costs

(2001, 516)

(1985, 232)

Source: Runzheimer International

60. The graph below shows the U.S. government's projected spending (in billions of dollars) on technology. Every 1 year, $7.7 billion more dollars are spent on technology.

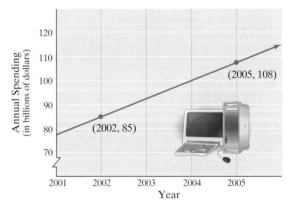

(2005, 108)

(2002, 85)

64. undefined slope; no y-intercept

(3.5) *Determine the slope and the y-intercept of the graph of each equation.*   **61.** $m = -3$; $(0, 7)$ **62.** $m = \frac{1}{6}$; $\left(0, \frac{1}{6}\right)$

61. $3x + y = 7$
62. $x - 6y = -1$
63. $y = 2$ $m = 0$; $(0, 2)$
64. $x = -5$

Determine whether the lines are parallel, perpendicular, or neither.

△ **65.** $x - y = -6$
        $x + y = 3$ perpendicular

△ **66.** $3x + y = 7$
        $-3x - y = 10$ parallel

△ **67.** $y = 4x + \dfrac{1}{2}$
        $4x + 2y = 1$ neither

Write an equation of each line in slope-intercept form.

68. slope -5; y-intercept $\dfrac{1}{2}$ $y = -5x + \dfrac{1}{2}$

69. slope $\dfrac{2}{3}$; y-intercept 6 $y = \dfrac{2}{3}x + 6$

70.–73. See graphing answer section.

Use the slope-intercept form to graph each equation.

70. $y = -3x$

71. $y = 3x - 1$

72. $-x + 2y = 8$

73. $5x - 3y = 15$

Match each equation with its graph.

74. $y = -2x + 1$ d

75. $y = -4x$ c

76. $y = 2x$ a

77. $y = 2x - 1$ b

a.
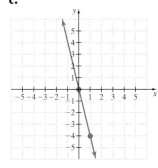

b.
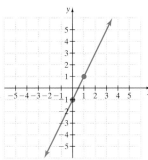

c.
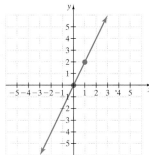

d.
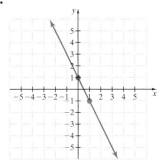

(3.6) Write an equation of each line in standard form.

78. With slope 4, through $(2, 0)$ $4x - y = 8$

79. With slope -3, through $(0, -5)$ $3x + y = -5$

80. With slope $\dfrac{1}{2}$, through $\left(0, -\dfrac{7}{2}\right)$ $x - 2y = 7$

81. With slope 0, through $(-2, -3)$ $y = -3$

82. With 0 slope, through the origin $y = 0$

83. With slope -6, through $(2, -1)$ $6x + y = 11$

84. With slope 12, through $\left(\dfrac{1}{2}, 5\right)$ $12x - y = 1$

85. Through $(0, 6)$ and $(6, 0)$ $x + y = 6$

86. Through $(0, -4)$ and $(-8, 0)$ $x + 2y = -8$

87. Vertical line, through $(5, 7)$ $x = 5$

88. Horizontal line, through $(-6, 8)$ $y = 8$

△ **89.** Through $(6, 0)$, perpendicular to $y = 8$ $x = 6$

△ **90.** Through $(10, 12)$, perpendicular to $x = -2$ $y = 12$

△ **91.** Write an equation in standard form of the line that contains $(5, 0)$ and is

   a. parallel to the line $y = -3x + 7$. $3x + y = 15$

   b. perpendicular to the line $y = -3x + 7$. $x - 3y = 5$

(3.7) Determine which of the following are functions

92. $\{(7, 1), (7, 5), (2, 6)\}$ no

93. $\{(0, -1), (5, -1), (2, 2)\}$ yes   **94.** $7x - 6y = 1$ yes

95. $y = 7$ yes   **96.** $x = 2$ no

97. $y = x^3$ yes   **98.** $x = y^2$ no

99. no
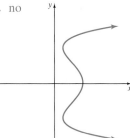

100. yes
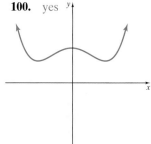

Given the following functions, find the indicated function values.

101. Given $f(x) = -2x + 6$, find

   a. $f(0)$ 6   **b.** $f(-2)$ 10   **c.** $f\left(\dfrac{1}{2}\right)$ 5

102. Given $h(x) = -5 - 3x$, find

   a. $h(2)$ -11 **b.** $h(-3)$ 4   **a.** $h(0)$ -5

103. Given $g(x) = x^2 + 12x$, find

   a. $g(3)$ 45   **b.** $g(-5)$ -35 **a.** $g(0)$ 0

104. Given $h(x) = 6 - |x|$, find

   a. $h(-1)$ 5   **b.** $h(1)$ 5   **c.** $h(-4)$ 2

Find the domain of each function.

105. $f(x) = 2x + 7$ $(-\infty, \infty)$ **106.** $g(x) = \dfrac{7}{x - 2}$

106. all real numbers except 2 or $(-\infty, 2) \cup (2, \infty)$

Find the domain and the range of each function graphed.

107.

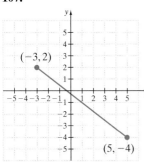

108.

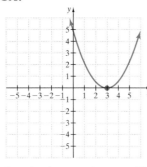

109.

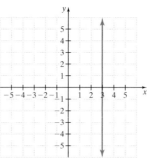

110.

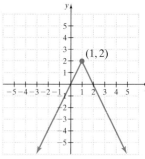

domain: $[-3, 5]$; range: $[-4, 2]$ domain: $(-\infty, \infty)$; range: $[0, \infty)$ domain: $\{3\}$; range: $(-\infty, \infty)$ domain: $(-\infty, \infty)$; range: $(-\infty, 2]$

CHAPTER 3 TEST

Remember to use your Chapter Test Prep Video CD to help you study and view solutions to the test questions you need help with.

1. a. $(1980, 38), (1984, 47), (1988, 51), (1992, 54), (1996, 59), (2000, 55)$

Graph the following.

1. The table gives the percent of new mothers who returned to work after having their child.

Year	*Percent of New Mothers Returning to Work*
1980	38
1984	47
1988	51
1992	54
1996	59
2000	55

(*Source*: Census Bureau)

  a. Write this data as a set of ordered pairs of the form (year, percent of new mothers returning to work).

  b. Create a scatter diagram of the data. Be sure to label the axes properly. See graphing answer section.

2. $2x + y = 8$ **3.** $5x - 7y = 10$

4. $y = -1$ **5.** $x - 3 = 0$

2.–5. See graphing answer section.

Find the slopes of the following lines.

6. $\dfrac{2}{5}$ **7.** 0

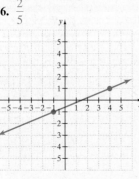

8. Through $(6, -5)$ and $(-1, 2)$ -1

9. $-3x + y = 5$ 3 **10.** $x = 6$ undefined

11. Determine the slope and the y-intercept of the graph of $7x - 3y = 2$. $\frac{7}{3}$; $\left(0, -\frac{2}{3}\right)$

△ **12.** Determine whether the graphs of $y = 2x - 6$ and $-4x = 2y$ are parallel lines, perpendicular lines, or neither.   neither

Find equations of the following lines. Write the equation in standard form.

13. With slope of $-\dfrac{1}{4}$, through $(2, 2)$ $x + 4y = 10$

14. Through the origin and $(6, -7)$ $7x + 6y = 0$

15. Through $(2, -5)$ and $(1, 3)$ $8x + y = 11$

△ **16.** Through $(-5, -1)$ and parallel to $x = 7$ $x = -5$

17. With slope $\dfrac{1}{8}$ and y-intercept $(0, 12)$ $x - 8y = -96$

Which of the following are functions?

18. yes

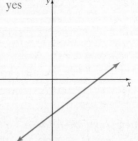

19. no

Given the following functions, find the indicated function values.

20. $h(x) = x^3 - x$

  a. $h(-1)$ 0 **b.** $h(0)$ 0 **c.** $h(4)$ 60

21. Find the domain of $y = \dfrac{1}{x + 1}$.

Find the domain and the range of each function graphed.

22.

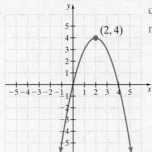

domain: $(-\infty, \infty)$

range: $(-\infty, 4]$

23.

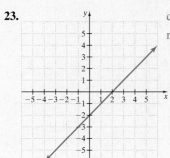

domain: $(-\infty, \infty)$

range: $(-\infty, \infty)$

This graph shows the sunset times for Seward, Alaska. Use this graph to answer Exercises 24 through 29.

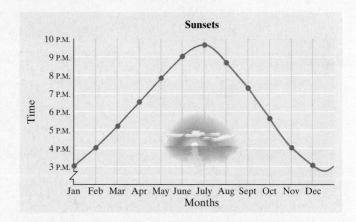

24. Approximate the time of sunset on June 1. 9 P.M.
25. Approximate the time of sunset on November 1. 4 P.M.
26. Approximate the date(s) when the sunset is 3 P.M.
27. Approximate the date(s) when the sunset is 9 P.M.
28. Is this graph the graph of a function? Why or why not?
29. Do you think a graph of sunset times for any location will always be a function? Why or why not?

26. January 1st and December 1st **27.** June 1st and end of July **28.** yes; it passes the vertical line test
29. yes; every location has exactly 1 sunset time per day.

3 CHAPTER CUMULATIVE REVIEW

3. $\dfrac{2}{39}$ (Sec. 1.3, Ex. 3)

1. Insert $<$, $>$, or $=$ in the space between each pair of numbers to make each statement true. (Sec. 1.2, Ex. 1)
  a. $2 < 3$ **b.** $7 > 4$ **c.** $72 > 27$

2. Write the fraction $\dfrac{56}{64}$ in lowest terms. $\dfrac{7}{8}$ (Sec. 1.3)

3. Multiply $\dfrac{2}{15}$ and $\dfrac{5}{13}$. Write the product in lowest terms

4. Add: $\dfrac{10}{3} + \dfrac{5}{21}$ $\dfrac{75}{21}$ or $3\dfrac{12}{21}$ (Sec. 1.3)

5. Simplify: $\dfrac{3 + |4 - 3| + 2^2}{6 - 3}$ $\dfrac{8}{3}$ (Sec. 1.4, Ex. 3)

6. Simplify: $16 - 3 \cdot 3 + 2^4$ 23 (Sec. 1.4)

7. Add.
  a. $-8 + (-11)$ -19 **b.** $-5 + 35$ 30
  c. $0.6 + (-1.1)$ -0.5 **d.** $-\dfrac{7}{10} + \left(-\dfrac{1}{10}\right)$ $-\dfrac{4}{5}$
  e. $11.4 + (-4.7)$ 6.7   **f.** $-\dfrac{3}{8} + \dfrac{2}{5}$ $\dfrac{1}{40}$ (Sec. 1.5, Ex. 6)

8. Simplify: $|9 + (-20)| + |-10|$ 21 (Sec. 1.5)

9. Simplify each expression.
  a. $-14 - 8 + 10 - (-6)$ -6 (Sec. 1.6, Ex. 4)
  b. $1.6 - (-10.3) + (-5.6)$ 6.3

10. Simplify: $-9 - (3 - 8)$ -4 (Sec. 1.6)

11. If $x = -2$ and $y = -4$, evaluate each expression.
  a. $5x - y$ -6 **b.** $x^4 - y^2$ 0
  c. $\dfrac{3x}{2y}$ $\dfrac{3}{4}$ (Sec. 1.7, Ex. 10)

12. Is -20 a solution of $\dfrac{x}{-10} = 2$? yes (Sec. 1.7)

13. Simplify each expression.   **b.** $-21x$ (Sec. 1.8, Ex. 3)
  a. $10 + (x + 12)$ $22 + x$ **b.** $-3(7x)$

14. Simplify: $(12 + x) - (4x - 7)$ $19 - 3x$ (Sec. 2.1)

15. Identify the numerical coefficient in each term.
  a. $-3y$ -3 **b.** $22z^4$ 22
  c. y 1   **d.** $-x$ -1
  e. $\dfrac{x}{7}$ $\dfrac{1}{7}$ (Sec. 2.1, Ex. 1)

16. Multiply: $-5(x - 7)$ $-5x + 35$ (Sec. 2.1)

17. Solve $y + 0.6 = -1.0$ for y. -1.6 (Sec. 2.2, Ex. 2)

18. Solve: $5(3 + z) - (8z + 9) = -4$ $\dfrac{10}{3}$ (Sec. 2.2)

19. Solve: $-\dfrac{2}{3}x = -\dfrac{5}{2}$ $\dfrac{15}{4}$ (Sec. 2.3, Ex. 5)

20. Solve: $\dfrac{x}{4} - 1 = -7$ -24 (Sec. 2.3)

21. If x is the first of three consecutive integers, express the sum of the three integers in terms of x. Simplify if possible.

22. Solve: $\dfrac{x}{3} - 2 = \dfrac{x}{3}$ no solution (Sec. 2.4)

23. Solve: $\dfrac{2(a + 3)}{3} = 6a + 2$ 0 (Sec. 2.4, Ex. 4)

24. Solve: $x + 2y = 6$ for y. $y = \dfrac{6 - x}{2}$ (Sec. 2.6)

25. In a recent year, the U.S. House of Representatives had a total of 431 Democrats and Republicans. There were 15 more Republican representatives than Democratic. Find the number of representatives from each party.

26. Solve $5(x + 4) \ge 4(2x + 3)$. Write the solution set in interval notation. $\left(-\infty, \dfrac{8}{3}\right]$ (Sec. 2.8)

27. Charles Pecot can afford enough fencing to enclose a rectangular garden with a perimeter of 140 feet. If the width of his garden is to be 30 feet, find the length.   40 ft (Sec. 2.6, Ex. 2)

28. Solve $-3 < 4x - 1 \le 2$. Write the solution set in interval notation. $\left(-\dfrac{1}{2}, \dfrac{3}{4}\right]$ (Sec. 2.8)

29. Solve $y = mx + b$ for x. $\dfrac{y - b}{m} = x$ (Sec. 2.6, Ex. 6)

30. Complete the table for $y = -5x$.   (Sec. 3.1)

x	y
0	0
-1	5
-2	-10

31. A chemist working on his doctoral degree at Massachusetts Institute of Technology needs 12 liters of a 50% acid solution for a lab experiment. The stockroom has only 40% and 70% solutions. How much of each solution should be mixed together to form 12 liters of a 50% solution?

32. Graph: $y = -3x + 5$ See graphing answer section. (Sec. 3.2)

33. Graph $x \ge -1$.

34. Find the x- and y-intercepts of $2x + 4y = -8$.

35. Solve $-1 \le 2x - 3 < 5$, and graph the solution set.

36. Graph $x = 2$ on a rectangular coordinate system.

37. Determine whether each ordered pair is a solution of the equation $x - 2y = 6$.

   a. $(6, 0)$ solution

   b. $(0, 3)$ not a solution

   c. $\left(1, -\dfrac{5}{2}\right)$ solution (Sec. 3.1, Ex. 3)

38. Find the slope of the line through $(0, 5)$ and $(-5, 4)$.

39. Identify the linear equations in two variables.

   a. $x - 1.5y = -1.6$ linear

   b. $y = -2x$ linear

   c. $x + y^2 = 9$ not linear

   d. $x = 5$ linear (Sec. 3.2, Ex. 1)

40. Find the slope of $x = -10$.   undefined slope (Sec. 3.4)

41. Find the slope of the line $y = -1$.   0 (Sec. 3.4, Ex. 3)

42. Find the slope and y-intercept of the line whose equation is $2x - 5y = 10$. $m = \dfrac{2}{5}$, y-intercept: $(0, -2)$ (Sec. 3.5)

43. Find the slope of a line perpendicular to the line passing through the points $(-1, 7)$ and $(2, 2)$. $\frac{3}{5}$ (Sec. 3.4, Ex. 6)

44. Write an equation of the line through $(2, 3)$ and $(0, 0)$. Write the equation in standard form. $3x - 2y = 0$ (Sec. 3.6)

21. $3x + 3$ (Sec. 2.3, Ex. 8)   **25.** 208 Democratic representatives 223 Republican representatives (Sec. 2.5, Ex. 4)

31. 40% solution: 8 l; 70% solution: 4 l (Sec. 2.7, Ex. 4)   **33.** $\longleftarrow$ $[-1, \infty)$; (Sec. 2.8, Ex. 1)   **34.** $(0, -2)$, $(-4, 0)$ (Sec. 3.3)

35. $\longleftarrow$ $\{x \mid 1 \le x < 4\}$ (Sec. 2.8, Ex. 10)   **36.** See graphing answer section. (Sec. 3.3)   **38.** $\dfrac{1}{5}$ (Sec. 3.4)

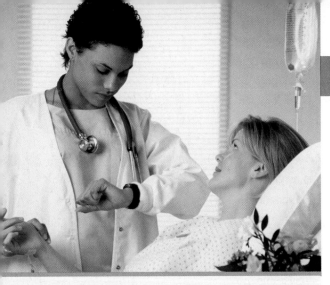

CHAPTER

Solving Systems of Linear Equations and Inequalities

In Chapter 3, we graphed equations containing two variables. Equations like these are often needed to represent relationships between two different values. For example, an economist attempts to predict what effects a price change will have on the sales prospects of calculators. There are many real-life opportunities to compare and contrast two such equations, called a system of equations. This chapter presents linear systems and ways we solve these systems and apply them to real-life situations.

LARGEST HEALTH CARE OCCUPATION IN THE UNITED STATES

Over 2.5 million people work as nurses in settings as varied as hospitals, private homes, corporate offices, nursing homes, overnight camps, doctors' offices, and community health centers. The U.S. Bureau of Labor Statistics predicts that the demand for nurses will continue to grow rapidly as the American population ages, requiring more long-term health care and home health care.

Registered nurses must be licensed in the state in which they work. About 25% of registered nurses hold a diploma from a hospital program, 35% hold an associate's degree, 30% hold a bachelor's degree, and 10% hold a higher degree. Nurses use math skills nearly every day in taking and comparing vital signs, administering medications, and tracking fluid intake and output.

In the Spotlight on Decision Making feature on page 286, you will have the opportunity, as a registered nurse, to make a decision concerning a patient's blood pressure.

4.1 SOLVING SYSTEMS OF LINEAR EQUATIONS BY GRAPHING

Objectives

1. Determine if an ordered pair is a solution of a system of equations in two variables.
2. Solve a system of linear equations by graphing.
3. Without graphing, determine the number of solutions of a system.

1 A **system of linear equations** consists of two or more linear equations. In this section, we focus on solving systems of linear equations containing two equations in two variables. Examples of such linear systems are

$$\begin{cases} 3x - 3y = 0 \\ x = 2y \end{cases} \qquad \begin{cases} x - y = 0 \\ 2x + y = 10 \end{cases} \qquad \begin{cases} y = 7x - 1 \\ y = 4 \end{cases}$$

A **solution** of a system of two equations in two variables is an ordered pair of numbers that is a solution of both equations in the system.

EXAMPLE 1

Which of the following ordered pairs is a solution of the given system?

$$\begin{cases} 2x - 3y = 6 & \text{First equation} \\ x = 2y & \text{Second equation} \end{cases}$$

a. $(12, 6)$ **b.** $(0, -2)$

Solution If an ordered pair is a solution of both equations, it is a solution of the system.

a. Replace x with 12 and y with 6 in both equations.

$$2x - 3y = 6 \quad \text{First equation} \qquad\qquad x = 2y \quad \text{Second equation}$$
$$2(12) - 3(6) \stackrel{?}{=} 6 \quad \text{Let } x = 12 \text{ and } y = 6. \qquad 12 \stackrel{?}{=} 2(6) \quad \text{Let } x = 12 \text{ and } y = 6.$$
$$24 - 18 \stackrel{?}{=} 6 \quad \text{Simplify.} \qquad\qquad 12 = 12 \quad \text{True}$$
$$6 = 6 \quad \text{True}$$

Since $(12, 6)$ is a solution of both equations, it is a solution of the system.

b. Start by replacing x with 0 and y with -2 in both equations.

$$2x - 3y = 6 \quad \text{First equation} \qquad\qquad x = 2y \quad \text{Second equation}$$
$$2(0) - 3(-2) \stackrel{?}{=} 6 \quad \text{Let } x = 0 \text{ and } y = -2. \qquad 0 \stackrel{?}{=} 2(-2) \quad \text{Let } x = 0 \text{ and } y = -2.$$
$$0 + 6 \stackrel{?}{=} 6 \quad \text{Simplify.} \qquad\qquad 0 = -4 \quad \text{False}$$
$$6 = 6 \quad \text{True}$$

While $(0, -2)$ is a solution of the first equation, it is not a solution of the second equation, so it is **not** a solution of the system.

2 Since a solution of a system of two equations in two variables is a solution common to both equations, it is also a point common to the graphs of both equations. Let's practice finding solutions of both equations in a system—that is, solutions of a system—by graphing and identifying points of intersection.

EXAMPLE 2

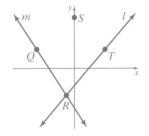
Solve the system of equations by graphing.

$$\begin{cases} -x + 3y = 10 \\ x + y = 2 \end{cases}$$

Solution   On a single set of axes, graph each linear equation.

$-x + 3y = 10$

x	y
0	$\dfrac{10}{3}$
-4	2
2	4

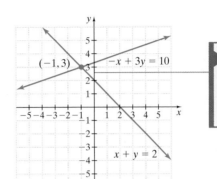

$x + y = 2$

x	y
0	2
2	0
1	1

> **Helpful Hint**
>
> The point of intersection gives the solution of the system.

The two lines appear to intersect at the point $(-1, 3)$. To check, we replace x with -1 and y with 3 in both equations.

$-x + 3y = 10$ First equation $x + y = 2$ Second equation

$-(-1) + 3(3) \stackrel{?}{=} 10$ Let $x = -1$ and $y = 3$. $-1 + 3 \stackrel{?}{=} 2$ Let $x = -1$ and $y = 3$.

$\qquad 1 + 9 \stackrel{?}{=} 10$ Simplify. $\qquad 2 = 2$ True

$\qquad\quad 10 = 10$ True

$(-1, 3)$ checks, so it is the solution of the system.

> **Helpful Hint**
>
> Neatly drawn graphs can help when you are estimating the solution of a system of linear equations by graphing.
>
> In the example above, notice that the two lines intersected in a point. This means that the system has 1 solution.

A system of equations that has at least one solution as in Example 2 is said to be a **consistent system**. A system that has no solution is said to be an **inconsistent system**.

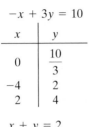
    EXAMPLE 3

Solve the following system of equations by graphing.

$$\begin{cases} 2x + y = 7 \\ 2y = -4x \end{cases}$$

Solution   Graph the two lines in the system.

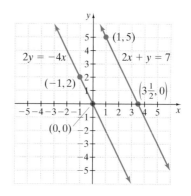

The lines **appear** to be parallel. To confirm this, write both equations in slope-intercept form by solving each equation for y.

$2x + y = 7$	First equation	$2y = -4x$	Second equation
$y = -2x + 7$	Subtract $2x$ from both sides.	$\dfrac{2y}{2} = \dfrac{-4x}{2}$	Divide both sides by 2.
		$y = -2x$	

Recall that when an equation is written in slope-intercept form, the coefficient of x is the slope. Since both equations have the same slope, -2, but different y-intercepts, the lines are parallel and have no points in common. Thus, there is no solution of the system and the system is inconsistent.

In Examples 2 and 3, the graphs of the two linear equations of each system are different. When this happens, we call these equations **independent equations**. If the graphs of the two equations in a system are identical, we call the equations **dependent equations**.

 EXAMPLE 4

Solve the system of equations by graphing.

$$\begin{cases} x - y = 3 \\ -x + y = -3 \end{cases}$$

Solution   Graph each line.

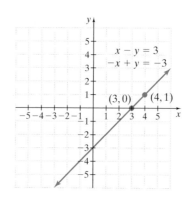

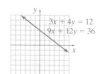

These graphs **appear** to be identical. To confirm this, write each equation in slope-intercept form.

$x - y = 3$ First equation $-x + y = -3$ Second equation

$-y = -x + 3$ Subtract x from both sides. $y = x - 3$ Add x to both sides.

$\dfrac{-y}{-1} = \dfrac{-x}{-1} + \dfrac{3}{-1}$ Divide both sides by -1.

$y = x - 3$

The equations are identical and so must be their graphs. The lines have an infinite number of points in common. Thus, there is an infinite number of solutions of the system and this is a consistent system. The equations are dependent equations.

As we have seen, three different situations can occur when graphing the two lines associated with the equations in a linear system:

One point of intersection: one solution

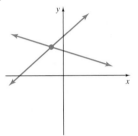

Consistent system
(at least one solution)
Independent equations
(graphs of equations differ)

Parallel lines: no solution

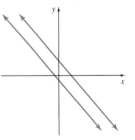

Inconsistent system
(no solution)
Independent equations
(graphs of equations differ)

Same line: infinite number of solutions

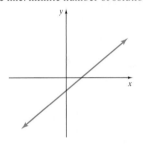

Consistent system
(at least one solution)
Dependent equations
(graphs of equations identical)

3   You may have suspected by now that graphing alone is not an accurate way to solve a system of linear equations. For example, a solution of $\left(\frac{1}{2}, \frac{2}{9}\right)$ is unlikely to be read correctly from a graph. The next two sections present two accurate methods of solving these systems. In the meantime, we can decide how many solutions a system has by writing each equation in the slope-intercept form.

EXAMPLE 5

Without graphing, determine the number of solutions of the system.

$$\begin{cases} \dfrac{1}{2}x - y = 2 \\ \quad x = 2y + 5 \end{cases}$$

Solution   First write each equation in slope-intercept form.

$\dfrac{1}{2}x - y = 2$ First equation $x = 2y + 5$ Second equation

$\dfrac{1}{2}x = y + 2$ Add y to both sides. $x - 5 = 2y$ Subtract 5 from both sides.

$\dfrac{x}{2} - \dfrac{5}{2} = \dfrac{2y}{2}$ Divide both sides by 2.

$$\frac{1}{2}x - 2 = y \qquad \text{Subtract 2 from} \qquad \frac{1}{2}x - \frac{5}{2} = y \qquad \text{Simplify.}$$
both sides.

The slope of each line is $\frac{1}{2}$, but they have different y-intercepts. This tells us that the lines representing these equations are parallel. Since the lines are parallel, the system has no solution and is inconsistent.

EXAMPLE 6

Determine the number of solutions of the system.

$$\begin{cases} 3x - y = 4 \\ x + 2y = 8 \end{cases}$$

Solution   Once again, the slope-intercept form helps determine how many solutions this system has.

CLASSROOM EXAMPLE

Determine the number of solutions of the system.

$$\begin{cases} x + 3y = -1 \\ 4x - y = 10 \end{cases}$$

answer: one solution

$3x - y = 4$	First equation	$x + 2y = 8$	Second equation
$3x = y + 4$	Add y to both sides.	$x = -2y + 8$	Subtract $2y$ from both sides.
$3x - 4 = y$	Subtract 4 from both sides.	$x - 8 = -2y$	Subtract 8 from both sides.

$$\frac{x}{-2} - \frac{8}{-2} = \frac{-2y}{-2} \qquad \text{Divide both sides by } -2.$$

$$-\frac{1}{2}x + 4 = y \qquad \text{Simplify.}$$

The slope of the second line is $-\frac{1}{2}$, whereas the slope of the first line is 3. Since the slopes are not equal, the two lines are neither parallel nor identical and must intersect. Therefore, this system has one solution and is consistent.

Graphing Calculator Explorations

A graphing calculator may be used to approximate solutions of systems of equations. For example, to approximate the solution of the system

$$\begin{cases} y = -3.14x - 1.35 \\ y = 4.88x + 5.25, \end{cases}$$

first graph each equation on the same set of axes. Then use the intersect feature of your calculator to approximate the point of intersection.

The approximate point of intersection is $(-0.82, 1.23)$.

Solve each system of equations. Approximate the solutions to two decimal places.

1. $\begin{cases} y = -2.68x + 1.21 \\ y = 5.22x - 1.68 \end{cases}$ $(0.37, 0.23)$ **2.** $\begin{cases} y = 4.25x + 3.89 \\ y = -1.88x + 3.21 \end{cases}$ $(-0.11, 3.42)$

3. $\begin{cases} 4.3x - 2.9y = 5.6 \\ 8.1x + 7.6y = -14.1 \end{cases}$ $(0.03, -1.89)$ **4.** $\begin{cases} -3.6x - 8.6y = 10 \\ -4.5x + 9.6y = -7.7 \end{cases}$ $(-0.41, -0.99)$

MENTAL MATH

Each rectangular coordinate system shows the graph of the equations in a system of equations. Use each graph to determine the number of solutions for each associated system. If the system has only one solution, give its coordinates. (The coordinates will be integers.)

1.

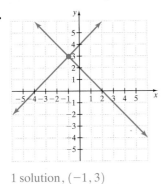

1 solution, $(-1, 3)$

2.

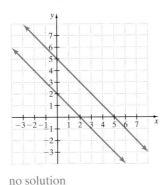

no solution

3.

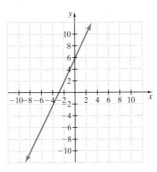

infinite number of solutions

4.

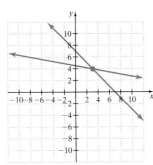

1 solution, $(3, 4)$

5.

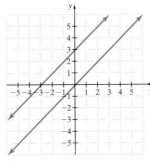

no solution

6.

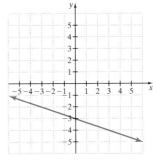

infinite number of solutions

7.

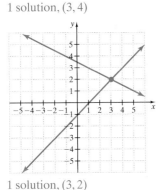

1 solution, $(3, 2)$

8.

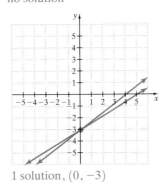

1 solution, $(0, -3)$

EXERCISE SET 4.1

STUDY GUIDE/SSM   CD/VIDEO   PH MATH TUTOR CENTER   MathXL®Tutorials ON CD   MathXL®   MyMathLab®

TEACHING TIP

Determine whether the given ordered pairs satisfy the system of the linear equations. See Example 1.

A Group Activity for this section is available in the Instructor's Resource Manual.

1. $\begin{cases} x + y = 8 \\ 3x + 2y = 21 \end{cases}$

  a. $(2, 4)$ no

  b. $(5, 3)$ yes

  c. $(1, 9)$ no

2. $\begin{cases} 2x + y = 5 \\ x + 3y = 5 \end{cases}$

  a. $(5, 0)$ no

  b. $(1, 2)$ no

  c. $(2, 1)$ yes

3. $\begin{cases} 3x - y = 5 \\ x + 2y = 11 \end{cases}$

  a. $(2, -1)$ no

  b. $(3, 4)$ yes

  c. $(0, -5)$ no

4. $\begin{cases} 2x - 3y = 8 \\ x - 2y = 6 \end{cases}$

  a. $(4, 0)$ no

  b. $(-2, -4)$ yes

  c. $(7, 2)$ no

5. $\begin{cases} 2y = 4x \\ 2x - y = 0 \end{cases}$

   a. $(-3, -6)$ yes

   b. $(0, 0)$ yes

   c. $(1, 2)$ yes

6. $\begin{cases} 4x = 1 - y \\ x - 3y = -8 \end{cases}$

   a. $(0, 1)$ no

   b. $(1, -3)$ no

   c. $(-2, 2)$ no

7. Construct a system of two linear equations that has $(2, 5)$ as a solution.   answers may vary

8. Construct a system of two linear equations that has $(0, 1)$ as a solution.   answers may vary

MIXED PRACTICE   **9.–28.** See graphing answer section.

Solve each system of equations by graphing the equations on the same set of axes. Tell whether the system is consistent or inconsistent and whether the equations are dependent or independent. See Examples 2 through 4.

9. $\begin{cases} y = x + 1 \\ y = 2x - 1 \end{cases}$ $(2, 3)$

10. $\begin{cases} y = 3x - 4 \\ y = x + 2 \end{cases}$ $(3, 5)$

11. $\begin{cases} 2x + y = 0 \\ 3x + y = 1 \end{cases}$ $(1, -2)$

12. $\begin{cases} 2x + y = 1 \\ 3x + y = 0 \end{cases}$ $(-1, 3)$

13. $\begin{cases} y = -x - 1 \\ y = 2x + 5 \end{cases}$ $(-2, 1)$

14. $\begin{cases} y = x - 1 \\ y = -3x - 5 \end{cases}$ $(-1, -2)$

15. $\begin{cases} 2x - y = 6 \\ \quad\quad y = 2 \end{cases}$ $(4, 2)$

16. $\begin{cases} x + y = 5 \\ \quad\quad x = 4 \end{cases}$ $(4, 1)$

17. $\begin{cases} x + y = 5 \\ x + y = 6 \end{cases}$

18. $\begin{cases} 2x + y = 4 \\ \quad x + y = 2 \end{cases}$ $(2, 0)$

19. $\begin{cases} y - 3x = -2 \\ 6x - 2y = 4 \end{cases}$

20. $\begin{cases} y + 2x = 3 \\ 4x = 2 - 2y \end{cases}$

21. $\begin{cases} x - 2y = 2 \\ 3x + 2y = -2 \end{cases}$ $(0, -1)$

22. $\begin{cases} x + 3y = 7 \\ 2x - 3y = -4 \end{cases}$ $(1, 2)$

23. $\begin{cases} \dfrac{1}{2}x + y = -1 \\ \quad\quad\quad x = 4 \end{cases}$ $(4, -3)$

24. $\begin{cases} x + \dfrac{3}{4}y = 2 \\ \quad\quad\quad x = -1 \end{cases}$ $(-1, 4)$

25. $\begin{cases} y = x - 2 \\ y = 2x + 3 \end{cases}$ $(-5, -7)$

26. $\begin{cases} y = x + 5 \\ y = -2x - 4 \end{cases}$ $(-3, 2)$

27. $\begin{cases} x + y = 7 \\ x - y = 3 \end{cases}$ $(5, 2)$

28. $\begin{cases} x + y = -4 \\ x - y = 2 \end{cases}$ $(-1, -3)$

29. Explain how to use a graph to determine the number of solutions of a system.   **29.–30.** answers may vary

30. The ordered pair $(-2, 3)$ is a solution of all three independent equations:

$$x + y = 1$$
$$2x - y = -7$$
$$x + 3y = 7$$

Describe the graph of all three equations on the same axes.

Without graphing, decide.

   a. Are the graphs of the equations identical lines, parallel lines, or lines intersecting at a single point?

   b. How many solutions does the system have? See Examples 5 and 6.

31. $\begin{cases} 4x + y = 24 \\ x + 2y = 2 \end{cases}$

32. $\begin{cases} 3x + y = 1 \\ 3x + 2y = 6 \end{cases}$

33. $\begin{cases} 2x + y = 0 \\ 2y = 6 - 4x \end{cases}$

34. $\begin{cases} 3x + y = 0 \\ 2y = -6x \end{cases}$

35. $\begin{cases} 6x - y = 4 \\ \dfrac{1}{2}y = -2 + 3x \end{cases}$

36. $\begin{cases} 3x - y = 2 \\ \dfrac{1}{3}y = -2 + 3x \end{cases}$

37. $\begin{cases} x = 5 \\ y = -2 \end{cases}$

38. $\begin{cases} y = 3 \\ x = -4 \end{cases}$

39. $\begin{cases} 3y - 2x = 3 \\ \quad x + 2y = 9 \end{cases}$

40. $\begin{cases} 2y = x + 2 \\ y + 2x = 3 \end{cases}$

41. $\begin{cases} 6y + 4x = 6 \\ 3y - 3 = -2x \end{cases}$

42. $\begin{cases} 8y + 6x = 4 \\ 4y - 2 = 3x \end{cases}$

43. $\begin{cases} x + y = 4 \\ x + y = 3 \end{cases}$

44. $\begin{cases} 2x + y = 0 \\ y = -2x + 1 \end{cases}$

REVIEW AND PREVIEW

Solve each equation. See Section 2.4.   **46.** -1 **48.** 3

45. $5(x - 3) + 3x = 1$ 2

46. $-2x + 3(x + 6) = 17$

47. $4\left(\dfrac{y + 1}{2}\right) + 3y = 0$ $-\dfrac{2}{5}$

48. $-y + 12\left(\dfrac{y - 1}{4}\right) = 3$

49. $8a - 2(3a - 1) = 6$ 2

50. $3z - (4z - 2) = 9$ -7

Concept Extensions   **51.–52.** answers may vary

51. Explain how writing each equation in a linear system in the point-slope form helps determine the number of solutions of a system.

52. Is it possible for a system of two linear equations in two variables to be inconsistent, but with dependent equations? Why or why not?

The double line graph below shows the number of pounds of fishery products from U.S. domestic catch and from imports. Use this graph for Exercises 53 and 54.

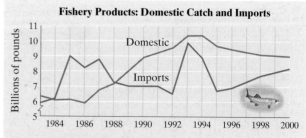

Fishery Products: Domestic Catch and Imports

Source: U.S. Bureau of the Census, Statistical Abstract of the United States: 1998, 115th ed., Washington, DC, 1995.

9.–16., consistent; independent   **17.** no solution; inconsistent; independent **18.** consistent; independent   **19.** infinite number of solutions; consistent; dependent **20.** no solution; inconsistent; independent **21.–28.** consistent; independent **31.–32.** intersecting; one solution **33.** parallel; no solution **34.–35.** identical lines; infinite number of solutions **36.–40.** intersecting; one solution **41.** identical lines; infinite number of solutions **42.** intersecting; one solution   **43.–44.** parallel; no solution

53. In what year(s) was the number of pounds of fishery products imported equal to the number of pounds of domestic catch?

54. In what year(s) was the number of pounds of fishery products imported greater than the number of pounds of domestic catch?   1984 through 1988

53. 1984, 1988

The double line graph below shows the number of Kmart stores vs. the number of Wal-Mart and Wal-Mart Supercenter stores. Use this graph to answer Exercises 55–58.

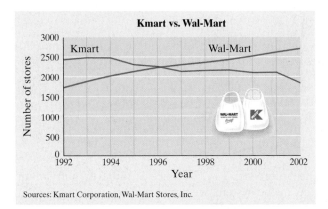

Kmart vs. Wal-Mart

Sources: Kmart Corporation, Wal-Mart Stores, Inc.

55. In what year was the number of Kmart stores approximately equal to the number of Wal-Mart stores?   1996

56. In what years was the number of Wal-Mart stores greater than the number of Kmart stores?   mid-1996 through 2002

57. Describe any trends you see in this graph.   answers may vary

58. Approximate how many more stores Wal-Mart had than Kmart in 2002.   885 stores

59. Construct a system of two linear equations that has $(1, 3)$ as a solution.   answers may vary

60. Construct a system of two linear equations that has $(0, 7)$ as a solution.   answers may vary

61. Below are two tables of values for two linear equations. Using the tables,

   a. find a solution of the corresponding system. $(4, 9)$

   b. graph several ordered pairs from each table and sketch the two lines.   See graphing answer section.

Does your graph confirm the solution from part a?   yes

x	y	x	y
1	3	1	6
2	5	2	7
3	7	3	8
4	9	4	9
5	11	5	10

4.2	SOLVING SYSTEMS OF LINEAR EQUATIONS BY SUBSTITUTION

Objective

  1 Use the substitution method to solve a system of linear equations.

  1   As we stated in the preceding section, graphing alone is not an accurate way to solve a system of linear equations. In this section, we discuss a second, more accurate method for solving systems of equations. This method is called the **substitution method** and is introduced in the next example.

EXAMPLE 1

Solve the system:

$$\begin{cases} 2x + y = 10 & \text{First equation} \\ x = y + 2 & \text{Second equation} \end{cases}$$

Solution   The second equation in this system is $x = y + 2$. This tells us that x and $y + 2$ have the same value. This means that we may substitute $y + 2$ for x in the first equation.

CLASSROOM EXAMPLE

Solve the system:
$$\begin{cases} 2x + 3y = 13 \\ x = y + 4 \end{cases}$$
answer: $(5, 1)$

$$2x + y = 10 \qquad \text{First equation}$$
$$2\,(y + 2) + y = 10 \qquad \text{Substitute } y + 2 \text{ for } x \text{ since } x = y + 2.$$

Notice that this equation now has one variable, y. Let's now solve this equation for y.

> **Helpful Hint**
>
> Don't forget the distributive property.

$$2(y + 2) + y = 10$$

$$2y + 4 + y = 10 \qquad \text{Use the distributive property.}$$

$$3y + 4 = 10 \qquad \text{Combine like terms.}$$

$$3y = 6 \qquad \text{Subtract 4 from both sides.}$$

$$y = 2 \qquad \text{Divide both sides by 3.}$$

Now we know that the y-value of the ordered pair solution of the system is 2. To find the corresponding x-value, we replace y with 2 in the equation $x = y + 2$ and solve for x.

$$x = y + 2$$

$$x = 2 + 2 \qquad \text{Let } y = 2.$$

$$x = 4$$

The solution of the system is the ordered pair $(4, 2)$. Since an ordered pair solution must satisfy both linear equations in the system, we could have chosen the equation $2x + y = 10$ to find the corresponding x-value. The resulting x-value is the same.

Check We check to see that $(4, 2)$ satisfies both equations of the original system.

First Equation	*Second Equation*
$2x + y = 10$	$x = y + 2$
$2(4) + 2 \stackrel{?}{=} 10$	$4 \stackrel{?}{=} 2 + 2 \qquad \text{Let } x = 4 \text{ and } y = 2.$
$10 = 10 \qquad \text{True}$	$4 = 4 \qquad \text{True}$

The solution of the system is $(4, 2)$.

A graph of the two equations shows the two lines intersecting at the point $(4, 2)$.

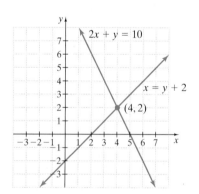

To solve a system of equations by substitution, we first need an equation solved for one of its variables.

EXAMPLE 2

Solve the system:

$$\begin{cases} x + 2y = 7 \\ 2x + 2y = 13 \end{cases}$$

Solution   We choose one of the equations and solve for x or y. We will solve the first equation for x by subtracting $2y$ from both sides.

$$x + 2y = 7 \qquad \text{First equation}$$
$$x = 7 - 2y \qquad \text{Subtract } \mathbf{2y} \text{ from both sides.}$$

Since $x = 7 - 2y$, we now substitute $7 - 2y$ for x in the second equation and solve for y.

$$2x + 2y = 13 \qquad \text{Second equation}$$
$$2(7 - 2y) + 2y = 13 \qquad \text{Let } x = \mathbf{7 - 2y}.$$

$$14 - 4y + 2y = 13 \qquad \text{Use the distributive property.}$$
$$14 - 2y = 13 \qquad \text{Simplify.}$$
$$-2y = -1 \qquad \text{Subtract } \mathbf{14} \text{ from both sides.}$$
$$y = \frac{1}{2} \qquad \text{Divide both sides by } \mathbf{-2}.$$

To find x, we let $y = \dfrac{1}{2}$ in the equation $x = 7 - 2y$.

$$x = 7 - 2y$$
$$x = 7 - 2\left(\frac{1}{2}\right) \qquad \text{Let } y = \frac{1}{2}.$$
$$x = 7 - 1$$
$$x = 6$$

The solution is $\left(6, \dfrac{1}{2}\right)$. Check the solution in both equations of the original system.

The following steps may be used to solve a system of equations by the substitution method.

Solving a System of Two Linear Equations by the Substitution Method
Step 1. Solve one of the equations for one of its variables.
Step 2. Substitute the expression for the variable found in step 1 into the other equation.
Step 3. Solve the equation from step 2 to find the value of one variable.
Step 4. Substitute the value found in step 3 in any equation containing both variables to find the value of the other variable.
Step 5. Check the proposed solution in the original system.

CLASSROOM EXAMPLE

Solve the system:

$$\begin{cases} 3x + y = 5 \\ 3x - 2y = -7 \end{cases}$$

answer: $\left(\frac{1}{3}, 4\right)$

Helpful Hint

Don't forget to insert parentheses when substituting $7 - 2y$ for x.

Concept Check Answer:
No, the solution will be an ordered pair.

✔ **CONCEPT CHECK**

As you solve the system $\begin{cases} 2x + y = -5 \\ x - y = 5 \end{cases}$ you find that $y = -5$. Is this the solution of the system?

EXAMPLE 3

Solve the system: $\begin{cases} 7x - 3y = -14 \\ -3x + y = 6 \end{cases}$

Solution   To avoid introducing fractions, we will solve the second equation for y.

$$-3x + y = 6 \qquad \textit{Second equation}$$
$$y = 3x + 6$$

Next, substitute $3x + 6$ for y in the first equation.

$$7x - 3y = -14 \qquad \textit{First equation}$$
$$7x - 3(3x + 6) = -14$$
$$7x - 9x - 18 = -14$$
$$-2x - 18 = -14$$
$$-2x = 4$$
$$\frac{-2x}{-2} = \frac{4}{-2}$$
$$x = -2$$

To find the corresponding y-value, substitute -2 for x in the equation $y = 3x + 6$. Then $y = 3(-2) + 6$ or $y = 0$. The solution of the system is $(-2, 0)$. Check this solution in both equations of the system.

> ### Helpful Hint
> When solving a system of equations by the substitution method, begin by solving an equation for one of its variables. If possible, solve for a variable that has a coefficient of 1 or -1. This way, we avoid working with time-consuming fractions.

EXAMPLE 4

Solve the system: $\begin{cases} \dfrac{1}{2}x - y = 3 \\ x = 6 + 2y \end{cases}$

Solution   The second equation is already solved for x in terms of y. Thus we substitute $6 + 2y$ for x in the first equation and solve for y.

$$\frac{1}{2}x - y = 3 \qquad \textit{First equation}$$
$$\frac{1}{2}(6 + 2y) - y = 3 \qquad \textit{Let } x = 6 + 2y.$$
$$3 + y - y = 3$$
$$3 = 3$$

Arriving at a true statement such as $3 = 3$ indicates that the two linear equations in the original system are equivalent. This means that their graphs are identical and there are an infinite number of solutions of the system. Any solution of one equation is also a solution of the other.

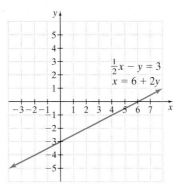

EXAMPLE 5

Use substitution to solve the system.

$$\begin{cases} 6x + 12y = 5 \\ -4x - 8y = 0 \end{cases}$$

Solution   Choose the second equation and solve for y.

$$-4x - 8y = 0 \qquad \text{Second equation}$$
$$-8y = 4x \qquad \text{Add } 4x \text{ to both sides.}$$
$$\frac{-8y}{-8} = \frac{4x}{-8} \qquad \text{Divide both sides by } -8.$$
$$y = -\frac{1}{2}x \qquad \text{Simplify.}$$

Now replace y with $-\dfrac{1}{2}x$ in the first equation.

$$6x + 12y = 5 \qquad \text{First equation}$$
$$6x + 12\left(-\frac{1}{2}x\right) = 5 \qquad \text{Let } y = -\frac{1}{2}x.$$
$$6x + (-6x) = 5 \qquad \text{Simplify.}$$
$$0 = 5 \qquad \text{Combine like terms.}$$

The false statement $0 = 5$ indicates that this system has no solution and is inconsistent. The graph of the linear equations in the system is a pair of parallel lines.

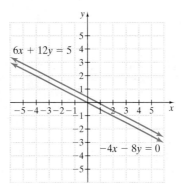

✔ **CONCEPT CHECK**

Describe how the graphs of the equations in a system appear if the system has

a. no solution   **b.** one solution   **c.** an infinite number of solutions

STUDY SKILLS REMINDER

Are You Satisfied with Your Performance on a Particular Quiz or Exam?

If not, don't forget to analyze your quiz or exam and look for common errors.

Were most of your errors a result of

▶ *Carelessness?* If your errors were careless, did you turn in your work before the allotted time expired? If so, resolve to use the entire time allotted next time. Any extra time can be spent checking your work.

▶ *Running out of time?* If so, make a point to better manage your time on your next exam. A few suggestions are to work any questions that you are unsure of last and to check your work after all questions have been answered.

▶ *Not understanding a concept?* If so, review that concept and correct your work. Remember next time to make sure that all concepts on a quiz or exam are understood before the exam.

MENTAL MATH

Give the solution of each system. If the system has no solution or an infinite number of solutions, say so. If the system as 1 solution, find it.

1. $\begin{cases} y = 4x \\ -3x + y = 1 \end{cases}$

When solving, you obtain $x = 1$ $(1,4)$

2. $\begin{cases} 4x - y = 17 \\ -8x + 2y = 0 \end{cases}$

When solving, you obtain $0 = 34$ no solution

3. $\begin{cases} 4x - y = 17 \\ -8x + 2y = -34 \end{cases}$

When solving, you obtain $0 = 0$ infinite number of solutions

4. $\begin{cases} 5x + 2y = 25 \\ x = y + 5 \end{cases}$

When solving, you obtain $y = 0$ $(5,0)$

5. $\begin{cases} x + y = 0 \\ 7x - 7y = 0 \end{cases}$

When solving, you obtain $x = 0$ $(0,0)$

6. $\begin{cases} y = -2x + 5 \\ 4x + 2y = 10 \end{cases}$

When solving, you obtain $0 = 0$ infinite number of solutions

EXERCISE SET 4.2

STUDY GUIDE/SSM   CD/VIDEO   PH MATH TUTOR CENTER   MathXL®Tutorials ON CD   MathXL®   MyMathLab®

MIXED PRACTICE

Solve each system of equations by the substitution method. See Examples 1 through 5.

1. $\begin{cases} x + y = 3 \\ x = 2y \end{cases}$ $(2,1)$

2. $\begin{cases} x + y = 20 \\ x = 3y \end{cases}$ $(15,5)$

 3. $\begin{cases} x + y = 6 \\ y = -3x \end{cases}$ $(-3,9)$

4. $\begin{cases} x + y = 6 \\ y = -4x \end{cases}$ $(-2,8)$

5. $\begin{cases} 3x + 2y = 16 \\ x = 3y - 2 \end{cases}$ $(4,2)$

6. $\begin{cases} 2x + 3y = 18 \\ x = 2y - 5 \end{cases}$ $(3,4)$

7. $\begin{cases} 3x - 4y = 10 \\ x = 2y \end{cases}$ $(10,5)$

8. $\begin{cases} 3x - 4y = 10 \\ y = 2x \end{cases}$ $(-2,-4)$

9. $\begin{cases} y = 3x + 1 \\ 4y - 8x = 12 \end{cases}$ $(2,7)$

10. $\begin{cases} y = 2x + 3 \\ 5y - 7x = 18 \end{cases}$ $(1,5)$

11. $\begin{cases} x + 2y = 6 \\ 2x + 3y = 8 \end{cases}$ $(-2,4)$

12. $\begin{cases} x + 3y = -5 \\ 2x + 2y = 6 \end{cases}$ $(7,-4)$

13. $\begin{cases} 2x - 5y = 1 \\ 3x + y = -7 \end{cases}$ $(-2, -1)$

14. $\begin{cases} 4x + 2y = 5 \\ 2x + y = -4 \end{cases}$

15. $\begin{cases} 2y = x + 2 \\ 6x - 12y = 0 \end{cases}$

16. $\begin{cases} 3y = x + 6 \\ 4x + 12y = 0 \end{cases}$ $(-3, 1)$

17. $\begin{cases} \dfrac{1}{3}x - y = 2 \\ x - 3y = 6 \end{cases}$

18. $\begin{cases} \dfrac{1}{4}x - 2y = 1 \\ x - 8y = 4 \end{cases}$

19. $\begin{cases} 4x + y = 11 \\ 2x + 5y = 1 \end{cases}$ $(3, -1)$

20. $\begin{cases} 3x + y = -14 \\ 4x + 3y = -22 \end{cases}$ $(-4, -2)$

21. $\begin{cases} 2x - 3y = -9 \\ 3x = y + 4 \end{cases}$ $(3, 5)$

22. $\begin{cases} 8x - 3y = -4 \\ 7x = y + 3 \end{cases}$ $(1, 4)$

23. $\begin{cases} 6x - 3y = 5 \\ x + 2y = 0 \end{cases}$ $\left(\frac{2}{3}, -\frac{1}{3}\right)$

24. $\begin{cases} 10x - 5y = -21 \\ x + 3y = 0 \end{cases}$ $\left(-\frac{9}{5}, \frac{3}{5}\right)$

🔒 25. $\begin{cases} 3x - y = 1 \\ 2x - 3y = 10 \end{cases}$ $(-1, -4)$

26. $\begin{cases} 2x - y = -7 \\ 4x - 3y = -11 \end{cases}$ $(-5, -3)$

27. $\begin{cases} -x + 2y = 10 \\ -2x + 3y = 18 \end{cases}$ $(-6, 2)$

28. $\begin{cases} -x + 3y = 18 \\ -3x + 2y = 19 \end{cases}$ $(-3, 5)$

29. $\begin{cases} 5x + 10y = 20 \\ 2x + 6y = 10 \end{cases}$ $(2, 1)$

30. $\begin{cases} 2x + 4y = 6 \\ 5x + 10y = 15 \end{cases}$

🔒 31. $\begin{cases} 3x + 6y = 9 \\ 4x + 8y = 16 \end{cases}$ no solution

32. $\begin{cases} 6x + 3y = 12 \\ 9x + 6y = 15 \end{cases}$ $(3, -2)$

33. $\begin{cases} y = 2x + 9 \\ y = 7x + 10 \end{cases}$ $\left(-\frac{1}{5}, \frac{43}{5}\right)$

34. $\begin{cases} y = 5x - 3 \\ y = 8x + 4 \end{cases}$ $\left(-\frac{7}{3}, -\frac{44}{3}\right)$

35. Explain how to identify an inconsistent system (no solution) when using the substitution method.   answers may vary

36. Occasionally, when using the substitution method, the equation $0 = 0$ is obtained. Explain how this result indicates that the equations are dependent.   answers may vary

Solve each system by the substitution method. First simplify each equation by combining like terms.

37. $\begin{cases} -5y + 6y = 3x + 2(x - 5) - 3x + 5 \\ 4(x + y) - x + y = -12 \end{cases}$ $(1, -3)$

38. $\begin{cases} 5x + 2y - 4x - 2y = 2(2y + 6) - 7 \\ 3(2x - y) - 4x = 1 + 9 \end{cases}$ $(5, 0)$

REVIEW AND PREVIEW

Write equivalent equations by multiplying both sides of the given equation by the given nonzero number. See Section 2.3.

39. $-6x - 4y = -12$
39. $3x + 2y = 6; -2$

40. $-5x + 5y = 50$
40. $-x + y = 10; 5$

41. $-4x + y = 3; 3$
41. $-12x + 3y = 9$

42. $5a - 7b = -4; -4$
42. $-20a + 28b = 16$

Add the binomials. See Section 4.2.

43. $\begin{aligned} 3n + 6m \\ 2n - 6m \end{aligned}$ $5n$

44. $\begin{aligned} -2x + 5y \\ 2x + 11y \end{aligned}$ $16y$

45. $\begin{aligned} -5a - 7b \\ 5a - 8b \end{aligned}$ $-15b$

46. $\begin{aligned} 9q + p \\ -9q - p \end{aligned}$ 0

Concept Extensions

📱 47. The number of men and women receiving bachelor's degrees each year has been steadily increasing. For the years 1970 through the projection of 2010, the number of men receiving degrees (in thousands) is given by the equation $y = 2.5x + 450$, and for women, the equation is $y = 11.85x + 337$ where x is the number of years after 1970. (*Source*: National Center for Education Statistics)

   a. Use the substitution method to solve this system of equations. (Round your final results to the nearest whole numbers.)   **a.** $(12, 480)$ **b.** answers may vary

   b. Explain the meaning of your answer to part (a).

   c. Sketch a graph of the system of equations. Write a sentence describing the trends for men and women receiving bachelor degrees.   See graphing answer section.

📱 48. The number of country music stations in the United States from 1995–2002 is given by the equation $y = -82.18x + 2612.39$ where x is the number of years after 1995. The number of Top 40 music stations is given by the equation $y = 27.89x + 306.46$ for the same time period. (*Source*: M Street Corporation)   **a.** $(20.9, 890.7)$ **b.** answers may vary

   a. Use the substitution method to solve this system of equations. (Round your final results to the nearest tenth.)

   b. Explain the meaning of your answer to part (a).

   c. Sketch a graph of the system of equations. Write a sentence describing the trends in the popularity of these two types of music formats.   See graphing answer section.

14.–15. no solution   **17.** infinite number of solutions   **18.** infinite number of solutions   **30.** infinite number of solutions

Solve each system by substitution. When necessary, round answers to the nearest hundredth.

49. $\begin{cases} y = 5.1x + 14.56 \\ y = -2x - 3.9 \end{cases}$
$(-2.6, 1.3)$

50. $\begin{cases} y = 3.1x - 16.35 \\ y = -9.7x + 28.45 \end{cases}$
$(3.5, -5.5)$

51. $\begin{cases} 3x + 2y = 14.05 \\ 5x + y = 18.5 \end{cases}$
$(3.28, 2.11)$

52. $\begin{cases} x + y = -15.2 \\ -2x + 5y = -19.3 \end{cases}$
$(-8.1, -7.1)$

4.3 SOLVING SYSTEMS OF LINEAR EQUATIONS BY ADDITION

Objective

1 Use the addition method to solve a system of linear equations.

1 We have seen that substitution is an accurate way to solve a linear system. Another method for solving a system of equations accurately is the **addition** or **elimination method**. The addition method is based on the addition property of equality: adding equal quantities to both sides of an equation does not change the solution of the equation. In symbols,

$$\text{if } A = B \text{ and } C = D, \text{ then } A + C = B + D.$$

EXAMPLE 1

Solve the system: $\begin{cases} x + y = 7 \\ x - y = 5 \end{cases}$

Solution Since the left side of each equation is equal to the right side, we add equal quantities by adding the left sides of the equations together and the right sides of the equations together. If we choose wisely, this adding gives us an equation in one variable, x, which we can solve for x.

$$
\begin{array}{ll}
x + y = 7 & \text{First equation} \\
\underline{x - y = 5} & \text{Second equation} \\
2x \quad\quad = 12 & \text{Add the equations.} \\
x = 6 & \text{Divide both sides by 2.}
\end{array}
$$

The x-value of the solution is 6. To find the corresponding y-value, let $x = 6$ in either equation of the system. We will use the first equation.

$$
\begin{array}{ll}
x + y = 7 & \text{First equation} \\
6 + y = 7 & \text{Let } x = 6. \\
y = 7 - 6 & \text{Solve for } y. \\
y = 1 & \text{Simplify.}
\end{array}
$$

The solution is $(6, 1)$. Check this in both equations.

First Equation	**Second Equation**	
$x + y = 7$	$x - y = 5$	
$6 + 1 \stackrel{?}{=} 7$	$6 - 1 \stackrel{?}{=} 5$	Let $x = 6$ and $y = 1$.
$7 = 7$ True	$5 = 5$ True	

Thus, the solution of the system is $(6, 1)$ and the graphs of the two equations intersect at the point $(6, 1)$ as shown on the top of the next page.

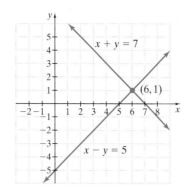

EXAMPLE 2

Solve the system: $\begin{cases} -2x + y = 2 \\ -x + 3y = -4 \end{cases}$

Solution

If we simply add the two equations, the result is still an equation in two variables. However, our goal is to eliminate one of the variables. Notice what happens if we multiply *both sides* of the first equation by -3, which we are allowed to do by the multiplication property of equality. The system

$$\begin{cases} -3(-2x + y) = -3(2) \\ -x + 3y = -4 \end{cases} \quad \text{simplifies to} \quad \begin{cases} 6x - 3y = -6 \\ -x + 3y = -4 \end{cases}$$

Now add the resulting equations and the y-variable is eliminated.

$$\begin{array}{r} 6x - 3y = -6 \\ -x + 3y = -4 \\ \hline 5x \qquad\quad = -10 \end{array} \quad \text{Add.}$$

$$x = -2 \qquad \text{Divide both sides by 5.}$$

To find the corresponding y-value, let $x = -2$ in any of the preceding equations containing both variables. We use the first equation of the original system.

$$-2x + y = 2 \qquad \text{First equation}$$
$$-2(-2) + y = 2 \qquad \text{Let } x = -2.$$
$$4 + y = 2$$
$$y = -2$$

The solution is $(-2, -2)$. Check this ordered pair in both equations of the original system.

In Example 2, the decision to multiply the first equation by -3 was no accident. **To eliminate a variable** when adding two equations, **the coefficient of the variable in one equation must be the opposite of its coefficient in the other equation.**

> **Helpful Hint**
>
> Be sure to multiply *both sides* of an equation by a chosen number when solving by the addition method. A common mistake is to multiply only the side containing the variables.

EXAMPLE 3

Solve the system: $\begin{cases} 2x - y = 7 \\ 8x - 4y = 1 \end{cases}$

Solution Multiply both sides of the first equation by -4 and the resulting coefficient of x is -8, the opposite of 8, the coefficient of x in the second equation. The system becomes

> **Helpful Hint**
> Don't forget to multiply both sides by -4.

$$\begin{cases} -4(2x - y) = -4(7) \\ 8x - 4y = 1 \end{cases} \quad \text{simplifies to} \quad \begin{cases} -8x + 4y = -28 \\ 8x - 4y = 1 \end{cases}$$

Now add the resulting equations.

$$\begin{array}{r} -8x + 4y = -28 \\ 8x - 4y = 1 \\ \hline 0 = -27 \quad \text{False} \end{array}$$

When we add the equations, both variables are eliminated and we have $0 = -27$, a false statement. This means that the system has no solution. The graphs of these equations are parallel lines.

CLASSROOM EXAMPLE

Solve the system: $\begin{cases} x - 3y = -2 \\ -3x + 9y = 5 \end{cases}$

answer: no solution

EXAMPLE 4

Solve the system: $\begin{cases} 3x - 2y = 2 \\ -9x + 6y = -6 \end{cases}$

Solution First we multiply both sides of the first equation by 3, then we add the resulting equations.

CLASSROOM EXAMPLE

Solve the system: $\begin{cases} 2x + 5y = 1 \\ -4x - 10y = -2 \end{cases}$

answer: infinite number of solutions

$$\begin{cases} 3(3x - 2y) = 3(2) \\ -9x + 6y = -6 \end{cases} \quad \text{simplifies to} \quad \begin{cases} 9x - 6y = 6 \\ -9x + 6y = -6 \quad \text{Add the equations.} \\ \hline 0 = 0 \quad \text{True} \end{cases}$$

Both variables are eliminated and we have $0 = 0$, a true statement. Whenever you eliminate a variable and get the equation $0 = 0$, the system has an infinite number of solutions. The graphs of these equations are identical.

TEACHING TIP

Continue to remind students to make sure that they multiply *both sides* of an equation by the same nonzero number. If they don't, the new equation is not equivalent to the old equation.

✔ **CONCEPT CHECK**

Suppose you are solving the system

$$\begin{cases} 3x + 8y = -5 \\ 2x - 4y = 3 \end{cases}$$

You decide to use the addition method by multiplying both sides of the second equation by 2. In which of the following was the multiplication performed correctly? Explain.

a. $4x - 8y = 3$ **b.** $4x - 8y = 6$

EXAMPLE 5

Solve the system: $\begin{cases} 3x + 4y = 13 \\ 5x - 9y = 6 \end{cases}$

Concept Check
Answer: b

Solution We can eliminate the variable y by multiplying the first equation by 9 and the second equation by 4.

CLASSROOM EXAMPLE

Solve the system: $\begin{cases} 4x + 5y = 14 \\ 3x - 2y = -1 \end{cases}$

answer: $(1, 2)$

$$\begin{cases} 9(3x + 4y) = 9(13) \\ 4(5x - 9y) = 4(6) \end{cases} \quad \text{simplifies to} \quad \begin{cases} 27x + 36y = 117 \\ \underline{20x - 36y = 24} \\ 47x = 141 \\ x = 3 \end{cases} \quad \text{Add the equations.}$$

To find the corresponding y-value, we let $x = 3$ in any equation in this example containing two variables. Doing so in any of these equations will give $y = 1$. The solution to this system is $(3, 1)$. Check to see that $(3, 1)$ satisfies each equation in the original system.

If we had decided to eliminate x instead of y in Example 5, the first equation could have been multiplied by 5 and the second by -3. Try solving the original system this way to check that the solution is $(3, 1)$.

The following steps summarize how to solve a system of linear equations by the addition method.

Solving a System of Two Linear Equations by the Addition Method

Step 1. Rewrite each equation in standard form $Ax + By = C$.

Step 2. If necessary, multiply one or both equations by a nonzero number so that the coefficients of a chosen variable in the system are opposites.

Step 3. Add the equations.

Step 4. Find the value of one variable by solving the resulting equation from Step 3.

Step 5. Find the value of the second variable by substituting the value found in Step 4 into either of the original equations.

Step 6. Check the proposed solution in the original system.

TEACHING TIP

Before attempting Example 6, remind students to use what they know to make this system as easy as possible to solve. Then ask how Example 6 could be written as a simpler system.

✔ **CONCEPT CHECK**

Suppose you are solving the system

$$\begin{cases} -4x + 7y = 6 \\ x + 2y = 5 \end{cases}$$

by the addition method.

a. What step(s) should you take if you wish to eliminate x when adding the equations?
b. What step(s) should you take if you wish to eliminate y when adding the equations?

EXAMPLE 6

Concept Check Answer:

a. multiply the second equation by 4
b. possible answer: multiply the first equation by -2 and the second equation by 7

Solve the system: $\begin{cases} -x - \dfrac{y}{2} = \dfrac{5}{2} \\ -\dfrac{x}{2} + \dfrac{y}{4} = 0 \end{cases}$

Solution   We begin by clearing each equation of fractions. To do so, we multiply both sides of the first equation by the LCD 2 and both sides of the second equation by the LCD 4. Then

the system

$$\begin{cases} 2\left(-x - \dfrac{y}{2}\right) = 2\left(\dfrac{5}{2}\right) \\ 4\left(-\dfrac{x}{2} + \dfrac{y}{4}\right) = 4(0) \end{cases} \quad \text{simplifies to} \quad \begin{cases} -2x - y = 5 \\ -2x + y = 0 \end{cases}$$

Now we add the resulting equations in the simplified system.

$$\begin{array}{rl} -2x - y &= 5 \\ \underline{-2x + y} &\underline{= 0} \\ -4x &= 5 \qquad \text{Add.} \\[4pt] x &= -\dfrac{5}{4} \end{array}$$

To find y, we could replace x with $-\dfrac{5}{4}$ in one of the equations with two variables.

Instead, let's go back to the simplified system and multiply by appropriate factors to eliminate the variable x and solve for y. To do this, we multiply the first equation in the simplified system by -1. Then the system

$$\begin{cases} -1(-2x - y) = -1(5) \\ -2x + y = 0 \end{cases} \quad \text{simplifies to} \quad \begin{array}{rl} 2x + y &= -5 \\ \underline{-2x + y} &\underline{= 0} \\ 2y &= -5 \qquad \text{Add.} \\[4pt] y &= -\dfrac{5}{2} \end{array}$$

Check the ordered pair $\left(-\dfrac{5}{4}, -\dfrac{5}{2}\right)$ in both equations of the original system. The solution is $\left(-\dfrac{5}{4}, -\dfrac{5}{2}\right)$.

Spotlight on

DECISION
& MAKING

Suppose you have been offered two similar positions as a sales associate. In one position, you would be paid a monthly salary of $1500 plus a 2% commission on all sales you make during the month. In the other position, you would be paid a monthly salary of $500 plus a 6% commission on all sales you make during the month. Which position would you choose? Explain your reasoning. Would knowing that the sales positions were at a car dealership affect your choice? What if the positions were at a shoe store?

EXERCISE SET 4.3

STUDY GUIDE/SSM   CD/VIDEO   PH MATH TUTOR CENTER   MathXL®Tutorials ON CD   MathXL®   MyMathLab®

10. $\left(4, -\dfrac{1}{5}\right)$

Solve each system of equations by the addition method. See Examples 1 through 5.   **4.** $(-3, 4)$ **8.** infinite number of solutions

5. $\begin{cases} x + y = 6 \\ x - y = 6 \end{cases}$ $(6, 0)$

6. $\begin{cases} x - y = 1 \\ -x + 2y = 0 \end{cases}$ $(2, 1)$

1. $\begin{cases} 3x + y = 5 \\ 6x - y = 4 \end{cases}$ $(1, 2)$

2. $\begin{cases} 4x + y = 13 \\ 2x - y = 5 \end{cases}$ $(3, 1)$

7. $\begin{cases} 3x + y = 4 \\ 9x + 3y = 6 \end{cases}$ no solution

8. $\begin{cases} 2x + y = 6 \\ 4x + 2y = 12 \end{cases}$

3. $\begin{cases} x - 2y = 8 \\ -x + 5y = -17 \end{cases}$ $(2, -3)$

4. $\begin{cases} x - 2y = -11 \\ -x + 5y = 23 \end{cases}$

9. $\begin{cases} 3x - 2y = 7 \\ 5x + 4y = 8 \end{cases}$ $\left(2, -\dfrac{1}{2}\right)$

10. $\begin{cases} 6x - 5y = 25 \\ 4x + 15y = 13 \end{cases}$

12. no solution   **13.** infinite number of solutions   **14.** no solution   **44.** $\left(-\frac{1}{4}, 2\right)$ **45.** infinite number of solutions

11. $\begin{cases} \frac{2}{3}x + 4y = -4 \\ 5x + 6y = 18 \end{cases}$ $(6, -2)$

12. $\begin{cases} \frac{3}{2}x + 4y = 1 \\ 9x + 24y = 5 \end{cases}$

13. $\begin{cases} 4x - 6y = 8 \\ 6x - 9y = 12 \end{cases}$

14. $\begin{cases} 9x - 3y = 12 \\ 12x - 4y = 18 \end{cases}$

15. $\begin{cases} 3x + y = -11 \\ 6x - 2y = -2 \end{cases}$ $(-2, -5)$

16. $\begin{cases} 4x + y = -13 \\ 6x - 3y = -15 \end{cases}$ $(-3, -1)$

17. $\begin{cases} 3x + 2y = 11 \\ 5x - 2y = 29 \end{cases}$ $(5, -2)$

18. $\begin{cases} 4x + 2y = 2 \\ 3x - 2y = 12 \end{cases}$ $(2, -3)$

19. $\begin{cases} x + 5y = 18 \\ 3x + 2y = -11 \end{cases}$ $(-7, 5)$

20. $\begin{cases} x + 4y = 14 \\ 5x + 3y = 2 \end{cases}$ $(-2, 4)$

21. $\begin{cases} 2x - 5y = 4 \\ 3x - 2y = 4 \end{cases}$ $\left(\frac{12}{11}, -\frac{4}{11}\right)$

22. $\begin{cases} 6x - 5y = 7 \\ 4x - 6y = 7 \end{cases}$ $\left(\frac{7}{16}, -\frac{7}{8}\right)$

23. $\begin{cases} 2x + 3y = 0 \\ 4x + 6y = 3 \end{cases}$ no solution

24. $\begin{cases} -x + 5y = -1 \\ 3x - 15y = 3 \end{cases}$

24. infinite number of solutions

Solve each system of equations by the addition method. See Example 6.   **29.** infinite number of solutions   **30.** no solution

25. $\begin{cases} \frac{x}{3} + \frac{y}{6} = 1 \\ \frac{x}{2} - \frac{y}{4} = 0 \end{cases}$ $\left(\frac{3}{2}, 3\right)$

26. $\begin{cases} \frac{x}{2} + \frac{y}{8} = 3 \\ x - \frac{y}{4} = 0 \end{cases}$ $(3, 12)$

27. $\begin{cases} x - \frac{y}{3} = -1 \\ -\frac{x}{2} + \frac{y}{8} = \frac{1}{4} \end{cases}$ $(1, 6)$

28. $\begin{cases} 2x - \frac{3y}{4} = -3 \\ x + \frac{y}{9} = \frac{13}{3} \end{cases}$ $(3, 12)$

29. $\begin{cases} \frac{x}{3} - y = 2 \\ -\frac{x}{2} + \frac{3y}{2} = -3 \end{cases}$

30. $\begin{cases} \frac{x}{2} + \frac{y}{4} = 1 \\ -\frac{x}{4} - \frac{y}{8} = 1 \end{cases}$

31. $\begin{cases} 8x = -11y - 16 \\ 2x + 3y = -4 \end{cases}$ $(-2, 0)$

32. $\begin{cases} 10x + 3y = -12 \\ 5x = -4y - 16 \end{cases}$ $(0, -4)$

33. When solving a system of equations by the addition method, how do we know when the system has no solution?

34. To solve the system $\begin{cases} 2x - 3y = 5 \\ 5x + 2y = 6 \end{cases}$, explain why the addition method might be preferred rather than the substitution method.   **33.–34.** answers may vary

MIXED PRACTICE

Solve each system by either the addition method or the substitution method.

35. $\begin{cases} 2x - 3y = -11 \\ y = 4x - 3 \end{cases}$ $(2, 5)$

36. $\begin{cases} 4x - 5y = 6 \\ y = 3x - 10 \end{cases}$ $(4, 2)$

37. $\begin{cases} x + 2y = 1 \\ 3x + 4y = -1 \end{cases}$ $(-3, 2)$

38. $\begin{cases} x + 3y = 5 \\ 5x + 6y = -2 \end{cases}$ $(-4, 3)$

39. $\begin{cases} 2y = x + 6 \\ 3x - 2y = -6 \end{cases}$ $(0, 3)$

40. $\begin{cases} 3y = x + 14 \\ 2x - 3y = -16 \end{cases}$ $(-2, 4)$

41. $\begin{cases} y = 2x - 3 \\ y = 5x - 18 \end{cases}$ $(5, 7)$

42. $\begin{cases} y = 6x - 5 \\ y = 4x - 11 \end{cases}$ $(-3, -23)$

43. $\begin{cases} x + \frac{1}{6}y = \frac{1}{2} \\ 3x + 2y = 3 \end{cases}$ $\left(\frac{1}{3}, 1\right)$

44. $\begin{cases} x + \frac{1}{3}y = \frac{5}{12} \\ 8x + 3y = 4 \end{cases}$

45. $\begin{cases} \frac{x+2}{2} = \frac{y+11}{3} \\ \frac{x}{2} = \frac{2y+16}{6} \end{cases}$

46. $\begin{cases} \frac{x+5}{2} = \frac{y+14}{4} \\ \frac{x}{3} = \frac{2y+2}{6} \end{cases}$ **46.** $(3, 2)$

47. $\begin{cases} 2x + 3y = 14 \\ 3x - 4y = -69.1 \end{cases}$ $(-8.9, 10.6)$

48. $\begin{cases} 5x - 2y = -19.8 \\ -3x + 5y = -3.7 \end{cases}$

48. $(-5.6, -4.1)$

REVIEW AND PREVIEW

Rewrite the following sentences using mathematical symbols. Do not solve the equations. See Sections 2.4 and 2.5.
49. $2x + 6 = x - 3$ **50.** $n + (n + 1) + (n + 2) = 66$
49. Twice a number, added to 6, is 3 less than the number.

50. The sum of three consecutive integers is 66.

51. Three times a number, subtracted from 20, is 2. $20 - 3x = 2$

52. Twice the sum of 8 and a number is the difference of the number and 20. $2(8 + x) = x - 20$

53. The product of 4 and the sum of a number and 6 is twice a number. $4(n + 6) = 2n$

54. The quotient of twice a number and 7 is subtracted from the reciprocal of the number. $\frac{1}{x} - \frac{2x}{7}$

Concept Extensions

55. In recent years, the number of Americans (in millions) who went downhill skiing more than once in a year has been decreasing, while the number who went snowboarding has been increasing. The number of downhill skiers who went more than once in a year from 1996 to 2001 is given by $0.35x + y = 9.3$ and the number of snowboarders can be given by the equation $0.56x - y = -2.5$. For both equations, x is the number of years after 1996. (*Source:* National Ski & Snowboard Association)

55. b. Skiers: 6.5 million Snowboarders: 6.98 million; rounding in part a caused a difference

  a. Suppose the trend continues. Use the addition method to predict the year in which the number of downhill skiers will equal the number of snowboarders. (Round up to the next whole number.)   2004

  b. Use the equations to determine the number of skiers and the number of snowboarders during the year found in part a. Are the numbers the same? Why or why not?

56. Commercial broadcast television stations can be divided into VHF stations (channels 2 through 13) and UHF stations (channels 14 through 83). The number y of VHF stations in the United States from 1980 through 2001 is given by the equation $3x - y = -514$, where x is the number of years after 1980. The number y of UHF stations in the United States from 1980 through 2001 is given by the equation $-23x + y = 244$, where x is the number of years after 1980. (*Source:* Based on data from the Television Bureau of Advertising, Inc.)

a. Use the addition method to solve this system of equations. $(13.5, 554.5)$

b. Interpret your solution from part (a). answers may vary

c. During which years were there more UHF commercial television stations than VHF stations? 1993 to 2001

57. Use the system of linear equations below to answer the questions. **57. b.** any real number except 15

$$\begin{cases} x + y = 5 \\ 3x + 3y = b \end{cases}$$

a. Find the value of b so that the equations are dependent and the system has an infinite number of solutions. $b = 15$

b. Find a value of b so that the system is inconsistent and there are no solutions to the system.

58. Use the system of linear equations below to answer the questions.

$$\begin{cases} x + y = 4 \\ 2x + by = 8 \end{cases}$$

a. Find the value of b so that the equations are dependent and the system has an infinite number of solutions. $b = 2$

b. Find a value of b so that the system is consistent and the system has a single solution. any real number except 2

59. Suppose you are solving the system

$$\begin{cases} -4x + 7y = 6 \\ x + 2y = 5 \end{cases}$$

by the addition method.

a. What step(s) should you take if you wish to eliminate x when adding the equations? answers may vary

b. What step(s) should you take if you wish to eliminate y when adding the equations? answers may vary

60. Suppose you are solving the system

$$\begin{cases} 3x + 8y = -5 \\ 2x - 4y = 3 \end{cases}$$

You decide to use the addition method by multiplying both sides of the second equation by 2. In which of the following was the multiplication performed correctly? Explain.

a. $4x - 8y = 3$

b. $4x - 8y = 6$ b, answers may vary

INTEGRATED REVIEW — SOLVING SYSTEMS OF EQUATIONS

Solve each system by either the addition method or the substitution method.

1. $\begin{cases} 2x - 3y = -11 \\ y = 4x - 3 \end{cases}$ $(2, 5)$

2. $\begin{cases} 4x - 5y = 6 \\ y = 3x - 10 \end{cases}$ $(4, 2)$

3. $\begin{cases} x + y = 3 \\ x - y = 7 \end{cases}$ $(5, -2)$

4. $\begin{cases} x - y = 20 \\ x + y = -8 \end{cases}$ $(6, -14)$

5. $\begin{cases} x + 2y = 1 \\ 3x + 4y = -1 \end{cases}$ $(-3, 2)$

6. $\begin{cases} x + 3y = 5 \\ 5x + 6y = -2 \end{cases}$ $(-4, 3)$

7. $\begin{cases} y = x + 3 \\ 3x - 2y = -6 \end{cases}$ $(0, 3)$

8. $\begin{cases} y = -2x \\ 2x - 3y = -16 \end{cases}$ $(-2, 4)$

9. $\begin{cases} y = 2x - 3 \\ y = 5x - 18 \end{cases}$ $(5, 7)$

10. $\begin{cases} y = 6x - 5 \\ y = 4x - 11 \end{cases}$ $(-3, -23)$

11. $\begin{cases} x + \dfrac{1}{6}y = \dfrac{1}{2} \\ 3x + 2y = 3 \end{cases}$ $\left(\dfrac{1}{3}, 1\right)$

12. $\begin{cases} x + \dfrac{1}{3}y = \dfrac{5}{12} \\ 8x + 3y = 4 \end{cases}$ $\left(-\dfrac{1}{4}, 2\right)$

13. $\begin{cases} x - 5y = 1 \\ -2x + 10y = 3 \end{cases}$ no solution   **14.** $\begin{cases} -x + 2y = 3 \\ 3x - 6y = -9 \end{cases}$ **15.** $\begin{cases} 0.2x - 0.3y = -0.95 \\ 0.4x + 0.1y = 0.55 \end{cases}$ (0.5, 3.5)

14. infinite number of solutions

16. $\begin{cases} 0.08x - 0.04y = -0.11 \\ 0.02x - 0.06y = -0.09 \end{cases}$ **17.** $\begin{cases} x = 3y - 7 \\ 2x - 6y = -14 \end{cases}$ **18.** $\begin{cases} y = \dfrac{x}{2} - 3 \\ 2x - 4y = 0 \end{cases}$ no solution

✎ **19.** Which method, substitution or addition, would you prefer to use to solve the system below? Explain your reasoning.   answers may vary

$$\begin{cases} 3x + 2y = -2 \\ y = -2x \end{cases}$$

✎ **20.** Which method, substitution or addition, would you prefer to use to solve the system below? Explain your reasoning.   answers may vary

16. $(-0.75, 1.25)$
17. infinite number of solutions

$$\begin{cases} 3x - 2y = -3 \\ 6x + 2y = 12 \end{cases}$$

4.4   SYSTEMS OF LINEAR EQUATIONS AND PROBLEM SOLVING

Objective

1   Use a system of equations to solve problems.

1   Many of the word problems solved earlier using one-variable equations can also be solved using two equations in **two** variables. We use the same problem-solving steps that have been used throughout this text. The only difference is that two variables are assigned to represent the two unknown quantities and that the problem is translated into **two** equations.

Problem-Solving Steps

1. **UNDERSTAND** the problem. During this step, become comfortable with the problem. Some ways of doing this are to

     Read and reread the problem.

     Choose two variables to represent the two unknowns.

     Construct a drawing, if possible.

     Propose a solution and check. Pay careful attention to how you check your proposed solution. This will help when writing equations to model the problem.

2. **TRANSLATE** the problem into two equations.

3. **SOLVE** the system of equations.

4. **INTERPRET** the results: **Check** the proposed solution in the stated problem and **state** your conclusion.

EXAMPLE 1

FINDING UNKNOWN NUMBERS

Find two numbers whose sum is 37 and whose difference is 21.

Solution

1. UNDERSTAND.   Read and reread the problem. Suppose that one number is 20. If their sum is 37, the other number is 17 because $20 + 17 = 37$. Is their difference 21? No; $20 - 17 = 3$. Our proposed solution is incorrect, but we now have a better understanding of the problem.

   Since we are looking for two numbers, we let

   x = first number

   y = second number

2. TRANSLATE.   Since we have assigned two variables to this problem, we translate our problem into two equations.

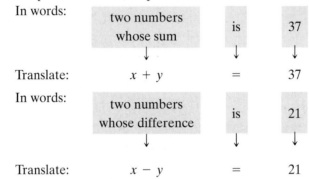

3. SOLVE.   Now we solve the system

$$\begin{cases} x + y = 37 \\ x - y = 21 \end{cases}$$

Notice that the coefficients of the variable y are opposites. Let's then solve by the addition method and begin by adding the equations.

$$\begin{array}{ll} x + y = 37 & \\ \underline{x - y = 21} & \\ 2x \quad = 58 & \text{Add the equations.} \\ x = \dfrac{58}{2} = 29 & \text{Divide both sides by 2.} \end{array}$$

Now we let $x = 29$ in the first equation to find y.

$$\begin{array}{ll} x + y = 37 & \text{First equation} \\ 29 + y = 37 & \\ y = 37 - 29 = 8 & \end{array}$$

4. INTERPRET.   The solution of the system is $(29, 8)$.
   Check: Notice that the sum of 29 and 8 is $29 + 8 = 37$, the required sum. Their difference is $29 - 8 = 21$, the required difference.
   State: The numbers are 29 and 8.

EXAMPLE 2

SOLVING A PROBLEM ABOUT PRICES

The Cirque du Soleil show Alegria is performing locally. Matinee admission for 4 adults and 2 children is $374, while admission for 2 adults and 3 children is $285.

a. What is the price of an adult's ticket?

b. What is the price of a child's ticket?

c. Suppose that a special rate of $1000 is offered for groups of 20 persons. Should a group of 4 adults and 16 children use the group rate? Why or why not?

Solution

CLASSROOM EXAMPLE

Admission prices at a local weekend fair were $5 for children and $7 for adults. The total money collected was $3379, and 587 people attended the fair. How many children and how many adults attended the fair?

answer:

365 children and 222 adults

1. UNDERSTAND.   Read and reread the problem and guess a solution. Let's suppose that the price of an adult's ticket is $50 and the price of a child's ticket is $40. To check our proposed solution, let's see if admission for 4 adults and 2 children is $374. Admission for 4 adults is 4($50) or $200 and admission for 2 children is 2($40) or $80. This gives a total admission of $200 + $80 = $280, not the required $374. Again though, we have accomplished the purpose of this process. We have a better understanding of the problem. To continue, we let

A = the price of an adult's ticket and

C = the price of a child's ticket

2. TRANSLATE.   We translate the problem into two equations using both variables.

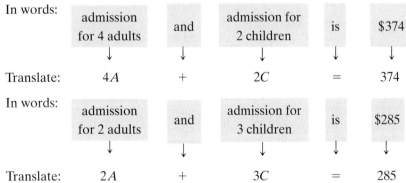

3. SOLVE.   We solve the system.

$$\begin{cases} 4A + 2C = 374 \\ 2A + 3C = 285 \end{cases}$$

Since both equations are written in standard form, we solve by the addition method. First we multiply the second equation by -2 so that when we add the equations we eliminate the variable A. Then the system

$$\begin{cases} 4A + 2C = 374 \\ -2(2A + 3C) = -2(285) \end{cases}$$ simplifies to $$\begin{cases} 4A + 2C = 374 \\ \underline{-4A - 6C = -570} \\ {-4C} = -196 \end{cases}$$ *Add the equations.*

$$-4C = -196$$
$$C = \frac{-196}{-4} = 49 \text{ or } \$49, \text{ the children's ticket price.}$$

To find A, we replace C with 49 in the first equation.

$$4A + 2C = 374 \qquad \qquad \text{First equation}$$
$$4A + 2(49) = 374 \qquad \qquad \text{Let } C = 49.$$
$$4A + 98 = 374$$
$$4A = 276$$
$$A = \frac{276}{4} = \begin{array}{l} 69 \text{ or } \$69, \\ \text{the adult's ticket price.} \end{array}$$

4. INTERPRET.

Check: Notice that 4 adults and 2 children will pay $4(\$69) + 2(\$49) = \$276 + \$98 = \$374$, the required amount. Also, the price for 2 adults and 3 children is $2(\$69) + 3(\$49) = \$138 + \$147 = \$285$, the required amount.

State: Answer the three original questions.

a. Since $A = 69$, the price of an adult's ticket is $69.

b. Since $C = 49$, the price of a child's ticket is $49.

c. The regular admission price for 4 adults and 16 children is

$$4(\$69) + 16(\$49) = \$276 + \$784$$
$$= \$1060$$

This is $60 more than the special group rate of $1000, so they should request the group rate. ⬭

(**EXAMPLE 3**)

FINDING RATES

As part of an exercise program, Albert and Louis started walking each morning. They live 15 miles away from each other and decided to meet one day by walking toward one another. After 2 hours they meet. If Louis walks one mile per hour faster than Albert, find both walking speeds.

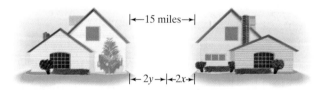

Solution **1. UNDERSTAND.** Read and reread the problem. Let's propose a solution and use the formula $d = r \cdot t$ to check. Suppose that Louis's rate is 4 miles per hour. Since Louis's rate is 1 mile per hour faster, Albert's rate is 3 miles per hour. To check, see if they can walk a total of 15 miles in 2 hours. Louis's distance is rate · time $= 4(2) = 8$ miles and Albert's distance is rate time $= 3(2) = 6$ miles. Their total distance is 8 miles + 6 miles = 14 miles, not the required 15 miles. Now that we have a better understanding of the problem, let's model it with a system of equations.

First, we let

x = Albert's rate in miles per hour

y = Louis's rate in miles per hour

Now we use the facts stated in the problem and the formula $d = rt$ to fill in the following chart.

	r	$\cdot$ t	$=$ d
Albert	x	2	$2x$
Louis	y	2	$2y$

2. TRANSLATE.   We translate the problem into two equations using both variables.

In words: | Albert's distance | + | Louis's distance | = | 15 |

Translate: $2x$ + $2y$ =   15

In words: | Louis's rate | is | 1 mile per hour faster than Albert's |

Translate: y = $x + 1$

3. SOLVE.   The system of equations we are solving is

$$\begin{cases} 2x + 2y = 15 \\ y = x + 1 \end{cases}$$

Let's use substitution to solve the system since the second equation is solved for y.

$$2x + 2y = 15 \qquad \text{\textit{First equation}}$$

$$2x + 2(x + 1) = 15 \qquad \text{\textit{Replace y with x + 1.}}$$
$$2x + 2x + 2 = 15$$
$$4x = 13$$
$$x = \frac{13}{4} = 3.25$$
$$y = x + 1 = 3.25 + 1 = 4.25$$

TEACHING TIP

Before beginning Example 4, you may want to have students make a table showing the amounts of salt and water in 1, 5, 10, 25, 50, and 100 liters of a 5% saline solution.

4. INTERPRET.   Albert's proposed rate is 3.25 miles per hour and Louis's proposed rate is 4.25 miles per hour.

Check: Use the formula $d = rt$ and find that in 2 hours, Albert's distance is (3.25)(2) miles or 6.5 miles. In 2 hours, Louis's distance is (4.25)(2) miles or 8.5 miles. The total distance walked is 6.5 miles + 8.5 miles or 15 miles, the given distance.

State: Albert walks at a rate of 3.25 miles per hour and Louis walks at a rate of 4.25 miles per hour.

EXAMPLE 4

FINDING AMOUNTS OF SOLUTIONS

Eric Daly, a chemistry teaching assistant, needs 10 liters of a 20% saline solution (salt water) for his 2 P.M. laboratory class. Unfortunately, the only mixtures on hand are a 5% saline solution and a 25% saline solution. How much of each solution should he mix to produce the 20% solution?

Solution

1. **UNDERSTAND.**   Read and reread the problem. Suppose that we need 4 liters of the 5% solution. Then we need $10 - 4 = 6$ liters of the 25% solution. To see if this gives us 10 liters of a 20% saline solution, let's find the amount of pure salt in each solution.

	concentration rate	×	amount of solution	=	amount of pure salt
	↓		↓		↓
5% solution:	0.05	×	4 liters	=	0.2 liters
25% solution:	0.25	×	6 liters	=	1.5 liters
20% solution:	0.20	×	10 liters	=	2 liters

Since 0.2 liters + 1.5 liters = 1.7 liters, not 2 liters, our proposed solution is incorrect. But we have gained some insight into how to model and check this problem. We let

x = number of liters of 5% solution

y = number of liters of 25% solution

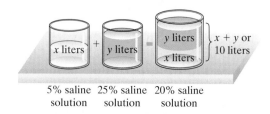

| | 5% saline solution | 25% saline solution | 20% saline solution |

Now we use a table to organize the given data.

	Concentration Rate	Liters of Solution	Liters of Pure Salt
First Solution	5%	x	$0.05x$
Second Solution	25%	y	$0.25y$
Mixture Needed	20%	10	$(0.20)(10)$

2. **TRANSLATE.**   We translate into two equations using both variables.

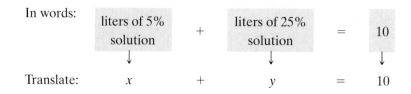

In words: liters of 5% solution + liters of 25% solution = 10

Translate: x + y = 10

In words:

salt in 5% solution	+	salt in 25% solution	=	salt in mixture
↓		↓		↓

Translate: $0.05x$ + $0.25y$ = $(0.20)(10)$

3. SOLVE.   Here we solve the system

$$\begin{cases} x + y = 10 \\ 0.05x + 0.25y = 2 \end{cases}$$

To solve by the addition method, we first multiply the first equation by -25 and the second equation by 100. Then the system

$$\begin{cases} -25(x + y) = -25(10) \\ 100(0.05x + 0.25y) = 100(2) \end{cases} \begin{matrix} \text{simplifies} \\ \text{to} \end{matrix} \begin{cases} -25x - 25y = -250 \\ \underline{5x + 25y = 200} \\ -20x = -50 \end{cases} \text{Add.}$$

$$x = 2.5$$

To find y, we let $x = 2.5$ in the first equation of the original system.

$$x + y = 10$$
$$2.5 + y = 10 \qquad \text{Let } x = \mathbf{2.5}.$$
$$y = 7.5$$

4. INTERPRET.   Thus, we propose that Eric needs to mix 2.5 liters of 5% saline solution with 7.5 liters of 25% saline solution.

Check: Notice that $2.5 + 7.5 = 10$, the required number of liters. Also, the sum of the liters of salt in the two solutions equals the liters of salt in the required mixture:

$$0.05(2.5) + 0.25(7.5) = 0.20(10)$$
$$0.125 + 1.875 = 2$$

State: Eric needs 2.5 liters of the 5% saline solution and 7.5 liters of the 25% solution.

✔ CONCEPT CHECK

Suppose you mix an amount of a 30% acid solution with an amount of a 50% acid solution. Which of the following acid strengths would be possible for the resulting acid mixture?

Concept Check Answer:
b

a. 22%   **b.** 44%   **c.** 63%

EXERCISE SET 4.4

STUDY GUIDE/SSM	CD/ VIDEO	PH MATH TUTOR CENTER	MathXL®Tutorials ON CD	MathXL®	MyMathLab®

Without actually solving each problem, choose each correct solution by deciding which choice satisfies the given conditions.

△ **1.** The length of a rectangle is 3 feet longer than the width. The perimeter is 30 feet. Find the dimensions of the rectangle.   c

  a. length = 8 feet; width = 5 feet

  b. length = 8 feet; width = 7 feet

  c. length = 9 feet; width = 6 feet

TEACHING TIP
A Group Activity for this section is available in the Instructor's Resource Manual.

△ **2.** An isosceles triangle, a triangle with two sides of equal length, has a perimeter of 20 inches. Each of the equal sides is one inch longer than the third side. Find the lengths of the three sides.

  a. 6 inches, 6 inches, and 7 inches   b

  b. 7 inches, 7 inches, and 6 inches

  c. 6 inches, 7 inches, and 8 inches

3. Two computer disks and three notebooks cost $17. However, five computer disks and four notebooks cost $32. Find the price of each.   b

   a. notebook = $4; computer disk = $3

   b. notebook = $3; computer disk = $4

   c. notebook = $5; computer disk = $2

4. Two music CDs and four music cassette tapes cost a total of $40. However, three music CDs and five cassette tapes cost $55. Find the price of each.   c

   a. CD = $12; cassette = $4

   b. CD = $15; cassette = $2

   c. CD = $10; cassette = $5

5. Kesha has a total of 100 coins, all of which are either dimes or quarters. The total value of the coins is $13.00. Find the number of each type of coin.   a

   a. 80 dimes; 20 quarters   **b.** 20 dimes; 44 quarters

   c. 60 dimes; 40 quarters

6. Yolanda has 28 gallons of saline solution available in two large containers at her pharmacy. One container holds three times as much as the other container. Find the capacity of each container.   c

   a. 15 gallons; 5 gallons   **b.** 20 gallons; 8 gallons

   c. 21 gallons; 7 gallons

Write a system of equations describing each situation. Do not solve the system. See Example 1.

7. Two numbers add up to 15 and have a difference of 7.

8. The total of two numbers is 16. The first number plus 2 more than 3 times the second equals 18.

9. Keiko has a total of $6500, which she has invested in two accounts. The larger account is $800 greater than the smaller account.

10. Dominique has four times as much money in his savings account as in his checking account. The total amount is $2300.

MIXED PRACTICE

Solve. See Examples 1 through 4.

11. Two numbers total 83 and have a difference of 17. Find the two numbers.   33 and 50

12. The sum of two numbers is 76 and their difference is 52. Find the two numbers.   64 and 12

13. A first number plus twice a second number is 8. Twice the first number plus the second totals 25. Find the numbers.   14 and −3

14. One number is 4 more than twice the second number. Their total is 25. Find the numbers.   18 and 7

15. The highest scorer during the WNBA 2003 regular season was Lauren Jackson of the Seattle Storm. Over the season, Jackson scored 12 more points than Katie Smith of the Minnesota Lynx. Together, Jackson and Smith scored 990 points during the 2003 regular season. How many points did each

player score over the course of the season? (*Source:* Women's National Basketball Association)
Jackson: 501 points; Smith 489 points

16. During the 2002–2003 regular NHL season, Markus Naslund of the Vancouver Vanucks scored 2 points less than Peter Forsberg of the Colorado Avalanche, the 2002–2003 point leader. Together, they scored 210 points. How many points did each score? (*Source*: National Hockey League)
Forsberg: 106 points; Naslund: 104 points

7. $\begin{cases} x + y = 15 \\ x - y = 7 \end{cases}$

8. $\begin{cases} x + y = 16 \\ x + 3y + 2 = 18 \end{cases}$

9. $\begin{cases} x + y = 6500 \\ x = y + 800 \end{cases}$

10. $\begin{cases} x + y = 2300 \\ y = 4x \end{cases}$

19. quarters: 53; nickels: 27   **21.** IBM: $84.92; GA Financial: $25.80

17. Ann Marie Jones has been pricing Amtrak train fares for a group trip to New York. Three adults and four children must pay $159. Two adults and three children must pay $112. Find the price of an adult's ticket, and find the price of a child's ticket.   child's ticket: $18; adult's ticket: $29

18. Last month, Jerry Papa purchased five cassettes and two compact discs at Wall-to-Wall Sound for $65. This month he bought three cassettes and four compact discs for $81. Find the price of each cassette, and find the price of each compact disc.   cassette: $7; compact disc: $15

19. Johnston and Betsy Waring have a jar containing 80 coins, all of which are either quarters or nickels. The total value of the coins is $14.60. How many of each type of coin do they have?

20. Bette Meish purchased 40 stamps, a mixture of 32¢ and 19¢ stamps. Find the number of each type of stamp if she spent $12.15.   32¢ stamps: 35; 19¢ stamps: 5

21. Brian and Abby Robinson own 50 shares of IBM stock and 40 shares of GA Financial stock. At the close of the markets on June 22, 2003, their stock portfolio was worth $5278. The closing price of GA Financial stock was $59.12 less than the closing price of IBM stock on that day. What was the price of each stock on June 22, 2003? (*Source*: New York Stock Exchange)

22. Kroger: 60 shares; General Motors: 120 shares

22. Sarah Stevenson has investments in Kroger and General Motors stock. On June 22, 2003, Kroger stock closed at $15.92 per share and General Motors stock closed at $39.49 per share. Sarah's portfolio was worth $5694 at the end of the day. If Sarah owns 60 more shares of General Motors stock than Kroger stock, how many of each type of stock does she own? (*Source*: New York Stock Exchange)

23. Pratap Puri rowed 18 miles down the Delaware River in 2 hours, but the return trip took him $4\frac{1}{2}$ hours. Find the rate Pratap could row in still water, and find the rate of the current.   still water: 6.5 mph; current: 2.5 mph

	d	$=$	r	$\cdot$	t
Downstream	18		$x + y$		2
Upstream	18		$x - y$		$4\frac{1}{2}$

24. The Jonathan Schultz family took a canoe 10 miles down the Allegheny River in 1 hour and 15 minutes. After lunch it took them 4 hours to return. Find the rate of the current.   2.75 mph

	d	$=$	r	$\cdot$	t
Downstream	10		$x + y$		$1\frac{1}{4}$
Upstream	10		$x - y$		4

25. Dave and Sandy Hartranft are frequent flyers with Delta Airlines. They often fly from Philadelphia to Chicago, a distance of 780 miles. On one particular trip they fly into the wind, and the flight takes 2 hours. The return trip, with the wind behind them, only takes $1\frac{1}{2}$ hours. Find the speed of the wind and find the speed of the plane in still air.   still air: 455 mph; wind: 65 mph

26. With a strong wind behind it, a United Airlines jet flies 2400 miles from Los Angeles to Orlando in 4 hours and 45 minutes. The return trip takes 6 hours, as the plane flies into the wind. Find the speed of the plane in still air, and find the wind speed to the nearest tenth mile per hour.
still air: 452.6 mph; wind: 52.6 mph

27. Alyssa Durheim is a chemist with Gemco Pharmaceutical. She needs to prepare 12 ounces of a 9% hydrochloric acid solution. Find the amount of 4% and the amount of 12% solution she should mix to get this solution.
12% solution: $7\frac{1}{2}$ oz; 4% solution: $4\frac{1}{2}$ oz

28. Diane Gray is preparing 15 liters of a 25% saline solution. Diane has two other saline solutions with strengths of 40% and 10%. Find the amount of 40% solution and the amount of 10% solution she should mix to get 15 liters of a 25% solution.   40% solution: $7\frac{1}{2}$ l; 10% solution: $7\frac{1}{2}$ l

29. Wayne Osby blends coffee for Maxwell House. He needs to prepare 200 pounds of blended coffee beans selling for $3.95 per pound. He intends to do this by blending together a high-quality bean costing $4.95 per pound and a cheaper bean costing $2.65 per pound. To the nearest pound, find how much high-quality coffee bean and how much cheaper coffee bean he should blend. $4.95 beans: 113 lb; $2.65 beans: 87 lb

30. Macadamia nuts cost an astounding $16.50 per pound, but research by Planter's Peanuts says that mixed nuts sell better if macadamias are included. The standard mix costs $9.25 per pound. Find how many pounds of macadamias and how many pounds of the standard mix should be combined to produce 40 pounds that will cost $10 per pound. Find the amounts to the nearest tenth of a pound.

△ **31.** Find the measures of two complementary angles if one angle is twice the other. (Recall that two angles are complementary if their sum is 90°.)   60°, 30°

30. macadamia: 4.1 lb; standard mix: 35.9 lb

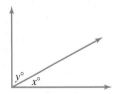

△ **32.** Find the measures of two supplementary angles if one angle is 20° more than four times the other. (Recall that two angles are supplementary if their sum is 180°.)   32°, 148°

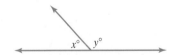

35. 20% solution: 10 *l*, 70%; solution: 40 *l*

△ **33.** Find the measures of two complementary angles if one angle is 10° more than three times the other.   20°, 70°

△ **34.** Find the measures of two supplementary angles if one angle is 18° more than twice the other.   54°, 126°

35. Barb Hayes, a pharmacist, needs 50 liters of a 60% alcohol solution. She currently has available a 20% solution and a 70% solution. How many liters of each does she need to make the needed 50 liters of 60% alcohol solution?

36. Two cars are 440 miles apart and traveling toward each other. They meet in 3 hours. If one car's speed is 10 miles per hour faster than the other car's speed, find the speed of each car. $68\frac{1}{3}$ mph and $78\frac{1}{3}$ mph

37. Kathi and Robert Haun had a pottery stand at the annual Skippack Craft Fair. They sold some of their pottery at the original price of $9.50 each, but later decreased the price of each by $2. If they sold all 90 pieces and took in $721, find
number sold at $9.50: 23; number sold at $7.50: 67

how many they sold at the original price and how many they sold at the reduced price.

38. A charity fund-raiser consisted of a spaghetti supper where a total of 387 people were fed. They charged $6.80 for adults and half-price for children. If they took in $2444.60, find how many adults and how many children attended the supper.

39. Traffic signs are regulated by the *Manual on Uniform Traffic Control Devices* (MUTCD). According to this manual, if the sign below is placed on a freeway, its perimeter must be 144 inches. Also, its length is 12 inches longer than its width. Find the dimensions of this sign.   length: 42 in.; width: 30 in.

38. adults: 332; children: 55

40. length: 18 in.; width: 12 in.

40. According to the MUTCD (see Exercise 39), this sign must have a perimeter of 60 inches. Also, its length must be 6 inches longer than its width. Find the length and width of this sign.

43. westbound: 80 mph; eastbound: 40 mph

41. Chris McShea began a 186-mile bicycle trip to build up stamina for a triathlete competition. Unfortunately, his bicycle chain broke, so he finished the trip walking. The whole trip took 6 hours. If Chris walks at a rate of 4 miles per hour and rides at 40 mph, find the amount of time he spent on the bicycle. $4\frac{1}{2}$ hr

42. Heather Johnson rented a car from Hertz, which rents its cars for a daily fee plus an additional charge per mile driven. Heather recalls that a car rented for 5 days and driven for 300 miles cost her $178, while a car rented for 4 days and driven for 500 miles cost $197. Find the daily fee, and find the mileage charge. $23 daily fee; $0.21 per mile

43. In Canada, eastbound and westbound trains travel along the same track, with sidings to pull onto to avoid accidents. Two trains are now 150 miles apart, with the westbound train traveling twice as fast as the eastbound train. A warning must be issued to pull one train onto a siding or else the trains will crash in $1\frac{1}{4}$ hours. Find the speed of the eastbound train and the speed of the westbound train.

44. Cyril and Anoa Nantambu operate a small construction and supply company. In July they charged the Shaffers $1702.50 for 65 hours of labor and 3 tons of material. In August the Shaffers paid $1349 for 49 hours of labor and $2\frac{1}{2}$ tons of material. Find the cost per hour of labor and the cost per ton of material.   labor: $13.50 per hr; material: $275 per ton

45. The Santa Fe National Historic Trail is approximately 1200 miles between Old Franklin, Missouri, and Santa Fe, New Mexico. Suppose that a group of hikers start from each town and walk the trail toward each other. They meet after a total hiking time of 240 hours. If one group travels $\frac{1}{2}$ mile per hour slower than the other group, find the rate of each group. (*Source*: National Park Service) $2\frac{1}{4}$ mph and $2\frac{3}{4}$ mph

46. California 1 South is a historic highway that stretches 123 miles along the coast from Monterey to Morro Bay. Suppose that two cars start driving this highway, one from each town. They meet after 3 hours. Find the rate of each car if one car travels 1 mile per hour faster than the other car. (*Source: National Geographic*)   20 mph, 21 mph

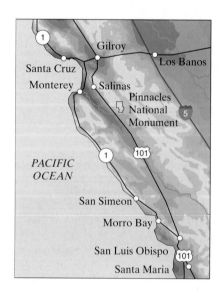

47. A 30% solution of fertilizer is to be mixed with a 60% solution of fertilizer in order to get 150 gallons of a 50% solution. How many gallons of the 30% solution and 60% solution should be mixed?   30%: 50 gal; 60%: 100 gal

48. A 10% acid solution is to be mixed with a 50% acid solution in order to get 120 ounces of a 20% acid solution. How many ounces of the 10% solution and 50% solution should be mixed?   10%: 90 oz; 50%: 30 oz

REVIEW AND PREVIEW

Solve each linear inequality. See Section 2.8.

49. $-3x < -9$ $(3, \infty)$ **50.** $2x - 7 \leq 5x + 11$ $[-6, \infty)$

51. $4(2x - 1) \geq 0$ $\left[\frac{1}{2}, \infty\right)$ **52.** $\frac{2}{3}x < \frac{1}{3}$ $\left(-\infty, \frac{1}{2}\right)$

Concept Extensions

△ **53.** David Lopez has decided to fence off a garden plot behind his house, using his house as the "fence" along one side of the garden. The length (which runs parallel to the house) is 3 feet less than twice the width. Find the dimensions if 33 feet of fencing is used.   width: 9 ft; length: 15 ft

55. c. increasing: cable; decreasing: networks

△ **54.** Judy McElroy plans to erect 152 feet of fencing around her rectangular horse pasture. A river bank serves as one side length of the rectangle. If each width is 4 feet longer than half the length, find the dimensions.   width: 40 ft; length: 72 ft

55. b. In 5.3 years after 1995, cable viewers equaled network viewers. This number of viewers was 38.2 million for each.

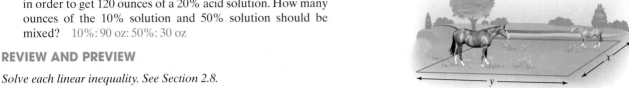

55. The number of viewers (in millions) who watch cable TV during the summer is approximated by the equation $y = 3.1x + 21.8$ where x is the number of years after 1995. The number of viewers (in millions) who watch the networks during the summer is approximated by the equation $y = -2.1x + 49.3$ where x is also the number of years after 1995.

a. Solve the system

$$\begin{cases} y = 3.1x + 21.8 \\ y = -2.1x + 49.3 \end{cases}$$

Round x and y to the nearest tenth. $(5.3, 38.2)$

b. Explain what the point of intersection means in terms of the context of the exercise.

c. Look at the slope of both equations of the system. What type of viewers is increasing and what type is decreasing?

56. In the triangle below, the measure of angle x is 6 times the measure of angle y. Find the measure of x and y by writing a system of two equations in two unknowns and solving the system. $x = 12\frac{6}{7}°$ $y = 77\frac{1}{7}°$

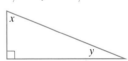

4.5   GRAPHING LINEAR INEQUALITIES

Objective

1 Graph a linear inequality in two variables.

1   In the next section, we continue our work with systems by solving systems of linear inequalities. Before that section, we first need to learn to graph a single linear inequality in two variables.

Recall that a linear equation in two variables is an equation that can be written in the form $Ax + By = C$ where A, B, and C are real numbers and A and B are not both 0. The definition of a linear inequality is the same except that the equal sign is replaced with an inequality sign.

A **linear inequality in two variables** is an inequality that can be written in one of the forms:

$$\begin{array}{ll} Ax + By < C & Ax + By \leq C \\ Ax + By > C & Ax + By \geq C \end{array}$$

where A, B, and C are real numbers and A and B are not both 0. Just as for linear equations in x and y, an ordered pair is a **solution** of an inequality in x and y if replacing the variables by coordinates of the ordered pair results in a true statement.

To graph a linear inequality in two variables, we will begin by graphing a related equation. For example, to graph $x - y \le 1$, we begin by graphing $x - y = 1$.

The linear equation $x - y = 1$ is graphed next. Recall that all points on the line correspond to ordered pairs that satisfy the equation $x - y = 1$. It can be shown that all the points above the line $x - y = 1$ have coordinates that satisfy the inequality $x - y < 1$. Similarly, all points below the line have coordinates that satisfy the inequality $x - y > 1$. To see this, a few points have been selected on one side of the line. These points all satisfy $x - y < 1$ as shown in the table below.

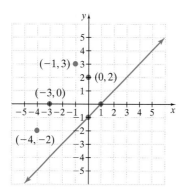

Point	$x - y < 1$	
Check $(0, 2)$	$0 - 2 < 1$	
	$-2 < 1$	True
Check $(-3, 0)$	$-3 - 0 < 1$	
	$-3 < 1$	True
Check $(-4, -2)$	$-4 - (-2) < 1$	
	$-2 < 1$	True
Check $(-1, 3)$	$-1 - 3 < 1$	
	$-4 < 1$	True

The graph of $x - y < 1$ is in blue, and the graph of $x - y > 1$ is in red.

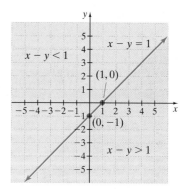

The region above the line and the region below the line are called **half-planes**. Every line divides the plane (similar to a sheet of paper extending indefinitely in all directions) into two half-planes; the line is called the **boundary**.

How do we graph $x - y \le 1$? Recall that the inequality $x - y \le 1$ means

$$x - y = 1 \quad \text{or} \quad x - y < 1$$

Thus, the graph of $x - y \le 1$ is the half-plane $x - y < 1$ along with the boundary line $x - y = 1$.

> ### Graphing a Linear Inequality in Two Variables
>
> **Step 1.**   Graph the boundary line found by replacing the inequality sign with an equal sign. If the inequality sign is $>$ or $<$, graph a dashed boundary line (indicating that the points on the line are not solutions of the inequality). If the inequality sign is $\geq$ or $\leq$, graph a solid boundary line (indicating that the points on the line are solutions of the inequality).
>
> **Step 2.**   Choose a point, not on the boundary line, as a test point. Substitute the coordinates of this test point into the original inequality.
>
> **Step 3.**   If a true statement is obtained in Step 2, shade the half-plane that contains the test point. If a false statement is obtained, shade the half-plane that does not contain the test point.

EXAMPLE 1

Graph: $x + y < 7$

Solution   First we graph the boundary line by graphing the equation $x + y = 7$. We graph this boundary as a dashed line because the inequality sign is $<$, and thus the points on the line are not solutions of the inequality $x + y < 7$.

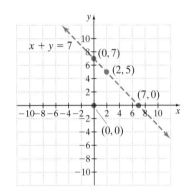

Next, choose a test point, being careful not to choose a point on the boundary line. We choose $(0, 0)$. Substitute the coordinates of $(0, 0)$ into $x + y < 7$.

$$x + y < 7 \qquad \text{Original inequality}$$
$$0 + 0 \overset{?}{<} 7 \qquad \text{Replace } x \text{ with } 0 \text{ and } y \text{ with } 0.$$
$$0 < 7 \qquad \text{True}$$

Since the result is a true statement, $(0, 0)$ is a solution of $x + y < 7$, and every point in the same half-plane as $(0, 0)$ is also a solution. To indicate this, shade the entire half-plane containing $(0, 0)$, as shown.

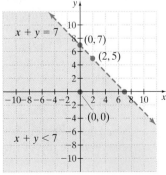

TEACHING TIP

For Examples 1 and 2, ask students what inequality describes the unshaded region of each graph.

EXAMPLE 2

Graph: $2x - y \geq 3$

Solution Graph the boundary line by graphing $2x - y = 3$. Draw this line as a solid line since the inequality sign is $\geq$, and thus the points on the line are solutions of $2x - y \geq 3$. Once again, $(0, 0)$ is a convenient test point since it is not on the boundary line.

Substitute 0 for x and 0 for y into the **original inequality**.

$$2x - y \geq 3$$
$$2(0) - 0 \overset{?}{\geq} 3 \qquad \text{Let } x = 0 \text{ and } y = 0.$$
$$0 \geq 3 \qquad \text{False}$$

Since the statement is false, no point in the half-plane containing $(0, 0)$ is a solution. Shade the half-plane that does not contain $(0, 0)$. Every point in the shaded half-plane and every point on the boundary line satisfies $2x - y \geq 3$.

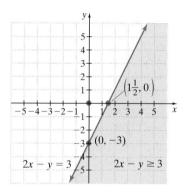

> ▶ **Helpful Hint**
>
> When graphing an inequality, make sure the test point is substituted into the **original inequality**. For Example 2, we substituted the test point $(0, 0)$ into the **original inequality** $2x - y \geq 3$, *not* $2x - y = 3$.

EXAMPLE 3

Graph: $x < 2y$

Solution We find the boundary line by graphing $x = 2y$. The boundary line is a dashed line since the inequality symbol is $<$. We cannot use $(0, 0)$ as a test point because it is a point on the boundary line. We choose instead $(0, 2)$.

$$x < 2y$$
$$0 \overset{?}{<} 2(2) \qquad \text{Let } x = 0 \text{ and } y = 2.$$
$$0 < 4 \qquad \text{True}$$

Since the statement is true, we shade the half-plane that contains the test point $(0, 2)$, as shown at the top of the next page.

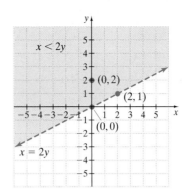

EXAMPLE 4

Graph: $5x + 4y \leq 20$

Solution

We graph the solid boundary line $5x + 4y = 20$ and choose $(0, 0)$ as the test point.

$$5x + 4y \leq 20$$
$$5(0) + 4(0) \overset{?}{\leq} 20 \qquad \text{Let } x = 0 \text{ and } y = 0.$$
$$0 \leq 20 \qquad \text{True}$$

We shade the half-plane that contains $(0, 0)$, as shown.

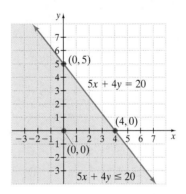

EXAMPLE 5

Graph: $y > 3$

Solution

We graph the dashed boundary line $y = 3$ and choose $(0, 0)$ as the test point. (Recall that the graph of $y = 3$ is a horizontal line with y-intercept 3.)

$$y > 3$$
$$0 \overset{?}{>} 3 \qquad \text{Let } y = 0.$$
$$0 > 3 \qquad \text{False}$$

We shade the half-plane that does not contain $(0, 0)$, as shown on the top of the next page.

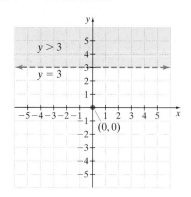

MENTAL MATH

State whether the graph of each inequality includes its corresponding boundary line.

1. $y \geq x + 4$ yes   **2.** $x - y > -7$ no   **3.** $y \geq x$ yes   **4.** $x > 0$ no

Decide whether $(0, 0)$ is a solution of each given inequality.

5. $x + y > -5$ yes   **6.** $2x + 3y < 10$ yes   **7.** $x - y \leq -1$ no   **8.** $\frac{2}{3}x + \frac{5}{6}y > 4$ no

TEACHING TIP
A Group Activity for this section is available in the Instructor's Resource Manual.

EXERCISE SET 4.5

STUDY GUIDE/SSM   CD/VIDEO   PH MATH TUTOR CENTER   MathXL®Tutorials ON CD   MathXL®   MyMathLab®

Determine which ordered pairs given are solutions of the linear inequality in two variables. See Example 1.

1. $x - y > 3$; $(2, -1), (5, 1)$ no; yes

2. $y - x < -2$; $(2, 1), (5, -1)$ no; yes

3. $3x - 5y \leq -4$; $(-1, -1), (4, 0)$ no; no

4. $2x + y \geq 10$; $(-1, -4), (5, 0)$ no; yes

5. $x < -y$; $(0, 2), (-5, 1)$ no; yes

6. $y > 3x$; $(0, 0), (-1, -4)$ no; no

MIXED PRACTICE   **7.–42.** See graphing answer section.

Graph each inequality. See Examples 2 through 6.

7. $x + y \leq 1$ **8.** $x + y \geq -2$ **9.** $2x + y > -4$

10. $x + 3y \leq 3$ **11.** $x + 6y \leq -6$ **12.** $7x + y > -14$

13. $2x + 5y > -10$ **14.** $5x + 2y \leq 10$ **15.** $x + 2y \leq 3$

16. $2x + 3y > -5$ **17.** $2x + 7y > 5$ **18.** $3x + 5y \leq -2$

19. $x - 2y \geq 3$ **20.** $4x + y \leq 2$ **21.** $5x + y < 3$

22. $x + 2y > -7$ **23.** $4x + y < 8$ **24.** $9x + 2y \geq -9$

25. $y \geq 2x$ **26.** $x < 5y$ **27.** $x \geq 0$

28. $y \leq 0$ **29.** $y \leq -3$ **30.** $x > -\frac{2}{3}$

31. $2x - 7y > 0$ **32.** $5x + 2y \leq 0$ **33.** $3x - 7y \geq 0$

34. $-2x - 9y > 0$ **35.** $x > y$ **36.** $x \leq -y$

37. $x - y \leq 6$ **38.** $x - y > 10$ **39.** $-\frac{1}{4}y + \frac{1}{3}x > 1$

40. $\frac{1}{2}x - \frac{1}{3}y \leq -1$ **41.** $-x < 0.4y$ **42.** $0.3x \geq 0.1y$

In Exercises 43–48, match each inequality with its graph.

a. $x > 2$ **b.** $y < 2$ **c.** $y < 2x$
d. $y \leq -3x$ **e.** $2x + 3y < 6$ **f.** $3x + 2y > 6$

43. e

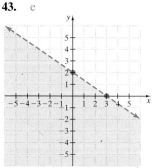

44. a

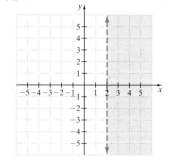

45. c

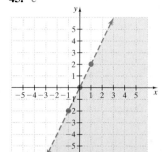

46. b

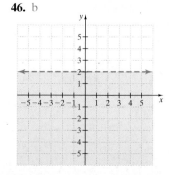

47. f 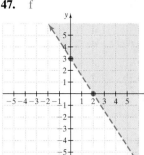   **48.** d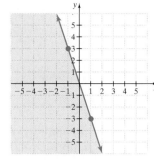

REVIEW AND PREVIEW

Evaluate. See Section 1.4.

49. 2^3 8   **50.** 3^4 81   **51.** $(-2)^5$ -32

52. -2^5 -32 **53.** $3 \cdot 4^2$ 48   **54.** $4 \cdot 3^3$ 108

Evaluate each expression for the given replacement value. See Section 1.4.

55. x^2 if x is -5 25   **56.** x^3 if x is -5 -125

57. $2x^3$ if x is -1 -2 **58.** $3x^2$ if x is -1 3

Concept Extensions

59. Write an inequality whose solutions are all pairs of numbers x and y whose sum is at least 13. Graph the inequality.

60. Write an inequality whose solutions are all the pairs of numbers x and y whose sum is at most -4. Graph the inequality.

61. Explain why a point on the boundary line should not be chosen as the test point.   answers may vary

62. Describe the graph of a linear inequality.   answers may vary

63. The price for a taxi cab in a small city is $2.50 per mile, x, while traveling, and $.25 every minute, y, while waiting. If you have $20 to spend on a cab ride, the inequality

$$2.5x + 0.25y \le 20.$$

represents your situation. Graph this inequality in the first quadrant only.

64. A word processor charges $22 per hour, x, for typing a first draft, and $15 per hour, y, for making changes and type a second draft. If you need a document typed and have $100, the inequality

$$22x + 15y \le 100$$

represents your situation. Graph the inequality in the first quadrant only.

59. $x + y \ge 13$; see graphing answer section.
60. $x + y \le -4$; see graphing answer section.
63.–64. See graphing answer section.

65. In Exercises 63 and 64, why were you instructed to graph each inequality in the first quadrant only?   answers may vary

66. Scott Sambracci and Sara Thygeson are planning their wedding. They have calculated that they want the cost of their wedding ceremony x plus the cost of their reception y to be no more than $5000.   **66. a.** $x + y \le 5000$

   a. Write an inequality describing this relationship.

   b. Graph this inequality.

   c. Why should we be interested in only quadrant I of this graph?   answers may vary

67. a. $30x + 0.15y \le 500$

67. It's the end of the budgeting period for Dennis Fernandes and he has $500 left in his budget for car rental expenses. He plans to spend this budget on a sales trip throughout southern Texas. He will rent a car that costs $30 per day and $0.15 per mile and he can spend no more than $500.

   a. Write an inequality describing this situation. Let x = number of days and let y = number of miles.

   b. Graph this inequality.

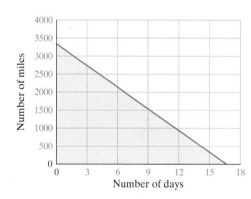

   c. Why should we be interested in only quadrant I of this graph?   answers may vary

4.6   SYSTEMS OF LINEAR INEQUALITIES

Objective

1   Solve a system of linear inequalities.

1   In Section 4.5, we graphed linear inequalities in two variables. Just as two linear equations make a system of linear equations, two linear inequalities make a **system of linear inequalities**. Systems of inequalities are very important in a process called linear programming. Many businesses use linear programming to find the most profitable way to use limited resources such as employees, machines, or buildings.

A **solution of a system of linear inequalities** is an ordered pair that satisfies each inequality in the system. The set of all such ordered pairs is the solution set of the system. Graphing this set gives us a picture of the solution set. We can graph a system of inequalities by graphing each inequality in the system and identifying the region of overlap.

EXAMPLE 1

Graph the solution of the system: $\begin{cases} 3x \ge y \\ x + 2y \le 8 \end{cases}$

Solution   We begin by graphing each inequality on the same set of axes. The graph of the solution of the system is the region contained in the graphs of both inequalities. It is their intersection.

First, graph $3x \ge y$. The boundary line is the graph of $3x = y$. Sketch a solid boundary line since the inequality $3x \ge y$ means $3x > y$ or $3x = y$. The test point $(1, 0)$ satisfies the inequality, so shade the half-plane that includes $(1, 0)$.

CLASSROOM EXAMPLE

Graph the solution of the system:
$\begin{cases} y - x \ge 4 \\ x + 3y \le -2 \end{cases}$

answer:

TEACHING TIP

Some students will have trouble finding the solution region even when both inequalities of the system have been graphed correctly. Here are two suggestions that may help.

1. Shade each inequality in the system with a different colored pencil.

2. If shading with pencil lead, try shading each inequality in the system at a different angle.

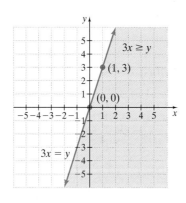

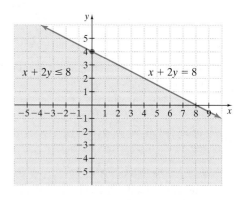

Next, sketch a solid boundary line $x + 2y = 8$ on the same set of axes. The test point $(0, 0)$ satisfies the inequality $x + 2y \le 8$, so shade the half-plane that includes $(0, 0)$. (For clarity, the graph of $x + 2y \le 8$ is shown on a separate set of axes.)

An ordered pair solution of the system must satisfy both inequalities. These solutions are points that lie in both shaded regions. The solution of the system is the purple shaded region as seen on the top of the next page. This solution includes parts of both boundary lines.

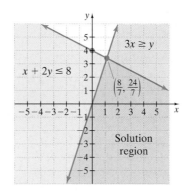

In linear programming, it is sometimes necessary to find the coordinates of the **corner point**: the point at which the two boundary lines intersect. To find the point of intersection, solve the related linear system

$$\begin{cases} 3x = y \\ x + 2y = 8 \end{cases}$$

by the substitution method or the addition method. The lines intersect at $\left(\frac{8}{7}, \frac{24}{7}\right)$, the corner point of the graph.

Graphing the Solution of a System of Linear Inequalities

Step 1.   Graph each inequality in the system on the same set of axes.

Step 2.   The solutions of the system are the points common to the graphs of all the inequalities in the system.

EXAMPLE 2

Graph the solution of the system: $\begin{cases} x - y < 2 \\ x + 2y > -1 \end{cases}$

Solution   Graph both inequalities on the same set of axes. Both boundary lines are dashed lines since the inequality symbols are $<$ and $>$. The solution of the system is the region shown by the purple shading. In this example, the boundary lines are not a part of the solution.

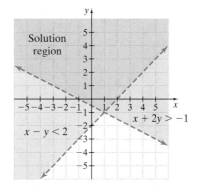

EXAMPLE 3

Graph the solution of the system: $\begin{cases} -3x + 4y < 12 \\ x \geq 2 \end{cases}$

Solution   Graph both inequalities on the same set of axes.

CLASSROOM EXAMPLE

Graph the solution of the system:
$$\begin{cases} -2x + 5y < 10 \\ y \leq 3 \end{cases}$$

answer:

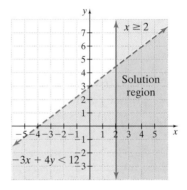

The solution of the system is the purple shaded region, including a portion of the line $x = 2$.

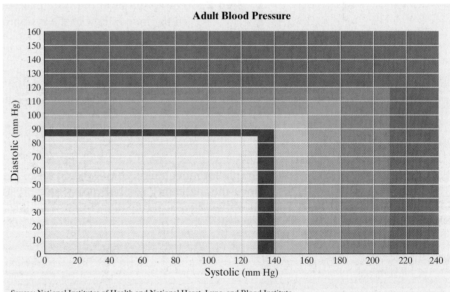

Spotlight on

DECISION
✢ MAKING

Suppose you are a registered nurse. Today you are working at a health fair providing free blood pressure screenings. You measure an attendee's blood pressure as 168/82 (read as "168 over 82," where the systolic blood pressure is listed first and the diastolic blood pressure is listed second). What would you recommend that this health fair attendee do? Explain.

Adult Blood Pressure

Source: National Institutes of Health and National Heart, Lung, and Blood Institute

Blood Pressure Category and Recommended Follow-up:
☐ Normal: recheck in 2 years
■ High normal: recheck in 1 year
■ Mild hypertension: confirm within 2 months
■ Moderate hypertension: see primary care physician within 1 month
■ Severe hypertension: see primary care physician within 1 week
■ Very severe hypertension: see primary care physician immediately

EXERCISE SET 4.6

| STUDY GUIDE/SSM | CD/ VIDEO | PH MATH TUTOR CENTER | MathXL®Tutorials ON CD | MathXL® | MyMathLab® |

MIXED PRACTICE

Graph the solution of each system of linear inequalities. See Examples 1 through 3.   **1.–24.** See graphing answer section.

1. $\begin{cases} y \geq x + 1 \\ y \geq 3 - x \end{cases}$

2. $\begin{cases} y \geq x - 3 \\ y \geq -1 - x \end{cases}$

3. $\begin{cases} y < 3x - 4 \\ y \leq x + 2 \end{cases}$

4. $\begin{cases} y \leq 2x + 1 \\ y > x + 2 \end{cases}$

5. $\begin{cases} y \leq -2x - 2 \\ y \geq x + 4 \end{cases}$

6. $\begin{cases} y \leq 2x + 4 \\ y \geq -x - 5 \end{cases}$

7. $\begin{cases} y \geq -x + 2 \\ y \leq 2x + 5 \end{cases}$

8. $\begin{cases} y \geq x - 5 \\ y \leq -3x + 3 \end{cases}$

9. $\begin{cases} x \geq 3y \\ x + 3y \leq 6 \end{cases}$

10. $\begin{cases} -2x < y \\ x + 2y < 3 \end{cases}$

11. $\begin{cases} y + 2x \geq 0 \\ 5x - 3y \leq 12 \end{cases}$

12. $\begin{cases} y + 2x \leq 0 \\ 5x + 3y \geq -2 \end{cases}$

13. $\begin{cases} 3x - 4y \geq -6 \\ 2x + y \leq 7 \end{cases}$

14. $\begin{cases} 4x - y \geq -2 \\ 2x + 3y \leq -8 \end{cases}$

15. $\begin{cases} x \leq 2 \\ y \geq -3 \end{cases}$

16. $\begin{cases} x \geq -3 \\ y \geq -2 \end{cases}$

17. $\begin{cases} y \geq 1 \\ x < -3 \end{cases}$

18. $\begin{cases} y > 2 \\ x \geq -1 \end{cases}$

19. $\begin{cases} 2x + 3y < -8 \\ x \geq -4 \end{cases}$

20. $\begin{cases} 3x + 2y \leq 6 \\ x < 2 \end{cases}$

21. $\begin{cases} 2x - 5y \leq 9 \\ y \leq -3 \end{cases}$

22. $\begin{cases} 2x + 5y \leq -10 \\ y \geq 1 \end{cases}$

23. $\begin{cases} y \geq \dfrac{1}{2}x + 2 \\ y \leq \dfrac{1}{2}x - 3 \end{cases}$

24. $\begin{cases} y \geq \dfrac{-3}{2}x + 3 \\ y < \dfrac{-3}{2}x + 6 \end{cases}$

REVIEW AND PREVIEW

Find the square of each expression. For example, the square of 7 is 7^2 or 49. The square of 5x is $(5x)^2$ or $25x^2$. See Section 1.4.

25. 4   16

26. 3   9

27. 6x $36x^2$

28. 11y $121y^2$

29. $10y^3$ $100y^6$

30. $8x^5$ $64x^{10}$

Concept Extensions

For each system of inequalities, choose the corresponding graph.

31. $\begin{cases} y < 5 \\ x > 3 \end{cases}$ C

32. $\begin{cases} y > 5 \\ x < 3 \end{cases}$ A

33. $\begin{cases} y \leq 5 \\ x < 3 \end{cases}$ D

34. $\begin{cases} y > 5 \\ x \geq 3 \end{cases}$ B

A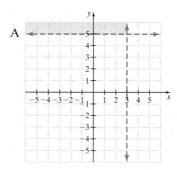

B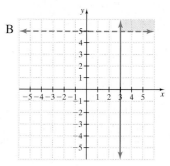

C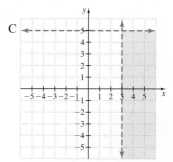

D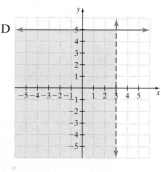

35. Explain how to decide which region to shade to show the solution region of the following system.

$\begin{cases} x \geq 3 \\ y \geq -2 \end{cases}$ answers may vary

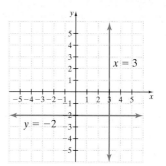

36. Describe the location of the solution region of the system

$\begin{cases} x > 0 \\ y > 0. \end{cases}$ answers may vary

37. Graph the solution of $\begin{cases} 2x - y \leq 6 \\ \quad x \geq 3 \\ \quad y > 2 \end{cases}$

38. Graph the solution of $\begin{cases} x + y < 5 \\ \quad y < 2x \\ \quad x \geq 0 \\ \quad y \geq 0 \end{cases}$

37.–38. See graphing answer section.

CHAPTER 4 PROJECT

Analyzing the Courses of Ships

From overhead photographs or satellite imagery of ships on the ocean, defense analysts can tell a lot about a ship's immediate course by looking at its wake. Assuming that two ships will maintain their present courses, it is possible to extend their paths, based on the wakes visible in the photograph. This can be used to find possible points of collision for the two ships.

In this project, you will investigate the courses and possibility of collision of the two ships shown in the figure. This project may be completed by working in groups or individually.

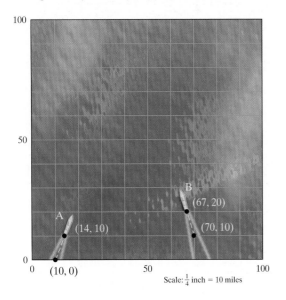

Scale: $\frac{1}{4}$ inch = 10 miles

1. Using each ship's wake as a guide, use a straight-edge to extend the paths of the ships on the figure. Estimate the coordinates of the point of intersection of the ships' courses from the grid. If the ships continue in these courses, they could possibly collide at the point of intersection of their paths.

2. What factor will govern whether or not the ships actually collide at the point found in Question 1?

3. Using the coordinates labeled on each ship's wake, find a linear equation that describes each path.

4. (Optional) Use a graphing calculator to graph both equations in the same window. Use the Intersect or Trace feature to estimate the point of intersection of the two paths. Compare this estimate to your estimate in Question 1.

5. Solve the system of two linear equations using one of the methods in this chapter. The solution is the point of intersection of the two paths. Compare your answer to your estimates from Questions 1 and 4.

6. Plot the point of intersection you found in Question 5 on the figure. Use the figure's scale to find each ship's distance from this point of collision by measuring from the bow (tip) of each ship with a ruler. Suppose that the speed of Ship A is r_1 and the speed of ship B is r_2. Given the present positions and courses of the two ships, find a relationship between their speeds that would ensure their collision.

4 CHAPTER VOCABULARY CHECK

Fill in each blank with one of the words or phrases listed below.

system of linear equations	solution	consistent	independent
dependent	inconsistent	substitution	addition
system of linear inequalities			

1. In a system of linear equations in two variables, if the graphs of the equations are the same, the equations are <u>dependent</u> equations.

2. Two or more linear equations are called a <u>system of linear equations</u>.

3. A system of equations that has at least one solution is called a(n) <u>consistent</u> system.

4. A <u>solution</u> of a system of two equations in two variables is an ordered pair of numbers that is a solution of both equations in the system.

5. Two algebraic methods for solving systems of equations are <u>addition</u> and <u>substitution</u>.

6. A system of equations that has no solution is called a(n) <u>inconsistent</u> system.

7. In a system of linear equations in two variables, if the graphs of the equations are different, the equations are <u>independent</u> equations.

8. Two or more linear inequalities are called a <u>system of linear inequalities</u>.

STUDY SKILLS REMINDER

Are You Preparing for a Test on Chapter 4?

Below I have listed some common trouble areas for topics covered in Chapter 4. After studying for your test—but before taking your test—read these.

▶ If you are having trouble drawing a neat graph, remember to ask your instructor if you can use graph paper on your test. This will save you time and keep your graphs neat.

▶ Do you remember how to check solutions of systems of equations? If $(-1, 5)$ is a solution of the system

$$\begin{cases} 3x - y = -8 \\ -x + y = 6 \end{cases},$$

then the ordered pair will make *both* equations a true statement.

$$
\begin{array}{ll}
3x - y = -8 & -x + y = 6 \\
3(-1) - 5 = -8 & -(-1) + 5 = 6 \quad \text{Let } x = -1 \\
& \qquad\qquad\qquad\quad \text{and } y = 5. \\
-8 = -8 \quad \text{True} & 6 = 6 \quad \text{True}
\end{array}
$$

Remember: This is simply a list of a few common trouble areas. For a review of Chapter 4, see the Highlights and Chapter Review at the end of this chapter.

CHAPTER 4 HIGHLIGHTS

Definitions and Concepts	Examples
Section 4.1 Solving Systems of Linear Equations by Graphing	
A **solution** of a system of two equations in two variables is an ordered pair of numbers that is a solution of both equations in the system.	Determine whether $(-1, 3)$ is a solution of the system: $$\begin{cases} 2x - y = -5 \\ x = 3y - 10 \end{cases}$$ *(continued)*

Definitions and Concepts	**Examples**

Section 4.1   Solving Systems of Linear Equations by Graphing

	Replace x with -1 and y with 3 in both equations. 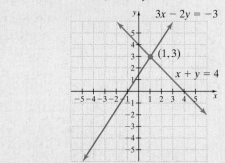 $(-1, 3)$ is a solution of the system.
Graphically, a solution of a system is a point common to the graphs of both equations.	Solve by graphing. $\begin{cases} 3x - 2y = -3 \\ x + y = 4 \end{cases}$ 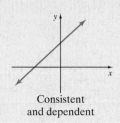
A system of equations with at least one solution is a **consistent system**. A system that has no solution is an **inconsistent system**. If the graphs of two linear equations are identical, the equations are **dependent**. If their graphs are different, the equations are **independent**.	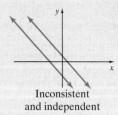

Section 4.2   Solving Systems of Linear Equations by Substitution

| To solve a system of linear equations by the substitution method
Step 1. Solve one equation for a variable.
Step 2. Substitute the expression for the variable into the other equation.
Step 3. Solve the equation from Step 2 to find the value of one variable.
Step 4. Substitute the value from Step 3 in either original equation to find the value of the other variable. | Solve by substitution.
$$\begin{cases} 3x + 2y = 1 \\ x = y - 3 \end{cases}$$
Substitute $y - 3$ for x in the first equation.
$$3x + 2y = 1$$
$$3(y - 3) + 2y = 1$$
$$3y - 9 + 2y = 1$$
$$5y = 10$$
$$y = 2 \quad \text{Divide by 5.}$$
(*continued*) |

Definitions and Concepts	Examples

Step 5. Check the solution in both equations.

To find x, substitute 2 for y in $x = y - 3$ so that $x = 2 - 3$ or -1. The solution $(-1, 2)$ checks.

To solve a system of linear equations by the addition method

Step 1. Rewrite each equation in standard form $Ax + By = C$.

Step 2. Multiply one or both equations by a nonzero number so that the coefficients of a variable are opposites.

Step 3. Add the equations.

Step 4. Find the value of one variable by solving the resulting equation.

Step 5. Substitute the value from Step 4 into either original equation to find the value of the other variable.

Step 6. Check the solution in both equations.

If solving a system of linear equations by substitution or addition yields a true statement such as $-2 = -2$, then the graphs of the equations in the system are identical and there is an infinite number of solutions of the system.

Solve by addition.

$$\begin{cases} x - 2y = 8 \\ 3x + y = -4 \end{cases}$$

Multiply both sides of the first equation by -3.

$$\begin{cases} -3x + 6y = -24 \\ \underline{3x + y = -4} \end{cases}$$
$$\begin{aligned} 7y &= -28 &&\text{Add.} \\ y &= -4 &&\text{Divide by 7.} \end{aligned}$$

To find x, let $y = -4$ in an original equation.

$$\begin{aligned} x - 2(-4) &= 8 &&\text{First equation} \\ x + 8 &= 8 \\ x &= 0 \end{aligned}$$

The solution $(0, -4)$ checks.

Solve: $\begin{cases} 2x - 6y = -2 \\ x = 3y - 1 \end{cases}$

Substitute $3y - 1$ for x in the first equation.

$$\begin{aligned} 2(3y - 1) - 6y &= -2 \\ 6y - 2 - 6y &= -2 \\ -2 &= -2 &&\text{True} \end{aligned}$$

The system has an infinite number of solutions.

Problem-solving steps
1. UNDERSTAND. Read and reread the problem.

Two angles are supplementary if their sum is 180°.
The larger of two supplementary angles is three times the smaller, decreased by twelve. Find the measure of each angle. Let

$$x = \text{measure of smaller angle}$$
$$y = \text{measure of larger angle}$$

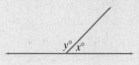

(continued)

Definitions and Concepts	Examples

Section 4.4 Systems of Linear Equations and Problem Solving

2. TRANSLATE.

In words:

the sum of supplementary angles	is	180°
↓	↓	↓

Translate: $x + y$ $=$ 180

In words:

larger angle	is	3 times smaller	decreased by	12
↓	↓	↓	↓	↓

Translate: y $=$ $3x$ $-$ 12

3. SOLVE.

Solve the system:

$$\begin{cases} x + y = 180 \\ y = 3x - 12 \end{cases}$$

Use the substitution method and replace y with $3x - 12$ in the first equation.

$$x + y = 180$$
$$x + (3x - 12) = 180$$
$$4x = 192$$
$$x = 48$$

Since $y = 3x - 12$, then $y = 3 \cdot 48 - 12$ or 132.

4. INTERPRET.

The solution checks. The smaller angle measures 48° and the larger angle measures 132°.

Section 4.5 Graphing Linear Inequalities

A **linear inequality in two variables** is an inequality that can be written in one of the forms:

$$Ax + By < C \qquad Ax + By \le C$$
$$Ax + By > C \qquad Ax + By \ge C$$

Linear Inequalities

$$2x - 5y < 6 \qquad x \ge -5$$
$$y > -8x \qquad y \le 2$$

To graph a linear inequality

1. Graph the boundary line by graphing the related equation. Draw the line solid if the inequality symbol is $\le$ or $\ge$. Draw the line dashed if the inequality symbol is $<$ or $>$.

2. Choose a test point not on the line. Substitute its coordinates into the original inequality.

Graph $2x - y \le 4$.

1. Graph $2x - y = 4$. Draw a solid line because the inequality symbol is $\le$.

2. Check the test point $(0, 0)$ in the inequality $2x - y \le 4$.

$2 \cdot 0 - 0 \le 4$ Let $x = 0$ and $y = 0$.

$0 \le 4$ True.   *(continued)*

Definitions and Concepts	**Examples**

Section 4.5   Graphing Linear Inequalities

3. If the resulting inequality is true, shade the half-plane that contains the test point. If the inequality is not true, shade the half-plane that does not contain the test point.

3. The inequality is true so we shade the half-plane containing $(0, 0)$.

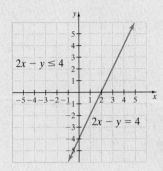

Section 4.6   Systems of Linear Inequalities

A system of linear inequalities consists of two or more linear inequalities.

To graph a system of inequalities, graph each inequality in the system. The overlapping region is the solution of the system.

System of Linear Inequalities

$$\begin{cases} x - y \geq 3 \\ y \leq -2x \end{cases}$$

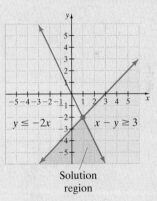

Solution region

CHAPTER REVIEW

(4.1) *Determine whether any of the following ordered pairs satisfy the system of linear equations.*

1. $\begin{cases} 2x - 3y = 12 \\ 3x + 4y = 1 \end{cases}$

   a. $(12, 4)$ no

   b. $(3, -2)$ yes

   c. $(-3, 6)$ no

2. $\begin{cases} 4x + y = 0 \\ -8x - 5y = 9 \end{cases}$

   a. $\left(\dfrac{3}{4}, -3\right)$ yes

   b. $(-2, 8)$ no

   c. $\left(\dfrac{1}{2}, -2\right)$ no

3. $\begin{cases} 5x - 6y = 18 \\ 2y - x = -4 \end{cases}$

   a. $(-6, -8)$ no

   b. $\left(3, \dfrac{5}{2}\right)$ no

   c. $\left(3, -\dfrac{1}{2}\right)$ yes

4. $\begin{cases} 2x + 3y = 1 \\ 3y - x = 4 \end{cases}$

   a. $(2, 2)$ no

   b. $(-1, 1)$ yes

   c. $(2, -1)$ no

Solve each system of equations by graphing.

5. $\begin{cases} 2x + y = 5 \\ \quad\; 3y = -x \end{cases}$ $(3, -1)$

6. $\begin{cases} 3x + y = -2 \\ 2x - y = -3 \end{cases}$ $(-1, 1)$

7. $\begin{cases} y - 2x = 4 \\ \; x + y = -5 \end{cases}$ $(-3, -2)$

8. $\begin{cases} y - 3x = 0 \\ 2y - 3 = 6x \end{cases}$ no solution

9. $\begin{cases} 3x + y = 2 \\ 3x - 6 = -9y \end{cases}$ $\left(\frac{1}{2}, \frac{1}{2}\right)$

10. $\begin{cases} 2y + x = 2 \\ \; x - y = 5 \end{cases}$ $(4, -1)$

5.–10. See graphing answer section.

Without graphing, (a) decide whether the graphs of the system are identical lines, parallel lines, or lines intersecting at a single point and (b) determine the number of solutions for each system.
11. intersecting, one solution   **12.** parallel, no solution

11. $\begin{cases} 2x - y = 3 \\ \quad\; y = 3x + 1 \end{cases}$

12. $\begin{cases} 3x + y = 4 \\ \quad\; y = -3x + 1 \end{cases}$

13. $\begin{cases} \dfrac{2}{3}x + \dfrac{1}{6}y = 0 \\ \quad\quad y = -4x \end{cases}$

14. $\begin{cases} \dfrac{1}{4}x + \dfrac{1}{8}y = 0 \\ \quad\quad y = -6x \end{cases}$

13. identical, infinite number of solutions
14. intersecting, one solution

(4.2) Solve the following systems of equations by the substitution method. If there is a single solution, give the ordered pair. If not, state whether the system is inconsistent or whether the equations are dependent.   **19.** no solution; inconsistent

15. $\begin{cases} \quad\quad y = 2x + 6 \\ 3x - 2y = -11 \end{cases}$ $(-1, 4)$

16. $\begin{cases} \quad\quad y = 3x - 7 \\ 2x - 3y = 7 \end{cases}$ $(2, -1)$

17. $\begin{cases} x + 3y = -3 \\ 2x + y = 4 \end{cases}$ $(3, -2)$

18. $\begin{cases} 3x + y = 11 \\ x + 2y = 12 \end{cases}$ $(2, 5)$

19. $\begin{cases} 4y = 2x - 3 \\ x - 2y = 4 \end{cases}$

20. $\begin{cases} 2x = 3y - 18 \\ x + 4y = 2 \end{cases}$ $(-6, 2)$

21. $\begin{cases} 2(3x - y) = 7x - 5 \\ 3(x - y) = 4x - 6 \end{cases}$ $(3, 1)$

22. $\begin{cases} 4(x - 3y) = 3x - 1 \\ 3(4y - 3x) = 1 - 8x \end{cases}$

23. $\begin{cases} \dfrac{3}{4}x + \dfrac{2}{3}y = 2 \\ \quad 3x + y = 18 \end{cases}$ $(8, -6)$

24. $\begin{cases} \dfrac{2}{5}x + \dfrac{3}{4}y = 1 \\ \quad x + 3y = -2 \end{cases}$ $(10, -4)$

22. infinite number of solutions; dependent

(4.3) Solve the following systems of equations by the addition method. If there is a single solution, give the ordered pair. If not, state whether the system is inconsistent or whether the equations are dependent.   **29.** infinite number of solutions; dependent

25. $\begin{cases} 2x + 3y = -6 \\ \; x - 3y = -12 \end{cases}$ $(-6, 2)$

26. $\begin{cases} \quad 4x + y = 15 \\ -4x + 3y = -19 \end{cases}$ $(4, -1)$

27. $\begin{cases} 2x - 3y = -15 \\ \; x + 4y = 31 \end{cases}$ $(3, 7)$

28. $\begin{cases} \; x - 5y = -22 \\ 4x + 3y = 4 \end{cases}$ $(-2, 4)$

29. $\begin{cases} 2x = 6y - 1 \\ \dfrac{1}{3}x - y = \dfrac{-1}{6} \end{cases}$

30. $\begin{cases} 8x = 3y - 2 \\ \dfrac{4}{7}x - y = \dfrac{-5}{2} \end{cases}$ $\left(\frac{7}{8}, 3\right)$

31. $\begin{cases} 5x = 6y + 25 \\ -2y = 7x - 9 \end{cases}$ $\left(2, -2\frac{1}{2}\right)$

32. $\begin{cases} -4x = 8 + 6y \\ -3y = 2x - 3 \end{cases}$

32. no solution; inconsistent

33. $\begin{cases} 3(x - 4) = -2y \\ \quad 2x = 3(y - 19) \end{cases}$ $(-6, 15)$

34. $\begin{cases} \quad\quad 4(x + 5) = -3y \\ 3x - 2(y + 18) = 0 \end{cases}$ $(4, -12)$

35. $\begin{cases} \dfrac{2x + 9}{3} = \dfrac{y + 1}{2} \\ \dfrac{x}{3} = \dfrac{y - 7}{6} \end{cases}$ $(-3, 1)$

36. $\begin{cases} \dfrac{2 - 5x}{4} = \dfrac{2y - 4}{2} \\ \dfrac{x + 5}{3} = \dfrac{y}{5} \end{cases}$ $(-2, 5)$

37. −6 and 22
(4.4) Solve by writing and solving a system of linear equations.

37. The sum of two numbers is 16. Three times the larger number decreased by the smaller number is 72. Find the two numbers.

38. The Forrest Theater can seat a total of 360 people. They take in $15,150 when every seat is sold. If orchestra section tickets cost $45 and balcony tickets cost $35, find the number of people that can be seated in the orchestra section.   255 people

39. A riverboat can head 340 miles upriver in 19 hours, but the return trip takes only 14 hours. Find the current of the river and find the speed of the ship in still water to the nearest tenth of a mile.   ship: 21.1 mph; current: 3.2 mph

42. 6% solution: 12.5 cc; 14% solution: 37.5 cc

	d	$=$ r	$\cdot$ t
Upriver	340	$x - y$	19
Downriver	340	$x + y$	14

43. one egg: $0.40; one strip of bacon: $0.65

40. Sam Abney invested $9000 one year ago. Part of the money was invested at 6%, the rest at 10%. If the total interest earned in one year was $652.80, find how much was invested at each rate. $6180 at 6%; $2820 at 10%

△ **41.** Ancient Greeks thought that the most pleasing dimensions for a picture are those where the length is approximately 1.6 times longer than the width. This ratio is known as the Golden Ratio. If Sandreka Walker has 6 feet of framing material, find the dimensions of the largest frame she can make that satisfies the Golden Ratio. Find the dimensions to the nearest hundredth of a foot.   width: 1.15 ft; length: 1.85 ft

42. Find the amount of 6% acid solution and the amount of 14% acid solution Pat should combine to prepare 50 cc (cubic centimeters) of a 12% solution.

43. The Deli charges $3.80 for a breakfast of 3 eggs and 4 strips of bacon. The charge is $2.75 for 2 eggs and 3 strips of bacon. Find the cost of each egg and the cost of each strip of bacon.

44. An exercise enthusiast alternates between jogging and walking. He traveled 15 miles during the past 3 hours. He jogs at a rate of 7.5 miles per hour and walks at a rate of 4 miles per hour. Find how much time, to the nearest hundredth of an hour, he actually spent jogging.   0.86 hr

(4.5) Graph the following inequalities.

45. $3x - 4y \le 0$

46. $3x - 4y \ge 0$

47. $x + 6y < 6$

48. $x + y > -2$

45.–48. See graphing answer section.

49.–52. See graphing answer section.

49. $y \geq -7$ **50.** $y \leq -4$

51. $-x \leq y$ **52.** $x \geq -y$

(4.6) *Graph the solutions of the following systems of linear inequalities.* **53.–60.** See graphing answer section.

53. $\begin{cases} y \geq 2x - 3 \\ y \leq -2x + 1 \end{cases}$ **54.** $\begin{cases} y \leq -3x - 3 \\ y \leq 2x + 7 \end{cases}$

55. $\begin{cases} x + 2y > 0 \\ x - y \leq 6 \end{cases}$ **56.** $\begin{cases} x - 2y \geq 7 \\ x + y \leq -5 \end{cases}$

57. $\begin{cases} 3x - 2y \leq 4 \\ 2x + y \geq 5 \end{cases}$ **58.** $\begin{cases} 4x - y \leq 0 \\ 3x - 2y \geq -5 \end{cases}$

59. $\begin{cases} -3x + 2y > -1 \\ y < -2 \end{cases}$ **60.** $\begin{cases} -2x + 3y > -7 \\ x \geq -2 \end{cases}$

CHAPTER 4 TEST

Remember to use your Chapter Test Prep Video CD to help you study and view solutions to the test questions you need help with.

Answer each question true or false. **3.** true **5.** no **6.** yes

1. A system of two linear equations in two variables can have exactly two solutions. false

2. Although (1, 4) is not a solution of $x + 2y = 6$, it can still be a solution of the system $\begin{cases} x + 2y = 6 \\ x + y = 5 \end{cases}$. false

3. If the two equations in a system of linear equations are added and the result is $3 = 0$, the system has no solution.

4. If the two equations in a system of linear equations are added and the result is $3x = 0$, the system has no solution. false

Is the ordered pair a solution of the given linear system?

5. $\begin{cases} 2x - 3y = 5 \\ 6x + y = 1 \end{cases}; (1, -1)$ **6.** $\begin{cases} 4x - 3y = 24 \\ 4x + 5y = -8 \end{cases}; (3, -4)$

7. Use graphing to find the solutions of the system $\begin{cases} y - x = 6 \\ y + 2x = -6 \end{cases}$ $(-4, 2)$, see graphing answer section.

8. Use the substitution method to solve the system $\begin{cases} 3x - 2y = -14 \\ x + 3y = -1 \end{cases}$ $(-4, 1)$

9. Use the substitution method to solve the system $\begin{cases} \dfrac{1}{2}x + 2y = -\dfrac{15}{4} \\ 4x = -y \end{cases}$ $\left(\dfrac{1}{2}, -2\right)$

10. Use the addition method to solve the system $\begin{cases} 3x + 5y = 2 \\ 2x - 3y = 14 \end{cases}$ $(4, -2)$

11. Use the addition method to solve the system $\begin{cases} 4x - 6y = 7 \\ -2x + 3y = 0 \end{cases}$ no solution

Solve each system using the substitution method or the addition method.

12. $\begin{cases} 3x + y = 7 \\ 4x + 3y = 1 \end{cases}$ $(4, -5)$

13. $\begin{cases} 3(2x + y) = 4x + 20 \\ x - 2y = 3 \end{cases}$ $(7, 2)$

14. $\begin{cases} \dfrac{x - 3}{2} = \dfrac{2 - y}{4} \\ \dfrac{7 - 2x}{3} = \dfrac{y}{2} \end{cases}$ $(5, -2)$

15. Find the amount of a 12% saline solution a lab assistant should add to 80 cc (cubic centimeters) of a 22% saline solution in order to have a 16% solution. 120 cc

16. Don has invested $4000, part at 5% simple annual interest and the rest at 9%. Find how much he invested at each rate if the total interest after 1 year is $311.

17. Although the number of farms in the U.S. is still decreasing, small farms are making a comeback. Texas and Missouri are the states with the most number of farms. Texas has 116 thousand more farms than Missouri and the total number of farms for these two states is 336 thousand. Find the number of farms for each state.

16. $1225 at 5%; $2775 at 9%

17. Texas: 226 thousand; Missouri: 110 thousand

Graph each linear equality.

18. $y \geq -4x$ See graphing answer section.

19. $2x - 3y > -6$ See graphing answer section.

Graph the solutions of the following systems of linear inequalities. **20.–21.** See graphing answer section.

20. $\begin{cases} y + 2x \leq 4 \\ y \geq 2 \end{cases}$ **21.** $\begin{cases} 2y - x \geq 1 \\ x + y \geq -4 \end{cases}$

CHAPTER CUMULATIVE REVIEW

1. Insert $<$, $>$, or $=$ in the space between the paired numbers to make each statement true. (Sec. 1.2, Ex. 6)
   a. $-1 < 0$ **b.** $7 = \frac{14}{2}$ **c.** $-5 > -6$

2. Evaluate.
   a. 5^2 25   **b.** 2^5 32 (Sec. 1.4)

3. Name the property or properties illustrated by each true statement.
   a. $3 \cdot y = y \cdot 3$ commutative property of multiplication
   b. $(x + 7) + 9 = x + (7 + 9)$
   c. $(b + 0) + 3 = b + 3$ identity element for addition
   d. $0.2 \cdot (z \cdot 5) = 0.2 \cdot (5 \cdot z)$ **e.** $-2 \cdot \left(-\frac{1}{2}\right) = 1$
   f. $-2 + 2 = 0$
   g. $-6 \cdot (y \cdot 2) = (-6 \cdot 2) \cdot y$

4. Evaluate $y^2 - 3x$ for $x = 8$ and $y = 5$.   1 (Sec. 1.4)

5. Subtract $4x - 2$ from $2x - 3$ $-2x - 1$ (Sec. 2.1, Ex. 7)

6. Simplify: $7 - 12 + (-5) - 2 + (-2)$ -14 (Sec. 1.6)

7. Solve $5t - 5 = 6t + 2$ for t. -7 (Sec. 2.2, Ex. 4)

8. Evaluate $2y^2 - x^2$ for $x = -7$ and $y = -3$. -31 (Sec. 1.7)

9. Solve: $\frac{5}{2}x = 15$ 6 (Sec. 2.3, Ex. 1)

10. Simplify: $0.4y - 6.7 + y - 0.3 - 2.6y$ $-1.2y - 7$ (Sec. 2.1)

11. Solve: $\frac{x}{2} - 1 = \frac{2}{3}x - 3$ 12 (Sec. 2.4, Ex. 3)

12. Solve: $7(x - 2) - 6(x + 1) = 20$ 40 (Sec. 2.2)

13. Twice the sum of a number and 4 is the same as four times the number, decreased 12. Find the number.   10 (Sec. 2.5, Ex. 1)

14. Solve: $5(y - 5) = 5y + 10$ no solution (Sec. 2.4)

15. Solve $y = mx + b$ for x. $x = \frac{y - b}{m}$ (Sec. 2.6, Ex. 6)

16. Five times the sum of a number and -1 is the same as 6 times the number. Find the number. -5 (Sec. 2.5)

17. Solve $-2x \le -4$. Write the solution set in interval notation.

18. Solve $P = a + b + c$ for b. $b = P - a - c$ (Sec. 2.6)

19. Graph $x = -2y$ by plotting intercepts.

20. Solve $3x + 7 \ge x - 9$. Write the solution set in interval notation. $[-8, \infty)$ (Sec. 2.8)

21. Find the slope of the line through $(-1, 5)$ and $(2, -3)$.

22. Complete the table of values for $x - 3y = 3$ (Sec. 3.1)

x	y
0	-1
3	0
9	2

23. $\frac{3}{4}$ (Sec. 3.5, Ex. 1)

23. Find the slope of the line whose equation is $y = \frac{3}{4}x + 6$.

24. Find the slope of the line parallel to the line passing through $(-1, 3)$ and $(2, -8)$. $-\frac{11}{3}$ (Sec. 3.4)

25. Find the slope and the y-intercept of the line whose equation is $3x - 4y = 4$.   slope: $\frac{3}{4}$; y-intercept: $(0, -1)$ (Sec. 3.5, Ex. 3)

26. Find the slope and y-intercept of the line whose equation is $y = 7x$. $m = 7$, y-intercept $(0, 0)$ (Sec. 3.5)

27. Find an equation of the line passing through $(-1, 5)$ with slope -2. Write the equation in standard form: $Ax + By = C$. $2x + y = 3$ (Sec. 3.6, Ex. 1)

28. Determine whether the lines are parallel, perpendicular or neither.   parallel (Sec. 3.5)

$$y = 4x - 5$$
$$-4x + y = 7$$

29. Find an equation of the vertical line through $(-1, 5)$.

30. Write an equation of the line with slope -5, through $(-2, 3)$.

31. Find the domain and the range of the relation $\{(0, 2), (3, 3), (-1, 0), (3, -2)\}$.

32. If $f(x) = 5x^2 - 6$, find $f(0)$ and $f(-2)$. $-6, 14$ (Sec. 3.7)

33. Which of the following relations are also functions?
   a. $\{(-1, 1), (2, 3), (7, 3), (8, 6)\}$ function
   b. $\{(0, -2), (1, 5), (0, 3), (7, 7)\}$

3. b. associative property of addition   **3. d.** commutative property of multiplication   **3. e.** multiplicative inverse property
3. f. additive inverse property   **3. g.** commutative and associative properties of multiplication (Sec. 1.8, Ex. 6)
17. $[2, \infty)$ (Sec. 2.8, Ex. 3) $(-\infty, -2)$ (Sec. 2.8, Ex. 4)   **19.** See graphing answer section. (Sec. 3.3, Ex. 3)   **21.** $-\frac{8}{3}$ (Sec. 3.4, Ex. 1)
29. $x = -1$ (Sec. 3.6, Ex. 3)   **30.** $y = -5x - 7$ (Sec. 3.6)   **31.** domain: $\{-1, 0, 3\}$; range: $\{-2, 0, 2, 3\}$ (Sec. 3.7, Ex. 1)
33. b. not a function (Sec. 3.7, Ex. 2)

34. Determine which graph(s) are graphs of functions.

a.   **b.**   **c.**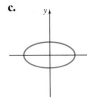

35. Determine the number of solutions of the system.

$$\begin{cases} 3x - y = 4 \\ x + 2y = 8 \end{cases}$$ one solution (Sec. 4.1, Ex. 6)

36. Determine whether any ordered pairs satisfy the given system.

$$\begin{cases} 2x - y = 6 \\ 3x + 2y = -5 \end{cases}$$

a. $(1, -4)$ **b.** $(0, 6)$ **c.** $(3, 0)$

Solve each system.

37. $\begin{cases} x + 2y = 7 \\ 2x + 2y = 13 \end{cases}$ **38.** $\begin{cases} 3x - 4y = 10 \\ y = 2x \end{cases}$

39. $\begin{cases} x + y = 7 \\ x - y = 5 \end{cases}$ **40.** $\begin{cases} x = 5y - 3 \\ x = 8y + 4 \end{cases}$

41. Find two numbers whose sum is 37 and whose difference is 21.

42. Graph: $x > 1$ See graphing answer section. (Sec. 4.5)

43. Graph: $2x - y \geq 3$

44. Graph the solution of the system: $\begin{cases} 2x + 3y < 6 \\ y < 2 \end{cases}$

34. a. no **b.** yes **c.** no (Sec. 3.7) **36. a.** yes **b.** no   **c.** no (Sec. 4.1)   **37.** $\left(6, \dfrac{1}{2}\right)$ (Sec. 4.2, Ex. 2)   **38.** $(-2, -4)$ (Sec. 4.2)

39. $(6, 1)$ (Sec. 4.3, Ex. 1) **40.** $\left(-\dfrac{44}{3}, -\dfrac{7}{3}\right)$ (Sec. 4.2) **41.** 29 and 8 (Sec. 4.4, Ex. 1) **43.–44.** See graphing answer section. (Sec. 4.6)

Exponents and Polynomials

Recall from Chapter 1 that an exponent is a shorthand notation for repeated factors. This chapter explores additional concepts about exponents and exponential expressions. An especially useful type of exponential expression is a polynomial. Polynomials model many real-world phenomena. This chapter will focus on operations on polynomials.

People who have been stuck in rush hour traffic know that there are millions of automobiles on the road today. Keeping those cars in good working order is the job of a professional auto technician. Auto technicians enjoy problem solving and, due to technological advances in automobile design, they dabble in computer science, electronics, and mathematics. Many technicians are also small business owners, and college courses in business and finance are helpful.

Auto technicians can be certified by the National Institute for Automobile Service Excellence (ASE). ASE offers certification exams in over 40 areas of specialization and requires its certified technicians to participate in continuing education programs. Employment in this field is expected to increase through the year 2010, according to the U.S. Bureau of Labor Statistics.

In the Spotlight on Decision Making feature on page 304, you will have the opportunity to make a decision concerning the modification of a car engine as an auto mechanic.

Link: ASE—www.asecert.org
Sources of text: (ASE) www.asecert.org
(BLS) http://www.bls.gov/oco/ocos181.htm

5.1 EXPONENTS

Objectives

1 Evaluate exponential expressions.

2 Use the product rule for exponents.

3 Use the power rule for exponents.

4 Use the power rules for products and quotients.

5 Use the quotient rule for exponents, and define a number raised to the 0 power.

1 As we reviewed in Section 1.4, an exponent is a shorthand notation for repeated factors. For example, $2 \cdot 2 \cdot 2 \cdot 2 \cdot 2$ can be written as 2^5. The expression 2^5 is called an **exponential expression**. It is also called the fifth **power** of 2, or we say that 2 is **raised** to the fifth power.

$$5^6 = \underbrace{5 \cdot 5 \cdot 5 \cdot 5 \cdot 5 \cdot 5}_{6 \text{ factors; each factor is } 5} \qquad \text{and} \qquad (-3)^4 = \underbrace{(-3) \cdot (-3) \cdot (-3) \cdot (-3)}_{4 \text{ factors; each factor is } -3}$$

The **base** of an exponential expression is the repeated factor. The **exponent** is the number of times that the base is used as a factor.

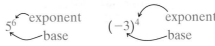

$$5^6 \overset{\text{exponent}}{\underset{\text{base}}{}} \qquad (-3)^4 \overset{\text{exponent}}{\underset{\text{base}}{}}$$

EXAMPLE 1

Evaluate each expression.

a. 2^3 **b.** 3^1 **c.** $(-4)^2$ **d.** -4^2 **e.** $\left(\dfrac{1}{2}\right)^4$ **f.** $(0.5)^3$ **g.** $4 \cdot 3^2$

Solution

a. $2^3 = 2 \cdot 2 \cdot 2 = 8$

b. To raise 3 to the first power means to use 3 as a factor only once. Therefore, $3^1 = 3$. Also, when no exponent is shown, the exponent is assumed to be 1.

c. $(-4)^2 = (-4)(-4) = 16$ **d.** $-4^2 = -(4 \cdot 4) = -16$

e. $\left(\dfrac{1}{2}\right)^4 = \dfrac{1}{2} \cdot \dfrac{1}{2} \cdot \dfrac{1}{2} \cdot \dfrac{1}{2} = \dfrac{1}{16}$ **f.** $(0.5)^3 = (0.5)(0.5)(0.5) = 0.125$

g. $4 \cdot 3^2 = 4 \cdot 9 = 36$

Notice how similar -4^2 is to $(-4)^2$ in the example above. The difference between the two is the parentheses. In $(-4)^2$, the parentheses tell us that the base, or repeated factor, is -4. In -4^2, only 4 is the base.

> ### Helpful Hint
>
> Be careful when identifying the base of an exponential expression. Pay close attention to the use of parentheses.
>
> $(-3)^2$ -3^2 $2 \cdot 3^2$
> The base is -3.   The base is 3.   The base is 3.
> $(-3)^2 = (-3)(-3) = 9$ $-3^2 = -(3 \cdot 3) = -9$ $2 \cdot 3^2 = 2 \cdot 3 \cdot 3 = 18$

2 An exponent has the same meaning whether the base is a number or a variable. If x is a real number and n is a positive integer, then x^n is the product of n factors, each of which is x.

$$x^n = \underbrace{x \cdot x \cdot x \cdot x \cdot x \cdot \ldots \cdot x}_{n \text{ factors of } x}$$

EXAMPLE 2

Evaluate each expression for the given value of x.

a. $2x^3$; x is 5

b. $\dfrac{9}{x^2}$; x is -3

Solution

a. If x is 5, $2x^3 = 2 \cdot (5)^3$

$$= 2 \cdot (5 \cdot 5 \cdot 5)$$
$$= 2 \cdot 125$$
$$= 250$$

b. If x is -3, $\dfrac{9}{x^2} = \dfrac{9}{(-3)^2}$

$$= \dfrac{9}{(-3)(-3)}$$
$$= \dfrac{9}{9} = 1$$

Exponential expressions can be multiplied, divided, added, subtracted, and themselves raised to powers. By our definition of an exponent,

$$5^4 \cdot 5^3 = \underbrace{(5 \cdot 5 \cdot 5 \cdot 5)}_{4 \text{ factors of } 5} \cdot \underbrace{(5 \cdot 5 \cdot 5)}_{3 \text{ factors of } 5}$$
$$= \underbrace{5 \cdot 5 \cdot 5 \cdot 5 \cdot 5 \cdot 5 \cdot 5}_{7 \text{ factors of } 5}$$
$$= 5^7$$

Also,

$$x^2 \cdot x^3 = (x \cdot x) \cdot (x \cdot x \cdot x)$$
$$= x \cdot x \cdot x \cdot x \cdot x$$
$$= x^5$$

In both cases, notice that the result is exactly the same if the exponents are added.

$$5^4 \cdot 5^3 = 5^{4+3} = 5^7 \quad \text{and} \quad x^2 \cdot x^3 = x^{2+3} = x^5$$

This suggests the following rule.

Product Rule for Exponents

If m and n are positive integers and a is a real number, then

$$a^m \cdot a^n = a^{m+n}$$

For example, $3^5 \cdot 3^7 = 3^{5+7} = 3^{12}$.

In other words, to multiply two exponential expressions with a **common base**, keep the base and add the exponents. We call this simplifying the exponential expression.

EXAMPLE 3

Use the product rule to simplify.

a. $4^2 \cdot 4^5$ **b.** $x^4 \cdot x^6$ **c.** $y^3 \cdot y$ **d.** $y^3 \cdot y^2 \cdot y^7$ **e.** $(-5)^7 \cdot (-5)^8$ **f.** $a^2 \cdot b^2$

Solution

a. $4^2 \cdot 4^5 = 4^{2+5} = 4^7$

b. $x^4 \cdot x^6 = x^{4+6} = x^{10}$

c. $y^3 \cdot y = y^3 \cdot y^1$

   $= y^{3+1}$

   $= y^4$

d. $y^3 \cdot y^2 \cdot y^7 = y^{3+2+7} = y^{12}$

e. $(-5)^7 \cdot (-5)^8 = (-5)^{7+8} = (-5)^{15}$

f. $a^2 \cdot b^2$ Cannot be simplified because a and b are different bases.

CLASSROOM EXAMPLE

Use the product rule to simplify.

a. $7^3 \cdot 7^2$ **b.** $x^4 \cdot x^9$

c. $r^5 \cdot r$ **d.** $s^6 \cdot s^2 \cdot s^3$

e. $(-3)^9 \cdot (-3)$

answer:

a. 7^5 **b.** x^{13} **c.** r^6

d. s^{11} **e.** $(-3)^{10}$

EXAMPLE 4

Use the product rule to simplify $(2x^2)(-3x^5)$.

Solution   Recall that $2x^2$ means $2 \cdot x^2$ and $-3x^5$ means $-3 \cdot x^5$.

$$(2x^2)(-3x^5) = 2 \cdot x^2 \cdot -3 \cdot x^5 \qquad \text{Remove parentheses.}$$
$$= 2 \cdot -3 \cdot x^2 \cdot x^5 \qquad \text{Group factors with common bases.}$$
$$= -6x^7 \qquad \text{Simplify.}$$

CLASSROOM EXAMPLE

Use the product rule to simplify $(6x^3)(-2x^9)$.

answer: $-12x^{12}$

Helpful Hint

These examples will remind you of the difference between adding and multiplying terms.

Addition

$$5x^3 + 3x^3 = (5 + 3)x^3 = 8x^3 \qquad \text{By the distributive property.}$$
$$7x + 4x^2 = 7x + 4x^2 \qquad \text{Cannot be combined.}$$

Multiplication

$$(5x^3)(3x^3) = 5 \cdot 3 \cdot x^3 \cdot x^3 = 15x^{3+3} = 15x^6 \qquad \text{By the product rule.}$$
$$(7x)(4x^2) = 7 \cdot 4 \cdot x \cdot x^2 = 28x^{1+2} = 28x^3 \qquad \text{By the product rule.}$$

3   Exponential expressions can themselves be raised to powers. Let's try to discover a rule that simplifies an expression like $(x^2)^3$. By definition,

$$(x^2)^3 = \underbrace{(x^2)(x^2)(x^2)}_{3 \text{ factors of } x^2}$$

which can be simplified by the product rule for exponents.

$$(x^2)^3 = (x^2)(x^2)(x^2) = x^{2+2+2} = x^6$$

Notice that the result is exactly the same if we multiply the exponents.

$$(x^2)^3 = x^{2 \cdot 3} = x^6$$

The following property states this result.

Power Rule for Exponents

If m and n are positive integers and a is a real number, then

$$(a^m)^n = a^{mn}$$

For example, $(7^2)^5 = 7^{2 \cdot 5} = 7^{10}$.

To raise a power to a power, keep the base and multiply the exponents.

EXAMPLE 5

Use the power rule to simplify.

a. $(x^2)^5$ **b.** $(y^8)^2$ **c.** $[(-5)^3]^7$

Solution **a.** $(x^2)^5 = x^{2 \cdot 5} = x^{10}$ **b.** $(y^8)^2 = y^{8 \cdot 2} = y^{16}$ **c.** $[(-5)^3]^7 = (-5)^{21}$

Helpful Hint

Take a moment to make sure that you understand when to apply the product rule and when to apply the power rule.

Product Rule → Add Exponents	Power Rule → Multiply Exponents
$x^5 \cdot x^7 = x^{5+7} = x^{12}$	$(x^5)^7 = x^{5 \cdot 7} = x^{35}$
$y^6 \cdot y^2 = y^{6+2} = y^8$	$(y^6)^2 = y^{6 \cdot 2} = y^{12}$

---

4 When the base of an exponential expression is a product, the definition of x^n still applies. To simplify $(xy)^3$, for example,

$$(xy)^3 = (xy)(xy)(xy) \quad (xy)^3 \text{ means 3 factors of } (xy).$$
$$= x \cdot x \cdot x \cdot y \cdot y \cdot y \quad \text{Group factors with common bases.}$$
$$= x^3 y^3 \quad \text{Simplify.}$$

Notice that to simplify the expression $(xy)^3$, we raise each factor within the parentheses to a power of 3.

$$(xy)^3 = x^3 y^3$$

In general, we have the following rule.

Power of a Product Rule

If n is a positive integer and a and b are real numbers, then

$$(ab)^n = a^n b^n$$

For example, $(3x)^5 = 3^5 x^5$.

In other words, to raise a product to a power, we raise each factor to the power.

EXAMPLE 6

Simplify each expression.

a. $(st)^4$ **b.** $(2a)^3$ **c.** $\left(\dfrac{1}{3}mn^3\right)^2$ **d.** $(-5x^2y^3z)^2$

Solution

a. $(st)^4 = s^4 \cdot t^4 = s^4 t^4$ Use the power of a product rule.

b. $(2a)^3 = 2^3 \cdot a^3 = 8a^3$ Use the power of a product rule.

c. $\left(\dfrac{1}{3}mn^3\right)^2 = \left(\dfrac{1}{3}\right)^2 \cdot (m)^2 \cdot (n^3)^2 = \dfrac{1}{9}m^2n^6$

d. $(-5x^2y^3z)^2 = (-5)^2 \cdot (x^2)^2 \cdot (y^3)^2 \cdot (z^1)^2$ Use the power of a product rule.

$= 25x^4y^6z^2$ Use the power rule for exponents.

CLASSROOM EXAMPLE

Simplify each expression.

a. $(xy)^7$ **b.** $(3y)^4$ **c.** $(-2p^4q^2r)^3$

answer:

a. x^7y^7 **b.** $81y^4$ **c.** $-8p^{12}q^6r^3$

Let's see what happens when we raise a quotient to a power. To simplify $\left(\dfrac{x}{y}\right)^3$, for example,

$$\left(\dfrac{x}{y}\right)^3 = \left(\dfrac{x}{y}\right)\left(\dfrac{x}{y}\right)\left(\dfrac{x}{y}\right) \qquad \left(\dfrac{x}{y}\right)^3 \text{ means 3 factors of } \left(\dfrac{x}{y}\right).$$

$$= \dfrac{x \cdot x \cdot x}{y \cdot y \cdot y} \qquad \text{Multiply fractions.}$$

$$= \dfrac{x^3}{y^3} \qquad \text{Simplify.}$$

Notice that to simplify the expression $\left(\dfrac{x}{y}\right)^3$, we raise both the numerator and the denominator to a power of 3.

$$\left(\dfrac{x}{y}\right)^3 = \dfrac{x^3}{y^3}$$

In general, we have the following.

Power of a Quotient Rule

If n is a positive integer and a and c are real numbers, then

$$\left(\dfrac{a}{c}\right)^n = \dfrac{a^n}{c^n}, \quad c \neq 0$$

For example, $\left(\dfrac{y}{7}\right)^4 = \dfrac{y^4}{7^4}$.

In other words, to raise a quotient to a power, we raise both the numerator and the denominator to the power.

EXAMPLE 7

Simplify each expression.

a. $\left(\dfrac{m}{n}\right)^7$ **b.** $\left(\dfrac{x^3}{3y^5}\right)^4$

Solution   **a.** $\left(\dfrac{m}{n}\right)^7 = \dfrac{m^7}{n^7}, n \neq 0$ Use the power of a quotient rule.

b. $\left(\dfrac{x^3}{3y^5}\right)^4 = \dfrac{(x^3)^4}{3^4 \cdot (y^5)^4}, y \neq 0$ Use the power of a product or quotient rule.

$= \dfrac{x^{12}}{81y^{20}}$ Use the power rule for exponents.

CLASSROOM EXAMPLE

Simplify each expression.

a. $\left(\dfrac{r}{s}\right)^6$ **b.** $\left(\dfrac{5x^6}{9y^3}\right)^2$

answer:

a. $\dfrac{r^6}{s^6}$ **b.** $\dfrac{25x^{12}}{81y^6}$

5   Another pattern for simplifying exponential expressions involves quotients. To simplify an expression like $\dfrac{x^5}{x^3}$, in which the numerator and the denominator have a common base, we can apply the fundamental principle of fractions and divide the numerator and the denominator by the common base factors. Assume for the remainder of this section that denominators are not 0.

TEACHING TIP

Before introducing the quotient rule, have students simplify the expressions $\dfrac{x^9}{x^5}$, $\dfrac{x^6}{x}$, and $\dfrac{x^3}{x^2}$. Then have them write their own quotient rule.

$$\dfrac{x^5}{x^3} = \dfrac{x \cdot x \cdot x \cdot x \cdot x}{x \cdot x \cdot x}$$
$$= \dfrac{\cancel{x} \cdot \cancel{x} \cdot \cancel{x} \cdot x \cdot x}{\cancel{x} \cdot \cancel{x} \cdot \cancel{x}}$$
$$= x \cdot x$$
$$= x^2$$

Notice that the result is exactly the same if we subtract exponents of the common bases.

$$\dfrac{x^5}{x^3} = x^{5-3} = x^2$$

The quotient rule for exponents states this result in a general way.

Quotient Rule for Exponents

If m and n are positive integers and a is a real number, then

$$\dfrac{a^m}{a^n} = a^{m-n}$$

as long as a is not 0.

For example, $\dfrac{x^6}{x^2} = x^{6-2} = x^4$.

In other words, to divide one exponential expression by another with a common base, keep the base and subtract exponents.

EXAMPLE 8

Simplify each quotient.

a. $\dfrac{x^5}{x^2}$ **b.** $\dfrac{4^7}{4^3}$ **c.** $\dfrac{(-3)^5}{(-3)^2}$ **d.** $\dfrac{s^2}{t^3}$ **e.** $\dfrac{2x^5y^2}{xy}$

Solution   **a.** $\dfrac{x^5}{x^2} = x^{5-2} = x^3$ Use the quotient rule.

b. $\dfrac{4^7}{4^3} = 4^{7-3} = 4^4 = 256$ Use the quotient rule.

c. $\dfrac{(-3)^5}{(-3)^2} = (-3)^3 = -27$

d. $\dfrac{s^2}{t^3}$ *Cannot be simplified because s and t are different bases.*

e. Begin by grouping common bases.

$$\frac{2x^5y^2}{xy} = 2 \cdot \frac{x^5}{x^1} \cdot \frac{y^2}{y^1}$$

$$= 2 \cdot (x^{5-1}) \cdot (y^{2-1}) \qquad \text{Use the quotient rule.}$$

$$= 2x^4y^1 \quad \text{or} \quad 2x^4y$$

✔ **CONCEPT CHECK**

Suppose you are simplifying each expression. Tell whether you would *add* the exponents, *subtract* the exponents, *multiply* the exponents, *divide* the exponents, or *none of these*.

a. $(x^{63})^{21}$ b. $\dfrac{y^{15}}{y^3}$ c. $z^{16} + z^8$ d. $w^{45} \cdot w^9$

Let's now give meaning to an expression such as x^0. To do so, we will simplify $\dfrac{x^3}{x^3}$ in two ways and compare the results.

$$\frac{x^3}{x^3} = x^{3-3} = x^0 \qquad \text{Apply the quotient rule.}$$

$$\frac{x^3}{x^3} = \frac{x \cdot x \cdot x}{x \cdot x \cdot x} = 1 \qquad \text{Apply the fundamental principle for fractions.}$$

Since $\dfrac{x^3}{x^3} = x^0$ and $\dfrac{x^3}{x^3} = 1$, we define that $x^0 = 1$ as long as x is not 0.

Zero Exponent

$a^0 = 1$, as long as a is not 0.

In other words, any base raised to the 0 power is 1, as long as the base is not 0.

EXAMPLE 9

Simplify the following expressions.

a. 3^0 b. $(ab)^0$ c. $(-5)^0$ d. -5^0 e. $\left(\dfrac{3}{100}\right)^0$

Solution   a. $3^0 = 1$

b. Assume that neither a nor b is zero.

$$(ab)^0 = a^0 \cdot b^0 = 1 \cdot 1 = 1$$

c. $(-5)^0 = 1$

d. $-5^0 = -1 \cdot 5^0 = -1 \cdot 1 = -1$

e. $\left(\dfrac{3}{100}\right)^0 = 1$

In the next example, exponential expressions are simplified using two or more of the exponent rules presented in this section.

EXAMPLE 10

Simplify the following.

a. $\left(\dfrac{-5x^2}{y^3}\right)^2$ **b.** $\dfrac{(x^3)^4 x}{x^7}$ **c.** $\dfrac{(2x)^5}{x^3}$ **d.** $\dfrac{(a^2 b)^3}{a^3 b^2}$

Solution **a.** Use the power of a product or quotient rule; then use the power rule for exponents.

$$\left(\frac{-5x^2}{y^3}\right)^2 = \frac{(-5)^2 (x^2)^2}{(y^3)^2} = \frac{25x^4}{y^6}$$

b. $\dfrac{(x^3)^4 x}{x^7} = \dfrac{x^{12} \cdot x}{x^7} = \dfrac{x^{12+1}}{x^7} = \dfrac{x^{13}}{x^7} = x^{13-7} = x^6$

c. Use the power of a product rule; then use the quotient rule.

$$\frac{(2x)^5}{x^3} = \frac{2^5 \cdot x^5}{x^3} = 2^5 \cdot x^{5-3} = 32x^2$$

d. Begin by applying the power of a product rule to the numerator.

$$\frac{(a^2 b)^3}{a^3 b^2} = \frac{(a^2)^3 \cdot b^3}{a^3 \cdot b^2}$$
$$= \frac{a^6 b^3}{a^3 b^2} \qquad \text{Use the power rule for exponents.}$$
$$= a^{6-3} b^{3-2} \qquad \text{Use the quotient rule.}$$
$$= a^3 b^1 \quad \text{or} \quad a^3 b$$

Spotlight on

DECISION
MAKING

Suppose you are an auto mechanic and amateur racing enthusiast. You have been modifying the engine in your car and would like to enter a local amateur race. The racing classes depend on the size, or displacement, of the engine.

Engine displacement can be calculated using the formula

$d = \dfrac{\pi}{4} b^2 sc$. In the formula,

d = the engine displacement in cubic centimeters (cc)

b = the bore or engine cylinder diameter in centimeters

s = the stroke or distance the piston travels up or down within the cylinder in centimeters

c = the number of cylinders the engine has.

You have made the following measurements on your modified engine: 8.4 cm bore, 7.6 cm stroke, and 6 cylinders. In which racing class would you enter your car? Explain.

BRENTWOOD AMATEUR RACING CLUB	
Racing Class	*Displacement Limit*
A	up to 2000 cc
B	up to 2400 cc
C	up to 2650 cc
D	up to 3000 cc

STUDY SKILLS REMINDER

What Should You Do on the Day of an Exam?

On the day of an exam, try the following:

▶ Allow yourself plenty of time to arrive at your classroom.

▶ Read the directions on the test carefully.

▶ Read each problem carefully as you take your test. Make sure that you answer the question asked.

▶ Watch your time and pace yourself. Work the problems that you are most confident with first.

▶ If you have time, check your work and answers.

▶ Do not turn your test in early. If you have extra time, spend it double-checking your work.

Good luck!

1. base: 3; exponent: 2 **2.** base: 5; exponent: 4 **3.** base: −3; exponent: 6 **4.** base: 3; exponent: 7 **5.** base: 4; exponent: 2
6. base: −4; exponent: 3 **7.** base: 5; exponent: 1; base: 3; exponent: 4 **8.** base: 9; exponent: 1; base: 7; exponent: 6
9. base: 5; exponent: 1; base: x; exponent: 2 **10.** base: $5x$; exponent: 2

MENTAL MATH

State the bases and the exponents for each of the following expressions.

1. 3^2 **2.** 5^4 **3.** $(-3)^6$ **4.** -3^7 **5.** -4^2

6. $(-4)^3$ **7.** $5 \cdot 3^4$ **8.** $9 \cdot 7^6$ **9.** $5x^2$ **10.** $(5x)^2$

EXERCISE SET 5.1

STUDY GUIDE/SSM   CD/VIDEO   PH MATH TUTOR CENTER   MathXL®Tutorials ON CD   MathXL®   MyMathLab®

Evaluate each expression. See Example 1.

 1. 7^2 49   **2.** -3^2 −9   **3.** $(-5)^1$ −5

4. $(-3)^2$ 9   **5.** -2^4 −16   **6.** -4^3 −64

7. $(-2)^4$ 16   **8.** $(-4)^3$ −64   **9.** $(0.1)^5$ 0.00001

10. $(0.2)^5$ 0.00032   **11.** $\left(\frac{1}{3}\right)^4$ $\frac{1}{81}$ **12.** $\left(-\frac{1}{9}\right)^2$ $\frac{1}{81}$

13. $7 \cdot 2^5$ 224   **14.** $9 \cdot 1^7$ 9   **15.** $-2 \cdot 5^3$ −250

16. $-4 \cdot 3^3$ −108

17. Explain why $(-5)^4 = 625$, while $-5^4 = -625$.

18. Explain why $5 \cdot 4^2 = 80$, while $(5 \cdot 4)^2 = 400$.

17.–18. answers may vary

Evaluate each expression given the replacement values for x. See Example 2.

19. x^2; $x = -2$ 4   **20.** x^3; $x = -2$ −8

21. $5x^3$; $x = 3$ 135   **22.** $4x^2$; $x = -1$ 4

23. $2xy^2$; $x = 3$ and $y = 5$ 150   **24.** $-4x^2y^3$; $x = 2$ and $y = -1$ 16

25. $\frac{2z^4}{5}$; $z = -2$ $\frac{32}{5}$ **26.** $\frac{10}{3y^3}$; $y = 5$ $\frac{2}{75}$

27. The formula $V = x^3$ can be used to find the volume V of a cube with side length x. Find the volume of a cube with side length 7 meters. (Volume is measured in cubic units.)   343 cu. m

28. The formula $S = 6x^2$ can be used to find the surface area S of a cube with side length x. Find the surface area of the cube with side length 5 meters. (Surface area is measured in square units.)   150 sq. m

29. To find the amount of water that a swimming pool in the shape of a cube can hold, do we use the formula for volume of the cube or surface area of the cube? (See Exercises 27 and 28.) volume

△ **30.** To find the amount of material needed to cover an ottoman in the shape of a cube, do we use the formula for volume of the cube or surface area of the cube? (See Exercises 27 and 28.)   surface area

Use the product rule to simplify each expression. Write the results using exponents. See Examples 3 and 4.

31. $x^2 \cdot x^8$ x^{10} **32.** $y^2 \cdot y$ y^3

33. $(-3)^3 \cdot (-3)^9$ $(-3)^{12}$ **34.** $(-5)^7 \cdot (-5)^6$ $(-5)^{13}$

35. $(5y^4)(3y)$ $15y^5$ **36.** $(-2z^3)(-2z^2)$ $4z^5$

37. $(4z^{10})(-6z^7)(z^3)$ $-24z^{20}$ **38.** $(12x^5)(-x^6)(x^4)$ $-12x^{15}$

Use the power rule and the power of a product or quotient rule to simplify each expression. See Examples 5 through 7.

39. $(pq)^7$ p^7q^7 **40.** $(4s)^3$ $64s^3$ **41.** $\left(\dfrac{m}{n}\right)^9$ $\dfrac{m^9}{n^9}$

42. $\left(\dfrac{xy}{7}\right)^2$ $\dfrac{x^2y^2}{49}$ **43.** $(x^2y^3)^5$ $x^{10}y^{15}$ **44.** $(a^4b)^7$ $a^{28}b^7$

45. $\left(\dfrac{-2xz}{y^5}\right)^2$ $\dfrac{4x^2z^2}{y^{10}}$ **46.** $\left(\dfrac{y^4}{-3z^3}\right)^3$ $-\dfrac{y^{12}}{27z^9}$

Use the quotient rule and simplify each expression. See Example 8.

47. $\dfrac{x^3}{x}$ x^2 **48.** $\dfrac{y^{10}}{y^9}$ y **49.** $\dfrac{(-2)^5}{(-2)^3}$ 4

50. $\dfrac{(-5)^{14}}{(-5)^{11}}$ -125 **51.** $\dfrac{p^7q^{20}}{pq^{15}}$ p^6q^5 **52.** $\dfrac{x^8y^6}{y^5}$ x^8y

53. $\dfrac{7x^2y^6}{14x^2y^3}$ $\dfrac{y^3}{2}$ **54.** $\dfrac{9a^4b^7}{3ab^2}$ $3a^3b^5$

Simplify the following. See Example 9.

55. $(2x)^0$ 1 **56.** $-4x^0$ -4 **57.** $-2x^0$ -2

58. $(4y)^0$ 1 **59.** $5^0 + y^0$ 2 **60.** $-3^0 + 4^0$ 0

Simplify the following. See Example 10.

61. $\left(\dfrac{-3a^2}{b^3}\right)^3$ $-\dfrac{27a^6}{b^9}$ **62.** $\left(\dfrac{q^7}{-2p^5}\right)^5$ $-\dfrac{q^{35}}{32p^{25}}$

63. $\dfrac{(x^5)^7 \cdot x^8}{x^4}$ x^{39} **64.** $\dfrac{y^{20}}{(y^2)^3 \cdot y^9}$ y^5

65. $\dfrac{(z^3)^6}{(5z)^4}$ $\dfrac{z^{14}}{625}$ **66.** $\dfrac{(3x)^4}{(x^2)^2}$ 81

67. $\dfrac{(6mn)^5}{mn^2}$ $7776m^4n^3$ **68.** $\dfrac{(6xy)^2}{9x^2y^2}$ 4

MIXED PRACTICE

Simplify the following.

69. -5^2 -25 **70.** $(-5)^2$ 25 **71.** $\left(\dfrac{1}{4}\right)^3$ $\dfrac{1}{64}$

72. $\left(\dfrac{2}{3}\right)^3$ $\dfrac{8}{27}$ **73.** $(9xy)^2$ $81x^2y^2$ **74.** $(2ab)^5$ $32a^5b^5$

75. $(6b)^0$ 1 **76.** $(5ab)^0$ 1 **77.** $2^3 + 2^5$ 40

78. $7^2 - 7^0$ 48 **79.** b^4b^2 b^6

80. y^4y^1 y^5 **81.** $a^2a^3a^4$ a^9

82. $x^2x^{15}x^9$ x^{26} **83.** $(2x^3)(-8x^4)$ $-16x^7$

84. $(3y^4)(-5y)$ $-15y^5$ **85.** $(4a)^3$ $64a^3$

86. $(2ab)^4$ $16a^4b^4$ **87.** $(-6xyz^3)^2$ $36x^2y^2z^6$

88. $(-3xy^2a^3b)^3$ $-27x^3y^6a^9b^3$ **89.** $\left(\dfrac{3y^5}{6x^4}\right)^3$ $\dfrac{y^{15}}{8x^{12}}$

90. $\left(\dfrac{2ab}{6yz}\right)^4$ $\dfrac{a^4b^4}{81y^4z^4}$ **91.** $\dfrac{x^5}{x^4}$ x

92. $\dfrac{5x^9}{x^3}$ $5x^6$ **93.** $\dfrac{2x^3y^2z}{xyz}$ $2x^2y$

94. $\dfrac{x^{12}y^{13}}{x^5y^7}$ x^7y^6 **95.** $\dfrac{(3x^2y^5)^5}{x^3y}$ $243x^7y^{24}$

96. $\dfrac{(4a^2)^4}{a^4b}$ $\dfrac{256a^4}{b}$

97. In your own words, explain why $5^0 = 1$.   answers may vary

98. In your own words, explain when $(-3)^n$ is positive and when it is negative.   answers may vary

REVIEW AND PREVIEW

Simplify each expression by combining any like terms. Use the distributive property to remove any parentheses. See Section 2.1.

99. $y - 10 + y$ $2y - 10$ **100.** $-6z + 20 - 3z$

101. $7x + 2 - 8x - 6$ $-x - 4$ **102.** $10y - 14 - y - 14$

103. $2(x - 5) + 3(5 - x)$ **104.** $-3(w + 7) + 5(w + 1)$

100. $-9z + 20$ **102.** $9y - 28$ **103.** $-x + 5$ **104.** $2w - 16$

Concept Extensions

Find the area of each figure. See the geometric formulas on the inside covers. If the formula contains π, do not approximate.

△ **105.**

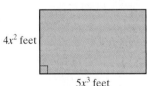

$20x^5$ sq. ft
$4x^2$ feet
$5x^3$ feet

△ **106.**

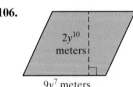

$18y^{17}$ sq. m
$2y^{10}$ meters
$9y^7$ meters

△ **107.**

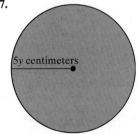

$25\pi y^2$ sq. cm
$5y$ centimeters

△ **108.**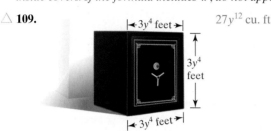
$8z^5$ decimeters

$64z^{10}$ sq. dm

Find the volume of each figure. See the geometric formulas in the inside covers. If the formula includes π, do not approximate.

△ **109.**

←$3y^4$ feet→

$3y^4$ feet

←$3y^4$ feet→

$27y^{12}$ cu. ft

△ **110.**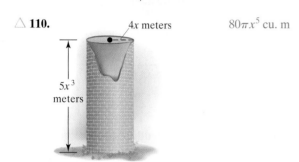

$4x$ meters

$5x^3$ meters

$80\pi x^5$ cu. m

Simplify each expression. Assume that variables represent positive integers. **116.** $9^{ab}a^{2ab}b^{3ab}c^{4ab}d^{5ab}$

111. $x^{5a}x^{4a}$ x^{9a}

112. $b^{9a}b^{4a}$ b^{13a}

113. $\dfrac{x^{9a}}{x^{4a}}$ x^{5a}

114. $\dfrac{y^{15b}}{y^{6b}}$ y^{9b}

115. $(x^a y^b z^c)^{5a}$ $x^{5a^2}y^{5ab}z^{5ac}$

116. $(9a^2 b^3 c^4 d^5)^{ab}$

117. Suppose you borrow money for 6 months. If the interest rate is compounded monthly, the formula $A = P\left(1 + \dfrac{r}{12}\right)^6$ gives the total amount A to be repaid at the end of 6 months. For a loan of $P = \$1000$ and interest rate of 9% $(r = 0.09)$, how much money will you need to pay off the loan? $\$1045.85$

118. On April 18, 2001, the Federal Reserve discount rate was set at 4%. (*Source:* Federal Reserve Board) The discount rate is the interest rate at which banks can borrow money from the Federal Reserve System. Suppose a bank needs to borrow money from the Federal Reserve System for 3 months. If the interest is compounded monthly, the formula $A = P\left(1 + \dfrac{r}{12}\right)^3$ gives the total amount A to be repaid at the end of 3 months. For a loan of $P = \$500,000$ and interest rate of $r = 0.04$, how much money will the bank repay to the Federal Reserve at the end of 3 months? Round to the nearest dollar. $\$505,017$

ADDING AND SUBTRACTING POLYNOMIALS

Objectives

1 Define monomial, binomial, trinomial, polynomial, and degree.

2 Find the value of a polynomial given replacement values for the variables.

3 Simplify a polynomial by combining like terms.

4 Add and subtract polynomials.

1   In this section, we introduce a special algebraic expression called a polynomial. Let's first review some definitions presented in Section 2.1.

Recall that a term is a number or the product of a number and variables raised to powers. The terms of the expression $4x^2 + 3x$ are $4x^2$ and $3x$. The terms of the expression $9x^4 - 7x - 1$ are $9x^4$, $-7x$, and -1.

Expression	Terms
$4x^2 + 3x$	$4x^2, 3x$
$9x^4 - 7x - 1$	$9x^4, -7x, -1$
$7y^3$	$7y^3$
5	5

The **numerical coefficient** of a term, or simply the **coefficient**, is the numerical factor of each term. If no numerical factor appears in the term, then the coefficient is understood to be 1. If the term is a number only, it is called a **constant** term or simply a constant.

Term	*Coefficient*
x^5	1
$3x^2$	3
$-4x$	-4
$-x^2y$	-1
3 (constant)	3

Polynomial

A **polynomial in x** is a finite sum of terms of the form ax^n, where a is a real number and n is a whole number.

For example,

$$x^5 - 3x^3 + 2x^2 - 5x + 1$$

is a polynomial. Notice that this polynomial is written in **descending powers** of x because the powers of x decrease from left to right. (Recall that the term 1 can be thought of as $1x^0$.)

On the other hand,

$$x^{-5} + 2x - 3$$

is **not** a polynomial because it contains an exponent, -5, that is not a whole number. (We study negative exponents in Section 5 of this chapter.)

Some polynomials are given special names.

A **monomial** is a polynomial with exactly one term.

A **binomial** is a polynomial with exactly two terms.

A **trinomial** is a polynomial with exactly three terms.

The following are examples of monomials, binomials, and trinomials. Each of these examples is also a polynomial.

POLYNOMIALS

Monomials	*Binomials*	*Trinomials*	*None of These*
ax^2	$x + y$	$x^2 + 4xy + y^2$	$5x^3 - 6x^2 + 3x - 6$
$-3z$	$3p + 2$	$x^5 + 7x^2 - x$	$-y^5 + y^4 - 3y^3 - y^2 + y$
4	$4x^2 - 7$	$-q^4 + q^3 - 2q$	$x^6 + x^4 - x^3 + 1$

Each term of a polynomial has a **degree**.

Degree of a Term

The degree of a term is the sum of the exponents on the variables contained in the term.

EXAMPLE 1

Find the degree of each term.

a. $-3x^2$ **b.** $5x^3yz$ **c.** 2

Solution   **a.** The exponent on x is 2, so the degree of the term is 2.

b. $5x^3yz$ can be written as $5x^3y^1z^1$. The degree of the term is the sum of its exponents, so the degree is $3 + 1 + 1$ or 5.

c. The constant, 2, can be written as $2x^0$ (since $x^0 = 1$). The degree of 2 or $2x^0$ is 0.

From the preceding, we can say that **the degree of a constant is 0**.
Each polynomial also has a degree.

> ## Degree of a Polynomial
>
> The degree of a polynomial is the greatest degree of any term of the polynomial.

EXAMPLE 2

Find the degree of each polynomial and tell whether the polynomial is a monomial, binomial, trinomial, or none of these.

a. $-2t^2 + 3t + 6$ **b.** $15x - 10$ **c.** $7x + 3x^3 + 2x^2 - 1$

Solution   **a.** The degree of the trinomial $-2t^2 + 3t + 6$ is 2, the greatest degree of any of its terms.

b. The degree of the binomial $15x - 10$ or $15x^1 - 10$ is 1.

c. The degree of the polynomial $7x + 3x^3 + 2x^2 - 1$ is 3.

EXAMPLE 3

Complete the table for the polynomial

$$7x^2y - 6xy + x^2 - 3y + 7$$

Use the table to give the degree of the polynomial.

Solution

Term	Numerical Coefficient	Degree of Term
$7x^2y$	7	3
$-6xy$	-6	2
x^2	1	2
$-3y$	-3	1
7	7	0

The degree of the polynomial is 3.

2   Polynomials have different values depending on replacement values for the variables.

EXAMPLE 4

Find the value of the polynomial $3x^2 - 2x + 1$ when $x = -2$.

Solution Replace x with -2 and simplify.

$$3x^2 - 2x + 1 = 3(-2)^2 - 2(-2) + 1$$
$$= 3(4) + 4 + 1$$
$$= 12 + 4 + 1$$
$$= 17$$

Many physical phenomena can be modeled by polynomials.

EXAMPLE 5

FINDING THE HEIGHT OF A DROPPED OBJECT

The CN Tower in Toronto, Ontario, is 1821 feet tall and is the world's tallest self-supporting structure. An object is dropped from the Skypod of the Tower which is at 1150 feet. Neglecting air resistance, the height of the object at time t seconds is given by the polynomial $-16t^2 + 1150$. Find the height of the object when $t = 1$ second and when $t = 7$ seconds.

Solution To find each height, we evaluate the polynomial when $t = 1$ and when $t = 7$.

$$-16t^2 + 1150 = -16(1)^2 + 1150 \quad \text{Replace } t \text{ with } 1.$$
$$= -16(1) + 1150$$
$$= -16 + 1150$$
$$= 1134$$

The height of the object at 1 second is 1134 feet.

$$-16t^2 + 1150 = -16(7)^2 + 1150 \quad \text{Replace } t \text{ with } 7.$$
$$= -16(49) + 1150$$
$$= -784 + 1150$$
$$= 366$$

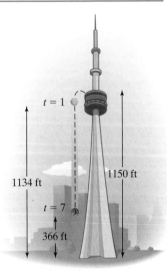

The height of the object at 7 seconds is 366 feet.

3 Polynomials with like terms can be simplified by combining the like terms. Recall that like terms are terms that contain exactly the same variables raised to exactly the same powers.

Like Terms	Unlike Terms
$5x^2, -7x^2$	$3x, 3y$
$y, 2y$	$-2x^2, -5x$
$\frac{1}{2}a^2b, -a^2b$	$6st^2, 4s^2t$

Only like terms can be combined. We combine like terms by applying the distributive property.

EXAMPLE 6

Simplify each polynomial by combining any like terms.

a. $-3x + 7x$ **b.** $x + 3x^2$ **c.** $11x^2 + 5 + 2x^2 - 7$ **d.** $\dfrac{2}{5}x^4 + \dfrac{2}{3}x^3 - x^2 + \dfrac{1}{10}x^4 - \dfrac{1}{6}x^3$

Solution

a. $-3x + 7x = (-3 + 7)x = 4x$

b. $x + 3x^2$ These terms cannot be combined because x and $3x^2$ are not like terms.

c. $11x^2 + 5 + 2x^2 - 7 = 11x^2 + 2x^2 + 5 - 7$

$$= 13x^2 - 2 \qquad \text{Combine like terms.}$$

d. $\dfrac{2}{5}x^4 + \dfrac{2}{3}x^3 - x^2 + \dfrac{1}{10}x^4 - \dfrac{1}{6}x^3$

$$= \left(\dfrac{2}{5} + \dfrac{1}{10}\right)x^4 + \left(\dfrac{2}{3} - \dfrac{1}{6}\right)x^3 - x^2$$

$$= \left(\dfrac{4}{10} + \dfrac{1}{10}\right)x^4 + \left(\dfrac{4}{6} - \dfrac{1}{6}\right)x^3 - x^2$$

$$= \dfrac{5}{10}x^4 + \dfrac{3}{6}x^3 - x^2$$

$$= \dfrac{1}{2}x^4 + \dfrac{1}{2}x^3 - x^2$$

CLASSROOM EXAMPLE

Simplify each polynomial by combining any like terms.

a. $14y^2 + 3 - 10y^2 - 9$
b. $7x^3 + x^3$
c. $23x^2 - 6x - x - 15$
d. $\frac{2}{7}x^3 - \frac{1}{4}x + 2 - \frac{1}{2}x^3 + \frac{3}{8}x$

answer:

a. $4y^2 - 6$ **b.** $8x^3$
c. $23x^2 - 7x - 15$
d. $-\frac{3}{14}x^3 + \frac{1}{8}x + 2$

✔ **CONCEPT CHECK**

When combining like terms in the expression $5x - 8x^2 - 8x$, which of the following is the proper result?

a. $-11x^2$ **b.** $-8x^2 - 3x$ **c.** $-11x$ **d.** $-11x^4$

EXAMPLE 7

CLASSROOM EXAMPLE

Combine like terms to simplify.
a. $11ab - 6a^2 - ba + 8b^2$
b. $7x^2y^2 + 2y^2 - 4y^2x^2$
     $+ x^2 - y^2 + 5x^2$

answer:
a. $10ab - 6a^2 + 8b^2$
b. $3x^2y^2 + y^2 + 6x^2$

Combine like terms to simplify.

$$-9x^2 + 3xy - 5y^2 + 7xy$$

Solution

$$-9x^2 + 3xy - 5y^2 + 7xy = -9x^2 + (3 + 7)xy - 5y^2$$
$$= -9x^2 + 10xy - 5y^2$$

> ▶ **Helpful Hint**
> This term can be written as $10xy$ or $10yx$.

EXAMPLE 8

Write a polynomial that describes the total area of the squares and rectangles shown below. Then simplify the polynomial.

Solution

Area: $x \cdot x \ + \ 3 \cdot x \ + \ 3 \cdot 3 \ + \ 4 \cdot x \ + \ x \cdot 2x$ Recall that the area of a rectangle is length times width.

$$= x^2 + 3x + 9 + 4x + 2x^2$$
$$= 3x^2 + 7x + 9 \qquad \text{Combine like terms.}$$

4   We now practice adding and subtracting polynomials.

> ## Adding Polynomials
>
> To add polynomials, combine all like terms.

EXAMPLE 9

Add $(-2x^2 + 5x - 1)$ and $(-2x^2 + x + 3)$.

Solution $(-2x^2 + 5x - 1) + (-2x^2 + x + 3) = -2x^2 + 5x - 1 - 2x^2 + x + 3$
$$= (-2x^2 - 2x^2) + (5x + 1x) + (-1 + 3)$$
$$= -4x^2 + 6x + 2$$

EXAMPLE 10

Add: $(4x^3 - 6x^2 + 2x + 7) + (5x^2 - 2x)$.

Solution $(4x^3 - 6x^2 + 2x + 7) + (5x^2 - 2x) = 4x^3 - 6x^2 + 2x + 7 + 5x^2 - 2x$
$$= 4x^3 + (-6x^2 + 5x^2) + (2x - 2x) + 7$$
$$= 4x^3 - x^2 + 7$$

Polynomials can be added vertically if we line up like terms underneath one another.

EXAMPLE 11

Add $(7y^3 - 2y^2 + 7)$ and $(6y^2 + 1)$ using the vertical format.

Solution   Vertically line up like terms and add.

$$\begin{array}{r} 7y^3 - 2y^2 + 7 \\ 6y^2 + 1 \\ \hline 7y^3 + 4y^2 + 8 \end{array}$$

To subtract one polynomial from another, recall the definition of subtraction. To subtract a number, we add its opposite: $a - b = a + (-b)$. To subtract a polynomial, we also add its opposite. Just as $-b$ is the opposite of b, $-(x^2 + 5)$ is the opposite of $(x^2 + 5)$.

EXAMPLE 12

Subtract: $(5x - 3) - (2x - 11)$.

Solution   From the definition of subtraction, we have

$$(5x - 3) - (2x - 11) = (5x - 3) + [-(2x - 11)] \qquad \text{Add the opposite.}$$
$$= (5x - 3) + (-2x + 11) \qquad \text{Apply the distributive property.}$$

$$= (5x - 2x) + (-3 + 11)$$
$$= 3x + 8$$

Subtracting Polynomials

To subtract two polynomials, change the signs of the terms of the polynomial being subtracted and then add.

EXAMPLE 13

Subtract: $(2x^3 + 8x^2 - 6x) - (2x^3 - x^2 + 1)$.

Solution   First, change the sign of each term of the second polynomial and then add.

> **Helpful Hint**
> Notice the sign of each term is changed.

$$(2x^3 + 8x^2 - 6x) - (2x^3 - x^2 + 1) = (2x^3 + 8x^2 - 6x) + (-2x^3 + x^2 - 1)$$
$$= 2x^3 - 2x^3 + 8x^2 + x^2 - 6x - 1$$
$$= 9x^2 - 6x - 1 \qquad \textit{Combine like terms.}$$

EXAMPLE 14

Subtract $(5y^2 + 2y - 6)$ from $(-3y^2 - 2y + 11)$ using the vertical format.

Solution   Arrange the polynomials in vertical format, lining up like terms.

$$
\begin{array}{rr}
-3y^2 - 2y + 11 & -3y^2 - 2y + 11 \\
-(5y^2 + 2y - 6) & -5y^2 - 2y + 6 \\
\hline
 & -8y^2 - 4y + 17
\end{array}
$$

EXAMPLE 15

Subtract $(5z - 7)$ from the sum of $(8z + 11)$ and $(9z - 2)$.

Solution   Notice that $(5z - 7)$ is to be subtracted **from** a sum. The translation is

$$[(8z + 11) + (9z - 2)] - (5z - 7)$$
$$= 8z + 11 + 9z - 2 - 5z + 7 \qquad \textit{Remove grouping symbols.}$$
$$= 8z + 9z - 5z + 11 - 2 + 7 \qquad \textit{Group like terms.}$$
$$= 12z + 16 \qquad \textit{Combine like terms.}$$

EXAMPLE 16

Add or subtract as indicated.

a. $(3x^2 - 6xy + 5y^2) + (-2x^2 + 8xy - y^2)$

b. $(9a^2b^2 + 6ab - 3ab^2) - (5b^2a + 2ab - 3 - 9b^2)$

Solution   **a.** $(3x^2 - 6xy + 5y^2) + (-2x^2 + 8xy - y^2)$
$$= 3x^2 - 6xy + 5y^2 - 2x^2 + 8xy - y^2$$
$$= x^2 + 2xy + 4y^2 \qquad \textit{Combine like terms.}$$

b. $(9a^2b^2 + 6ab - 3ab^2) - (5b^2a + 2ab - 3 - 9b^2)$

    $= 9a^2b^2 + 6ab - 3ab^2 - 5b^2a - 2ab + 3 + 9b^2$ Change the sign of each term of the polynomial being subtracted.

    $= 9a^2b^2 + 4ab - 8ab^2 + 3 + 9b^2$ Combine like terms.

MENTAL MATH

Simplify by combining like terms if possible.

1. $-9y - 5y$ $-14y$ **2.** $6m^5 + 7m^5$ $13m^5$ **3.** $4y^3 + 3y^3$ $7y^3$ **4.** $21y^5 - 19y^5$ $2y^5$

5. $x + 6x$ $7x$ **6.** $7z - z$ $6z$ **7.** $5m^2 + 2m$ $5m^2 + 2m$ **8.** $8p^3 + 3p^2$ $8p^3 + 3p^2$

EXERCISE SET 5.2

STUDY GUIDE/SSM   CD/VIDEO   PH MATH TUTOR CENTER   MathXL®Tutorials ON CD   MathXL®   MyMathLab®

Find the degree of each of the following polynomials and determine whether it is a monomial, binomial, trinomial, or none of these. See Examples 1 through 3.

1. $x + 2$ 1; binomial

2. $-6y + y^2 + 4$

3. $9m^3 - 5m^2 + 4m - 8$

4. $5a^2 + 3a^3 - 4a^4$

5. $12x^4y - x^2y^2 - 12x^2y^4$

6. $7r^2s^2 + 2r - 3s^5$

7. $3zx - 5x^2$ 2; binomial

8. $5y + 2$ 1; binomial

Match each polynomial in the first column with its degree in the second column. See Examples 1 through 3.

Polynomial	Degree	
9. $3xy^2 - 4$	4	3
10. $8x^2y^2$	2	4
11. $5a^2 - 2a + 1$	6	2
12. $4z^6 + 3z^2$	3	6

13. Describe how to find the degree of a term.

14. Describe how to find the degree of a polynomial.

15. Explain why xyz is a monomial while $x + y + z$ is a trinomial.

16. Explain why the degree of the term $5y^3$ is 3 and the degree of the polynomial $2y + y + 2y$ is 1.

Find the value of each polynomial when **(a)** $x = 0$ *and* **(b)** $x = -1$. *See Examples 4 and 5.*

17. $x + 6$ **(a)** 6;   **(b)** 5   **18.** $2x - 10$

19. $x^2 - 5x - 2$ **(a)** -2; **(b)** 4   **20.** $x^2 - 4$

21. $x^3 - 15$ **(a)** -15; **(b)** -16 **22.** $-2x^3 + 3x^2 - 6$

23. Find the height of the object in Example 5 when t is 9 seconds. Explain your result.

24. Find the height of the object in Example 5 when t is 8 seconds. Use the results of this exercise and the previous one and tell what happens between 8 and 9 seconds.

Simplify each of the following by combining like terms. See Examples 6 and 7.

25. $14x^2 + 9x^2$ $23x^2$ **26.** $18x^3 - 4x^3$ $14x^3$

27. $15x^2 - 3x^2 - y$ $12x^2 - y$ **28.** $12k^3 - 9k^3 + 11$

29. $8s - 5s + 4s$ $7s$ **30.** $5y + 7y - 6y$ $6y$

31. $0.1y^2 - 1.2y^2 + 6.7 - 1.9$ $-1.1y^2 + 4.8$

32. $7.6y + 3.2y^2 - 8y - 2.5y^2$ $0.7y^2 - 0.4y$

33. $\dfrac{2}{5}x^2 - \dfrac{1}{3}x^3 + x^2 - \dfrac{1}{4}x^3 + 6$ $-\dfrac{7}{12}x^3 + \dfrac{7}{5}x^2 + 6$

34. $\dfrac{1}{6}x^4 - \dfrac{1}{7}x^2 + 5 - \dfrac{3}{2}x^4 - \dfrac{3}{7}x^2 + \dfrac{1}{3}$ $-\dfrac{1}{3}x^4 - \dfrac{4}{7}x^2 + \dfrac{16}{3}$

35. $6a^2 - 4ab + 7b^2 - a^2 - 5ab + 9b^2$ $5a^2 - 9ab + 16b^2$

36. $x^2y + xy - y + 10x^2y - 2y + xy$ $11x^2y + 2xy - 3y$

Perform the indicated operations. See Examples 9 through 13.

37. $(3x + 7) + (9x + 5)$ $12x + 12$

38. $(3x^2 + 7) + (3x^2 + 9)$ $6x^2 + 16$

39. $(-7x + 5) + (-3x^2 + 7x + 5)$ $-3x^2 + 10$

40. $(3x - 8) + (4x^2 - 3x + 3)$ $4x^2 - 5$

41. $(2x^2 + 5) - (3x^2 - 9)$ $-x^2 + 14$

42. $(5x^2 + 4) - (-2y^2 + 4)$ $5x^2 + 2y^2$

43. $3x - (5x - 9)$ $-2x + 9$ **44.** $4 - (-y - 4)$ $y + 8$

45. $(2x^2 + 3x - 9) - (-4x + 7)$ $2x^2 + 7x - 16$

46. $(-7x^2 + 4x + 7) - (-8x + 2)$ $-7x^2 + 12x + 5$

47. Given the following triangle, find its perimeter.

$(x^2 + 7x + 4)$ ft

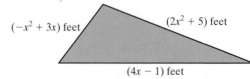

$(-x^2 + 3x)$ feet $(2x^2 + 5)$ feet

$(4x - 1)$ feet

2. 2; trinomial   **3.** 3; none of these   **4.** 4; trinomial   **5.** 6; trinomial   **6.** 5; trinomial   **13.–16.** answers may vary

18. (a) -10; **(b)** -12 **20. (a)** -4; **(b)** -3 **22. (a)** -6; **(b)** -1 **23.** -146 ft; the object has reached the ground

24. 126 ft; answers may vary   **28.** $3k^3 + 11$

△ **48.** Given the following quadrilateral, find its perimeter.
$(2x^2 - 2x + 2)$ cm

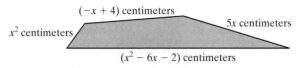

(−x + 4) centimeters

5x centimeters

x^2 centimeters

$(x^2 - 6x - 2)$ centimeters

△ **49.** A wooden beam is $(4y^2 + 4y + 1)$ meters long. If a piece $(y^2 - 10)$ meters is cut, express the length of the remaining piece of beam as a polynomial in y. $(3y^2 + 4y + 11)$ m

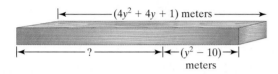

$(4y^2 + 4y + 1)$ meters

? $(y^2 - 10)$

meters

△ **50.** A piece of quarter-round molding is $(13x - 7)$ inches long. If a piece $(2x + 2)$ inches is removed, express the length of the remaining piece of molding as a polynomial in x. $(11x - 9)$ in.

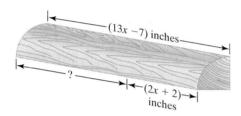

$(13x - 7)$ inches

? $(2x + 2)$

inches

Perform the indicated operations. See Examples 11 and 14.

53. $-2z^2 - 16z + 6$ **54.** $2u^5 - 10u^2 + 11u - 9$

51.
$$3t^2 + 4$$
$$+5t^2 - 8$$
$8t^2 - 4$

52.
$$7x^3 + 3$$
$$+2x^3 + 1$$
$9x^3 + 4$

53.
$$4z^2 - 8z + 3$$
$$-(6z^2 + 8z - 3)$$

54.
$$5u^5 - 4u^2 + 3u - 7$$
$$-(3u^5 + 6u^2 - 8u + 2)$$

55.
$$5x^3 - 4x^2 + 6x - 2$$
$$-(3x^3 - 2x^2 - x - 4)$$

56.
$$7a^2 - 9a + 6$$
$$-(11a^2 - 4a + 2)$$

55. $2x^3 - 2x^2 + 7x + 2$ **56.** $-4a^2 - 5a + 4$ **59.** $12x + 2$

Perform the indicated operations. See Examples 14 and 15.

57. Subtract $(19x^2 + 5)$ from $(81x^2 + 10)$. $62x^2 + 5$

58. Subtract $(2x + xy)$ from $(3x - 9xy)$. $x - 10xy$

59. Subtract $(2x + 2)$ from the sum of $(8x + 1)$ and $(6x + 3)$.

60. Subtract $(-12x - 3)$ from the sum of $(-5x - 7)$ and $(12x + 3)$. $19x - 1$

MIXED PRACTICE

Perform the indicated operations.

61. $-15x - (-4x)$ $-11x$ **62.** $16y - (-4y)$ $20y$

63. $2x - 5 + 5x - 8$ $7x - 13$ **64.** $x - 3 + 8x + 10$ $9x + 7$

65. $(-3y^2 - 4y) + (2y^2 + y - 1)$ $-y^2 - 3y - 1$

66. $(7x^2 + 2x - 9) + (-3x^2 + 5)$ $4x^2 + 2x - 4$

67. $(5x + 8) - (-2x^2 - 6x + 8)$ $2x^2 + 11x$

71. $7x^2 + 14x + 18$ **76.** $2y^2 + y - 10$

68. $(-6y^2 + 3y - 4) - (9y^2 - 3y)$ $-15y^2 + 6y - 4$

69. $(-8x^4 + 7x) + (-8x^4 + x + 9)$ $-16x^4 + 8x + 9$

70. $(6y^5 - 6y^3 + 4) + (-2y^5 - 8y^3 - 7)$ $4y^5 - 14y^3 - 3$

71. $(3x^2 + 5x - 8) + (5x^2 + 9x + 12) - (x^2 - 14)$

72. $(-a^2 + 1) - (a^2 - 3) + (5a^2 - 6a + 7)$ $3a^2 - 6a + 11$

73. Subtract $4x$ from $7x - 3$. $3x - 3$

74. Subtract y from $y^2 - 4y + 1$. $y^2 - 5y + 1$

75. Subtract $(5x + 7)$ from $(7x^2 + 3x + 9)$. $7x^2 - 2x + 2$

76. Subtract $(5y^2 + 8y + 2)$ from $(7y^2 + 9y - 8)$.

77. Subtract $(4y^2 - 6y - 3)$ from the sum of $(8y^2 + 7)$ and $(6y + 9)$. $4y^2 + 12y + 19$

78. Subtract $(5y + 7x^2)$ from the sum of $(8y - x)$ and $(3 + 8x^2)$. $x^2 + 3y - x + 3$

79. Subtract $(-2x^2 + 4x - 12)$ from the sum of $(-x^2 - 2x)$ and $(5x^2 + x + 9)$. $6x^2 - 5x + 21$

80. Subtract $(4x^2 - 2x + 2)$ from the sum of $(x^2 + 7x + 1)$ and $(7x + 5)$. $-3x^2 + 16x + 4$

81. Subtract $(3x^3 - x + 4)$ from the sum of $(x^3 + x^2 + 1)$ and $(5x^3 - 2x^2 + 9)$. $3x^3 - x^2 + x + 6$

82. Subtract $(5x^3 + 2x^2 - 7)$ from the sum of $(2x^3 - 7x + 2)$ and $(-x^3 - 6x^2 - 19)$. $-4x^3 - 8x^2 - 7x - 10$

REVIEW AND PREVIEW

Multiply. See Section 5.1.

83. $3x(2x)$ $6x^2$ **84.** $-7x(x)$ $-7x^2$

85. $(12x^3)(-x^5)$ $-12x^8$ **86.** $6r^3(7r^{10})$ $42r^{13}$

87. $10x^2(20xy^2)$ $200x^3y^2$ **88.** $-z^2y(11zy)$ $-11y^2z^3$

Find the area of each figure. Write a polynomial that describes the total area of the rectangles and squares shown in Exercises 89–90. Then simplify the polynomial. See Example 8.

△ **89.**

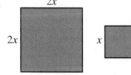

2x

2x

7

x

x

x

5

x

$4x^2 + 7x + x^2 + 5x; 5x^2 + 12x$

△ **90.**

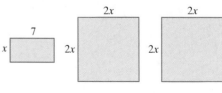

2x   2x

7

x   2x   2x

6   4

3   4

$7x + 4x^2 + 4x^2 + 18 + 16; 8x^2 + 7x + 34$

Recall that the perimeter of a figure such as the ones shown in Exercises 91 and 92 is the sum of the lengths of its sides. Write each perimeter as a polynomial. Then simplify the polynomial.

△ **91.**

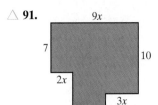

△ **92.**

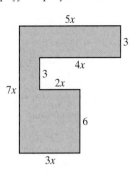

91. $9x + 10 + 3x + 12 + 4x + 15 + 2x + 7; 18x + 44$
92. $5x + 3 + 4x + 3 + 2x + 6 + 3x + 7x; 21x + 12$

△ *Write a polynomial that describes the surface area of each figure. (Recall that the surface area of a solid is the sum of the areas of the faces or sides of the solid.)* **93.** $2x^2 + 4xy$ **94.** $28x + 90$

93.

94.

Add or subtract as indicated. See Example 16.

95. $(9a + 6b - 5) + (-11a - 7b + 6)$ $-2a - b + 1$

96. $(3x - 2 + 6y) + (7x - 2 - y)$ $10x - 4 + 5y$

97. $(4x^2 + y^2 + 3) - (x^2 + y^2 - 2)$ $3x^2 + 5$

98. $(7a^2 - 3b^2 + 10) - (-2a^2 + b^2 - 12)$ $9a^2 - 4b^2 + 22$

99. $(x^2 + 2xy - y^2) + (5x^2 - 4xy + 20y^2)$ $6x^2 - 2xy + 19y^2$

100. $(a^2 - ab + 4b^2) + (6a^2 + 8ab - b^2)$ $7a^2 + 7ab + 3b^2$

101. $(11r^2s + 16rs - 3 - 2r^2s^2) - (3sr^2 + 5 - 9r^2s^2)$

102. $(3x^2y - 6xy + x^2y^2 - 5) - (11x^2y^2 - 1 + 5yx^2)$
101. $8r^2s + 16rs - 8 + 7r^2s^2$ **102.** $-2x^2y - 6xy - 10x^2y^2 - 4$
Simplify each polynomial by combining like terms.
103. $-5.42x^2 + 7.75x - 19.61$
103. $7.75x + 9.16x^2 - 1.27 - 14.58x^2 - 18.34$

104. $1.85x^2 - 3.76x + 9.25x^2 + 10.76 - 4.21x$
104. $11.1x^2 - 7.97x + 10.76$
Perform each indicated operation.

105. $[(7.9y^4 - 6.8y^3 + 3.3y) + (6.1y^3 - 5)]$
     $- (4.2y^4 + 1.1y - 1)$ $3.7y^4 - 0.7y^3 + 2.2y - 4$

106. $[(1.2x^2 - 3x + 9.1) - (7.8x^2 - 3.1 + 8)] + (1.2x - 6)$
106. $-6.6x^2 - 1.8x - 1.8$

Concept Extensions

107. A rocket is fired upward from the ground with an initial velocity of 200 feet per second. Neglecting air resistance, the height of the rocket at any time t can be described by the polynomial $-16t^2 + 200t$. Find the height of the rocket at the given times.

a. $t = 1$ second   184 ft   **b.** $t = 5$ seconds   600 ft
c. $t = 7.6$ seconds   595.84 ft **d.** $t = 10.3$ seconds 362.56 ft

108. Explain why the height of the rocket in Exercise 107 increases and then decreases as time passes.   answers may vary

109. The number of U.S. boating deaths is approximated by the polynomial $270.9x^2 - 3240.2x + 10,084$ from 1997 to 2001. Use the model to predict how many boating deaths there will be in 2005 ($x = 8$). (*Source:* National Safe Boating Council, U.S. Coast Guard)   1500

110. The projected high school enrollments (in millions) from 1991 to 2011 can be given by the equation $-0.018x^2 + 0.526x + 15.092$. Use this model to predict how many students will be enrolled in high school in 2007 ($x = 17$). (*Source:* National Center for Education Statistics)   18.832 million

111. The percentages of fatal motor vehicle accidents caused by improper driving for the years 1999 to 2001 are given by the polynomial $4.45x^2 - 15.45x + 72.6$. The percentages of fatal accidents caused by speeding are given by the polynomial $4.4x^2 - 8.8x + 23$. Find a polynomial for the percentage of nonspeeding-related fatal traffic accidents caused by improper driving. In each polynomial, x represents the number of years after 1999. $0.05x^2 - 6.65x + 49.6$

112. The number of wireless telephone subscribers (in millions) x years after 1990 is given by the polynomial $0.97x^2 - 0.91x + 7.46$ for 1993 through 2000. Use this model to predict the number of wireless telephone subscribers in 2005 ($x = 15$). (*Source:* Based on data from Cellular Telecommunications & Internet Association)   212.06 million wireless subscribers

113. The annual per capita consumption of chicken in pounds in the United States x years after 1990 is given by the polynomial $0.08x^3 - 1.19x^2 + 6.45x + 69.93$ for 1991 through 2000. Use this model to predict the per capita consumption of chicken in 2003 ($x = 13$). (*Source:* Based on data from U.S. Department of Agriculture, Economic Research Service)   128.43 lb

114. The polynomial $588x^2 + 983x + 26,493$ represents the annual U.S. production of beef (in millions of pounds) during 1999–2001. The polynomial $280.5x^2 + 660.5x + 19,791$ represents the combined annual U.S. production of veal, lamb, and pork (in millions of pounds) during 1999–2001. In both polynomials, x represents the number of years after 1999. Find a polynomial for the *total* U.S. meat production in millions of pounds during this period. (*Source:* Based on data from the U.S. Department of Agriculture) $868.5x^2 + 1643.5x + 46,284$

5.3   *MULTIPLYING POLYNOMIALS*

Objectives

1 Use the distributive property to multiply polynomials.

2 Multiply polynomials vertically.

1   To multiply polynomials, we apply our knowledge of the rules and definitions of exponents.

To multiply two monomials such as $(-5x^3)$ and $(-2x^4)$, use the associative and commutative properties and regroup. Remember that to multiply exponential expressions with a common base we add exponents.

$$(-5x^3)(-2x^4) = (-5)(-2)(x^3)(x^4) = 10x^7$$

EXAMPLES

Multiply.

1. $6x \cdot 4x = (6 \cdot 4)(x \cdot x)$ Use the commutative and associative properties.

$$= 24x^2$$ Multiply.

2. $-7x^2 \cdot 2x^5 = (-7 \cdot 2)(x^2 \cdot x^5)$

$$= -14x^7$$

3. $(-12x^5)(-x) = (-12x^5)(-1x)$

$$= (-12)(-1)(x^5 \cdot x)$$

$$= 12x^6$$

To multiply polynomials that are not monomials, use the distributive property.

EXAMPLE 4

Use the distributive property to find each product.

a. $5x(2x^3 + 6)$ **b.** $-3x^2(5x^2 + 6x - 1)$

Solution   **a.** $5x(2x^3 + 6) = 5x(2x^3) + 5x(6)$ Use the distributive property.

$$= 10x^4 + 30x$$ Multiply.

b. $-3x^2(5x^2 + 6x - 1)$

$$= (-3x^2)(5x^2) + (-3x^2)(6x) + (-3x^2)(-1)$$ Use the distributive property.

$$= -15x^4 - 18x^3 + 3x^2$$ Multiply.

We also use the distributive property to multiply two binomials. To multiply $(x + 3)$ by $(x + 1)$, distribute the factor $(x + 3)$ first.

$$(x + 3)(x + 1) = x(x + 1) + 3(x + 1)$$ Distribute $(x + 3)$.

$$= x(x) + x(1) + 3(x) + 3(1)$$ Apply distributive property a second time.

$$= x^2 + x + 3x + 3$$ Multiply.

$$= x^2 + 4x + 3$$ Combine like terms.

This idea can be expanded so that we can multiply any two polynomials.

CLASSROOM EXAMPLE

Multiply.

a. $7y \cdot 8y$ **b.** $(-9y^4)(-y)$

answer: **a.** $56y^2$ **b.** $9y^5$

CLASSROOM EXAMPLE

Use the distributive property to find each product.

a. $8x(7x^4 + 1)$

b. $-2x^3(3x^2 - x + 2)$

answer:

a. $56x^5 + 8x$

b. $-6x^5 + 2x^4 - 4x^3$

TEACHING TIP

Example 4b can be illustrated with an area diagram. Note that the result of the multiplication is written inside the rectangles.

	$5x^2$	$+6x$	-1
$-3x^2$	$-15x^4$	$-18x^3$	$+3x^2$

> ## To Multiply Two Polynomials
>
> Multiply each term of the first polynomial by each term of the second polynomial, and then combine like terms.

EXAMPLE 5

Multiply $(3x + 2)(2x - 5)$.

Solution Multiply each term of the first binomial by each term of the second.

$$(3x + 2)(2x - 5) = 3x(2x) + 3x(-5) + 2(2x) + 2(-5)$$

$$= 6x^2 - 15x + 4x - 10 \qquad \text{Multiply.}$$

$$= 6x^2 - 11x - 10 \qquad \text{Combine like terms.}$$

EXAMPLE 6

Multiply $(2x - y)^2$.

Solution Recall that $a^2 = a \cdot a$, so $(2x - y)^2 = (2x - y)(2x - y)$. Multiply each term of the first polynomial by each term of the second.

$$(2x - y)(2x - y) = 2x(2x) + 2x(-y) + (-y)(2x) + (-y)(-y)$$

$$= 4x^2 - 2xy - 2xy + y^2 \qquad \text{Multiply.}$$

$$= 4x^2 - 4xy + y^2 \qquad \text{Combine like terms.}$$

EXAMPLE 7

Multiply $(t + 2)$ by $(3t^2 - 4t + 2)$.

Solution Multiply each term of the first polynomial by each term of the second.

$$(t + 2)(3t^2 - 4t + 2) = t(3t^2) + t(-4t) + t(2) + 2(3t^2) + 2(-4t) + 2(2)$$

$$= 3t^3 - 4t^2 + 2t + 6t^2 - 8t + 4$$

$$= 3t^3 + 2t^2 - 6t + 4 \qquad \text{Combine like terms.}$$

EXAMPLE 8

Multiply $(3a + b)^3$.

Solution Write $(3a + b)^3$ as $(3a + b)(3a + b)(3a + b)$.

$$(3a + b)(3a + b)(3a + b) = (9a^2 + 3ab + 3ab + b^2)(3a + b)$$

$$= (9a^2 + 6ab + b^2)(3a + b)$$

$$= (9a^2 + 6ab + b^2)3a + (9a^2 + 6ab + b^2)b$$

$$= 27a^3 + 18a^2b + 3ab^2 + 9a^2b + 6ab^2 + b^3$$

$$= 27a^3 + 27a^2b + 9ab^2 + b^3$$

2 Another convenient method for multiplying polynomials is to use a vertical format similar to the format used to multiply real numbers. We demonstrate this method by multiplying $(3y^2 - 4y + 1)$ by $(y + 2)$.

EXAMPLE 9

Multiply $(3y^2 - 4y + 1)(y + 2)$. Use a vertical format.

Solution

Step 1. Multiply 2 by each term of the top polynomial. Write the first **partial product** below the line.

$$
\begin{array}{r}
3y^2 - 4y + 1 \\
\times \qquad y + 2 \\
\hline
6y^2 - 8y + 2 \qquad \text{Partial product}
\end{array}
$$

Step 2. Multiply y by each term of the top polynomial. Write this partial product underneath the previous one, being careful to line up like terms.

$$
\begin{array}{r}
3y^2 - 4y + 1 \\
\times \qquad y + 2 \\
\hline
6y^2 - 8y + 2 \qquad \text{Partial product} \\
3y^3 - 4y^2 + \; y \qquad \text{Partial product}
\end{array}
$$

Step 3. Combine like terms of the partial products.

$$
\begin{array}{r}
3y^2 - 4y + 1 \\
\times \qquad y + 2 \\
\hline
6y^2 - 8y + 2 \\
3y^3 - 4y^2 + \; y \\
\hline
3y^3 + 2y^2 - 7y + 2 \qquad \text{Combine like terms.}
\end{array}
$$

Thus, $(y + 2)(3y^2 - 4y + 1) = 3y^3 + 2y^2 - 7y + 2$.

When multiplying vertically, be careful if a power is missing, you may want to leave space in the partial products and take care that like terms are lined up.

EXAMPLE 10

Multiply $(2x^3 - 3x + 4)(x^2 + 1)$. Use a vertical format.

Solution

$$
\begin{array}{r}
2x^3 - 3x + 4 \\
\times \qquad\qquad x^2 + 1 \\
\hline
2x^3 \qquad - 3x + 4 \qquad \text{Leave space for missing powers of } x. \\
2x^5 - 3x^3 + 4x^2 \qquad\qquad\qquad \\
\hline
2x^5 - \; x^3 + 4x^2 - 3x + 4 \qquad \text{Combine like terms.}
\end{array}
$$

MENTAL MATH

Find the following products mentally. See Examples 1 through 3.

1. $5x(2y)$ $10xy$

2. $7a(4b)$ $28ab$

3. $x^2 \cdot x^5$ x^7

4. $z \cdot z^4$ z^5

5. $6x(3x^2)$ $18x^3$

6. $5a^2(3a^2)$ $15a^4$

7. $-9x^3 \cdot 3x^2$ $-27x^5$

8. $-8x(-4x^7)$ $32x^8$

Simplify, if possible.

9. $a^2 \cdot a^5$ a^7

10. $(a^2)^5$ a^{10}

11. $a^2 + a^5$

12. $\dfrac{a^5}{a^2}$ a^3 **11.** cannot simplify

EXERCISE SET 5.3

3. $7x^3 + 14x^2 - 7x$ **4.** $-5y^3 - 5y^2 + 50y$ STUDY GUIDE/SSM · CD/ VIDEO · PH MATH TUTOR CENTER · MathXL®Tutorials ON CD · MathXL® · MyMathLab®
6. $-20y^3 + 24y^4$

Find the following products. See Example 4.

1. $2a(2a - 4)$ $4a^2 - 8a$
2. $3a(2a + 7)$ $6a^2 + 21a$
 3. $7x(x^2 + 2x - 1)$
4. $-5y(y^2 + y - 10)$
5. $3x^2(2x^2 - x)$ $6x^4 - 3x^3$
6. $-4y^2(5y - 6y^2)$

△ **7.** The area of the larger rectangle below is $x(x + 3)$. Find another expression for this area by finding the sum of the areas of the smaller rectangles. $x^2 + 3x$

10. $y^2 + 12y + 35$
12. $36x^2 - 84x + 49$
13. $30x^2 - 79xy + 45y^2$
14. $21x^2 - 43xy - 14y^2$

△ **8.** Write an expression for the area of the larger rectangle below in two different ways. $x(1 + 2x); x + 2x^2$

Find the following products. See Examples 5 and 6.

 9. $(a + 7)(a - 2)$ $a^2 + 5a - 14$ **10.** $(y + 5)(y + 7)$
11. $(2y - 4)^2$ $4y^2 - 16y + 16$ **12.** $(6x - 7)^2$
13. $(5x - 9y)(6x - 5y)$ **14.** $(3x - 7y)(7x + 2y)$
15. $(2x^2 - 5)^2$ $4x^4 - 20x^2 + 25$ **16.** $(x^2 - 4)^2$ $x^4 - 8x^2 + 16$

△ **17.** The area of the figure below is $(x + 2)(x + 3)$. Find another expression for this area by finding the sum of the areas of the smaller rectangles. $x^2 + 5x + 6$

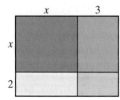

△ **18.** Write an expression for the area of the figure below in two different ways. $(3x + 1)(3x + 1); 9x^2 + 6x + 1$

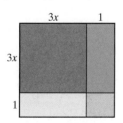

Find the following products. See Example 7.

19. $(x - 2)(x^2 - 3x + 7)$
20. $(x + 3)(x^2 + 5x - 8)$
21. $(x + 5)(x^3 - 3x + 4)$
22. $(a + 2)(a^3 - 3a^2 + 7)$
23. $(2a - 3)(5a^2 - 6a + 4)$
24. $(3 + b)(2 - 5b - 3b^2)$

Find the following products. See Example 8.

25. $(x + 2)^3$ $x^3 + 6x^2 + 12x + 8$ **26.** $(y - 1)^3$
27. $(2y - 3)^3$
28. $(3x + 4)^3$

Find the following products. Use the vertical multiplication method. See Examples 9 and 10.

 29. $(x + 3)(2x^2 + 4x - 1)$ **30.** $(2x - 5)(3x^2 - 4x + 7)$
31. $(x^3 + 5x - 7)(x^2 - 9)$ **32.** $(3x^4 - x^2 + 2)(2x^3 + 1)$

33. Evaluate each of the following. **33. b.** 324; 164
a. $(2 + 3)^2; 2^2 + 3^2$ 25; 13 **b.** $(8 + 10)^2; 8^2 + 10^2$
c. Does $(a + b)^2 = a^2 + b^2$ no matter what the values of a and b are? Why or why not? no; answers may vary

34. Perform the indicated operations. Explain the difference between the two expressions. answers may vary
a. $(3x + 5) + (3x + 7)$ **b.** $(3x + 5)(3x + 7)$
34. a. $6x + 12$; **34. b.** $9x^2 + 36x + 35$

MIXED PRACTICE

Find the following products.

35. $2a(a + 4)$ $2a^2 + 8a$ **36.** $-3a(2a + 7)$
37. $3x(2x^2 - 3x + 4)$ **38.** $-4x(5x^2 - 6x - 10)$
39. $(5x + 9y)(3x + 2y)$ **40.** $(5x - 5y)(2x - y)$
41. $(x + 2)(x^2 + 5x + 6)$ **42.** $(x - 7)(x^2 - 15x + 56)$
43. $(7x + 4)^2$ $49x^2 + 56x + 16$ **44.** $(3x - 2)^2$
45. $-2a^2(3a^2 - 2a + 3)$ **46.** $-4b^2(3b^3 - 12b^2 - 6)$
47. $(x + 3)(x^2 + 7x + 12)$ **48.** $(n + 1)(n^2 - 7n - 9)$
49. $(a + 1)^3$ $a^3 + 3a^2 + 3a + 1$ **50.** $(x - y)^3$
51. $(x + y)(x + y)$ **52.** $(x + 3)(7x + 1)$
53. $(x - 7)(x - 6)$ **54.** $(4x + 5)(-3x + 2)$
55. $3a(a^2 + 2)$ $3a^3 + 6a$ **56.** $x^3(x + 12)$ $x^4 + 12x^3$

19. $x^3 - 5x^2 + 13x - 14$ **20.** $x^3 + 8x^2 + 7x - 24$
21. $x^4 + 5x^3 - 3x^2 - 11x + 20$ **22.** $a^4 - a^3 - 6a^2 + 7a + 14$
23. $10a^3 - 27a^2 + 26a - 12$ **24.** $-3b^3 - 14b^2 - 13b + 6$
26. $y^3 - 3y^2 + 3y - 1$ **27.** $8y^3 - 36y^2 + 54y - 27$
28. $27x^3 + 108x^2 + 144x + 64$ **29.** $2x^3 + 10x^2 + 11x - 3$
30. $6x^3 - 23x^2 + 34x - 35$ **31.** $x^5 - 4x^3 - 7x^2 - 45x + 63$

32. $6x^7 - 2x^5 + 3x^4 + 4x^3 - x^2 + 2$ **36.** $-6a^2 - 21a$ **37.** $6x^3 - 9x^2 + 12x$ **38.** $-20x^3 + 24x^2 + 40x$ **39.** $15x^2 + 37xy + 18y^2$
40. $10x^2 - 15xy + 5y^2$ **41.** $x^3 + 7x^2 + 16x + 12$ **42.** $x^3 - 22x^2 + 161x - 392$ **44.** $9x^2 - 12x + 4$ **45.** $-6a^4 + 4a^3 - 6a^2$
46. $-12b^5 + 48b^4 + 24b^2$ **47.** $x^3 + 10x^2 + 33x + 36$ **48.** $n^3 - 6n^2 - 16n - 9$ **50.** $x^3 - 3x^2y + 3xy^2 - y^3$ **51.** $x^2 + 2xy + y^2$
52. $7x^2 + 22x + 3$ **53.** $x^2 - 13x + 42$ **54.** $-12x^2 - 7x + 10$

57. $-4y^3 - 12y^2 + 44y$

57. $-4y(y^2 + 3y - 11)$

58. $-10x^3 + 12x^2 - 2x$

58. $-2x(5x^2 - 6x + 1)$

59. $(5x + 1)(5x - 1)$ $25x^2 - 1$ **60.** $(2x + y)(3x - y)$

61. $(5x + 4)(x^2 - x + 4)$

62. $(x - 2)(x^2 - x + 3)$

63. $(2x - 5)^3$

64. $(3y - 1)^3$

65. $(4x + 5)(8x^2 + 2x - 4)$

66. $(x + 7)(x^2 - 7x - 8)$

67. $(7xy - y)^2$

68. $(x + y)^2$

69. $(5y^2 - y + 3)(y^2 - 3y - 2)$ $5y^4 - 16y^3 - 4y^2 - 7y - 6$

70. $(2x^2 + x - 1)(x^2 + 3x - 4)$ $2x^4 + 7x^3 - 6x^2 - 7x + 4$

71. $(3x^2 + 2x - 4)(2x^2 - 4x + 3)$ $6x^4 - 8x^3 - 7x^2 + 22x - 12$

72. $(a^2 + 3a - 2)(2a^2 - 5a - 1)$ $2a^4 + a^3 - 20a^2 + 7a + 2$

REVIEW AND PREVIEW

Perform the indicated operation. See Section 5.1.

73. $(5x)^2$ $25x^2$

74. $(4p)^2$ $16p^2$

75. $(-3y^3)^2$ $9y^6$

76. $(-7m^2)^2$ $49m^4$

*For income tax purposes, Rob Calcutta, the owner of Copy Services, uses a method called **straight-line depreciation** to show the depreciated (or decreased) value of a copy machine he recently purchased. Rob assumes that he can use the machine for 7 years. The graph below shows the depreciated value of the machine over the years. Use this graph to answer Exercises 77 through 82. See Sections 1.9 and 3.1.*

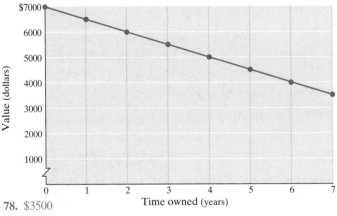

78. $3500

77. What was the purchase price of the copy machine? $7000

78. What is the depreciated value of the machine in 7 years?

79. What loss in value occurred during the first year? $500

80. What loss in value occurred during the second year? $500

81. Why do you think this method of depreciating is called straight-line depreciation?   answers may vary

82. Why is the line tilted downward?   There is a loss in value each year.

Concept Extensions

Express each of the following as polynomials.

△ **83.** Find the area of the rectangle. $(4x^2 - 25)$ sq. yd

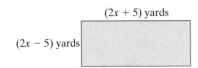

(2x + 5) yards

(2x − 5) yards

△ **84.** Find the area of the square field. $(x^2 + 8x + 16)$ sq. ft

(x + 4) feet

△ **85.** Find the area of the triangle. $(6x^2 - 4x)$ sq. in.

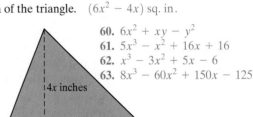

60. $6x^2 + xy - y^2$
61. $5x^3 - x^2 + 16x + 16$
62. $x^3 - 3x^2 + 5x - 6$
63. $8x^3 - 60x^2 + 150x - 125$

4x inches

(3x − 2) inches

△ **86.** Find the volume of the cube. $(y^3 - 3y^2 + 3y - 1)$ cu. m

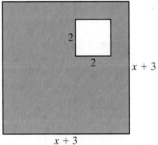

(y − 1) meters

64. $27y^3 - 27y^2 + 9y - 1$ **65.** $32x^3 + 48x^2 - 6x - 20$

△ **87.** Write a polynomial that describes the area of the shaded region. $(x^2 + 6x + 5)$ sq. units

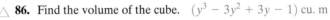

2

2

x + 3

x + 3

66. $x^3 - 57x - 56$ **67.** $49x^2y^2 - 14xy^2 + y^2$ **68.** $x^2 + 2xy + y^2$

△ **88.** Write a polynomial that describes the area of the shaded region. $(5x + 4)$ sq. units

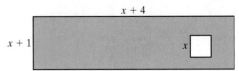

x + 4

x + 1

x

89. Multiply the following polynomials.   **89. b.** $4x^2 - 9y^2$

a. $(a + b)(a - b)$ $a^2 - b^2$ **b.** $(2x + 3y)(2x - 3y)$

c. $(4x + 7)(4x - 7)$ $16x^2 - 49$

d. Can you make a general statement about all products of the form $(x + y)(x - y)$?   answers may vary

5.4   SPECIAL PRODUCTS

Objectives

1. Multiply two binomials using the FOIL method.
2. Square a binomial.
3. Multiply the sum and difference of two terms.

1   In this section, we multiply binomials using special products. First, a special order for multiplying binomials called the FOIL order or method is introduced. This method is demonstrated by multiplying $(3x + 1)$ by $(2x + 5)$.

TEACHING TIP

Point out that the special products in this section are shortcuts for multiplying binomials. They can all be worked out by the method in the previous section in which each term in the first binomial is multiplied by every term in the second binomial.

F stands for the product of the **First** terms. $(3x + 1)(2x + 5)$
$$(3x)(2x) = 6x^2 \quad \mathbf{F}$$

O stands for the product of the **Outer** terms. $(3x + 1)(2x + 5)$
$$(3x)(5) = 15x \quad \mathbf{O}$$

I stands for the product of the **Inner** terms. $(3x + 1)(2x + 5)$
$$(1)(2x) = 2x \quad \mathbf{I}$$

L stands for the product of the **Last** terms. $(3x + 1)(3x + 5)$
$$(1)(5) = 5 \quad \mathbf{L}$$

$$
\begin{array}{ccccc}
 & \mathbf{F} & \mathbf{O} & \mathbf{I} & \mathbf{L} \\
(3x + 1)(2x + 5) &=& 6x^2 + 15x + 2x + 5 & &
\end{array}
$$
$$= 6x^2 + 17x + 5 \qquad \text{Combine like terms.}$$

✔ **CONCEPT CHECK**

Multiply $(3x + 1)(2x + 5)$ using methods from the last section. Show that the product is still $6x^2 + 17x + 5$.

EXAMPLE 1

Multiply $(x - 3)(x + 4)$ by the FOIL method.

Solution

CLASSROOM EXAMPLE
Multiply $(x + 7)(x - 5)$.
answer: $x^2 + 2x - 35$

$$
\begin{array}{ccccc}
 & \mathbf{F} & \mathbf{O} & \mathbf{I} & \mathbf{L} \\
(x - 3)(x + 4) &=& (x)(x) + (x)(4) + (-3)(x) + (-3)(4)
\end{array}
$$

$$= x^2 + 4x - 3x - 12$$
$$= x^2 + x - 12 \qquad \text{Combine like terms.}$$

EXAMPLE 2

Multiply $(5x - 7)(x - 2)$ by the FOIL method.

Solution

CLASSROOM EXAMPLE
Multiply $(6x - 1)(x - 4)$.
answer: $6x^2 - 25x + 4$

$$
\begin{array}{ccccc}
 & \mathbf{F} & \mathbf{O} & \mathbf{I} & \mathbf{L} \\
(5x - 7)(x - 2) &=& 5x(x) + 5x(-2) + (-7)(x) + (-7)(-2)
\end{array}
$$

$$= 5x^2 - 10x - 7x + 14$$
$$= 5x^2 - 17x + 14 \qquad \text{Combine like terms.}$$

Concept Check Answer:
Multiply and simplify:
$3x(2x + 5) + 1(2x + 5)$

EXAMPLE 3

Multiply $2(y + 6)(2y - 1)$.

Solution

$$\begin{array}{c} \qquad\qquad \text{F} \quad\ \ \text{O} \quad\ \text{I} \quad\ \ \text{L} \\ 2(y + 6)(2y - 1) = 2(2y^2 - 1y + 12y - 6) \\ = 2(2y^2 + 11y - 6) \\ = 4y^2 + 22y - 12 \qquad \textit{Now use the distributive property.} \end{array}$$

CLASSROOM EXAMPLE
Multiply $3(x - 4)(3x + 1)$.
answer: $9x^2 - 33x - 12$

2   Now, try squaring a binomial using the FOIL method.

EXAMPLE 4

Multiply $(3y + 1)^2$.

Solution

$$\begin{array}{c} (3y + 1)^2 = (3y + 1)(3y + 1) \\ \qquad\qquad \text{F} \qquad\ \ \ \text{O} \qquad\ \ \text{I} \qquad \text{L} \\ = (3y)(3y) + (3y)(1) + 1(3y) + 1(1) \\ = 9y^2 + 3y + 3y + 1 \\ = 9y^2 + 6y + 1 \end{array}$$

CLASSROOM EXAMPLE
Multiply $(2x + 5)^2$.
answer: $4x^2 + 20x + 25$

TEACHING TIP
Consider having students discover patterns for squaring binomials themselves. Before doing Example 4, you may want to have students multiply $(4x + 3)^2$. Then have them multiply $(4x - 3)^2$. Ask if they notice any relationship between the problem and its solution.

Notice the pattern that appears in Example 4.

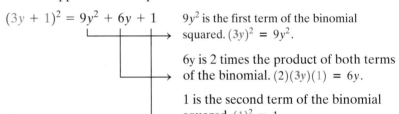

$(3y + 1)^2 = 9y^2 + 6y + 1$ $9y^2$ is the first term of the binomial squared. $(3y)^2 = 9y^2$.

6y is 2 times the product of both terms of the binomial. $(2)(3y)(1) = 6y$.

1 is the second term of the binomial squared. $(1)^2 = 1$.

This pattern leads to the following, which can be used when squaring a binomial. We call these **special products**.

> ### Squaring a Binomial
>
> A binomial squared is equal to the square of the first term plus or minus twice the product of both terms plus the square of the second term.
>
> $$(a + b)^2 = a^2 + 2ab + b^2$$
> $$(a - b)^2 = a^2 - 2ab + b^2$$

This product can be visualized geometrically.

TEACHING TIP
Now is probably a good time to once again remind students that $(x + y)^2 \neq x^2 + y^2$. Point out the partitioned square illustration in this section. It may help students if they visually see the product.

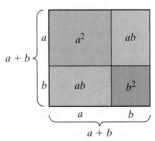

The area of the large square is side · side.

Area $= (a + b)(a + b) = (a + b)^2$

The area of the large square is also the sum of the areas of the smaller rectangles.

Area $= a^2 + ab + ab + b^2 = a^2 + 2ab + b^2$

Thus, $(a + b)^2 = a^2 + 2ab + b^2$.

EXAMPLE 5

Use a special product to square each binomial.

a. $(t + 2)^2$ **b.** $(p - q)^2$ **c.** $(2x + 5)^2$ **d.** $(x^2 - 7y)^2$

Solution

first term squared	plus or minus	twice the product of the terms	plus	second term squared

a. $(t + 2)^2 = t^2 + 2(t)(2) + 2^2 = t^2 + 4t + 4$

b. $(p - q)^2 = p^2 - 2(p)(q) + q^2 = p^2 - 2pq + q^2$

c. $(2x + 5)^2 = (2x)^2 + 2(2x)(5) + 5^2 = 4x^2 + 20x + 25$

d. $(x^2 - 7y)^2 = (x^2)^2 - 2(x^2)(7y) + (7y^2) = x^4 - 14x^2y + 49y^2$

Helpful Hint

Notice that

$(a + b)^2 \neq a^2 + b^2$ The middle term $2ab$ is missing.

$(a + b)^2 = (a + b)(a + b) = a^2 + 2ab + b^2$

Likewise,

$(a - b)^2 \neq a^2 - b^2$

$(a - b)^2 = (a - b)(a - b) = a^2 - 2ab + b^2$

Another special product is the product of the sum and difference of the same two terms, such as $(x + y)(x - y)$. Finding this product by the FOIL method, we see a pattern emerge.

$$(x + y)(x - y) = x^2 - xy + xy - y^2$$

$$= x^2 - y^2$$

Notice that the middle two terms subtract out. This is because the **O**uter product is the opposite of the **I**nner product. Only the **difference of squares** remains.

3

Multiplying the Sum and Difference of Two Terms

The product of the sum and difference of two terms is the square of the first term minus the square of the second term.

$$(a + b)(a - b) = a^2 - b^2$$

EXAMPLE 6

Use a special product to multiply.

a. $4(x + 4)(x - 4)$ **b.** $(6t + 7)(6t - 7)$ **c.** $\left(x - \dfrac{1}{4}\right)\left(x + \dfrac{1}{4}\right)$

d. $(2p - q)(2p + q)$ **e.** $(3x^2 - 5y)(3x^2 + 5y)$

Solution

first term squared ↓   minus ↓   second term squared ↙

a. $4(x + 4)(x - 4) = 4(x^2 - 4^2) = 4(x^2 - 16) = 4x^2 - 64$

b. $(6t + 7)(6t - 7) = (6t)^2 - 7^2 = 36t^2 - 49$

c. $\left(x - \dfrac{1}{4}\right)\left(x + \dfrac{1}{4}\right) = x^2 - \left(\dfrac{1}{4}\right)^2 = x^2 - \dfrac{1}{16}$

d. $(2p - q)(2p + q) = (2p)^2 - q^2 = 4p^2 - q^2$

e. $(3x^2 - 5y)(3x^2 + 5y) = (3x^2)^2 - (5y)^2 = 9x^4 - 25y^2$

✔ **CONCEPT CHECK**

Match each expression on the left to the equivalent expression or expressions in the list below.

$(a + b)^2$

$(a + b)(a - b)$

a. $(a + b)(a + b)$ **b.** $a^2 - b^2$ **c.** $a^2 + b^2$ **d.** $a^2 - 2ab + b^2$ **e.** $a^2 + 2ab + b^2$

Let's now practice multiplying polynomials in general. If possible, use a special product.

EXAMPLE 7

Multiply.

a. $(x - 5)(3x + 4)$ **b.** $(7x + 4)^2$ **c.** $(y - 0.6)(y + 0.6)$

d. $(y^4 + 2)(3y^2 - 1)$ **e.** $(a - 3)(a^2 + 2a - 1)$

Solution   **a.** $(x - 5)(3x + 4) = 3x^2 + 4x - 15x - 20$ FOIL.

$= 3x^2 - 11x - 20$

b. $(7x + 4)^2 = (7x)^2 + 2(7x)(4) + 4^2$ Squaring a binomial.

$= 49x^2 + 56x + 16$

c. $(y - 0.6)(y + 0.6) = y^2 - (0.6)^2 = y^2 - 0.36$ Multiplying the sum and difference of 2 terms.

d. $(y^4 + 2)(3y^2 - 1) = 3y^6 - y^4 + 6y^2 - 2$

e. I've inserted this product as a reminder that since it is not a binomial times a binomial, the FOIL order may not be used.

$(a - 3)(a^2 + 2a - 1) = a(a^2 + 2a - 1) - 3(a^2 + 2a - 1)$ Multiplying each term of the binomial by

$= a^3 + 2a^2 - a - 3a^2 - 6a + 3$ each term of the

$= a^3 - a^2 - 7a + 3$ trinomial.

MENTAL MATH

Answer each exercise true or false.

1. $(x + 4)^2 = x^2 + 16$ false   **2.** For $(x + 6)(2x - 1)$ the product of the first terms is $2x^2$.   true

3. $(x + 4)(x - 4) = x^2 + 16$ false   **4.** The product $(x - 1)(x^3 + 3x - 1)$ is a polynomial of degree 5.   false

EXERCISE SET 5.4

STUDY GUIDE/SSM   CD/VIDEO   PH MATH TUTOR CENTER   MathXL®Tutorials ON CD   MathXL®   MyMathLab®

Find each product using the FOIL method. See Examples 1 through 3.

1. $(x + 3)(x + 4)$ $x^2 + 7x + 12$ **2.** $(x + 5)(x - 1)$

3. $(x - 5)(x + 10)$ $x^2 + 5x - 50$ **4.** $(y - 12)(y + 4)$

5. $(5x - 6)(x + 2)$ **6.** $(3y - 5)(2y - 7)$

7. $(y - 6)(4y - 1)$ **8.** $(2x - 9)(x - 11)$

9. $(2x + 5)(3x - 1)$ **10.** $(6x + 2)(x - 2)$

Find each product. See Examples 4 and 5.

11. $(x - 2)^2$ $x^2 - 4x + 4$ **12.** $(x + 7)^2$

13. $(2x - 1)^2$ $4x^2 - 4x + 1$ **14.** $(7x - 3)^2$

15. $(3a - 5)^2$ $9a^2 - 30a + 25$ **16.** $(5a + 2)^2$

17. $(5x + 9)^2$ $25x^2 + 90x + 81$ **18.** $(6s - 2)^2$

19. Using your own words, explain how to square a binomial such as $(a + b)^2$.   answers may vary

20. Explain how to find the product of two binomials using the FOIL method.   answers may vary

Find each product. See Example 6.

21. $(a - 7)(a + 7)$ $a^2 - 49$ **22.** $(b + 3)(b - 3)$ $b^2 - 9$

23. $(3x - 1)(3x + 1)$ $9x^2 - 1$ **24.** $(4x - 5)(4x + 5)$

25. $\left(3x - \dfrac{1}{2}\right)\left(3x + \dfrac{1}{2}\right)$ $9x^2 - \dfrac{1}{4}$ **26.** $\left(10x + \dfrac{2}{7}\right)\left(10x - \dfrac{2}{7}\right)$

27. $(9x + y)(9x - y)$ $81x^2 - y^2$ **28.** $(2x - y)(2x + y)$

29. $(2x + 0.1)(2x - 0.1)$ **30.** $(5x - 1.3)(5x + 1.3)$

MIXED PRACTICE

Find each product. See Example 7.

31. $(a + 5)(a + 4)$ $a^2 + 9a + 20$ **32.** $(a - 5)(a - 7)$

33. $(a + 7)^2$ $a^2 + 14a + 49$ **34.** $(b - 2)^2$ $b^2 - 4b + 4$

35. $(4a + 1)(3a - 1)$ **36.** $(6a + 7)(6a + 5)$

37. $(x + 2)(x - 2)$ $x^2 - 4$ **38.** $(x - 10)(x + 10)$

39. $(3a + 1)^2$ $9a^2 + 6a + 1$ **40.** $(4a - 2)^2$

41. $(x^2 + y)(4x - y^4)$ **42.** $(x^3 - 2)(5x + y)$

43. $(x + 3)(x^2 - 6x + 1)$ **44.** $(x - 2)(x^2 - 4x + 2)$

45. $(2a - 3)^2$ $4a^2 - 12a + 9$ **46.** $(5b - 4x)^2$

47. $(5x - 6z)(5x + 6z)$ **48.** $(11x - 7y)(11x + 7y)$

49. $(x^5 - 3)(x^5 - 5)$ **50.** $(a^4 + 5)(a^4 + 6)$

51. $\left(x - \dfrac{1}{3}\right)\left(x + \dfrac{1}{3}\right)$ $x^2 - \dfrac{1}{9}$ **52.** $\left(3x + \dfrac{1}{5}\right)\left(3x - \dfrac{1}{5}\right)$

53. $(a^3 + 11)(a^4 - 3)$ **54.** $(x^5 + 5)(x^2 - 8)$

55. $3(x - 2)^2$ $3x^2 - 12x + 12$ **56.** $2(3b + 7)^2$

57. $(3b + 7)(2b - 5)$ **58.** $(3y - 13)(y - 3)$

59. $(7p - 8)(7p + 8)$ $49p^2 - 64$ **60.** $(3s - 4)(3s + 4)$

61. $\left(\dfrac{1}{3}a^2 - 7\right)\left(\dfrac{1}{3}a^2 + 7\right)$ **62.** $\left(\dfrac{2}{3}a - b^2\right)\left(\dfrac{2}{3}a - b^2\right)$

63. $5x^2(3x^2 - x + 2)$ **64.** $4x^3(2x^2 + 5x - 1)$

65. $(2r - 3s)(2r + 3s)$ $4r^2 - 9s^2$ **66.** $(6r - 2x)(6r + 2x)$

67. $(3x - 7y)^2$ **68.** $(4s - 2y)^2$

69. $(4x + 5)(4x - 5)$ $16x^2 - 25$ **70.** $(3x + 5)(3x - 5)$

71. $(8x + 4)^2$ $64x^2 + 64x + 16$ **72.** $(3x + 2)^2$

73. $\left(a - \dfrac{1}{2}y\right)\left(a + \dfrac{1}{2}y\right)$ $a^2 - \dfrac{1}{4}y^2$ **74.** $\left(\dfrac{a}{2} + 4y\right)\left(\dfrac{a}{2} - 4y\right)$

75. $\left(\dfrac{1}{5}x - y\right)\left(\dfrac{1}{5}x + y\right)$ **76.** $\left(\dfrac{y}{6} - 8\right)\left(\dfrac{y}{6} + 8\right)$

77. $(a + 1)(3a^2 - a + 1)$ **78.** $(b + 3)(2b^2 + b - 3)$

REVIEW AND PREVIEW

Simplify each expression. See Section 5.1.

79. $\dfrac{50b^{10}}{70b^5}$ $\dfrac{5b^5}{7}$ **80.** $\dfrac{x^3y^6}{xy^2}$ x^2y^4 **81.** $\dfrac{8a^{17}b^{15}}{-4a^7b^{10}}$ $-2a^{10}b^5$

82. $\dfrac{-6a^8y}{3a^4y}$ $-2a^4$ **83.** $\dfrac{2x^4y^{12}}{3x^4y^4}$ $\dfrac{2y^8}{3}$ **84.** $\dfrac{-48ab^6}{32ab^3}$ $-\dfrac{3b^3}{2}$

Find the slope of each line. See Section 3.4.   **86.** $-\dfrac{1}{2}$

85. $\dfrac{1}{3}$ **86.**

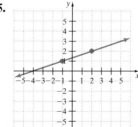
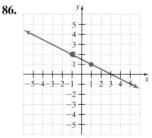

87.

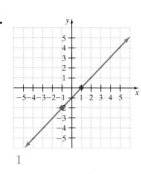

1

88.

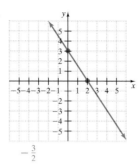

$-\frac{3}{2}$

Express each of the following as a polynomial in x.

△ **89.** Find the area of the square rug shown if its side is $(2x + 1)$ feet. $(4x^2 + 4x + 1)$ sq. ft

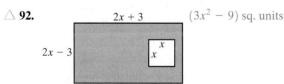

$(2x + 1)$ feet

$(2x + 1)$ feet

△ **90.** Find the area of the rectangular canvas if its length is $(3x - 2)$ inches and its width is $(x - 4)$ inches.
$(3x^2 - 14x + 8)$ sq. in.

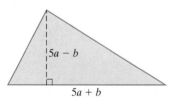

$(x - 4)$ inches

$(3x - 2)$ inches

Concept Extensions

Find the area of each shaded region.

△ **91.** $\left(\frac{25}{2}a^2 - \frac{1}{2}b^2\right)$ sq. units

$5a - b$

$5a + b$

△ **92.** $2x + 3$ $(3x^2 - 9)$ sq. units

$2x - 3$ x
                              x

△ **93.** $(24x^2 - 32x + 8)$ sq. m

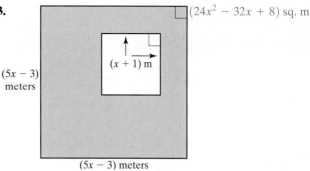

$(x + 1)$ m

$(5x - 3)$ meters

$(5x - 3)$ meters

△ **94.**

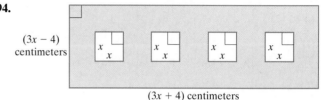

$(3x - 4)$ centimeters

x x x x
x x x x

$(3x + 4)$ centimeters

$(5x^2 - 16)$ sq. cm

△ **95.** x 5 △ **96.** $2y$ 11

x $2y$

5 11

$x^2 + 10x + 25$ $4y^2 + 44y + 121$

97. When is the product of two binomials a binomial? To help you answer this question, find the product $(x + 2)(x + 2)$ and then find the product $(x + 2)(x - 2)$.   answers may vary

Find each product. For example,

$$[(a + b) - 2][(a + b) + 2] = (a + b)^2 - 2^2$$
$$= a^2 + 2ab + b^2 - 4$$

98. $[(x + y) - 3][(x + y) + 3]$ $x^2 + 2xy + y^2 - 9$

99. $[(a + c) - 5][(a + c) + 5]$ $a^2 + 2ac + c^2 - 25$

100. $[(a - 3) + b][(a - 3) - b]$ $a^2 - 6a + 9 - b^2$

101. $[(x - 2) + y][(x - 2) - y]$ $x^2 - 4x + 4 - y^2$

2. $x^2 + 4x - 5$ **4.** $y^2 - 8y - 48$ **5.** $5x^2 + 4x - 12$
6. $6y^2 - 31y + 35$ **7.** $4y^2 - 25y + 6$ **8.** $2x^2 - 31x + 99$
9. $6x^2 + 13x - 5$ **10.** $6x^2 - 10x - 4$ **12.** $x^2 + 14x + 49$
14. $49x^2 - 42x + 9$ **16.** $25a^2 + 20a + 4$ **18.** $36s^2 - 24s + 4$
24. $16x^2 - 25$ **26.** $100x^2 - \frac{4}{49}$ **28.** $4x^2 - y^2$ **29.** $4x^2 - 0.01$

30. $25x^2 - 1.69$ **32.** $a^2 - 12a + 35$ **35.** $12a^2 - a - 1$ **36.** $36a^2 + 72a + 35$ **38.** $x^2 - 100$ **40.** $16a^2 - 16a + 4$
41. $4x^3 - x^2y^4 + 4xy - y^5$ **42.** $5x^4 + x^3y - 10x - 2y$ **43.** $x^3 - 3x^2 - 17x + 3$ **44.** $x^3 - 6x^2 + 10x - 4$
46. $25b^2 - 40bx + 16x^2$ **47.** $25x^2 - 36z^2$ **48.** $121x^2 - 49y^2$ **49.** $x^{10} - 8x^5 + 15$ **50.** $a^8 + 11a^4 + 30$ **52.** $9x^2 - \frac{1}{25}$
53. $a^7 - 3a^3 + 11a^4 - 33$ **54.** $x^7 - 8x^5 + 5x^2 - 40$ **56.** $18b^2 + 84b + 98$ **57.** $6b^2 - b - 35$ **58.** $3y^2 - 22y + 39$ **60.** $9s^2 - 16$
61. $\frac{1}{9}a^4 - 49$ **62.** $\frac{4}{9}a^2 - \frac{4}{3}ab^2 + b^4$ **63.** $15x^4 - 5x^3 + 10x^2$ **64.** $8x^5 + 20x^4 - 4x^3$ **66.** $36r^2 - 4x^2$ **67.** $9x^2 - 42xy + 49y^2$
68. $16s^2 - 16sy + 4y^2$ **70.** $9x^2 - 25$ **72.** $9x^2 + 12x + 4$ **74.** $\frac{a^2}{4} - 16y^2$ **75.** $\frac{1}{25}x^2 - y^2$ **76.** $\frac{y^2}{36} - 64$ **77.** $3a^3 + 2a^2 + 1$
78. $2b^3 + 7b^2 - 9$

INTEGRATED REVIEW | EXPONENTS AND OPERATIONS ON POLYNOMIALS

Perform the indicated operations and simplify.

18. $6x^2 + 13x - 11$

20. $8.4x^2 - 6.8x - 5.7$

26. $-a^2 - 3ab + 6b^2$

30. $20x^7 + 25x^3 - 4x^4 - 5$

32. $5x^3 + 9x^2 - 17x + 3$

1. $(5x^2)(7x^3)$ $35x^5$

2. $(4y^2)(8y^7)$ $32y^9$

3. -4^2 -16

4. $(-4)^2$ 16

5. $(x - 5)(2x + 1)$ $2x^2 - 9x - 5$

6. $(3x - 2)(x + 5)$ $3x^2 + 13x - 10$

7. $(x - 5) + (2x + 1)$ $3x - 4$

8. $(3x - 2) + (x + 5)$ $4x + 3$

9. $\dfrac{7x^9y^{12}}{x^3y^{10}}$ $7x^6y^2$

10. $\dfrac{20a^2b^8}{14a^2b^2}$ $\dfrac{10b^6}{7}$

11. $(12m^7n^6)^2$ $144m^{14}n^{12}$

12. $(4y^9z^{10})^3$ $64y^{27}z^{30}$

13. $3(4y - 3)(4y + 3)$ $48y^2 - 27$

14. $2(7x - 1)(7x + 1)$ $98x^2 - 2$

15. $(x^7y^5)^9$ $x^{63}y^{45}$

16. $(3^1x^9)^3$ $27x^{27}$

17. $(7x^2 - 2x + 3) - (5x^2 + 9)$ $2x^2 - 2x - 6$

18. $(10x^2 + 7x - 9) - (4x^2 - 6x + 2)$

19. $0.7y^2 - 1.2 + 1.8y^2 - 6y + 1$ $2.5y^2 - 6y - 0.2$

20. $7.8x^2 - 6.8x + 3.3 + 0.6x^2 - 9$

21. $(x + 4y)^2$ $x^2 + 8xy + 16y^2$

22. $(y - 9z)^2$ $y^2 - 18yz + 81z^2$

23. $(x + 4y) + (x + 4y)$ $2x + 8y$

24. $(y - 9z) + (y - 9z)$ $2y - 18z$

25. $7x^2 - 6xy + 4(y^2 - xy)$ $7x^2 - 10xy + 4y^2$

26. $5a^2 - 3ab + 6(b^2 - a^2)$

27. $(x - 3)(x^2 + 5x - 1)$ $x^3 + 2x^2 - 16x + 3$

28. $(x + 1)(x^2 - 3x - 2)$ $x^3 - 2x^2 - 5x - 2$

29. $(2x^3 - 7)(3x^2 + 10)$ $6x^5 + 20x^3 - 21x^2 - 70$

30. $(5x^3 - 1)(4x^4 + 5)$

31. $(2x - 7)(x^2 - 6x + 1)$ $2x^3 - 19x^2 + 44x - 7$

32. $(5x - 1)(x^2 + 2x - 3)$

Perform exercises and simplify, if possible.

33. $5x^3 + 5y^3$ cannot simplify

34. $(5x^3)(5y^3)$ $25x^3y^3$

35. $(5x^3)^3$ $125x^9$

36. $\dfrac{5x^3}{5y^3}$ $\dfrac{x^3}{y^3}$

37. $x + x$ $2x$

38. $x \cdot x$ x^2

5.5   *NEGATIVE EXPONENTS AND SCIENTIFIC NOTATION*

Objectives

1   Evaluate numbers raised to negative integer powers.

2   Use all the rules and definitions for exponents to simplify exponential expressions.

3   Write numbers in scientific notation.

4   Convert numbers from scientific notation to standard form.

1   Our work with exponential expressions so far has been limited to exponents that are positive integers or 0. Here we expand to give meaning to an expression like x^{-3}.

Suppose that we wish to simplify the expression $\dfrac{x^2}{x^5}$. If we use the quotient rule for exponents, we subtract exponents:

$$\frac{x^2}{x^5} = x^{2-5} = x^{-3}, \quad x \neq 0$$

But what does x^{-3} mean? Let's simplify $\dfrac{x^2}{x^5}$ using the definition of x^n.

$$\frac{x^2}{x^5} = \frac{x \cdot x}{x \cdot x \cdot x \cdot x \cdot x}$$

$$= \frac{\cancel{x} \cdot \cancel{x}}{\cancel{x} \cdot \cancel{x} \cdot x \cdot x \cdot x} \qquad \text{Divide numerator and denominator by common factors by applying the fundamental principle for fractions.}$$

$$= \frac{1}{x^3}$$

If the quotient rule is to hold true for negative exponents, then x^{-3} must equal $\dfrac{1}{x^3}$.

From this example, we state the definition for negative exponents.

Negative Exponents

If a is a real number other than 0 and n is an integer, then

$$a^{-n} = \frac{1}{a^n}$$

For example, $x^{-3} = \dfrac{1}{x^3}$.

In other words, another way to write a^{-n} is to take its reciprocal and change the sign of its exponent.

EXAMPLE 1

Simplify by writing each expression with positive exponents only.

a. 3^{-2} **b.** $2x^{-3}$ **c.** $2^{-1} + 4^{-1}$ **d.** $(-2)^{-4}$ **e.** $\dfrac{1}{y^{-4}}$ **f.** $\dfrac{1}{7^{-2}}$

Solution

a. $3^{-2} = \dfrac{1}{3^2} = \dfrac{1}{9}$ *Use the definition of negative exponents.*

b. $2x^{-3} = 2 \cdot \dfrac{1}{x^3} = \dfrac{2}{x^3}$ *Use the definition of negative exponents.*

> **Helpful Hint**
> Don't forget that since there are no parentheses, only x is the base for the exponent -3.

c. $2^{-1} + 4^{-1} = \dfrac{1}{2} + \dfrac{1}{4} = \dfrac{2}{4} + \dfrac{1}{4} = \dfrac{3}{4}$

d. $(-2)^{-4} = \dfrac{1}{(-2)^4} = \dfrac{1}{(-2)(-2)(-2)(-2)} = \dfrac{1}{16}$

e. $\dfrac{1}{y^{-4}} = \dfrac{1}{\dfrac{1}{y^4}} = y^4$ **f.** $\dfrac{1}{7^{-2}} = \dfrac{1}{\dfrac{1}{7^2}} = \dfrac{7^2}{1}$ or 49

> **Helpful Hint**
>
> From Example 1, we see that
>
> $$\frac{1}{y^{-4}} = \frac{y^4}{1} \text{ or } y^4.$$
>
> Also, notice that a negative exponent *does not affect* the sign of its base. The key word to remember when working with negative exponents is reciprocal.
>
> $$x^{-2} = \frac{1}{x^2} \qquad 2^{-3} = \frac{1}{2^3} = \frac{1}{8}$$
>
> $$\frac{1}{5^{-2}} = \frac{5^2}{1} \text{ or } 25 \qquad \frac{y^{-7}}{x^{-11}} = \frac{x^{11}}{y^7}$$

EXAMPLE 2

CLASSROOM EXAMPLE

Simplify.

a. $\dfrac{1}{y^{-7}}$ b. $\dfrac{1}{2^{-5}}$ c. $\dfrac{x^{-5}}{y^{-2}}$

answer:

a. y^7 b. 32   c. $\dfrac{y^2}{x^5}$

Simplify each expression. Write results using positive exponents only.

a. $\dfrac{1}{x^{-3}}$ b. $\dfrac{1}{3^{-4}}$ c. $\dfrac{p^{-4}}{q^{-9}}$ d. $\dfrac{5^{-3}}{2^{-5}}$

Solution

a. $\dfrac{1}{x^{-3}} = \dfrac{x^3}{1} = x^3$ b. $\dfrac{1}{3^{-4}} = \dfrac{3^4}{1} = 81$ c. $\dfrac{p^{-4}}{q^{-9}} = \dfrac{q^9}{p^4}$ d. $\dfrac{5^{-3}}{2^{-5}} = \dfrac{2^5}{5^3} = \dfrac{32}{125}$

EXAMPLE 3

Simplify each expression. Write answers with positive exponents.

a. $\dfrac{y}{y^{-2}}$ b. $\dfrac{3}{x^{-4}}$ c. $\dfrac{x^{-5}}{x^7}$

Solution

a. $\dfrac{y}{y^{-2}} = \dfrac{y^1}{y^{-2}} = y^{1-(-2)} = y^3$ Remember that $\dfrac{a^m}{a^n} = a^{m-n}$.

CLASSROOM EXAMPLE

Simplify.

a. $\dfrac{x}{x^{-5}}$ b. $\dfrac{7z^{-2}}{z^4}$

answer:

a. x^6 b. $\dfrac{7}{z^6}$

b. $\dfrac{3}{x^{-4}} = 3 \cdot \dfrac{1}{x^{-4}} = 3 \cdot x^4$ or $3x^4$ c. $\dfrac{x^{-5}}{x^7} = x^{-5-7} = x^{-12} = \dfrac{1}{x^{12}}$

2   All the previously stated rules for exponents apply for negative exponents also. Here is a summary of the rules and definitions for exponents.

TEACHING TIP

Consider asking students if they see another approach to Example 4a. For example.

$$\left(\frac{2}{3}\right)^{-3} = \left(\frac{3}{2}\right)^3$$

$$= \frac{3^3}{2^3}$$

$$= \frac{27}{8}$$

> ## Summary of Exponent Rules
>
> If m and n are integers and a, b, and c are real numbers, then:
>
> Product rule for exponents: $a^m \cdot a^n = a^{m+n}$
>
> Power rule for exponents: $\left(a^m\right)^n = a^{m \cdot n}$
>
> Power of a product: $(ab)^n = a^n b^n$
>
> Power of a quotient: $\left(\dfrac{a}{c}\right)^n = \dfrac{a^n}{c^n}, \quad c \neq 0$
>
> Quotient rule for exponents: $\dfrac{a^m}{a^n} = a^{m-n}, \quad a \neq 0$
>
> Zero exponent: $a^0 = 1, \quad a \neq 0$
>
> Negative exponent: $a^{-n} = \dfrac{1}{a^n}, \quad a \neq 0$

EXAMPLE 4

Simplify the following expressions. Write each result using positive exponents only.

a. $\left(\dfrac{2}{3}\right)^{-3}$ **b.** $\dfrac{(x^3)^4 x}{x^7}$ **c.** $\left(\dfrac{3a^2}{b}\right)^{-3}$

d. $\dfrac{4^{-1}x^{-3}y}{4^{-3}x^2 y^{-6}}$ **e.** $(y^{-3}z^6)^{-6}$ **f.** $\left(\dfrac{-2x^3 y}{xy^{-1}}\right)^3$

Solution

a. $\left(\dfrac{2}{3}\right)^{-3} = \dfrac{2^{-3}}{3^{-3}} = \dfrac{3^3}{2^3} = \dfrac{27}{8}$

b. $\dfrac{(x^3)^4 x}{x^7} = \dfrac{x^{12} \cdot x}{x^7} = \dfrac{x^{12+1}}{x^7} = \dfrac{x^{13}}{x^7} = x^{13-7} = x^6$ Use the power rule.

c. $\left(\dfrac{3a^2}{b}\right)^{-3} = \dfrac{3^{-3}(a^2)^{-3}}{b^{-3}}$ Raise each factor in the numerator and the denominator to the -3 power.

$= \dfrac{3^{-3}a^{-6}}{b^{-3}}$ Use the power rule.

$= \dfrac{b^3}{3^3 a^6}$ Use the negative exponent rule.

$= \dfrac{b^3}{27a^6}$ Write 3^3 as **27**.

d. $\dfrac{4^{-1}x^{-3}y}{4^{-3}x^2 y^{-6}} = 4^{-1-(-3)}x^{-3-2}y^{1-(-6)} = 4^2 x^{-5}y^7 = \dfrac{4^2 y^7}{x^5} = \dfrac{16y^7}{x^5}$

e. $(y^{-3}z^6)^{-6} = y^{18} \cdot z^{-36} = \dfrac{y^{18}}{z^{36}}$

f. $\left(\dfrac{-2x^3 y}{xy^{-1}}\right)^3 = \dfrac{(-2)^3 x^9 y^3}{x^3 y^{-3}} = \dfrac{-8x^9 y^3}{x^3 y^{-3}} = -8x^{9-3}y^{3-(-3)} = -8x^6 y^6$

3 Both very large and very small numbers frequently occur in many fields of science. For example, the distance between the sun and the planet Pluto is approximately 5,906,000,000 kilometers, and the mass of a proton is approximately 0.000000000000000000000165 gram. It can be tedious to write these numbers in this standard decimal notation, so **scientific notation** is used as a convenient shorthand for expressing very large and very small numbers.

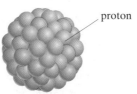

Mass of proton is approximately
0.000 000 000 000 000 000 000 001 65 gram

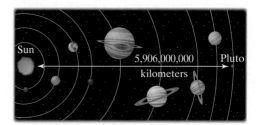

Scientific Notation

A positive number is written in scientific notation if it is written as the product of a number a, where $1 \le a < 10$, and an integer power r of 10:

$$a \times 10^r$$

The numbers below are written in scientific notation. The $\times$ sign for multiplication is used as part of the notation.

$$2.03 \times 10^2 \quad 7.362 \times 10^7 \quad 5.906 \times 10^9 \quad \text{(Distance between the sun and Pluto)}$$
$$1 \times 10^{-3} \quad 8.1 \times 10^{-5} \quad 1.65 \times 10^{-24} \quad \text{(Mass of a proton)}$$

The following steps are useful when writing numbers in scientific notation.

To Write a Number in Scientific Notation

Step 1. Move the decimal point in the original number to the left or right so that the new number has a value between 1 and 10.

Step 2. Count the number of decimal places the decimal point is moved in Step 1. If the original number is 10 or greater, the count is positive. If the original number is less than 1, the count is negative.

Step 3. Multiply the new number in Step 1 by 10 raised to an exponent equal to the count found in Step 2.

EXAMPLE 5

Write each number in scientific notation.

a. 367,000,000 **b.** 0.000003 **c.** 20,520,000,000 **d.** 0.00085

Solution

a. Step 1. Move the decimal point until the number is between 1 and 10.
367,000,000
8 places

Step 2. The decimal point is moved 8 places, and the original number is 10 or greater, so the count is positive 8.

Step 3. $367,000,000 = 3.67 \times 10^8$.

b. Step 1. Move the decimal point until the number is between 1 and 10.
0.000003
6 places

Step 2. The decimal point is moved 6 places, and the original number is less than 1, so the count is -6.

Step 3. $0.000003 = 3.0 \times 10^{-6}$

c. $20,520,000,000 = 2.052 \times 10^{10}$

d. $0.00085 = 8.5 \times 10^{-4}$

4   A number written in scientific notation can be rewritten in standard form. For example, to write 8.63×10^3 in standard form, recall that $10^3 = 1000$.

$$8.63 \times 10^3 = 8.63(1000) = 8630$$

Notice that the exponent on the 10 is positive 3, and we moved the decimal point 3 places to the right.

To write 7.29×10^{-3} in standard form, recall that $10^{-3} = \dfrac{1}{10^3} = \dfrac{1}{1000}$.

$$7.29 \times 10^{-3} = 7.29\left(\frac{1}{1000}\right) = \frac{7.29}{1000} = 0.00729$$

The exponent on the 10 is negative 3, and we moved the decimal to the left 3 places.

In general, **to write a scientific notation number in standard form**, move the decimal point the same number of places as the exponent on 10. If the exponent is positive, move the decimal point to the right; if the exponent is negative, move the decimal point to the left.

EXAMPLE 6

Write each number in standard notation, without exponents.

a. 1.02×10^5 **b.** 7.358×10^{-3} **c.** 8.4×10^7 **d.** 3.007×10^{-5}

Solution   **a.** Move the decimal point 5 places to the right.

$$1.02 \times 10^5 = 102,000.$$

b. Move the decimal point 3 places to the left.

$$7.358 \times 10^{-3} = 0.007358.$$

c. $8.4 \times 10^7 = 84,000,000.$ *7 places to the right*

d. $3.007 \times 10^{-5} = 0.00003007$ *5 places to the left*

✔ **CONCEPT CHECK**

Which number in each pair is larger?

a. 7.8×10^3 or 2.1×10^5 **b.** 9.2×10^{-2} or 2.7×10^4 **c.** 5.6×10^{-4} or 6.3×10^{-5}

Performing operations on numbers written in scientific notation makes use of the rules and definitions for exponents.

EXAMPLE 7

Perform each indicated operation. Write each result in standard decimal notation.

a. $(8 \times 10^{-6})(7 \times 10^3)$ **b.** $\dfrac{12 \times 10^2}{6 \times 10^{-3}}$

Solution   **a.** $(8 \times 10^{-6})(7 \times 10^3) = (8 \cdot 7) \times (10^{-6} \cdot 10^3)$

$$= 56 \times 10^{-3}$$

$$= 0.056$$

b. $\dfrac{12 \times 10^2}{6 \times 10^{-3}} = \dfrac{12}{6} \times 10^{2-(-3)} = 2 \times 10^5 = 200,000$

Calculator Explorations

Scientific Notation

To enter a number written in scientific notation on a scientific calculator, locate the scientific notation key, which may be marked [EE] or [EXP]. To enter 3.1×10^7, press [3.1] [EE] [7]. The display should read [3.1   07].

(continued)

Enter each number written in scientific notation on your calculator. **8.** 1.083612×10^6

1. 5.31×10^3 5.31 EE 3

2. -4.8×10^{14} −4.8 EE 14

3. 6.6×10^{-9} 6.6 EE −9

4. -9.9811×10^{-2} −9.9811 EE −2

Multiply each of the following on your calculator. Notice the form of the result.

5. $3,000,000 \times 5,000,000$ 1.5×10^{13}

6. $230,000 \times 1000$ 2.3×10^8

Multiply each of the following on your calculator. Write the product in scientific notation.

7. $(3.26 \times 10^6)(2.5 \times 10^{13})$ 8.15×10^{19}

8. $(8.76 \times 10^{-4})(1.237 \times 10^9)$

Spotlight on DECISION & MAKING

Suppose you are a paralegal for a law firm. You are investigating the facts of a mineral rights case. Drilling on property adjacent to the client's has struck a natural gas reserve. The client believes that a portion of this reserve lies within her own property boundaries and she is, therefore, entitled to a portion of the proceeds from selling the natural gas. As part of your investigation, you contact two different experts, who fax you the following estimates for the size of the natural gas reserve:

Expert A	***Expert B***
Estimate of entire reserve: 4.6×10^7 cubic feet	Estimate of entire reserve: 6.7×10^6 cubic feet
Estimate of size of reserve on client's property 1.84×10^7 cubic feet	Estimate of size of reserve on client's property: 2.68×10^6 cubic feet

How, if at all, would you use these estimates in the case? Explain.

MENTAL MATH

State each expression using positive exponents only.

1. $5x^{-2}$ $\dfrac{5}{x^2}$

2. $3x^{-3}$ $\dfrac{3}{x^3}$

3. $\dfrac{1}{y^{-6}}$ y^6

4. $\dfrac{1}{x^{-3}}$ x^3

5. $\dfrac{4}{y^{-3}}$ $4y^3$

6. $\dfrac{16}{y^{-7}}$ $16y^7$

TEACHING TIP

A Group Activity for this section is available in the Instructor's Resource Manual.

EXERCISE SET 5.5

STUDY GUIDE/SSM CD/VIDEO PH MATH TUTOR CENTER MathXL®Tutorials ON CD MathXL® MyMathLab®

Simplify each expression. Write each result using positive exponents only. See Examples 1 through 3.

1. 4^{-3} $\dfrac{1}{64}$

2. 6^{-2} $\dfrac{1}{36}$

3. $(-2)^{-4}$ $\dfrac{1}{16}$

4. $(-3)^{-5}$ $-\dfrac{1}{243}$

5. $7x^{-3}$ $\dfrac{7}{x^3}$

6. $(7x)^{-3}$ $\dfrac{1}{343x^3}$

7. $\left(\dfrac{1}{2}\right)^{-5}$ 32

8. $\left(\dfrac{1}{8}\right)^{-2}$ 64

9. $\left(-\dfrac{1}{4}\right)^{-3}$ -64

10. $\left(-\dfrac{1}{8}\right)^{-2}$ 64

11. $3^{-1} + 2^{-1}$ $\dfrac{5}{6}$

12. $4^{-1} + 4^{-2}$ $\dfrac{5}{16}$

13. $\dfrac{1}{p^{-3}}$ p^3

14. $\dfrac{1}{q^{-5}}$ q^5

15. $\dfrac{p^{-5}}{q^{-4}}$ $\dfrac{q^4}{p^5}$

16. $\dfrac{r^{-5}}{s^{-2}}$ $\dfrac{s^2}{r^5}$

17. $\dfrac{x^{-2}}{x}$ $\dfrac{1}{x^3}$

18. $\dfrac{y}{y^{-3}}$ y^4

19. $2^0 + 3^{-1}$ $\frac{4}{3}$ **20.** $4^{-2} - 4^{-3}$ $\frac{3}{64}$ **21.** $\frac{-1}{p^{-4}}$ $-p^4$

22. $\frac{-1}{y^{-6}}$ $-y^6$ **23.** $-2^0 - 3^0$ -2 **24.** $5^0 + (-5)^0$ 2

Simplify each expression. Write each result using positive exponents only. See Example 4.

25. $\frac{x^2 x^5}{x^3}$ x^4 **26.** $\frac{y^4 y^5}{y^6}$ y^3 **27.** $\frac{p^2 p}{p^{-1}}$ p^4

28. $\frac{y^3 y}{y^{-2}}$ y^6 **29.** $\frac{(m^5)^4 m}{m^{10}}$ m^{11} **30.** $\frac{(x^2)^8 x}{x^9}$ x^8

31. $\frac{r}{r^{-3} r^{-2}}$ r^6 **32.** $\frac{p}{p^{-3} q^{-5}}$ $p^4 q^5$ **33.** $(x^5 y^3)^{-3}$ $\frac{1}{x^{15} y^9}$

34. $(z^5 x^5)^{-3}$ $\frac{1}{z^{15} x^{15}}$ **35.** $\frac{(x^2)^3}{x^{10}}$ $\frac{1}{x^4}$ **36.** $\frac{(y^4)^2}{y^{12}}$ $\frac{1}{y^4}$

37. $\frac{(a^5)^2}{(a^3)^4}$ $\frac{1}{a^2}$ **38.** $\frac{(x^2)^5}{(x^4)^3}$ $\frac{1}{x^2}$ **39.** $\frac{8k^4}{2k}$ $4k^3$

40. $\frac{27r^4}{3r^6}$ $\frac{9}{r^2}$ **41.** $\frac{-6m^4}{-2m^3}$ $3m$ **42.** $\frac{15a^4}{-15a^5}$ $-\frac{1}{a}$

43. $\frac{-24a^6 b}{6ab^2}$ $-\frac{4a^5}{b}$ **44.** $\frac{-5x^4 y^5}{15x^4 y^2}$ $-\frac{y^3}{3}$

45. $(-2x^3 y^{-4})(3x^{-1} y)$ $-\frac{6x^2}{y^3}$

46. $(-5a^4 b^{-7})(-a^{-4} b^3)$ $\frac{5}{b^4}$

47. $(a^{-5} b^2)^{-6}$ $\frac{a^{30}}{b^{12}}$ **48.** $(4^{-1} x^5)^{-2}$ $\frac{16}{x^{10}}$

49. $\left(\frac{x^{-2} y^4}{x^3 y^7}\right)^2$ $\frac{1}{x^{10} y^6}$ **50.** $\left(\frac{a^5 b}{a^7 b^{-2}}\right)^{-3}$ $\frac{a^6}{b^9}$

51. $\frac{4^2 z^{-3}}{4^3 z^{-5}}$ $\frac{z^2}{4}$ **52.** $\frac{3^{-1} x^4}{3^3 x^{-7}}$ $\frac{x^{11}}{81}$

53. $\frac{2^{-3} x^{-4}}{2^2 x}$ $\frac{1}{32x^5}$ **54.** $\frac{5^{-1} z^7}{5^{-2} z^9}$ $\frac{5}{z^2}$

55. $\frac{7ab^{-4}}{7^{-1} a^{-3} b^2}$ $\frac{49a^4}{b^6}$ **56.** $\frac{6^{-5} x^{-1} y^2}{6^{-2} x^{-4} y^4}$ $\frac{x^3}{216y^2}$

57. $\left(\frac{a^{-5} b}{ab^3}\right)^{-4}$ $a^{24} b^8$ **58.** $\left(\frac{r^{-2} s^{-3}}{r^{-4} s^{-3}}\right)^{-3}$ $\frac{1}{r^6}$

59. $\frac{(xy^3)^5}{(xy)^{-4}}$ $x^9 y^{19}$ **60.** $\frac{(rs)^{-3}}{(r^2 s^3)^2}$ $\frac{1}{r^7 s^9}$

61. $\frac{(-2xy^{-3})^{-3}}{(xy^{-1})^{-1}}$ $-\frac{y^8}{8x^2}$ **62.** $\frac{(-3x^2 y^2)^{-2}}{(xyz)^{-2}}$ $\frac{z^2}{9x^2 y^2}$

63. $\frac{6x^2 y^3}{-7xy^5}$ $-\frac{6x}{7y^2}$ **64.** $\frac{-8xa^2 b}{-5xa^5 b}$ $\frac{8}{5a^3}$

Write each number in scientific notation. See Example 5.

65. $78{,}000$ 7.8×10^4 **66.** $9{,}300{,}000{,}000$ 9.3×10^9

67. 0.00000167 1.67×10^{-6} **68.** 0.00000017 1.7×10^{-7}

69. 0.00635 6.35×10^{-3} **70.** 0.00194 1.94×10^{-3}

71. $1{,}160{,}000$ 1.16×10^6 **72.** $700{,}000$ 7.0×10^5

73. The temperature at the interior of the Earth is 20,000,000 degrees Celsius. Write 20,000,000 in scientific notation. 2.0×10^7

75. 1.56×10^7

74. The half-life of a carbon isotope is 5000 years. Write 5000 in scientific notation. 5.0×10^3

75. The temperature of the Sun at its core is 15,600,000 degrees Kelvin. Write 15,600,000 in scientific notation. (*Source:* Students for the Exploration and Development of Space)

76. Google.com is an Internet search engine that allows users to search over 1,346,966,000 Web pages. Write 1,346,966,000 in scientific notation. (*Source:* Google, Inc.) 1.346966×10^9

77. At this writing, the world's largest optical telescopes are the twin Keck Telescopes located near the summit of Mauna Kea in Hawaii. The elevation of the Keck Telescopes is about 13,600 feet above sea level. Write 13,600 in scientific notation. (*Source:* W.M. Keck Observatory) 1.36×10^4

78. More than 2,000,000,000 pencils are manufactured in the United States annually. Write this number in scientific notation. (*Source:* AbsoluteTrivia.com) 2×10^9

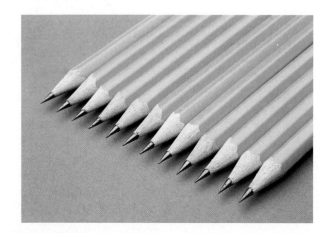

79. In 2003, the population of the United States was roughly 292,000,000. Write 292,000,000 in scientific notation. (*Source:* U.S. Census Bureau) 2.92×10^8

80. Pioneer 10 became the first spacecraft to leave the solar system eleven years after it was launched in 1972. When it was contacted in April 2001 to verify that it could still transmit a radio signal, Pioneer 10 was 7,290,000,000 miles from Earth. Write 7,290,000,000 in scientific notation. (*Source:* NASA Ames Research Center) 7.29×10^9

Write each number in standard notation. See Example 6.

81. 8.673×10^{-10} 0.0000000008673 **82.** 9.056×10^{-4}

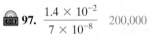 **83.** 3.3×10^{-2} 0.033   **84.** 4.8×10^{-6} 0.0000048

85. 2.032×10^4 20,320   **86.** 9.07×10^{10}

87. One coulomb of electricity is 6.25×10^{18} electrons. Write this number in standard notation.

88. The mass of a hydrogen atom is 1.7×10^{-24} grams. Write this number in standard notation.

89. The distance light travels in 1 year is 9.460×10^{12} kilometers. Write this number in standard notation.

90. The population of the world is 6.067×10^9. Write this number in standard notation. (*Source:* U.S. Bureau of the Census)

Evaluate each expression using exponential rules. Write each result in standard notation. See Example 7.
91. 0.000036 **93.** 0.0000000000000000028

91. $(1.2 \times 10^{-3})(3 \times 10^{-2})$ **92.** $(2.5 \times 10^6)(2 \times 10^{-6})$ 5

93. $(4 \times 10^{-10})(7 \times 10^{-9})$ **94.** $(5 \times 10^6)(4 \times 10^{-8})$ 0.2

95. $\dfrac{8 \times 10^{-1}}{16 \times 10^5}$ 0.0000005   **96.** $\dfrac{25 \times 10^{-4}}{5 \times 10^{-9}}$ 500,000

97. $\dfrac{1.4 \times 10^{-2}}{7 \times 10^{-8}}$ 200,000   **98.** $\dfrac{0.4 \times 10^5}{0.2 \times 10^{11}}$ 0.000002

REVIEW AND PREVIEW

Simplify the following. See Section 5.1.

99. $\dfrac{5x^7}{3x^4}$ **100.** $\dfrac{27y^{14}}{3y^7}$ **101.** $\dfrac{15z^4y^3}{21zy}$ **102.** $\dfrac{18a^7b^{17}}{30a^7b}$

Use the distributive property and multiply. See Sections 5.3 and 5.5.

103. $\dfrac{1}{y}(5y^2 - 6y + 5)$ **104.** $\dfrac{2}{x}(3x^5 + x^4 - 2)$

105. Find the volume of the cube. $\dfrac{27}{x^6z^3}$ cu. in.

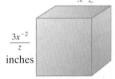

$\dfrac{3x^{-2}}{z}$ inches

106. Find the area of the triangle. $\dfrac{10}{7x^4}$ sq. m

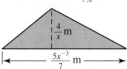

$\dfrac{4}{x}$ m

$\dfrac{5x^{-3}}{7}$ m

Concept Extensions

Simplify each expression. Write each result in standard notation.

107. $(2.63 \times 10^{12})(-1.5 \times 10^{-10})$ -394.5

108. $(6.785 \times 10^{-4})(4.68 \times 10^{10})$ 31,753,800

Light travels at a rate of 1.86×10^5 miles per second. Use this information and the distance formula $d = r \cdot t$ to answer Exercises 109 and 110.

109. If the distance from the moon to the Earth is 238,857 miles, find how long it takes the reflected light of the moon to reach the Earth. (Round to the nearest tenth of a second.) 1.3 sec

110. If the distance from the sun to the Earth is 93,000,000 miles, find how long it takes the light of the sun to reach the Earth. (Round to the nearest tenth of a second.) 500 sec

111. It was stated earlier that for an integer n,
$$x^{-n} = \dfrac{1}{x^n}, \quad x \neq 0$$
Explain why x may not equal 0. answers may vary

112. Determine whether each statement is true or false.

a. $5^{-1} < 5^{-2}$ false   **b.** $\left(\dfrac{1}{5}\right)^{-1} < \left(\dfrac{1}{5}\right)^{-2}$ true

c. $a^{-1} < a^{-2}$ for all nonzero numbers. false

113. 2.7×10^9 gal
113. Although the actual amount varies by season and time of day, the average volume of water that flows over Niagara Falls (the American and Canadian falls combined) each second is 7.5×10^5 gallons. How much water flows over Niagara Falls in an hour? Write the result in scientific notation. (*Hint:* 1 hour equals 3600 seconds) (*Source:* niagarafallslive.com)

114. A beam of light travels 9.460×10^{12} kilometers per year. How far does light travel in 10,000 years? Write the result in scientific notation. 9.46×10^{16} km

115. Explain why $(a^{-1})^3$ has the same value as $(a^3)^{-1}$.

116. If $a = \dfrac{1}{10}$, then find the value of a^{-2}. 100

115. answers may vary
Simplify each expression. Assume that variables represent positive integers.

117. $a^{-4m} \cdot a^{5m}$ a^m **118.** $(x^{-3s})^3$ $\dfrac{1}{x^{9s}}$

119. $\dfrac{y^{4a}}{y^{-a}}$ y^{5a} **120.** $\dfrac{y^{-6a}}{zy^{6a}}$ $\dfrac{1}{y^{12a}z}$

121. $(z^{3a+2})^{-2}$ $\dfrac{1}{z^{6a+4}}$ **122.** $(a^{4x-1})^{-1}$ $\dfrac{1}{a^{4x-1}}$

82. 0.0009056 **86.** 90,700,000,000 **87.** 6,250,000,000,000,000,000 **88.** 0.0000000000000000000000017 **89.** 9,460,000,000,000
90. 6,067,000,000 **99.** $\dfrac{5x^3}{3}$ **100.** $9y^7$ **101.** $\dfrac{5z^3y^2}{7}$ **102.** $\dfrac{3b^{16}}{5}$ **103.** $5y - 6 + \dfrac{5}{y}$ **104.** $6x^4 + 2x^3 - \dfrac{4}{x}$

5.6   *DIVISION OF POLYNOMIALS*

Objectives

1   Divide a polynomial by a monomial.

2   Use long division to divide a polynomial by another polynomial.

1   Now that we know how to add, subtract, and multiply polynomials, we practice dividing polynomials.

To divide a polynomial by a monomial, recall addition of fractions. Fractions that have a common denominator are added by adding the numerators:

$$\frac{a}{c} + \frac{b}{c} = \frac{a + b}{c}$$

If we read this equation from right to left and let a, b, and c be monomials, $c \neq 0$, we have the following:

Dividing a Polynomial By a Monomial

Divide each term of the polynomial by the monomial.

$$\frac{a + b}{c} = \frac{a}{c} + \frac{b}{c}, \quad c \neq 0$$

Throughout this section, we assume that denominators are not 0.

EXAMPLE 1

Divide $6m^2 + 2m$ by $2m$.

Solution   We begin by writing the quotient in fraction form. Then we divide each term of the polynomial $6m^2 + 2m$ by the monomial $2m$.

$$\frac{6m^2 + 2m}{2m} = \frac{6m^2}{2m} + \frac{2m}{2m}$$
$$= 3m + 1 \qquad \text{Simplify.}$$

Check   We know that if $\dfrac{6m^2 + 2m}{2m} = 3m + 1$, then $2m \cdot (3m + 1)$ must equal $6m^2 + 2m$. Thus, to check, we multiply.

TEACHING TIP

Ask students whether the answer to Example 1 would be true for all values of m. Have them verify that it is not true for $m = 0$.

$$2m(3m + 1) = 2m(3m) + 2m(1) = 6m^2 + 2m$$

The quotient $3m + 1$ checks.

EXAMPLE 2

Divide $\dfrac{9x^5 - 12x^2 + 3x}{3x^2}$.

Solution

$$\frac{9x^5 - 12x^2 + 3x}{3x^2} = \frac{9x^5}{3x^2} - \frac{12x^2}{3x^2} + \frac{3x}{3x^2} \qquad \text{Divide each term by } 3x^2.$$
$$= 3x^3 - 4 + \frac{1}{x} \qquad \text{Simplify.}$$

Notice that the quotient is not a polynomial because of the term $\dfrac{1}{x}$. This expression is called a rational expression—we will study rational expressions further in Chapter 7. Although the quotient of two polynomials is not always a polynomial, we may still check by multiplying.

Check

$$3x^2\left(3x^3 - 4 + \frac{1}{x}\right) = 3x^2(3x^3) - 3x^2(4) + 3x^2\left(\frac{1}{x}\right)$$

$$= 9x^5 - 12x^2 + 3x$$

EXAMPLE 3

Divide $\dfrac{8x^2y^2 - 16xy + 2x}{4xy}$.

Solution

$$\frac{8x^2y^2 - 16xy + 2x}{4xy} = \frac{8x^2y^2}{4xy} - \frac{16xy}{4xy} + \frac{2x}{4xy} \qquad \text{Divide each term by } 4xy.$$

$$= 2xy - 4 + \frac{1}{2y} \qquad \text{Simplify.}$$

Check

$$4xy\left(2xy - 4 + \frac{1}{2y}\right) = 4xy(2xy) - 4xy(4) + 4xy\left(\frac{1}{2y}\right)$$

$$= 8x^2y^2 - 16xy + 2x$$

✔ CONCEPT CHECK

In which of the following is $\dfrac{x+5}{5}$ simplified correctly?

a. $\dfrac{x}{5} + 1$ **b.** x **c.** $x + 1$

2 To divide a polynomial by a polynomial other than a monomial, we use a process known as long division. Polynomial long division is similar to number long division, so we review long division by dividing 13 into 3660.

> **Helpful Hint**
> Recall that 3660 is called the dividend.

$$
\begin{array}{r}
281 \\
13\overline{)3660} \\
\end{array}
$$

$\underline{26}\!\downarrow$ $\quad 2 \cdot 13 = 26$

106 $\quad$ Subtract and bring down the next digit in the dividend.

$\underline{104}\!\downarrow$ $\quad 8 \cdot 13 = 104$

20 $\quad$ Subtract and bring down the next digit in the dividend.

$\underline{13}$ $\quad 1 \cdot 13 = 13$

7 $\quad$ Subtract. There are no more digits to bring down, so the remainder is 7.

The quotient is 281 R 7, which can be written as $281\dfrac{7}{13}$ $\begin{array}{l} \leftarrow \text{remainder} \\ \leftarrow \text{divisor} \end{array}$

Recall that division can be checked by multiplication. To check a division problem such as this one, we see that

$$13 \cdot 281 + 7 = 3660$$

Now we demonstrate long division of polynomials.

EXAMPLE 4

Divide $x^2 + 7x + 12$ by $x + 3$ using long division.

Solution

To subtract, change the signs of these terms and add.

$$x + 3 \overline{) x^2 + 7x + 12}$$

How many times does x divide x^2? $\dfrac{x^2}{x} = x$.

$x^2 + 3x$ Multiply: $x(x + 3)$.

$4x + 12$ Subtract and bring down the next term.

Now we repeat this process.

How many times does x divide $4x$? $\dfrac{4x}{x} = 4$.

$$\begin{array}{r} x + 4 \\ x + 3 \overline{) x^2 + 7x + 12} \\ x^2 + 3x \\ \hline 4x + 12 \\ 4x + 12 \\ \hline 0 \end{array}$$

To subtract, change the signs of these terms and add.

Multiply: $4(x + 3)$.

Subtract. The remainder is 0.

The quotient is $x + 4$.

Check   We check by multiplying.

	divisor	·	quotient	+	remainder	=	dividend
	↓		↓		↓		↓
or	$(x + 3)$	·	$(x + 4)$	+	0	=	$x^2 + 7x + 12$

The quotient checks.

EXAMPLE 5

Divide $6x^2 + 10x - 5$ by $3x - 1$ using long division.

Solution

$$\begin{array}{r} 2x + 4 \\ 3x - 1 \overline{) 6x^2 + 10x - 5} \\ 6x^2 - 2x \\ \hline 12x - 5 \\ 12x - 4 \\ \hline -1 \end{array}$$

$\dfrac{6x^2}{3x} = 2x$, so $2x$ is a term of the quotient.

Multiply $2x(3x - 1)$.

Subtract and bring down the next term.

$\dfrac{12x}{3x} = 4,\ 4(3x - 1)$

Subtract. The remainder is -1.

Thus $(6x^2 + 10x - 5)$ divided by $(3x - 1)$ is $(2x + 4)$ with a remainder of -1. This can be written as

$$\frac{6x^2 + 10x - 5}{3x - 1} = 2x + 4 + \frac{-1}{3x - 1} \begin{array}{l} \leftarrow \text{remainder} \\ \leftarrow \text{divisor} \end{array}$$

Check   To check, we multiply $(3x - 1)(2x + 4)$. Then we add the remainder, -1, to this product.

$$(3x - 1)(2x + 4) + (-1) = (6x^2 + 12x - 2x - 4) - 1$$
$$= 6x^2 + 10x - 5$$

The quotient checks.

   In Example 5, the degree of the divisor, $3x - 1$, is 1 and the degree of the remainder, -1, is 0. The division process is continued until the degree of the remainder polynomial is less than the degree of the divisor polynomial.

EXAMPLE 6

Divide $\dfrac{4x^2 + 7 + 8x^3}{2x + 3}$.

Solution

Before we begin the division process, we rewrite $4x^2 + 7 + 8x^3$ as $8x^3 + 4x^2 + 0x + 7$. Notice that we have written the polynomial in descending order and have represented the missing x term by $0x$.

$$
\begin{array}{r}
4x^2 - 4x + 6 \\
2x + 3 \overline{)8x^3 + 4x^2 + 0x + 7} \\
\underline{8x^3 + 12x^2} \\
-8x^2 + 0x \\
\underline{8x^2 + 12x} \\
12x + 7 \\
\underline{12x + 18} \\
-11 \quad \text{Remainder.}
\end{array}
$$

Thus, $\dfrac{4x^2 + 7 + 8x^3}{2x + 3} = 4x^2 - 4x + 6 + \dfrac{-11}{2x + 3}$.

EXAMPLE 7

Divide $\dfrac{2x^4 - x^3 + 3x^2 + x - 1}{x^2 + 1}$.

Solution

Before dividing, rewrite the divisor polynomial $(x^2 + 1)$ as $(x^2 + 0x + 1)$. The $0x$ term represents the missing x^1 term in the divisor.

$$
\begin{array}{r}
2x^2 - x + 1 \\
x^2 + 0x + 1 \overline{)2x^4 - x^3 + 3x^2 + x - 1} \\
\underline{2x^4 + 0x^3 + 2x^2} \\
-x^3 + x^2 + x \\
\underline{x^3 + 0x^2 + x} \\
x^2 + 2x - 1 \\
\underline{x^2 + 0x + 1} \\
2x - 2 \quad \text{Remainder.}
\end{array}
$$

Thus, $\dfrac{2x^4 - x^3 + 3x^2 + x - 1}{x^2 + 1} = 2x^2 - x + 1 + \dfrac{2x - 2}{x^2 + 1}$.

MENTAL MATH

Simplify each expression mentally.

1. $\dfrac{a^6}{a^4}$ a^2 **2.** $\dfrac{y^2}{y}$ y **3.** $\dfrac{a^3}{a}$ a^2 **4.** $\dfrac{p^8}{p^3}$ p^5 **5.** $\dfrac{k^5}{k^2}$ k^3

6. $\dfrac{k^7}{k^5}$ k^2 **7.** $\dfrac{p^8}{p^3}$ p^5 **8.** $\dfrac{k^5}{k^2}$ k^3 **9.** $\dfrac{k^7}{k^5}$ k^2

EXERCISE SET 5.6

STUDY GUIDE/SSM   CD/ VIDEO   PH MATH TUTOR CENTER   MathXL®Tutorials ON CD   MathXL®   MyMathLab®

Perform each division. See Examples 1 through 3.

1. $\dfrac{15p^3 + 18p^2}{3p}$ **2.** $\dfrac{14m^2 - 27m^3}{7m}$ **3.** $\dfrac{-9x^4 + 18x^5}{6x^5}$

4. $\dfrac{6x^5 + 3x^4}{3x^4}$ **5.** $\dfrac{-9x^5 + 3x^4 - 12}{3x^3}$ **6.** $\dfrac{6a^2 - 4a + 12}{2a^2}$

7. $\dfrac{4x^4 - 6x^3 + 7}{-4x^4}$ **8.** $\dfrac{-12a^3 + 36a - 15}{3a}$

9. $\dfrac{25x^5 - 15x^3 + 5}{5x^2}$ **10.** $\dfrac{-4y^2 + 4y + 6}{2y}$

△ **11.** The perimeter of a square is $(12x^3 + 4x - 16)$ feet. Find the length of its side. $(3x^3 + x - 4)$ ft

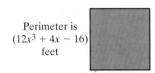

Perimeter is $(12x^3 + 4x - 16)$ feet

△ **12.** The volume of the swimming pool shown is $(36x^5 - 12x^3 + 6x^2)$ cubic feet. If its height is $2x$ feet and its width is $3x$ feet, find its length. $(6x^3 - 2x + 1)$ ft

3x feet
2x feet

Perform each division. See Examples 4 through 7.

13. $\dfrac{x^2 + 4x + 3}{x + 3}$ $x + 1$ **14.** $\dfrac{x^2 + 7x + 10}{x + 5}$ $x + 2$

15. $\dfrac{2x^2 + 13x + 15}{x + 5}$ $2x + 3$ **16.** $\dfrac{3x^2 + 8x + 4}{x + 2}$ $3x + 2$

17. $\dfrac{2x^2 - 7x + 3}{x - 4}$ $2x + 1 + \dfrac{7}{x - 4}$ **18.** $\dfrac{3x^2 - x - 4}{x - 1}$

19. $\dfrac{8x^2 + 6x - 27}{2x - 3}$ $4x + 9$ **20.** $\dfrac{18w^2 + 18w - 8}{3w + 4}$ $6w - 2$

21. $\dfrac{9a^3 - 3a^2 - 3a + 4}{3a + 2}$ **22.** $\dfrac{-x^3 - 6x^2 + 2x - 3}{x - 1}$

23. $\dfrac{2b^3 + 9b^2 + 6b - 4}{b + 4}$ **24.** $\dfrac{2x^3 + 3x^2 - 3x + 4}{x + 2}$

25. Explain how to check a polynomial long division result when the remainder is 0.   answers may vary

26. Explain how to check a polynomial long division result when the remainder is not 0.   answers may vary

△ **27.** The area of the following parallelogram is $(10x^2 + 31x + 15)$ square meters. If its base is $(5x + 3)$ meters, find its height. $(2x + 5)$ m

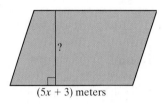

$(5x + 3)$ meters

△ **28.** The area of the top of the Ping-Pong table is $(49x^2 + 70x - 200)$ square inches. If its length is $(7x + 20)$ inches, find its width. $(7x - 10)$ in.

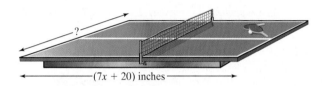

$(7x + 20)$ inches

MIXED PRACTICE

Perform each division.

29. $\dfrac{20x^2 + 5x + 9}{5x^3}$ $\dfrac{4}{x} + \dfrac{1}{x^2} + \dfrac{9}{5x^3}$ **30.** $\dfrac{8x^3 - 4x^2 + 6x + 2}{2x^2}$

31. $\dfrac{5x^2 + 28x - 10}{x + 6}$ **32.** $\dfrac{2x^2 + x - 15}{x + 3}$ $2x - 5$

33. $\dfrac{10x^3 - 24x^2 - 10x}{10x}$ **34.** $\dfrac{2x^3 + 12x^2 + 16}{4x^2}$

35. $\dfrac{6x^2 + 17x - 4}{x + 3}$ $6x - 1 - \dfrac{1}{x + 3}$ **36.** $\dfrac{2x^2 - 9x + 15}{x - 6}$

37. $\dfrac{12x^4 + 3x^2}{3x^2}$ $4x^2 + 1$ **38.** $\dfrac{15x^2 - 9x^5}{9x^5}$ $\dfrac{5}{3x^3} - 1$

39. $\dfrac{2x^3 + 2x^2 - 17x + 8}{x - 2}$ **40.** $\dfrac{4x^3 + 11x^2 - 8x - 10}{x + 3}$

41. $\dfrac{30x^2 - 17x + 2}{5x - 2}$ $6x - 1$ **42.** $\dfrac{4x^2 - 13x - 12}{4x + 3}$ $x - 4$

43. $\dfrac{3x^4 - 9x^3 + 12}{-3x}$ **44.** $\dfrac{8y^6 - 3y^2 - 4y}{4y}$

1. $5p^2 + 6p$ **2.** $2m - \frac{27m^2}{7}$ **3.** $-\frac{3}{2x} + 3$ **4.** $2x + 1$ **5.** $-3x^2 + x - \frac{4}{x}$ **6.** $3 - \frac{2}{a} + \frac{6}{a}$ **7.** $-1 + \frac{3}{2x} - \frac{7}{4x}$ **8.** $-4a^2 + 12 - \frac{5}{a}$

9. $5x^3 - 3x + \frac{1}{x}$ **10.** $-2y + 2 + \frac{3}{y}$ **18.** $3x + 2 - \frac{2}{x - 1}$ **21.** $3a^2 - 3a + 1 + \frac{2}{3a + 2}$ **22.** $-x^2 - 7x - 5 - \frac{8}{x - 1}$

23. $2b^2 + b + 2 - \frac{12}{b + 4}$ **24.** $2x^2 - x - 1 + \frac{6}{x + 2}$ **30.** $4x - 2 + \frac{3}{x} + \frac{1}{x^2}$ **31.** $5x - 2 + \frac{2}{x + 6}$ **33.** $x^2 - \frac{12x}{5} - 1$ **34.** $\frac{x}{2} + 3 + \frac{4}{x^2}$

36. $2x + 3 + \frac{33}{x - 6}$ **39.** $2x^2 + 6x - 5 - \frac{2}{x - 2}$ **40.** $4x^2 - x - 5 + \frac{5}{x + 3}$ **43.** $-x^3 + 3x^2 - \frac{4}{x}$ **44.** $2y^5 - \frac{3y}{4} - 1$

45. $\dfrac{8x^2 + 10x + 1}{2x + 1}$

46. $\dfrac{3x^2 + 17x + 7}{3x + 2}$

47. $\dfrac{4x^2 - 81}{2x - 9}$ $2x + 9$

48. $\dfrac{16x^2 - 36}{4x + 6}$ $4x - 6$

49. $\dfrac{4x^3 + 12x^2 + x - 12}{2x + 3}$

50. $\dfrac{6x^2 + 11x - 10}{3x - 2}$

51. $\dfrac{x^3 - 27}{x - 3}$

52. $\dfrac{x^3 + 64}{x + 4}$

53. $\dfrac{x^3 + 1}{x + 1}$

54. $\dfrac{x^5 + x^2}{x^2 + x}$

55. $\dfrac{1 - 3x^2}{x + 2}$

56. $\dfrac{7 - 5x^2}{x + 3}$

57. $\dfrac{-4b + 4b^2 - 5}{2b - 1}$

58. $\dfrac{-3y + 2y^2 - 15}{2y + 5}$

REVIEW AND PREVIEW

Multiply each expression. See Section 5.3.

59. $2a(a^2 + 1)$ $2a^3 + 2a$

60. $-4a(3a^2 - 4)$ $-12a^3 + 16a$

61. $2x(x^2 + 7x - 5)$

62. $4y(y^2 - 8y - 4)$

63. $-3xy(xy^2 + 7x^2y + 8)$

64. $-9xy(4xyz + 7xy^2z + 2)$

65. $9ab(ab^2c + 4bc - 8)$

66. $-7sr(6s^2r + 9sr^2 + 9rs + 8)$

Use the bar graph below to answer Exercises 67 through 70. See Section 1.9.

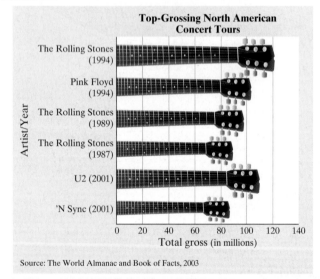

Top-Grossing North American Concert Tours

Source: The World Almanac and Book of Facts, 2003

67. Which artist has grossed the most money on one tour?

68. Estimate the amount of money made by the 1994 concert tour of Pink Floyd. $103 million

69. Estimate the amount of money made by the 2001 concert tour of U2. $110 million

70. Which artist shown has grossed the least amount of money on a tour?   'NSync (2001)

Concept Extensions

Divide.

71. $\dfrac{x^7 + x^4}{x^4 + x^3}$ $x^3 - x^2 + x$

72. $\dfrac{x^6 - x^4}{x^3 + 1}$ $x^3 - x - 1 + \dfrac{x + 1}{x^3 + 1}$

73. $(18x^{10a} - 12x^{8a} + 14x^{5a} - 2x^{3a}) \div 2x^{3a}$

74. $(25y^{11b} + 5y^{6b} - 20y^{3b} + 100y^b) \div 5y^b$

45. $4x + 3 - \dfrac{2}{2x + 1}$ **46.** $x + 5 - \dfrac{3}{3x + 2}$ **49.** $2x^2 + 3x - 4$

50. $2x + 5$ **51.** $x^2 + 3x + 9$ **52.** $x^2 - 4x + 16$

53. $x^2 - x + 1$ **54.** $x^3 - x^2 + x$ **55.** $-3x + 6 - \dfrac{11}{x + 2}$

56. $-5x + 15 - \dfrac{38}{x + 3}$ **57.** $2b - 1 - \dfrac{6}{2b - 1}$ **58.** $y - 4 + \dfrac{5}{2y + 5}$

61. $2x^3 + 14x^2 - 10x$ **62.** $4y^3 - 32y^2 - 16y$

63. $-3x^2y^3 - 21x^3y^2 - 24xy$ **64.** $-36x^2y^2z - 63x^2y^3z - 18xy$

65. $9a^2b^3c + 36ab^2c - 72ab$ **66.** $-42s^3r^2 - 63s^2r^3 - 63s^2r^2 - 56sr$

67. The Rolling Stones (1994)   **73.** $9x^{7a} - 6x^{5a} + 7x^{2a} - 1$

74. $5y^{10b} + y^{5b} - 4y^{2b} + 20$

CHAPTER 5 PROJECT

Modeling with Polynomials

The polynomial model $1.65x^2 + 14.77x + 364.27$ dollars represents consumer spending per person per year on all U.S. media from 1990 to 1996. This includes spending on subscription TV services, recorded music, newspapers, magazines, books, home video, theater movies, video games, and educational software. The polynomial model $0.81x^2 + 3.41x + 88.98$ dollars represents consumer spending per person per year on subscription TV services alone during this same period. In both models, x is the number of years after 1990. (*Source:* Based on data from Veronis,

Suhler & Associates, New York, NY, *Communications Industry Report*, annual)

In this project, you will have the opportunity to investigate these polynomial models numerically, algebraically, and graphically. This project may be completed by working in groups or individually.

1. Use the polynomials to complete the following table showing the annual consumer spending per person over the period 1990–1996 by evaluating each polynomial at the given values of x. Then subtract each value in the fourth column from the corresponding value in the third column. Record the result in the last column, "Difference." What do you think these values represent? What trends do you notice in the data?

Year	x	Consumer Spending per Person per Year on ALL U.S. Media	Consumer spending per Person per Year on Subscription TV	Difference
1990	0			
1992	2			
1994	4			
1996	6			

2. Use the polynomial models to find a new polynomial model representing the amount of consumer spending per person on U.S. media other than subscription TV services (such as recorded music, newspapers, magazines, books, home video, theater movies, video games, and educational software). Then use this new polynomial model to complete the following table.

Year	x	Consumer Spending per Person per Year on Media Other Than Subscription TV
1990	0	
1992	2	
1994	4	
1996	6	

3. Compare the values in the last column of the table in Question 1 to the values in the last column of the table in Question 2. What do you notice? What can you conclude?

4. Use the polynomial models to estimate consumer spending on (a) all U.S. media, (b) subscription TV, and (c) media other than subscription TV for the year 1998.

5. Use the polynomial models to estimate consumer spending on (a) all U.S. media, (b) subscription TV, and (c) media other than subscription TV for the year 2000.

6. Create a bar graph that represents the data for consumer spending on all U.S. media in the years 1990, 1992, 1994, and 1996 along with your estimates for 1998 and 2000. Study your bar graph. Discuss what the graph implies about the future.

5 CHAPTER VOCABULARY CHECK

Fill in each blank with one of the words or phrases listed below.

term   coefficient   monomial   binomial   trinomial
polynomials   degree of a term   degree of a polynomial   FOIL

1. A term is a number or the product of numbers and variables raised to powers.

2. The FOIL method may be used when multiplying two binomials.

3. A polynomial with exactly 3 terms is called a trinomial.

4. The degree of a polynomial is the greatest degree of any term of the polynomial.

5. A polynomial with exactly 2 terms is called a binomial.

6. The coefficient of a term is its numerical factor.

7. The degree of a term is the sum of the exponents on the variables in the term.

8. A polynomial with exactly 1 term is called a monomial.

9. Monomials, binomials, and trinomials are all examples of polynomials.

STUDY SKILLS REMINDER

Are You Preparing for a Test on Chapter 5?

Below is a list of some *common trouble areas* for topics covered in Chapter 5. After studying for your test—but before taking your test—read these.

▶ Do you know that a negative exponent does not make the base a negative number? For example,

$$3^{-2} = \frac{1}{3^2} = \frac{1}{9}$$

▶ Make sure you remember that x has an understood coefficient of 1 and an understood exponent of 1. For example,

$$2x + x = 2x + 1x = 3x; \quad x^5 \cdot x = x^5 \cdot x^1 = x^6$$

▶ Do you know the difference between $5x^2$ and $(5x)^2$?

$$5x^2 \text{ is } 5 \cdot x^2; \quad (5x)^2 = 5^2 \cdot x^2 \text{ or } 25 \cdot x^2$$

▶ Can you evaluate $x^2 - x$ when $x = -2$?

$$x^2 - x = (-2)^2 - (-2) = 4 - (-2) = 4 + 2 = 6$$

▶ Can you subtract $5x^2 + 1$ from $3x^2 - 6$?

$$(3x^2 - 6) - (5x^2 + 1) = 3x^2 - 6 - 5x^2 - 1 = -2x^2 - 7$$

▶ Make sure you are familiar with squaring a binomial and other special products.

$$(3x - 4)^2 = (3x)^2 - 2(3x)(4) + 4^2 = 9x^2 - 24x + 16$$

or

$$(3x - 4)^2 = (3x - 4)(3x - 4) = 9x^2 - 24x + 16$$
$$(2x^2 + 1)(2x^2 - 1) = (2x^2)^2 - 1^2 = 4x^4 - 1$$

Remember: This is simply a checklist of common trouble areas. For a review of Chapter 5, see the Highlights and Chapter Review.

CHAPTER 5 HIGHLIGHTS

Definitions and Concepts	Examples
Section 5.1 Exponents	
a^n means the product of n factors, each of which is a.	$3^2 = 3 \cdot 3 = 9$ $(-5)^3 = (-5)(-5)(-5) = -125$ $\left(\dfrac{1}{2}\right)^4 = \dfrac{1}{2} \cdot \dfrac{1}{2} \cdot \dfrac{1}{2} \cdot \dfrac{1}{2} = \dfrac{1}{16}$

Definitions and Concepts	**Examples**

Section 5.1 Exponents

If m and n are integers and no denominators are 0,

Product Rule: $a^m \cdot a^n = a^{m+n}$

Power Rule: $(a^m)^n = a^{mn}$

Power of a Product Rule: $(ab)^n = a^n b^n$

Power of a Quotient Rule: $\left(\dfrac{a}{b}\right)^n = \dfrac{a^n}{b^n}$

Quotient Rule: $\dfrac{a^m}{a^n} = a^{m-n}$

Zero Exponent: $a^0 = 1, a \neq 0$.

$x^2 \cdot x^7 = x^{2+7} = x^9$

$(5^3)^8 = 5^{3 \cdot 8} = 5^{24}$

$(7y)^4 = 7^4 y^4$

$\left(\dfrac{x}{8}\right)^3 = \dfrac{x^3}{8^3}$

$\dfrac{x^9}{x^4} = x^{9-4} = x^5$

$5^0 = 1, x^0 = 1, x \neq 0$

Section 5.2 Adding and Subtracting Polynomials

A **term** is a number or the product of numbers and variables raised to powers.

The **numerical coefficient** or **coefficient** of a term is its numerical factor.

A **polynomial** is a term or a finite sum of terms in which variables may appear in the numerator raised to whole number powers only.

A **monomial** is a polynomial with exactly 1 term.
A **binomial** is a polynomial with exactly 2 terms.
A **trinomial** is a polynomial with exactly 3 terms.

The **degree of a term** is the sum of the exponents on the variables in the term.

The **degree of a polynomial** is the greatest degree of any term of the polynomial.

To add polynomials, add or combine like terms.

Terms

$-5x, 7a^2b, \dfrac{1}{4}y^4, 0.2$

Term	*Coefficient*
$7x^2$	7
y	1
$-a^2b$	-1

Polynomials

$3x^2 - 2x + 1$	(Trinomial)
$-0.2a^2b - 5b^2$	(Binomial)
$\dfrac{5}{6}y^3$	(Monomial)

Term	*Degree*
$-5x^3$	3
3 (or $3x^0$)	0
$2a^2b^2c$	5

Polynomial	*Degree*
$5x^2 - 3x + 2$	2
$7y + 8y^2z^3 - 12$	$2 + 3 = 5$

Add:

$(7x^2 - 3x + 2) + (-5x - 6) = 7x^2 - 3x + 2 - 5x - 6$

$$= 7x^2 - 8x - 4$$

Definitions and Concepts	**Examples**

Section 5.2 Adding and Subtracting Polynomials

To subtract two polynomials, change the signs of the terms of the second polynomial, then add.	Subtract: $$(17y^2 - 2y + 1) - (-3y^3 + 5y - 6)$$ $$= (17y^2 - 2y + 1) + (3y^3 - 5y + 6)$$ $$= 17y^2 - 2y + 1 + 3y^3 - 5y + 6$$ $$= 3y^3 + 17y^2 - 7y + 7$$

Section 5.3 Multiplying Polynomials

To multiply two polynomials, multiply each term of one polynomial by each term of the other polynomial, and then combine like terms.	Multiply: $$(2x + 1)(5x^2 - 6x + 2)$$ $$= 2x(5x^2 - 6x + 2) + 1(5x^2 - 6x + 2)$$ $$= 10x^3 - 12x^2 + 4x + 5x^2 - 6x + 2$$ $$= 10x^3 - 7x^2 - 2x + 2$$

Section 5.4 Special Products

The **FOIL method** may be used when multiplying two binomials.	Multiply: $(5x - 3)(2x + 3)$ $$\begin{array}{cccc} & F & O & I & L \\ (5x - 3)(2x + 3) = (5x)(2x) + (5x)(3) + (-3)(2x) + (-3)(3) \end{array}$$ $$= 10x^2 + 15x - 6x - 9$$ $$= 10x^2 + 9x - 9$$
Squaring a Binomial $$(a + b)^2 = a^2 + 2ab + b^2$$	Square each binomial. $$(x + 5)^2 = x^2 + 2(x)(5) + 5^2$$ $$= x^2 + 10x + 25$$
$$(a - b)^2 = a^2 - 2ab + b^2$$	$$(3x - 2y)^2 = (3x)^2 - 2(3x)(2y) + (2y)^2$$ $$= 9x^2 - 12xy + 4y^2$$
Multiplying the Sum and Difference of Two Terms $$(a + b)(a - b) = a^2 - b^2$$	Multiply. $$(6y + 5)(6y - 5) = (6y)^2 - 5^2$$ $$= 36y^2 - 25$$

Section 5.5 Negative Exponents and Scientific Notation

If $a \neq 0$ and n is an integer, $$a^{-n} = \frac{1}{a^n}$$	$$3^{-2} = \frac{1}{3^2} = \frac{1}{9}; 5x^{-2} = \frac{5}{x^2}$$

Definitions and Concepts	**Examples**

Section 5.5 Negative Exponents and Scientific Notation

Rules for exponents are true for positive and negative integers.

Simplify:
$$\left(\frac{x^{-2}y}{x^5}\right)^{-2} = \frac{x^4 y^{-2}}{x^{-10}}$$
$$= x^{4-(-10)}y^{-2}$$
$$= \frac{x^{14}}{y^2}$$

A positive number is written in scientific notation if it is as the product of a number a, $1 \le a < 10$, and an integer power r of 10.

$$a \times 10^r$$

Write each number in scientific notation.

$$12{,}000 = 1.2 \times 10^4$$
$$0.00000568 = 5.68 \times 10^{-6}$$

Section 5.6 Division of Polynomials

To divide a polynomial by a monomial:

$$\frac{a+b}{c} = \frac{a}{c} + \frac{b}{c}$$

To divide a polynomial by a polynomial other than a monomial, use long division.

Divide:

$$\frac{15x^5 - 10x^3 + 5x^2 - 2x}{5x^2} = \frac{15x^5}{5x^2} - \frac{10x^3}{5x^2} + \frac{5x^2}{5x^2} - \frac{2x}{5x^2}$$

$$= 3x^3 - 2x + 1 - \frac{2}{5x}$$

$$5x - 1 + \frac{-4}{2x+3}$$
$$2x+3 \overline{)10x^2 + 13x - 7}$$
$$\underline{10x^2 + 15x}$$
$$-2x - 7$$
$$\underline{-2x - 3}$$
$$-4$$

5 CHAPTER REVIEW

(5.1) *State the base and the exponent for each expression.*

1. 3^2 base: 3; exponent: 2

2. $(-5)^4$

3. -5^4 base 5; exponent: 4

Evaluate each expression.   **2.** base: -5; exponent: 4

4. 8^3 512

5. $(-6)^2$ 36

6. -6^2 -36

7. $-4^3 - 4^0$ -65

8. $(3b)^0$ 1

9. $\frac{8b}{8b}$ 1

Simplify each expression.

10. $5b^3 b^5 a^6$ $5a^6 b^8$

11. $2^3 \cdot x^0$ 8

12. $[(-3)^2]^3$ 729

13. $(2x^3)(-5x^2)$ $-10x^5$

14. $\left(\frac{mn}{q}\right)^2 \cdot \left(\frac{mn}{q}\right)$ $\frac{m^3 n^3}{q^3}$

15. $\left(\frac{3ab^2}{6ab}\right)^4$ $\frac{b^4}{16}$

16. $\frac{x^9}{x^4}$ x^5

17. $\frac{2x^7 y^8}{8xy^2}$ $\frac{x^6 y^6}{4}$

18. $\frac{3x^4 y^{10}}{12xy^6}$ $\frac{x^3 y^4}{4}$

19. $5a^7(2a^4)^3$ $40a^{19}$

20. $(2x)^2(9x)$ $36x^3$

21. $\frac{(-4)^2(3^3)}{(4^5)(3^2)}$ $\frac{3}{64}$

47. $2s^5 + 3s^4 + 4s^3 + s^2 - 7s - 6$ **48.** $-6m^7 - 3x^4 + 7m^6 - 4m^2$
64. $-2x^3 + 18x^2 - 2x$ **65.** $-3a^3b - 3a^2b - 3ab^2$ **66.** $-6a^4 + 8a^2 - 2a$

22. $\dfrac{(-7)^2(3^5)}{(-7)^3(3^4)}$ $-\dfrac{3}{7}$

23. $\dfrac{(2x)^0(-4)^2}{16x}$ $\dfrac{1}{x}$

24. $\dfrac{(8xy)(3xy)}{18x^2y^2}$ $\dfrac{4}{3}$

25. $m^0 + p^0 + 3q^0$ 5

26. $(-5a)^0 + 7^0 + 8^0$ 3

27. $(3xy^2 + 8x + 9)^0$ 1

28. $8x^0 + 9^0$ 9

29. $6(a^2b^3)^3$ $6a^6b^9$

30. $\dfrac{(x^3z)^a}{x^2z^2}$ $x^{3a-2}z^{a-2}$

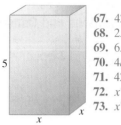

5

x x x

67. $42b^4 - 28b^2 + 14b$
68. $2x^2 - 9x - 35$
69. $6x^2 - 11x - 10$
70. $4a^2 + 27a - 7$
71. $42a^2 + 11a - 3$
72. $x^4 + 7x^3 + 4x^2 + 23x - 3$
73. $x^6 + 2x^5 + x^2 + 3x + 2$

(5.2) *Find the degree of each term.*

31. $-5x^4y^3$ 7

32. $10x^3y^2z$ 6

33. $35a^5bc^2$ 8

34. $95xyz$ 3

Find the degree of each polynomial.

35. $y^5 + 7x - 8x^4$ 5

36. $9y^2 + 30y + 25$ 2

37. $-14x^2y - 28x^2y^3 - 42x^2y^2$ 5

38. $6x^2y^2z^2 + 5x^2y^3 - 12xyz$ 6

39. a. Complete the table for the polynomial
$3a^2b - 2a^2 + ab - b^2 - 6$.

Term	Numerical Coefficient	Degree of Term
$3a^2b$	3	3
$-2a^2$	-2	2
ab	1	2
$-b^2$	-1	2
-6	-6	0

b. What is the degree of the polynomial?   3

40. a. Complete the table for the polynomial
$x^2y^2 + 5x^2 - 7y^2 + 11xy - 1$.

Term	Numerical Coefficient	Degree of Term
x^2y^2	1	4
$5x^2$	5	2
$-7y^2$	-7	2
$11xy$	11	2
-1	-1	0

b. What is the degree of the polynomial?   4

△ **41.** The surface area of a box with a square base and a height of 5 units is given by the polynomial $2x^2 + 20x$. Fill in the table below by evaluating $2x^2 + 20x$ for the given values of x.

x	1	3	5.1	10
$2x^2 + 20x$	22	78	154.02	400

Combine like terms.

42. $6a^2b^2 + 4ab + 9a^2b^2$ $15a^2b^2 + 4ab$

43. $21x^2y^3 + 3xy + x^2y^3 + 6$ $22x^2y^3 + 3xy + 6$

44. $4a^2b - 3b^2 - 8q^2 - 10a^2b + 7q^2$ $-6a^2b - 3b^2 - q^2$

45. $2s^{14} + 3s^{13} + 12s^{12} - s^{10}$ cannot be combined

Add or subtract as indicated.

46. $(3k^2 + 2k + 6) + (5k^2 + k)$ $8k^2 + 3k + 6$

47. $(2s^5 + 3s^4 + 4s^3 + 5s^2) - (4s^2 + 7s + 6)$

48. $(2m^7 + 3x^4 + 7m^6) - (8m^7 + 4m^2 + 6x^4)$

49. $(11r^2 + 16rs - 2s^2) - (3r^2 + 5rs - 9s^2)$ $8r^2 + 11rs + 7s^2$

50. $(3x^2 - 6xy + y^2) - (11x^2 - xy + 5y^2)$ $-8x^2 - 5xy - 4y^2$

51. Subtract $(3x - y)$ from $(7x - 14y)$. $4x - 13y$

52. Subtract $(4x^2 + 8x - 7)$ from the sum of $(x^2 + 7x + 9)$ and $(x^2 + 4)$. $-2x^2 - x + 20$

53. With the popularity of the Internet growing rapidly, the number of new trademarks registered each year has also been increasing. The number of trademark registrations (in thousands) from 1990 to 2000 can be given by the polynomial $72.5x^2 - 17.5x + 120$, where x is the number of years since 1990. Use this model to predict the number of trademark registrations in 2010. (*Source*: International Trademark Association) 28,770,000 trademark registrations

(5.3) *Multiply each expression.*

54. $9x(x^2y)$ $9x^3y$

55. $-7(8xz^2)$ $-56xz^2$

56. $(6xa^2)(xya^3)$ $6x^2a^5y$

57. $(4xy)(-3xa^2y^3)$ $-12x^2a^2y^4$

58. $6(x + 5)$ $6x + 30$

59. $9(x - 7)$ $9x - 63$

60. $4(2a + 7)$ $8a + 28$

61. $9(6a - 3)$ $54a - 27$

62. $-7x(x^2 + 5)$ $-7x^3 - 35x$

63. $-8y(4y^2 - 6)$ $-32y^3 + 48y$

64. $-2(x^3 - 9x^2 + x)$

65. $-3a(a^2b + ab + b^2)$

66. $(3a^3 - 4a + 1)(-2a)$

67. $(6b^3 - 4b + 2)(7b)$

68. $(2x + 5)(x - 7)$

69. $(2x - 5)(3x + 2)$

70. $(4a - 1)(a + 7)$

71. $(6a - 1)(7a + 3)$

72. $(x + 7)(x^3 + 4x - 5)$

73. $(x + 2)(x^5 + x + 1)$

74. $(x^2 + 2x + 4)(x^2 + 2x - 4)$ $x^4 + 4x^3 + 4x^2 - 16$

75. $(x^3 + 4x + 4)(x^3 + 4x - 4)$ $x^6 + 8x^4 + 16x^2 - 16$

76. $(x + 7)^3$

76. $x^3 + 21x^2 + 147x + 343$

77. $(2x - 5)^3$

77. $8x^3 - 60x^2 + 150x - 125$

(5.4) Multiply.

78. $2x(3x^2 - 7x + 1)$

79. $3y(5y^2 - y + 2)$

80. $(6x^5 - 1)(4x^2 + 3)$

81. $(4a^3 - 1)(3a^2 + 7)$

82. $(x^2 + 7y)^2$

83. $(x^3 - 5y)^2$ $x^6 - 10x^3y + 25y^2$

84. $(3x - 7)^2$ $9x^2 - 42x + 49$

85. $(4x + 2)^2$ $16x^2 + 16x + 4$

86. $(y + 1)(y^2 - 6y - 5)$

87. $(x - 2)(x^2 - x - 2)$

88. $(5x - 9)^2$ $25x^2 - 90x + 81$

89. $(5x + 1)(5x - 1)$ $25x^2 - 1$

90. $(7x + 4)(7x - 4)$ $49x^2 - 16$

91. $(a + 2b)(a - 2b)$ $a^2 - 4b^2$

92. $(2x - 6)(2x + 6)$ $4x^2 - 36$

93. $(4a^2 - 2b)(4a^2 + 2b)$

93. $16a^4 - 4b^2$

(5.5) Simplify each expression.

94. 7^{-2} $\frac{1}{49}$

95. -7^{-2} $-\frac{1}{49}$

96. $2x^{-4}$ $\frac{2}{x^4}$

97. $(2x)^{-4}$ $\frac{1}{16x^4}$

98. $\left(\frac{1}{5}\right)^{-3}$ 125

99. $\left(\frac{-2}{3}\right)^{-2}$ $\frac{9}{4}$

100. $2^0 + 2^{-4}$ $\frac{17}{16}$

101. $6^{-1} - 7^{-1}$ $\frac{1}{42}$

Simplify each expression. Assume that variables in an exponent represent positive integers only. Write each answer using positive exponents.

102. $\frac{1}{(2q)^{-3}}$ $8q^3$

103. $\frac{-1}{(qr)^{-3}}$ $-q^3r^3$

104. $\frac{r^{-3}}{s^{-4}}$ $\frac{s^4}{r^3}$

105. $\frac{rs^{-3}}{r^{-4}}$ $\frac{r^5}{s^3}$

106. $(-x^{-3}y^5)(5xy^{-2})$ $-\frac{5y^3}{x^2}$

107. $(-3x^5y^{-2})(-4x^{-5}y)$ $\frac{12}{y}$

108. $(2x^{-5})^{-3}$ $\frac{x^{15}}{8}$

109. $(3y^{-6})^{-1}$ $\frac{y^6}{3}$

110. $(3a^{-1}b^{-1}c^{-2})^{-2}$ $\frac{a^2b^2c^4}{9}$

111. $(4x^{-2}y^{-3}z)^{-3}$ $\frac{x^6y^9}{64z^3}$

112. $\frac{5^{-2}x^8}{5^{-3}x^{11}}$ $\frac{5}{x^3}$

113. $\frac{7^5y^{-2}}{7^7y^{-10}}$ $\frac{y^8}{49}$

114. $\left(\frac{bc^{-2}}{bc^{-3}}\right)^4$ c^4

115. $\left(\frac{x^{-3}y^{-4}}{x^{-2}y^{-5}}\right)^{-3}$ $\frac{x^3}{y^3}$

116. $\frac{x^{-4}y^{-6}z^3}{x^2y^7z^3}$ $\frac{1}{x^6y^{13}}$

117. $\frac{a^5b^{-5}c^4}{a^{-5}b^5c^4}$ $\frac{a^{10}}{b^{10}}$

118. $-2^0 + 2^{-4}$ $-\frac{15}{16}$

119. $-3^{-2} - 3^{-3}$ $-\frac{4}{27}$

120. $a^{6m}a^{5m}$ a^{11m}

121. $\frac{(x^{5+h})^3}{x^5}$ x^{10+3h}

122. $(3xy^{2z})^3$ $27x^3y^{6z}$

123. $a^{m+2}a^{m+3}$ a^{2m+5}

Write each number in scientific notation.

124. 0.00027 2.7×10^{-4}

125. 0.8868 8.868×10^{-1}

126. $80,800,000$ 8.08×10^7

127. $868,000$ 8.68×10^5

128. In a recent year, the United States imported approximately 109,379,000 kilograms of coffee. Write this number in scientific notation. (*Source:* International Coffee Organization) 1.09379×10^8

129. The approximate diameter of the Milky Way galaxy is 150,000 light years. Write this number in scientific notation. (*Source:* NASA IMAGE/POETRY Education and Public Outreach Program) 1.5×10^5

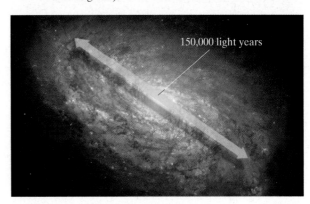

150,000 light years

Write each number in standard form.

130. 8.67×10^5 $867,000$

131. 3.86×10^{-3} 0.00386

132. 8.6×10^{-4} 0.00086

133. 8.936×10^5 $893,600$

134. The volume of the planet Jupiter is 1.43128×10^{15} cubic kilometers. Write this number in standard notation. (*Source:* National Space Science Data Center) $1,431,280,000,000,000$

78. $6x^3 - 14x^2 + 2x$ **79.** $15y^3 - 3y^2 + 6y$ **80.** $24x^7 + 18x^5 - 4x^2 - 3$ **81.** $12a^5 + 28a^3 - 3a^2 - 7$ **82.** $x^4 + 14x^2y + 49y^2$
86. $y^3 - 5y^2 - 11y - 5$ **87.** $x^3 - 3x^2 + 4$

135. An angstrom is a unit of measure, equal to 1×10^{-10} meter, used for measuring wavelengths or the diameters of atoms. Write this number in standard notation. (*Source:* National Institute of Standards and Technology) 0.0000000001 m

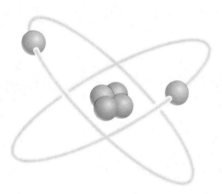

Simplify. Express each result in standard form.

136. $(8 \times 10^4)(2 \times 10^{-7})$ 0.016 **137.** $\dfrac{8 \times 10^4}{2 \times 10^{-7}}$

137. 400,000,000,000   **139.** $-a^2 + 3b - 4$

(5.6) *Perform each division.*

138. $\dfrac{x^2 + 21x + 49}{7x^2}$ $\frac{1}{7} + \frac{3}{x} + \frac{7}{x^2}$ **139.** $\dfrac{5a^3b - 15ab^2 + 20ab}{-5ab}$

140. $(a^2 - a + 4) \div (a - 2)$ $a + 1 + \dfrac{6}{a - 2}$

141. $(4x^2 + 20x + 7) \div (x + 5)$ $4x + \dfrac{7}{x + 5}$

142. $\dfrac{a^3 + a^2 + 2a + 6}{a - 2}$ **143.** $\dfrac{9b^3 - 18b^2 + 8b - 1}{3b - 2}$

144. $\dfrac{4x^4 - 4x^3 + x^2 + 4x - 3}{2x - 1}$ **145.** $\dfrac{-10x^2 - x^3 - 21x + 18}{x - 6}$

142. $a^2 + 3a + 8 + \dfrac{22}{a - 2}$ **143.** $3b^2 - 4b - \dfrac{1}{3b - 2}$

144. $2x^3 - x^2 + 2 - \dfrac{1}{2x - 1}$ **145.** $-x^2 - 16x - 117 - \dfrac{684}{x - 6}$

 CHAPTER 5 TEST

Remember to use your Chapter Test Prep Video CD to help you study and view solutions to the test questions you need help with.

Evaluate each expression.

1. 2^5 32 **2.** $(-3)^4$ 81 **3.** -3^4 -81 **4.** 4^{-3} $\frac{1}{64}$

Simplify each exponential expression. Write the result using only positive exponents.

5. $(3x^2)(-5x^9)$ $-15x^{11}$ **6.** $\dfrac{y^7}{y^2}$ y^5

7. $\dfrac{r^{-8}}{r^{-3}}$ $\dfrac{1}{r^5}$ **8.** $\left(\dfrac{x^2y^3}{x^3y^{-4}}\right)^2$ $\dfrac{y^{14}}{x^2}$ **9.** $\dfrac{6^2x^{-4}y^{-1}}{6^3x^{-3}y^7}$ $\dfrac{1}{6xy^8}$

Express each number in scientific notation.

10. 563,000 5.63×10^5 **11.** 0.0000863 8.63×10^{-5}

Write each number in standard form.

12. 1.5×10^{-3} 0.0015 **13.** 6.23×10^4 62,300

14. Simplify. Write the answer in standard form.
$(1.2 \times 10^5)(3 \times 10^{-7})$ 0.036

15. a. Complete the table for the polynomial $4xy^2 + 7xyz + x^3y - 2$.

Term	Numerical Coefficient	Degree of Term
$4xy^2$	4	3
$7xyz$	7	3
x^3y	1	4
-2	-2	0

b. What is the degree of the polynomial? 4

16. Simplify by combining like terms.
$5x^2 + 4xy - 7x^2 + 11 + 8xy$ $-2x^2 + 12xy + 11$

Perform each indicated operation. **17.** $16x^3 + 7x^2 - 3x - 13$

17. $(8x^3 + 7x^2 + 4x - 7) + (8x^3 - 7x - 6)$

18. $\begin{aligned} 5x^3 + \ \ x^2 + 5x - 2 \\ -(8x^3 - 4x^2 + \ x - 7) \end{aligned}$ $-3x^3 + 5x^2 + 4x + 5$

19. Subtract $(4x + 2)$ from the sum of $(8x^2 + 7x + 5)$ and $(x^3 - 8)$. $x^3 + 8x^2 + 3x - 5$

Multiply.

20. $(3x + 7)(x^2 + 5x + 2)$ $3x^3 + 22x^2 + 41x + 14$

21. $3x^2(2x^2 - 3x + 7)$ $6x^4 - 9x^3 + 21x^2$

22. $(x + 7)(3x - 5)$ $3x^2 + 16x - 35$

23. $(4x - 2)^2$ $16x^2 - 16x + 4$

24. $(x^2 - 9b)(x^2 + 9b)$ $x^4 - 81b^2$

25. The height of the Bank of China in Hong Kong is 1001 feet. Neglecting air resistance, the height of an object dropped from this building at time t seconds is given by the polynomial $-16t^2 + 1001$. Find the height of the object at the given times below.

t	0 seconds	1 second	3 seconds	5 seconds
$-16t^2 + 1001$	1001 ft	985 ft	857 ft	601 ft

Divide. **26.** $\dfrac{x}{2y} + 3 - \dfrac{7}{8y}$ **27.** $x + 2$

26. $\dfrac{4x^2 + 24xy - 7x}{8xy}$ **27.** $(x^2 + 7x + 10) \div (x + 5)$

28. $\dfrac{27x^3 - 8}{3x + 2}$ $9x^2 - 6x + 4 - \dfrac{16}{3x + 2}$

5 CHAPTER CUMULATIVE REVIEW

28. 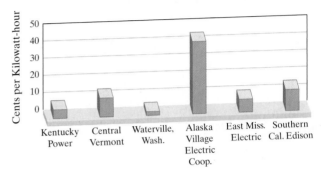 $\xrightarrow{\qquad 5 \qquad}$ $(-\infty, 5)$ (Sec. 2.8)

1. Tell whether each statement is true or false.

   a. $8 \geq 8$ true   **b.** $8 \leq 8$ true

   c. $23 \leq 0$ false   **d.** $23 \geq 0$ true (Sec. 1.2, Ex. 2)

2. Find the absolute value of each number.

   a. $|-7.2|$ 7.2   **b.** $|0|$ 0   **c.** $\left|-\dfrac{1}{2}\right|$ $\dfrac{1}{2}$ (Sec. 1.2)

3. Divide. Write all quotients in lowest terms.

   a. $\dfrac{4}{5} \div \dfrac{5}{16}$ $\dfrac{64}{25}$ **b.** $\dfrac{7}{10} \div 14$ $\dfrac{1}{20}$

   c. $\dfrac{3}{8} \div \dfrac{3}{10}$ $\dfrac{5}{4}$ (Sec. 1.3, Ex. 4)

4. Multiply. Write products in lowest terms.

   a. $\dfrac{3}{4} \cdot \dfrac{7}{21}$ $\dfrac{1}{4}$ **b.** $\dfrac{1}{2} \cdot 4\dfrac{5}{6}$ $2\dfrac{5}{12}$ (Sec. 1.3)

5. Evaluate the following:

   a. 3^2 9   **b.** 5^3 125

   c. 2^4 16   **d.** 7^1 7

   e. $\left(\dfrac{3}{7}\right)^2$ $\dfrac{9}{49}$ (Sec. 1.4, Ex. 1)

6. Evaluate $\dfrac{2x - 7y}{x^2}$ for $x = 5$ and $y = 1$. $\dfrac{3}{25}$ (Sec. 1.4)

7. Add.

   a. $-3 + (-7)$ -10 **b.** $-1 + (-20)$ -21

   c. $-2 + (-10)$ -12 (Sec. 1.5, Ex. 3)

8. Simplify: $8 + 3(2 \cdot 6 - 1)$ 41 (Sec 1.4)

9. Subtract 8 from -4. -12 (Sec. 1.6, Ex. 3)

10. Is $x = 1$ a solution of $5x^2 + 2 = x - 8$.   no (Sec. 1.4)

11. Find the reciprocal of each number.

    a. 22 $\frac{1}{22}$ **b.** $\dfrac{3}{16}$ $\frac{16}{3}$

    c. -10 $-\frac{1}{10}$ **d.** $-\dfrac{9}{13}$ $-\frac{13}{9}$ (Sec. 1.7, Ex. 5)

12. Subtract:

    a. $7 - 40$ -33 **b.** $-5 - (-10)$ 5 (Sec. 1.6)

13. Use an associative property to complete each statement.

    a. $5 + (4 + 6) =$ _____ $(5 + 4) + 6$

    b. $(-1 \cdot 2) \cdot 5 =$ _____ $-1 \cdot (2 \cdot 5)$ (Sec. 1.8, Ex. 2)

14. Simplify: $\dfrac{4(-3) + (-8)}{5 + (-5)}$ undefined (Sec. 1.7)

15. The bar graph in the next column shows the cents charged per kilowatt-hour for selected electricity companies.

    a. Which company charges the highest rate?

    b. Which company charges the lowest rate?

    c. Approximate the electricity rate charged by the first four companies listed.   15. **a.** Alaska Village Electric

d. Approximate the difference in the rates charged by the companies in parts (a) and (b).   37¢ (Sec. 1.9, Ex. 1)

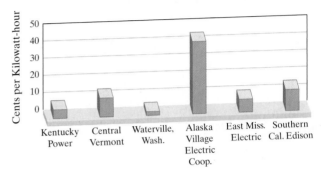

Source: *USA Today* Research

16. $-2x - 6y + 2z$ (Sec. 1.8)

16. Use the distributive property to write $-2(x + 3y - z)$ without parentheses.

17. Find each product by using the distributive property to remove parentheses.

    a. $5(x + 2)$ $5x + 10$

    b. $-2(y + 0.3z - 1)$ $-2y - 0.6z + 2$

    c. $-(x + y - 2z + 6)$ $-x - y + 2z - 6$ (Sec. 2.1, Ex. 5)

18. Simplify: $2(6x - 1) - (x - 7)$ $11x + 5$ (Sec. 2.1)

19. Solve $x - 7 = 10$ for x.   17 (Sec. 2.2, Ex. 1)

20. Write the phrase as an algebraic expression: double a number, subtracted from the sum of a number and seven

21. Solve: $\dfrac{5}{2}x = 15$ 6 (Sec. 2.3, Ex. 1)

22. Solve: $2x + \dfrac{1}{8} = x - \dfrac{3}{8}$ $-\dfrac{1}{2}$ (Sec. 2.2)

23. Twice a number, added to seven, is the same as three subtracted from the number. Find the number. -10 (Sec. 2.4, Ex. 8)

24. Solve: $10 = 5j - 2$ $\dfrac{12}{5}$ (Sec. 2.3)

25. Twice the sum of a number and 4 is the same as four times the number decreased by 12. Find the number.   10 (Sec. 2.5, Ex. 1)

26. Solve: $\dfrac{7x + 5}{3} = x + 3$ 1 (Sec. 2.4)

27. The length of a rectangular road sign is 2 feet less than three times its width. Find the dimensions if the perimeter is 28 feet.   width: 4 ft; length: 10 ft (Sec. 2.6, Ex. 4)

28. Graph $x < 5$ and write in interval notation.

29. Solve $F = \dfrac{9}{5}C + 32$ for C. $\dfrac{5F - 160}{9} = C$ (Sec. 2.6, Ex. 8)

30. Find the slope of each line.   20. $(x + 7) - 2x$ (Sec. 2.1)

    a. $x = -1$ undefined slope   **b.** $y = 7$ $m = 0$ (Sec. 3.4)

15. **b.** American Electric Power. **c.** American Electric Power: 5¢ per kilowatt-hour; Green Mountain Power: 11¢ per kilowatt-hour; Montana Power Co.: 6¢ per kilowatt-hour; Alaska Village Electric: 42¢ per kilowatt-hour

31. Graph $2 < x \le 4$. 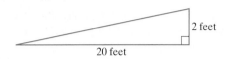 (Sec. 2.8, Ex. 9)

32. Recall that the grade of a road is its slope written as a percent. Find the grade of the road shown.

2 feet

20 feet

33. Complete the following ordered-pair solutions for the equation $3x + y = 12$.

    a. $(0,\ )$ $(0, 12)$ **b.** $(\ , 6)$ $(2, 6)$

    c. $(-1,\ )$ $(-1, 15)$ (Sec. 3.1, Ex. 4)

34. Solve the system: $\begin{cases} 3x + 2y = -8 \\ 2x - 6y = -9 \end{cases}$ $\left(-3, \dfrac{1}{2}\right)$ (Sec. 4.2)

35. Graph the linear equation $2x + y = 5$.

36. Solve the system: $\begin{cases} x = -3y + 3 \\ 2x + 9y = 5 \end{cases}$ $\left(4, -\dfrac{1}{3}\right)$ (Sec 4.3)

37. Graph $x = 2$.   See graphing answer section. (Sec. 3.3, Ex. 5)

38. Evaluate.

    a. $(-5)^2$ 25   **b.** -5^2 -25 **c.** $2 \cdot 5^2$ 50 (Sec. 5.1)

39. Find the slope of the line $x = 5$.

40. Simplify: $\dfrac{(z^2)^3 \cdot z^7}{z^9}$ z^4 (Sec. 5.1)

41. Graph $x + y < 7$.

42. Subtract $(5y^2 - 6) - (y^2 + 2)$. $4y^2 - 8$ (Sec. 5.2)

43. Use the product rule to simplify $(2x^2)(-3x^5)$.

44. Find the value of $-x^2$ when

    a. $x = 2$ -4 **b.** $x = -2$ -4 (Sec. 5.2)

45. Add $(-2x^2 + 5x - 1)$ and $(-2x^2 + x + 3)$.

46. Multiply $(10x^2 - 3)(10x^2 + 3)$. $100x^4 - 9$ (Sec. 5.4)

47. Multiply $(2x - y)^2$. $4x^2 - 4xy + y^2$ (Sec. 5.3, Ex. 6)

48. Multiply $(10x^2 + 3)^2$. $100x^4 + 60x^2 + 9$ (Sec. 5.4)

49. Divide $6m^2 + 2m$ by $2m$. $3m + 1$ (Sec. 5.6, Ex. 1)

50. Evaluate.

    a. 5^{-1} $\dfrac{1}{5}$ **b.** 7^{-2} $\dfrac{1}{49}$ (Sec. 5.5)

32. 10% (Sec. 3.4)

35. See graphing answer section. (Sec. 3.2, Ex. 2)

39. undefined slope (Sec. 3.4, Ex. 4)   **41.** See graphing answer section. (Sec. 4.5, Ex. 1)   **43.** $-6x^7$ (Sec. 5.1, Ex. 4)

45. $-4x^2 + 6x + 2$ (Sec. 5.2, Ex. 9)

Factoring Polynomials

In Chapter 5, you learned how to multiply polynomials. This chapter deals with an operation that is the reverse process of multiplying, called *factoring*. Factoring is an important algebraic skill because this process allows us to write a sum as a product.

At the end of this chapter, we use factoring to help us solve equations other than linear equations, and in Chapter 7 we use factoring to simplify and perform arithmetic operations on rational expressions.

Results from a national survey conducted by the Nellie Mae Corporation show that graduating college students in 2002 had an average credit card debt of $3000. The National Foundation for Credit Counseling (NFCC) offers college students the following advice on the responsible use of credit cards:

- Compare credit cards and shop around for the best interest rates.
- Be careful of hidden credit card fees.
- Avoid stress by shopping wisely and buying only what you can afford to repay immediately.
- Closely guard your credit card to protect yourself from identity theft.

For further help, contact a certified credit counselor through the NFCC. These counselors offer their services at little or no cost to the public.

In the Spotlight on Decision Making feature on page 381, you will have the opportunity to make a decision about the best choice of three credit card offers.

Link: National Foundation for Credit Counseling: www.nfcc.org
Source of text: (same)

6.1 THE GREATEST COMMON FACTOR AND FACTORING BY GROUPING

Objectives

1. Find the greatest common factor of a list of integers.
2. Find the greatest common factor of a list of terms.
3. Factor out the greatest common factor from a polynomial.
4. Factor a polynomial by grouping.

When an integer is written as the product of two or more other integers, each of these integers is called a **factor** of the product. This is true for polynomials, also. When a polynomial is written as the product of two or more other polynomials, each of these polynomials is called a factor of the product.

The process of writing a polynomial as a product is called **factoring** the polynomial.

$$
\underset{\text{factor}}{2} \cdot \underset{\text{factor}}{3} = \underset{\text{product}}{6}
$$

$$
\underset{\text{factor}}{x^2} \cdot \underset{\text{factor}}{x^3} = \underset{\text{product}}{x^5}
$$

$$
\underset{\text{factor}}{(x + 2)}\underset{\text{factor}}{(x + 3)} = \underset{\text{product}}{x^2 + 5x + 6}
$$

Notice that factoring is the reverse process of multiplying.

$$
x^2 + 5x + 6 = (x + 2)(x + 3)
$$

factoring / multiplying

✔ **CONCEPT CHECK**

Multiply: $2(x - 4)$

What do you think the result of factoring $2x - 8$ would be? Why?

The first step in factoring a polynomial is to see whether the terms of the polynomial have a common factor. If there is one, we can write the polynomial as a product by **factoring out** the common factor. We will usually factor out the **greatest common factor (GCF)**.

1 The GCF of a list of integers is the largest integer that is a factor of all the integers in the list. For example, the GCF of 12 and 20 is 4 because 4 is the largest integer that is a factor of both 12 and 20. With large integers, the GCF may not be easily found by inspection. When this happens, use the following steps.

Finding the GCF of a List of Integers

Step 1. Write each number as a product of prime numbers.

Step 2. Identify the common prime factors.

Step 3. The product of all common prime factors found in step 2 is the greatest common factor. If there are no common prime factors, the greatest common factor is 1.

Recall from Section 1.3 that a prime number is a whole number other than 1, whose only factors are 1 and itself.

EXAMPLE 1

Find the GCF of each list of numbers.

a. 28 and 40   **b.** 55 and 21   **c.** 15, 18, and 66

Solution   **a.** Write each number as a product of primes.

$$28 = 2 \cdot 2 \cdot 7 = 2^2 \cdot 7$$
$$40 = 2 \cdot 2 \cdot 2 \cdot 5 = 2^3 \cdot 5$$

There are two common factors, each of which is 2, so the GCF is

$$\text{GCF} = 2 \cdot 2 = 4$$

b. $55 = 5 \cdot 11$
    $21 = 3 \cdot 7$

There are no common prime factors; thus, the GCF is 1.

c. $15 = 3 \cdot 5$
    $18 = 2 \cdot 3 \cdot 3 = 2 \cdot 3^2$
    $66 = 2 \cdot 3 \cdot 11$

The only prime factor common to all three numbers is 3, so the GCF is

$$\text{GCF} = 3$$

2   The greatest common factor of a list of variables raised to powers is found in a similar way. For example, the GCF of x^2, x^3, and x^5 is x^2 because each term contains a factor of x^2 and no higher power of x is a factor of each term.

$$x^2 = x \cdot x$$
$$x^3 = x \cdot x \cdot x$$
$$x^5 = x \cdot x \cdot x \cdot x \cdot x$$

There are two common factors, each of which is x, so the GCF $= x \cdot x$ or x^2. From this example, we see that **the GCF of a list of common variables raised to powers is the variable raised to the smallest exponent in the list.**

EXAMPLE 2

Find the GCF of each list of terms.

a. x^3, x^7, and x^5 **b.** y, y^4, and y^7

Solution   **a.** The GCF is x^3, since 3 is the smallest exponent to which x is raised.

b. The GCF is y^1 or y, since 1 is the smallest exponent on y.

In general, the **greatest common factor (GCF) of a list of terms** is the product of the GCF of the numerical coefficients and the GCF of the variable factors.

EXAMPLE 3

Find the GCF of each list of terms.

a. $6x^2$, $10x^3$, and $-8x$ **b.** $8y^2$, y^3, and y^5 **c.** a^3b^2, a^5b, and a^6b^2

Solution   **a.** The GCF of the numerical coefficients 6, 10, and -8 is 2.

The GCF of variable factors x^2, x^3, and x is x.

Thus, the GCF of $6x^2$, $10x^3$, and $-8x$ is $2x$.

CLASSROOM EXAMPLE
Find the GCF of each list of terms.
a. $-9x^2, 15x^4$, and $6x$
b. $11ab^3, a^2b^3$, and ab^2
answer: a. $3x$ b. ab^2

b. The GCF of the numerical coefficients $8, 1$, and 1 is 1.
The GCF of variable factors y^2, y^3, and y^5 is y^2.
Thus, the GCF of $8y^2$, y^3, and y^5 is $1y^2$ or y^2.

c. The GCF of a^3, a^5, and a^6 is a^3.

The GCF of b^2, b, and b^2 is b. Thus, the GCF of the terms is a^3b.

3   The first step in factoring a polynomial is to find the GCF of its terms. Once we do so, we can write the polynomial as a product by **factoring out** the GCF.

The polynomial $8x + 14$, for example, contains two terms: $8x$ and 14. The GCF of these terms is 2. We factor out 2 from each term by writing each term as a product of 2 and the term's remaining factors.

$$8x + 14 = 2 \cdot 4x + 2 \cdot 7$$

Using the distributive property, we can write

$$8x + 14 = 2 \cdot 4x + 2 \cdot 7$$
$$= 2(4x + 7)$$

Thus, a factored form of $8x + 14$ is $2(4x + 7)$.

> **Helpful Hint**
> A factored form of $8x + 14$ is *not*
> $$2 \cdot 4x + 2 \cdot 7$$
> Although the *terms* have been factored (written as a product), the *polynomial $8x + 14$* has not been factored (written as a product). A factored form of $8x + 14$ is the *product $2(4x + 7)$*.

✔ **CONCEPT CHECK**

Which of the following is/are factored form(s) of $7t + 21$?

   a. 7 **b.** $7 \cdot t + 7 \cdot 3$ **c.** $7(t + 3)$ **d.** $7(t + 21)$

EXAMPLE 4

Factor each polynomial by factoring out the GCF.

   a. $6t + 18$ **b.** $y^5 - y^7$

Solution   **a.** The GCF of terms $6t$ and 18 is 6.

$$6t + 18 = 6 \cdot t + 6 \cdot 3$$
$$= 6(t + 3)$$ Apply the distributive property.

Our work can be checked by multiplying 6 and $(t + 3)$.

$$6(t + 3) = 6 \cdot t + 6 \cdot 3 = 6t + 18, \text{ the original polynomial.}$$

b. The GCF of y^5 and y^7 is y^5. Thus,

$$y^5 - y^7 = (y^5)1 - (y^5)y^2$$
$$= y^5(1 - y^2)$$

CLASSROOM EXAMPLE
Factor.
a. $49x - 35$ b. $z^3 - z^2$
answer:
a. $7(7x - 5)$ b. $z^2(z - 1)$

> **Helpful Hint**
> Don't forget the 1.

Concept Check Answer:
c

EXAMPLE 5

Factor $-9a^5 + 18a^2 - 3a$.

Solution

$$-9a^5 + 18a^2 - 3a = (3a)(-3a^4) + (3a)(6a) + (3a)(-1)$$
$$= 3a(-3a^4 + 6a - 1)$$

> **Helpful Hint**
> Don't forget the -1.

In Example 5 we could have chosen to factor out a $-3a$ instead of $3a$. If we factor out a $-3a$, we have

$$-9a^5 + 18a^2 - 3a = (-3a)(3a^4) + (-3a)(-6a) + (-3a)(1)$$
$$= -3a(3a^4 - 6a + 1)$$

> **Helpful Hint**
> Notice the changes in signs when factoring out $-3a$.

EXAMPLE 6

Factor $25x^4z + 15x^3z + 5x^2z$.

Solution

The greatest common factor is $5x^2z$.

$$25x^4z + 15x^3z + 5x^2z = 5x^2z(5x^2 + 3x + 1)$$

> **Helpful Hint**
> Be careful when the GCF of the terms is the same as one of the terms in the polynomial. The greatest common factor of the terms of $8x^2 - 6x^3 + 2x$ is $2x$. When factoring out $2x$ from the terms of $8x^2 - 6x^3 + 2x$, don't forget a term of 1.
> $$8x^2 - 6x^3 + 2x = 2x(4x) - 2x(3x^2) + 2x(1)$$
> $$= 2x(4x - 3x^2 + 1)$$
> Check by multiplying.
> $$2x(4x - 3x^2 + 1) = 8x^2 - 6x^3 + 2x$$

TEACHING TIP

A common mistake is for students to factor $5(x + 3) + y(x + 3)$ as $(x + 3)^2(5 + y)$. Before working Example 7, make sure that students understand the concept of *factoring out a common factor*.

EXAMPLE 7

Factor $5(x + 3) + y(x + 3)$.

Solution

The binomial $(x + 3)$ is the greatest common factor. Use the distributive property to factor out $(x + 3)$.

$$5(x + 3) + y(x + 3) = (x + 3)(5 + y)$$

EXAMPLE 8

Factor $3m^2n(a + b) - (a + b)$.

Solution

The greatest common factor is $(a + b)$.

$$3m^2n(a + b) - 1(a + b) = (a + b)(3m^2n - 1)$$

4   Once the GCF is factored out, we can often continue to factor the polynomial, using a variety of techniques. We discuss here a technique for factoring polynomials called **grouping**.

EXAMPLE 9

Factor $xy + 2x + 3y + 6$ by grouping. Check by multiplying.

Solution   The GCF of the first two terms is x, and the GCF of the last two terms is 3.

$$xy + 2x + 3y + 6 = \underline{x(y + 2) + 3(y + 2)}$$

> ▶ **Helpful Hint**
>
> Notice that this is *not* a factored form of the original polynomial. It is a sum, not a product.

Next, factor out the common binomial factor of $(y + 2)$.

$$x(y + 2) + 3(y + 2) = (y + 2)(x + 3)$$

To check, multiply $(y + 2)$ by $(x + 3)$.

$$(y + 2)(x + 3) = xy + 2x + 3y + 6, \text{ the original polynomial.}$$

Thus, the factored form of $xy + 2x + 3y + 6$ is $(y + 2)(x + 3)$.

Factoring a Four-term Polynomial by Grouping

Step 1. Arrange the terms so that the first two terms have a common factor and the last two terms have a common factor.

Step 2. For each pair of terms, use the distributive property to factor out the pair's greatest common factor.

Step 3. If there is now a common binomial factor, factor it out.

Step 4. If there is no common binomial factor in step 3, begin again, rearranging the terms differently. If no rearrangement leads to a common binomial factor, the polynomial cannot be factored.

EXAMPLE 10

Factor $3x^2 + 4xy - 3x - 4y$ by grouping.

Solution   The first two terms have a common factor x. Factor -1 from the last two terms so that the common binomial factor of $(3x + 4y)$ appears.

$$3x^2 + 4xy - 3x - 4y = x(3x + 4y) - 1(3x + 4y)$$

Next, factor out the common factor $(3x + 4y)$.

$$= (3x + 4y)(x - 1)$$

> ### Helpful Hint
> One more reminder: When **factoring** a polynomial, make sure the polynomial is written as a **product**. For example, it is true that
>
> $$3x^2 + 4xy - 3x - 4y = x(3x + 4y) - 1(3x + 4y)$$
>
> but $x(3x + 4y) - 1(3x + 4y)$ is not a **factored form** of the original polynomial since it is a **sum (difference)**, not a **product**. The factored form of $3x^2 + 4xy - 3x - 4y$ is $(3x + 4y)(x - 1)$.

Factoring out a greatest common factor first makes factoring by any method easier, as we see in the next example.

EXAMPLE 11

Factor $4ax - 4ab - 2bx + 2b^2$.

Solution   First, factor out the common factor 2 from all four terms.

$$
\begin{aligned}
4ax - 4ab &- 2bx + 2b^2 \\
&= 2(2ax - 2ab - bx + b^2) &&\text{Factor out } \mathbf{2} \text{ from all four terms.} \\
&= 2[2a(x - b) - b(x - b)] &&\text{Factor out common factors from each pair of terms.} \\
&= 2(x - b)(2a - b) &&\text{Factor out the common binomial.}
\end{aligned}
$$

Notice that we factored out $-b$ instead of b from the second pair of terms so that the binomial factor of each pair is the same.

CLASSROOM EXAMPLE
Factor $15xz + 15yz - 5xy - 5y^2$.
answer: $5(x + y)(3z - y)$

STUDY SKILLS REMINDER

How Well Do You Know Your Textbook?

See if you can answer the questions below.

1. What does the 🔒 icon mean?
2. What does the ⟍ icon mean?
3. What does the △ icon mean?
4. Where can you find a review for each chapter? What answers to this review can be found in the back of your text?
5. Each chapter contains an overview of the chapter along with examples. What is this feature called?
6. Does this text contain any solutions to exercises? If so, where?

MENTAL MATH

Find the prime factorization of the following integers.

1. 14 $2 \cdot 7$ **2.** 15 $3 \cdot 5$ **3.** 10 $2 \cdot 5$ **4.** 70 $2 \cdot 5 \cdot 7$

Find the GCF of the following pairs of integers.

5. 6, 15   3   **6.** 20, 15   5   **7.** 3, 18   3   **8.** 14, 35   7

20. $5x^2y^2(5x^2y - 3)$ **22.** $3ax(-5a + 3)$ **23.** $4x(3x^2 + 4x - 2)$ **24.** $3x(2x^2 - 3x + 4)$ **25.** $5xy(x^2 - 3x + 2)$ **26.** $7xy(2x^2 + x - 1)$

EXERCISE SET 6.1

STUDY GUIDE/SSM | CD/ VIDEO | PH MATH TUTOR CENTER | MathXL®Tutorials ON CD | MathXL® | MyMathLab®

Find the GCF for each list. See Examples 1 through 3.

1. $32, 36$ 4

2. $36, 90$ 18

3. $12, 18, 36$ 6

4. $24, 14, 21$ 1

5. y^2, y^4, y^7 y^2

6. x^3, x^2, x^3 x^2

7. $x^{10}y^2, xy^2, x^3y^3$ xy^2

8. p^7q, p^8q^2, p^9q^3 p^7q

9. $8x, 4$ 4

10. $9y, y$ y

 11. $12y^4, 20y^3$ $4y^3$

12. $32x, 18x^2$ $2x$

13. $12x^3, 6x^4, 3x^5$ $3x^3$

14. $15y^2, 5y^7, 20y^3$ $5y^2$

15. $18x^2y, 9x^3y^3, 36x^3y$ $9x^2y$

16. $7x, 21x^2y^2, 14xy$ $7x$

Factor out the GCF from each polynomial. See Examples 4 through 6. **21.** $-6a^3x(4a - 3)$

17. $30x - 15$ $15(2x - 1)$

18. $42x - 7$ $7(6x - 1)$

19. $24cd^3 - 18c^2d$ $6cd(4d^2 - 3c)$ **20.** $25x^4y^3 - 15x^2y^2$

21. $-24a^4x + 18a^3x$

22. $-15a^2x + 9ax$

23. $12x^3 + 16x^2 - 8x$

24. $6x^3 - 9x^2 + 12x$

25. $5x^3y - 15x^2y + 10xy$

26. $14x^3y + 7x^2y - 7xy$

27. Construct a binomial whose greatest common factor is $5a^3$.

28. Construct a trinomial whose greatest common factor is $2x^2$.

27.–28. answers may vary

Factor out the GCF from each polynomial. See Examples 7 and 8.
29. $(x + 2)(y + 3)$ **30.** $(y + 4)(z + 3)$ **31.** $(y - 3)(x - 4)$

29. $y(x + 2) + 3(x + 2)$

30. $z(y + 4) + 3(y + 4)$

31. $x(y - 3) - 4(y - 3)$

32. $6(x + 2) - y(x + 2)$

33. $2x(x + y) - (x + y)$

34. $xy(y + 1) - (y + 1)$

32. $(x + 2)(6 - y)$ **33.** $(x + y)(2x - 1)$ **34.** $(y + 1)(xy - 1)$

Factor the following four-term polynomials by grouping. See Examples 9 through 11. **35.** $(x + 3)(5 + y)$ **36.** $(x + 1)(y + 2)$
37. $(y - 4)(2 + x)$ **38.** $(x - 7)(6 + y)$ **39.** $(y - 2)(3x + 8)$

35. $5x + 15 + xy + 3y$

36. $xy + y + 2x + 2$

37. $2y - 8 + xy - 4x$

38. $6x - 42 + xy - 7y$

39. $3xy - 6x + 8y - 16$

40. $xy - 2yz + 5x - 10z$

41. $y^3 + 3y^2 + y + 3$

42. $x^3 + 4x + x^2 + 4$

40. $(x - 2z)(y + 5)$ **41.** $(y + 3)(y^2 + 1)$ **42.** $(x^2 + 4)(x + 1)$

Write an expression for the area of each shaded region. Then write the expression as a factored polynomial.

△ **43.**

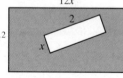

$12x^3 - 2x; 2x(6x^2 - 1)$

△ **44.**

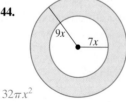

$32\pi x^2$

△ **45.**

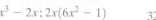
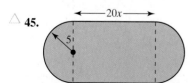

$200x + 25\pi; 25(8x + \pi)$

△ **46.**

$28x^2 - \pi x^2; x^2(28 - \pi)$

MIXED PRACTICE

Factor the following polynomials.

47. $3x - 6$ $3(x - 2)$

48. $4x - 16$ $4(x - 4)$

49. $32xy - 18x^2$ $2x(16y - 9x)$ **50.** $10xy - 15x^2$

51. $4x - 8y + 4$ $4(x - 2y + 1)$ **52.** $7x + 21y - 7$

53. $8(x + 2) - y(x + 2)$ $(x + 2)(8 - y)$

54. $x(y^2 + 1) - 3(y^2 + 1)$ $(y^2 + 1)(x - 3)$

55. $-40x^8y^6 - 16x^9y^5$

56. $-21x^3y - 49x^2y^2$

57. $-3x + 12$ $-3(x - 4)$

58. $-10x + 20$

59. $18x^3y^3 - 12x^3y^2 + 6x^5y^2$ $6x^3y^2(3y - 2 + x^2)$

60. $32x^3y^3 - 24x^2y^3 + 8x^2y^4$ $8x^2y^3(4x - 3 + y)$

61. $(x - 2)(y^2 + 1)$ **62.** $(y + 4)(x + 1)$ **63.** $(y + 3)(5x + 6)$

Concept Extensions

Factor. **64.** $(2x + 1)(x^2 + 4)$ **65.** $(x - 2y)(4x - 3)$

61. $y^2(x - 2) + (x - 2)$

62. $x(y + 4) + (y + 4)$

63. $5xy + 15x + 6y + 18$

64. $2x^3 + x^2 + 8x + 4$

65. $4x^2 - 8xy - 3x + 6y$

66. $2x^3 - x^2 - 10x + 5$

67. $126x^3yz + 210y^4z^3$

68. $231x^3y^2z - 143yz^2$

69. $3y - 5x + 15 - xy$

70. $2x - 9y + 18 - xy$

71. $12x^2y - 42x^2 - 4y + 14$ $2(3x^2 - 1)(2y - 7)$

72. $90 + 15y^2 - 18x - 3xy^2$ $3(6 + y^2)(5 - x)$

Fill in the chart by finding two numbers that have the given product and sum. The first row is filled in for you.

	Two Numbers	Their Product	Their Sum
	$4, 7$	28	11
73.	$2, 6$	12	8
74.	$4, 5$	20	9
75.	$-1, -8$	8	-9
76.	$-2, -8$	16	-10
77.	$-2, 5$	-10	3
78.	$-3, 3$	-9	0
79.	$-8, 3$	-24	-5
80.	$-9, 4$	-36	-5

REVIEW AND PREVIEW

Multiply. See Section 5.4.

81. $(x + 2)(x + 5)$ $x^2 + 7x + 10$ **82.** $(y + 3)(y + 6)$ $y^2 + 9y + 18$

83. $(a - 7)(a - 8)$ $a^2 - 15a + 56$ **84.** $(z - 4)(z - 4)$ $z^2 - 8z + 16$

85. Explain how you can tell whether a polynomial is written in factored form.   answers may vary

86. Construct a 4-term polynomial that can be factored by grouping.   answers may vary

Which of the following expressions are factored?

87. $(a + 6)(b - 2)$ **87.–88.** factored

88. $(x + 5)(x + y)$

89. $5(2y + z) - b(2y + z)$ **89.–90.** not factored

90. $3x(a + 2b) + 2(a + 2b)$

Write an expression for the length of each rectangle.

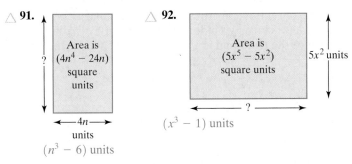

△ **91.**   Area is $(4n^4 - 24n)$ square units

?

◄—$4n$—►
units

$(n^3 - 6)$ units

△ **92.**   Area is $(5x^5 - 5x^2)$ square units

$5x^2$ units

◄—— ? ——►

$(x^3 - 1)$ units

Factor each polynomial by grouping.

93. $x^{2n} + 2x^n + 3x^n + 6$ (*Hint:* Don't forget that $x^{2n} = x^n \cdot x^n$.)

94. $x^{2n} + 6x^n + 10x^n + 60$ **95.** $3x^{2n} + 21x^n - 5x^n - 35$

96. $12x^{2n} - 10x^n - 30x^n + 25$ $(2x^n - 5)(6x^n - 5)$

97. The number (in millions) of CDs shipped annually in the United States each year during 1999–2001 can be modeled by the polynomial $8x^2 - 56x + 124$, where x is the number of years since 1999. (*Source:* Recording Industry Association of America)

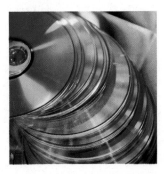

a. Find the number of CDs sold in 2001. To do so, let $x = 2$ and evaluate $8x^2 - 56x + 124$.   44 million

b. Use this expression to predict the number of CDs sold in 2006.   124 million

c. Factor the polynomial $8x^2 - 56x + 124$. $4(2x^2 - 14x + 31)$

98. The number (in thousands) of students who graduated from U.S. high schools each year during 1998–2000 can be modeled by $12x^2 + 21x + 2457$, where x is the number of years since 1998. (*Source:* U.S. Bureau of the Census)

a. Find the number of students who graduated from U.S. high schools in 2000. To do so, let $x = 2$ and evaluate $12x^2 + 21x + 2457$.   2547 (in thousands), or 2,547,000

b. Use this expression to predict the number of students who will graduate from U.S. high schools in 2007.

c. Factor the polynomial $12x^2 + 21x + 2457$.

50. $5x(2y - 3x)$ **52.** $7(x + 3y - 1)$ **55.** $-8x^8y^5(5y + 2x)$ **56.** $-7x^2y(3x + 7y)$ **58.** $-10(x - 2)$ **66.** $(2x - 1)(x^2 - 5)$
67. $42yz(3x^3 + 5y^3z^2)$ **68.** $11yz(21x^3y - 13z)$ **69.** $(3 - x)(5 + y)$ **70.** $(x + 9)(2 - y)$ **93.** $(x^n + 2)(x^n + 3)$ **94.** $(x^n + 6)(x^n + 10)$
95. $(3x^n - 5)(x^n + 7)$ **98.b.** 3618 (in thousands), or 3,618,000   **98.c.** $3(4x^2 + 7x + 819)$

6.2   FACTORING TRINOMIALS OF THE FORM $x^2 + bx + c$

Objectives

1 Factor trinomials of the form $x^2 + bx + c$.

2 Factor out the greatest common factor and then factor a trinomial of the form $x^2 + bx + c$.

1   In this section, we factor trinomials of the form $x^2 + bx + c$, such as

$$x^2 + 4x + 3, \qquad x^2 - 8x + 15, \qquad x^2 + 4x - 12, \qquad r^2 - r - 42$$

Notice that for these trinomials, the coefficient of the squared variable is 1.

Recall that factoring means to write as a product and that factoring and multiplying are reverse processes. Using the FOIL method of multiplying binomials, we have that

$$(x + 3)(x + 1) = \overset{F}{x^2} + \overset{O}{1x} + \overset{I}{3x} + \overset{L}{3}$$
$$= x^2 + 4x + 3$$

Thus, a factored form of $x^2 + 4x + 3$ is $(x + 3)(x + 1)$.

Notice that the product of the first terms of the binomials is $x \cdot x = x^2$, the first term of the trinomial. Also, the product of the last two terms of the binomials is $3 \cdot 1 = 3$, the third term of the trinomial. The sum of these same terms is $3 + 1 = 4$, the coefficient of the middle term, x, of the trinomial.

The product of these numbers is **3**.

$$x^2 + 4x + 3 = (x + 3)(x + 1)$$

The sum of these numbers is **4**.

Many trinomials, such as the one above, factor into two binomials. To factor $x^2 + 7x + 10$, let's assume that it factors into two binomials and begin by writing two pairs of parentheses. The first term of the trinomial is x^2, so we use x and x as the first terms of the binomial factors.

$$x^2 + 7x + 10 = (x + \quad)(x + \quad)$$

To determine the last term of each binomial factor, we look for two integers whose product is 10 and whose sum is 7. Since our numbers must have a positive product and a positive sum, we list pairs of positive integer factors of 10 only.

Positive Factors of 10	Sum of Factors
1, 10	$1 + 10 = 11$
2, 5	$2 + 5 = 7$

The correct pair of numbers is 2 and 5 because their product is 10 and their sum is 7. Now we can fill in the last terms of the binomial factors.

$$x^2 + 7x + 10 = (x + 2)(x + 5)$$

To see if we have factored correctly, multiply.

$$(x + 2)(x + 5) = x^2 + 5x + 2x + 10$$
$$= x^2 + 7x + 10 \qquad \text{Combine like terms.}$$

> **Helpful Hint**
>
> Since multiplication is commutative, the factored form of $x^2 + 7x + 10$ can be written as either $(x + 2)(x + 5)$ or $(x + 5)(x + 2)$.

Factoring a Trinomial of the Form $x^2 + bx + c$

To factor a trinomial of the form $x^2 + bx + c$, look for two numbers whose product is c and whose sum is b. The factored form of $x^2 + bx + c$ is

product is c

$$(x + \text{one number})(x + \text{other number})$$

sum is b

EXAMPLE 1

Factor $x^2 + 7x + 12$.

Solution   Begin by writing the first terms of the binomial factors.

$$(x +\ \ )(x +\ \ )$$

Next, look for two numbers whose product is 12 and whose sum is 7. Since our numbers must have a positive product and a positive sum, we look at positive pairs of factors of 12 only.

Positive Factors of 12	*Sum of Factors*
1, 12	$1 + 12 = 13$
2, 6	$2 + 6 = 8$
3, 4	$3 + 4 = 7$

The correct pair of numbers is 3 and 4 because their product is 12 and their sum is 7. Use these numbers as the last terms of the binomial factors.

$$x^2 + 7x + 12 = (x + 3)(x + 4)$$

To check, multiply $(x + 3)$ by $(x + 4)$.

EXAMPLE 2

Factor $x^2 - 8x + 15$.

Solution   Begin by writing the first terms of the binomials.

$$(x +\ \ )(x +\ \ )$$

Now look for two numbers whose product is 15 and whose sum is -8. Since our numbers must have a positive product and a negative sum, we look at negative factors of 15 only.

Negative Factors of 15	*Sum of Factors*
$-1, -15$	$-1 + (-15) = -16$
$-3, -5$	$-3 + (-5) = -8$

The correct pair of numbers is -3 and -5 because their product is 15 and their sum is -8. Then

$$x^2 - 8x + 15 = (x - 3)(x - 5)$$

EXAMPLE 3

Factor $x^2 + 4x - 12$.

Solution $x^2 + 4x - 12 = (x +\ \ )(x +\ \ )$

Look for two numbers whose product is -12 and whose sum is 4. Since our numbers have a negative product, their signs must be different.

Factors of -12	*Sum of Factors*
$-1, 12$	$-1 + 12 = 11$
$1, -12$	$1 + (-12) = -11$
$-2, 6$	$-2 + 6 = 4$
$2, -6$	$2 + (-6) = -4$
$-3, 4$	$-3 + 4 = 1$
$3, -4$	$3 + (-4) = -1$

The correct pair of numbers is -2 and 6 since their product is -12 and their sum is 4. Hence

$$x^2 + 4x - 12 = (x - 2)(x + 6)$$

EXAMPLE 4

Factor $r^2 - r - 42$.

Solution Because the variable in this trinomial is r, the first term of each binomial factor is r.

$$r^2 - r - 42 = (r +)(r +)$$

Find two numbers whose product is -42 and whose sum is -1, the numerical coefficient of r. The numbers are 6 and -7. Therefore,

$$r^2 - r - 42 = (r + 6)(r - 7)$$

CLASSROOM EXAMPLE
Factor $q^2 - 3q - 40$.
answer: $(q + 5)(q - 8)$

EXAMPLE 5

Factor $a^2 + 2a + 10$.

CLASSROOM EXAMPLE
Factor $y^2 + 6y + 15$.
answer: prime polynomial

Solution Look for two numbers whose product is 10 and whose sum is 2. Neither 1 and 10 nor 2 and 5 give the required sum, 2. We conclude that $a^2 + 2a + 10$ is not factorable with integers. The polynomial $a^2 + 2a + 10$ is called a **prime polynomial**.

EXAMPLE 6

Factor $x^2 + 5xy + 6y^2$.

Solution $x^2 + 5xy + 6y^2 = (x +)(x +)$

CLASSROOM EXAMPLE
Factor $a^2 - 13ab + 30b^2$.
answer: $(a - 3b)(a - 10b)$

Look for two terms whose product is $6y^2$ and whose sum is $5y$, the coefficient of x in the middle term of the trinomial. The terms are $2y$ and $3y$ because $2y \cdot 3y = 6y^2$ and $2y + 3y = 5y$. Therefore,

$$x^2 + 5xy + 6y^2 = (x + 2y)(x + 3y)$$

The following sign patterns may be useful when factoring trinomials.

> ### Helpful Hint—Sign Patterns
>
> A positive constant in a trinomial tells us to look for two numbers with the same sign. The sign of the coefficient of the middle term tells us whether the signs are both positive or both negative.
>
> $$\begin{array}{cc} \text{both} & \text{same} \\ \text{positive} & \text{sign} \\ \downarrow & \downarrow \end{array}$$
> $$x^2 + 10x + 16 = (x + 2)(x + 8)$$
>
> $$\begin{array}{cc} \text{both} & \text{same} \\ \text{negative} & \text{sign} \\ \downarrow & \downarrow \end{array}$$
> $$x^2 - 10x + 16 = (x - 2)(x - 8)$$ *(continued)*

A negative constant in a trinomial tells us to look for two numbers with opposite signs.

opposite
signs
↓

opposite
signs
↓

$$x^2 + 6x - 16 = (x + 8)(x - 2) \qquad x^2 - 6x - 16 = (x - 8)(x + 2)$$

2   Remember that the first step in factoring any polynomial is to factor out the greatest common factor (if there is one other than 1 or -1).

EXAMPLE 7

Factor $3m^2 - 24m - 60$.

Solution   First factor out the greatest common factor, 3, from each term.

$$3m^2 - 24m - 60 = 3(m^2 - 8m - 20)$$

Next, factor $m^2 - 8m - 20$ by looking for two factors of -20 whose sum is -8. The factors are -10 and 2.

$$3m^2 - 24m - 60 = 3(m + 2)(m - 10)$$

Remember to write the common factor 3 as part of the answer. Check by multiplying.

$$3(m + 2)(m - 10) = 3(m^2 - 8m - 20)$$
$$= 3m^2 - 24m - 60$$

Helpful Hint

When factoring a polynomial, remember that factored out common factors are part of the final factored form. For example,

$$5x^2 - 15x - 50 = 5(x^2 - 3x - 10)$$
$$= 5(x + 2)(x - 5)$$

Thus, $5x^2 - 15x - 50$ **factored completely** is $5(x + 2)(x - 5)$.

MENTAL MATH

Complete the following.

1. $x^2 + 9x + 20 = (x + 4)(x + 5)$ **2.** $x^2 + 12x + 35 = (x + 5)(x + 7)$ **3.** $x^2 - 7x + 12 = (x - 4)(x - 3)$

4. $x^2 - 13x + 22 = (x - 2)(x - 11)$ **5.** $x^2 + 4x + 4 = (x + 2)(x + 2)$ **6.** $x^2 + 10x + 24 = (x + 6)(x + 4)$

EXERCISE SET 6.2

STUDY GUIDE/SSM | CD/ VIDEO | PH MATH TUTOR CENTER | MathXL®Tutorials ON CD | MathXL® | MyMathLab®

Factor each trinomial completely. If the polynomial can't be factored, write prime. See Examples 1 through 5.

 1. $x^2 + 7x + 6$ $(x + 6)(x + 1)$ **2.** $x^2 + 6x + 8$
3. $x^2 + x - 20$ $(x + 5)(x - 4)$ **4.** $x^2 + 7x - 30$
5. $x^2 - 8x + 15$ $(x - 5)(x - 3)$ **6.** $x^2 - 9x + 14$
7. $x^2 - 10x + 9$ $(x - 9)(x - 1)$ **8.** $x^2 - 6x + 9$
9. $x^2 - 15x + 5$ prime   **10.** $x^2 - 13x + 30$
11. $x^2 - 3x - 18$ $(x - 6)(x + 3)$ **12.** $x^2 - x - 30$
13. $x^2 + 5x + 2$ prime   **14.** $x^2 - 7x + 5$ prime

Factor each trinomial completely. See Example 6.

15. $x^2 + 8xy + 15y^2$ **16.** $x^2 + 6xy + 8y^2$
17. $x^2 - 2xy + y^2$ **18.** $x^2 - 11xy + 30y^2$
19. $x^2 - 3xy - 4y^2$ **20.** $x^2 - 4xy - 77y^2$

Factor each trinomial completely. See Example 7.

21. $2z^2 + 20z + 32$ **22.** $3x^2 + 30x + 63$
23. $2x^3 - 18x^2 + 40x$ **24.** $x^3 - x^2 - 56x$
25. $7x^2 + 14xy - 21y^2$ **26.** $3r^2 - 3rs - 6s^2$

27. To factor $x^2 + 13x + 42$, think of two numbers whose <u>product</u> is 42 and whose <u>sum</u> is 13.

28. Write a polynomial that factors as $(x - 3)(x + 8)$.
$x^2 + 5x - 24$

MIXED PRACTICE

Factor each trinomial completely.

29. $x^2 + 15x + 36$ **30.** $x^2 + 19x + 60$
31. $x^2 - x - 2$ $(x - 2)(x + 1)$ **32.** $x^2 - 5x - 14$
33. $r^2 - 16r + 48$ $(r - 12)(r - 4)$ **34.** $r^2 - 10r + 21$ $(r - 7)(r - 3)$
35. $x^2 - 4x - 21$ $(x - 7)(x + 3)$ **36.** $x^2 - 4x - 32$ $(x - 8)(x + 4)$
37. $x^2 + 7xy + 10y^2$ **38.** $x^2 + 5xy + 4y^2$
39. $r^2 - 3r + 6$ prime   **40.** $x^2 + 4x - 10$ prime
41. $2t^2 + 24t + 64$ $2(t + 8)(t + 4)$ **42.** $2t^2 + 20t + 50$
43. $x^3 - 2x^2 - 24x$ **44.** $x^3 - 3x^2 - 28x$
45. $x^2 - 16x + 63$ $(x - 9)(x - 7)$ **46.** $x^2 - 19x + 88$
47. $x^2 + xy - 2y^2$ **48.** $x^2 - xy - 6y^2$
49. $3x^2 - 60x + 108$ **50.** $2x^2 - 24x + 70$
51. $x^2 - 18x - 144$ **52.** $x^2 + x - 42$
53. $6x^3 + 54x^2 + 120x$ **54.** $3x^3 + 3x^2 - 126x$
55. $2t^5 - 14t^4 + 24t^3$ **56.** $3x^6 + 30x^5 + 72x^4$
57. $5x^3y - 25x^2y^2 - 120xy^3$ **58.** $3x^2 - 6xy - 72y^2$
59. $4x^2y + 4xy - 12y$ **60.** $3x^2y - 9xy + 45y$
61. $2a^2b - 20ab^2 + 42b^3$ **62.** $-x^2z + 14xz^2 - 28z^3$

REVIEW AND PREVIEW

Multiply. See Section 5.4.

63. $(2x + 1)(x + 5)$ **64.** $(3x + 2)(x + 4)$

65. $(5y - 4)(3y - 1)$ **66.** $(4z - 7)(7z - 1)$
67. $(a + 3)(9a - 4)$ **68.** $(y - 5)(6y + 5)$

Concept Extensions

Find a positive value of b so that each trinomial is factorable.

69. $x^2 + bx + 15$ 8; 16   **70.** $y^2 + by + 20$ 9; 12; 21
71. $m^2 + bm - 27$ 6; 26   **72.** $x^2 + bx - 14$ 5; 13

Find a positive value of c so that each trinomial is factorable.

73. $x^2 + 6x + c$ 5; 8; 9   **74.** $t^2 + 8t + c$ 7; 12; 15; 16
75. $y^2 - 4y + c$ 3; 4   **76.** $n^2 - 16n + c$
                     76. 15; 28; 39; 48; 55; 60; 63; 64

Write the perimeter of each rectangle as a simplified polynomial. Then factor the polynomial.

△ **77.**

$4x + 33$

$x^2 + 10x$

$2x^2 + 28x + 66$: $2(x + 3)(x + 11)$

△ **78.**

$4x^3 + 24x^2 + 32x$; $4x(x + 4)(x + 2)$

$12x^2$

$2x^3 + 16x$

Complete the following sentences in your own words.

79. If $x^2 + bx + c$ is factorable and c is negative, then the signs of the last term factors of the binomial are opposite because. . . .   **79.–80.** answers may vary

80. If $x^2 + bx + c$ is factorable and c is positive, then the signs of the last term factors of the binomials are the same because. . . .

Factor each trinomial completely. Don't forget to first factor out the greatest common factor.

81. $2x^2y + 30xy + 100y$ **82.** $3x^2z^2 + 9xz^2 + 6z^2$
83. $-12x^2y^3 - 24xy^3 - 36y^3$ **84.** $-4x^2t^4 + 4xt^4 + 24t^4$
85. $y^2(x + 1) - 2y(x + 1) - 15(x + 1)$
86. $z^2(x + 1) - 3z(x + 1) - 70(x + 1)$

Factor each trinomial. (Hint: Notice that $x^{2n} + 4x^n + 3$ factors as $(x^n + 1)(x^n + 3)$.)

87. $x^{2n} + 5x^n + 6$ **88.** $x^{2n} + 8x^n - 20$

2. $(x + 4)(x + 2)$ **4.** $(x - 3)(x + 10)$ **6.** $(x - 7)(x - 2)$ **8.** $(x - 3)(x - 3)$ **10.** $(x - 3)(x - 10)$ **12.** $(x - 6)(x + 5)$
15. $(x + 5y)(x + 3y)$ **16.** $(x + 4y)(x + 2y)$ **17.** $(x - y)(x - y)$ **18.** $(x - 6y)(x - 5y)$ **19.** $(x - 4y)(x + y)$

6.3   FACTORING TRINOMIALS OF THE FORM $ax^2 + bx + c$

20. $(x - 11y)(x + 7y)$
21. $2(z + 8)(z + 2)$
22. $3(x + 7)(x + 3)$
23. $2x(x - 5)(x - 4)$
24. $x(x - 8)(x + 7)$
25. $7(x + 3y)(x - y)$
26. $3(r + s)(r - 2s)$
29. $(x + 12)(x + 3)$
30. $(x + 4)(x + 15)$
32. $(x - 7)(x + 2)$
37. $(x + 5y)(x + 2y)$
38. $(x + 4y)(x + y)$
42. $2(t + 5)(t + 5)$
43. $x(x - 6)(x + 4)$
44. $x(x - 7)(x + 4)$
46. $(x - 8)(x - 11)$
47. $(x + 2y)(x - y)$
48. $(x - 3y)(x + 2y)$
49. $3(x - 18)(x - 2)$
50. $2(x - 7)(x - 5)$
51. $(x - 24)(x + 6)$
52. $(x + 7)(x - 6)$
53. $6x(x + 4)(x + 5)$
54. $3x(x + 7)(x - 6)$
55. $2t^3(t - 4)(t - 3)$
56. $3x^4(x + 6)(x + 4)$
57. $5xy(x - 8y)(x + 3y)$
58. $3(x - 6y)(x + 4y)$
59. $4y(x^2 + x - 3)$
60. $3y(x^2 - 3x + 15)$
61. $2b(a - 7b)(a - 3b)$
62. $-z(x^2 - 14xz + 28z^2)$
63. $2x^2 + 11x + 5$
64. $3x^2 + 14x + 8$
65. $15y^2 - 17y + 4$
66. $28z^2 - 53z + 7$
67. $9a^2 + 23a - 12$
68. $6y^2 - 25y - 25$
81. $2y(x + 5)(x + 10)$
82. $3z^2(x + 2)(x + 1)$
83. $-12y^3(x^2 + 2x + 3)$
84. $-4t^4(x - 3)(x + 2)$
85. $(x + 1)(y - 5)(y + 3)$
86. $(x + 1)(z - 10)(z + 7)$
87. $(x^n + 2)(x^n + 3)$
88. $(x^n + 10)(x^n - 2)$

Objectives

1 Factor trinomials of the form $ax^2 + bx + c$.

2 Factor out a GCF before factoring a trinomial of the form $ax^2 + bx + c$.

3 Factor perfect square trinomials.

4 Factor trinomials of the form $ax^2 + bx + c$ by grouping.

1   In this section, we factor trinomials of the form $ax^2 + bx + c$, such as

$$3x^2 + 11x + 6, \qquad 8x^2 - 22x + 5, \qquad 2x^2 + 13x - 7$$

Notice that the coefficient of the squared variable in these trinomials is a number other than 1. We will factor these trinomials using a trial-and-check method based on FOIL and our work in the last section.

To begin, let's review the relationship between the numerical coefficients of the trinomial and the numerical coefficients of its factored form. For example, since $(2x + 1)(x + 6) = 2x^2 + 13x + 6$, the factored form of $2x^2 + 13x + 6$ is

$$2x^2 + 13x + 6 = (2x + 1)(x + 6)$$

Notice that $2x$ and x are factors of $2x^2$, the first term of the trinomial. Also, 6 and 1 are factors of 6, the last term of the trinomial, as shown:

$$2x^2 + 13x + 6 = (2x + 1)(x + 6)$$

with $2x \cdot x$ and $1 \cdot 6$ indicated.

Also notice that $13x$, the middle term, is the sum of the following products:

$$2x^2 + 13x + 6 = (2x + 1)(x + 6)$$

$$\begin{array}{c} 1x \\ +12x \\ \hline 13x \end{array} \quad \text{Middle term}$$

Let's use this pattern to factor $5x^2 + 7x + 2$. First, we find factors of $5x^2$. Since all numerical coefficients in this trinomial are positive, we will use factors with positive numerical coefficients only. Thus, the factors of $5x^2$ are $5x$ and x. Let's try these factors as first terms of the binomials. Thus far, we have

$$5x^2 + 7x + 2 = (5x + \quad)(x + \quad)$$

Next, we need to find positive factors of 2. Positive factors of 2 are 1 and 2. Now we try possible combinations of these factors as second terms of the binomials until we obtain a middle term of $7x$.

$$(5x + 1)(x + 2) = 5x^2 + 11x + 2$$

$$\begin{array}{c} 1x \\ +10x \\ \hline 11x \end{array} \longrightarrow \text{Incorrect middle term}$$

Let's try switching factors 2 and 1.

$$(5x + 2)(x + 1) = 5x^2 + 7x + 2$$

$$\underbrace{}$$

$$\frac{2x}{+5x}$$
$$7x \longrightarrow \text{Correct middle term}$$

Thus the factored form of $5x^2 + 7x + 2$ is $(5x + 2)(x + 1)$. To check, we multiply $(5x + 2)$ and $(x + 1)$. The product is $5x^2 + 7x + 2$.

EXAMPLE 1

Factor $3x^2 + 11x + 6$.

Solution Since all numerical coefficients are positive, we use factors with positive numerical coefficients. We first find factors of $3x^2$.

Factors of $3x^2$: $3x^2 = 3x \cdot x$

If factorable, the trinomial will be of the form

$$3x^2 + 11x + 6 = (3x + \quad)(x + \quad)$$

Next we factor 6.

Factors of 6: $6 = 1 \cdot 6$, $6 = 2 \cdot 3$

Now we try combinations of factors of 6 until a middle term of $11x$ is obtained. Let's try 1 and 6 first.

$$(3x + 1)(x + 6) = 3x^2 + 19x + 6$$

$$\frac{1x}{+18x}$$
$$19x \longrightarrow \text{Incorrect middle term}$$

Now let's next try 6 and 1.

$$(3x + 6)(x + 1)$$

Before multiplying, notice that the terms of the factor $3x + 6$ have a common factor of 3. The terms of the original trinomial $3x^2 + 11x + 6$ have no common factor other than 1, so the terms of the factored form of $3x^2 + 11x + 6$ can contain no common factor other than 1. This means that $(3x + 6)(x + 1)$ is not a factored form.

Next let's try 2 and 3 as last terms.

$$(3x + 2)(x + 3) = 3x^2 + 11x + 6$$

$$\frac{2x}{+9x}$$
$$11x \longrightarrow \text{Correct middle term}$$

Thus the factored form of $3x^2 + 11x + 6$ is $(3x + 2)(x + 3)$.

> **Helpful Hint**
>
> If the terms of a trinomial have no common factor (other than 1), then the terms of neither of its binomial factors will contain a common factor (other than 1).

For example,

$$3x^2 + 11x + 6 \qquad (3x + 3)(x + 2)$$

$\underbrace{}$
no common factor

↑   ↑
Common factor of 3,
so cannot be the
factorization of
$3x^2 + 11x + 6$

✔ **CONCEPT CHECK**

Do the terms of $3x^2 + 29x + 18$ have a common factor? Without multiplying, decide which of the following factored forms could not be a factored form of $3x^2 + 29x + 18$.

a. $(3x + 18)(x + 1)$ **b.** $(3x + 2)(x + 9)$ **c.** $(3x + 6)(x + 3)$ **d.** $(3x + 9)(x + 2)$

EXAMPLE 2

Factor $8x^2 - 22x + 5$.

Solution   Factors of $8x^2$: $8x^2 = 8x \cdot x$, $8x^2 = 4x \cdot 2x$

We'll try $8x$ and x.

$$8x^2 - 22x + 5 = (8x + \quad)(x + \quad)$$

Since the middle term, $-22x$, has a negative numerical coefficient, we factor 5 into negative factors.

Factors of 5: $5 = -1 \cdot -5$

Let's try -1 and -5.

$$(8x - 1)(x - 5) = 8x^2 - 41x + 5$$

$\underbrace{\quad -1x \quad}$
$+(-40x)$
$-41x \longrightarrow$ Incorrect middle term

Now let's try -5 and -1.

$$(8x - 5)(x - 1) = 8x^2 - 13x + 5$$

$\underbrace{\quad -5x \quad}$
$+(-8x)$
$-13x \longrightarrow$ Incorrect middle term

Don't give up yet! We can still try other factors of $8x^2$. Let's try $4x$ and $2x$ with -1 and -5.

$$(4x - 1)(2x - 5) = 8x^2 - 22x + 5$$

$\underbrace{\quad -2x \quad}$
$+(-20x)$
$-22x \longrightarrow$ Correct middle term

The factored form of $8x^2 - 22x + 5$ is $(4x - 1)(2x - 5)$.

EXAMPLE 3

Factor: $2x^2 + 13x - 7$

Solution   Factors of $2x^2$: $2x^2 = 2x \cdot x$

Factors of -7: $-7 = -1 \cdot 7$, $-7 = 1 \cdot -7$

We try possible combinations of these factors:

Concept Check Answer:
no; a, c, d

$$(2x + 1)(x - 7) = 2x^2 - 13x - 7 \quad \text{Incorrect middle term}$$
$$(2x - 1)(x + 7) = 2x^2 + 13x - 7 \quad \text{Correct middle term}$$

The factored form of $2x^2 + 13x - 7$ is $(2x - 1)(x + 7)$.

EXAMPLE 4

Factor $10x^2 - 13xy - 3y^2$.

Solution Factors of $10x^2$: $\quad 10x^2 = 10x \cdot x, \quad 10x^2 = 2x \cdot 5x$

Factors of $-3y^2$: $\quad -3y^2 = -3y \cdot y, \quad -3y^2 = 3y \cdot -y$

We try some combinations of these factors:

$$(10x - 3y)(x + y) = 10x^2 + 7xy - 3y^2$$
$$(x + 3y)(10x - y) = 10x^2 + 29xy - 3y^2$$
$$(5x + 3y)(2x - y) = 10x^2 + xy - 3y^2$$
$$(2x - 3y)(5x + y) = 10x^2 - 13xy - 3y^2 \quad \text{Correct middle term}$$

The factored form of $10x^2 - 13xy - 3y^2$ is $(2x - 3y)(5x + y)$.

2 Don't forget that the best first step in factoring any polynomial is to look for a common factor to factor out.

EXAMPLE 5

Factor $24x^4 + 40x^3 + 6x^2$.

Solution Notice that all three terms have a common factor of $2x^2$. First, factor out $2x^2$.

$$24x^4 + 40x^3 + 6x^2 = 2x^2(12x^2 + 20x + 3)$$

Next, factor $12x^2 + 20x + 3$.

Factors of $12x^2$: $\quad 12x^2 = 6x \cdot 2x, \quad 12x^2 = 4x \cdot 3x, \quad 12x^2 = 12x \cdot x$

Since all terms in the trinomial have positive numerical coefficients, factor 3 using positive factors only.

Factors of 3: $\quad 3 = 1 \cdot 3$

We try some combinations of the factors.

$$2x^2(4x + 3)(3x + 1) = 2x^2(12x^2 + 13x + 3)$$
$$2x^2(12x + 1)(x + 3) = 2x^2(12x^2 + 37x + 3)$$
$$2x^2(2x + 3)(6x + 1) = 2x^2(12x^2 + 20x + 3) \quad \text{Correct middle term}$$

The factored form of $24x^4 + 40x^3 + 6x^2$ is $2x^2(2x + 3)(6x + 1)$.

> **Helpful Hint**
>
> Don't forget to include the common factor in the factored form.

EXAMPLE 6

Factor $4x^2 - 12x + 9$.

Solution Factors of $4x^2$: $\quad 4x^2 = 2x \cdot 2x, \quad 4x^2 = 4x \cdot x$

Since the middle term $-12x$ has a negative numerical coefficient, factor 9 into negative factors only.

Factors of 9: $9 = -3 \cdot -3$, $9 = -1 \cdot -9$

The correct combination is

$$(2x - 3)(2x - 3) = 4x^2 - 12x + 9$$

$-6x$
$+(-6x)$
$-12x$ ⟶ *Correct middle term*

Thus, $4x^2 - 12x + 9 = (2x - 3)(2x - 3)$, which can also be written as $(2x - 3)^2$.

3   Notice in Example 6 that $4x^2 - 12x + 9 = (2x - 3)^2$. The trinomial $4x^2 - 12x + 9$ is called a **perfect square trinomial** since it is the square of the binomial $2x - 3$.

In the last chapter, we learned a shortcut special product for squaring a binomial, recognizing that

$$(a + b)^2 = a^2 + 2ab + b^2$$

The trinomial $a^2 + 2ab + b^2$ is a perfect square trinomial, since it is the square of the binomial $a + b$. We can use this pattern to help us factor perfect square trinomials. To use this pattern, we must first be able to recognize a perfect square trinomial. A trinomial is a perfect square when its first term is the square of some expression a, its last term is the square of some expression b, and its middle term is twice the product of the expressions a and b. When a trinomial fits this description, its factored form is $(a + b)^2$.

Perfect Square Trinomials

$$a^2 + 2ab + b^2 = (a + b)^2$$
$$a^2 - 2ab + b^2 = (a - b)^2$$

EXAMPLE 7

Factor $x^2 + 12x + 36$.

Solution   This trinomial is a perfect square trinomial since:

1. The first term is the square of x: $x^2 = (x)^2$.
2. The last term is the square of 6: $36 = (6)^2$.
3. The middle term is twice the product of x and 6: $12x = 2 \cdot x \cdot 6$.
Thus, $x^2 + 12x + 36 = (x + 6)^2$.

EXAMPLE 8

Factor $25x^2 + 25xy + 4y^2$.

Solution   Determine whether or not this trinomial is a perfect square by considering the same three questions.

1. Is the first term a square? Yes, $25x^2 = (5x)^2$.
2. Is the last term a square? Yes, $4y^2 = (2y)^2$.
3. Is the middle term twice the product of $5x$ and $2y$? **No.** $2 \cdot 5x \cdot 2y = 20xy$, not $25xy$.

Therefore, $25x^2 + 25xy + 4y^2$ is not a perfect square trinomial. It is factorable, though. Using earlier techniques, we find that $25x^2 + 25xy + 4y^2 = (5x + 4y)(5x + y)$.

> **Helpful Hint**
>
> A perfect square trinomial that is not recognized as such can be factored by other methods.

EXAMPLE 9

Factor $4m^2 - 4m + 1$.

Solution   This is a perfect square trinomial since $4m^2 = (2m)^2$, $1 = (1)^2$, and $4m = 2 \cdot 2m \cdot 1$.

$$4m^2 - 4m + 1 = (2m - 1)^2$$

CLASSROOM EXAMPLE
Factor $25x^2 - 20x + 4$.
answer: $(5x - 2)^2$

4   If we extend our work from Section 6.1, grouping can also be used to factor trinomials of the form $ax^2 + bx + c$. To use this method, write the trinomial as a four-term polynomial. For example, to factor $2x^2 + 11x + 12$ using grouping, find two numbers whose product is $2 \cdot 12 = 24$ and whose sum is 11. Since we want a positive product and a positive sum, we consider positive pairs of factors of 24 only.

Factors of 24	*Sum of Factors*
1, 24	$1 + 24 = 25$
2, 12	$2 + 12 = 14$
3, 8	$3 + 8 = 11$

The factors are 3 and 8. Use these factors to write the middle term $11x$ as $3x + 8x$. Replace $11x$ with $3x + 8x$ in the original trinomial and factor by grouping.

$$
\begin{aligned}
2x^2 + 11x + 12 &= 2x^2 + 3x + 8x + 12 \\
&= (2x^2 + 3x) + (8x + 12) \\
&= x(2x + 3) + 4(2x + 3) \\
&= (2x + 3)(x + 4)
\end{aligned}
$$

In general, we have the following:

> ### Factoring Trinomials of the Form $ax^2 + bx + c$ by Grouping
>
> **Step 1.**   Find two numbers whose product is $a \cdot c$ and whose sum is b.
>
> **Step 2.**   Write the middle term, bx, using the factors found in Step 1.
>
> **Step 3.**   Factor by grouping.

EXAMPLE 10

Factor $8x^2 - 14x + 5$ by grouping.

Solution   This trinomial is of the form $ax^2 + bx + c$ with $a = 8$, $b = -14$, and $c = 5$.

Step 1.   Find two numbers whose product is $a \cdot c$ or $8 \cdot 5 = 40$, and whose sum is b or -14. The numbers are -4 and -10.

CLASSROOM EXAMPLE
Factor $3x^2 + 14x + 8$ by grouping.

answer: $(x + 4)(3x + 2)$

Step 2.   Write $-14x$ as $-4x - 10x$ so that

$$8x^2 - 14x + 5 = 8x^2 - 4x - 10x + 5$$

Step 3.   Factor by grouping.

$$
\begin{aligned}
8x^2 - 4x - 10x + 5 &= 4x(2x - 1) - 5(2x - 1) \\
&= (2x - 1)(4x - 5)
\end{aligned}
$$

EXAMPLE 11

Factor $3x^2 - x - 10$ by grouping.

Solution   In $3x^2 - x - 10$, $a = 3$, $b = -1$, and $c = -10$.

Step 1. Find two numbers whose product is $a \cdot c$ or $3(-10) = -30$ and whose sum is b or -1. The numbers are -6 and 5.

Step 2. $3x^2 - x - 10 = 3x^2 - 6x + 5x - 10$

Step 3. $\qquad\qquad = 3x(x - 2) + 5(x - 2)$

$\qquad\qquad\qquad = (x - 2)(3x + 5)$

EXAMPLE 12

Factor $4x^2 + 11x - 3$ by grouping.

Solution   In $4x^2 + 11x - 3$, $a = 4$, $b = 11$, and $c = -3$.

Step 1. Find two numbers whose product is $a \cdot c$ or $4(-3) = -12$ and whose sum is b or 11. The numbers are -1 and 12.

Step 2. $4x^2 + 11x - 3 = 4x^2 - 1x + 12x - 3$

$\qquad\qquad\qquad = x(4x - 1) + 3(4x - 1)$

$\qquad\qquad\qquad = (4x - 1)(x + 3)$

STUDY SKILLS REMINDER

Are You Satisfied with Your Performance in This Course Thus Far?

If not, ask yourself the following questions:

▶ Am I attending all class periods and arriving on time?

▶ Am I working and checking my homework assignments?

▶ Am I getting help when I need it?

▶ In addition to my instructor, am I using the supplements to this text that could help me? For example, the tutorial video lessons? MathPro, the tutorial software?

▶ Am I satisfied with my performance on quizzes and tests?

If you answered no to *any* of these questions, read or reread Section 1.1 for suggestions in these areas. Also, you may want to contact your instructor for additional feedback.

MENTAL MATH

State whether or not each trinomial is a perfect trinomial square.

1. $x^2 + 14x + 49$ yes

2. $9x^2 - 12x + 4$ yes

3. $y^2 + 2y + 4$ no

4. $x^2 - 4x + 2$ no

5. $9y^2 + 6y + 1$ yes

6. $y^2 - 16y + 64$ yes

EXERCISE SET 6.3

STUDY GUIDE/SSM   CD/VIDEO   PH MATH TUTOR CENTER   MathXL®Tutorials ON CD   MathXL®   MyMathLab®

Factor completely. See Examples 1 through 5 or 10 through 12.

1. $2x^2 + 13x + 15$ **2.** $3x^2 + 8x + 4$

3. $2x^2 - 9x - 5$ $(2x + 1)(x - 5)$ **4.** $3x^2 + 20x - 63$ $(3x - 7)(x + 9)$

5. $2y^2 - y - 6$ $(2y + 3)(y - 2)$ **6.** $8y^2 - 17y + 9$ $(y - 1)(8y - 9)$

7. $16a^2 - 24a + 9$ $(4a - 3)^2$ **8.** $25x^2 + 20x + 4$ $(5x + 2)^2$

9. $36r^2 - 5r - 24$ **10.** $20r^2 + 27r - 8$ $(4r - 1)(5r + 8)$

11. $10x^2 + 17x + 3$ **12.** $21x^2 - 41x + 10$

13. $21x^2 - 48x - 45$ **14.** $12x^2 - 14x - 10$

15. $12x^2 - 14x - 6$ **16.** $20x^2 - 2x + 6$

17. $4x^3 - 9x^2 - 9x$ **18.** $6x^3 - 31x^2 + 5x$

Factor the following perfect square trinomials. See Examples 6 through 9.   **26.** $(2xy - 7)^2$

19. $x^2 + 22x + 121$ $(x + 11)^2$ **20.** $x^2 + 18x + 81$ $(x + 9)^2$

21. $x^2 - 16x + 64$ $(x - 8)^2$ **22.** $x^2 - 12x + 36$ $(x - 6)^2$

23. $16y^2 - 40y + 25$ $(4y - 5)^2$ **24.** $9y^2 + 48y + 64$ $(3y + 8)^2$

25. $x^2y^2 - 10xy + 25$ $(xy - 5)^2$ **26.** $4x^2y^2 - 28xy + 49$

27. Describe a perfect square trinomial. **27.–28.** answers may vary

28. Write a perfect square trinomial that factors as $(x + 3y)^2$.

MIXED PRACTICE   See page 375 for answers.
Factor the following completely.

29. $2x^2 - 7x - 99$ **30.** $2x^2 + 7x - 72$

31. $4x^2 - 8x - 21$ **32.** $6x^2 - 11x - 10$

33. $30x^2 - 53x + 21$ **34.** $21x^2 - 6x - 30$

35. $24x^2 - 58x + 9$ **36.** $36x^2 + 55x - 14$

37. $9x^2 - 24xy + 16y^2$ $(3x - 4y)^2$ **38.** $25x^2 + 60xy + 36y^2$ $(5x + 6y)^2$

39. $x^2 - 14xy + 49y^2$ $(x - 7y)^2$ **40.** $x^2 + 10xy + 25y^2$ $(x + 5y)^2$

41. $2x^2 + 7x + 5$ $(2x + 5)(x + 1)$ **42.** $2x^2 + 7x + 3$ $(2x + 1)(x + 3)$

43. $3x^2 - 5x + 1$ prime   **44.** $3x^2 - 7x + 6$ prime

45. $-2y^2 + y + 10$ **46.** $-4x^2 - 23x + 6$

47. $16x^2 + 24xy + 9y^2$ $(4x + 3y)^2$ **48.** $4x^2 - 36xy + 81y^2$ $(2x - 9y)^2$

49. $8x^2y + 34xy - 84y$ **50.** $6x^2y^2 - 2xy^2 - 60y^2$

51. $3x^2 + x - 2$ $(3x - 2)(x + 1)$ **52.** $8y^2 + y - 9$ $(y - 1)(8y + 9)$

53. $x^2y^2 + 4xy + 4$ $(xy + 2)^2$ **54.** $x^2y^2 - 6xy + 9$ $(xy - 3)^2$

55. $49y^2 + 42xy + 9x^2$ $(7y + 3x)^2$ **56.** $16x^2 - 8xy + y^2$ $(4x - y)^2$

57. $3x^2 - 42x + 63$ **58.** $5x^2 - 75x + 60$

59. $42a^2 - 43a + 6$ **60.** $54a^2 + 39ab - 8b^2$

61. $18x^2 - 9x - 14$ **62.** $8x^2 + 6x - 27$

63. $25p^2 - 70pq + 49q^2$ **64.** $36p^2 - 18pq + 9q^2$

65. $15x^2 - 16x - 15$ **66.** $12x^2 + 7x - 12$

67. $-27t + 7t^2 - 4$ **68.** $4t^2 - 7 - 3t$

The area of the largest square in the figure is $(a + b)^2$. Use this figure to answer Exercises 69–70.

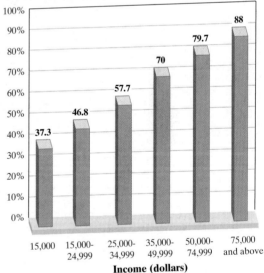

△ **69.** Write the area of the largest square as the sum of the areas of the smaller squares and rectangles. $a^2 + 2ab + b^2$

△ **70.** What factoring formula from this section is visually represented by this square?   perfect square trinomial

REVIEW AND REVIEW 73. $a^3 + 27$ **74.** $z^3 - 8$ **75.** $y^3 - 125$
Multiply the following. See Section 5.4.   **76.** $m^3 + 343$

71. $(x - 2)(x + 2)$ $x^2 - 4$ **72.** $(y^2 + 3)(y^2 - 3)$ $y^4 - 9$

73. $(a + 3)(a^2 - 3a + 9)$ **74.** $(z - 2)(z^2 + 2z + 4)$

75. $(y - 5)(y^2 + 5y + 25)$ **76.** $(m + 7)(m^2 - 7m + 49)$

As of 2001, approximately 66% of U.S. households had a personal computer. The following graph shows the percent of selected households having a computer grouped according to household income. See Section 1.9.

Percent of households with computer by income:
- 15,000: 37.3
- 15,000–24,999: 46.8
- 25,000–34,999: 57.7
- 35,000–49,999: 70
- 50,000–74,999: 79.7
- 75,000 and above: 88

Income (dollars)

       77. $75,000 and above
77. Which range of household income corresponds to the greatest percent of households having a personal computer?

78. Which range of household income corresponds to the greatest increase in percent of households having a personal computer? $35,000–$49,999   **79.–80.** answers may vary

79. Describe any trend you notice from this graph.

80. Why don't the percents shown in the graph add to 100%?

1. $(2x + 3)(x + 5)$ **2.** $(3x + 2)(x + 2)$ **9.** $(9r - 8)(4r + 3)$ **11.** $(5x + 1)(2x + 3)$ **12.** $(7x - 2)(3x - 5)$ **13.** $3(7x + 5)(x - 3)$
14. $2(3x - 5)(2x + 1)$ **15.** $2(2x - 3)(3x + 1)$ **16.** $2(10x^2 - x + 3)$ **17.** $x(4x + 3)(x - 3)$ **18.** $x(x - 5)(6x - 1)$

Concept Extensions

Find a positive value of b so that each trinomial is factorable.

81. $3x^2 + bx - 5$ $2; 14$ **82.** $2y^2 + by + 3$ $5; 7$

83. $2z^2 + bz - 7$ $5; 13$ **84.** $5x^2 + bx - 1$ 4

Find a positive value of c so that each trinomial is factorable.

85. $5x^2 + 7x + c$ 2 **86.** $7x^2 + 22x + c$ $3; 15; 16$

87. $3x^2 - 8x + c$ $4; 5$ **88.** $11y^2 - 40y + c$ $21; 29; 36$

Factor completely. Don't forget to first factor out the greatest common factor.

89. $-12x^3y^2 + 3x^2y^2 + 15xy^2$ **90.** $-12r^3x^2 + 38r^2x^2 + 14rx^2$

91. $-30p^3q + 88p^2q^2 + 6pq^3$ **92.** $3x^3y^2 + 3x^2y^3 - 18xy^4$

93. $4x^2(y - 1)^2 + 10x(y - 1)^2 + 25(y - 1)^2$

94. $3x^2(a + 3)^3 - 28x(a + 3)^3 + 25(a + 3)^3$

Factor.

95. $3x^{2n} + 17x^n + 10$ **96.** $2x^{2n} + 5x^n - 12$
95. $(3x^n + 2)(x^n + 5)$ **96.** $(2x^n - 3)(x^n + 4)$

6.4   FACTORING BINOMIALS

29. $(2x + 11)(x - 9)$
30. $(2x - 9)(x + 8)$
31. $(2x - 7)(2x + 3)$
32. $(3x + 2)(2x - 5)$
33. $(6x - 7)(5x - 3)$
34. $3(7x^2 - 2x - 10)$
35. $(4x - 9)(6x - 1)$
36. $(4x + 7)(9x - 2)$
45. $(-2y + 5)(y + 2)$
46. $(-4x + 1)(x + 6)$
49. $2y(4x - 7)(x + 6)$
50. $2y^2(3x - 10)(x + 3)$
57. $3(x^2 - 14x + 21)$
58. $5(x^2 - 15x + 12)$
59. $(7a - 6)(6a - 1)$
60. $(9a + 8b)(6a - b)$
61. $(6x - 7)(3x + 2)$
62. $(4x + 9)(2x - 3)$
63. $(5p - 7q)^2$
64. $9(4p^2 - 2pq + q^2)$
65. $(5x + 3)(3x - 5)$
66. $(3x + 4)(4x - 3)$
67. $(7t + 1)(t - 4)$
68. $(4t - 7)(t + 1)$
89. $-3xy^2(4x - 5)(x + 1)$
90. $-2rx^2(2r - 7)(3r + 1)$
91. $-2pq(p - 3q)(15p + q)$
92. $3xy^2(x + 3y)(x - 2y)$
93. $(y - 1)^2(4x^2 + 10x + 25)$
94. $(a + 3)^3(x - 1)(3x - 25)$

Objectives

1 Factor the difference of two squares.

2 Factor the sum or difference of two cubes.

1   When learning to multiply binomials in Chapter 5, we studied a special product, the product of the sum and difference of two terms, a and b:

$$(a + b)(a - b) = a^2 - b^2$$

For example, the product of $x + 3$ and $x - 3$ is

$$(x + 3)(x - 3) = x^2 - 9$$

The binomial $x^2 - 9$ is called a **difference of squares**. In this section, we use the pattern for the product of a sum and difference to factor the binomial difference of squares.

To use this pattern to help us factor, we must be able to recognize a difference of squares. A binomial is a difference of squares when it is the difference of the square of some expression a and the square of some expression b.

Difference of Two Squares

$$a^2 - b^2 = (a + b)(a - b)$$

EXAMPLE 1

Factor $x^2 - 25$.

Solution $x^2 - 25$ is the difference of two squares since $x^2 - 25 = x^2 - 5^2$. Therefore,

$$x^2 - 25 = x^2 - 5^2 = (x + 5)(x - 5)$$

Multiply to check.

EXAMPLE 2

Factor each difference of squares.

a. $4x^2 - 1$ **b.** $25a^2 - 9b^2$

Solution   **a.** $4x^2 - 1 = (2x)^2 - 1^2 = (2x + 1)(2x - 1)$
b. $25a^2 - 9b^2 = (5a)^2 - (3b)^2 = (5a + 3b)(5a - 3b)$

EXAMPLE 3

Factor $x^4 - y^6$.

Solution   Write x^4 as $(x^2)^2$ and y^6 as $(y^3)^2$.

$$x^4 - y^6 = (x^2)^2 - (y^3)^2 = (x^2 + y^3)(x^2 - y^3)$$

EXAMPLE 4

Factor $9x^2 - 36$.

Solution   Remember when factoring always to check first for common factors. If there are common factors, factor out the GCF and then factor the resulting polynomial.

$$\begin{aligned} 9x^2 - 36 &= 9(x^2 - 4) \qquad \text{Factor out the GCF 9.} \\ &= 9(x^2 - 2^2) \\ &= 9(x + 2)(x - 2) \end{aligned}$$

In this example, if we forget to factor out the GCF first, we still have the difference of two squares.

$$9x^2 - 36 = (3x)^2 - (6)^2 = (3x + 6)(3x - 6)$$

This binomial has not been factored completely since both terms of both binomial factors have a common factor of 3.

$$3x + 6 = 3(x + 2) \quad \text{and} \quad 3x - 6 = 3(x - 2)$$

Then

$$9x^2 - 36 = (3x + 6)(3x - 6) = 3(x + 2)3(x - 2) = 9(x + 2)(x - 2)$$

Factoring is easier if the GCF is factored out first before using other methods.

EXAMPLE 5

Factor $x^2 + 4$.

Solution   The binomial $x^2 + 4$ is the **sum** of squares since we can write $x^2 + 4$ as $x^2 + 2^2$. We might try to factor using $(x + 2)(x + 2)$ or $(x - 2)(x - 2)$. But when multiplying to check, neither factoring is correct.

$$(x + 2)(x + 2) = x^2 + 4x + 4$$
$$(x - 2)(x - 2) = x^2 - 4x + 4$$

In both cases, the product is a trinomial, not the required binomial. In fact, $x^2 + 4$ is a **prime polynomial**.

> **Helpful Hint**
> After the greatest common factor has been removed, the *sum* of two squares cannot be factored further using real numbers.

2   Although the sum of two squares usually does not factor, the sum or difference of two cubes can be factored and reveals factoring patterns. The pattern for the sum of cubes is illustrated by multiplying the binomial $x + y$ and the trinomial $x^2 - xy + y^2$.

$$
\begin{array}{r}
x^2 - xy + y^2 \\
x + y \\
\hline
x^2y - xy^2 + y^3 \\
x^3 - x^2y + xy^2 \\
\hline
x^3 \qquad\qquad + y^3
\end{array}
$$

$$(x + y)(x^2 - xy + y^2) = x^3 + y^3 \qquad \textit{Sum of cubes.}$$

The pattern for the difference of two cubes is illustrated by multiplying the binomial $x - y$ by the trinomial $x^2 + xy + y^2$. The result is

$$(x - y)(x^2 + xy + y^2) = x^3 - y^3 \qquad \textit{Difference of cubes.}$$

> **Sum or Difference of Two Cubes**
> $$a^3 + b^3 = (a + b)(a^2 - ab + b^2)$$
> $$a^3 - b^3 = (a - b)(a^2 + ab + b^2)$$

EXAMPLE 6

Factor $x^3 + 8$.

Solution   First, write the binomial in the form $a^3 + b^3$.

$$x^3 + 8 = x^3 + 2^3 \qquad \textit{Write in the form } a^3 + b^3.$$

If we replace a with x and b with 2 in the formula above, we have

$$x^3 + 2^3 = (x + 2)[x^2 - (x)(2) + 2^2]$$
$$= (x + 2)(x^2 - 2x + 4)$$

CLASSROOM EXAMPLE
Factor $x^3 + 27$.
answer: $(x + 3)(x^2 - 3x + 9)$

> **Helpful Hint**
> When factoring sums or differences of cubes, notice the sign patterns.
>
> same sign
> $$x^3 + y^3 = (x + y)(x^2 - xy + y^2)$$
> opposite signs   always positive
>
> same sign
> $$x^3 - y^3 = (x - y)(x^2 + xy + y^2)$$
> opposite signs   always positive

EXAMPLE 7

Factor $y^3 - 27$.

Solution

$$y^3 - 27 = y^3 - 3^3 \qquad \text{Write in the form } a^3 - b^3.$$
$$= (y - 3)[y^2 + (y)(3) + 3^2]$$
$$= (y - 3)(y^2 + 3y + 9)$$

CLASSROOM EXAMPLE

Factor $z^3 - 8$.

answer: $(z - 2)(z^2 + 2z + 4)$

EXAMPLE 8

Factor $64x^3 + 1$.

Solution

$$64x^3 + 1 = (4x)^3 + 1^3$$
$$= (4x + 1)[(4x)^2 - (4x)(1) + 1^2]$$
$$= (4x + 1)(16x^2 - 4x + 1)$$

CLASSROOM EXAMPLE

Factor $125a^3 - 1$.

answer: $(5a - 1)(25a^2 + 5a + 1)$

 EXAMPLE 9

Factor $54a^3 - 16b^3$.

Solution Remember to factor out common factors first before using other factoring methods.

$$54a^3 - 16b^3 = 2(27a^3 - 8b^3) \qquad \text{Factor out the GCF 2.}$$
$$= 2[(3a)^3 - (2b)^3] \qquad \text{Difference of two cubes.}$$
$$= 2(3a - 2b)[(3a)^2 + (3a)(2b) + (2b)^2]$$
$$= 2(3a - 2b)(9a^2 + 6ab + 4b^2)$$

CLASSROOM EXAMPLE

Factor $16x^3 + 250y^3$.

answer:
$2(2x + 5y)(4x^2 - 10xy + 25y^2)$

Graphing Calculator Explorations

$2 \to X$	
	2
$X^2 - 6X$	
	-8

A graphing calculator is a convenient tool for evaluating an expression at a given replacement value. For example, let's evaluate $x^2 - 6x$ when $x = 2$. To do so, store the value 2 in the variable x and then enter and evaluate the algebraic expression.

The value of $x^2 - 6x$ when $x = 2$ is -8. You may want to use this method for evaluating expressions as you explore the following.

We can use a graphing calculator to explore factoring patterns numerically. Use your calculator to evaluate $x^2 - 2x + 1$, $x^2 - 2x - 1$, and $(x - 1)^2$ for each value of x given in the table. What do you observe?

	$x^2 - 2x + 1$	$x^2 - 2x - 1$	$(x - 1)^2$
$x = 5$	16	14	16
$x = -3$	16	14	16
$x = 2.7$	2.89	0.89	2.89
$x = -12.1$	171.61	169.61	171.61
$x = 0$	1	-1	1

(continued)

Notice in each case that $x^2 - 2x - 1 \neq (x - 1)^2$. Because for each x in the table the value of $x^2 - 2x + 1$ and the value of $(x - 1)^2$ are the same, we might guess that $x^2 - 2x + 1 = (x - 1)^2$. We can verify our guess algebraically with multiplication:

$$(x - 1)(x - 1) = x^2 - x - x + 1 = x^2 - 2x + 1$$

MENTAL MATH

Write each number as a square.

1. 1 1^2 **2.** 25 5^2 **3.** 81 9^2 **4.** 64 8^2 **5.** 9 3^2 **6.** 100 10^2

Write each number as a cube.

7. 1 1^3 **8.** 64 4^3 **9.** 8 2^3 **10.** 27 3^3

EXERCISE SET 6.4

STUDY GUIDE/SSM | CD/ VIDEO | PH MATH TUTOR CENTER | MathXL®Tutorials ON CD | MathXL® | MyMathLab®

Factor the difference of two squares. See Examples 1 through 4.

1. $x^2 - 4$ $(x + 2)(x - 2)$ **2.** $y^2 - 81$ $(y + 9)(y - 9)$

3. $y^2 - 49$ $(y + 7)(y - 7)$ **4.** $x^2 - 100$

5. $25y^2 - 9$ $(5y + 3)(5y - 3)$ **6.** $49a^2 - 16$

7. $121 - 100x^2$ **8.** $144 - 81x^2$

9. $12x^2 - 27$ $3(2x + 3)(2x - 3)$ **10.** $36x^2 - 64$

11. $169a^2 - 49b^2$ **12.** $225a^2 - 81b^2$

13. $x^2y^2 - 1$ $(xy + 1)(xy - 1)$ **14.** $16 - a^2b^2$

15. $x^4 - 9$ $(x^2 + 3)(x^2 - 3)$ **16.** $y^4 - 25$ $(y^2 + 5)(y^2 - 5)$

17. $49a^4 - 16$ **18.** $49b^4 - 1$

19. $x^4 - y^{10}$ $(x^2 + y^5)(x^2 - y^5)$ **20.** $x^{14} - y^4$

21. What binomial multiplied by $(x - 6)$ gives the difference of two squares? $(x + 6)$

22. What binomial multiplied by $(5 + y)$ gives the difference of two squares? $(5 - y)$

Factor the sum or difference of two cubes. See Examples 6 through 9.

23. $a^3 + 27$ **24.** $b^3 - 8$

25. $8a^3 + 1$ **26.** $64x^3 - 1$

27. $5k^3 + 40$ $5(k + 2)(k^2 - 2k + 4)$ **28.** $6r^3 - 162$

29. $x^3y^3 - 64$ **30.** $8x^3 - y^3$

31. $x^3 + 125$ **32.** $a^3 - 216$

33. $24x^4 - 81xy^3$ **34.** $375y^6 - 24y^3$

35. What binomial multiplied by $(4x^2 - 2xy + y^2)$ gives the sum or difference of two cubes? $(2x + y)$

36. What binomial multiplied by $(1 + 4y + 16y^2)$ gives the sum or difference of two cubes? $(1 - 4y)$

MIXED PRACTICE

Factor the binomials completely.

37. $x^2 - 121$ $(x + 11)(x - 11)$ **38.** $x^2 - 36$ $(x + 6)(x - 6)$

39. $81 - p^2$ $(9 - p)(9 + p)$ **40.** $100 - t^2$

41. $4r^2 - 1$ $(2r + 1)(2r - 1)$ **42.** $9t^2 - 1$

43. $9x^2 - 16$ $(3x + 4)(3x - 4)$ **44.** $36y^2 - 25$

45. $16r^2 + 1$ prime   **46.** $49y^2 + 1$ prime

47. $27 - t^3$ $(3 - t)(9 + 3t + t^2)$ **48.** $125 + r^3$

49. $8r^3 - 64$ **50.** $54r^3 + 2$

51. $t^3 - 343$ **52.** $s^3 + 216$

4. $(x + 10)(x - 10)$ **6.** $(7a + 4)(7a - 4)$ **7.** $(11 + 10x)(11 - 10x)$ **8.** $9(4 + 3x)(4 - 3x)$ **10.** $4(3x + 4)(3x - 4)$
11. $(13a + 7b)(13a - 7b)$ **12.** $9(5a + 3b)(5a - 3b)$ **14.** $(4 + ab)(4 - ab)$ **17.** $(7a^2 + 4)(7a^2 - 4)$ **18.** $(7b^2 + 1)(7b^2 - 1)$
20. $(x^7 + y^2)(x^7 - y^2)$ **23.** $(a + 3)(a^2 - 3a + 9)$ **24.** $(b - 2)(b^2 + 2b + 4)$ **25.** $(2a + 1)(4a^2 - 2a + 1)$
26. $(4x - 1)(16x^2 + 4x + 1)$ **28.** $6(r - 3)(r^2 + 3r + 9)$ **29.** $(xy - 4)(x^2y^2 + 4xy + 16)$ **30.** $(2x - y)(4x^2 + 2xy + y^2)$
31. $(x + 5)(x^2 - 5x + 25)$ **32.** $(a - 6)(a^2 + 6a + 36)$ **33.** $3x(2x - 3y)(4x^2 + 6xy + 9y^2)$ **34.** $3y^3(5y - 2)(25y^2 + 10y + 4)$
40. $(10 + t)(10 - t)$ **42.** $(3t + 1)(3t - 1)$ **44.** $(6y + 5)(6y - 5)$ **48.** $(5 + r)(25 - 5r + r^2)$ **49.** $8(r - 2)(r^2 + 2r + 4)$
50. $2(3r + 1)(9r^2 - 3r + 1)$ **51.** $(t - 7)(t^2 + 7t + 49)$ **52.** $(s + 6)(s^2 - 6s + 36)$

53. $(x + 13y)(x - 13y)$ **54.** $(x + 15y)(x - 15y)$ **56.** $(xy - z)(x^2y^2 + xyz + z^2)$ **57.** $(xy + 1)(x^2y^2 - xy + 1)$

53. $x^2 - 169y^2$

54. $x^2 - 225y^2$

55. $x^2y^2 - z^2$ $(xy + z)(xy - z)$ **56.** $x^3y^3 - z^3$

57. $x^3y^3 + 1$

58. $x^2y^2 + z^2$ prime

59. $s^3 - 64t^3$

60. $8t^3 + s^3$

61. $18r^2 - 8$ $2(3r + 2)(3r - 2)$ **62.** $32t^2 - 50$ $2(4t + 5)(4t - 5)$

63. $9xy^2 - 4x$ $x(3y + 2)(3y - 2)$ **64.** $16xy^2 - 64x$

65. $25y^4 - 100y^2$ **66.** $xy^3 - 9xyz^2$

67. $x^3y - 4xy^3$ $xy(x - 2y)(x + 2y)$**68.** $12s^3t^3 + 192s^5t$

69. $8s^6t^3 + 100s^3t^6$ $4s^3t^3(2s^3 + 25t^3)$**70.** $25x^5y + 121x^3y$

71. $27x^2y^3 - xy^2$ $xy^2(27xy - 1)$ **72.** $8x^3y^3 + x^3y$

REVIEW AND PREVIEW

73. $4x^3 + 2x^2 - 1 + \frac{3}{x}$ **74.** $y^2 + 3 - \frac{2}{y} + \frac{1}{3y^2}$

Divide the following. See Section 5.6.

73. $\dfrac{8x^4 + 4x^3 - 2x + 6}{2x}$ **74.** $\dfrac{3y^4 + 9y^2 - 6y + 1}{3y^2}$

Use long division to divide the following. See Section 5.6.

75. $\dfrac{2x^2 - 3x - 2}{x - 2}$ $2x + 1$ **76.** $\dfrac{4x^2 - 21x + 21}{x - 3}$

77. $\dfrac{3x^2 + 13x + 10}{x + 3}$ **78.** $\dfrac{5x^2 + 14x + 12}{x + 2}$

76. $4x - 9 - \frac{6}{x - 3}$ **77.** $3x + 4 - \frac{2}{x + 3}$ **78.** $5x + 4 + \frac{4}{x + 2}$

Concept Extensions   **79.** $(x + 2 + y)(x + 2 - y)$

Factor each expression completely.   **80.** $(y - 6 + z)(y - 6 - z)$

79. $(x + 2)^2 - y^2$ **80.** $(y - 6)^2 - z^2$

81. $a^2(b - 4) - 16(b - 4)$ **82.** $m^2(n + 8) - 9(n + 8)$

83. $(x^2 + 6x + 9) - 4y^2$ (*Hint:* Factor the trinomial in parentheses first.) $(x + 3 + 2y)(x + 3 - 2y)$

84. $(x^2 + 2x + 1) - 36y^2$ **85.** $x^{2n} - 100$

86. $x^{2n} - 81$ $(x^n + 9)(x^n - 9)$

87. An object is dropped from the top of Pittsburgh's USX Towers, which is 841 feet tall. (*Source: World Almanac* research) The height of the object after t seconds is given by the expression $841 - 16t^2$.

   a. Find the height of the object after 2 seconds.   777 ft

   b. Find the height of the object after 5 seconds.   441 ft

   c. To the nearest whole second, estimate when the object hits the ground.   7 sec

   d. Factor $841 - 16t^2$. $(29 + 4t)(29 - 4t)$

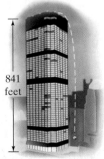

841 feet

88. A worker on the top of the Aetna Life Building in San Francisco accidentally drops a bolt. The Aetna Life Building is 529 feet tall. (*Source: World Almanac* research) The height of the bolt after t seconds is given by the expression $529 - 16t^2$.

   a. Find the height of the bolt after 1 second.   513 ft

   b. Find the height of the bolt after 4 seconds.   273 ft

   c. To the nearest whole second, estimate when the bolt hits the ground.   6 sec

   d. Factor $529 - 16t^2$. $(23 + 4t)(23 - 4t)$

59. $(s - 4t)(s^2 + 4st + 16t^2)$
60. $(2t + s)(4t^2 - 2st + s^2)$
64. $16x(y + 2)(y - 2)$
65. $25y^2(y - 2)(y + 2)$
66. $xy(y + 3z)(y - 3z)$
68. $12s^3t(t^2 + 16s^2)$
70. $x^3y(25x^2 + 121)$
72. $x^3y(8y^2 + 1)$

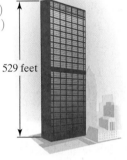

529 feet

89. At this writing, the world's tallest building is the Petronas Twin Towers in Kuala Lumpur, Malaysia, at a height of 1483 feet. (*Source:* Council on Tall Buildings and Urban Habitat) Suppose a worker is suspended 39 feet below the tip of the pinnacle atop one of the towers, at a height of 1444 feet above the ground. If the worker accidentally drops a bolt, the height of the bolt after t seconds is given by the expression $1444 - 16t^2$.

   a. Find the height of the bolt after 3 seconds.   1300 ft

   b. Find the height of the bolt after 7 seconds.   660 ft

   c. To the nearest whole second, estimate when the bolt hits the ground.   10 sec

   d. Factor $1444 - 16t^2$. $4(19 + 2t)(19 - 2t)$

90. A performer with the Moscow Circus is planning a stunt involving a free fall from the top of the Moscow State University building, which is 784 feet tall. (*Source:* Council on Tall Buildings and Urban Habitat) Neglecting air resistance, the performer's height above gigantic cushions positioned at ground level after t seconds is given by the expression $784 - 16t^2$.

81. $(a + 4)(a - 4)(b - 4)$ **82.** $(m + 3)(m - 3)(n + 8)$ **84.** $(x + 1 + 6y)(x + 1 - 6y)$ **85.** $(x^n + 10)(x^n - 10)$

c. To the nearest whole second, estimate when the performer reaches the cushions positioned at ground level.   7 sec

d. Factor $784 - 16t^2$. $16(7 + t)(7 - t)$

91. In your own words, explain how to tell whether a binomial is a difference of squares. Then explain how to factor a difference of squares.   answers may vary

a. Find the performer's height after 2 seconds.   720 ft

b. Find the performer's height after 5 seconds.   384 ft

92. In your own words, explain how to tell whether a binomial is a sum of cubes. Then explain how to factor a sum of cubes.   answers may vary

Spotlight on DECISION MAKING

Suppose you are shopping for a credit card and have received the following credit card offers in the mail.

Credit Card Offer No. 1

We offer a low 9.8% APR interest rate, coupled with a generous 20-day grace period and low $35 annual fee. We can offer you a $2000 credit limit.

In addition, you'll earn one frequent flier mile for every dollar charged on your card. And every time you use your card to pay for airline tickets, we'll even give you $250,000 in flight insurance—absolutely free!

Credit Card Offer No. 2

With our card, you'll get a 14.5% APR interest rate, absolutely *no* annual fee, a 25-day grace period, and a $2500 credit limit.

And, only with our card, you can receive a year-end cash bonus of up to 1% of the total value of purchases made to your card during the year! That's our way of saying "Thank you" for choosing our card.

Credit Card Offer No. 3

Just say "yes" to this offer, and you can get our lowest 17% APR interest rate, along with a 30-day grace period and easy-to-swallow $20 annual fee. You also qualify for a $3000 credit limit.

That's not all: you earn gift certificates for a local mall with every purchase! Plus, you'll receive $100,000 of flight insurance whenever you charge airline tickets.

You construct a decision grid to help make your choice. In the decision grid, give each of the decision criteria a rank reflecting its importance to you, with 1 being not important to 10 being very important. Then for each card offer, decide how well the criteria are supported, assigning a rating of 1 for poor support to a rating of 10 for excellent support. For Credit Card Offer #1, fill in the Score column by multiplying rank by rating for each criteria. Repeat for each credit card offer. Finally, total the scores for each credit card offer. The offer with the highest score is likely to be the best choice for you.

Based on your decision-grid analysis, which credit card would you choose? Explain.   *(continued)*

Spotlight on
DECISION
ꙮ MAKING

Criteria	Rank	Credit Card Offer #1		Credit Card Offer #2		Credit Card Offer #3	
		Rating	*Score*	*Rating*	*Score*	*Rating*	*Score*
Interest rate							
Annual fee							
Grace period							
Credit limit							
Automatic flight insurance							
Rebates/ incentives/bonuses							
TOTAL							

INTEGRATED REVIEW — CHOOSING A FACTORING STRATEGY

The following steps may be helpful when factoring polynomials.

Factoring a Polynomial

Step 1. Are there any common factors? If so, factor out the GCF.

Step 2. How many terms are in the polynomial?

  a. If there are **two** terms, decide if one of the following can be applied.

    i. Difference of two squares: $a^2 - b^2 = (a + b)(a - b)$.

    ii. Difference of two cubes: $a^3 - b^3 = (a - b)(a^2 + ab + b^2)$.

    iii. Sum of two cubes: $a^3 + b^3 = (a + b)(a^2 - ab + b^2)$.

  b. If there are **three** terms, try one of the following.

    i. Perfect square trinomial: $a^2 + 2ab + b^2 = (a + b)^2$.

    ii. If not a perfect square trinomial, factor using the methods presented in Sections 6.2 and 6.3.

  c. If there are **four** or more terms, try factoring by grouping.

Step 3. See if any factors in the factored polynomial can be factored further.

Step 4. Check by multiplying.

Study the next five examples to help you use the steps on the previous page.

EXAMPLE 1

Factor $10t^2 - 17t + 3$.

Solution **Step 1.** The terms of this polynomial have no common factor (other than 1).

Step 2. There are three terms, so this polynomial is a trinomial. This trinomial is not a perfect square trinomial, so factor using methods from earlier sections.

Factors of $10t^2$: $\quad 10t^2 = 2t \cdot 5t, \qquad 10t^2 = t \cdot 10t$

Since the middle term, $-17t$, has a negative numerical coefficient, find negative factors of 3.

Factors of 3: $\quad 3 = -1 \cdot -3$

Try different combinations of these factors. The correct combination is

$$(2t - 3)(5t - 1) = 10t^2 - 17t + 3$$

$$\underbrace{}_{-15t}$$
$$\frac{-2t}{-17t} \qquad \textit{Correct middle term.}$$

Step 3. No factor can be factored further, so we have factored completely.

Step 4. To check, multiply $2t - 3$ and $5t - 1$.

$$(2t - 3)(5t - 1) = 10t^2 - 2t - 15t + 3 = 10t^2 - 17t + 3$$

The factored form of $10t^2 - 17t + 3$ is $(2t - 3)(5t - 1)$.

EXAMPLE 2

Factor $2x^3 + 3x^2 - 2x - 3$.

Solution **Step 1.** There are no factors common to all terms.

Step 2. Try factoring by grouping since this polynomial has four terms.

$$2x^3 + 3x^2 - 2x - 3 = x^2(2x + 3) - 1(2x + 3) \quad \textit{Factor out the greatest common factor for each pair of terms.}$$

$$= (2x + 3)(x^2 - 1) \quad \textit{Factor out } 2x + 3.$$

Step 3. The binomial $x^2 - 1$ can be factored further. It is the difference of two squares.

$$= (2x + 3)(x + 1)(x - 1) \quad \textit{Factor } x^2 - 1 \textit{ as a difference of squares.}$$

Step 4. Check by finding the product of the three binomials.
The polynomial factored completely is $(2x + 3)(x + 1)(x - 1)$.

EXAMPLE 3

Factor $12m^2 - 3n^2$.

Solution **Step 1.** The terms of this binomial contain a greatest common factor of 3.

$$12m^2 - 3n^2 = 3(4m^2 - n^2) \quad \textit{Factor out the greatest common factor.}$$

Step 2. The binomial $4m^2 - n^2$ is a difference of squares.

$$= 3(2m + n)(2m - n) \quad \text{Factor the difference of squares.}$$

Step 3. No factor can be factored further.

Step 4. We check by multiplying.

$$3(2m + n)(2m - n) = 3(4m^2 - n^2) = 12m^2 - 3n^2$$

The factored form of $12m^2 - 3n^2$ is $3(2m + n)(2m - n)$.

EXAMPLE 4

Factor $x^3 + 27y^3$.

Solution **Step 1.** The terms of this binomial contain no common factor (other than 1).

Step 2. This binomial is the sum of two cubes.

$$x^3 + 27y^3 = (x)^3 + (3y)^3$$
$$= (x + 3y)[x^2 - x(3y) + (3y)^2]$$
$$= (x + 3y)(x^2 - 3xy + 9y^2)$$

Step 3. No factor can be factored further.

Step 4. We check by multiplying.

$$(x + 3y)(x^2 - 3xy + 9y^2) = x(x^2 - 3xy + 9y^2) + 3y(x^2 - 3xy + 9y^2)$$
$$= x^3 - 3x^2y + 9xy^2 + 3x^2y - 9xy^2 + 27y^3$$
$$= x^3 + 27y^3$$

Thus, $x^3 + 27y^3$ factored completely is $(x + 3y)(x^2 - 3xy + 9y^2)$.

EXAMPLE 5

Factor $30a^2b^3 + 55a^2b^2 - 35a^2b$.

Solution **Step 1.** $30a^2b^3 + 55a^2b^2 - 35a^2b = 5a^2b(6b^2 + 11b - 7)$ *Factor out the GCF.*

Step 2. $= 5a^2b(2b - 1)(3b + 7)$ *Factor the resulting trinomial.*

Step 3. No factor can be factored further.

Step 4. Check by multiplying.

The trinomial factored completely is $5a^2b(2b - 1)(3b + 7)$.

11. $(x + 3)(x + 4)$
14. $(x - 5)(x - 2)$
16. $(x + 6)(x + 5)$
17. $(x - 6)(x + 5)$
18. $(x + 8)(x + 3)$
20. $3(x + 5)(x - 5)$
21. $(x + 3)(x + y)$
22. $(y - 7)(3 + x)$
23. $(x + 8)(x - 2)$
24. $(x - 7)(x + 4)$
25. $4x(x + 7)(x - 2)$
26. $6x(x - 5)(x + 4)$
27. $2(3x + 4)(2x + 3)$
28. $(2a - b)(4a + 5b)$
29. $(2a + b)(2a - b)$

Factor the following completely.

1. $a^2 + 2ab + b^2$ $(a + b)^2$
2. $a^2 - 2ab + b^2$ $(a - b)^2$
3. $a^2 + a - 12$ $(a - 3)(a + 4)$
4. $a^2 - 7a + 10$ $(a - 5)(a - 2)$
5. $a^2 - a - 6$ $(a + 2)(a - 3)$
6. $a^2 + 2a + 1$ $(a + 1)^2$
7. $x^2 + 2x + 1$ $(x + 1)^2$
8. $x^2 + x - 2$ $(x + 2)(x - 1)$
9. $x^2 + 4x + 3$ $(x + 1)(x + 3)$
10. $x^2 + x - 6$ $(x + 3)(x - 2)$
11. $x^2 + 7x + 12$
12. $x^2 + x - 12$ $(x + 4)(x - 3)$
13. $x^2 + 3x - 4$ $(x + 4)(x - 1)$
14. $x^2 - 7x + 10$
15. $x^2 + 2x - 15$ $(x + 5)(x - 3)$
16. $x^2 + 11x + 30$
17. $x^2 - x - 30$
18. $x^2 + 11x + 24$
19. $2x^2 - 98$ $2(x + 7)(x - 7)$
20. $3x^2 - 75$
21. $x^2 + 3x + xy + 3y$
22. $3y - 21 + xy - 7x$
23. $x^2 + 6x - 16$
24. $x^2 - 3x - 28$
25. $4x^3 + 20x^2 - 56x$
26. $6x^3 - 6x^2 - 120x$
27. $12x^2 + 34x + 24$
28. $8a^2 + 6ab - 5b^2$
29. $4a^2 - b^2$
30. $28 - 13x - 6x^2$ $(4 - 3x)(7 + 2x)$

31. $20 - 3x - 2x^2$ $(5 - 2x)(4 + x)$ **32.** $x^2 - 2x + 4$ prime   **33.** $a^2 + a - 3$ prime

34. $6y^2 + y - 15$ $(3y + 5)(2y - 3)$ **35.** $4x^2 - x - 5$ **36.** $x^2y - y^3$ $y(x + y)(x - y)$

37. $4t^2 + 36$ $4(t^2 + 9)$ **38.** $x^2 + x + xy + y$ **39.** $ax + 2x + a + 2$ $(x + 1)(a + 2)$

40. $18x^3 - 63x^2 + 9x$ **41.** $12a^3 - 24a^2 + 4a$ **42.** $x^2 + 14x - 32$ $(x + 16)(x - 2)$

43. $x^2 - 14x - 48$ prime   **44.** $16a^2 - 56ab + 49b^2$ **45.** $25p^2 - 70pq + 49q^2$ $(5p - 7q)^2$

46. $7x^2 + 24xy + 9y^2$ 🔒 **47.** $125 - 8y^3$ **48.** $64x^3 + 27$

49. $-x^2 - x + 30$ **50.** $-x^2 + 6x - 8$ **51.** $14 + 5x - x^2$ $(7 - x)(2 + x)$

52. $3 - 2x - x^2$ $(3 + x)(1 - x)$ **53.** $3x^4y + 6x^3y - 72x^2y$ **54.** $2x^3y + 8x^2y^2 - 10xy^3$

55. $5x^3y^2 - 40x^2y^3 + 35xy^4$ **56.** $4x^4y - 8x^3y - 60x^2y$ **57.** $12x^3y + 243xy$ $3xy(4x^2 + 81)$

58. $6x^3y^2 + 8xy^2$ $2xy^2(3x^2 + 4)$ **59.** $4 - x^2$ $(2 + x)(2 - x)$ **60.** $9 - y^2$ $(3 + y)(3 - y)$

61. $3rs - s + 12r - 4$ **62.** $x^3 - 2x^2 + 3x - 6$ **63.** $4x^2 - 8xy - 3x + 6y$

64. $4x^2 - 2xy - 7yz + 14xz$ **65.** $6x^2 + 18xy + 12y^2$ **66.** $12x^2 + 46xy - 8y^2$

67. $xy^2 - 4x + 3y^2 - 12$ **68.** $x^2y^2 - 9x^2 + 3y^2 - 27$ **69.** $5(x + y) + x(x + y)$

70. $7(x - y) + y(x - y)$ **71.** $14t^2 - 9t + 1$ **72.** $3t^2 - 5t + 1$ prime

73. $3x^2 + 2x - 5$ **74.** $7x^2 + 19x - 6$ **75.** $x^2 + 9xy - 36y^2$

76. $3x^2 + 10xy - 8y^2$ **77.** $1 - 8ab - 20a^2b^2$ **78.** $1 - 7ab - 60a^2b^2$

79. $9 - 10x^2 + x^4$ **80.** $36 - 13x^2 + x^4$ **81.** $x^4 - 14x^2 - 32$

82. $x^4 - 22x^2 - 75$ **83.** $x^2 - 23x + 120$ **84.** $y^2 + 22y + 96$ $(y + 16)(y + 6)$

85. $6x^3 - 28x^2 + 16x$ **86.** $6y^3 - 8y^2 - 30y$ **87.** $27x^3 - 125y^3$

88. $216y^3 - z^3$ **89.** $x^3y^3 + 8z^3$ **90.** $27a^3b^3 + 8$

91. $2xy - 72x^3y$ **92.** $2x^3 - 18x$ **93.** $x^3 + 6x^2 - 4x - 24$

94. $x^3 - 2x^2 - 36x + 72$ **95.** $6a^3 + 10a^2$ $2a^2(3a + 5)$ **96.** $4n^2 - 6n$ $2n(2n - 3)$

97. $a^2(a + 2) + 2(a + 2)$ **98.** $a - b + x(a - b)$ **99.** $x^3 - 28 + 7x^2 - 4x$

100. $a^3 - 45 - 9a + 5a^2$ $(a + 3)(a - 3)(a + 5)$

Concept Extensions

Factor.

101. $(x - y)^2 - z^2$ $(x - y + z)(x - y - z)$ **102.** $(x + 2y)^2 - 9$ $(x + 2y + 3)(x + 2y - 3)$

103. $81 - (5x + 1)^2$ $(9 + 5x + 1)(9 - 5x - 1)$ **104.** $b^2 - (4a + c)^2$ $(b + 4a + c)(b - 4a - c)$

105. Explain why it makes good sense to factor out the GCF first, before using other methods of factoring.

106. The sum of two squares usually does not factor. Is the sum of two squares $9x^2 + 81y^2$ factorable?

107. Which of the following are equivalent to $(x + 10)(x - 7)$? a, c

  a. $(x - 7)(x + 10)$ **b.** $-1(x + 10)(x - 7)$ **c.** $-1(x + 10)(7 - x)$ **d.** $-1(-x - 10)(7 - x)$

35. $(4x - 5)(x + 1)$
38. $(x + 1)(x + y)$
40. $9x(2x^2 - 7x + 1)$
41. $4a(3a^2 - 6a + 1)$
44. $(4a - 7b)^2$
46. $(7x + 3y)(x + 3y)$
47. $(5 - 2y)(25 + 10y + 4y^2)$
48. $(4x + 3)(16x^2 - 12x + 9)$
49. $-(x - 5)(x + 6)$
50. $-(x - 2)(x - 4)$
53. $3x^2y(x + 6)(x - 4)$
54. $2xy(x + 5y)(x - y)$
55. $5xy^2(x - 7y)(x - y)$
56. $4x^2y(x - 5)(x + 3)$
61. $(s + 4)(3r - 1)$
62. $(x - 2)(x^2 + 3)$
63. $(4x - 3)(x - 2y)$
64. $(2x - y)(2x + 7z)$
65. $6(x + 2y)(x + y)$
66. $2(x + 4y)(6x - y)$
67. $(x + 3)(y + 2)(y - 2)$
68. $(y + 3)(y - 3)(x^2 + 3)$
69. $(5 + x)(x + y)$
70. $(x - y)(7 + y)$
71. $(7t - 1)(2t - 1)$
73. $(3x + 5)(x - 1)$
74. $(7x - 2)(x + 3)$
75. $(x + 12y)(x - 3y)$
76. $(3x - 2y)(x + 4y)$
77. $(1 - 10ab)(1 + 2ab)$
78. $(1 + 5ab)(1 - 12ab)$

79. $(3 - x)(3 + x)(1 - x)(1 + x)$ **80.** $(3 - x)(3 + x)(2 - x)(2 + x)$ **81.** $(x + 4)(x - 4)(x^2 + 2)$

82. $(x + 5)(x - 5)(x^2 + 3)$ **83.** $(x - 15)(x - 8)$ **85.** $2x(3x - 2)(x - 4)$ **86.** $2y(3y + 5)(y - 3)$ **87.** $(3x - 5y)(9x^2 + 15xy + 25y^2)$

88. $(6y - z)(36y^2 + 6yz + z^2)$ **89.** $(xy + 2z)(x^2y^2 - 2xyz + 4z^2)$ **90.** $(3ab + 2)(9a^2b^2 - 6ab + 4)$ **91.** $2xy(1 + 6x)(1 - 6x)$

92. $2x(x + 3)(x - 3)$ **93.** $(x + 2)(x - 2)(x + 6)$ **94.** $(x - 2)(x + 6)(x - 6)$ **97.** $(a^2 + 2)(a + 2)$ **98.** $(a - b)(1 + x)$

99. $(x + 2)(x - 2)(x + 7)$ **105.** answers may vary **106.** yes; $9(x^2 + 9y^2)$

6.5   SOLVING QUADRATIC EQUATIONS BY FACTORING

Objectives

1 Define quadratic equation.

2 Solve quadratic equations by factoring.

3 Solve equations with degree greater than 2 by factoring.

1   Linear equations, while versatile, are not versatile enough to model many real-life phenomena. For example, let's suppose an object is dropped from the top of a 256-foot cliff and we want to know how long before the object strikes the ground. The answer to this question is found by solving the equation $-16t^2 + 256 = 0$. (See Example 1 in the next section.) This equation is called a **quadratic equation** because it contains a variable with an exponent of 2, and no other variable in the equation contains an exponent greater than 2. In this section, we solve quadratic equations by factoring.

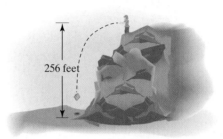

256 feet

Quadratic Equation

A quadratic equation is one that can be written in the form

$$ax^2 + bx + c = 0,$$

where a, b, and c are real numbers, and $a \neq 0$.

Notice that the degree of the polynomial $ax^2 + bx + c$ is 2. Here are a few more examples of quadratic equations.

Quadratic Equations

$$3x^2 + 5x + 6 = 0 \qquad x^2 = 9 \qquad y^2 + y = 1$$

The form $ax^2 + bx + c = 0$ is called the **standard form** of a quadratic equation. The quadratic equations $3x^2 + 5x + 6 = 0$ and $-16t^2 + 256 = 0$ are in standard form. One side of the equation is 0 and the other side is a polynomial of degree 2 written in descending powers of the variable.

2   Some quadratic equations can be solved by making use of factoring and the **zero factor theorem**.

Zero Factor Theorem

If a and b are real numbers and if $ab = 0$, then $a = 0$ or $b = 0$.

This theorem states that if the product of two numbers is 0 then at least one of the numbers must be 0.

EXAMPLE 1

Solve $(x - 3)(x + 1) = 0$.

Solution   If this equation is to be a true statement, then either the factor $x - 3$ must be 0 or the factor $x + 1$ must be 0. In other words, either

$$x - 3 = 0 \qquad \text{or} \qquad x + 1 = 0$$

If we solve these two linear equations, we have

$$x = 3 \quad \text{or} \quad x = -1$$

Thus, 3 and -1 are both solutions of the equation $(x - 3)(x + 1) = 0$. To check, we replace x with 3 in the original equation. Then we replace x with -1 in the original equation.

Check

$$(x - 3)(x + 1) = 0 \qquad\qquad (x - 3)(x + 1) = 0$$
$$(3 - 3)(3 + 1) \stackrel{?}{=} 0 \quad \text{Replace } x \text{ with 3.} \quad (-1 - 3)(-1 + 1) \stackrel{?}{=} 0 \quad \text{Replace } x \text{ with } -1.$$
$$0(4) = 0 \quad \text{True} \qquad\qquad (-4)(0) = 0 \quad \text{True}$$

The solutions are 3 and -1, or we say that the solution set is $\{-1, 3\}$.

> **Helpful Hint**
>
> The zero factor property says that *if a product is 0, then a factor is 0.*
> If $a \cdot b = 0$, then $a = 0$ or $b = 0$.
> If $x(x + 5) = 0$, then $x = 0$ or $x + 5 = 0$.
> If $(x + 7)(2x - 3) = 0$, then $x + 7 = 0$ or $2x - 3 = 0$.
>
> Use this property only when the product is 0. For example, if $a \cdot b = 8$, we do not know the value of a or b. The values may be $a = 2, b = 4$ or $a = 8, b = 1$, or any other two numbers whose product is 8.

EXAMPLE 2

Solve $x^2 - 9x = -20$.

Solution
First, write the equation in standard form; then factor.

$$x^2 - 9x = -20$$
$$x^2 - 9x + 20 = 0 \qquad \text{Write in standard form by adding 20 to both sides.}$$
$$(x - 4)(x - 5) = 0 \qquad \text{Factor.}$$

Next, use the zero factor theorem and set each factor equal to 0.

$$x - 4 = 0 \quad \text{or} \quad x - 5 = 0 \qquad \text{Set each factor equal to 0.}$$
$$x = 4 \quad \text{or} \quad x = 5 \qquad \text{Solve.}$$

Check the solutions by replacing x with each value in the original equation. The solutions are 4 and 5.

The following steps may be used to solve a quadratic equation by factoring.

Solving Quadratic Equations by Factoring

Step 1. Write the equation in standard form: $ax^2 + bx + c = 0$.

Step 2. Factor the quadratic completely.

Step 3. Set each factor containing a variable equal to 0.

Step 4. Solve the resulting equations.

Step 5. Check each solution in the original equation.

Since it is not always possible to factor a quadratic polynomial, not all quadratic equations can be solved by factoring. Other methods of solving quadratic equations are presented in Chapter 9.

EXAMPLE 3

Solve $x(2x - 7) = 4$.

Solution   First, write the equation in standard form; then factor.

$$x(2x - 7) = 4$$
$$2x^2 - 7x = 4 \qquad \text{Multiply.}$$
$$2x^2 - 7x - 4 = 0 \qquad \text{Write in standard form.}$$
$$(2x + 1)(x - 4) = 0 \qquad \text{Factor.}$$

$$2x + 1 = 0 \quad \text{or} \quad x - 4 = 0 \quad \text{Set each factor equal to zero.}$$
$$2x = -1 \quad \text{or} \quad x = 4 \quad \text{Solve.}$$
$$x = -\frac{1}{2}$$

Check both solutions $-\dfrac{1}{2}$ and 4.

> **Helpful Hint**
>
> To solve the equation $x(2x - 7) = 4$, above do **not** set each factor equal to 4. Remember that to apply the zero factor property, one side of the equation *must be* 0 and the other side of the equation must be in factored form.

EXAMPLE 4

Solve $-2x^2 - 4x + 30 = 0$.

Solution   The equation is in standard form so we begin by factoring out a common factor of -2.

$$-2x^2 - 4x + 30 = 0$$
$$-2(x^2 + 2x - 15) = 0 \qquad \text{Factor out } -2.$$
$$-2(x + 5)(x - 3) = 0 \qquad \text{Factor the quadratic.}$$

Next, set each factor **containing a variable** equal to 0.

$$x + 5 = 0 \quad \text{or} \quad x - 3 = 0 \qquad \text{Set each factor containing a variable equal to 0.}$$
$$x = -5 \quad \text{or} \quad x = 3 \qquad \text{Solve.}$$

Note that the factor -2 is a constant term containing no variables and can never equal 0. The solutions are -5 and 3.

3   Some equations involving polynomials of degree higher than 2 may also be solved by factoring and then applying the zero factor theorem.

EXAMPLE 5

Solve $3x^3 - 12x = 0$.

Solution   Factor the left side of the equation. Begin by factoring out the common factor of $3x$.

$$3x^3 - 12x = 0$$
$$3x(x^2 - 4) = 0 \qquad \text{Factor out the GCF } 3x.$$
$$3x(x + 2)(x - 2) = 0 \qquad \text{Factor } x^2 - 4, \text{ a difference of squares.}$$

$$3x = 0 \quad \text{or} \quad x + 2 = 0 \quad \text{or} \quad x - 2 = 0 \qquad \textit{Set each factor}$$
$$\textit{equal to 0.}$$

$$x = 0 \quad \text{or} \qquad x = -2 \quad \text{or} \qquad x = 2 \qquad \textit{Solve.}$$

Thus, the equation $3x^3 - 12x = 0$ has three solutions: $0, -2$, and 2. To check, replace x with each solution in the original equation.

<table>
<tr><td align="center">***Let x = 0.***</td><td align="center">***Let x = -2.***</td><td align="center">***Let x = 2.***</td></tr>
<tr><td align="center">$3(0)^3 - 12(0) \stackrel{?}{=} 0$</td><td align="center">$3(-2)^3 - 12(-2) \stackrel{?}{=} 0$</td><td align="center">$3(2)^3 - 12(2) \stackrel{?}{=} 0$</td></tr>
<tr><td align="center">$0 = 0$</td><td align="center">$3(-8) + 24 \stackrel{?}{=} 0$</td><td align="center">$3(8) - 24 \stackrel{?}{=} 0$</td></tr>
<tr><td></td><td align="center">$0 = 0$</td><td align="center">$0 = 0$</td></tr>
</table>

Substituting $0, -2$, or 2 into the original equation results each time in a true equation. The solutions are $0, -2$, and 2.

EXAMPLE 6

Solve $(5x - 1)(2x^2 + 15x + 18) = 0$.

Solution

$$(5x - 1)(2x^2 + 15x + 18) = 0$$
$$(5x - 1)(2x + 3)(x + 6) = 0 \qquad \textit{Factor the trinomial.}$$

$$5x - 1 = 0 \quad \text{or} \quad 2x + 3 = 0 \quad \text{or} \quad x + 6 = 0 \qquad \textit{Set each factor equal to 0.}$$
$$5x = 1 \quad \text{or} \qquad 2x = -3 \quad \text{or} \qquad x = -6 \qquad \textit{Solve.}$$
$$x = \frac{1}{5} \quad \text{or} \qquad x = -\frac{3}{2}$$

The solutions are $\dfrac{1}{5}, -\dfrac{3}{2}$, and -6. Check by replacing x with each solution in the original equation. The solutions are $-6, -\dfrac{3}{2}$, and $\dfrac{1}{5}$.

EXAMPLE 7

Solve $2x^3 - 4x^2 - 30x = 0$.

Solution   Begin by factoring out the GCF $2x$.

$$2x^3 - 4x^2 - 30x = 0$$
$$2x(x^2 - 2x - 15) = 0 \qquad \textit{Factor out the GCF 2x.}$$
$$2x(x - 5)(x + 3) = 0 \qquad \textit{Factor the quadratic.}$$

$$2x = 0 \quad \text{or} \quad x - 5 = 0 \quad \text{or} \quad x + 3 = 0 \qquad \textit{Set each factor containing a variable equal to 0.}$$
$$x = 0 \quad \text{or} \qquad x = 5 \quad \text{or} \qquad x = -3 \qquad \textit{Solve.}$$

Check by replacing x with each solution in the cubic equation. The solutions are $-3, 0$, and 5.

In Chapter 3, we graphed linear equations in two variables, such as $y = 5x - 6$. Recall that to find the x-intercept of the graph of a linear equation, let $y = 0$ and solve for x. This is also how to find the x-intercepts of the graph of a **quadratic equation in two variables**, such as $y = x^2 - 5x + 4$.

EXAMPLE 8

Find the x-intercepts of the graph of $y = x^2 - 5x + 4$.

Solution Let $y = 0$ and solve for x.

$$y = x^2 - 5x + 4$$
$$0 = x^2 - 5x + 4 \qquad \text{Let } y = 0.$$
$$0 = (x - 1)(x - 4) \qquad \text{Factor.}$$
$$x - 1 = 0 \quad \text{or} \quad x - 4 = 0 \qquad \text{Set each factor equal to 0.}$$
$$x = 1 \quad \text{or} \qquad x = 4 \qquad \text{Solve.}$$

The x-intercepts of the graph of $y = x^2 - 5x + 4$ are $(1, 0)$ and $(4, 0)$.

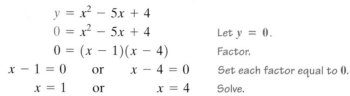

The graph of $y = x^2 - 5x + 4$ is shown in the margin.

In general, a quadratic equation in two variables is one that can be written in the form $y = ax^2 + bx + c$ where $a \neq 0$. The graph of such an equation is called a **parabola** and will open up or down depending on the value of a.

Notice that the x-intercepts of the graph of $y = ax^2 + bx + c$ are the real number solutions of $0 = ax^2 + bx + c$. Also, the real number solutions of $0 = ax^2 + bx + c$ are the x-intercepts of the graph of $y = ax^2 + bx + c$. We study more about graphs of quadratic equations in two variables in Chapter 9.

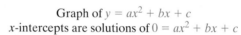

Graph of $y = ax^2 + bx + c$
x-intercepts are solutions of $0 = ax^2 + bx + c$

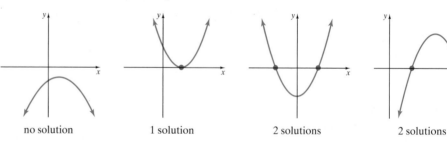

no solution   1 solution   2 solutions   2 solutions

Graphing Calculator Explorations

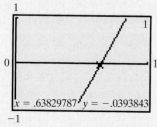

A grapher may be used to find solutions of a quadratic equation whether the related quadratic polynomial is factorable or not. For example, let's use a grapher to approximate the solutions of $0 = x^2 + 4x - 3$. To do so, graph $y_1 = x^2 + 4x - 3$. Recall that the x-intercepts of this graph are the solutions of $0 = x^2 + 4x - 3$.

Notice that the graph appears to have an x-intercept between -5 and -4 and one between 0 and 1. Many graphers contain a TRACE feature. This feature activates a graph cursor that can be used to *trace* along a graph while the corresponding x- and y-coordinates are shown on the screen. Use the TRACE feature to confirm that x-intercepts lie between -5 and -4 and also 0 and 1. To approximate the x-intercepts to the nearest tenth, use a ROOT or a ZOOM feature on your grapher or redefine the viewing window. (A ROOT feature calculates the x-intercept. A ZOOM feature magnifies the viewing window around a specific location such as the graph cursor.) If we redefine the window to $[0, 1]$ on the x-axis and $[-1, 1]$ on the y-axis, the following graph is generated.

By using the TRACE feature, we can conclude that one x-intercept is approximately 0.6 to the nearest tenth. By repeating these steps for the other x-intercept, we find that it is approximately -4.6. *(continued)*

1. $-0.9, 2.2$
2. $-2.5, 3.5$
3. no real solution
4. no real solution
5. $-1.8, 2.8$
6. $-0.9, 0.3$

Use a grapher to approximate the real number solutions to the nearest tenth. If an equation has no real number solution, state so.

1. $3x^2 - 4x - 6 = 0$

2. $x^2 - x - 9 = 0$

3. $2x^2 + x + 2 = 0$

4. $-4x^2 - 5x - 4 = 0$

5. $-x^2 + x + 5 = 0$

6. $10x^2 + 6x - 3 = 0$

MENTAL MATH

Solve each equation by inspection.

1. $(a - 3)(a - 7) = 0$ $3, 7$

2. $(a - 5)(a - 2) = 0$ $5, 2$

3. $(x + 8)(x + 6) = 0$ $-8, -6$

4. $(x + 2)(x + 3) = 0$ $-2, -3$

5. $(x + 1)(x - 3) = 0$ $-1, 3$

6. $(x - 1)(x + 2) = 0$ $1, -2$

EXERCISE SET 6.5

STUDY GUIDE/SSM   CD/ VIDEO   PH MATH TUTOR CENTER   MathXL®Tutorials ON CD   MathXL®   MyMathLab®

2. $-3, -2$ **6.** $\frac{2}{3}, -\frac{1}{5}$ **8.** $-\frac{1}{9}, \frac{3}{4}$

45. $-\frac{9}{2}, \frac{8}{3}$

Solve each equation. See Example 1.   **9.** $(x - 6)(x + 1) = 0$

MIXED PRACTICE   **46.** $\frac{1}{4}, 7$ **50.** $\frac{5}{4}, -\frac{2}{7}$ **56.** no real solution

1. $(x - 2)(x + 1) = 0$ $2, -1$

2. $(x + 3)(x + 2) = 0$

 3. $x(x + 6) = 0$ $0, -6$

4. $2x(x - 7) = 0$ $0, 7$

Solve each equation. Be careful. Some of the equations are quadratic and higher degree and some are linear.

5. $(2x + 3)(4x - 5) = 0$ $-\frac{3}{2}, \frac{5}{4}$

6. $(3x - 2)(5x + 1) = 0$

7. $(2x - 7)(7x + 2) = 0$ $\frac{7}{2}, -\frac{2}{7}$

8. $(9x + 1)(4x - 3) = 0$

31. $x(x + 7) = 0$ $0, -7$

32. $y(6 - y) = 0$ $0, 6$

9. Write a quadratic equation that has two solutions, 6 and -1. Leave the polynomial in the equation in factored form.

33. $(x + 5)(x - 4) = 0$ $-5, 4$

34. $(x - 8)(x - 1) = 0$ $8, 1$

35. $x^2 - x = 30$ $-5, 6$

36. $x^2 + 13x = -36$ $-9, -4$

10. Write a quadratic equation that has two solutions, 0 and -2. Leave the polynomial in the equation in factored form.
$x(x + 2) = 0$

37. $6y^2 - 22y - 40 = 0$ $-\frac{4}{3}, 5$

38. $3x^2 - 6x - 9 = 0$ $3, -1$

39. $(2x + 3)(2x^2 - 5x - 3) = 0$ $-\frac{3}{2}, -\frac{1}{2}, 3$

40. $(2x - 9)(x^2 + 5x - 36) = 0$ $\frac{9}{2}, -9, 4$

Solve each equation. See Examples 2 through 4.   **20.** $\frac{3}{4}, -\frac{7}{9}$

41. $x^2 - 15 = -2x$ $-5, 3$

42. $x^2 - 26 = -11x$ $-13, 2$

 11. $x^2 - 13x + 36 = 0$ $9, 4$

12. $x^2 + 2x - 63 = 0$ $-9, 7$

43. $x^2 - 16x = 0$ $0, 16$

44. $x^2 + 5x = 0$ $0, -5$

13. $x^2 + 2x - 8 = 0$ $-4, 2$

14. $x^2 - 5x + 6 = 0$ $3, 2$

45. $-18y^2 - 33y + 216 = 0$

46. $-20y^2 + 145y - 35 = 0$

15. $x^2 - 4x = 32$ $8, -4$

16. $x^2 - 5x = 24$ $8, -3$

47. $12x^2 - 59x + 55 = 0$ $\frac{5}{4}, \frac{11}{3}$

48. $30x^2 - 97x + 60 = 0$ $\frac{5}{6}, \frac{12}{5}$

17. $x(3x - 1) = 14$ $\frac{7}{3}, -2$

18. $x(4x - 11) = 3$ $-\frac{1}{4}, 3$

49. $18x^2 + 9x - 2 = 0$ $-\frac{2}{3}, \frac{1}{6}$

50. $28x^2 - 27x - 10 = 0$

19. $3x^2 + 19x - 72 = 0$ $\frac{8}{3}, -9$

20. $36x^2 + x - 21 = 0$

51. $x(6x + 7) = 5$ $-\frac{5}{3}, \frac{1}{2}$

52. $4x(8x + 9) = 5$ $\frac{1}{8}, -\frac{5}{4}$

21. Write a quadratic equation in standard form that has two solutions, 5 and 7. $x^2 - 12x + 35 = 0$

53. $4(x - 7) = 6$ $\frac{17}{2}$

54. $5(3 - 4x) = 9$ $\frac{3}{10}$

22. Write an equation that has three solutions, 0, 1, and 2.
$x^3 - 3x^2 + 2x = 0$

55. $5x^2 - 6x - 8 = 0$ $2, -\frac{4}{5}$

56. $9x^2 + 6x + 2 = 0$

Solve each equation. See Examples 5 through 7.

57. $(y - 2)(y + 3) = 6$ $-4, 3$

58. $(y - 5)(y - 2) = 28$ $9, -2$

23. $x^3 - 12x^2 + 32x = 0$ $0, 8, 4$

24. $x^3 - 14x^2 + 49x = 0$ $0, 7$

59. $4y^2 - 1 = 0$ $\frac{1}{2}, -\frac{1}{2}$

60. $4y^2 - 81 = 0$ $\frac{9}{2}, -\frac{9}{2}$

25. $(4x - 3)(16x^2 - 24x + 9) = 0$ $\frac{3}{4}$

61. $t^2 + 13t + 22 = 0$ $-2, -11$

62. $x^2 - 9x + 18 = 0$ $6, 3$

26. $(2x + 5)(4x^2 - 10x + 25) = 0$ $-\frac{5}{2}$

63. $5t - 3 = 12$ 3

64. $9 - t = -1$ 10

27. $4x^3 - x = 0$ $0, \frac{1}{2}, -\frac{1}{2}$

28. $4y^3 - 36y = 0$ $0, 3, -3$

65. $x^2 + 6x - 17 = -26$ -3

66. $x^2 - 8x - 4 = -20$ 4

29. $32x^3 - 4x^2 - 6x = 0$
$0, \frac{1}{2}, -\frac{3}{8}$

30. $15x^3 + 24x^2 - 63x = 0$
$0, \frac{7}{5}, -3$

67. $12x^2 + 7x - 12 = 0$ $\frac{3}{4}, -\frac{4}{3}$

68. $30x^2 - 11x - 30 = 0$ $-\frac{5}{6}, \frac{6}{5}$

69. $10t^3 - 25t - 15t^2 = 0$
$0, \frac{5}{2}, -1$

70. $36t^3 - 48t - 12t^2 = 0$
$0, \frac{4}{3}, -1$

Find the x-intercepts of the graph of each equation. See Example 8.

71. $y = (3x + 4)(x - 1)$ $-\frac{4}{3}, 1$ **72.** $y = (5x - 3)(x - 4)$

73. $y = x^2 - 3x - 10$ $-2, 5$ **74.** $y = x^2 + 7x + 6$ $-1, -6$

75. $y = 2x^2 + 11x - 6$ $-6, \frac{1}{2}$ **76.** $y = 4x^2 + 11x + 6$ $-2, -\frac{3}{4}$

72. $\frac{3}{5}, 4$

For Exercises 77 through 82, match each equation with its graph. See Example 8.

a.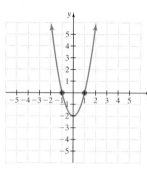

b.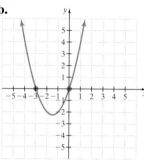

c.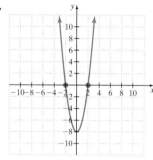

d.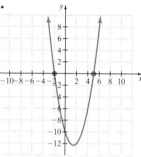

e.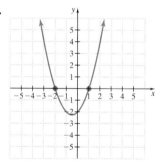

f.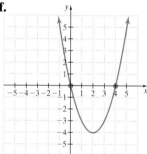

77. $y = (x + 2)(x - 1)$ e **78.** $y = (x - 5)(x + 2)$ d

79. $y = x(x + 3)$ b **80.** $y = x(x - 4)$ f

81. $y = 2x^2 - 8$ c **82.** $y = 2x^2 - 2$ a

REVIEW AND PREVIEW **86.** $\frac{5}{36}$ **90.** $\frac{36}{119}$

Perform the following operations. Write all results in lowest terms. See Section 1.3.

83. $\frac{3}{5} + \frac{4}{9}$ $\frac{47}{45}$ **84.** $\frac{2}{3} + \frac{3}{7}$ $\frac{23}{21}$ **85.** $\frac{7}{10} - \frac{5}{12}$ $\frac{17}{60}$ **86.** $\frac{5}{9} - \frac{5}{12}$

87. $\frac{7}{8} \div \frac{7}{15}$ $\frac{15}{8}$ **88.** $\frac{5}{12} - \frac{3}{10}$ $\frac{7}{60}$ **89.** $\frac{4}{5} \cdot \frac{7}{8}$ $\frac{7}{10}$ **90.** $\frac{3}{7} \cdot \frac{12}{17}$

Concept Extensions

91. A compass is accidentally thrown upward and out of an air balloon at a height of 300 feet. The height, y, of the compass at time x in seconds is given by the equation

$$y = -16x^2 + 20x + 300$$

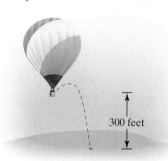

300 feet

a. Find the height of the compass at the given times by filling in the table below.

time x	0	1	2	3	4	5	6
height y	300	304	276	216	124	0	−156

b. Use the table to determine when the compass strikes the ground. 5 sec

c. Use the table to approximate the maximum height of the compass. 304 ft

d. Plot the points (x, y) on a rectangular coordinate system and connect them with a smooth curve. Explain your results. See graphing answer section.

92. A rocket is fired upward from the ground with an initial velocity of 100 feet per second. The height, y, of the rocket at any time x is given by the equation

$$y = -16x^2 + 100x$$

a. Find the height of the rocket at the given times by filling in the table below.

time x	0	1	2	3	4	5	6	7
height y	0	84	136	156	144	100	24	−84

b. Use the table to approximate when the rocket strikes the ground to the nearest second.   6 sec

c. Use the table to approximate the maximum height of the rocket.   156 ft

✎ **d.** Plot the points (x, y) on a rectangular coordinate system and connect them with a smooth curve. Explain your results.   See graphing answer section.

93. $(x - 3)(3x + 4) = (x + 2)(x - 6)$ $0, \frac{1}{2}$

94. $(2x - 3)(x + 6) = (x - 9)(x + 2)$ $0, -16$

95. $(2x - 3)(x + 8) = (x - 6)(x + 4)$ $0, -15$

96. $(x + 6)(x - 6) = (2x - 9)(x + 4)$ $0, 1$

Solve each equation. First, multiply the binomial.

To solve $(x - 6)(2x - 3) = (x + 2)(x + 9)$, see below.

$$(x - 6)(2x - 3) = (x + 2)(x + 9)$$
$$2x^2 - 15x + 18 = x^2 + 11x + 18$$
$$x^2 - 26x = 0$$
$$x(x - 26) = 0$$
$$x = 0 \quad \text{or} \quad x - 26 = 0$$
$$x = 26$$

6.6   QUADRATIC EQUATIONS AND PROBLEM SOLVING

Objective

1 Solve problems that can be modeled by quadratic equations.

1   Some problems may be modeled by quadratic equations. To solve these problems, we use the same problem-solving steps that were introduced in Section 2.5. When solving these problems, keep in mind that a solution of an equation that models a problem may not be a solution to the problem. For example, a person's age or the length of a rectangle is always a positive number. Discard solutions that do not make sense as solutions of the problem.

EXAMPLE 1

FINDING THE LENGTH OF TIME

For a TV commercial, a piece of luggage is dropped from a cliff 256 feet above the ground to show the durability of the luggage. Neglecting air resistance, the height h in feet of the luggage above the ground after t seconds is given by the quadratic equation

$$h = -16t^2 + 256$$

Find how long it takes for the luggage to hit the ground.

Solution   **1. UNDERSTAND.** Read and reread the problem. Then draw a picture of the problem.

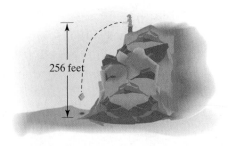

256 feet

The equation $h = -16t^2 + 256$ models the height of the falling luggage at time t. Familiarize yourself with this equation by finding the height of the luggage at $t = 1$ second and $t = 2$ seconds.

When $t = 1$ second, the height of the suitcase is

$$h = -16(1)^2 + 256 = 240 \text{ feet.}$$

When $t = 2$ seconds, the height of the suitcase is

$$h = -16(2)^2 + 256 = 192 \text{ feet.}$$

2. TRANSLATE. To find how long it takes the luggage to hit the ground, we want to know the value of t for which the height $h = 0$.

$$0 = -16t^2 + 256$$

3. SOLVE. We solve the quadratic equation by factoring.

$$0 = -16t^2 + 256$$
$$0 = -16(t^2 - 16)$$
$$0 = -16(t - 4)(t + 4)$$
$$t - 4 = 0 \quad \text{or} \quad t + 4 = 0$$
$$t = 4 \quad\quad\quad t = -4$$

4. INTERPRET. Since the time t cannot be negative, the proposed solution is 4 seconds.

Check: Verify that the height of the luggage when t is 4 seconds is 0.
When $t = 4$ seconds, $h = -16(4)^2 + 256 = -256 + 256 = 0$ feet.

State: The solution checks and the luggage hits the ground 4 seconds after it is dropped.

EXAMPLE 2

FINDING A NUMBER

The square of a number plus three times the number is 70. Find the number.

Solution   1. UNDERSTAND. Read and reread the problem. Suppose that the number is 5. The square of 5 is 5^2 or 25. Three times 5 is 15. Then $25 + 15 = 40$, not 70, so the number must be greater than 5. Remember, the purpose of proposing a number, such as 5, is to better understand the problem. Now that we do, we will let $x =$ the number.

2. TRANSLATE.

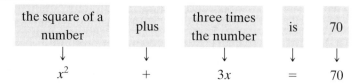

the square of a number	plus	three times the number	is	70
↓	↓	↓	↓	↓
x^2	$+$	$3x$	$=$	70

3. SOLVE.

$$x^2 + 3x = 70$$
$$x^2 + 3x - 70 = 0 \qquad \text{Subtract 70 from both sides.}$$
$$(x + 10)(x - 7) = 0 \qquad \text{Factor.}$$
$$x + 10 = 0 \quad \text{or} \quad x - 7 = 0 \qquad \text{Set each factor equal to 0.}$$
$$x = -10 \quad\quad\quad x = 7 \qquad \text{Solve.}$$

4. INTERPRET.

Check: The square of -10 is $(-10)^2$, or 100. Three times -10 is $3(-10)$ or -30. Then $100 + (-30) = 70$, the correct sum, so -10 checks.

The square of 7 is 7^2 or 49. Three times 7 is 3(7), or 21. Then $49 + 21 = 70$, the correct sum, so 7 checks.

State: There are two numbers. They are -10 and 7.

△ (**EXAMPLE 3**)

FINDING THE BASE AND HEIGHT OF A SAIL

The height of a triangular sail is 2 meters less than twice the length of the base. If the sail has an area of 30 square meters, find the length of its base and the height.

Solution

1. UNDERSTAND. Read and reread the problem. Since we are finding the length of the base and the height, we let

$$x = \text{the length of the base}$$

and since the height is 2 meters less than twice the base,

$$2x - 2 = \text{the height}$$

An illustration is shown to the left.

2. TRANSLATE. We are given that the area of the triangle is 30 square meters, so we use the formula for area of a triangle.

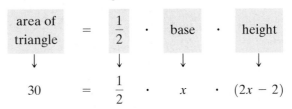

area of triangle	=	$\frac{1}{2}$	·	base	·	height
↓		↓		↓		↓
30	=	$\frac{1}{2}$	·	x	·	$(2x - 2)$

3. SOLVE. Now we solve the quadratic equation.

$$30 = \frac{1}{2}x(2x - 2)$$

$$30 = x^2 - x \qquad \text{Multiply.}$$

$$x^2 - x - 30 = 0 \qquad \text{Write in standard form.}$$

$$(x - 6)(x + 5) = 0 \qquad \text{Factor.}$$

$$x - 6 = 0 \quad \text{or} \quad x + 5 = 0 \qquad \text{Set each factor equal to 0.}$$

$$x = 6 \qquad\qquad x = -5$$

4. INTERPRET. Since x represents the length of the base, we discard the solution -5. The base of a triangle cannot be negative. The base is then 6 meters and the height is $2(6) - 2 = 10$ meters.

Check: To check this problem, we recall that $\frac{1}{2}$base · height = area, or

$$\frac{1}{2}(6)(10) = 30 \qquad \text{The required area}$$

State: The base of the triangular sail is 6 meters and the height is 10 meters.

The next example makes use of the **Pythagorean theorem** and consecutive integers. Before we review this theorem, recall that a **right triangle** is a triangle that contains a 90° or right angle. The **hypotenuse** of a right triangle is the side opposite the right angle and is the longest side of the triangle. The **legs** of a right triangle are the other sides of the triangle.

Helpful Hint

If you use this formula, don't forget that c represents the length of the hypotenuse.

Pythagorean Theorem

In a right triangle, the sum of the squares of the lengths of the two legs is equal to the square of the length of the hypotenuse.

$$(\text{leg})^2 + (\text{leg})^2 = (\text{hypotenuse})^2 \quad \text{or} \quad a^2 + b^2 = c^2$$

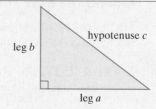

hypotenuse c

leg b

leg a

Study the following diagrams for a review of consecutive integers.

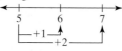

Consecutive integers:

If x is the first integer: $x, x + 1, x + 2$

Consecutive even integers:

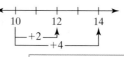

If x is the first even integer: $x, x + 2, x + 4$

Helpful Hint

This 2 means that even numbers are 2 units between each other.

Consecutive odd integers:

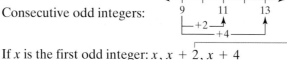

If x is the first odd integer: $x, x + 2, x + 4$

Helpful Hint

This 2 means that odd numbers are 2 units between each other.

EXAMPLE 4

FINDING THE DIMENSIONS OF A TRIANGLE

Find the lengths of the sides of a right triangle if the lengths can be expressed as three consecutive even integers.

Solution

1. UNDERSTAND. Read and reread the problem. Let's suppose that the length of one leg of the right triangle is 4 units. Then the other leg is the next even integer, or 6 units, and the hypotenuse of the triangle is the next even integer, or 8 units. Remember that the hypotenuse is the longest side. Let's see if a triangle with sides of these lengths forms a right triangle. To do this, we check to see whether the Pythagorean theorem holds true.

$$4^2 + 6^2 \overset{?}{=} 8^2$$
$$16 + 36 \overset{?}{=} 64$$
$$52 = 64 \quad \text{False}$$

Our proposed numbers do not check, but we now have a better understanding of the problem.

We let x, $x + 2$, and $x + 4$ be three consecutive even integers. Since these integers represent lengths of the sides of a right triangle, we have

$$x = \text{one leg}$$
$$x + 2 = \text{other leg}$$
$$x + 4 = \text{hypotenuse (longest side)}$$

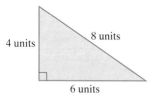

4 units
8 units
6 units

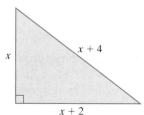

x
$x + 4$
$x + 2$

2. TRANSLATE. By the Pythagorean theorem, we have that

$$(\text{hypotenuse})^2 = (\text{leg})^2 + (\text{leg})^2$$
$$(x + 4)^2 = (x)^2 + (x + 2)^2$$

3. SOLVE. Now we solve the equation.

$(x + 4)^2 = x^2 + (x + 2)^2$	
$x^2 + 8x + 16 = x^2 + x^2 + 4x + 4$	Multiply.
$x^2 + 8x + 16 = 2x^2 + 4x + 4$	Combine like terms.
$x^2 - 4x - 12 = 0$	Write in standard form.
$(x - 6)(x + 2) = 0$	Factor.
$x - 6 = 0 \quad \text{or} \quad x + 2 = 0$	Set each factor equal to 0.
$x = 6 \qquad\qquad x = -2$	

4. INTERPRET. We discard $x = -2$ since length cannot be negative. If $x = 6$, then $x + 2 = 8$ and $x + 4 = 10$.

Check: Verify that $(\text{hypotenuse})^2 = (\text{leg})^2 + (\text{leg})^2$, or $10^2 = 6^2 + 8^2$, or $100 = 36 + 64$.

State: The sides of the right triangle have lengths 6 units, 8 units, and 10 units.

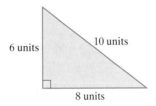

6 units   10 units   8 units

Spotlight on

DECISION
✕ MAKING

Suppose you are a landscaper. You are landscaping a public park and have just put in a flower bed measuring 8 feet by 12 feet. You would also like to surround the bed with a decorative floral border consisting of low-growing, spreading plants. Each plant will cover approximately 1 square foot when mature, and you have 224 plants to use. How wide of a strip of ground should you prepare around the flower bed for the border? Explain.

Grows to cover 1 sq ft

1 foot

1 foot

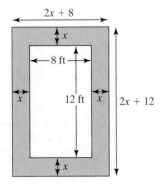

2x + 8

x

8 ft

x   12 ft   x   2x + 12

x

TEACHING TIP A Group Activity is available in the Instructor's Resource Manual.

EXERCISE SET 6.6

STUDY GUIDE/SSM CD/VIDEO PH MATH TUTOR CENTER MathXL®Tutorials ON CD MathXL® MyMathLab®

MIXED PRACTICE

See Examples 1 through 4 for all exercises. Represent each given condition using a single variable, x.

△ **1.** The length and width of a rectangle whose length is 4 centimeters more than its width   width = x; length = $x + 4$

△ **2.** The length and width of a rectangle whose length is twice its width   width = x; length = $2x$

3. Two consecutive odd integers x and $x + 2$ if x is an odd integer

4. Two consecutive even integers x and $x + 2$ if x is an even integer

△ **5.** The base and height of a triangle whose height is one more than four times its base   base = x; height = $4x + 1$

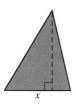

△ **6.** The base and height of a trapezoid whose base is three less than five times its height   height = x; base = $5x - 3$

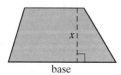

base

Use the information given to find the dimensions of each figure.

△ **7.** The *area* of the square is 121 square units. Find the length of its sides.   11 units

△ **8.** The *area* of the rectangle is 84 square inches. Find its length and width.   length = 12 in.; width = 7 in.

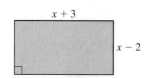

x + 3

x − 2

△ **9.** The *perimeter* of the quadrilateral is 120 centimeters. Find the lengths of the sides.   15 cm, 13 cm, 70 cm, 22 cm

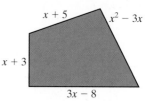

$x + 5$ $x^2 - 3x$

$x + 3$

$3x - 8$

△ **10.** The *perimeter* of the triangle is 85 feet. Find the lengths of its sides.   14 ft, 19 ft, 52 ft

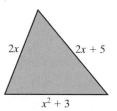

$2x$ $2x + 5$

$x^2 + 3$

△ **11.** The *area* of the parallelogram is 96 square miles. Find its base and height.   base = 16 mi; height = 6 mi

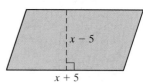

$x - 5$

$x + 5$

△ **12.** The *area* of the circle is 25π square kilometers. Find its radius.   radius = 5 km

x

Solve.

▢ **13.** An object is thrown upward from the top of an 80-foot building with an initial velocity of 64 feet per second. The height h of the object after t seconds is given by the quadratic equation $h = -16t^2 + 64t + 80$. When will the object hit the ground?   5 sec

14. A hang glider pilot accidentally drops her compass from the top of a 400-foot cliff. The height h of the compass after t seconds is given by the quadratic equation $h = -16t^2 + 400$. When will the compass hit the ground?   5 sec

16. length = 16 in.; width = 7 in.

△ **15.** The length of a rectangle is 7 centimeters less than twice its width. Its area is 30 square centimeters. Find the dimensions of the rectangle.   length = 5 cm; width = 6 cm

△ **16.** The length of a rectangle is 9 inches more than its width. Its area is 112 square inches. Find the dimensions of the rectangle.

The equation $D = \frac{1}{2}n(n - 3)$ gives the number of diagonals D for a polygon with n sides. For example, a polygon with 6 sides has $D = \frac{1}{2} \cdot 6(6 - 3)$ or $D = 9$ diagonals. (See if you can count all 9 diagonals. Some are shown in the figure.) Use this equation, $D = \frac{1}{2}n(n - 3)$, for Exercises 17–20.

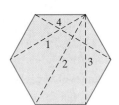

17. 54 diagonals
18. 90 diagonals
19. 10 sides
20. 7 sides

△ **17.** Find the number of diagonals for a polygon that has 12 sides.

△ **18.** Find the number of diagonals for a polygon that has 15 sides.

△ **19.** Find the number of sides n for a polygon that has 35 diagonals.

△ **20.** Find the number of sides n for a polygon that has 14 diagonals.

Solve.   **21.** −12 or 11   **22.** −14 or 13

21. The sum of a number and its square is 132. Find the number(s).

22. The sum of a number and its square is 182. Find the number(s).

△ **23.** Two boats travel at a right angle to each other after leaving the same dock at the same time. One hour later the boats are 17 miles apart. If one boat travels 7 miles per hour faster than the other boat, find the rate of each boat.
slow boat: 8 mph; fast boat: 15 mph

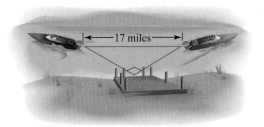

△ **24.** The side of a square equals the width of a rectangle. The length of the rectangle is 6 meters longer than its width. The sum of the areas of the square and the rectangle is 176 square meters. Find the side of the square.   8 m

25. The sum of two numbers is 20, and the sum of their squares is 218. Find the numbers.   13 and 7

26. The sum of two numbers is 25, and the sum of their squares is 325. Find the numbers.   10 and 15

△ **27.** If the sides of a square are increased by 3 inches, the area becomes 64 square inches. Find the length of the sides of the original square.   5 in.

△ **28.** If the sides of a square are increased by 5 meters, the area becomes 100 square meters. Find the length of the sides of the original square.   5 m

29. 12 mm, 16 mm, 20 mm

🔒 **29.** One leg of a right triangle is 4 millimeters longer than the smaller leg and the hypotenuse is 8 millimeters longer than the smaller leg. Find the lengths of the sides of the triangle.

△ **30.** One leg of a right triangle is 9 centimeters longer than the other leg and the hypotenuse is 45 centimeters. Find the lengths of the legs of the triangle.   27 cm and 36 cm

△ **31.** The length of the base of a triangle is twice its height. If the area of the triangle is 100 square kilometers, find the height.   10 km

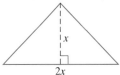

△ **32.** The height of a triangle is 2 millimeters less than the base. If the area is 60 square millimeters, find the base.   12 mm

△ **33.** Find the length of the shorter leg of a right triangle if the longer leg is 12 feet more than the shorter leg and the hypotenuse is 12 feet less than twice the shorter leg.   36 ft

△ **34.** Find the length of the shorter leg of a right triangle if the longer leg is 10 miles more than the shorter leg and the hypotenuse is 10 miles less than twice the shorter leg.   30 mi

35. An object is dropped from the top of the 625-foot-tall Waldorf-Astoria Hotel on Park Avenue in New York City. (*Source: World Almanac* research) The height h of the object after t seconds is given by the equation $h = -16t^2 + 625$. Find how many seconds pass before the object reaches the ground.   6.25 sec

36. A 6-foot-tall person drops an object from the top of the Westin Peachtree Plaza in Atlanta, Georgia. The Westin building is 723 feet tall. (*Source: World Almanac* research) The height h of the object after t seconds is given by the equation $h = -16t^2 + 729$. Find how many seconds pass before the object reaches the ground.   6.75 sec

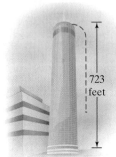

39. length: 15 mi; width: 8 mi   **40.** width: 8 in.; length: 20 in.
37. At the end of 2 years, P dollars invested at an interest rate r compounded annually increases to an amount, A dollars, given by

$$A = P(1 + r)^2$$

Find the interest rate if $100 increased to $144 in 2 years.   20%

38. At the end of 2 years, P dollars invested at an interest rate r compounded annually increases to an amount, A dollars, given by   10%

$$A = P(1 + r)^2$$

Find the interest rate if $2000 increased to $2420 in 2 years.

△ **39.** Find the dimensions of a rectangle whose width is 7 miles less than its length and whose area is 120 square miles.

△ **40.** Find the dimensions of a rectangle whose width is 2 inches less than half its length and whose area is 160 square inches.

41. If the cost, C, for manufacturing x units of a certain product is given by $C = x^2 - 15x + 50$, find the number of units manufactured at a cost of $9500.   105 units

42. If a switchboard handles n telephones, the number C of telephone connections it can make simultaneously is given by the equation $C = \dfrac{n(n - 1)}{2}$. Find how many telephones are handled by a switchboard making 120 telephone connections simultaneously.   16 telephones

REVIEW AND PREVIEW

The following double line graph shows a comparison of the number of farms in the United States and the size of the average farm. Use this graph to answer Exercises 43–49. See Section 1.9.

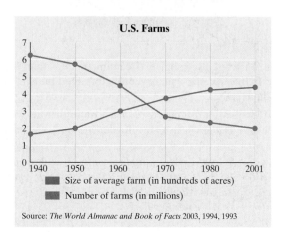

Source: *The World Almanac and Book of Facts* 2003, 1994, 1993

△ **43.** Approximate the size of the average farm in 1940.   175 acres

△ **44.** Approximate the size of the average farm in 2001.   435 acres

45. Approximate the number of farms in 1940.   6.25 million

46. Approximate the number of farms in 2001.   2 million

47. Approximate the year that the colored lines in this graph intersect.   1966

48. In your own words, explain the meaning of the point of intersection in the graph.   answers may vary

49. Describe the trends shown in this graph and speculate as to why these trends have occurred.   answers may vary

Write each fraction in simplest form. See Section 1.3.

50. $\dfrac{20}{35}$ $\dfrac{4}{7}$ **51.** $\dfrac{24}{32}$ $\dfrac{3}{4}$

52. $\dfrac{27}{18}$ $\dfrac{3}{2}$ **53.** $\dfrac{15}{27}$ $\dfrac{5}{9}$

54. $\dfrac{14}{42}$ $\dfrac{1}{3}$ **55.** $\dfrac{45}{50}$ $\dfrac{9}{10}$

Concept Extensions

△ **56.** According to the International America's Cup Class (IACC) rule, a sailboat competing in the America's Cup match must have a 110-foot-tall mast and a combined mainsail and jib sail area of 3000 square feet. (*Source:* America's Cup Organizing Committee) A design for an IACC-class sailboat calls for the mainsail to be 60% of the combined sail area. If the height of the triangular mainsail is 28 feet more than twice the length of the boom, find the length of the boom and the height of the mainsail.   boom length: 36 ft; height of mainsail: 100 ft

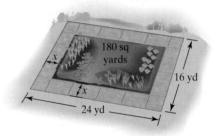

110 feet

mainsail | jib sail

boom

57. Write down two numbers whose sum is 10. Square each number and find the sum of the squares. Use this work to write a word problem like Exercise 21. Then give the word problem to a classmate to solve.

△ **58.** A rectangular pool is surrounded by a walk 4 meters wide. The pool is 6 meters longer than its width. If the total area is 576

square meters more than the area of the pool, find the dimensions of the pool.   width of pool: 29 m; length of pool: 35 m

△ **59.** A rectangular garden is surrounded by a walk of uniform width. The area of the garden is 180 square yards. If the dimensions of the garden plus the walk are 16 yards by 24 yards, find the width of the walk.   3 yd

180 sq yards
16 yd
24 yd

CHAPTER 6 PROJECT

Choosing Among Building Options

Whether putting in a new floor, hanging new wallpaper, or retiling a bathroom, it may be necessary to choose among several different materials with different pricing schemes. If a fixed amount of money is available for projects like these, it can be helpful to compare the choices by calculating how much area can be covered by a fixed dollar-value of material.

In this project, you will have the opportunity to choose among three different choices of materials for building a patio around a swimming pool. This project may be completed by working in groups or individually.

Situation: Suppose you have just had a 10-foot-by-15-foot in-ground swimming pool installed in your backyard. You have $3000 left from the building project that you would like to spend on surrounding the pool with a patio, equally wide on all sides (see figure). You have talked to several local suppliers about options for building this patio and must choose among the following.

Option	Material	Price
A	Poured cement	$5 per square foot
B	Brick	$7.50 per square foot plus a $30 flat fee for delivering the bricks
C	Outdoor carpeting	$4.50 per square foot plus $10.86 per foot of the pool's perimeter to install edging

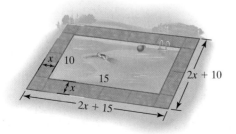

x | 10
15
x
2x + 10
2x + 15

△ **1.** Find the area of the swimming pool.

△ **2.** Write an algebraic expression for the total area of the region containing both the pool and the patio.

3. Use subtraction to find an algebraic expression for the area of just the patio (not including the area of the pool).

 4. Find the perimeter of the swimming pool alone.

5. For each patio material option, write an algebraic expression for the total cost of installing the patio based on its area and the given price information.

6. If you plan to spend the entire $3000 on the patio, how wide would the patio in option A be?

7. If you plan to spend the entire $3000 on the patio, how wide would the patio in option B be?

8. If you plan to spend the entire $3000 on the patio, how wide would the patio in option C be?

9. Which option would you choose? Why? Discuss the pros and cons of each option.

6 CHAPTER VOCABULARY CHECK

Fill in each blank with one of the words or phrases listed below.

factoring   quadratic equation   perfect square trinomial   greatest common factor

1. An equation that can be written in the form $ax^2 + bx + c = 0$ (with a not 0) is called a <u>quadratic equation</u>.

2. <u>Factoring</u> is the process of writing an expression as a product.

3. The <u>greatest common factor</u> of a list of terms is the product of all common factors.

4. A trinomial that is the square of some binomial is called a <u>perfect square trinomial</u>.

CHAPTER 6 HIGHLIGHTS

Definitions and Concepts	Examples
Section 6.1 The Greatest Common Factor and Factoring by Grouping	

Definitions and Concepts	Examples
Factoring is the process of writing an expression as a product.	Factor: $6 = 2 \cdot 3$ $x^2 + 5x + 6 = (x + 2)(x + 3)$
To Find the GCF of a List of Integers	Find the GCF of 12, 36, and 48.
Step 1. Write each number as a product of primes.	$12 = 2 \cdot 2 \cdot 3$
Step 2. Identify the common prime factors.	$36 = 2 \cdot 2 \cdot 3 \cdot 3$
Step 3. The product of all common factors is the greatest common factor. If there are no common prime factors, the GCF is 1.	$48 = 2 \cdot 2 \cdot 2 \cdot 2 \cdot 3$ $\text{GCF} = 2 \cdot 2 \cdot 3 = 12$
The GCF of a list of common variables raised to powers is the variable raised to the smallest exponent in the list.	The GCF of z^5, z^3, and z^{10} is z^3.
The GCF of a list of terms is the product of all common factors.	Find the GCF of $8x^2y$, $10x^3y^2$, and $26x^2y^3$. The GCF of 8, 10, and 26 is 2. The GCF of x^2, x^3, and x^2 is x^2. The GCF of y, y^2, and y^3 is y. The GCF of the terms is $2x^2y$.

Definitions and Concepts	**Examples**

Section 6.1 The Greatest Common Factor and Factoring by Grouping

To Factor by Grouping

Step 1. Arrange the terms so that the first two terms have a common factor and the last two have a common factor.

Step 2. For each pair of terms, factor out the pair's GCF.

Step 3. If there is now a common binomial factor, factor it out.

Step 4. If there is no common binomial factor, begin again, rearranging the terms differently. If no rearrangement leads to a common binomial factor, the polynomial cannot be factored.

Factor $10ax + 15a - 6xy - 9y$.

Step 1. $10ax + 15a - 6xy - 9y$

Step 2. $5a(2x + 3) - 3y(2x + 3)$

Step 3. $(2x + 3)(5a - 3y)$

Section 6.2 Factoring Trinomials of the Form $x^2 + bx + c$

To factor a trinomial of the form $x^2 + bx + c$, look for two numbers whose product is c and whose sum is b. The factored form is

$$(x + \text{one number})(x + \text{other number})$$

Factor: $x^2 + 7x + 12$

$$3 + 4 = 7 \qquad 3 \cdot 4 = 12$$
$$(x + 3)(x + 4)$$

Section 6.3 Factoring Trinomials of the Form $ax^2 + bx + c$

Method 1: To factor $ax^2 + bx + c$, try various combinations of factors of ax^2 and c until a middle term of bx is obtained when checking.

Factor: $3x^2 + 14x - 5$

Factors of $3x^2$: $3x, x$

Factors of -5: $-1, 5$ and $1, -5$.

$$(3x - \underline{1})(x + 5)$$
$$\underbrace{\qquad\qquad}_{-1x}$$
$$\frac{15x}{14x}$$
Correct middle term

Method 2: Factor $ax^2 + bx + c$ by grouping.

Step 1. Find two numbers whose product is $a \cdot c$ and and whose sum is b.

Step 2. Rewrite bx, using the factors found in step 1.

Step 3. Factor by grouping.

Factor: $3x^2 + 14x - 5$

Step 1. Find two numbers whose product is $3 \cdot (-5)$ or -15 whose sum is 14. They are 15 and -1.

Step 2. $3x^2 + 14x - 5$
$$= 3x^2 + 15x - 1x - 5$$

Step 3. $= 3x(x + 5) - 1(x + 5)$
$$= (x + 5)(3x - 1)$$

A **perfect square trinomial** is a trinomial that is the square of some binomial.

Perfect square trinomial = square of binomial
$$x^2 + 4x + 4 = (x + 2)^2$$
$$25x^2 - 10x + 1 = (5x - 1)^2$$

Definitions and Concepts	**Examples**

Section 6.3 Factoring Trinomials of the Form $ax^2 + bx + c$

Factoring Perfect Square Trinomials: $$a^2 + 2ab + b^2 = (a + b)^2$$ $$a^2 - 2ab + b^2 = (a - b)^2$$	Factor: $$x^2 + 6x + 9 = x^2 + 2 \cdot x \cdot 3 + 3^2 = (x + 3)^2$$ $$4x^2 - 12x + 9 = (2x)^2 - 2 \cdot 2x \cdot 3 + 3^2 = (2x - 3)^2$$

Section 6.4 Factoring Binomials

Difference of Squares $$a^2 - b^2 = (a + b)(a - b)$$ ***Sum or Difference of Cubes*** $$a^3 + b^3 = (a + b)(a^2 - ab + b^2)$$ $$a^3 - b^3 = (a - b)(a^2 + ab + b^2)$$	Factor: $$x^2 - 9 = x^2 - 3^2 = (x + 3)(x - 3)$$ $$y^3 + 8 = y^3 + 2^3 = (y + 2)(y^2 - 2y + 4)$$ $$125z^3 - 1 = (5z)^3 - 1^3 = (5z - 1)(25z^2 + 5z + 1)$$

Integrated Review—Choosing a Factoring Strategy

To Factor a Polynomial, **Step 1.** Factor out the GCF. **Step 2.** **a.** If two terms, **i.** $a^2 - b^2 = (a + b)(a - b)$ **ii.** $a^3 - b^3 = (a - b)(a^2 + ab + b^2)$ **iii.** $a^3 + b^3 = (a + b)(a^2 - ab + b^2)$ **b.** If three terms, **i.** $a^2 + 2ab + b^2 = (a + b)^2$ **ii.** Methods in Sections 6.2 and 6.3 **c.** If four or more terms, try factoring by grouping. **Step 3.** See if any factors can be factored further. **Step 4.** Check by multiplying.	Factor: $2x^4 - 6x^2 - 8$ **Step 1.** $2x^4 - 6x^2 - 8 = 2(x^4 - 3x^2 - 4)$ **Step 2. b. ii.** $= 2(x^2 + 1)(x^2 - 4)$ **Step 3.** $= 2(x^2 + 1)(x + 2)(x - 2)$ **Step 4.** Check by multiplying. $$2(x^2 + 1)(x + 2)(x - 2) = 2(x^2 + 1)(x^2 - 4)$$ $$= 2(x^4 - 3x^2 - 4)$$ $$= 2x^4 - 6x^2 - 8$$

Section 6.5 Solving Quadratic Equations by Factoring

A **quadratic equation** is an equation that can be written in the form $ax^2 + bx + c = 0$ with a not 0. The form $ax^2 + bx + c = 0$ is called the **standard form** of a quadratic equation.	**Quadratic Equation** **Standard Form** $x^2 = 16$ $x^2 - 16 = 0$ $y = -2y^2 + 5$ $2y^2 + y - 5 = 0$

Definitions and Concepts	**Examples**

Section 6.5 Solving Quadratic Equations by Factoring

Zero Factor Theorem If a and b are real numbers and if $ab = 0$, then $a = 0$ or $b = 0$.	If $(x + 3)(x - 1) = 0$, then $x + 3 = 0$ or $x - 1 = 0$
To solve quadratic equations by factoring,	Solve: $3x^2 = 13x - 4$
Step 1. Write the equation in standard form: $ax^2 + bx + c = 0$.	**Step 1.** $3x^2 - 13x + 4 = 0$
Step 2. Factor the quadratic.	**Step 2.** $(3x - 1)(x - 4) = 0$
Step 3. Set each factor containing a variable equal to 0.	**Step 3.** $3x - 1 = 0$ or $x - 4 = 0$
Step 4. Solve the equations.	**Step 4.** $3x = 1$ or $x = 4$ $x = \dfrac{1}{3}$
Step 5. Check in the original equation.	**Step 5.** Check both $\frac{1}{3}$ and 4 in the original equation.

Section 6.6 Quadratic Equations and Problem Solving

Problem-Solving Steps	A garden is in the shape of a rectangle whose length is two feet more than its width. If the area of the garden is 35 square feet, find its dimensions.
1. UNDERSTAND the problem.	**1.** Read and reread the problem. Guess a solution and check your guess. Let x be the width of the rectangular garden. Then $x + 2$ is the length. 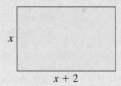 $x + 2$
2. TRANSLATE.	**2.** In words: length · width = area Translate: $(x + 2)$ · x = 35
3. SOLVE.	**3.** $\qquad (x + 2)x = 35$ $\qquad x^2 + 2x - 35 = 0$ $\qquad (x - 5)(x + 7) = 0$ $x - 5 = 0$ or $x + 7 = 0$ $x = 5$ or $x = -7$
4. INTERPRET.	**4.** Discard the solution of -7 since x represents width. *Check:* If x is 5 feet then $x + 2 = 5 + 2 = 7$ feet. The area of a rectangle whose width is 5 feet and whose length is 7 feet is (5 feet)(7 feet) or 35 square feet. *State:* The garden is 5 feet by 7 feet.

6 CHAPTER REVIEW

(6.1) *Complete the factoring.*

1. $6x^2 - 15x = 3x(\quad)$ $2x - 5$

2. $2x^3y - 6x^2y^2 - 8xy^3 = 2xy(\quad)$
$x^2 - 3xy - 4y^2$ or $(x - 4y)(x + y)$

Factor the GCF from each polynomial.

3. $20x^2 + 12x$ $4x(5x + 3)$ **4.** $6x^2y^2 - 3xy^3$

5. $-8x^3y + 6x^2y^2$ $-2x^2y(4x - 3y)$ **4.** $3xy^2(2x - y)$

6. $3x(2x + 3) - 5(2x + 3)$ **7.** $5x(x + 1) - (x + 1)$
$(2x + 3)(3x - 5)$ $(x + 1)(5x - 1)$

Factor. **8.** $(x - 1)(3x + 2)$ **9.** $(2x - 1)(3x + 5)$

8. $3x^2 - 3x + 2x - 2$ **9.** $6x^2 + 10x - 3x - 5$

10. $3a^2 + 9ab + 3b^2 + ab$ $(a + 3b)(3a + b)$

(6.2) *Factor each trinomial.*

11. $x^2 + 6x + 8$ $(x + 4)(x + 2)$ **12.** $x^2 - 11x + 24$ $(x - 8)(x - 3)$

13. $x^2 + x + 2$ prime **14.** $x^2 - 5x - 6$ $(x - 6)(x + 1)$

15. $x^2 + 2x - 8$ $(x + 4)(x - 2)$ **16.** $x^2 + 4xy - 12y^2$

17. $x^2 + 8xy + 15y^2$ **18.** $3x^2y + 6xy^2 + 3y^3$

19. $72 - 18x - 2x^2$ **20.** $32 + 12x - 4x^2$

(6.3) *Factor each trinomial.* **16.** $(x + 6y)(x - 2y)$ **17.** $(x + 5y)(x + 3y)$

21. $2x^2 + 11x - 6$ **22.** $4x^2 - 7x + 4$ prime

23. $4x^2 + 4x - 3$ **24.** $6x^2 + 5xy - 4y^2$

25. $6x^2 - 25xy + 4y^2$ **26.** $18x^2 - 60x + 50$ $2(3x - 5)^2$

27. $2x^2 - 23xy - 39y^2$ **28.** $4x^2 - 28xy + 49y^2$ $(2x - 7y)^2$

29. $18x^2 - 9xy - 20y^2$ **30.** $36x^3y + 24x^2y^2 - 45xy^3$

(6.4) *Factor each binomial.* **32.** $(3t + 5s)(3t - 5s)$

31. $4x^2 - 9$ $(2x + 3)(2x - 3)$ **32.** $9t^2 - 25s^2$

33. $16x^2 + y^2$ prime **34.** $x^3 - 8y^3$

35. $8x^3 + 27$ **36.** $2x^3 + 8x$ $2x(x^2 + 4)$

37. $54 - 2x^3y^3$ **38.** $9x^2 - 4y^2$ $(3x + 2y)(3x - 2y)$

39. $16x^4 - 1$ **40.** $x^4 + 16$ prime

39. $(4x^2 + 1)(2x + 1)(2x - 1)$

INTEGRATED REVIEW EXERCISES

Factor.

41. $2x^2 + 5x - 12$ **42.** $3x^2 - 12$ $3(x + 2)(x - 2)$

43. $x(x - 1) + 3(x - 1)$ **44.** $x^2 + xy - 3x - 3y$ $(x + y)(x - 3)$

45. $4x^2y - 6xy^2$ $2xy(2x - 3y)$ **46.** $8x^2 - 15x - x^3$ $-x(x - 5)(x - 3)$

47. $125x^3 + 27$ **48.** $24x^2 - 3x - 18$ $3(8x^2 - x - 6)$

49. $(x + 7)^2 - y^2$ **50.** $x^2(x + 3) - 4(x + 3)$

(6.5) *Solve the following equations.*

51. $(x + 6)(x - 2) = 0$ $-6, 2$

52. $3x(x + 1)(7x - 2) = 0$ **53.** $4(5x + 1)(x + 3) = 0$

54. $x^2 + 8x + 7 = 0$ $-7, -1$ **55.** $x^2 - 2x - 24 = 0$ $-4, 6$

56. $x^2 + 10x = -25$ -5 **57.** $x(x - 10) = -16$ $2, 8$

58. $(3x - 1)(9x^2 + 3x + 1) = 0$ **59.** $56x^2 - 5x - 6 = 0$

60. $20x^2 - 7x - 6 = 0$ $\frac{3}{4}, -\frac{2}{5}$ **61.** $5(3x + 2) = 4$ $-\frac{2}{5}$

62. $6x^2 - 3x + 8 = 0$ **63.** $12 - 5t = -3$ 3

64. $5x^3 + 20x^2 + 20x = 0$ $-2, 0$ **65.** $4t^3 - 5t^2 - 21t = 0$
$0, -\frac{7}{4}, 3$

58. $\frac{1}{3}$

(6.6) *Solve the following problems.*

△ **66.** A flag for a local organization is in the shape of a rectangle whose length is 15 inches less than twice its width. If the area of the flag is 500 square inches, find its dimensions.
width: 20 in.; length: 25 in.

△ **67.** The base of a triangular sail is four times its height. If the area of the triangle is 162 square yards, find the base. 36 yd

height

base

68. 19 and 20

68. Find two consecutive positive integers whose product is 380.

69. A rocket is fired from the ground with an initial velocity of 440 feet per second. Its height h after t seconds is given by the equation

$$h = -16t^2 + 440t$$

69. a. 17.5 sec and 10 sec; The rocket reaches a height of 2800 ft on its way up and on its way back down.

a. Find how many seconds pass before the rocket reaches a height of 2800 feet. Explain why two answers are obtained.

b. Find how many seconds pass before the rocket reaches the ground again.   27.5 sec

△ **70.** An architect's squaring instrument is in the shape of a right triangle. Find the length of the long leg of the right triangle if the hypotenuse is 8 centimeters longer than the long leg and the short leg is 8 centimeters shorter than the long leg.   32 cm

Answers to Chapter Review exercises on p. 406.

18. $3y(x + y)^2$

19. $2(3 - x)(12 + x)$

20. $4(8 + 3x - x^2)$

21. $(2x - 1)(x + 6)$

23. $(2x + 3)(2x - 1)$

24. $(3x + 4y)(2x - y)$

25. $(6x - y)(x - 4y)$

27. $(2x + 3y)(x - 13y)$

29. $(6x + 5y)(3x - 4y)$

30. $3xy(2x + 3y)(6x - 5y)$

34. $(x - 2y)(x^2 + 2xy + 4y^2)$

35. $(2x + 3)(4x^2 - 6x + 9)$

37. $2(3 - xy)(9 + 3xy + x^2y^2)$

41. $(2x - 3)(x + 4)$

43. $(x - 1)(x + 3)$

47. $(5x + 3)(25x^2 - 15x + 9)$

49. $(x + 7 + y)(x + 7 - y)$

50. $(x + 3)(x + 2)(x - 2)$

52. $0, -1, \frac{2}{7}$

53. $-\frac{1}{5}, -3$

59. $-\frac{2}{7}, \frac{3}{8}$

62. no real solution

STUDY SKILLS REMINDER

Are You Prepared for a Test on Chapter 6?

Below is a list of some *common trouble areas* for topics covered in Chapter 6. After studying for your test—but before taking your test—read these.

▶ Don't forget that the first step to factor any polynomial is to first factor out any common factors.

$$9x^2 - 36 = 9(x^2 - 4) = 9(x + 2)(x - 2)$$

▶ Can you completely factor $x^4 - 24x^2 - 25$?

$$x^4 - 24x^2 - 25 = (x^2 - 25)(x^2 + 1)$$
$$= (x + 5)(x - 5)(x^2 + 1)$$

▶ Remember that to use the zero factor property to solve a quadratic equation, one side of the equation must be 0 and the other side must be a factored polynomial.

$$x(x - 2) = 3 \quad \text{Cannot use zero factor property.}$$
$$x^2 - 2x - 3 = 0$$
$$(x - 3)(x + 1) = 0 \quad \text{Now we can use zero factor property.}$$
$$x - 3 = 0 \quad \text{or} \quad x + 1 = 0$$
$$x = 3 \quad \text{or} \quad x = -1$$

Remember: This is simply a sampling of selected topics given to check your understanding. For a review of Chapter 6 in your text, see the material at the end of this chapter.

CHAPTER 6 TEST

Remember to use your Chapter Test Prep Video CD to help you study and view solutions to the test questions you need help with.

Factor each polynomial completely. If a polynomial cannot be factored, write "prime."

1. $y^2 - 8y - 48$ $(y - 12)(y + 4)$ **2.** $x^2 + x - 10$ prime

3. $9x^3 + 39x^2 + 12x$

4. $3a^2 + 3ab - 7a - 7b$

5. $3x^2 - 5x + 2$ $(3x - 2)(x - 1)$ **6.** $x^2 + 14xy + 24y^2$

7. $180 - 5x^2$ $5(6 + x)(6 - x)$ **8.** $6t^2 - t - 5$

9. $xy^2 - 7y^2 - 4x + 28$

10. $x - x^5$

11. $-xy^3 - x^3y$ $-xy(y^2 + x^2)$ **12.** $64x^3 - 1$

13. $8y^3 - 64$ $8(y - 2)(y^2 + 2y + 4)$

3. $3x(3x + 1)(x + 4)$ **4.** $(3a - 7)(a + b)$ **6.** $(x + 12y)(x + 2y)$ **8.** $(6t + 5)(t - 1)$ **9.** $(y + 2)(y - 2)(x - 7)$

10. $x(1 + x^2)(1 + x)(1 - x)$ **12.** $(4x - 1)(16x^2 + 4x + 1)$

Solve each equation.

14. $x^2 + 5x = 14$ $-7, 2$ **15.** $x(x + 6) = 7$ $-7, 1$

16. $3x(2x - 3)(3x + 4) = 0$ **17.** $5t^3 - 45t = 0$

18. $t^2 - 2t - 15 = 0$ $-3, 5$ **19.** $6x^2 = 15x$ $0, \frac{5}{2}$

16. $0, \frac{3}{2}, -\frac{4}{3}$

17. $0, 3, -3$

Solve each problem.

△ **20.** A deck for a home is in the shape of a triangle. The length of the base of the triangle is 9 feet longer than its altitude. If the area of the triangle is 68 square feet, find the length of the base.   17 ft

21. The sum of two numbers is 17 and the sum of their squares is 145. Find the numbers.   8 and 9

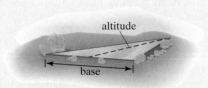

altitude

base

22. An object is dropped from the top of the Woolworth Building on Broadway in New York City. The height h of the object after t seconds is given by the equation

$$h = -16t^2 + 784$$

Find how many seconds pass before the object reaches the ground.   7 sec

6 CHAPTER CUMULATIVE REVIEW

1. Translate each sentence into a mathematical statement.
  a. Nine is less than or equal to eleven. $9 \le 11$
  b. Eight is greater than one. $8 > 1$
  c. Three is not equal to four. $3 \ne 4$ (Sec. 1.2, Ex. 3)

2. Insert $<$ or $>$ in the space to make each statement true.
  a. $|-5| \; > \; |-3|$
  b. $|0| \; < \; |-2|$ (Sec. 1.2)

3. Write each fraction in lowest terms.
  a. $\frac{42}{49}$ $\frac{6}{7}$ **b.** $\frac{11}{27}$ $\frac{11}{27}$ **c.** $\frac{88}{20}$ $\frac{22}{5}$ (Sec. 1.3, Ex. 2)

4. Evaluate $\frac{x}{y} + 5x$ if $x = 20$ and $y = 10$.   102 (Sec. 1.7)

5. Simplify: $\frac{8 + 2 \cdot 3}{2^2 - 1}$ $\frac{14}{3}$ (Sec. 1.4, Ex. 5)

6. Evaluate $\frac{x}{y} + 5x$ if $x = -20$ and $y = 10$. -102 (Sec. 1.7)

7. Add.
  a. $3 + (-7) + (-8)$ -12
  b. $[7 + (-10)] + [-2 + |-4|]$ -1 (Sec. 1.5, Ex. 7)

8. Evaluate $\frac{x}{y} + 5x$ if $x = -20$ and $y = -10$. -98 (Sec. 1.7)

9. Multiply.
  a. $(-6)(4)$ -24 **b.** $2(-1)$ -2
  c. $(-5)(-10)$ 50 (Sec. 1.7, Ex. 1)

10. Simplify: $5 - 2(3x - 7)$ $19 - 6x$ (Sec. 2.1)

11. Simplify each expression by combining like terms.
  a. $7x - 3x$ $4x$ **b.** $10y^2 + y^2$ $11y^2$
  c. $8x^2 + 2x - 3x$ $8x^2 - x$ (Sec. 2.1, Ex. 3)

12. Solve: $0.8y + 0.2(y - 1) = 1.8$ 2 (Sec. 2.2)

Solve.

13. $3 - x = 7$ -4 (Sec. 2.2, Ex. 7)   **14.** $\frac{x}{-7} = -4$

15. $-3x = 33$ **16.** $-\frac{2}{3}x = -22$

17. $8(2 - t) = -5t$ **18.** $-z = \frac{7z + 3}{5}$

19. A 10-foot board is to be cut into two pieces so that the longer piece is 4 times the shorter. Find the length of each piece.   shorter: 2 ft; longer: 8 ft (Sec. 2.5, Ex. 3)

20. Solve $3x + 9 \le 5(x - 1)$. Write the solution set using interval notation. $[7, \infty)$ (Sec. 2.8)

21. Graph the linear equation $y = -\frac{1}{3}x$.

22. Is the ordered pair $(-1, 2)$ a solution of $-7x - 8y = -9$?

23. Is the line passing through the points $(-6, 0)$ and $(-2, 3)$ parallel to the line passing through the points $(5, 4)$ and $(9, 7)$?

24. Find the slope of the line through $(5, -6)$ and $(5, 2)$.

25. Evaluate each expression for the given value of x.
  a. $2x^3$; x is 5   250   **b.** $\frac{9}{x^2}$; x is -3 1 (Sec. 5.1, Ex. 2)

26. Find the slope and y-intercept of the line whose equation is $7x - 3y = 2$. $m = \frac{7}{3}$, y-intercept $\left(0, -\frac{2}{3}\right)$ (Sec. 3.5)

14. 28 (Sec. 2.3)   **15.** -11 (Sec. 2.3, Ex. 2)   **16.** 33 (Sec. 2.3)   **17.** $\frac{16}{3}$ (Sec. 2.4, Ex. 2)   **18.** $-\frac{1}{4}$ (Sec. 2.4)

21. See graphing answer section. (Sec. 3.2, Ex. 5)   **22.** yes (Sec. 3.1)   **23.** yes (Sec. 3.4, Ex. 5)   **24.** undefined (Sec. 3.4)

27. Find the degree of each term.
    a. $-3x^2$ 2   b. $5x^3yz$ 5   c. 2   0 (Sec. 5.2, Ex. 1)

28. Find an equation of the vertical line through $(0, 7)$.

29. Subtract: $(2x^3 + 8x^2 - 6x) - (2x^3 - x^2 + 1)$

30. Find an equation of the line with slope 4 and y-intercept $\left(0, \frac{1}{2}\right)$. Write the equation in standard form.

31. Multiply $(3x + 2)(2x - 5)$.

32. Write an equation of the line through $(-4, 0)$ and $(6, -1)$. Write the equation in standard form.

33. Multiply $(3y + 1)^2$. $9y^2 + 6y + 1$ (Sec. 5.4, Ex. 4)

34. Solve the system: $\begin{cases} -x + 3y = 18 \\ -3x + 2y = 19 \end{cases}$ $(-3, 5)$ (Sec. 4.2)

35. Simplify by writing each expression with positive exponents only.
    a. 3^{-2} $\frac{1}{9}$ b. $2x^{-3}$ $\frac{2}{x^3}$
    c. $2^{-1} + 4^{-1}$ $\frac{3}{4}$ d. $(-2)^{-4}$ $\frac{1}{16}$ (Sec. 5.5, Ex. 1)

36. Simplify: $\dfrac{(5a^7)^2}{a^5}$ $25a^9$ (Sec. 5.1)

37. Write each number in scientific notation.
    a. 367,000,000 3.67×10^8
    b. 0.000003 3.0×10^{-6}
    c. 20,520,000,000 2.052×10^{10}
    d. 0.00085 8.5×10^{-4} (Sec. 5.5, Ex. 5)

38. Multiply: $(3x - 7y)^2$ $9x^2 - 42xy + 49y^2$ (Sec. 5.4)

39. Divide $x^2 + 7x + 12$ by $x + 3$ using long division.

40. Simplify: $\dfrac{(xy)^{-3}}{(x^5y^6)^3}$ $\dfrac{1}{x^{18}y^{21}}$ (Sec. 5.5)

41. Find the GCF of each list of terms.
    a. x^3, x^7, and x^5 x^3
    b. y, y^4, and y^7 y (Sec. 6.1, Ex. 2)

Factor.

42. $z^3 + 7z + z^2 + 7$ $(z + 1)(z^2 + 7)$ (Sec. 6.1)

43. $x^2 + 7x + 12$ $(x + 3)(x + 4)$ (Sec. 6.2, Ex. 1)

44. $2x^3 + 2x^2 - 84x$ $2x(x + 7)(x - 6)$ (Sec. 6.2)

45. $8x^2 - 22x + 5$ $(4x - 1)(2x - 5)$ (Sec. 6.3, Ex. 2)

46. $-4x^2 - 23x + 6$

47. $25a^2 - 9b^2$ $(5a + 3b)(5a - 3b)$ (Sec. 6.4, Ex. 2b)

48. $9xy^2 - 16x$ $x(3y + 4)(3y - 4)$ (Sec. 6.4)

49. Solve $(x - 3)(x + 1) = 0$. $3, -1$ (Sec. 6.5, Ex. 1)

50. Solve $x^2 - 13x = -36$. $9, 4$ (Sec. 6.5)

28. $x = 0$ (Sec. 3.6)   29. $9x^2 - 6x - 1$ (Sec. 5.2, Ex. 13)   30. $8x - 2y = -1$ (Sec. 3.6)   31. $6x^2 - 11x - 10$ (Sec. 5.3, Ex. 5)
32. $x + 10y = -4$ (Sec. 3.6)   39. $x + 4$ (Sec. 5.6, Ex. 4)   46. $(-4x + 1)(x + 6)$ or $-1(4x - 1)(x + 6)$ (Sec. 6.3)

CHAPTER  7

Rational Expressions

In this chapter, we expand our knowledge of algebraic expressions to include another category called rational expressions, such as $\frac{x+1}{x}$. We explore the operations of addition, subtraction, multiplication, and division for these algebraic fractions, using principles similar to the principles for number fractions. Thus, the material in this chapter will make full use of your knowledge of number fractions.

A WIDELY KNOWN ASPECT OF FORENSIC SCIENCE

Forensic science is the application of science and technology to the resolution of criminal and civil issues. One of the most widely known aspects of forensic science is criminalistics—the analysis of crime scenes and the examination of physical evidence. Criminalistic investigations are often carried out in what is popularly known as a crime laboratory.

Crime lab workers might be expected to analyze and identify chemical compounds, human hair and tissues, or materials such as metals, glass, or wood. They might also make measurements on items of physical evidence and compare footprints, firearms, or bullets. Anyone wishing to work in criminalistics, or the forensic sciences in general, should have excellent problem-solving and communication skills and also have a good mathematical background.

In the Spotlight on Decision Making feature on page 415, you will have the opportunity, as a forensic lab technician, to decide what material a piece of physical evidence is made of.

7.1 SIMPLIFYING RATIONAL EXPRESSIONS

Objectives

1 Find the value of a rational expression given a replacement number.

2 Identify values for which a rational expression is undefined.

3 Write rational expressions in lowest terms.

1   As we reviewed in Chapter 1, a rational number is a number that can be written as a quotient of integers. A **rational expression** is also a quotient; it is a quotient of polynomials.

> **Rational Expression**
>
> A rational expression is an expression that can be written in the form $\dfrac{P}{Q}$, where P and Q are polynomials and Q does not equal 0.

Rational Expressions

$$\frac{3y^3}{8} \qquad \frac{-4p}{p^3 + 2p + 1} \qquad \frac{5x^2 - 3x + 2}{3x + 7}$$

Rational expressions have different values depending on what value replaces the variable. Next, we review the standard order of operations by finding values of rational expressions for given replacement values of the variable.

EXAMPLE 1

Find the value of $\dfrac{x + 4}{2x - 3}$ for the given replacement values.

a. $x = 5$ **b.** $x = -2$

Solution **a.** Replace each x in the expression with 5 and then simplify.

$$\frac{x + 4}{2x - 3} = \frac{5 + 4}{2(5) - 3} = \frac{9}{10 - 3} = \frac{9}{7}$$

b. Replace each x in the expression with -2 and then simplify.

$$\frac{x + 4}{2x - 3} = \frac{-2 + 4}{2(-2) - 3} = \frac{2}{-7} \quad \text{or} \quad -\frac{2}{7}$$

For a negative fraction such as $\dfrac{2}{-7}$, recall from Chapter 1 that

$$\frac{2}{-7} = \frac{-2}{7} = -\frac{2}{7}$$

In general, for any fraction

$$\frac{-a}{b} = \frac{a}{-b} = -\frac{a}{b}, \qquad b \neq 0$$

This is also true for rational expressions. For example,

Notice the parentheses.
↓

$$\frac{-(x+2)}{x} = \frac{x+2}{-x} = -\frac{x+2}{x}$$

2   In the definition of rational expression on the previous page, notice that we wrote Q is not 0 for the denominator Q. This is because the denominator of a rational expression must not equal 0 since division by 0 is not defined. This means we must be careful when replacing the variable in a rational expression by a number. For example, suppose we replace x with 5 in the rational expression $\frac{2+x}{x-5}$. The expression becomes

$$\frac{2+x}{x-5} = \frac{2+5}{5-5} = \frac{7}{0}$$

But division by 0 is undefined. Therefore, in this rational expression we can allow x to be any real number *except* 5. A rational expression is undefined for values that make the denominator 0.

EXAMPLE 2

Are there any values for x for which each rational expression is undefined?

a. $\dfrac{x}{x-3}$ **b.** $\dfrac{x^2+2}{x^2-3x+2}$ **c.** $\dfrac{x^3-6x^2-10x}{3}$ **d.** $\dfrac{2}{x^2+1}$

Solution   To find values for which a rational expression is undefined, find values that make the *denominator* 0.

a. The denominator of $\frac{x}{x-3}$ is 0 when $x-3=0$ or when $x=3$. Thus, when $x=3$, the expression $\frac{x}{x-3}$ is undefined.

b. Set the denominator equal to zero.

$$x^2-3x+2=0$$
$$(x-2)(x-1)=0 \qquad \text{Factor.}$$
$$x-2=0 \quad \text{or} \quad x-1=0 \qquad \text{Set each factor equal to zero.}$$
$$x=2 \quad \text{or} \quad x=1 \qquad \text{Solve.}$$

Thus, when $x=2$ or $x=1$, the denominator x^2-3x+2 is 0. So the rational expression $\dfrac{x^2+2}{x^2-3x+2}$ is undefined when $x=2$ or when $x=1$.

c. The denominator of $\dfrac{x^3-6x^2-10x}{3}$ is never zero, so there are no values of x for which this expression is undefined.

d. No matter which real number x is replaced by, the denominator x^2+1 does not equal 0, so there are no real numbers for which this expression is undefined. ⬭

3   A fraction is said to be written in lowest terms or simplest form when the numerator and denominator have no common factors other than 1 (or -1). For example, the fraction $\frac{7}{10}$ is in lowest terms since the numerator and denominator have no common factors other than 1 (or -1).

The process of writing a rational expression in lowest terms or simplest form is called **simplifying** a rational expression. The following fundamental principle of rational expressions is used to simplify a rational expression.

> **Fundamental Principle of Rational Expressions**
>
> If P, Q, and R are polynomials, and Q and R are not 0,
>
> $$\frac{PR}{QR} = \frac{P}{Q}$$

Simplifying a rational expression is similar to simplifying a fraction. To simplify the fraction $\frac{15}{20}$, we factor the numerator and the denominator, look for common factors in both, and then use the fundamental principle.

$$\frac{15}{20} = \frac{3 \cdot 5}{2 \cdot 2 \cdot 5} = \frac{3}{2 \cdot 2} = \frac{3}{4}$$

To simplify the rational expression $\frac{x^2 - 9}{x^2 + x - 6}$, we also factor the numerator and denominator, look for common factors in both, and then use the fundamental principle of rational expressions.

$$\frac{x^2 - 9}{x^2 + x - 6} = \frac{(x - 3)(x + 3)}{(x - 2)(x + 3)} = \frac{x - 3}{x - 2}$$

This means that the rational expression $\frac{x^2 - 9}{x^2 + x - 6}$ has the same value as the rational expression $\frac{x-3}{x-2}$ for all values of x except 2 and -3. (Remember that when x is 2, the denominator of both rational expressions is 0 and when x is -3, the original rational expression has a denominator of 0.)

As we simplify rational expressions, we will assume that the simplified rational expression is equal to the original rational expression for all real numbers except those for which either denominator is 0. The following steps may be used to simplify rational expressions.

> **Simplifying a Rational Expression**
>
> **Step 1.** Completely factor the numerator and denominator.
>
> **Step 2.** Apply the fundamental principle of rational expressions to divide out common factors.

EXAMPLE 3

Write $\frac{21a^2b}{3a^5b}$ in simplest form.

Solution   Factor the numerator and denominator. Then apply the fundamental principle.

$$\frac{21a^2b}{3a^5b} = \frac{7 \cdot 3 \cdot a^2 \cdot b}{3 \cdot a^3 \cdot a^2 \cdot b} = \frac{7}{a^3}$$

EXAMPLE 4

Simplify: $\dfrac{6a - 33}{15}$

Solution Factor the numerator and denominator. Then apply the fundamental principle.

$$\frac{6a - 33}{15} = \frac{3(2a - 11)}{3 \cdot 5} = \frac{2a - 11}{5}$$

EXAMPLE 5

Simplify: $\dfrac{5x - 5}{x^3 - x^2}$

Solution Factor the numerator and denominator, if possible, and then apply the fundamental principle.

$$\frac{5x - 5}{x^3 - x^2} = \frac{5(x - 1)}{x^2(x - 1)} = \frac{5}{x^2}$$

EXAMPLE 6

Simplify: $\dfrac{x^2 + 8x + 7}{x^2 - 4x - 5}$

Solution Factor the numerator and denominator and apply the fundamental principle.

$$\frac{x^2 + 8x + 7}{x^2 - 4x - 5} = \frac{(x + 7)(x + 1)}{(x - 5)(x + 1)} = \frac{x + 7}{x - 5}$$

TEACHING TIP
Give students a concrete example that illustrates why $2x/x$ can be simplified to 2, and $(x + 2)/x$ cannot be simplified to 2. For example, evaluate each expression for $x = 1$ and then $x = 5$. Have students notice that the first expression evaluates to 2 for both values of x but the second expression does not.

> **Helpful Hint**
>
> When simplifying a rational expression, the fundamental principle applies to common *factors*, **not common *terms***.
>
> $\dfrac{x \cdot (x + 2)}{x \cdot x} = \dfrac{x + 2}{x}$ | $\dfrac{x + 2}{x}$
>
> Common factors. These can be divided out. | Common terms. Fundamental principle does not apply. This is in simplest form.

✔ **CONCEPT CHECK**

Recall that the fundamental principle applies to common factors only. Which of the following are *not* true? Explain why.

a. $\dfrac{3 - 1}{3 + 5} = -\dfrac{1}{5}$ **b.** $\dfrac{2x + 10}{2} = x + 5$ **c.** $\dfrac{37}{72} = \dfrac{3}{2}$ **d.** $\dfrac{2x + 3}{2} = x + 3$

EXAMPLE 7

Simplify each rational expression.

Concept Check Answer:
a, c, d

a. $\dfrac{x + y}{y + x}$ **b.** $\dfrac{x - y}{y - x}$

Solution   **a.** The expression $\frac{x + y}{y + x}$ can be simplified by using the commutative property of addition to rewrite the denominator $y + x$ as $x + y$.

$$\frac{x + y}{y + x} = \frac{x + y}{x + y} = 1$$

b. The expression $\frac{x - y}{y - x}$ can be simplified by recognizing that $y - x$ and $x - y$ are opposites. In other words, $y - x = -1(x - y)$. Proceed as follows:

$$\frac{x - y}{y - x} = \frac{1 \cdot (x - y)}{(-1)(x - y)} = \frac{1}{-1} = -1$$

EXAMPLE 8

Simplify: $\dfrac{4 - x^2}{3x^2 - 5x - 2}$

Solution

$$\frac{4 - x^2}{3x^2 - 5x - 2} = \frac{(2 - x)(2 + x)}{(x - 2)(3x + 1)} \qquad \text{Factor.}$$

$$= \frac{(-1)(x - 2)(2 + x)}{(x - 2)(3x + 1)} \qquad \text{Write } 2 - x \text{ as } -1(x - 2).$$

$$= \frac{(-1)(2 + x)}{3x + 1} \quad \text{or} \quad \frac{-2 - x}{3x + 1} \qquad \text{Simplify.}$$

Spotlight on DECISION MAKING

Suppose you are a forensic lab technician. You have been asked to try to identify a piece of metal found at a crime scene. You know that one way to analyze the piece of metal is to find its density (mass per unit volume), using the formula $density = \dfrac{mass}{volume}$. After weighing the metal, you find its mass as 36.2 grams. You have also determined the volume of the metal to be 4.5 milliliters. Use this information to decide which type of metal this piece is most likely made of, and explain your reasoning. What other characteristics might help the identification?

Densities	
Metal	**Density (g/ml)**
Aluminum	2.7
Iron	7.8
Lead	11.5
Silver	10.5

STUDY SKILLS REMINDER

How Are You Doing?

If you haven't done so yet, take a few moments and think about how you are doing in this course. Are you working toward your goal of successfully completing this course? Is your performance on homework, quizzes, and tests satisfactory? If not, you might want to see your instructor to see if he/she has any suggestions on how you can improve your performance. Let me once again re-mind you that, in addition to your instructor, there are many places to get help with your mathematics course. A few suggestions are below.

▶ This text has an accompanying video lesson for every section in this text.

▶ The back of this book contains answers to odd-numbered exercises and selected solutions.

▶ MathPro is available with this text. It is a tutorial software program with lessons corresponding to each section in the text.

▶ There is a student solutions manual available that contains worked-out solutions to odd-numbered exercises as well as solutions to every exercise in the Integrated Reviews, Chapter Reviews, Chapter Tests, and Cumulative Reviews.

▶ Don't forget to check with your instructor for other local resources available to you, such as a tutoring center.

MENTAL MATH

Find any real numbers for which each rational expression is undefined. See Example 2.

1. $\dfrac{x+5}{x}$ $x = 0$

2. $\dfrac{x^2 - 5x}{x - 3}$ $x = 3$

3. $\dfrac{x^2 + 4x - 2}{x(x - 1)}$ $x = 0, x = 1$

4. $\dfrac{x+2}{(x-5)(x-6)}$ $x = 5, x = 6$

Decide which rational expression can be simplified. (Do not actually simplify.)

5. $\dfrac{x}{x+7}$ no

6. $\dfrac{3+x}{x+3}$ yes

7. $\dfrac{5-x}{x-5}$ yes

8. $\dfrac{x+2}{x+8}$ no

EXERCISE SET 7.1

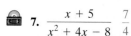

STUDY GUIDE/SSM · CD/VIDEO · PH MATH TUTOR CENTER · MathXL®Tutorials ON CD · MathXL® · MyMathLab®

19. $x = -1, x = 0, x = 1$ **20.** $x = 0, x = 2, x = -2$

Find the value of the following expressions when $x = 2$, $y = -2$, and $z = -5$. See Example 1.

Find any real numbers for which each rational expression is undefined. See Example 2.

1. $\dfrac{x+5}{x+2}$ $\dfrac{7}{4}$

2. $\dfrac{x+8}{2x+5}$ $\dfrac{10}{9}$

3. $\dfrac{z-8}{z+2}$ $\dfrac{13}{3}$

4. $\dfrac{y-2}{-5+y}$ $\dfrac{4}{7}$

5. $\dfrac{x^2 + 8x + 2}{x^2 - x - 6}$ $-\dfrac{11}{2}$

6. $\dfrac{z^2 + 8}{z^3 - 25z}$ undefined

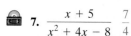 **7.** $\dfrac{x+5}{x^2 + 4x - 8}$ $\dfrac{7}{4}$

8. $\dfrac{z^3 + 1}{z^2 + 1}$ $-\dfrac{62}{13}$

9. $\dfrac{y^3}{y^2 - 1}$ $-\dfrac{8}{3}$

10. $\dfrac{z}{z^2 - 5}$ $-\dfrac{1}{4}$

11. $\dfrac{5}{3y}$ $y = 0$

12. $-\dfrac{7}{9x}$ $x = 0$

 13. $\dfrac{x+3}{x+2}$ $x = -2$

14. $\dfrac{5x+1}{x-3}$ $x = 3$

15. $\dfrac{4x^2 + 9}{2x - 8}$ $x = 4$

16. $\dfrac{x^2}{3x - 18}$ $x = 6$

17. $\dfrac{9x^3 + 4}{15x + 30}$ $x = -2$

18. $\dfrac{9x^3 + 4x}{15x + 45}$ $x = -3$

19. $\dfrac{19x^3 + 2}{x^3 - x}$

20. $\dfrac{20x^3 - 5}{x^3 - 4x}$

21. $\dfrac{x^2 - 5x - 2}{x^2 + 4}$ none

22. $\dfrac{9y^5 + y^3}{x^2 + 9}$ none

23. Explain why the denominator of a fraction or a rational expression must not equal zero. answers may vary

✎ **24.** Does $\dfrac{(x-3)(x+3)}{x-3}$ have the same value as $x+3$ for all real numbers? Explain why or why not.   no; answers may vary

Simplify each expression. See Examples 3 through 6. **28.** $3(x+7)$

25. $\dfrac{8x^5}{4x^9}$ $\dfrac{2}{x^4}$

26. $\dfrac{12y^7}{-2y^6}$ $-6y$

27. $\dfrac{5(x-2)}{(x-2)(x+1)}$ $\dfrac{5}{x+1}$

28. $\dfrac{9(x-7)(x+7)}{3(x-7)}$

🔒 **29.** $\dfrac{-5a-5b}{a+b}$ -5

30. $\dfrac{7x+35}{x^2+5x}$ $\dfrac{7}{x}$

31. $\dfrac{x+5}{x^2-4x-45}$ $\dfrac{1}{x-9}$

32. $\dfrac{x-3}{x^2-6x+9}$ $\dfrac{1}{x-3}$

33. $\dfrac{5x^2+11x+2}{x+2}$ $5x+1$

34. $\dfrac{12x^2+4x-1}{2x+1}$ $6x-1$

35. $\dfrac{x^2+x-12}{2x^2-5x-3}$ $\dfrac{x+4}{2x+1}$

36. $\dfrac{x^2+3x-4}{x^2-x-20}$ $\dfrac{x-1}{x-5}$

✎ **37.** Explain how to write a fraction in lowest terms.

✎ **38.** Explain how to write a rational expression in lowest terms.

37.–38. answers may vary

Simplify each expression. See Examples 7 and 8.

🔒 **39.** $\dfrac{x-7}{7-x}$ -1 **44.** $-\dfrac{x+3}{2}$

40. $\dfrac{y-z}{z-y}$ -1

41. $\dfrac{y^2-2y}{4-2y}$ $-\dfrac{y}{2}$

42. $\dfrac{x^2+5x}{20+4x}$ $\dfrac{x}{4}$

43. $\dfrac{x^2-4x+4}{4-x^2}$ $\dfrac{2-x}{x+2}$

44. $\dfrac{x^2+10x+21}{-2x-14}$

MIXED PRACTICE

Simplify each expression.

45. $\dfrac{15x^4y^8}{-5x^8y^3}$ $-\dfrac{3y^5}{x^4}$

46. $\dfrac{24a^3b^3}{6a^2b^4}$ $\dfrac{4a}{b}$

47. $\dfrac{(x-2)(x+3)}{5(x+3)}$ $\dfrac{x-2}{5}$

48. $\dfrac{-2(y-9)}{(y-9)^2}$ $-\dfrac{2}{y-9}$

49. $\dfrac{-6a-6b}{a+b}$ -6

50. $\dfrac{4a-4y}{4y-4a}$ -1

🔒 **51.** $\dfrac{2x^2-8}{4x-8}$ $\dfrac{x+2}{2}$

52. $\dfrac{5x^2-500}{35x+350}$ $\dfrac{x-10}{7}$

53. $\dfrac{11x^2-22x^3}{6x-12x^2}$ $\dfrac{11x}{6}$

54. $\dfrac{16r^2-4s^2}{4r-2s}$ $2(2r+s)$

55. $\dfrac{x+7}{x^2+5x-14}$ $\dfrac{1}{x-2}$

56. $\dfrac{x-10}{x^2-17x+70}$ $\dfrac{1}{x-7}$

57. $\dfrac{2x^2+3x-2}{2x-1}$ $x+2$

58. $\dfrac{4x^2+24x}{x+6}$ $4x$

59. $\dfrac{x^2-1}{x^2-2x+1}$ $\dfrac{x+1}{x-1}$

60. $\dfrac{x^2-16}{x^2-8x+16}$ $\dfrac{x+4}{x-4}$

61. $\dfrac{m^2-6m+9}{m^2-9}$ $\dfrac{m-3}{m+3}$

62. $\dfrac{m^2-4m+4}{m^2+m-6}$ $\dfrac{m-2}{m+3}$

63. $\dfrac{-2a^2+12a-18}{9-a^2}$ $\dfrac{2(a-3)}{a+3}$

64. $\dfrac{-4a^2+8a-4}{2a^2-2}$

65. $\dfrac{2-x}{x-2}$ -1 **64.** $-\dfrac{2(a-1)}{a+1}$

66. $\dfrac{7-y}{y-7}$ -1

67. $\dfrac{x^2-1}{1-x}$ $-x-1$

68. $\dfrac{x^2-xy}{2y-2x}$ $-\dfrac{x}{2}$

🔒 **69.** $\dfrac{x^2+7x+10}{x^2-3x-10}$ $\dfrac{x+5}{x-5}$

70. $\dfrac{2x^2+7x-4}{x^2+3x-4}$ $\dfrac{2x-1}{x-1}$

71. $\dfrac{3x^2+7x+2}{3x^2+13x+4}$ $\dfrac{x+2}{x+4}$

72. $\dfrac{4x^2-4x+1}{2x^2+9x-5}$ $\dfrac{2x-1}{x+5}$

REVIEW AND PREVIEW   **82.** $a+b$ **84.** $\dfrac{x+2}{y}$

Perform the indicated operations. See Section 1.3.

73. $\dfrac{1}{3}\cdot\dfrac{9}{11}$ $\dfrac{3}{11}$

74. $\dfrac{5}{27}\cdot\dfrac{2}{5}$ $\dfrac{2}{27}$

75. $\dfrac{1}{3}\div\dfrac{1}{4}$ $\dfrac{4}{3}$

76. $\dfrac{7}{8}\div\dfrac{1}{2}$ $\dfrac{7}{4}$

77. $\dfrac{5}{6}\cdot\dfrac{10}{11}\cdot\dfrac{2}{3}$ $\dfrac{50}{99}$

78. $\dfrac{4}{3}\cdot\dfrac{1}{7}\cdot\dfrac{10}{13}$ $\dfrac{40}{273}$

79. $\dfrac{13}{20}\div\dfrac{2}{9}$ $\dfrac{117}{40}$

80. $\dfrac{8}{15}\div\dfrac{5}{8}$ $\dfrac{64}{75}$

Concept Extensions   **88.** $-\dfrac{1}{x^2+3x+9}$

Simplify. These expressions contain 4-term polynomials and sums and differences of cubes.

81. $\dfrac{x^2+xy+2x+2y}{x+2}$ $x+y$

82. $\dfrac{ab+ac+b^2+bc}{b+c}$

83. $\dfrac{5x+15-xy-3y}{2x+6}$ $\dfrac{5-y}{2}$

84. $\dfrac{xy-6x+2y-12}{y^2-6y}$

85. $\dfrac{x^3+8}{x+2}$ x^2-2x+4

86. $\dfrac{x^3+64}{x+4}$ $x^2-4x+16$

87. $\dfrac{x^3-1}{1-x}$ $-x^2-x-1$

88. $\dfrac{3-x}{x^3-27}$

89. The total revenue R from the sale of a popular music compact disc is approximately given by the equation

$$R=\dfrac{150x^2}{x^2+3}$$

where x is the number of years since the CD has been released and revenue R is in millions of dollars.

a. Find the total revenue generated by the end of the first year. $\$37.5$ million   **89. c.** $\approx\$48.2$ million

b. Find the total revenue generated by the end of the second year. $\approx\$85.7$ million

c. Find the total revenue generated in the second year only.

90. For a certain model fax machine, the manufacturing cost C per machine is given by the equation

$$C=\dfrac{250x+10{,}000}{x}$$

where x is the number of fax machines manufactured and cost C is in dollars per machine.

a. Find the cost per fax machine when manufacturing 100 fax machines. $350

b. Find the cost per fax machine when manufacturing 1000 fax machines. $260

c. Does the cost per machine decrease or increase when more machines are manufactured? Explain why this is so.
decrease; answers may vary

Solve.

91. The dose of medicine prescribed for a child depends on the child's age A in years and the adult dose D for the medication. Young's Rule is a formula used by pediatricians that gives a child's dose C as

$$C = \frac{DA}{A + 12}$$

Suppose that an 8-year-old child needs medication, and the normal adult dose is 1000 mg. What size dose should the child receive?   400 mg

92. During a storm, water treatment engineers monitor how quickly rain is falling. If too much rain comes too fast, there is a danger of sewers backing up. A formula that gives the rainfall intensity i in millimeters per hour for a certain strength storm in eastern Virginia is

$$i = \frac{5840}{t + 29}$$

where t is the duration of the storm in minutes. What rainfall intensity should engineers expect for a storm of this strength in eastern Virginia that lasts for 80 minutes? Round your answer to one decimal place.   53.6 millimeters per hour

93. Calculating body-mass index is a way to gauge whether a person should lose weight. Doctors recommend that body-mass index values fall between 19 and 25. The formula for body-mass index B is   no; $B \approx 24$

$$B = \frac{705w}{h^2}$$

where w is weight in pounds and h is height in inches. Should a 148-pound person who is 5 feet 6 inches tall lose weight?

94. Anthropologists and forensic scientists use a measure called the cephalic index to help classify skulls. The cephalic index of a skull with width W and length L from front to back is given by the formula

$$C = \frac{100W}{L}$$

A long skull has an index value less than 75, a medium skull has an index value between 75 and 85, and a broad skull has an index value over 85. Find the cephalic index of a skull that is 5 inches wide and 6.4 inches long. Classify the skull.
$C = 78.125$; medium

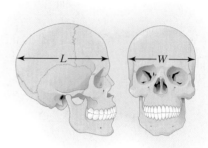

95. Recall that the fundamental principle applies to common *factors* only. Which of the following are *not* true? Explain why.   a, c; answers may vary

a. $\dfrac{3 - 1}{3 + 5} = -\dfrac{1}{5}$

b. $\dfrac{2x + 10}{2} = x + 5$

c. $\dfrac{37}{72} = \dfrac{3}{2}$

d. $\dfrac{2x + 6}{2} = x + 3$

96. A company's gross profit margin P can be computed with the formula $P = \frac{R - C}{R}$, where $R =$ the company's revenue and $C =$ the cost of goods sold. For fiscal year 2003, consumer electronics retailer Best Buy had revenues of $20.9 billion and cost of goods sold of $15.7 billion. (*Source:* Best Buy Annual Report) What was Best Buy's gross profit margin in 2003? Express the answer as a percent rounded to the nearest tenth of a percent.   24.9%

97. A baseball player's slugging percent S can be calculated with the following formula:

$$S = \frac{h + d + 2t + 3r}{b}, \text{ where } h = \text{ number of hits}, d = \text{number}$$

of doubles, $t =$ number of triples, $r =$ number of home runs, and $b =$ number of at bats. During the 2003 season, Manny Ramirez of the Boston Red Sox had 472 at bats, 150 hits, 30 doubles, 1 triple, and 31 home runs. (*Source:* Major League Baseball) Calculate Ramirez's 2003 slugging percent. Round to the nearest tenth of a percent.   58.3%

How does the graph of $y = \frac{x^2 - 9}{x - 3}$ compare to the graph of $y = x + 3$? Recall that $\frac{x^2 - 9}{x - 3} = \frac{(x + 3)(x - 3)}{x - 3} = x + 3$ as long as x is not 3. This means that the graph of $y = \frac{x^2 - 9}{x - 3}$ is the same as the graph of $y = x + 3$ with $x \neq 3$. To graph $y = \frac{x^2 - 9}{x - 3}$, then, graph the linear equation $y = x + 3$ and place an open dot on the graph at 3. This open dot or interruption of the line at 3 means $x \neq 3$.
See graphing answer section.

98. Graph $y = \dfrac{x^2 - 25}{x + 5}$.

99. Graph $y = \dfrac{x^2 - 16}{x - 4}$.

100. Graph $y = \dfrac{x^2 + x - 12}{x + 4}$.

101. Graph $y = \dfrac{x^2 - 6x + 8}{x - 2}$.

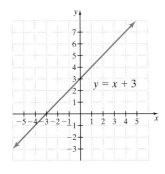

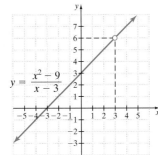

7.2

MULTIPLYING AND DIVIDING RATIONAL EXPRESSIONS

Objectives

1. Multiply rational expressions.
2. Divide rational expressions.

TEACHING TIP
Continually remind students throughout the next few sections that operations on rational expressions are the same as on fractions.

1   Just as simplifying rational expressions is similar to simplifying number fractions, multiplying and dividing rational expressions is similar to multiplying and dividing number fractions. To find the product of fractions and rational expressions, multiply the numerators and multiply the denominators.

$$\frac{3}{5} \cdot \frac{1}{4} = \frac{3 \cdot 1}{5 \cdot 4} = \frac{3}{20} \quad \text{and} \quad \frac{x}{y + 1} \cdot \frac{x + 3}{y - 1} = \frac{x(x + 3)}{(y + 1)(y - 1)}$$

Multiplying Rational Expressions

Let P, Q, R, and S be polynomials. Then

$$\frac{P}{Q} \cdot \frac{R}{S} = \frac{PR}{QS}$$

as long as $Q \neq 0$ and $S \neq 0$.

EXAMPLE 1

Multiply.

a. $\dfrac{25x}{2} \cdot \dfrac{1}{y^3}$

b. $\dfrac{-7x^2}{5y} \cdot \dfrac{3y^5}{14x^2}$

Solution To multiply rational expressions, multiply the numerators and then multiply the denominators of both expressions. Then simplify if possible.

a. $\dfrac{25x}{2} \cdot \dfrac{1}{y^3} = \dfrac{25x \cdot 1}{2 \cdot y^3} = \dfrac{25x}{2y^3}$

The expression $\dfrac{25x}{2y^3}$ is in simplest form.

b. $\dfrac{-7x^2}{5y} \cdot \dfrac{3y^5}{14x^2} = \dfrac{-7x^2 \cdot 3y^5}{5y \cdot 14x^2}$ Multiply.

The expression $\dfrac{-7x^2 \cdot 3y^5}{5y \cdot 14x^2}$ is not in simplest form, so we factor the numerator and the denominator and apply the fundamental principle.

$$= \dfrac{-1 \cdot 7 \cdot 3 \cdot x^2 \cdot y \cdot y^4}{5 \cdot 2 \cdot 7 \cdot x^2 \cdot y}$$

$$= -\dfrac{3y^4}{10}$$

When multiplying rational expressions, it is usually best to factor each numerator and denominator. This will help us when we apply the fundamental principle to write the product in lowest terms.

EXAMPLE 2

Multiply: $\dfrac{x^2 + x}{3x} \cdot \dfrac{6}{5x + 5}$

Solution

$$\dfrac{x^2 + x}{3x} \cdot \dfrac{6}{5x + 5} = \dfrac{x(x + 1)}{3x} \cdot \dfrac{2 \cdot 3}{5(x + 1)} \quad \text{Factor numerators and denominators.}$$

$$= \dfrac{x(x + 1) \cdot 2 \cdot 3}{3x \cdot 5(x + 1)} \quad \text{Multiply.}$$

$$= \dfrac{2}{5} \quad \text{Simplify.}$$

The following steps may be used to multiply rational expressions.

Multiplying Rational Expressions

Step 1. Completely factor numerators and denominators.

Step 2. Multiply numerators and multiply denominators.

Step 3. Simplify or write the product in lowest terms by applying the fundamental principle to all common factors.

✔ **CONCEPT CHECK**

Which of the following is a true statement?

a. $\dfrac{1}{3} \cdot \dfrac{1}{2} = \dfrac{1}{5}$ **b.** $\dfrac{2}{x} \cdot \dfrac{5}{x} = \dfrac{10}{x}$ **c.** $\dfrac{3}{x} \cdot \dfrac{1}{2} = \dfrac{3}{2x}$ **d.** $\dfrac{x}{7} \cdot \dfrac{x + 5}{4} = \dfrac{2x + 5}{28}$

EXAMPLE 3

Multiply: $\dfrac{3x + 3}{5x - 5x^2} \cdot \dfrac{2x^2 + x - 3}{4x^2 - 9}$

Solution

$\dfrac{3x + 3}{5x - 5x^2} \cdot \dfrac{2x^2 + x - 3}{4x^2 - 9} = \dfrac{3(x + 1)}{5x(1 - x)} \cdot \dfrac{(2x + 3)(x - 1)}{(2x - 3)(2x + 3)}$ Factor.

$= \dfrac{3(x + 1)(2x + 3)(x - 1)}{5x(1 - x)(2x - 3)(2x + 3)}$ Multiply.

$= \dfrac{3(x + 1)(x - 1)}{5x(1 - x)(2x - 3)}$ Apply the fundamental principle.

Next, recall that $x - 1$ and $1 - x$ are opposites so that $x - 1 = -1(1 - x)$.

$= \dfrac{3(x + 1)(-1)(1 - x)}{5x(1 - x)(2x - 3)}$ Write $x - 1$ as $-1(1 - x)$.

$= \dfrac{-3(x + 1)}{5x(2x - 3)}$ or $-\dfrac{3(x + 1)}{5x(2x - 3)}$ Apply the fundamental principle.

2 We can divide by a rational expression in the same way we divide by a fraction. To divide by a fraction, multiply by its reciprocal.

> **Helpful Hint**
> Don't forget how to find reciprocals. The reciprocal of $\dfrac{a}{b}$ is $\dfrac{b}{a}$, $a \neq 0$, $b \neq 0$.

For example, to divide $\frac{3}{2}$ by $\frac{7}{8}$, multiply $\frac{3}{2}$ by $\frac{8}{7}$.

$$\dfrac{3}{2} \div \dfrac{7}{8} = \dfrac{3}{2} \cdot \dfrac{8}{7} = \dfrac{3 \cdot 4 \cdot 2}{2 \cdot 7} = \dfrac{12}{7}$$

Dividing Rational Expressions

Let P, Q, R, and S be polynomials. Then,

$$\dfrac{P}{Q} \div \dfrac{R}{S} = \dfrac{P}{Q} \cdot \dfrac{S}{R} = \dfrac{PS}{QR}$$

as long as $Q \neq 0$, $S \neq 0$, and $R \neq 0$.

EXAMPLE 4

Divide: $\dfrac{3x^3 y^7}{40} \div \dfrac{4x^3}{y^2}$

Solution

$$\dfrac{3x^3 y^7}{40} \div \dfrac{4x^3}{y^2} = \dfrac{3x^3 y^7}{40} \cdot \dfrac{y^2}{4x^3} \qquad \text{Multiply by the reciprocal of } \dfrac{4x^3}{y^2}.$$

$$= \dfrac{3x^3 y^9}{160 x^3}$$

$$= \dfrac{3y^9}{160} \qquad \text{Simplify.} \quad \text{⬭}$$

EXAMPLE 5

Divide $\dfrac{(x-1)(x+2)}{10}$ by $\dfrac{2x+4}{5}$

Solution

$$\dfrac{(x-1)(x+2)}{10} \div \dfrac{2x+4}{5} = \dfrac{(x-1)(x+2)}{10} \cdot \dfrac{5}{2x+4} \qquad \begin{array}{l}\text{Multiply by the reciprocal of}\\[4pt] \dfrac{2x+4}{5}.\end{array}$$

$$= \dfrac{(x-1)(x+2)\cdot 5}{5\cdot 2\cdot 2\cdot(x+2)} \qquad \text{Factor and multiply.}$$

$$= \dfrac{x-1}{4} \qquad \text{Simplify.} \quad \text{⬭}$$

The following may be used to divide by a rational expression.

Dividing by a Rational Expression

Multiply by its reciprocal.

EXAMPLE 6

Divide: $\dfrac{6x+2}{x^2-1} \div \dfrac{3x^2+x}{x-1}$

Solution

$$\dfrac{6x+2}{x^2-1} \div \dfrac{3x^2+x}{x-1} = \dfrac{6x+2}{x^2-1} \cdot \dfrac{x-1}{3x^2+x} \qquad \text{Multiply by the reciprocal.}$$

$$= \dfrac{2(3x+1)(x-1)}{(x+1)(x-1)\cdot x(3x+1)} \qquad \text{Factor and multiply.}$$

$$= \dfrac{2}{x(x+1)} \qquad \text{Simplify.} \quad \text{⬭}$$

EXAMPLE 7

Divide: $\dfrac{2x^2-11x+5}{5x-25} \div \dfrac{4x-2}{10}$

Solution

$$\dfrac{2x^2-11x+5}{5x-25} \div \dfrac{4x-2}{10} = \dfrac{2x^2-11x+5}{5x-25} \cdot \dfrac{10}{4x-2} \qquad \text{Multiply by the reciprocal.}$$

$$= \dfrac{(2x-1)(x-5)\cdot 2\cdot 5}{5(x-5)\cdot 2(2x-1)} \qquad \text{Factor and multiply.}$$

$$= \dfrac{1}{1} \quad \text{or} \quad 1 \qquad \text{Simplify.} \quad \text{⬭}$$

Now that we know how to multiply fractions and rational expressions, we can use this knowledge to help us convert between units of measure. To do so, we will use **unit fractions**. A unit fraction is a fraction that equals 1. For example, since 12 in. = 1 ft, we have the unit fractions

$$\frac{12 \text{ in.}}{1 \text{ ft}} = 1 \quad \text{and} \quad \frac{1 \text{ ft}}{12 \text{ in.}} = 1$$

EXAMPLE 8

CONVERTING FROM SQUARE YARDS TO SQUARE FEET

The largest casino in the world is the Foxwoods Resort Casino in Ledyard, CT. The gaming area for this casino is 21,444 *square yards*. Find the size of the gaming area in *square feet*. (*Source: The Guinness Book of Records*)

CLASSROOM EXAMPLE

The largest building in the world is the Boeing Company's Everett, Washington, factory complex where the 747, 767, and 777 jets are built. The volume of this building is 472,370,319 cubic feet. Convert this to cubic yards.

answer: 17,495,197 cu yd

Solution   There are 9 square feet in 1 square yard.

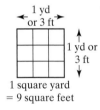

1 square yard
= 9 square feet

Unit fraction

$$21{,}444 \text{ square yards} = 21{,}444 \text{ sq. yd} \cdot \frac{9 \text{ sq. ft}}{1 \text{ sq. yd}}$$
$$= 192{,}996 \text{ square feet}$$

> **Helpful Hint**
>
> When converting a unit of measurement, if possible, write the unit fraction so that **the numerator contains the units we are converting to** and **the denominator contains the original units.** For example, suppose we want to convert 48 inches to feet.
>
> Unit fraction
>
> $$48 \text{ in.} = \frac{48 \text{ in.}}{1} \cdot \frac{1 \text{ ft}}{12 \text{ in.}} \quad \begin{array}{l} \leftarrow \textit{Units converting to} \\ \leftarrow \textit{Original units} \end{array}$$
> $$= \frac{48}{12} \text{ ft} = 4 \text{ ft}$$

MENTAL MATH

Find the following products. See Example 1.

1. $\dfrac{2}{y} \cdot \dfrac{x}{3}$ $\dfrac{2x}{3y}$

2. $\dfrac{3x}{4} \cdot \dfrac{1}{y}$ $\dfrac{3x}{4y}$

3. $\dfrac{5}{7} \cdot \dfrac{y^2}{x^2}$ $\dfrac{5y^2}{7x^2}$

4. $\dfrac{x^5}{11} \cdot \dfrac{4}{z^3}$ $\dfrac{4x^5}{11z^3}$

5. $\dfrac{9}{x} \cdot \dfrac{x}{5}$ $\dfrac{9}{5}$

6. $\dfrac{y}{7} \cdot \dfrac{3}{y}$ $\dfrac{3}{7}$

EXERCISE SET 7.2

STUDY GUIDE/SSM · CD/VIDEO · PH MATH TUTOR CENTER · MathXL®Tutorials ON CD · MathXL® · MyMathLab®

Multiply. Simplify if possible. See Examples 1 through 3.

 1. $\dfrac{8x}{2} \cdot \dfrac{x^5}{4x^2}$ x^4

2. $\dfrac{6x^2}{10x^3} \cdot \dfrac{5x}{12}$ $\dfrac{1}{4}$

3. $-\dfrac{5a^2b}{30a^2b^2} \cdot b^3$ $-\dfrac{b^2}{6}$

4. $-\dfrac{9x^3y^2}{18xy^5} \cdot y^3$ $-\dfrac{x^2}{2}$

5. $\dfrac{4}{x+2} \cdot \dfrac{x}{7}$ $\dfrac{4x}{7(x+2)}$

6. $\dfrac{9}{y-6} \cdot \dfrac{y}{5}$ $\dfrac{9y}{5(y-6)}$

7. $\dfrac{x}{2x-14} \cdot \dfrac{x^2-7x}{5}$ $\dfrac{x^2}{10}$

8. $\dfrac{4x-24}{20x} \cdot \dfrac{5}{x-6}$ $\dfrac{1}{x}$

9. $\dfrac{6x+6}{5} \cdot \dfrac{10}{36x+36}$ $\dfrac{1}{3}$

10. $\dfrac{x^2+x}{8} \cdot \dfrac{16}{x+1}$ $2x$

11. $\dfrac{m^2-n^2}{m+n} \cdot \dfrac{m}{m^2-mn}$ 1

12. $\dfrac{(m-n)^2}{m+n} \cdot \dfrac{m}{m^2-mn}$

13. $\dfrac{x^2-25}{x^2-3x-10} \cdot \dfrac{x+2}{x}$ $\dfrac{x+5}{x}$

14. $\dfrac{a^2+6a+9}{a^2-4} \cdot \dfrac{a+3}{a-2}$

 15. Find the area of the following rectangle. $\dfrac{2}{9x^2(x-5)}$ sq. ft

12. $\dfrac{m-n}{m+n}$ $\dfrac{2x}{x^2-25}$ feet

14. $\dfrac{(a+3)^3}{(a-2)^2(a+2)}$

$\dfrac{x+5}{9x^3}$ feet

16. Find the area of the following square. $\dfrac{4}{(5x+3)^2}$ sq. m

21. $x(x+4)$

22. $x+3$

24. $\dfrac{1}{20}$

25. m^2-n^2

26. $\dfrac{m-n}{m+n}$

$\dfrac{2x}{5x^2+3x}$ meters

Divide. Simplify if possible. See Examples 4 through 7.

17. $\dfrac{5x^7}{2x^5} \div \dfrac{10x}{4x^3}$ x^4

18. $\dfrac{9y^4}{6y} \div \dfrac{y^2}{3}$ $\dfrac{9y}{2}$

19. $\dfrac{8x^2}{y^3} \div \dfrac{4x^2y^3}{6}$ $\dfrac{12}{y^6}$

20. $\dfrac{7a^2b}{3ab^2} \div \dfrac{21a^2b^2}{14ab}$ $\dfrac{14}{9b^2}$

21. $\dfrac{(x-6)(x+4)}{4x} \div \dfrac{2x-12}{8x^2}$

22. $\dfrac{(x+3)^2}{5} \div \dfrac{5x+15}{25}$

23. $\dfrac{3x^2}{x^2-1} \div \dfrac{x^5}{(x+1)^2}$ $\dfrac{3(x+1)}{x^3(x-1)}$

24. $\dfrac{(x+1)}{(x+1)(2x+3)} \div \dfrac{20}{2x+3}$

25. $\dfrac{m^2-n^2}{m+n} \div \dfrac{m}{m^2+nm}$

43. $\dfrac{3x+4y}{2(x+2y)}$ **47.** $\dfrac{2(x+2)}{x-2}$ **48.** $-3x(x+2)$

27. $-\dfrac{x+2}{x-3}$

26. $\dfrac{(m-n)^2}{m+n} \div \dfrac{m^2-mn}{m}$

 27. $\dfrac{x+2}{7-x} \div \dfrac{x^2-5x+6}{x^2-9x+14}$

28. $(x-3) \div \dfrac{x^2+3x-18}{x}$ $\dfrac{x}{x+6}$

29. $\dfrac{x^2+7x+10}{1-x} \div \dfrac{x^2+2x-15}{x-1}$ $-\dfrac{x+2}{x-3}$

30. $\dfrac{a^2-b^2}{9} \div \dfrac{3b-3a}{27x^2}$ $-x^2(a+b)$

31. Explain how to multiply rational expressions.

32. Explain how to divide rational expressions.

31.–32. answers may vary

MIXED PRACTICE

Perform the indicated operations.

33. $\dfrac{5a^2b}{30a^2b^2} \cdot \dfrac{1}{b^3} \cdot \dfrac{1}{6b^4}$

34. $\dfrac{9x^2y^2}{42xy^5} \cdot \dfrac{6}{x^5} \cdot \dfrac{9}{7x^4y^3}$

35. $\dfrac{12x^3y}{8xy^7} \div \dfrac{7x^5y}{6x}$ $\dfrac{9}{7x^2y^7}$

36. $\dfrac{4y^2z}{3y^7z^7} \div \dfrac{12y}{6z}$ $\dfrac{2}{3y^6z^5}$

37. $\dfrac{5x-10}{12} \div \dfrac{4x-8}{8}$ $\dfrac{5}{6}$

38. $\dfrac{6x+6}{5} \div \dfrac{3x+3}{10}$ 4

39. $\dfrac{x^2+5x}{8} \cdot \dfrac{9}{3x+15}$ $\dfrac{3x}{8}$

40. $\dfrac{3x^2+12x}{6} \cdot \dfrac{9}{2x+8}$ $\dfrac{9x}{4}$

41. $\dfrac{7}{6p^2+q} \div \dfrac{14}{18p^2+3q}$ $\dfrac{3}{2}$

42. $\dfrac{5x-10}{12} \div \dfrac{4x-8}{8}$ $\dfrac{5}{6}$

 43. $\dfrac{3x+4y}{x^2+4xy+4y^2} \cdot \dfrac{x+2y}{2}$

44. $\dfrac{2a+2b}{3} \div \dfrac{a^2-b^2}{a-b}$ $\dfrac{2}{3}$

45. $\dfrac{x^2-9}{x^2+8} \div \dfrac{3-x}{2x^2+16}$ $-2(x+3)$

46. $\dfrac{x^2-y^2}{3x^2+3xy} \cdot \dfrac{3x^2+6x}{3x^2-2xy-y^2}$ $\dfrac{x+2}{3x+y}$

47. $\dfrac{(x+2)^2}{x-2} \div \dfrac{x^2-4}{2x-4}$

48. $\dfrac{x^2-4}{2y} \div \dfrac{2-x}{6xy}$

49. $\dfrac{a^2+7a+12}{a^2+5a+6} \cdot \dfrac{a^2+8a+15}{a^2+5a+4}$ $\dfrac{(a+5)(a+3)}{(a+2)(a+1)}$

50. $\dfrac{b^2+2b-3}{b^2+b-2} \cdot \dfrac{b^2-4}{b^2+6b+8}$ $\dfrac{(b+3)(b-2)}{(b+2)(b+4)}$

51. $\dfrac{1}{-x-4} \div \dfrac{x^2-7x}{x^2-3x-28}$ $-\dfrac{1}{x}$

52. $\dfrac{x^2-10x+21}{7-x} \div (x+3)$ $-\dfrac{x-3}{x+3}$

53. $\dfrac{x^2-5x-24}{2x^2-2x-24} \cdot \dfrac{4x^2+4x-24}{x^2-10x+16}$ $\dfrac{2(x+3)}{x-4}$

54. $\dfrac{a^2-b^2}{a} \cdot \dfrac{a+b}{a^2+ab}$

55. $(x-5) \div \dfrac{5-x}{x^2+2}$

54. $\dfrac{(a-b)(a+b)}{a^2}$ **55.** $-(x^2+2)$

56. $\dfrac{2x^2 + 3xy + y^2}{x^2 - y^2} \div \dfrac{1}{2x + 2y}$ $\dfrac{2(x + y)(2x + y)}{x - y}$

57. $\dfrac{x^2 - y^2}{x^2 - 2xy + y^2} \cdot \dfrac{y - x}{x + y}$ -1 **58.** $\dfrac{x + 3}{x^2 - 9} \cdot \dfrac{x^2 - 8x + 15}{5x}$

59. Find the quotient of $\dfrac{x^2 - 9}{2x}$ and $\dfrac{x + 3}{8x^4}$. $4x^3(x - 3)$

60. Find the quotient of $\dfrac{4x^2 + 4x + 1}{4x + 2}$ and $\dfrac{4x + 2}{16}$. 4

Convert as indicated. See Example 8.   **58.** $\dfrac{x - 5}{5x}$

61. 10 square feet = $\underline{}$ square inches.   1440

62. 1008 square inches = $\underline{}$ square feet.   7

63. The Pentagon, headquarters for the Department of Defense, contains 3,707,745 square feet of office space. Convert this to square yards. (Round to the nearest square yard.) (*Source: World Almanac*, 2003)   411,972 sq. yd

64. The Empire State building in Manhattan contains approximately 137,300 square yards of office space. Convert this to square feet.   1,235,700 sq. ft

65. 50 miles per hour = $\underline{}$ feet per second (round to the nearest whole).   73

66. 10 feet per second = $\underline{}$ miles per hour (round to the nearest tenth).   6.8

67. The speed of sound is 5023 feet per second in ocean water whose temperature is 77°F. Convert this speed of sound to miles per hour. Round to the nearest tenth. (*Source: CRC Handbook of Chemistry and Physics*, 65th edition)

68. On October 15, 1997, Andy Green broke the world land speed record. He reached a speed of approximately 763 mph in his car Thrust SSC in the Black Rock Desert, Nevada. Find this speed in feet per second. Round to the nearest whole. (*Source: The World Almanac and Book of Facts*, 2003)

67. 3424.8 mph   **68.** 1119 ft per sec

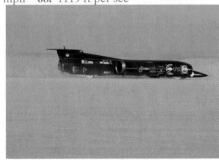

REVIEW AND PREVIEW

Perform each operation. See Section 1.3.

69. $\dfrac{1}{5} + \dfrac{4}{5}$ 1

70. $\dfrac{3}{15} + \dfrac{6}{15}$ $\dfrac{3}{5}$

71. $\dfrac{9}{9} - \dfrac{19}{9}$ $-\dfrac{10}{9}$

72. $\dfrac{4}{3} - \dfrac{8}{3}$ $-\dfrac{4}{3}$

73. $\dfrac{6}{5} + \left(\dfrac{1}{5} - \dfrac{8}{5}\right)$ $-\dfrac{1}{5}$

74. $-\dfrac{3}{2} + \left(\dfrac{1}{2} - \dfrac{3}{2}\right)$ $-\dfrac{5}{2}$

See Section 3.2.   **75.–76.** See graphing answer section.

75. Graph the linear equation $x - 2y = 6$.

76. Graph the linear equation $5x + y = 10$.

Concept Extensions   **83.** $\dfrac{a - b}{6(a^2 + ab + b^2)}$

Simplify. These expressions contain 4-term polynomials and sums and differences of cubes.

77. $\dfrac{a^2 + ac + ba + bc}{a - b} \div \dfrac{a + c}{a + b}$ $\dfrac{(a + b)^2}{a - b}$

78. $\dfrac{x^2 + 2x - xy - 2y}{x^2 - y^2} \div \dfrac{2x + 4}{x + y}$ $\dfrac{1}{2}$

79. $\dfrac{3x^2 + 8x + 5}{x^2 + 8x + 7} \cdot \dfrac{x + 7}{x^2 + 4}$ $\dfrac{3x + 5}{x^2 + 4}$

80. $\dfrac{16x^2 + 2x}{16x^2 + 10x + 1} \cdot \dfrac{1}{4x^2 + 2x}$ $\dfrac{1}{(2x + 1)^2}$

81. $\dfrac{x^3 + 8}{x^2 - 2x + 4} \cdot \dfrac{4}{x^2 - 4}$ $\dfrac{4}{x - 2}$

82. $\dfrac{9y}{3y - 3} \cdot \dfrac{y^3 - 1}{y^3 + y^2 + y}$ 3 **83.** $\dfrac{a^2 - ab}{6a^2 + 6ab} \div \dfrac{a^3 - b^3}{a^2 - b^2}$

84. $\dfrac{x^3 + 27y^3}{6x} \div \dfrac{x^2 - 9y^2}{x^2 - 3xy}$ $\dfrac{x^2 - 3xy + 9y^2}{6}$

Perform the following operations.   **87.** $\dfrac{5a(2a + b)(3a - 2b)}{b^2(a - b)(a + 2b)}$

85. $\left(\dfrac{x^2 - y^2}{x^2 + y^2} \div \dfrac{x^2 - y^2}{3x}\right) \cdot \dfrac{x^2 + y^2}{6}$ $\dfrac{x}{2}$

86. $\left(\dfrac{x^2 - 9}{x^2 - 1} \cdot \dfrac{x^2 + 2x + 1}{2x^2 + 9x + 9}\right) \div \dfrac{2x + 3}{1 - x}$ $-\dfrac{(x - 3)(x + 1)}{(2x + 3)^2}$

87. $\left(\dfrac{2a + b}{b^2} \cdot \dfrac{3a^2 - 2ab}{ab + 2b^2}\right) \div \dfrac{a^2 - 3ab + 2b^2}{5ab - 10b^2}$

88. $\left(\dfrac{x^2y^2 - xy}{4x - 4y} \div \dfrac{3y - 3x}{8x - 8y}\right) \cdot \dfrac{y - x}{8}$ $\dfrac{xy(xy - 1)}{12}$

89. On a recent day, 1 euro was equivalent to 1.09 U.S. dollars. If you had wanted to exchange $2000 U.S. to euros on that day for a European vacation, how much would you have received? Round to the nearest hundredth. (*Source: International Monetary Fund*)   1834.86 euros   **90.** no

90. An environmental technician finds that warm water from an industrial process is being discharged into a nearby pond at a rate of 30 gallons per minute. Plant regulations state that the flow rate should be no more than 0.1 cubic feet per second. Is the flow rate of 30 gallons per minute in violation of the plant regulations? (*Hint: 1 cubic foot is equivalent to 7.48 gallons.*)

7.3

ADDING AND SUBTRACTING RATIONAL EXPRESSIONS WITH COMMON DENOMINATORS AND LEAST COMMON DENOMINATOR

Objectives

1. Add and subtract rational expressions with the same denominator.
2. Find the least common denominator of a list of rational expressions.
3. Write a rational expression as an equivalent expression whose denominator is given.

1   Like multiplication and division, addition and subtraction of rational expressions is similar to addition and subtraction of rational numbers. In this section, we add and subtract rational expressions with a common (or the same) denominator.

Add: $\dfrac{6}{5} + \dfrac{2}{5}$ Add: $\dfrac{9}{x+2} + \dfrac{3}{x+2}$

Add the numerators and place the sum over the common denominator.

$$\dfrac{6}{5} + \dfrac{2}{5} = \dfrac{6+2}{5}$$

$$= \dfrac{8}{5} \quad \text{Simplify.}$$

$$\dfrac{9}{x+2} + \dfrac{3}{x+2} = \dfrac{9+3}{x+2}$$

$$= \dfrac{12}{x+2} \quad \text{Simplify.}$$

Adding and Subtracting Rational Expressions with Common Denominators

If $\dfrac{P}{R}$ and $\dfrac{Q}{R}$ are rational expressions, then

$$\dfrac{P}{R} + \dfrac{Q}{R} = \dfrac{P+Q}{R} \quad \text{and} \quad \dfrac{P}{R} - \dfrac{Q}{R} = \dfrac{P-Q}{R}$$

To add or subtract rational expressions, add or subtract numerators and place the sum or difference over the common denominator.

TEACHING TIP

Remind students throughout this section that to add or subtract when the denominators are the same, add or subtract numerators and *keep* the common denominator.

EXAMPLE 1

Add: $\dfrac{5m}{2n} + \dfrac{m}{2n}$

Solution

$$\dfrac{5m}{2n} + \dfrac{m}{2n} = \dfrac{5m+m}{2n} \quad \text{Add the numerators.}$$

$$= \dfrac{6m}{2n} \quad \text{Simplify the numerator by combining like terms.}$$

$$= \dfrac{3m}{n} \quad \text{Simplify by applying the fundamental principle.}$$

CLASSROOM EXAMPLE

Add: $\dfrac{8x}{3y} + \dfrac{x}{3y}$

answer: $\dfrac{3x}{y}$

EXAMPLE 2

Subtract: $\dfrac{2y}{2y-7} - \dfrac{7}{2y-7}$

Solution

$$\dfrac{2y}{2y-7} - \dfrac{7}{2y-7} = \dfrac{2y-7}{2y-7} \qquad \text{Subtract the numerators.}$$

$$= \dfrac{1}{1} \quad \text{or} \quad 1 \qquad \text{Simplify.}$$

EXAMPLE 3

Subtract: $\dfrac{3x^2 + 2x}{x-1} - \dfrac{10x-5}{x-1}$.

Solution

$$\dfrac{3x^2+2x}{x-1} - \dfrac{10x-5}{x-1} = \dfrac{(3x^2+2x)-(10x-5)}{x-1} \qquad \begin{array}{l}\text{Subtract the numerators}\\\text{Notice the parentheses.}\end{array}$$

$$= \dfrac{3x^2 + 2x - 10x + 5}{x-1} \qquad \text{Use the distributive property.}$$

$$= \dfrac{3x^2 - 8x + 5}{x-1} \qquad \text{Combine like terms.}$$

$$= \dfrac{(x-1)(3x-5)}{x-1} \qquad \text{Factor.}$$

$$= 3x - 5 \qquad \text{Simplify.}$$

> **Helpful Hint**
>
> Parentheses are inserted so that the entire numerator, $10x - 5$, is subtracted.

> **Helpful Hint**
>
> Notice how the numerator $10x - 5$ has been subtracted in Example 3.
>
> This − sign applies to the entire numerator of $10x - 5$.
>
> So parentheses are inserted here to indicate this.
>
> $$\dfrac{3x^2+2x}{x-1} - \dfrac{10x-5}{x-1} = \dfrac{3x^2+2x-(10x-5)}{x-1}$$

2   To add and subtract fractions with **unlike** denominators, first find a least common denominator (LCD), and then write all fractions as equivalent fractions with the LCD.

For example, suppose we add $\frac{8}{3}$ and $\frac{2}{5}$. The LCD of denominators 3 and 5 is 15, since 15 is the least common multiple (LCM) of 3 and 5. That is, 15 is the smallest number that both 3 and 5 divide into evenly. Rewrite each fraction so that its denominator is 15. (Notice how we apply the fundamental principle of rational expressions.)

$$\dfrac{8}{3} + \dfrac{2}{5} = \dfrac{8(5)}{3(5)} + \dfrac{2(3)}{5(3)} = \dfrac{40}{15} + \dfrac{6}{15} = \dfrac{40+6}{15} = \dfrac{46}{15}$$

We are multiplying by **1**.

To add or subtract rational expressions with unlike denominators, we also first find an LCD and then write all rational expressions as equivalent expressions with the LCD. The **least common denominator (LCD) of a list of rational expressions** is a polynomial of least degree whose factors include all the factors of the denominators in the list.

> **Finding the Least Common Denominator (LCD)**
>
> **Step 1.** Factor each denominator completely.
>
> **Step 2.** The least common denominator (LCD) is the product of all unique factors found in Step 1, each raised to a power equal to the greatest number of times that the factor appears in any one factored denominator.

EXAMPLE 4

Find the LCD for each pair.

a. $\dfrac{1}{8}, \dfrac{3}{22}$
  b. $\dfrac{7}{5x}, \dfrac{6}{15x^2}$

CLASSROOM EXAMPLE

Find the LCD for each pair.

a. $\dfrac{2}{9}, \dfrac{7}{15}$ **b.** $\dfrac{5}{6x^3}, \dfrac{11}{8x^5}$

answer: **a.** 45 **b.** $24x^5$

Solution **a.** Start by finding the prime factorization of each denominator.

$$8 = 2 \cdot 2 \cdot 2 = 2^3 \quad \text{and} \quad 22 = 2 \cdot 11$$

Next, write the product of all the unique factors, each raised to a power equal to the greatest number of times that the factor appears in any denominator.

The greatest number of times that the factor 2 appears is 3.

The greatest number of times that the factor 11 appears is 1.

$$\text{LCD} = 2^3 \cdot 11^1 = 8 \cdot 11 = 88$$

b. Factor each denominator.

$$5x = 5 \cdot x \quad \text{and} \quad 15x^2 = 3 \cdot 5 \cdot x^2$$

The greatest number of times that the factor 5 appears is 1.

The greatest number of times that the factor 3 appears is 1.

The greatest number of times that the factor x appears is 2.

$$\text{LCD} = 3^1 \cdot 5^1 \cdot x^2 = 15x^2$$

EXAMPLE 5

Find the LCD of

CLASSROOM EXAMPLE

Find the LCD of

a. $\dfrac{3a}{a+5}$ and $\dfrac{7a}{a-5}$

b. $\dfrac{-2}{y}$ and $\dfrac{5y}{y-3}$

answer: **a.** $(a+5)(a-5)$
b. $y(y-3)$

a. $\dfrac{7x}{x+2}$ and $\dfrac{5x^2}{x-2}$
  b. $\dfrac{3}{x}$ and $\dfrac{6}{x+4}$

Solution **a.** The denominators $x+2$ and $x-2$ are completely factored already. The factor $x+2$ appears once and the factor $x-2$ appears once.

$$\text{LCD} = (x+2)(x-2)$$

b. The denominators x and $x+4$ cannot be factored further. The factor x appeared once and the factor $x+4$ appears once.

$$\text{LCD} = x(x+4)$$

EXAMPLE 6

Find the LCD of $\dfrac{6m^2}{3m+15}$ and $\dfrac{2}{(m+5)^2}$.

Solution Factor each denominator.

$$3m + 15 = 3(m+5)$$
$$(m+5)^2 \text{ is already factored.}$$

The greatest number of times that the factor 3 appears is 1.
The greatest number of times that the factor $m + 5$ appears *in any one denominator* is 2.

$$\text{LCD} = 3(m+5)^2$$

✔ **CONCEPT CHECK**

Choose the correct LCD of $\dfrac{x}{(x+1)^2}$ and $\dfrac{5}{x+1}$.

a. $x + 1$ **b.** $(x+1)^2$ **c.** $(x+1)^3$ **d.** $5x(x+1)^2$

EXAMPLE 7

Find the LCD of $\dfrac{t-10}{t^2-t-6}$ and $\dfrac{t+5}{t^2+3t+2}$.

Solution   Start by factoring each denominator.

$$t^2 - t - 6 = (t-3)(t+2)$$
$$t^2 + 3t + 2 = (t+1)(t+2)$$
$$\text{LCD} = (t-3)(t+2)(t+1)$$

EXAMPLE 8

Find the LCD of $\dfrac{2}{x-2}$ and $\dfrac{10}{2-x}$.

Solution   The denominators $x - 2$ and $2 - x$ are opposites. That is, $2 - x = -1(x - 2)$. Use $x - 2$ or $2 - x$ as the LCD.

$$\text{LCD} = x - 2 \qquad \text{or} \qquad \text{LCD} = 2 - x$$

3   Next we practice writing a rational expression as an equivalent rational expression with a given denominator. To do this, we apply the fundamental principle, which says that $\frac{PR}{QR} = \frac{P}{Q}$, or equivalently that $\frac{P}{Q} = \frac{PR}{QR}$. This can be seen by recalling that multiplying an expression by 1 produces an equivalent expression. In other words,

$$\frac{P}{Q} = \frac{P}{Q} \cdot 1 = \frac{P}{Q} \cdot \frac{R}{R} = \frac{PR}{QR}.$$

EXAMPLE 9

Write $\dfrac{4b}{9a}$ as an equivalent fraction with the given denominator.

$$\frac{4b}{9a} = \frac{}{27a^2b}$$

Solution   Ask yourself: "What do we multiply $9a$ by to get $27a^2b$?" The answer is $3ab$, since $9a(3ab) = 27a^2b$. Multiply the numerator and denominator by $3ab$.

$$\frac{4b}{9a} = \frac{4b(3ab)}{9a(3ab)} = \frac{12ab^2}{27a^2b}$$

EXAMPLE 10

Write the rational expression as an equivalent rational expression with the given denominator.

$$\frac{5}{x^2 - 4} = \frac{}{(x - 2)(x + 2)(x - 4)}$$

Solution   First, factor the denominator $x^2 - 4$ as $(x - 2)(x + 2)$.

If we multiply the original denominator $(x - 2)(x + 2)$ by $x - 4$, the result is the new denominator $(x + 2)(x - 2)(x - 4)$. Thus, multiply the numerator and the denominator by $x - 4$.

$$\frac{5}{x^2 - 4} = \underbrace{\frac{5}{(x - 2)(x + 2)}}_{\substack{\text{Factored} \\ \text{denominator}}} = \frac{5(x - 4)}{(x - 2)(x + 2)(x - 4)}$$

$$= \frac{5x - 20}{(x - 2)(x + 2)(x - 4)}$$

CLASSROOM EXAMPLE

Write as an equivalent rational expression with the given denominator.

$$\frac{3}{x^2 - 25} = \frac{}{(x + 5)(x - 5)(x - 3)}$$

answer: $\dfrac{3x - 9}{(x + 5)(x - 5)(x - 3)}$

MENTAL MATH

Perform the indicated operations.

1. $\dfrac{2}{3} + \dfrac{1}{3}$ 1

2. $\dfrac{5}{11} + \dfrac{1}{11}$ $\dfrac{6}{11}$

3. $\dfrac{3x}{9} + \dfrac{4x}{9}$ $\dfrac{7x}{9}$

4. $\dfrac{3y}{8} + \dfrac{2y}{8}$ $\dfrac{5y}{8}$

5. $\dfrac{8}{9} - \dfrac{7}{9}$ $\dfrac{1}{9}$

6. $-\dfrac{4}{12} - \dfrac{3}{12}$ $-\dfrac{7}{12}$

7. $\dfrac{7}{5} - \dfrac{10y}{5}$ $\dfrac{7 - 10y}{5}$

8. $\dfrac{12x}{7} - \dfrac{4x}{7}$ $\dfrac{8x}{7}$

8. $\dfrac{7p + 5}{2p + 7}$ **14.** $2x + 5$ **16.** $2x + 5$ **22.** $-\dfrac{4}{2x + 1}$ **24.** $\dfrac{y - 2}{y + 2}$

EXERCISE SET 7.3

STUDY GUIDE/SSM	CD/ VIDEO	PH MATH TUTOR CENTER	MathXL®Tutorials ON CD	MathXL®	MyMathLab®

Add or subtract as indicated. Simplify the result if possible. See Examples 1 through 3.

1. $\dfrac{a}{13} + \dfrac{9}{13}$ $\dfrac{a + 9}{13}$

2. $\dfrac{x + 1}{7} + \dfrac{6}{7}$ $\dfrac{x + 7}{7}$

 3. $\dfrac{9}{3 + y} + \dfrac{y + 1}{3 + y}$ $\dfrac{y + 10}{3 + y}$

4. $\dfrac{9}{y + 9} + \dfrac{y}{y + 9}$ 1

5. $\dfrac{4m}{3n} + \dfrac{5m}{3n}$ $\dfrac{3m}{n}$

6. $\dfrac{3p}{2} + \dfrac{11p}{2}$ $7p$

7. $\dfrac{2x + 1}{x - 3} + \dfrac{3x + 6}{x - 3}$ $\dfrac{5x + 7}{x - 3}$

8. $\dfrac{4p - 3}{2p + 7} + \dfrac{3p + 8}{2p + 7}$

9. $\dfrac{7}{8} - \dfrac{3}{8}$ $\dfrac{1}{2}$

10. $\dfrac{4}{5} - \dfrac{13}{5}$ $-\dfrac{9}{5}$

11. $\dfrac{4m}{m - 6} - \dfrac{24}{m - 6}$ 4

12. $\dfrac{8y}{y - 2} - \dfrac{16}{y - 2}$ 8

13. $\dfrac{2x^2}{x - 5} - \dfrac{25 + x^2}{x - 5}$ $x + 5$

14. $\dfrac{6x^2}{2x - 5} - \dfrac{25 + 2x^2}{2x - 5}$

15. $\dfrac{-3x^2 - 4}{x - 4} - \dfrac{12 - 4x^2}{x - 4}$ $x + 4$

16. $\dfrac{7x^2 - 9}{2x - 5} - \dfrac{16 + 3x^2}{2x - 5}$

17. $\dfrac{2x + 3}{x + 1} - \dfrac{x + 2}{x + 1}$ 1

18. $\dfrac{1}{x^2 - 2x - 15} - \dfrac{4 - x}{x^2 - 2x - 15}$ $\dfrac{x - 3}{x^2 - 2x - 15}$

19. $\dfrac{3}{x^3} + \dfrac{9}{x^3}$ $\dfrac{12}{x^3}$

20. $\dfrac{5}{xy} + \dfrac{8}{xy}$ $\dfrac{13}{xy}$

21. $\dfrac{5}{x + 4} - \dfrac{10}{x + 4}$ $-\dfrac{5}{x + 4}$

22. $\dfrac{4}{2x + 1} - \dfrac{8}{2x + 1}$

23. $\dfrac{x}{x + y} - \dfrac{2}{x + y}$ $\dfrac{x - 2}{x + y}$

24. $\dfrac{y + 1}{y + 2} - \dfrac{3}{y + 2}$

25. $\dfrac{8x}{2x + 5} + \dfrac{20}{2x + 5}$ 4

26. $\dfrac{12y - 5}{3y - 1} + \dfrac{1}{3y - 1}$ 4

27. $\dfrac{5x + 4}{x - 1} - \dfrac{2x + 7}{x - 1}$ 3

28. $\dfrac{x^2 + 9x}{x + 7} - \dfrac{4x + 14}{x + 7}$

29. $\dfrac{a}{a^2 + 2a - 15} - \dfrac{3}{a^2 + 2a - 15}$ $\dfrac{1}{a + 5}$

30. $\dfrac{3y}{y^2 + 3y - 10} - \dfrac{6}{y^2 + 3y - 10}$ $\dfrac{3}{y + 5}$

31. $\dfrac{2x + 3}{x^2 - x - 30} - \dfrac{x - 2}{x^2 - x - 30}$ $\dfrac{1}{x - 6}$

32. $\dfrac{3x - 1}{x^2 + 5x - 6} - \dfrac{2x - 7}{x^2 + 5x - 6}$ $\dfrac{1}{x - 1}$

△ **33.** A square-shaped pasture has a side of length $\dfrac{5}{x - 2}$ meters. Express its perimeter as a rational expression. $\dfrac{20}{x - 2}$ m

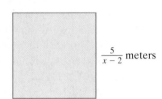
$\dfrac{5}{x - 2}$ meters

△ **34.** The following trapezoid has sides of indicated length. Find its perimeter. $\dfrac{2x + 15}{x + 3}$ in.

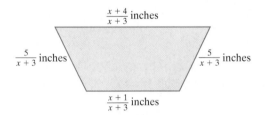
$\dfrac{x + 4}{x + 3}$ inches

$\dfrac{5}{x + 3}$ inches $\dfrac{5}{x + 3}$ inches

$\dfrac{x + 1}{x + 3}$ inches

35. Describe the process for adding and subtracting two rational expressions with the same denominators.   answers may vary

36. Explain the similarities between subtracting $\dfrac{3}{8}$ from $\dfrac{7}{8}$ and subtracting $\dfrac{6}{x + 3}$ from $\dfrac{9}{x + 3}$.   answers may vary

Find the LCD for the following lists of rational expressions. See Examples 4 through 8.

37. $\dfrac{2}{3}, \dfrac{4}{33}$ 33

38. $\dfrac{8}{20}, \dfrac{4}{15}$ 60

39. $\dfrac{19}{2x}, \dfrac{5}{4x^3}$ $4x^3$

40. $\dfrac{17x}{4y^5}, \dfrac{2}{8y}$ $8y^5$

41. $\dfrac{9}{8x}, \dfrac{3}{2x + 4}$ $8x(x + 2)$

42. $\dfrac{1}{6y}, \dfrac{3x}{4y + 12}$

43. $\dfrac{1}{3x + 3}, \dfrac{8}{2x^2 + 4x + 2}$ $6(x + 1)^2$

44. $\dfrac{19x + 5}{4x - 12}, \dfrac{3}{2x^2 - 12x + 18}$ $4(x - 3)^2$

45. $\dfrac{5}{x - 8}, \dfrac{3}{8 - x}$ $8 - x$ or $x - 8$

46. $\dfrac{2x + 5}{3x - 7}, \dfrac{5}{7 - 3x}$

47. $\dfrac{4 + x}{8x^2(x - 1)^2}, \dfrac{17}{10x^3(x - 1)}$

48. $\dfrac{2x + 3}{9x(x + 2)}, \dfrac{9x + 5}{12(x + 2)^2}$

49. $\dfrac{9x + 1}{2x + 1}, \dfrac{3x - 5}{2x - 1}$

50. $\dfrac{5}{4x - 2}, \dfrac{7}{4x + 2}$

51. $\dfrac{5x + 1}{2x^2 + 7x - 4}, \dfrac{3x}{2x^2 + 5x - 3}$ $(2x - 1)(x + 4)(x + 3)$

52. $\dfrac{4}{x^2 + 4x + 3}, \dfrac{4x - 2}{x^2 + 10x + 21}$ $(x + 3)(x + 1)(x + 7)$

53. Write some instructions to help a friend who is having difficulty finding the LCD of two rational expressions.

54. Explain why the LCD of rational expressions $\dfrac{7}{x + 1}$ and $\dfrac{9x}{(x + 1)^2}$ is $(x + 1)^2$ and not $(x + 1)^3$.

53–54. answers may vary

Rewrite each rational expression as an equivalent rational expression whose denominator is the given polynomial. See Examples 9 and 10.

55. $\dfrac{3}{2x} = \dfrac{6x}{4x^2}$

56. $\dfrac{3}{9y^5} = \dfrac{24y^4}{72y^9}$

57. $\dfrac{6}{3a} = \dfrac{24b^2}{12ab^2}$

58. $\dfrac{17a}{4y^2x} = \dfrac{136ayxz}{32y^3x^2z}$

59. $\dfrac{9}{x + 3} = \dfrac{18}{2(x + 3)}$

60. $\dfrac{4x + 1}{3x + 6} = \dfrac{4xy + y}{3y(x + 2)}$

61. $\dfrac{9a + 2}{5a + 10} = \dfrac{9ab + 2b}{5b(a + 2)}$

62. $\dfrac{5 + y}{2x^2 + 10} = \dfrac{10 + 2y}{4(x^2 + 5)}$

63. $\dfrac{x}{x^2 + 6x + 8} = \dfrac{x^2 + x}{(x + 4)(x + 2)(x + 1)}$

64. $\dfrac{5x}{x^2 + 2x - 3} = \dfrac{5x^2 - 25x}{(x - 1)(x - 5)(x + 3)}$

65. $\dfrac{9y - 1}{15x^2 - 30} = \dfrac{18y - 2}{30x^2 - 60}$

66. $\dfrac{8x + 3}{7y^2 - 21} = \dfrac{24x + 9}{21y^2 - 63}$

67. $\dfrac{5}{2x^2 - 9x - 5} = \dfrac{15x\,(x - 7)}{3x(2x + 1)(x - 7)(x - 5)}$

68. $\dfrac{x - 9}{3x^2 + 10x + 3} = \dfrac{x(x - 9)\,(x + 5)}{x(x + 3)(x + 5)(3x + 1)}$

Perform each operation. See Section 1.3.

69. $\dfrac{2}{3} + \dfrac{5}{7}$ $\dfrac{29}{21}$

70. $\dfrac{9}{10} - \dfrac{3}{5}$ $\dfrac{3}{10}$

71. $\dfrac{2}{6} - \dfrac{3}{4}$ $-\dfrac{5}{12}$

72. $\dfrac{11}{15} + \dfrac{5}{9}$ $\dfrac{58}{45}$

REVIEW AND PREVIEW

Solve the following quadratic equations by factoring. See Section 6.5.

73. $x(x - 3) = 0$ 0, 3

74. $2x(x + 5) = 0$ 0, −5

75. $x^2 + 6x + 5 = 0$ −5, −1

76. $x^2 - 6x + 5 = 0$ 1, 5

28. $x - 2$ **42.** $12y(y + 3)$ **46.** $7 - 3x$ or $3x - 7$ **47.** $40x^3(x - 1)^2$ **48.** $36x(x + 2)^2$ **49.** $(2x + 1)(2x - 1)$ **50.** $2(2x - 1)(2x + 1)$

Concept Extensions

Multiple choice. Select the correct result.

77. $\dfrac{3}{x} + \dfrac{y}{x} =$ c

 a. $\dfrac{3+y}{x^2}$ **b.** $\dfrac{3+y}{2x}$ **c.** $\dfrac{3+y}{x}$

78. $\dfrac{3}{x} - \dfrac{y}{x} =$ c

 a. $\dfrac{3-y}{x^2}$ **b.** $\dfrac{3-y}{2x}$ **c.** $\dfrac{3-y}{x}$

79. $\dfrac{3}{x} \cdot \dfrac{y}{x} =$ b

 a. $\dfrac{3y}{x}$ **b.** $\dfrac{3y}{x^2}$ **c.** $3y$

80. $\dfrac{3}{x} \div \dfrac{y}{x} =$ a

 a. $\dfrac{3}{y}$ **b.** $\dfrac{y}{3}$ **c.** $\dfrac{3}{x^2 y}$

Write each rational expression as an equivalent expression with a denominator of $x - 2$.

81. $\dfrac{5}{2-x} - \dfrac{5}{x-2}$ **82.** $\dfrac{8y}{2-x} - \dfrac{8y}{x-2}$

83. $-\dfrac{7+x}{2-x}\ \dfrac{7+x}{x-2}$ **84.** $\dfrac{x-3}{-(x-2)}\ \dfrac{3-x}{x-2}$

85. The planet Mercury revolves around the sun in 88 Earth days. It takes Jupiter 4332 Earth days to make one revolution

around the sun. (*Source:* National Space Science Data Center) If the two planets are aligned as shown in the figure, how long will it take for them to align again?   95,304 Earth days

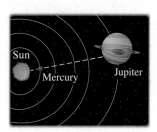

86. You are throwing a barbecue and you want to make sure that you purchase the same number of hot dogs as hot dog buns. Hot dogs come 8 to a package and hot dog buns come 12 to a package. What is the least number of each type of package you should buy?   3 pkg hot dogs and 2 pkg buns

87. An algebra student approaches you with a problem. He's tried to subtract two rational expressions, but his result does not match the book's. Check to see if the student has made an error. If so, correct his work shown below.   answers may vary

$$\dfrac{2x-6}{x-5} - \dfrac{x+4}{x-5}$$
$$= \dfrac{2x-6-x+4}{x-5}$$
$$= \dfrac{x-2}{x-5}$$

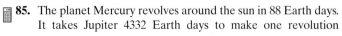

ADDING AND SUBTRACTING RATIONAL EXPRESSIONS WITH UNLIKE DENOMINATORS

Objective

1 Add and subtract rational expressions with unlike denominators.

1   In the previous section, we practiced all the skills we need to add and subtract rational expressions with unlike or different denominators. The steps are as follows:

Adding or Subtracting Rational Expressions with Unlike Denominators

Step 1. Find the LCD of the rational expressions.

Step 2. Rewrite each rational expression as an equivalent expression whose denominator is the LCD found in Step 1.

Step 3. Add or subtract numerators and write the sum or difference over the common denominator.

Step 4. Simplify or write the rational expression in simplest form.

EXAMPLE 1

Perform each indicated operation.

a. $\dfrac{a}{4} - \dfrac{2a}{8}$ **b.** $\dfrac{3}{10x^2} + \dfrac{7}{25x}$

Solution

a. First, we must find the LCD. Since $4 = 2^2$ and $8 = 2^3$, the LCD $= 2^3 = 8$. Next we write each fraction as an equivalent fraction with the denominator 8, then we subtract.

$$\frac{a}{4} - \frac{2a}{8} = \frac{a(2)}{4(2)} - \frac{2a}{8} = \frac{2a}{8} - \frac{2a}{8} = \frac{2a - 2a}{8} = \frac{0}{8} = 0$$

b. Since $10x^2 = 2 \cdot 5 \cdot x \cdot x$ and $25x = 5 \cdot 5 \cdot x$, the LCD $= 2 \cdot 5^2 \cdot x^2 = 50x^2$. We write each fraction as an equivalent fraction with a denominator of $50x^2$.

$$\frac{3}{10x^2} + \frac{7}{25x} = \frac{3(5)}{10x^2(5)} + \frac{7(2x)}{25x(2x)}$$

$$= \frac{15}{50x^2} + \frac{14x}{50x^2}$$

$$= \frac{15 + 14x}{50x^2} \qquad \text{Add numerators. Write the sum over the common denominator.}$$

EXAMPLE 2

Subtract: $\dfrac{6x}{x^2 - 4} - \dfrac{3}{x + 2}$

Solution

Since $x^2 - 4 = (x + 2)(x - 2)$, the LCD $= (x - 2)(x + 2)$. We write equivalent expressions with the LCD as denominators.

$$\frac{6x}{x^2 - 4} - \frac{3}{x + 2} = \frac{6x}{(x - 2)(x + 2)} - \frac{3(x - 2)}{(x + 2)(x - 2)}$$

$$= \frac{6x - 3(x - 2)}{(x + 2)(x - 2)} \qquad \text{Subtract numerators. Write the difference over the common denominator.}$$

$$= \frac{6x - 3x + 6}{(x + 2)(x - 2)} \qquad \text{Apply the distributive property in the numerator.}$$

$$= \frac{3x + 6}{(x + 2)(x - 2)} \qquad \text{Combine like terms in the numerator.}$$

Next we factor the numerator to see if this rational expression can be simplified.

$$= \frac{3(x + 2)}{(x + 2)(x - 2)} \qquad \text{Factor.}$$

$$= \frac{3}{x - 2} \qquad \text{Apply the fundamental principle to simplify.}$$

EXAMPLE 3

Add: $\dfrac{2}{3t} + \dfrac{5}{t+1}$

Solution

The LCD is $3t(t+1)$. We write each rational expression as an equivalent rational expression with a denominator of $3t(t+1)$.

$$\dfrac{2}{3t} + \dfrac{5}{t+1} = \dfrac{2(t+1)}{3t(t+1)} + \dfrac{5(3t)}{(t+1)(3t)}$$

$$= \dfrac{2(t+1) + 5(3t)}{3t(t+1)} \qquad \text{Add numerators. Write the sum over the common denominator.}$$

$$= \dfrac{2t + 2 + 15t}{3t(t+1)} \qquad \text{Apply the distributive property in the numerator.}$$

$$= \dfrac{17t + 2}{3t(t+1)} \qquad \text{Combine like terms in the numerator.} \;\; \ominus$$

CLASSROOM EXAMPLE

Add: $\dfrac{5}{7x} + \dfrac{2}{x+1}$

answer: $\dfrac{19x+5}{7x(x+1)}$

TEACHING TIP

Continue to remind students that they may not add or subtract numerators until the denominators are the same.

EXAMPLE 4

Subtract: $\dfrac{7}{x-3} - \dfrac{9}{3-x}$

Solution

To find a common denominator, we notice that $x-3$ and $3-x$ are opposites. That is, $3 - x = -(x-3)$. We write the denominator $3-x$ as $-(x-3)$ and simplify.

$$\dfrac{7}{x-3} - \dfrac{9}{3-x} = \dfrac{7}{x-3} - \dfrac{9}{-(x-3)}$$

$$= \dfrac{7}{x-3} - \dfrac{-9}{x-3} \qquad \text{Apply } \dfrac{a}{-b} = \dfrac{-a}{b}.$$

$$= \dfrac{7 - (-9)}{x-3} \qquad \text{Subtract numerators. Write the difference over the common denominator.}$$

$$= \dfrac{16}{x-3} \qquad\qquad \ominus$$

CLASSROOM EXAMPLE

Subtract: $\dfrac{10}{x-6} - \dfrac{15}{6-x}$

answer: $\dfrac{25}{x-6}$

TEACHING TIP

In Example 4, to verify for students that $x-3$ and $3-x$ are opposites, replace x with several values and have students notice the results. To verify that $3-x$ and $-(x-3)$ are equal, have students replace x with several values and notice the results.

EXAMPLE 5

Add: $1 + \dfrac{m}{m+1}$

Solution

Recall that 1 is the same as $\dfrac{1}{1}$. The LCD of and $\dfrac{1}{1}$ and $\dfrac{m}{m+1}$ is $m+1$.

$$1 + \dfrac{m}{m+1} = \dfrac{1}{1} + \dfrac{m}{m+1} \qquad \text{Write } \mathbf{1} \text{ as } \tfrac{1}{1}.$$

$$= \dfrac{1(m+1)}{1(m+1)} + \dfrac{m}{m+1} \qquad \text{Multiply both the numerator and the denominator of } \tfrac{1}{1} \text{ by } \boldsymbol{m+1}.$$

CLASSROOM EXAMPLE

Add: $2 + \dfrac{x}{x+5}$

answer: $\dfrac{3x+10}{x+5}$

$$= \frac{m + 1 + m}{m + 1}$$ Add numerators. Write the sum over the common denominator.

$$= \frac{2m + 1}{m + 1}$$ Combine like terms in the numerator.

EXAMPLE 6

Subtract: $\dfrac{3}{2x^2 + x} - \dfrac{2x}{6x + 3}$

Solution First, we factor the denominators.

$$\frac{3}{2x^2 + x} - \frac{2x}{6x + 3} = \frac{3}{x(2x + 1)} - \frac{2x}{3(2x + 1)}$$

The LCD is $3x(2x + 1)$. We write equivalent expressions with denominators of $3x(2x + 1)$.

$$= \frac{3(3)}{x(2x + 1)(3)} - \frac{2x(x)}{3(2x + 1)(x)}$$

$$= \frac{9 - 2x^2}{3x(2x + 1)}$$ Subtract numerators. Write the difference over the common denominator.

EXAMPLE 7

Add: $\dfrac{2x}{x^2 + 2x + 1} + \dfrac{x}{x^2 - 1}$

Solution First we factor the denominators.

$$\frac{2x}{x^2 + 2x + 1} + \frac{x}{x^2 - 1} = \frac{2x}{(x + 1)(x + 1)} + \frac{x}{(x + 1)(x - 1)}$$

Now we write the rational expressions as equivalent expressions with denominators of $(x + 1)(x + 1)(x - 1)$, the LCD.

$$= \frac{2x(x - 1)}{(x + 1)(x + 1)(x - 1)} + \frac{x(x + 1)}{(x + 1)(x - 1)(x + 1)}$$

$$= \frac{2x(x - 1) + x(x + 1)}{(x + 1)^2(x - 1)}$$ Add numerators. Write the sum over the common denominator.

$$= \frac{2x^2 - 2x + x^2 + x}{(x + 1)^2(x - 1)}$$ Apply the distributive property in the numerator.

$$= \frac{3x^2 - x}{(x + 1)^2(x - 1)} \quad \text{or} \quad \frac{x(3x - 1)}{(x + 1)^2(x - 1)}$$

The numerator was factored as a last step to see if the rational expression could be simplified further. Since there are no factors common to the numerator and the denominator, we can't simplify further.

MENTAL MATH

Match each exercise with the *first* step needed to perform the operation. Do not actually perform the operation.

1. $\dfrac{3}{4} - \dfrac{y}{4}$ D   **2.** $\dfrac{2}{a} \cdot \dfrac{3}{(a+6)}$ C   **3.** $\dfrac{x+1}{x} \div \dfrac{x-1}{x}$ A   **4.** $\dfrac{9}{x-2} - \dfrac{x}{x+2}$ B

A. Multiply the first rational expression by the reciprocal of the second rational expression.
B. Find the LCD. Write each expression as an equivalent expression with the LCD as denominator.
C. Multiply numerators, then multiply denominators.
D. Subtract numerators. Place the difference over a common denominator.

TEACHING TIP

A Group Activity for this section is available in the Instructor's Resource Manual.

8. $\dfrac{7}{x+4}$ **9.** $\dfrac{17x+30}{2(x-2)(x+2)}$ **10.** $\dfrac{3x-7}{(x-2)(x+2)}$ **12.** $\dfrac{x^2+2x+3}{(x+1)(x-1)}$ **14.** $\dfrac{2x-3}{2(4x-3)}$ **20.** $-\dfrac{16}{25x^2-1}$ **22.** $\dfrac{4}{x-3}$ **36.** $\dfrac{2x+6}{(x-2)^2}$

EXERCISE SET 7.4

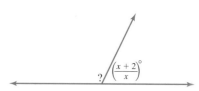

STUDY GUIDE/SSM   CD/ VIDEO   PH MATH TUTOR CENTER   MathXL®Tutorials ON CD   MathXL®   MyMathLab®

Perform the indicated operations. See Example 1.

1. $\dfrac{4}{2x} + \dfrac{9}{3x}$ $\dfrac{5}{x}$

2. $\dfrac{15}{7a} + \dfrac{8}{6a}$ $\dfrac{73}{21a}$

3. $\dfrac{15a}{b} + \dfrac{6b}{5}$ $\dfrac{75a+6b^2}{5b}$

4. $\dfrac{4c}{d} + \dfrac{8x}{5}$ $\dfrac{20c+8dx}{5d}$

5. $\dfrac{3}{x} + \dfrac{5}{2x^2}$ $\dfrac{6x+5}{2x^2}$

6. $\dfrac{14}{3x^2} + \dfrac{6}{x}$ $\dfrac{14+18x}{3x^2}$

Perform the indicated operations. See Examples 2 and 3.

7. $\dfrac{6}{x+1} + \dfrac{9}{2x+2}$ $\dfrac{21}{2(x+1)}$

8. $\dfrac{8}{x+4} - \dfrac{3}{3x+12}$

9. $\dfrac{15}{2x-4} + \dfrac{x}{x^2-4}$

10. $\dfrac{3}{x+2} - \dfrac{1}{x^2-4}$

11. $\dfrac{3}{4x} + \dfrac{8}{x-2}$ $\dfrac{35x-6}{4x(x-2)}$

12. $\dfrac{x}{x+1} + \dfrac{3}{x-1}$

13. $\dfrac{5}{y^2} - \dfrac{y}{2y+1}$ $\dfrac{5+10y-y^3}{y^2(2y+1)}$

14. $\dfrac{x}{4x-3} - \dfrac{3}{8x-6}$

15. In your own words, explain how to add two rational expressions with unlike denominators.   answers may vary

16. In your own words, explain how to subtract two rational expressions with unlike denominators.   answers may vary

Add or subtract as indicated. See Example 4.

17. $\dfrac{6}{x-3} + \dfrac{8}{3-x} - \dfrac{2}{x-3}$

18. $\dfrac{9}{x-3} + \dfrac{9}{3-x}$ 0

19. $\dfrac{-8}{x^2-1} - \dfrac{7}{1-x^2} - \dfrac{1}{x^2-1}$

20. $\dfrac{-9}{25x^2-1} + \dfrac{7}{1-25x^2}$

21. $\dfrac{x}{x^2-4} - \dfrac{2}{4-x^2} - \dfrac{1}{x-2}$

22. $\dfrac{5}{2x-6} - \dfrac{3}{6-2x}$

Add or subtract as indicated. See Example 5.

23. $\dfrac{5}{x} + 2$ $\dfrac{5+2x}{x}$

24. $\dfrac{7}{x^2} - 5x$ $\dfrac{7-5x^3}{x^2}$

25. $\dfrac{5}{x-2} + 6$ $\dfrac{6x-7}{x-2}$

26. $\dfrac{6y}{y+5} + 1$ $\dfrac{7y+5}{y+5}$

27. $\dfrac{y+2}{y+3} - 2$ $-\dfrac{y+4}{y+3}$

28. $\dfrac{7}{2x-3} - 3$ $\dfrac{16-6x}{2x-3}$

△ **29.** Two angles are said to be complementary if their sum is 90°. If one angle measures $\dfrac{40}{x}$ degrees, find the measure of its complement. $\left(\dfrac{90x-40}{x}\right)°$

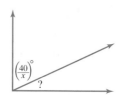

△ **30.** Two angles are said to be supplementary if their sum is 180°. If one angle measures $\dfrac{x+2}{x}$ degrees, find the measure of its supplement. $\left(\dfrac{179x-2}{x}\right)°$

MIXED PRACTICE

Perform the indicated operations. See Examples 1 through 7.

31. $\dfrac{5x}{x+2} - \dfrac{3x-4}{x+2}$ 2

32. $\dfrac{7x}{x-3} - \dfrac{4x+9}{x-3}$ 3

33. $\dfrac{3x^4}{x} - \dfrac{4x^2}{x^2}$ $3x^3-4$

34. $\dfrac{5x}{6} + \dfrac{15x^2}{2}$ $\dfrac{5x+45x^2}{6}$

35. $\dfrac{1}{x+3} - \dfrac{1}{(x+3)^2}$ $\dfrac{x+2}{(x+3)^2}$

36. $\dfrac{5x}{(x-2)^2} - \dfrac{3}{x-2}$

37. $\dfrac{4}{5b} + \dfrac{1}{b-1}$ $\dfrac{9b-4}{5b(b-1)}$

38. $\dfrac{1}{y+5} + \dfrac{2}{3y}$ $\dfrac{5y+10}{3y(y+5)}$

39. $\dfrac{2}{m} + 1$ $\dfrac{2+m}{m}$

40. $\dfrac{6}{x} - 1$ $\dfrac{6-x}{x}$

41. $\dfrac{6}{1-2x} - \dfrac{4}{2x-1}$ $\dfrac{10}{1-2x}$

42. $\dfrac{10}{3n-4} - \dfrac{5}{4-3n}$ $\dfrac{15}{3n-4}$

43. $\dfrac{7}{(x+1)(x-1)} + \dfrac{8}{(x+1)^2}$ $\dfrac{15x-1}{(x+1)^2(x-1)}$

44. $\dfrac{5x+2}{(x+1)(x+5)} - \dfrac{2}{x+5}$ $\dfrac{3x}{(x+1)(x+5)}$

45. $\dfrac{x}{x^2-1} - \dfrac{2}{x^2-2x+1}$ $\dfrac{x^2-3x-2}{(x-1)^2(x+1)}$

46. $\dfrac{x}{x^2-4} - \dfrac{5}{x^2-4x+4}$ $\dfrac{x^2-7x-10}{(x+2)(x-2)^2}$

🔒 **47.** $\dfrac{3a}{2a+6} - \dfrac{a-1}{a+3}$ $\dfrac{a+2}{2(a+3)}$

48. $\dfrac{1}{x+y} - \dfrac{y}{x^2-y^2}$

49. $\dfrac{5}{2-x} + \dfrac{x}{2x-4}$ $\dfrac{x-10}{2(x-2)}$

50. $\dfrac{-1}{a-2} + \dfrac{4}{4-2a} - \dfrac{3}{a-2}$

51. $\dfrac{-7}{y^2-3y+2} - \dfrac{2}{y-1}$

52. $\dfrac{2}{x^2+4x+4} + \dfrac{1}{x+2}$

53. $\dfrac{13}{x^2-5x+6} - \dfrac{5}{x-3}$

54. $\dfrac{27}{y^2-81} + \dfrac{3}{2(y+9)}$

55. $\dfrac{8}{(x+2)(x-2)} + \dfrac{4}{(x+2)(x-3)}$ $\dfrac{12x-32}{(x+2)(x-2)(x-3)}$

56. $\dfrac{5}{6x^2(x+2)} + \dfrac{4x}{x(x+2)^2}$

48. $\dfrac{x-2y}{(x+y)(x-y)}$

57. $\dfrac{5}{9x^2-4} + \dfrac{2}{3x-2}$

51. $\dfrac{-3-2y}{(y-1)(y-2)}$

58. $\dfrac{4}{x^2-x-6} + \dfrac{x}{x^2+5x+6}$ $\dfrac{x^2+x+12}{(x-3)(x+3)(x+2)}$

🔒 **59.** $\dfrac{x+8}{x^2-5x-6} + \dfrac{x+1}{x^2-4x-5}$ $\dfrac{2x^2-2x-46}{(x+1)(x-6)(x-5)}$

60. $\dfrac{x}{x^2+12x+20} - \dfrac{1}{x^2+8x-20}$ $\dfrac{x^2-3x-2}{(x+10)(x+2)(x-2)}$

REVIEW AND PREVIEW

Solve the following linear and quadratic equations. See Sections 2.4 and 6.5.

61. $3x+5=7$ $\tfrac{2}{3}$

62. $5x-1=8$ $\tfrac{9}{5}$

63. $2x^2-x-1=0$ $-\tfrac{1}{2}, 1$

64. $4x^2-9=0$ $\tfrac{3}{2}, -\tfrac{3}{2}$

Simplify the following rational expressions. See Section 7.1.

65. $\dfrac{2+x}{x+2}$ 1

66. $\dfrac{y^2+y}{y+y^2}$ 1

67. $\dfrac{2-x}{x-2}$ -1

68. $\dfrac{z-4}{4-z}$ -1

52. $\dfrac{x+4}{(x+2)^2}$

53. $\dfrac{-5x+23}{(x-3)(x-2)}$

54. $\dfrac{3}{2(y-9)}$

56. $\dfrac{24x^2+5x+10}{6x^2(x+2)^2}$

57. $\dfrac{6x+9}{(3x-2)(3x+2)}$

Concept Extensions   **69.** $\dfrac{10y-20}{y(y-5)}$ ft; $\dfrac{6}{y^2-5y}$ sq.ft

△ **69.** The length of the rectangle is $\dfrac{3}{y-5}$ feet, while its width is $\dfrac{2}{y}$ feet. Find its perimeter and then find its area.

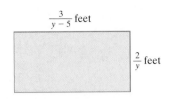

$\dfrac{3}{y-5}$ feet

$\dfrac{2}{y}$ feet

70. A board of length $\dfrac{3}{x+4}$ inches was cut into two pieces. If one piece is $\dfrac{1}{x-4}$ inches, express the length of the other board as a rational expression. $\dfrac{2x-16}{(x-4)(x+4)}$ in.

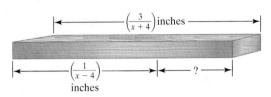

$\left(\dfrac{3}{x+4}\right)$ inches

$\left(\dfrac{1}{x-4}\right)$ inches

$?$

MIXED PRACTICE   **72.** $\dfrac{z}{3(9z-5)}$ **76.** $\dfrac{9}{4(x-1)}$

Perform the indicated operations. Addition, subtraction, multiplication, and division of rational expressions are included here.

71. $\dfrac{15x}{x+8} \cdot \dfrac{2x+16}{3x}$ 10

72. $\dfrac{9z+5}{15} \cdot \dfrac{5z}{81z^2-25}$

73. $\dfrac{8x+7}{3x+5} - \dfrac{2x-3}{3x+5}$ 2

74. $\dfrac{2z^2}{4z-1} - \dfrac{z-2z^2}{4z-1}$ z

75. $\dfrac{5a+10}{18} \div \dfrac{a^2-4}{10a}$ $\dfrac{25a}{9(a-2)}$

76. $\dfrac{9}{x^2-1} \div \dfrac{12}{3x+3}$

77. $\dfrac{5}{x^2-3x+2} + \dfrac{1}{x-2}$ $\dfrac{x+4}{(x-2)(x-1)}$

78. $\dfrac{4}{2x^2+5x-3} + \dfrac{2}{x+3}$ $\dfrac{4x+2}{(2x-1)(x+3)}$

79. Explain when the LCD is the product of the denominators.

80. Explain when the LCD is the same as one of the denominators of a rational expression to be added or subtracted.

79.–80. answers may vary

Add or subtract as indicated.   **81.** $\dfrac{4x^2-15x+6}{(x-2)^2(x+2)(x-3)}$

81. $\dfrac{5}{x^2-4} + \dfrac{2}{x^2-4x+4} - \dfrac{3}{x^2-x-6}$

82. $\dfrac{8}{x^2+6x+5} - \dfrac{3x}{x^2+4x-5} + \dfrac{2}{x^2-1}$ $\dfrac{-3x^2+7x+2}{(x+5)(x+1)(x-1)}$

83. $\dfrac{5+x}{x^3-27} + \dfrac{x}{x^3+3x^2+9x}$ $\dfrac{2(x+1)}{(x-3)(x^2+3x+9)}$

84. $\dfrac{x+5}{x^3+1} - \dfrac{3}{2x^2-2x+2}$ $\dfrac{-x+7}{2(x+1)(x^2-x+1)}$

85. The dose of medicine prescribed for a child depends on the child's age A in years and the adult dose D for the medication. Two expressions that give a child's dose are Young's

Rule, $\dfrac{DA}{A + 12}$, and Cowling's Rule, $\dfrac{D(A + 1)}{24}$. Find an expression for the difference in the doses given by these expressions. $\dfrac{11DA - DA^2 - 12D}{24(A + 12)}$

86. In ice hockey, penalty killing percentage is a statistic calculated as $1 - \dfrac{G}{P}$, where G = opponent's power play goals and P = opponent's power play opportunities. Subtract these terms. $\dfrac{P - G}{P}$

7.5 SOLVING EQUATIONS CONTAINING RATIONAL EXPRESSIONS

Objectives

1 Solve equations containing rational expressions.

2 Solve equations containing rational expressions for a specified variable.

1 In Chapter 2, we solved equations containing fractions. In this section, we continue the work we began in Chapter 2 by solving equations containing rational expressions.

TEACHING TIP
You may want to begin this section with a review of solving linear equations, such as $3x + 16 = 1$.

Examples of Equations Containing Rational Expressions

$$\frac{x}{5} + \frac{x + 2}{9} = 8 \quad \text{and} \quad \frac{x + 1}{9x - 5} = \frac{2}{3x}$$

To solve equations such as these, use the multiplication property of equality to clear the equation of fractions by multiplying both sides of the equation by the LCD.

EXAMPLE 1

Solve: $\dfrac{x}{2} + \dfrac{8}{3} = \dfrac{1}{6}$

Solution The LCD of denominators 2, 3, and 6 is 6, so we multiply both sides of the equation by 6.

CLASSROOM EXAMPLE
Solve: $\dfrac{x}{4} + \dfrac{4}{5} = \dfrac{1}{20}$
answer: -3

Helpful Hint
Make sure that *each* term is multiplied by the LCD, 6.

$$6\left(\frac{x}{2} + \frac{8}{3}\right) = 6\left(\frac{1}{6}\right)$$

$$6\left(\frac{x}{2}\right) + 6\left(\frac{8}{3}\right) = 6\left(\frac{1}{6}\right) \quad \text{Use the distributive property.}$$

$$3 \cdot x + 16 = 1 \quad \text{Multiply and simplify.}$$
$$3x = -15 \quad \text{Subtract 16 from both sides.}$$
$$x = -5 \quad \text{Divide both sides by 3.}$$

Check To check, we replace x with -5 in the original equation.

$$\frac{x}{2} + \frac{8}{3} = \frac{1}{6}$$

$$\frac{-5}{2} + \frac{8}{3} \stackrel{?}{=} \frac{1}{6} \quad \text{Replace } x \text{ with } -5.$$

$$\frac{1}{6} = \frac{1}{6} \quad \text{True}$$

This number checks, so the solution is -5.

EXAMPLE 2

Solve: $\dfrac{t-4}{2} - \dfrac{t-3}{9} = \dfrac{5}{18}$

Solution   The LCD of denominators 2, 9, and 18 is 18, so we multiply both sides of the equation by 18.

$$18\left(\dfrac{t-4}{2} - \dfrac{t-3}{9}\right) = 18\left(\dfrac{5}{18}\right)$$

$$18\left(\dfrac{t-4}{2}\right) - 18\left(\dfrac{t-3}{9}\right) = 18\left(\dfrac{5}{18}\right) \qquad \text{Use the distributive property.}$$

$$9(t-4) - 2(t-3) = 5 \qquad \text{Simplify.}$$

$$9t - 36 - 2t + 6 = 5 \qquad \text{Use the distributive property.}$$

$$7t - 30 = 5 \qquad \text{Combine like terms.}$$

$$7t = 35$$

$$t = 5 \qquad \text{Solve for } t.$$

Check

$$\dfrac{t-4}{2} - \dfrac{t-3}{9} = \dfrac{5}{18}$$

$$\dfrac{5-4}{2} - \dfrac{5-3}{9} \stackrel{?}{=} \dfrac{5}{18} \qquad \text{Replace } t \text{ with 5.}$$

$$\dfrac{1}{2} - \dfrac{2}{9} \stackrel{?}{=} \dfrac{5}{18} \qquad \text{Simplify.}$$

$$\dfrac{5}{18} = \dfrac{5}{18} \qquad \text{True}$$

The solution is 5.

Recall from Section 7.1 that a rational expression is defined for all real numbers except those that make the denominator of the expression 0. This means that if an equation contains *rational expressions with variables in the denominator*, we must be certain that the proposed solution does not make the denominator 0. If replacing the variable with the proposed solution makes the denominator 0, the rational expression is undefined and this proposed solution must be rejected.

EXAMPLE 3

Solve: $3 - \dfrac{6}{x} = x + 8$

Solution   In this equation, 0 cannot be a solution because if x is 0, the rational expression $\dfrac{6}{x}$ is undefined. The LCD is x, so we multiply both sides of the equation by x.

$$x\left(3 - \dfrac{6}{x}\right) = x(x+8)$$

$$x(3) - x\left(\dfrac{6}{x}\right) = x \cdot x + x \cdot 8 \qquad \text{Use the distributive property.}$$

$$3x - 6 = x^2 + 8x \qquad \text{Simplify.}$$

Now we write the quadratic equation in standard form and solve for x.

▶ **Helpful Hint**
Multiply *each* term by 18.

▶ **Helpful Hint**
Multiply *each* term by x.

$$0 = x^2 + 5x + 6$$

$$0 = (x + 3)(x + 2) \qquad \text{Factor.}$$

$$x + 3 = 0 \qquad \text{or} \qquad x + 2 = 0 \qquad \text{Set each factor equal to 0 and solve.}$$

$$x = -3 \qquad\qquad x = -2$$

Notice that neither -3 nor -2 makes the denominator in the original equation equal to 0.

Check   To check these solutions, we replace x in the original equation by -3, and then by -2.

If $x = -3$:

$$3 - \frac{6}{x} = x + 8$$

$$3 - \frac{6}{-3} \overset{?}{=} -3 + 8$$

$$3 - (-2) \overset{?}{=} 5$$

$$5 = 5 \qquad \text{True}$$

If $x = -2$:

$$3 - \frac{6}{x} = x + 8$$

$$3 - \frac{6}{-2} \overset{?}{=} -2 + 8$$

$$3 - (-3) \overset{?}{=} 6$$

$$6 = 6 \qquad \text{True}$$

Both -3 and -2 are solutions.

The following steps may be used to solve an equation containing rational expressions.

Solving an Equation Containing Rational Expressions

Step 1. Multiply both sides of the equation by the LCD of all rational expressions in the equation.

Step 2. Remove any grouping symbols and solve the resulting equation.

Step 3. Check the solution in the original equation.

EXAMPLE 4

Solve: $\dfrac{4x}{x^2 - 25} + \dfrac{2}{x - 5} = \dfrac{1}{x + 5}$

Solution   The denominator $x^2 - 25$ factors as $(x + 5)(x - 5)$. The LCD is then $(x + 5)(x - 5)$, so we multiply both sides of the equation by this LCD.

CLASSROOM EXAMPLE

Solve: $\dfrac{2}{x + 3} + \dfrac{3}{x - 3} = \dfrac{-2}{x^2 - 9}$

answer: -1

$$(x + 5)(x - 5)\left(\frac{4x}{(x + 5)(x - 5)} + \frac{2}{x - 5} \right) \qquad \text{Multiply by the LCD. Notice that } -5 \text{ and } 5 \text{ cannot be solutions.}$$

$$= (x + 5)(x - 5)\left(\frac{1}{x + 5} \right)$$

$$(x + 5)(x - 5) \cdot \frac{4x}{x^2 - 25} + (x + 5)(x - 5) \cdot \frac{2}{x - 5} \qquad \text{Use the distributive property.}$$

$$= (x + 5)(x - 5) \cdot \frac{1}{x + 5}$$

$$4x + 2(x + 5) = x - 5 \qquad \text{Simplify.}$$
$$4x + 2x + 10 = x - 5 \qquad \text{Use the distributive property.}$$
$$6x + 10 = x - 5 \qquad \text{Combine like terms.}$$
$$5x = -15$$
$$x = -3 \qquad \text{Divide both sides by 5.}$$

Check   Check by replacing x with -3 in the original equation. The solution is -3.

EXAMPLE 5

Solve: $\dfrac{2x}{x - 4} = \dfrac{8}{x - 4} + 1$

Solution   Multiply both sides by the LCD, $x - 4$.

CLASSROOM EXAMPLE

Solve: $\dfrac{5x}{x - 1} = \dfrac{5}{x - 1} + 3$

answer: no solution

$$(x - 4)\left(\dfrac{2x}{x - 4}\right) = (x - 4)\left(\dfrac{8}{x - 4} + 1\right) \qquad \begin{array}{l} \text{Multiply by the LCD.} \\ \text{Notice that 4 cannot be a solution.} \end{array}$$

$$(x - 4) \cdot \dfrac{2x}{x - 4} = (x - 4) \cdot \dfrac{8}{x - 4} + (x - 4) \cdot 1 \qquad \text{Use the distributive property.}$$

$$2x = 8 + (x - 4) \qquad \text{Simplify.}$$

$$2x = 4 + x$$

$$x = 4$$

Notice that 4 makes the denominator 0 in the original equation. Therefore, 4 is *not* a solution and this equation has *no solution*.

TEACHING TIP

One of the most important concepts that you can help students with is the difference between an equation and an expression. This is a great time to reinforce that difference.

> **Helpful Hint**
>
> As we can see from Example 5, it is important to check the proposed solution(s) in the *original* equation.

 CONCEPT CHECK

When can we clear fractions by multiplying through by the LCD?

  a. When adding or subtracting rational expressions
  b. When solving an equation containing rational expressions
  c. Both of these   **d.** Neither of these

EXAMPLE 6

Solve: $x + \dfrac{14}{x - 2} = \dfrac{7x}{x - 2} + 1$

Solution   Notice the denominators in this equation. We can see that 2 can't be a solution. The LCD is $x - 2$, so we multiply both sides of the equation by $x - 2$.

CLASSROOM EXAMPLE

Solve: $x - \dfrac{6}{x + 3} = \dfrac{2x}{x + 3} + 2$

answer: 4

Concept Check Answer:

b

$$(x - 2)\left(x + \dfrac{14}{x - 2}\right) = (x - 2)\left(\dfrac{7x}{x - 2} + 1\right)$$

$$(x - 2)(x) + (x - 2)\left(\dfrac{14}{x - 2}\right) = (x - 2)\left(\dfrac{7x}{x - 2}\right) + (x - 2)(1)$$

$$x^2 - 2x + 14 = 7x + x - 2 \quad \text{Simplify.}$$
$$x^2 - 2x + 14 = 8x - 2 \quad \text{Combine like terms.}$$
$$x^2 - 10x + 16 = 0 \quad \begin{array}{l}\text{Write the quadratic} \\ \text{equation in standard form.}\end{array}$$
$$(x - 8)(x - 2) = 0 \quad \text{Factor.}$$
$$x - 8 = 0 \quad \text{or} \quad x - 2 = 0 \quad \text{Set each factor equal to 0.}$$
$$x = 8 \qquad\qquad x = 2 \quad \text{Solve.}$$

As we have already noted, 2 can't be a solution of the original equation. So we need only replace x with 8 in the original equation. We find that 8 is a solution; the only solution is 8.

2 The last example in this section is an equation containing several variables. We are directed to solve for one of them. The steps used in the preceeding examples can be applied to solve equations for a specified variable as well.

EXAMPLE 7

Solve $\dfrac{1}{a} + \dfrac{1}{b} = \dfrac{1}{x}$ for x.

Solution

(This type of equation often models a work problem, as we shall see in Section 7.6.) The LCD is abx, so we multiply both sides by abx.

$$abx\left(\frac{1}{a} + \frac{1}{b}\right) = abx\left(\frac{1}{x}\right)$$
$$abx\left(\frac{1}{a}\right) + abx\left(\frac{1}{b}\right) = abx \cdot \frac{1}{x}$$
$$bx + ax = ab \quad \text{Simplify.}$$
$$x(b + a) = ab \quad \text{Factor out } x \text{ from each term on the left side.}$$
$$\frac{x(b + a)}{b + a} = \frac{ab}{b + a} \quad \text{Divide both sides by } b + a.$$
$$x = \frac{ab}{b + a} \quad \text{Simplify.}$$

This equation is now solved for x.

Graphing Calculator Explorations

A grapher may be used to check solutions of equations containing rational expressions. For example, to check the solution of Example 1, $\frac{x}{2} + \frac{8}{3} = \frac{1}{6}$, graph $y_1 = x/2 + 8/3$ and $y_2 = 1/6$.
Use TRACE and ZOOM, or use INTERSECT, to find the point of intersection. The point of intersection has an x-value of -5, so the solution of the equation is -5.

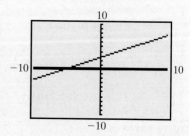

Use a grapher to check the examples of this section.

1. Example 2   2. Example 3   3. Example 4
4. Example 5   5. Example 6

STUDY SKILLS REMINDER

Is Your Notebook Still Organized?

Is your notebook still organized? If it's not, it's not too late to start organizing it. Start writing your notes and completing your homework assignment in a notebook with pockets (spiral or ring binder). Take class notes in this notebook, and then follow the notes with your completed homework assignment. When you receive graded papers or handouts, place them in the notebook pocket so that you will not lose them.

Remember to mark (possibly with an exclamation point) any note(s) that seems extra important to you. Also remember to mark (possibly with a question mark) any notes or homework that you are having trouble with. Don't forget to see your instructor or a math tutor to help you with the concepts or exercises that you are having trouble understanding.

Also—don't forget to write neatly.

MENTAL MATH

Solve each equation for the variable.

1. $\dfrac{x}{5} = 2$ 10

2. $\dfrac{x}{8} = 4$ 32

3. $\dfrac{z}{6} = 6$ 36

4. $\dfrac{y}{7} = 8$ 56

12. -6 **17.** no solution   **20.** $-4, 6$ **22.** -1

EXERCISE SET 7.5

STUDY GUIDE/SSM   CD/ VIDEO   PH MATH TUTOR CENTER   MathXL®Tutorials ON CD   MathXL®   MyMathLab®

Solve each equation. See Examples 1 through 3.

1. $\dfrac{x}{5} + 3 = 9$ 30

2. $\dfrac{x}{5} - 2 = 9$ 55

3. $\dfrac{x}{2} + \dfrac{5x}{4} = \dfrac{x}{12}$ 0

4. $\dfrac{x}{6} + \dfrac{4x}{3} = \dfrac{x}{18}$ 0

5. $\dfrac{y}{7} - \dfrac{3y}{2} = 1$ $-\dfrac{14}{19}$

6. $\dfrac{a}{9} - \dfrac{4a}{3} = 2$ $-\dfrac{18}{11}$

7. $2 + \dfrac{10}{x} = x + 5$ $-5, 2$

8. $6 + \dfrac{5}{y} = y - \dfrac{2}{y}$ $-1, 7$

9. $\dfrac{a}{5} = \dfrac{a - 3}{2}$ 5

10. $\dfrac{2b}{5} = \dfrac{b + 2}{6}$ $\dfrac{10}{7}$

 11. $\dfrac{x - 3}{5} + \dfrac{x - 2}{2} = \dfrac{1}{2}$ 3

12. $\dfrac{a + 5}{4} + \dfrac{a + 5}{2} = \dfrac{a}{8}$

13. $\dfrac{x + 1}{3} - \dfrac{x - 2}{4} = \dfrac{5}{6}$ 0

14. $\dfrac{x + 2}{15} - \dfrac{x - 1}{3} = \dfrac{7}{15}$ 0

Solve each equation. See Examples 4 through 6.

15. $\dfrac{9}{2a - 5} = -2$ $\dfrac{1}{4}$

16. $\dfrac{6}{4 - 3x} = 3$ $\dfrac{2}{3}$

17. $\dfrac{y}{y + 4} + \dfrac{4}{y + 4} = 3$

18. $\dfrac{5y}{y + 1} - \dfrac{3}{y + 1} = 4$ 7

19. $\dfrac{4y}{y - 3} - 3 = \dfrac{3y - 1}{y + 3}$ $-1, 12$

20. $\dfrac{2x}{x + 2} - 2 = \dfrac{x - 8}{x - 2}$

21. $\dfrac{4y}{y - 4} + 5 = \dfrac{5y}{y - 4}$ 5

22. $\dfrac{2a}{a + 2} - 5 = \dfrac{7a}{a + 2}$

 23. $\dfrac{7}{x - 2} + 1 = \dfrac{x}{x + 2}$ $-\dfrac{10}{9}$

24. $1 + \dfrac{3}{x + 1} = \dfrac{x}{x - 1}$ 2

25. $\dfrac{x + 1}{x + 3} = \dfrac{2x^2 - 15x}{x^2 + x - 6} - \dfrac{x - 3}{x - 2}$ $\dfrac{11}{14}$

26. $\dfrac{3}{x + 3} = \dfrac{12x + 19}{x^2 + 7x + 12} - \dfrac{5}{x + 4}$ 2

27. $\dfrac{y}{2y+2} + \dfrac{2y-16}{4y+4} = \dfrac{2y-3}{y+1}$ no solution

28. $\dfrac{1}{x+2} = \dfrac{4}{x^2-4} - \dfrac{1}{x-2}$ no solution

MIXED PRACTICE

Solve each equation.   **35.–36.** no solution

29. $\dfrac{2x}{7} - 5x = 9$ $-\dfrac{21}{11}$

30. $\dfrac{4x}{8} - 5x = 10$ $-\dfrac{20}{9}$

🔒 **31.** $\dfrac{2}{y} + \dfrac{1}{2} = \dfrac{5}{2y}$ 1

32. $\dfrac{6}{3y} + \dfrac{3}{y} = 1$ 5

33. $\dfrac{4x+10}{7} = \dfrac{8}{2}$ $\dfrac{9}{2}$

34. $\dfrac{1}{2} = \dfrac{x+1}{8}$ 3

🔒 **35.** $2 + \dfrac{3}{a-3} = \dfrac{a}{a-3}$

36. $\dfrac{2y}{y-2} - \dfrac{4}{y-2} = 4$

37. $\dfrac{5}{x} + \dfrac{2}{3} = \dfrac{7}{2x}$ $-\dfrac{9}{4}$

38. $\dfrac{5}{3} - \dfrac{3}{2x} = \dfrac{5}{4}$ $\dfrac{18}{5}$

39. $\dfrac{2a}{a+4} = \dfrac{3}{a-1}$ $-\dfrac{3}{2}, 4$

40. $\dfrac{5}{3x-8} = \dfrac{x}{x-2}$ $1, \dfrac{10}{3}$

41. $\dfrac{x+1}{3} - \dfrac{x-1}{6} = \dfrac{1}{6}$ -2

42. $\dfrac{3x}{5} - \dfrac{x-6}{3} = \dfrac{1}{5}$ $-\dfrac{27}{4}$

43. $\dfrac{4r-1}{r^2+5r-14} + \dfrac{2}{r+7} = \dfrac{1}{r-2}$ $\dfrac{12}{5}$

44. $\dfrac{2t+3}{t-1} - \dfrac{2}{t+3} = \dfrac{5-6t}{t^2+2t-3}$ $-6, -\dfrac{1}{2}$

🔒 **45.** $\dfrac{t}{t-4} = \dfrac{t+4}{6}$ $-2, 8$

46. $\dfrac{15}{x+4} = \dfrac{x-4}{x}$ $-1, 16$

47. $\dfrac{x}{2x+6} + \dfrac{x+1}{3x+9} = \dfrac{2}{4x+12}$ $\dfrac{1}{5}$

48. $\dfrac{a}{5a-5} - \dfrac{a-2}{2a-2} = \dfrac{5}{4a-4}$ $-\dfrac{5}{6}$

Solve each equation for the indicated variable. See Example 7.

49. $\dfrac{D}{R} = T$; for R $R = \dfrac{D}{T}$

△ **50.** $\dfrac{A}{W} = L$; for W $W = \dfrac{A}{L}$

51. $\dfrac{3}{x} = \dfrac{5y}{x+2}$; for y $y = \dfrac{3x+6}{5x}$

52. $\dfrac{7x-1}{2x} = \dfrac{5}{y}$; for y

53. $\dfrac{3a+2}{3b-2} = -\dfrac{4}{2a}$; for b

54. $\dfrac{6x+y}{7x} = \dfrac{3x}{h}$; for h

△ **55.** $\dfrac{A}{BH} = \dfrac{1}{2}$; for B $B = \dfrac{2A}{H}$

△ **56.** $\dfrac{V}{\pi r^2 h} = 1$; for h

△ **57.** $\dfrac{C}{\pi r} = 2$; for r $r = \dfrac{C}{2\pi}$

△ **58.** $\dfrac{3V}{A} = H$; for V

59. $\dfrac{1}{a} = \dfrac{1}{b} + \dfrac{1}{c}$; for a $a = \dfrac{bc}{c+b}$

60. $\dfrac{1}{2} - \dfrac{1}{x} = \dfrac{1}{y}$; for x

61. $\dfrac{m^2}{6} - \dfrac{n}{3} = \dfrac{p}{2}$; for n

62. $\dfrac{x^2}{r} + \dfrac{y^2}{t} = 1$; for r

52. $y = \dfrac{10x}{7x-1}$ **53.** $b = -\dfrac{3a^2+2a-4}{6}$ **54.** $h = \dfrac{21x^2}{6x+y}$ **56.** $h = \dfrac{V}{\pi r^2}$ **58.** $V = \dfrac{AH}{3}$ **60.** $x = \dfrac{2y}{y-2}$ **61.** $n = \dfrac{m^2-3p}{2}$ **62.** $r = \dfrac{x^2t}{t-y^2}$

REVIEW AND PREVIEW

Identify the x- and y-intercepts. See Section 3.3.

63.

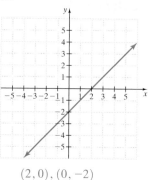

$(2, 0), (0, -2)$

64.

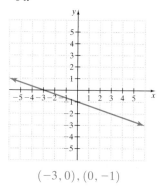

$(-3, 0), (0, -1)$

65.

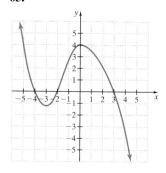

$(-4, 0), (-2, 0), (3, 0), (0, 4)$

66.

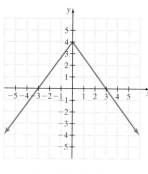

$(-3, 0), (3, 0), (0, 4)$

Concept Extensions

67. Explain the difference between solving an equation such as $\dfrac{x}{2} + \dfrac{3}{4} = \dfrac{x}{4}$ for x and performing an operation such as adding $\dfrac{x}{2} + \dfrac{3}{4}$.   answers may vary

68. When solving an equation such as $\dfrac{y}{4} = \dfrac{y}{2} - \dfrac{1}{4}$, we may multiply all terms by 4. When subtracting two rational expressions such as $\dfrac{y}{2} - \dfrac{1}{4}$, we may not. Explain why.   answers may vary

Determine whether each of the following is an equation or an expression. If it is an equation, then solve it for its variable. If it is an expression, perform the indicated operation.

△ **69.** $\dfrac{1}{x} + \dfrac{5}{9}$ $\dfrac{5x+9}{9x}$

70. $\dfrac{1}{x} + \dfrac{5}{9} = \dfrac{2}{3}$ 9

71. $\dfrac{5}{x-1} - \dfrac{2}{x} = \dfrac{5}{x(x-1)}$ no solution

72. $\dfrac{5}{x-1} - \dfrac{2}{x}$ $\dfrac{3x+2}{x(x-1)}$

Recall that two angles are supplementary if the sum of their measures is 180°. Find the measures of the following supplementary angles.

△ **73.** $100°, 80°$

$\left(\dfrac{20x}{3}\right)° \left(\dfrac{32x}{6}\right)°$

△ **74.** $30°, 150°$

$\left(\dfrac{5x}{2}\right)° \left(\dfrac{25x}{2}\right)°$

Recall that two angles are complementary if the sum of their measures is 90°. Find the measures of the following complementary angles.

△ **75.** 22.5°, 67.5°

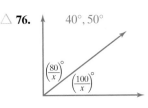

△ **76.** 40°, 50°

Solve each equation.

77. $\dfrac{5}{a^2 + 4a + 3} + \dfrac{2}{a^2 + a - 6} - \dfrac{3}{a^2 - a - 2} = 0$ $\quad\dfrac{17}{4}$

78. $-\dfrac{2}{a^2 + 2a - 8} + \dfrac{1}{a^2 + 9a + 20} = \dfrac{-4}{a^2 + 3a - 10}$ $\quad-\dfrac{4}{3}$

SUMMARY ON RATIONAL EXPRESSIONS

INTEGRATED REVIEW

It is important to know the difference between performing operations with rational expressions and solving an equation containing rational expressions. Study the examples below.

Performing Operations with Rational Expressions

Adding: $\dfrac{1}{x} + \dfrac{1}{x + 5} = \dfrac{1 \cdot (x + 5)}{x(x + 5)} + \dfrac{1 \cdot x}{x(x + 5)} = \dfrac{x + 5 + x}{x(x + 5)} = \dfrac{2x + 5}{x(x + 5)}$

Subtracting: $\dfrac{3}{x} - \dfrac{5}{x^2 y} = \dfrac{3 \cdot xy}{x \cdot xy} - \dfrac{5}{x^2 y} = \dfrac{3xy - 5}{x^2 y}$

Multiplying: $\dfrac{2}{x} \cdot \dfrac{5}{x - 1} = \dfrac{2 \cdot 5}{x(x - 1)} = \dfrac{10}{x(x - 1)}$

Dividing: $\dfrac{4}{2x + 1} \div \dfrac{x - 3}{x} = \dfrac{4}{2x + 1} \cdot \dfrac{x}{x - 3} = \dfrac{4x}{(2x + 1)(x - 3)}$

Solving an Equation Containing Rational Expressions

To solve an equation containing rational expressions, we clear the equation of fractions by multiplying both sides by the LCD.

$$\dfrac{3}{x} - \dfrac{5}{x - 1} = \dfrac{1}{x(x - 1)} \qquad \text{*Note that x can't be 0 or 1.*}$$

$$x(x - 1)\left(\dfrac{3}{x}\right) - x(x - 1)\left(\dfrac{5}{x - 1}\right) = x(x - 1) \cdot \dfrac{1}{x(x - 1)} \qquad \text{*Multiply both sides by the LCD.*}$$

$$3(x - 1) - 5x = 1 \qquad \text{*Simplify.*}$$

$$3x - 3 - 5x = 1 \qquad \text{*Use the distributive property.*}$$

$$-2x - 3 = 1 \qquad \text{*Combine like terms.*}$$

$$-2x = 4 \qquad \text{*Add 3 to both sides.*}$$

$$x = -2 \qquad \text{*Divide both sides by −2.*}$$

Determine whether each of the following is an equation or an expression. If it is an equation, solve it for its variable. If it is an expression, perform the indicated operation.

1. $\dfrac{1}{x} + \dfrac{2}{3}$ expression: $\dfrac{3 + 2x}{3x}$

2. $\dfrac{3}{a} + \dfrac{5}{6}$ expression: $\dfrac{18 + 5a}{6a}$

3. $\dfrac{1}{x} + \dfrac{2}{3} = \dfrac{3}{x}$ equation; 3

4. $\dfrac{3}{a} + \dfrac{5}{6} = 1$ equation; 18

5. $\dfrac{2}{x - 1} - \dfrac{1}{x}$ expression; $\dfrac{x + 1}{x(x - 1)}$

6. $\dfrac{4}{x - 3} - \dfrac{1}{x}$

6. expression; $\dfrac{3(x + 1)}{x(x - 3)}$

7. equation; no solution

8. equation; 1

9. expression; 10

7. $\dfrac{2}{x + 1} - \dfrac{1}{x} = 1$

8. $\dfrac{4}{x - 3} - \dfrac{1}{x} = \dfrac{6}{x(x - 3)}$

9. $\dfrac{15x}{x + 8} \cdot \dfrac{2x + 16}{3x}$

10. $\dfrac{9z + 5}{15} \cdot \dfrac{5z}{81z^2 - 25}$

11. $\dfrac{2x + 1}{x - 3} + \dfrac{3x + 6}{x - 3}$

12. $\dfrac{4p - 3}{2p + 7} + \dfrac{3p + 8}{2p + 7}$

13. $\dfrac{x + 5}{7} = \dfrac{8}{2}$ equation; 23

14. $\dfrac{1}{2} = \dfrac{x - 1}{8}$ equation; 5

15. $\dfrac{5a + 10}{18} \div \dfrac{a^2 - 4}{10a}$

16. $\dfrac{9}{x^2 - 1} + \dfrac{12}{3x + 3}$

17. $\dfrac{x + 2}{3x - 1} + \dfrac{5}{(3x - 1)^2}$

18. $\dfrac{4}{(2x - 5)^2} + \dfrac{x + 1}{2x - 5}$

19. $\dfrac{x - 7}{x} - \dfrac{x + 2}{5x}$

20. $\dfrac{9}{x^2 - 4} + \dfrac{2}{x + 2} = \dfrac{-1}{x - 2}$ equation; $-\dfrac{7}{3}$

21. $\dfrac{3}{x + 3} = \dfrac{5}{x^2 - 9} - \dfrac{2}{x - 3}$

22. $\dfrac{10x - 9}{x} - \dfrac{x - 4}{3x}$ expression; $\dfrac{29x - 23}{3x}$

7.6   PROPORTION AND PROBLEM SOLVING WITH RATIONAL EQUATIONS

10. expression; $\dfrac{z}{3(9z - 5)}$

11. expression; $\dfrac{5x + 7}{x - 3}$

12. expression; $\dfrac{7p + 5}{2p + 7}$

15. expression; $\dfrac{25a}{9(a - 2)}$

16. expression; $\dfrac{4x + 5}{(x + 1)(x - 1)}$

17. expression; $\dfrac{3x^2 + 5x + 3}{(3x - 1)^2}$

18. expression; $\dfrac{2x^2 - 3x - 1}{(2x - 5)^2}$

19. expression; $\dfrac{4x - 37}{5x}$

21. equation; $\dfrac{8}{5}$

TEACHING TIP

Ask students to mentally calculate whether the statements below are true or false by using cross products.

$\dfrac{2}{3} = \dfrac{6}{10}$

$\dfrac{4}{6} = \dfrac{10}{15}$

$\dfrac{5}{7} = \dfrac{7}{10}$

$\dfrac{3}{12} = \dfrac{2}{8}$

Objectives

1 Use proportions to solve problems.

2 Solve problems about numbers.

3 Solve problems about work.

4 Solve problems about distance.

1 A **ratio** is the quotient of two numbers or two quantities. For example, the ratio of 2 to 5 can be written as $\frac{2}{5}$, the quotient of 2 and 5.

If two ratios are equal, we say the ratios are **in proportion** to each other. A **proportion** is a mathematical statement that two ratios are equal.

For example, the equation $\frac{1}{2} = \frac{4}{8}$ is a proportion, as is $\frac{x}{5} = \frac{8}{10}$, because both sides of the equations are ratios. When we want to emphasize the equation as a proportion, we

$$\text{read the proportion } \frac{1}{2} = \frac{4}{8} \text{ as "one is to two as four is to eight"}$$

In a proportion, cross products are equal. To understand cross products, let's start with the proportion

$$\frac{a}{b} = \frac{c}{d}$$

and multiply both sides by the LCD, bd.

$$bd\left(\frac{a}{b}\right) = bd\left(\frac{c}{d}\right) \quad \text{Multiply both sides by the LCD, } bd.$$

$$ad = bc \quad \text{Simplify.}$$

Cross product   Cross product

Notice why ad and bc are called cross products.

$$\frac{a}{b} = \frac{c}{d}$$

bc

ad

> **Cross Products**
>
> If $\dfrac{a}{b} = \dfrac{c}{d}$, then $ad = bc$.

For example, if

$$\frac{1}{2} = \frac{4}{8}, \quad \text{then} \quad \begin{array}{c} 1 \cdot 8 = 2 \cdot 4 \quad \text{or} \\ 8 = 8 \end{array}$$

EXAMPLE 1

Solve for x: $\dfrac{45}{x} = \dfrac{5}{7}$

Solution

This is an equation with rational expressions, and also a proportion. Below are two ways to solve.

Since this is a rational equation, we can use the methods of the previous section.

$$\frac{45}{x} = \frac{5}{7}$$

$7x \cdot \dfrac{45}{x} = 7x \cdot \dfrac{5}{7}$ Multiply both sides by LCD **7x**.

$7 \cdot 45 = x \cdot 5$ Divide out common factors.

$315 = 5x$ Multiply.

$\dfrac{315}{5} = \dfrac{5x}{5}$ Divide both sides by **5**.

$63 = x$ Simplify.

Since this is also a proportion, we may set cross products equal.

$$\frac{45}{x} = \frac{5}{7}$$

$45 \cdot 7 = x \cdot 5$ Set cross products equal.

$315 = 5x$ Multiply.

$\dfrac{315}{5} = \dfrac{5x}{5}$ Divide both sides by **5**.

$63 = x$ Simplify.

Check

Both methods give us a solution of 63. To check, substitute 63 for x in the original proportion. The solution is 63.

In this section, if the rational equation is a proportion, we will use cross products to solve.

Proportions can be used to model and solve many real-life problems. When using proportions in this way, it is important to judge whether the solution is reasonable. Doing so helps us to decide if the proportion has been formed correctly. We use the same problem-solving steps that were introduced in Section 2.5.

EXAMPLE 2

CALCULATING THE COST OF RECORDABLE COMPACT DISCS

Three boxes of CD-Rs (recordable compact discs) cost $37.47. How much should 5 boxes cost?

Solution

1. UNDERSTAND.   Read and reread the problem. We know that the cost of 5 boxes is more than the cost of 3 boxes, or $37.47, and less than the cost of 6 boxes, which is double the cost of 3 boxes, or 2($37.47) = $74.94. Let's suppose that 5 boxes cost $60.00. To check, we see if 3 boxes is to 5 boxes as the *price* of 3 boxes is to the *price*

of 5 boxes. In other words, we see if

$$\frac{3 \text{ boxes}}{5 \text{ boxes}} = \frac{\text{price of 3 boxes}}{\text{price of 5 boxes}}$$

or

$$\frac{3}{5} = \frac{37.47}{60.00}$$

$$3(60.00) = 5(37.47) \qquad \textit{Set cross products equal.}$$

or

$$180.00 = 187.35 \qquad \textit{Not a true statement.}$$

Thus, $60 is not correct, but we now have a better understanding of the problem.

Let x = price of 5 boxes of CD-Rs.

2. TRANSLATE.

$$\frac{3 \text{ boxes}}{5 \text{ boxes}} = \frac{\text{price of 3 boxes}}{\text{price of 5 boxes}}$$

$$\frac{3}{5} = \frac{37.47}{x}$$

3. SOLVE.

$$\frac{3}{5} = \frac{37.47}{x}$$

$$3x = 5(37.47) \qquad \textit{Set cross products equal.}$$

$$3x = 187.35$$

$$x = 62.45 \qquad \textit{Divide both sides by 3.}$$

4. INTERPRET.

Check: Verify that 3 boxes is to 5 boxes as $37.47 is to $62.45. Also, notice that our solution is a reasonable one as discussed in Step 1.

State: Five boxes of CD-Rs cost $62.45.

Helpful Hint

The proportion $\dfrac{5 \text{ boxes}}{3 \text{ boxes}} = \dfrac{\text{price of 5 boxes}}{\text{price of 3 boxes}}$ could also have been used to solve the problem above. Notice that the cross products are the same.

Similar triangles have the same shape but not necessarily the same size. In similar triangles, the measures of corresponding angles are equal, and corresponding sides are in proportion.

If triangle ABC and triangle XYZ shown are similar, then we know that the measure of angle A = the measure of angle X, the measure of angle B = the measure of angle Y, and the measure of angle C = the measure of angle Z. We also know that corresponding sides are in proportion: $\frac{a}{x} = \frac{b}{y} = \frac{c}{z}$.

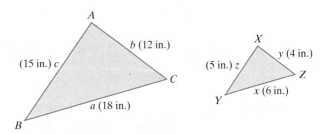

In this section, we will position similar triangles so that they have the same orientation.

To show that corresponding sides are in proportion for the triangles above, we write the ratios of the corresponding sides.

$$\frac{a}{x} = \frac{18}{6} = 3 \quad \frac{b}{y} = \frac{12}{4} = 3 \quad \frac{c}{z} = \frac{15}{5} = 3$$

EXAMPLE 3

FINDING THE LENGTH OF A SIDE OF A TRIANGLE

If the following two triangles are similar, find the missing length x.

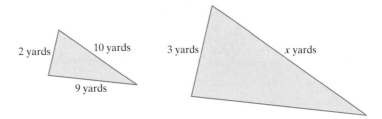

Solution

1. UNDERSTAND.   Read the problem and study the figure.
2. TRANSLATE.   Since the triangles are similar, their corresponding sides are in proportion and we have

$$\frac{2}{3} = \frac{10}{x}$$

3. SOLVE.   To solve, we multiply both sides by the LCD, $3x$, or cross multiply.

$$2x = 30$$
$$x = 15 \quad \text{Divide both sides by 2.}$$

4. INTERPRET.
   Check: To check, replace x with 15 in the original proportion and see that a true statement results.
   State: The missing length is 15 yards.

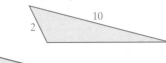
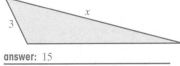

Spotlight on

DECISION ✗ MAKING

Suppose you coach your company's softball team. Two new employees are interested in playing on the company team, and you have only one open position. Employee A reports that last season he had 32 hits in 122 times at bat. Employee B reports that last season she had 19 hits in 56 times at bat. Which would you try to recruit first? Why? What other factors would you want to consider?

2 Let's continue to solve problems. The remaining problems are all modeled by rational equations.

EXAMPLE 4

Solution

FINDING AN UNKNOWN NUMBER

The quotient of a number and 6, minus $\frac{5}{3}$, is the quotient of the number and 2. Find the number.

1. UNDERSTAND. Read and reread the problem. Suppose that the unknown number is 2, then we see if the quotient of 2 and 6, or $\frac{2}{6}$, minus $\frac{5}{3}$ is equal to the quotient of 2 and 2, or $\frac{2}{2}$.

$$\frac{2}{6} - \frac{5}{3} = \frac{1}{3} - \frac{5}{3} = -\frac{4}{3}, \text{not } \frac{2}{2}$$

Don't forget that the purpose of a proposed solution is to better understand the problem.

Let x = the unknown number.

2. TRANSLATE.

In words: the quotient of x and 6   minus $\frac{5}{3}$ is   the quotient of x and 2

Translate: $\frac{x}{6}$ $-$ $\frac{5}{3}$ $=$ $\frac{x}{2}$

3. SOLVE. Here, we solve the equation $\frac{x}{6} - \frac{5}{3} = \frac{x}{2}$. We begin by multiplying both sides of the equation by the LCD, 6.

$$6\left(\frac{x}{6} - \frac{5}{3}\right) = 6\left(\frac{x}{2}\right)$$

$$6\left(\frac{x}{6}\right) - 6\left(\frac{5}{3}\right) = 6\left(\frac{x}{2}\right) \quad \text{Apply the distributive property.}$$

$$x - 10 = 3x \quad \text{Simplify.}$$

$$-10 = 2x \quad \text{Subtract } x \text{ from both sides.}$$

$$-\frac{10}{2} = \frac{2x}{2} \quad \text{Divide both sides by 2.}$$

$$-5 = x \quad \text{Simplify.}$$

TEACHING TIP
Take some time to review the concept of a "work" problem. Ask students a few questions like the following: If it takes you 2 hours to complete a job, what part of the job has been completed in 1 hour? After a few of these, see if you can successfully insert the variable x.

4. INTERPRET.

Check: To check, we verify that "the quotient of -5 and 6 minus $\frac{5}{3}$ is the quotient of -5 and 2," or $-\frac{5}{6} - \frac{5}{3} = -\frac{5}{2}$.

State: The unknown number is -5.

3   The next example is often called a work problem. Work problems usually involve people or machines doing a certain task.

EXAMPLE 5

FINDING WORK RATES

Sam Waterton and Frank Schaffer work in a plant that manufactures automobiles. Sam can complete a quality control tour of the plant in 3 hours while his assistant, Frank, needs 7 hours to complete the same job. The regional manager is coming to inspect the plant facilities, so both Sam and Frank are directed to complete a quality control tour together. How long will this take?

Solution   **1. UNDERSTAND.**   Read and reread the problem. The key idea here is the relationship between the **time** (hours) it takes to complete the job and the **part of the job** completed in 1 unit of time (hour). For example, if the **time** it takes Sam to complete the job is 3 hours, the **part of the job** he can complete in 1 hour is $\frac{1}{3}$. Similarly, Frank can complete $\frac{1}{7}$ of the job in 1 hour.

Let $x =$ the **time** in hours it takes Sam and Frank to complete the job together.

Then $\dfrac{1}{x} =$ the **part of the job** they complete in 1 hour.

	Hours to Complete Total Job	*Part of Job Completed in 1 Hour*
Sam	3	$\frac{1}{3}$
Frank	7	$\frac{1}{7}$
Together	x	$\frac{1}{x}$

2. TRANSLATE.

In words:

part of job Sam completed in 1 hour	added to	part of job Frank completed in 1 hour	is equal to	part of job they completed together in 1 hour
↓	↓	↓	↓	↓

Translate: $\dfrac{1}{3}$ $+$ $\dfrac{1}{7}$ $=$ $\dfrac{1}{x}$

3. SOLVE.   Here, we solve the equation $\dfrac{1}{3} + \dfrac{1}{7} = \dfrac{1}{x}$. We begin by multiplying both sides of the equation by the LCD, $21x$.

$$21x\left(\frac{1}{3}\right) + 21x\left(\frac{1}{7}\right) = 21x\left(\frac{1}{x}\right)$$
$$7x + 3x = 21 \qquad \text{Simplify.}$$
$$10x = 21$$
$$x = \frac{21}{10} \quad \text{or} \quad 2\frac{1}{10} \text{ hours}$$

4. INTERPRET.

Check: Our proposed solution is $2\frac{1}{10}$ hours. This proposed solution is reasonable since $2\frac{1}{10}$ hours is more than half of Sam's time and less than half of Frank's time. Check this solution in the originally *stated* problem.

State: Sam and Frank can complete the quality control tour in $2\frac{1}{10}$ hours. ⬭

4 Next we look at a problem solved by the distance formula.

EXAMPLE 6

FINDING SPEEDS OF VEHICLES

A car travels 180 miles in the same time that a truck travels 120 miles. If the car's speed is 20 miles per hour faster than the truck's, find the car's speed and the truck's speed.

Solution

1. UNDERSTAND. Read and reread the problem. Suppose that the truck's speed is 45 miles per hour. Then the car's speed is 20 miles per hour more, or 65 miles per hour.

We are given that the car travels 180 miles in the same time that the truck travels 120 miles. To find the time it takes the car to travel 180 miles, remember that since $d = rt$, we know that $\frac{d}{r} = t$.

Car's Time	*Truck's Time*
$t = \dfrac{d}{r} = \dfrac{180}{65} = 2\dfrac{50}{65} = 2\dfrac{10}{13}$ hours	$t = \dfrac{d}{r} = \dfrac{120}{45} = 2\dfrac{30}{45} = 2\dfrac{2}{3}$ hours

Since the times are not the same, our proposed solution is not correct. But we have a better understanding of the problem.

Let x = the speed of the truck.

Since the car's speed is 20 miles per hour faster than the truck's, then

$$x + 20 = \text{the speed of the car}$$

Use the formula $d = r \cdot t$ or **d**istance = **r**ate · **t**ime. Prepare a chart to organize the information in the problem.

	Distance	=	Rate	·	Time
Truck	120		x		$\dfrac{120}{x}$ ← distance ← rate
Car	180		$x + 20$		$\dfrac{180}{x + 20}$ ← distance ← rate

2. TRANSLATE. Since the car and the truck traveled the same amount of time, we have that

In words:

car's time	=	truck's time
↓		↓

Translate: $\dfrac{180}{x + 20} = \dfrac{120}{x}$

3. SOLVE. We begin by multiplying both sides of the equation by the LCD, $x(x + 20)$, or cross multiplying.

$$\frac{180}{x + 20} = \frac{120}{x}$$
$$180x = 120(x + 20)$$
$$180x = 120x + 2400 \qquad \text{Use the distributive property.}$$
$$60x = 2400 \qquad \text{Subtract } \mathbf{120x} \text{ from both sides.}$$
$$x = 40 \qquad \text{Divide both sides by } \mathbf{60}.$$

4. INTERPRET. The speed of the truck is 40 miles per hour. The speed of the car must then be $x + 20$ or 60 miles per hour.

Check: Find the time it takes the car to travel 180 miles and the time it takes the truck to travel 120 miles.

Car's Time	*Truck's Time*
$t = \dfrac{d}{r} = \dfrac{180}{60} = 3$ hours	$t = \dfrac{d}{r} = \dfrac{120}{40} = 3$ hours

Since both travel the same amount of time, the proposed solution is correct.

State: The car's speed is 60 miles per hour and the truck's speed is 40 miles per hour.

Spotlight on

DECISION
✗ MAKING

Suppose you must select a child care center for your two-year-old daughter. You have compiled information on two possible choices in the table shown at the right. Which child care center would you choose? Why?

	Center A	Center B
Weekly cost	$130	$110
Number of 2-year-olds	7	13
Number of adults for 2-year-olds	2	3
Distance from home	6 miles	11 miles

EXERCISE SET 7.6

STUDY GUIDE/SSM   CD/VIDEO   PH MATH TUTOR CENTER   MathXL®Tutorials ON CD   MathXL®   MyMathLab®

Solve each proportion. See Example 1.

1. $\dfrac{2}{3} = \dfrac{x}{6}$ 4

2. $\dfrac{x}{2} = \dfrac{16}{6}$ $\dfrac{16}{3}$

3. $\dfrac{x}{10} = \dfrac{5}{9}$ $\dfrac{50}{9}$

4. $\dfrac{9}{4x} = \dfrac{6}{2}$ $\dfrac{3}{4}$

5. $\dfrac{x + 1}{2x + 3} = \dfrac{2}{3}$ -3

6. $\dfrac{x + 1}{x + 2} = \dfrac{5}{3}$ $-\dfrac{7}{2}$

7. $\dfrac{9}{5} = \dfrac{12}{3x + 2}$ $\dfrac{14}{9}$

8. $\dfrac{6}{11} = \dfrac{27}{3x - 2}$ $\dfrac{103}{6}$

9. For which of the following equations can we immediately use cross products to solve for x?   a

a. $\dfrac{2-x}{5} = \dfrac{1+x}{3}$ **b.** $\dfrac{2}{5} - x = \dfrac{1+x}{3}$

10. For what value of x is $\dfrac{x}{x-1}$ in proportion to $\dfrac{x+1}{x}$? Explain your result.   none; answers may vary

Solve. See Example 2.

11. The ratio of the weight of an object on Earth to the weight of the same object on Pluto is 100 to 3. If an elephant weighs 4100 pounds on Earth, find the elephant's weight on Pluto.   123 lb

12. If a 170-pound person weighs approximately 65 pounds on Mars, how much does a 9000-pound satellite weigh?   3441 lb

 13. There are 110 calories per 28.4 grams of Crispy Rice cereal. Find how many calories are in 42.6 grams of this cereal.   165 cal

14. On an architect's blueprint, 1 inch corresponds to 4 feet. Find the length of a wall represented by a line that is $3\frac{7}{8}$ inches long on the blueprint. $15\frac{1}{2}$ ft

Find the unknown length x or y in the following pairs of similar triangles. See Example 3.

△ **15.**

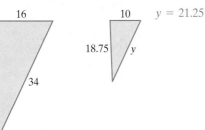

$y = 21.25$

△ **16.**

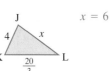

$x = 6$

△ **17.**

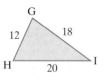

$y = 5\frac{5}{7}$ ft

△ **18.**

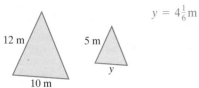

$y = 4\frac{1}{6}$ m

28. upstream time: $\dfrac{9}{r-3}$; downstream time: $\dfrac{11}{r+3}$; $r = 30$ mph

Solve the following. See Example 4.

19. Three times the reciprocal of a number equals 9 times the reciprocal of 6. Find the number.   2

27. trip to park rate: r; to park time: $\dfrac{12}{r}$; return trip rate: r; return time: $\dfrac{18}{r} - 1$; $r = 6$ mph

20. Twelve divided by the sum of x and 2 equals the quotient of 4 and the difference of x and 2. Find x.   4

21. If twice a number added to 3 is divided by the number plus 1, the result is three halves. Find the number. -3

22. A number added to the product of 6 and the reciprocal of the number equals -5. Find the number. -3 or -2

See Example 5.

23. Smith Engineering found that an experienced surveyor surveys a roadbed in 4 hours. An apprentice surveyor needs 5 hours to survey the same stretch of road. If the two work together, find how long it takes them to complete the job. $2\frac{2}{9}$ hr

24. An experienced bricklayer constructs a small wall in 3 hours. The apprentice completes the job in 6 hours. Find how long it takes if they work together.   2 hr

25. In 2 minutes, a conveyor belt moves 300 pounds of recyclable aluminum from the delivery truck to a storage area. A smaller belt moves the same quantity of cans the same distance in 6 minutes. If both belts are used, find how long it takes to move the cans to the storage area. $1\frac{1}{2}$ min

26. Find how long it takes the conveyor belts described in Exercise 25 to move 1200 pounds of cans. (*Hint:* Think of 1200 pounds as four 300-pound jobs.)   6 min

See Example 6.

 27. A jogger begins her workout by jogging to the park, a distance of 12 miles. She then jogs home at the same speed but along a different route. This return trip is 18 miles and her time is one hour longer. Find her jogging speed. Complete the accompanying chart and use it to find her jogging speed.

	distance	=	rate	•	time
Trip to park	12				
Return trip	18				

28. A boat can travel 9 miles upstream in the same amount of time it takes to travel 11 miles downstream. If the current of the river is 3 miles per hour, complete the chart below and use it to find the speed of the boat in still water.

	distance	=	rate	•	time
Upstream	9		$r-3$		
Downstream	11		$r+3$		

29. A cyclist rode the first 20-mile portion of his workout at a constant speed. For the 16-mile cooldown portion of his workout, he reduced his speed by 2 miles per hour. Each portion of the workout took the same time. Find the cyclist's speed during the first portion and find his speed during the cooldown portion.   1st portion: 10 mph; cooldown: 8 mph

30. A semi-truck travels 300 miles through the flatland in the same amount of time that it travels 180 miles through mountains.

The rate of the truck is 20 miles per hour slower in the mountains than in the flatland. Find both the flatland rate and mountain rate.   flatland: 50 mph; mountains: 30 mph

31. A human factors expert recommends that there be at least 9 square feet of floor space in a college classroom for every student in the class. Find the minimum floor space that 40 students need.   360 sq. ft

32. Due to space problems at a local university, a 20-foot by 12-foot conference room is converted into a classroom. Find the maximum number of students the room can accommodate. (See Exercise 31.)   26 students

MIXED PRACTICE

Solve the following.   **46.** car: 4 hr; jet: 3 hr

33. One-fourth equals the quotient of a number and 8. Find the number.   2

34. Four times a number added to 5 is divided by 6. The result is $\frac{7}{2}$. Find the number.   4

35. Marcus and Tony work for Lombardo's Pipe and Concrete. Mr. Lombardo is preparing an estimate for a customer. He knows that Marcus lays a slab of concrete in 6 hours. Tony lays the same size slab in 4 hours. If both work on the job and the cost of labor is $45.00 per hour, decide what the labor estimate should be. $108.00

36. Mr. Dodson can paint his house by himself in 4 days. His son needs an additional day to complete the job if he works by himself. If they work together, find how long it takes to paint the house. $2\frac{2}{9}$ days

37. While road testing a new make of car, the editor of a consumer magazine finds that he can go 10 miles into a 3-mile-per-hour wind in the same amount of time he can go 11 miles with a 3-mile-per-hour wind behind him. Find the speed of the car in still air.   63 mph

38. A fisherman on Pearl River rows 9 miles downstream in the same amount of time he rows 3 miles upstream. If the current is 6 miles per hour, find how long it takes him to cover the 12 miles.   1 hr

△ **39.** Find the unknown length y. $y = 37\frac{1}{2}$ ft

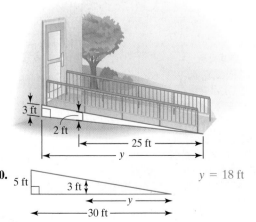

△ **40.** $y = 18$ ft

41. Ken Hall, a tailback, holds the high school sports record for total yards rushed in a season. In 1953, he rushed for 4045 total yards in 12 games. Find his average rushing yards per game.

42. To estimate the number of people in Jackson, population 50,000, who have no health insurance, 250 people were polled. Of those polled, 39 had no insurance. How many people in the city might we expect to be uninsured?   7800 people

43. Two divided by the difference of a number and 3 minus 4 divided by a number plus 3, equals 8 times the reciprocal of the difference of the number squared and 9. What is the number?   5   **41.** 337 yd/game

44. If 15 times the reciprocal of a number is added to the ratio of 9 times a number minus 7 and the number plus 2, the result is 9. What is the number?   3

45. A pilot flies 630 miles with a tail wind of 35 miles per hour. Against the wind, he flies only 455 miles in the same amount of time. Find the rate of the plane in still air.   217 mph

46. A marketing manager travels 1080 miles in a corporate jet and then an additional 240 miles by car. If the car ride takes one hour longer than the jet ride takes, and if the rate of the jet is 6 times the rate of the car, find the time the manager travels by jet and find the time the manager travels by car.

47. To mix weed killer with water correctly, it is necessary to mix 8 teaspoons of weed killer with 2 gallons of water. Find how many gallons of water are needed to mix with the entire box if it contains 36 teaspoons of weed killer.   9 gal

48. A recent headline read, "Women Earn Bigger Checks in 1 of Every 6 Couples." If there are 23,000 couples in a nearby metropolitan area, how many women would you expect to earn bigger paychecks?   3833 women

49. A cyclist rides 16 miles per hour on level ground on a still day. He finds that he rides 48 miles with the wind behind him in the same amount of time that he rides 16 miles into the wind. Find the rate of the wind.   8 mph

50. The current on a portion of the Mississippi River is 3 miles per hour. A barge can go 6 miles upstream in the same amount of time it takes to go 10 miles downstream. Find the speed of the boat in still water.   12 mph

51. One custodian cleans a suite of offices in 3 hours. When a second worker is asked to join the regular custodian, the job takes only $1\frac{1}{2}$ hours. How long does it take the second worker to do the same job alone?   3 hr

52. One person proofreads a copy for a small newspaper in 4 hours. If a second proofreader is also employed, the job can be done in $2\frac{1}{2}$ hours. How long does it take for the second proofreader to do the same job alone? $6\frac{2}{3}$ hr

53. An architect is completing the plans for a triangular deck. Use the diagram below to find the missing dimension. $26\frac{2}{3}$ ft

20 feet x

6 inches   8 inches

54. A student wishes to make a small model of a triangular mainsail in order to study the effects of wind on the sail. The smaller model will be the same shape as a regular-size sailboat's mainsail. Use the following diagram to find the missing dimensions. $x = 4.4$ ft; $y = 5.6$ ft

11 ft   14 ft   5 ft

x y

2 ft

58. first pump: 27 min; second pump: 54 min

55. The manufacturers of cans of salted mixed nuts state that the ratio of peanuts to other nuts is 3 to 2. If 324 peanuts are in a can, find how many other nuts should also be in the can.

56. There are 1280 calories in a 14-ounce portion of Eagle Brand Milk. Find how many calories are in 2 ounces of Eagle Brand Milk. $182\frac{6}{7}$ cal   **55.** 216 nuts

57. One pipe fills a storage pool in 20 hours. A second pipe fills the same pool in 15 hours. When a third pipe is added and all three are used to fill the pool, it takes only 6 hours. Find how long it takes the third pipe to do the job.   20 hr

58. One pump fills a tank 2 times as fast as another pump. If the pumps work together, they fill the tank in 18 minutes. How long does it take for each pump to fill the tank?

59. Andrew and Timothy Larson volunteer at a local recycling plant. Andrew can sort a batch of recyclables in 2 hours alone while his brother Timothy needs 3 hours to complete the same job. If they work together, how long will it take them to sort one batch? $1\frac{1}{5}$ hours

60. A car travels 280 miles in the same time that a motorcycle travels 240 miles. If the car's speed is 10 miles per hour more than the motorcycle's, find the speed of the car and the speed of the motorcycle.   car: 70 mph; motorcycle: 60 mph

61. In 6 hours, an experienced cook prepares enough pies to supply a local restaurant's daily order. Another cook prepares the same number of pies in 7 hours. Together with a third cook, they prepare the pies in 2 hours. Find the work rate of the third cook. $5\frac{1}{4}$ hr

62. Mrs. Smith balances the company books in 8 hours. It takes her assistant half again as long to do the same job. If they work together, find how long it takes them to balance the books. $4\frac{4}{5}$ hr

63. One pump fills a tank 3 times as fast as another pump. If the pumps work together, they fill the tank in 21 minutes. How long does it take for each pump to fill the tank?   first pump: 28 min; second pump: 84 min

REVIEW AND PREVIEW

Find the slope of the line through each pair of points. Use the slope to determine whether the line is vertical, horizontal, or moves upward or downward from left to right. See Section 3.4.

64. $(-2, 5), (4, -3)$

65. $(0, 4), (2, 10)$

66. $(-3, -6), (1, 5)$

67. $(-2, 7), (3, -2)$

68. $(3, 7), (3, -2)$

69. $(0, -4), (2, -4)$

64. $m = -\frac{4}{3}$; downward
65. $m = 3$; upward
66. $m = \frac{11}{4}$; upward
67. $m = -\frac{9}{5}$; downward
68. undefined slope; vertical
69. $m = 0$; horizontal

Concept Extensions

The following bar graph shows the capacity of the United States to generate electricity from the wind in the years shown. Use this graph for Exercises 70 and 71.

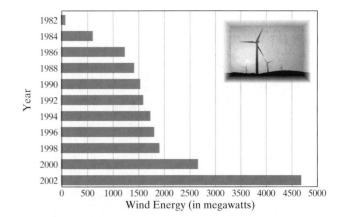

70. Find the approximate increase in megawatt capacity during the 2-year period from 1982 to 1984.   530 megawatts

71. Find the approximate increase in megawatt capacity during the 2-year period from 2000 to 2002.   2035 megawatts

In general, 1000 megawatts will serve the average electricity needs of 560,000 people. Use this fact and the preceding graph to answer Exercises 72 and 73.   **72.** 1,484,000 people

72. In 2000, the number of megawatts that were generated from wind would serve the electricity needs of how many people?

73. How many megawatts of electricity are needed to serve the city or town in which you live?   answers may vary

74. If x is 10, is $\frac{2}{x}$ in proportion to $\frac{x}{50}$? Explain why or why not.
yes; answers may vary

One of the great algebraists of ancient times was a man named Diophantus. Little is known of his life other than that he lived and worked in Alexandria. Some historians believe he lived during the first century of the Christian era, about the time of Nero. The only clue to his personal life is the following epigram found in a collection called the Palatine Anthology.

God granted him youth for a sixth of his life and added a twelfth part to this. He clothed his cheeks in down. He lit him the light of wedlock after a seventh part and five years after his marriage, He granted him a son. Alas, lateborn wretched child. After attaining the measure of half his father's life, cruel fate overtook him, thus leaving Diophantus during the last four years of his life only such consolation as the science of numbers. How old was Diophantus at his death?*

We are looking for Diophantus' age when he died, so let x represent that age. If we sum the parts of his life, we should get the total age.

Parts of his life $\begin{cases} \frac{1}{6}x + \frac{1}{12}x \text{ is the time of his youth.} \\ \frac{1}{7}x \text{ is the time between his youth and when he married.} \\ 5 \text{ years is the time between his marriage and the birth of his son.} \\ \frac{1}{2}x \text{ is the time Diophantus had with his son.} \\ 4 \text{ years is the time between his son's death and his own.} \end{cases}$

The sum of these parts should equal Diophantus' age when he died.

$$\frac{1}{6}\cdot x + \frac{1}{12}\cdot x + \frac{1}{7}\cdot x + 5 + \frac{1}{2}\cdot x + 4 = x$$

75. Solve the epigram.   84 yr

76. How old was Diophantus when his son was born? How old was the son when he died?   38 yr; 42 yr

77. Solve the following epigram:

I was four when my mother packed my lunch and sent me off to school. Half my life was spent in school and another sixth was spent on a farm. Alas, hard times befell me. My crops and cattle fared poorly and my land was sold. I returned to school for 3 years and have spent one tenth of my life teaching. How old am I?   30 yr

78. Write an epigram describing your life. Be sure that none of the time periods in your epigram overlap.   answers may vary

79. A hyena spots a giraffe 0.5 mile away and begins running toward it. The giraffe starts running away from the hyena just as the hyena begins running toward it. A hyena can run at a speed of 40 mph and a giraffe can run at 32 mph. How long will it take for the hyena to overtake the giraffe? (*Source: World Almanac* and *Book of Facts*)   3.75 min

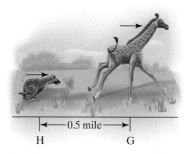

H ← 0.5 mile → G

*From *The Nature and Growth of Modern Mathematics*, Edna Kramer, 1970, Fawcett Premier Books, Vol. 1, pages 107–108.

VARIATION AND PROBLEM SOLVING

Objectives

1 Solve problems involving direct variation.

2 Solve problems involving inverse variation.

3 Other types of direct and inverse variation.

4 Variation and problem solving.

In Chapter 3, we studied linear equations in two variables. Recall that such an equation can be written in the form $Ax + By = C$, where A and B are not both 0.

Also recall that the graph of a linear equation in two variables is a line. In this section, we begin by looking at a particular family of linear equations—those that can be written in the form

$$y = kx,$$

where k is a constant. This family of equations is called *direct variation*.

 Let's suppose that you are earning \$7.25 per hour at a part-time job. The amount of money you earn depends on the number of hours you work. This is illustrated by the following table:

Hours Worked	0	1	2	3	4	
Money Earned (before deductions)	0	7.25	14.50	21.75	29.00	and so on

In general, to calculate your earnings (before deductions) multiply the constant \$7.25 by the number of hours you work. If we let y represent the amount of money earned and x represent the number of hours worked, we get the direct variation equation

$$y = 7.25 \cdot x$$

earnings $= \$7.25 \cdot$ hours worked

Notice that in this direct variation equation, as the number of hours increases, the pay increases as well.

Direct Variation

y varies directly as x, or **y is directly proportional to x**, if there is a nonzero constant k such that

$$y = kx$$

The number k is called the **constant of variation** or the **constant of proportionality**.

In our direct variation example: $y = 7.25x$, the constant of variation is 7.25.

Let's use the previous table to graph $y = 7.25x$. We begin our graph at the ordered-pair solution $(0, 0)$. Why? We assume that the least amount of hours worked is 0. If 0 hours are worked, then the pay is \$0.

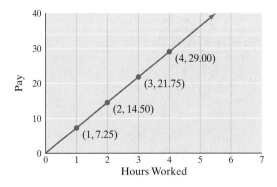

As illustrated in this graph, a direct variation equation $y = kx$ is linear. Also notice that $y = 7.25x$ is a function since its graph passes the vertical line test.

(**EXAMPLE 1**)

Write a direct variation equation of the form $y = kx$ that satisfies the ordered pairs in the table below.

x	2	9	1.5	-1
y	6	27	4.5	-3

Solution   We are given that there is a direct variation relationship between x and y. This means that

$$y = kx$$

By studying the given values, you may be able to mentally calculate k. If not, to find k, we simply substitute one given ordered pair into this equation and solve for k. We'll use the given pair $(2, 6)$.

$$y = kx$$
$$6 = k \cdot 2$$
$$\frac{6}{2} = \frac{k \cdot 2}{2}$$
$$3 = k \qquad \text{Solve for } k.$$

Since $k = 3$, we have the equation $y = 3x$.

To check, see that each given y is 3 times the given x.

Let's try another type of direct variation example.

CLASSROOM EXAMPLE

Write a direct variation equation that satisfies:

x	4	$\frac{1}{2}$	1.5	6
y	8	1	3	12

answer: $y = 2x$

(**EXAMPLE 2**)

Suppose that y varies directly as x. If y is 17 when x is 34, find the constant of variation and the direct variation equation. Then find y when x is 12.

Solution   Let's use the same method as in Example 1 to find x. Since we are told that y varies directly as x, we know the relationship is of the form

$$y = kx.$$

Let $y = 17$ and $x = 34$ and solve for k.

$$17 = k \cdot 34$$
$$\frac{17}{34} = \frac{k \cdot 34}{34}$$
$$\frac{1}{2} = k \qquad \text{Solve for } k.$$

Thus, the constant of variation is $\frac{1}{2}$ and the equation is $y = \frac{1}{2}x$.

To find y when $x = 12$, use $y = \frac{1}{2}x$ and replace x with 12.

$$y = \frac{1}{2}x$$
$$y = \frac{1}{2} \cdot 12 \qquad \text{Replace } x \text{ with } \mathbf{12}.$$
$$y = 6$$

CLASSROOM EXAMPLE

If y varies directly as x and y is 15 when x is 45, find the constant of variation and the direct variation equation. Then find y when x is 3.

answer: $y = \frac{1}{3}x; y = 1$

Thus, when x is 12, y is 6.

Let's review a few facts about linear equations of the form $y = kx$.

> ## Direct Variation: $y = kx$
>
> - There is a direct variation relationship between x and y.
> - The graph is a line.
> - The line will always go through the origin $(0, 0)$. Why?
>   Let $x = 0$. Then $y = k \cdot 0$ or $y = 0$.
> - The slope of the graph of $y = kx$ is k, the constant of variation. Why?
>   Remember that the slope of an equation of the form $y = mx + b$ is m, the coefficient of x.
> - The equation $y = kx$ describes a function. Each x has a unique y and its graph passes the vertical line test.

EXAMPLE 3

The line is the graph of a direct variation equation. Find the constant of variation and the direct variation equation.

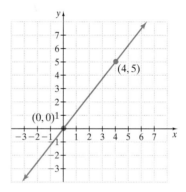

Solution Recall that k, the constant of variation is the same as the slope of the line. Thus, to find k, we use the slope formula and find slope.

Using the given points $(0, 0)$, and $(4, 5)$, we have

$$\text{slope} = \frac{5 - 0}{4 - 0} = \frac{5}{4}.$$

Thus, $k = \frac{5}{4}$ and the variation equation is $y = \frac{5}{4}x$.

2 In this section, we will introduce another type of variation, called inverse variation.

Let's suppose you need to drive a distance of 40 miles. You know that the faster you drive the distance, the sooner you arrive at your destination. Recall that there is a mathematical relationship between distance, rate, and time. It is $d = r \cdot t$. In our example, distance is a constant 40 miles, so we have $40 = r \cdot t$ or $t = \frac{40}{r}$.

For example, if you drive 10 mph, the time to drive the 40 miles is

$$t = \frac{40}{r} = \frac{40}{10} = 4 \text{ hours}$$

If you drive 20 mph, the time is

$$t = \frac{40}{r} = \frac{40}{20} = 2 \text{ hours}$$

Again, notice that as speed increases, time decreases. Below are some ordered-pair solutions of $t = \frac{40}{r}$ and its graph.

rate (mph)	r	5	10	20	40	60	80
time (hr)	t	8	4	2	1	$\frac{2}{3}$	$\frac{1}{2}$

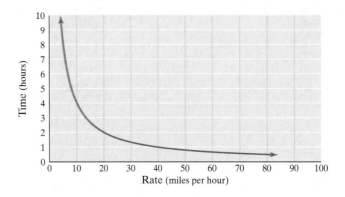

Notice that the graph of this variation is not a line, but it passes the vertical line test so $t = \frac{40}{r}$ does describe a function. This is an example of inverse variation.

Inverse Variation

***y* varies inversely as *x*,** or ***y* is inversely proportional to *x*,** if there is a nonzero constant k such that

$$y = \frac{k}{x}$$

The number k is called the **constant of variation** or the **constant of proportionality**.

In our inverse variation example, $t = \frac{40}{r}$ or $y = \frac{40}{x}$, the constant of variation is 40. We can immediately see differences and similarities in direct variation and inverse variation.

Direct variation	$y = kx$	linear equation	both functions
Inverse variation	$y = \dfrac{k}{x}$	rational equation	

Remember that $y = \frac{k}{x}$ is a rational equation and not a linear equation. Also notice that because x is in the denominator, that x can be any value except 0.

We can still derive an inverse variation equation from a table of values.

EXAMPLE 4

Write an inverse variation equation of the form $y = \frac{k}{x}$ that satisfies the ordered pairs in the table below.

x	2	4	$\frac{1}{2}$
y	6	3	24

Solution Since there is an inverse variation relationship between x and y, we know that $y = \frac{k}{x}$. To find k, choose one given ordered pair and substitute the values into the equation. We'll use $(2, 6)$.

$$y = \frac{k}{x}$$

$$6 = \frac{k}{2}$$

$$2 \cdot 6 = 2 \cdot \frac{k}{2} \qquad \text{Multiply both sides by 2.}$$

$$12 = k \qquad \text{Solve.}$$

Since $k = 12$, we have the equation $y = \frac{12}{x}$.

> **Helpful Hint**
>
> Multiply both sides of the inverse variation relationship equation $y = \frac{k}{x}$ by x (as long as x is not 0), and we have $xy = k$. This means that if y varies inversely as x, their product is always the constant of variation k. For an example of this, check the table below.
>
> | x | 2 | 4 | $\frac{1}{2}$ |
> |---|---|---|---|
> | y | 6 | 3 | 24 |
>
> $2 \cdot 6 = 12 \qquad 4 \cdot 3 = 12 \qquad \frac{1}{2} \cdot 24 = 12$

EXAMPLE 5

Suppose that y varies inversely as x. If $y = 0.02$ when $x = 75$, find the constant of variation and the inverse variation equation. Then find y when x is 30.

Solution Since y varies inversely as x, the constant of variation may be found by simply finding the product of the given x and y.

$$k = xy = 75(0.02) = 1.5$$

To check, we will use the inverse variation equation

$$y = \frac{k}{x}.$$

Let $y = 0.02$ and $x = 75$ and solve for k.

$$0.02 = \frac{k}{75}$$

$$75(0.02) = 75 \cdot \frac{k}{75} \qquad \text{Multiply both sides by 75.}$$

$$1.5 = k \qquad \text{Solve for } k.$$

Thus, the constant of variation is 1.5 and the equation is $y = \frac{1.5}{x}$.

To find y when $x = 30$ use $y = \frac{1.5}{x}$ and replace x with 30.

$$y = \frac{1.5}{x}$$

$$y = \frac{1.5}{30} \qquad \text{Replace } x \text{ with 30.}$$

$$y = 0.05$$

Thus, when x is 30, y is 0.05.

3   It is possible for y to vary directly or inversely as powers of x. In this section, our powers of x will be natural numbers only.

> ### Direct and Inverse
>
> Variation as nth Powers of x
>
> **y varies directly as a power of x** if there is a nonzero constant k and a natural number n such that
>
> $$y = kx^n$$
>
> **y varies inversely as a power of x** if there is a nonzero constant k and a natural number n such that
>
> $$y = \frac{k}{x^n}$$

EXAMPLE 6

The surface area of a cube A varies directly as the square of a length of its side s. If A is 54 when s is 3, find A when $s = 4.2$.

Solution   Since the surface area A varies directly as the square of side s, we have

$$A = ks^2.$$

To find k, let $A = 54$ and $s = 3$.

$$A = k \cdot s^2$$
$$54 = k \cdot 3^2 \qquad \text{Let } A = 54 \text{ and } s = 3.$$
$$54 = 9k \qquad 3^2 = 9.$$
$$6 = k \qquad \text{Divide by 9.}$$

The formula for surface area of a cube is then

$$A = 6s^2 \text{ where } s \text{ is the length of a side}.$$

To find the surface area when $s = 4.2$, substitute.

$$A = 6s^2$$
$$A = 6 \cdot (4.2)^2$$
$$A = 105.84$$

The surface area of a cube whose side measures 4.2 units is 105.84 square units.

4   There are many real-life applications of direct and inverse variation.

EXAMPLE 7

The weight of a body w varies inversely with the square of its distance from the center of Earth d. If a person weighs 160 pounds on the surface of Earth, what is the person's weight 200 miles above the surface? (Assume that the radius of Earth is 4000 miles.)

Solution

1. UNDERSTAND. Make sure you read and reread the problem.

2. TRANSLATE. Since we are told that weight w varies inversely with the square of its distance from the center of Earth, d, we have

$$w = \frac{k}{d^2}.$$

3. SOLVE. To solve the problem, we first find k. To do so, use the fact that the person weighs 160 pounds on Earth's surface, which is a distance of 4000 miles from Earth's center.

$$w = \frac{k}{d^2}$$
$$160 = \frac{k}{(4000)^2}$$
$$2{,}560{,}000{,}000 = k$$

Thus, we have $w = \dfrac{2{,}560{,}000{,}000}{d^2}$

Since we want to know the person's weight 200 miles above the Earth's surface, we let $d = 4200$ and find w.

$$w = \frac{2{,}560{,}000{,}000}{d^2}$$
$$w = \frac{2{,}560{,}000{,}000}{(4200)^2}$$

A person **200** miles above the Earth's surface is **4200** miles from the Earth's center.

$$w \approx 145$$

Simplify.

4. INTERPRET. Check: Your answer is reasonable since the further a person is from Earth, the less the person weighs.

State: Thus, 200 miles above the surface of the Earth, a 160-pound person weighs approximately 145 pounds.

1. direct **2.** inverse **3.** inverse **4.** direct **5.** inverse **6.** direct **7.** direct **8.** inverse

MENTAL MATH

State whether each equation represents direct or indirect variation.

1. $y = 5x$ **2.** $y = \dfrac{5}{x}$ **3.** $y = \dfrac{7}{x^2}$ **4.** $y = 6.5x^4$ **5.** $y = \dfrac{11}{x}$ **6.** $y = 18x$ **7.** $y = 12x^2$ **8.** $y = \dfrac{20}{x^3}$

EXERCISE SET 7.7

STUDY GUIDE/SSM · CD/VIDEO · PH MATH TUTOR CENTER · MathXL®Tutorials ON CD · MathXL® · MyMathLab®

Write a direct variation equation, $y = kx$, that satisfies the ordered pairs in each table. See Example 1.

2. $y = 7x$ **3.** $y = 6x$ **4.** $y = \frac{1}{3}x$

1.

x	0	6	10
y	0	3	5

$y = \dfrac{1}{2}x$

2.

x	0	2	−1	3
y	0	14	−7	21

3.

x	−2	2	4	5
y	−12	12	24	30

4.

x	3	9	−2	12
y	1	3	$-\frac{2}{3}$	4

Write a direct variation equation, y = kx, that describes each graph. See Example 3.

5. $y = 3x$

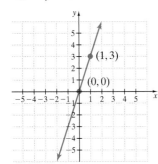

6. $y = \frac{1}{4}x$

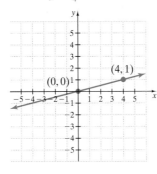

7. $y = \frac{2}{3}x$

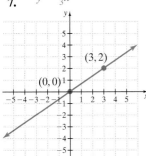

8. $y = \frac{5}{2}x$

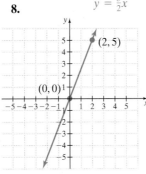

Write an inverse variation equation, $y = \frac{k}{x}$, that satisfies the ordered pairs in each table. See Example 4.

9.

x	1	−7	3.5	−2
y	7	−1	2	−3.5

$y = \frac{7}{x}$

10.

x	2	−11	4	−4
y	11	−2	5.5	−5.5

$y = \frac{22}{x}$

11.

x	10	$\frac{1}{2}$	$-\frac{3}{2}$
y	0.05	1	$-\frac{1}{3}$

$y = \frac{0.5}{x}$

12.

x	4	$\frac{1}{5}$	−8
y	0.1	2	−0.05

$y = \frac{0.4}{x}$

Write an equation to describe each variation. Use k for the constant of proportionality. See Examples 1 through 6.   **14.** $a = kb$

13. y varies directly as x $y = kx$ **14.** a varies directly as b

15. h varies inversely as t $h = \frac{k}{t}$ **16.** s varies inversely as t

17. z varies directly as x^2 $z = kx^2$ **18.** p varies inversely as x^2

19. y varies inversely as z^3 $y = \frac{k}{z^3}$ **20.** x varies directly as y^4

21. x varies inversely as $\sqrt{y}$ **22.** y varies directly as d^2

MIXED PRACTICE   **16.** $s = \frac{k}{t}$ **18.** $p = \frac{k}{x^2}$ **20.** $x = ky^4$ **21.** $x = \frac{k}{\sqrt{y}}$

Solve. See Examples 2, 5, and 6.   **22.** $y = kd^2$ **23.** $y = 40$ **24.** $y = 18$

23. y varies directly as x. If $y = 20$ when $x = 5$, find y when x is 10.

24. y varies directly as x. If $y = 27$ when $x = 3$, find y when x is 2.

25. y varies inversely as x. If $y = 5$ when $x = 60$, find y when x is 100. $y = 3$

26. y varies inversely as x. If $y = 200$ when $x = 5$, find y when x is 4. $y = 250$

27. z varies directly as x^2. If $z = 96$ when $x = 4$, find z when $x = 3$. $z = 54$

28. s varies directly as t^3. If $s = 270$ when $t = 3$, find s when $x = 1$. $s = 10$

29. a varies inversely as b^3. If $a = \frac{3}{2}$ when $b = 2$, find a when b is 3.

30. p varies inversely as q^2. If $p = \frac{5}{16}$ when $q = 8$, find p when $q = \frac{1}{2}$. $p = 80$

MIXED PRACTICE   **29.** $a = \frac{4}{9}$

Solve. See Examples 1 through 7.

31. Your paycheck (before deductions) varies directly as the number of hours you work. If your paycheck is $112.50 for 18 hours, find your pay for 10 hours. $62.50

32. If your paycheck (before deductions) is $244.50 for 30 hours, find your pay for 34 hours. See Exercise 31. $277.10

33. The cost of manufacturing a certain type of headphone varies inversely as the number of headphones increases. If 5000 headphones can be manufactured for $9.00 each, find the cost to manufacture 7500 headphones. $6

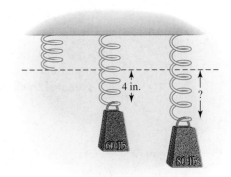

34. The cost of manufacturing a certain composition notebook varies inversely as the number of notebooks increases. If 10,000 notebooks can be manufactured for $0.50 each, find the cost to manufacture 18,000 notebooks. $0.28

35. The distance a spring stretches varies directly with the weight attached to the spring. If a 60-pound weight stretches the spring 4 inches, find the distance that an 80-pound weight stretches the spring. $5\frac{1}{3}$ inches

36. If a 30-pound weight stretches a spring 10 inches, find the distance a 20-pound weight stretches the spring. (See Exercise 35.)

37. The weight of an object varies inversely as the square of its distance from the *center* of the Earth. If a person weighs 180 pounds on Earth's surface, what is his weight 10 miles above the surface of the Earth? (Assume that the Earth's radius is 4000 miles.)   179.1 lb

38. For a constant distance, the rate of travel varies inversely as the time traveled. If a family travels 55 mph and arrives at a destination in 4 hours, how long will the return trip take traveling at 60 mph? $3\frac{2}{3}$ hr   **36.** $6\frac{2}{3}$ in.

39. The distance d that an object falls is directly proportional to the square of the time of the fall, t. A person who is parachuting for the first time is told to wait 10 seconds before opening the parachute. If the person falls 64 feet in 2 seconds, find how far he falls in 10 seconds.   1600 feet

40. The distance needed for a car to stop, d is directly proportional to the square of its rate of travel, r. Under certain driving conditions, a car traveling 60 mph needs 300 feet to stop.

With these same driving conditions, how long does it take a car to stop if the car is traveling 30 mph when the brakes are applied?   75 feet

REVIEW AND PREVIEW   **41.** $\frac{1}{2}$ **42.** $\frac{3}{4}$ **43.** $\frac{3}{7}$ **44.** $\frac{6}{5}$

Simplify. Follow the circled steps in the order shown.

41. $\dfrac{\frac{3}{4}+\frac{1}{4}}{\frac{3}{8}+\frac{13}{8}}$ ← ①Add. ③Divide. ②Add.

42. $\dfrac{\frac{9}{5}+\frac{6}{5}}{\frac{17}{6}+\frac{7}{6}}$ ← ① Add. ③ Divide. ② Add.

43. $\dfrac{\frac{2}{5}+\frac{1}{5}}{\frac{7}{10}+\frac{7}{10}}$ ← ①Add. ③Divide. ②Add.

44. $\dfrac{\frac{1}{4}+\frac{5}{4}}{\frac{3}{8}+\frac{7}{8}}$ ← ①Add. ③Divide. ②Add.

Concept Extensions

45. Suppose that y varies directly as x. If x is tripled, what is the effect on y?   multiplied by 3

46. Suppose that y varies directly as x^2. If x is tripled, what is the effect on y?   multiplied by 9

47. The period, P, of a pendulum (the time of one complete back and forth swing) varies directly with the square root of its length, l. If the length of the pendulum is quadrupled, what is the effect on the period, P?   it is doubled

48. For a constant distance, the rate of travel r varies inversely with the time traveled, t. If a car traveling 100 mph completes a test track in 6 minutes, find the rate needed to complete the same test track in 4 minutes. (Hint: Convert minutes to hours.)   150 mph

---

7.8   ## SIMPLIFYING COMPLEX FRACTIONS

Objectives

1   Simplify complex fractions using method 1.

2   Simplify complex fractions using method 2.

1   A rational expression whose numerator or denominator or both numerator and denominator contain fractions is called a **complex rational expression** or a **complex fraction**. Some examples are

$$\dfrac{4}{2-\frac{1}{2}}, \qquad \dfrac{\frac{3}{2}}{\frac{4}{7}-x}, \qquad \left.\dfrac{\frac{1}{x+2}}{x+2-\frac{1}{x}}\right\}$$

← Numerator of complex fraction
← Main fraction bar
← Denominator of complex fraction

Our goal in this section is to write complex fractions in simplest form. A complex fraction is in simplest form when it is in the form $\frac{P}{Q}$, where P and Q are polynomials that have no common factors.

In this section, two methods of simplifying complex fractions are represented. The first method presented uses the fact that the main fraction bar indicates division.

Method 1: Simplifying a Complex Fraction

Step 1. Add or subtract fractions in the numerator or denominator so that the numerator is a single fraction and the denominator is a single fraction.

Step 2. Perform the indicated division by multiplying the numerator of the complex fraction by the reciprocal of the denominator of the complex fraction.

Step 3. Write the rational expression in simplest form.

EXAMPLE 1

Simplify the complex fraction $\dfrac{\frac{5}{8}}{\frac{2}{3}}$.

Solution Since the numerator and denominator of the complex fraction are already single fractions, we proceed to step 2: perform the indicated division by multiplying the numerator $\frac{5}{8}$ by the reciprocal of the denominator $\frac{2}{3}$.

$$\frac{\frac{5}{8}}{\frac{2}{3}} = \frac{5}{8} \cdot \frac{3}{2} = \frac{15}{16}$$

The reciprocal of $\frac{2}{3}$ is $\frac{3}{2}$.

EXAMPLE 2

Simplify: $\dfrac{\frac{2}{3} + \frac{1}{5}}{\frac{2}{3} - \frac{2}{9}}$.

Solution Simplify above and below the main fraction bar separately. First, add $\frac{2}{3}$ and $\frac{1}{5}$ to obtain a single fraction in the numerator; then subtract $\frac{2}{9}$ from $\frac{2}{3}$ to obtain a single fraction in the denominator.

$$\frac{\frac{2}{3} + \frac{1}{5}}{\frac{2}{3} - \frac{2}{9}} = \frac{\frac{2(5)}{3(5)} + \frac{1(3)}{5(3)}}{\frac{2(3)}{3(3)} - \frac{2}{9}}$$

The LCD of the numerator's fractions is **15**.

The LCD of the denominator's fractions is **9**.

$$= \frac{\frac{10}{15} + \frac{3}{15}}{\frac{6}{9} - \frac{2}{9}}$$

Simplify.

$$= \dfrac{\dfrac{13}{15}}{\dfrac{4}{9}}$$ Add the numerator's fractions.

Subtract the denominator's fractions.

Next, perform the indicated division by multiplying the numerator of the complex fraction by the reciprocal of the denominator of the complex fraction.

$$\dfrac{\dfrac{13}{15}}{\dfrac{4}{9}} = \dfrac{13}{15} \cdot \dfrac{9}{4}$$ The reciprocal of $\dfrac{4}{9}$ is $\dfrac{9}{4}$.

$$= \dfrac{13 \cdot 3 \cdot 3}{3 \cdot 5 \cdot 4} = \dfrac{39}{20}$$

EXAMPLE 3

Simplify: $\dfrac{\dfrac{1}{z} - \dfrac{1}{2}}{\dfrac{1}{3} - \dfrac{z}{6}}$

Solution   Subtract to get a single fraction in the numerator and a single fraction in the denominator of the complex fraction.

$$\dfrac{\dfrac{1}{z} - \dfrac{1}{2}}{\dfrac{1}{3} - \dfrac{z}{6}} = \dfrac{\dfrac{2}{2z} - \dfrac{z}{2z}}{\dfrac{2}{6} - \dfrac{z}{6}}$$ The LCD of the numerator's fractions is $2z$.

The LCD of the denominator's fractions is 6.

$$= \dfrac{\dfrac{2 - z}{2z}}{\dfrac{2 - z}{6}}$$

$$= \dfrac{2 - z}{2z} \cdot \dfrac{6}{2 - z}$$ Multiply by the reciprocal of $\dfrac{2 - z}{6}$.

$$= \dfrac{2 \cdot 3 \cdot (2 - z)}{2 \cdot z \cdot (2 - z)}$$ Factor.

$$= \dfrac{3}{z}$$ Write in simplest form.

2   Next we study a second method for simplifying complex fractions. In this method, we multiply the numerator and the denominator of the complex fraction by the LCD of all fractions in the complex fraction.

> **Method 2: Simplifying a Complex Fraction**
>
> **Step 1.** Find the LCD of all the fractions in the complex fraction.
>
> **Step 2.** Multiply both the numerator and the denominator of the complex fraction by the LCD from Step 1.
>
> **Step 3.** Perform the indicated operations and write the result in simplest form.

We use method 2 to rework Example 2.

EXAMPLE 4

Simplify: $\dfrac{\dfrac{2}{3} + \dfrac{1}{5}}{\dfrac{2}{3} - \dfrac{2}{9}}$

Solution   The LCD of $\frac{2}{3}, \frac{1}{5}, \frac{2}{3}$ and $\frac{2}{9}$ is 45, so we multiply the numerator and the denominator of the complex fraction by 45. Then we perform the indicated operations, and write in simplest form.

$$\dfrac{\dfrac{2}{3} + \dfrac{1}{5}}{\dfrac{2}{3} - \dfrac{2}{9}} = \dfrac{45\left(\dfrac{2}{3} + \dfrac{1}{5}\right)}{45\left(\dfrac{2}{3} - \dfrac{2}{9}\right)}$$

$$= \dfrac{45\left(\dfrac{2}{3}\right) + 45\left(\dfrac{1}{5}\right)}{45\left(\dfrac{2}{3}\right) - 45\left(\dfrac{2}{9}\right)} \qquad \text{Apply the distributive property.}$$

$$= \dfrac{30 + 9}{30 - 10} = \dfrac{39}{20} \qquad \text{Simplify.}$$

TEACHING TIP

For Method 2, remind students that they must multiply the numerator and the denominator of the complex fraction by the **same** number or expression.

> ▶ **Helpful Hint**
>
> The same complex fraction was simplified using two different methods in Examples 2 and 4. Notice that each time the simplified result is the same.

EXAMPLE 5

Simplify: $\dfrac{\dfrac{x+1}{y}}{\dfrac{x}{y} + 2}$

Solution   The LCD of $\dfrac{x+1}{y}, \dfrac{x}{y}$, and $\dfrac{2}{1}$ is y, so we multiply the numerator and the denominator of the complex fraction by y.

$$\dfrac{\dfrac{x+1}{y}}{\dfrac{x}{y} + 2} = \dfrac{y\left(\dfrac{x+1}{y}\right)}{y\left(\dfrac{x}{y} + 2\right)}$$

$$= \dfrac{y\left(\dfrac{x+1}{y}\right)}{y\left(\dfrac{x}{y}\right) + y \cdot 2} \qquad \text{Apply the distributive property in the denominator.}$$

$$= \dfrac{x+1}{x+2y} \qquad \text{Simplify.}$$

EXAMPLE 6

Simplify: $\dfrac{\dfrac{x}{y} + \dfrac{3}{2x}}{\dfrac{x}{2} + y}$

Solution The LCD of $\dfrac{x}{y}, \dfrac{3}{2x}, \dfrac{x}{2}$, and $\dfrac{y}{1}$ is $2xy$, so we multiply both the numerator and the denominator of the complex fraction by $2xy$.

$$\frac{\dfrac{x}{y} + \dfrac{3}{2x}}{\dfrac{x}{2} + y} = \frac{2xy\left(\dfrac{x}{y} + \dfrac{3}{2x}\right)}{2xy\left(\dfrac{x}{2} + y\right)}$$

$$= \frac{2xy\left(\dfrac{x}{y}\right) + 2xy\left(\dfrac{3}{2x}\right)}{2xy\left(\dfrac{x}{2}\right) + 2xy(y)}$$ Apply the distributive property.

$$= \frac{2x^2 + 3y}{x^2 y + 2xy^2}$$

$$\text{or } \frac{2x^2 + 3y}{xy(x + 2y)}$$

MENTAL MATH

Complete the steps by stating the simplified complex fraction.

1. $\dfrac{\dfrac{y}{2}}{\dfrac{5x}{2}} = \dfrac{2\left(\dfrac{y}{2}\right)}{2\left(\dfrac{5x}{2}\right)} = \dfrac{?}{?} \quad \dfrac{y}{5x}$

2. $\dfrac{\dfrac{10}{x}}{\dfrac{z}{x}} = \dfrac{x\left(\dfrac{10}{x}\right)}{x\left(\dfrac{z}{x}\right)} = \dfrac{?}{?} \quad \dfrac{10}{z}$

3. $\dfrac{\dfrac{3}{x}}{\dfrac{5}{x^2}} = \dfrac{x^2\left(\dfrac{3}{x}\right)}{x^2\left(\dfrac{5}{x^2}\right)} = \dfrac{?}{?} \quad \dfrac{3x}{5}$

4. $\dfrac{\dfrac{a}{10}}{\dfrac{b}{20}} = \dfrac{20\left(\dfrac{a}{10}\right)}{20\left(\dfrac{b}{20}\right)} = \dfrac{?}{?} \quad \dfrac{2a}{b}$

EXERCISE SET 7.8

STUDY GUIDE/SSM   CD/VIDEO   PH MATH TUTOR CENTER   MathXL®Tutorials ON CD   MathXL®   MyMathLab®

MIXED PRACTICE

Simplify each complex fraction. See Examples 1 through 6.

1. $\dfrac{\dfrac{1}{2}}{\dfrac{3}{4}} \quad \dfrac{2}{3}$

2. $\dfrac{\dfrac{1}{8}}{-\dfrac{5}{12}} \quad -\dfrac{3}{10}$

7. $\dfrac{\dfrac{1}{2} + \dfrac{2}{3}}{\dfrac{5}{9} - \dfrac{5}{6}} \quad -\dfrac{21}{5}$

8. $\dfrac{\dfrac{3}{4} - \dfrac{1}{2}}{\dfrac{3}{8} + \dfrac{1}{6}} \quad \dfrac{6}{13}$

3. $\dfrac{-\dfrac{4x}{9}}{-\dfrac{2x}{3}} \quad \dfrac{2}{3}$

4. $\dfrac{-\dfrac{6y}{11}}{\dfrac{4y}{9}} \quad -\dfrac{27}{22}$

9. $\dfrac{2 + \dfrac{7}{10}}{1 + \dfrac{3}{5}} \quad \dfrac{27}{16}$

10. $\dfrac{4 - \dfrac{11}{12}}{5 + \dfrac{1}{4}} \quad \dfrac{37}{63}$

5. $\dfrac{\dfrac{1 + x}{6}}{\dfrac{1 + x}{3}} \quad \dfrac{1}{2}$

6. $\dfrac{\dfrac{6x - 3}{5x^2}}{\dfrac{2x - 1}{10x}} \quad \dfrac{6}{x}$

11. $\dfrac{\dfrac{1}{3}}{\dfrac{1}{2} - \dfrac{1}{4}} \quad \dfrac{4}{3}$

12. $\dfrac{\dfrac{7}{10} - \dfrac{3}{5}}{\dfrac{1}{2}} \quad \dfrac{1}{5}$

13. $\dfrac{-\dfrac{2}{9}}{-\dfrac{14}{3}}$ $\dfrac{1}{21}$

14. $\dfrac{\dfrac{3}{8}}{\dfrac{4}{15}}$ $\dfrac{45}{32}$

15. $\dfrac{-\dfrac{5}{12x^2}}{\dfrac{25}{16x^3}}$ $-\dfrac{4x}{15}$

16. $\dfrac{-\dfrac{7}{8y}}{\dfrac{21}{4y}}$ $-\dfrac{1}{6}$

17. $\dfrac{\dfrac{m}{n}-1}{\dfrac{m}{n}+1}$ $\dfrac{m-n}{m+n}$

18. $\dfrac{\dfrac{x}{2}+2}{\dfrac{x}{2}-2}$ $\dfrac{x+4}{x-4}$

🔒 **19.** $\dfrac{\dfrac{1}{5}-\dfrac{1}{x}}{\dfrac{7}{10}+\dfrac{1}{x^2}}$ $\dfrac{2x(x-5)}{7x^2+10}$

20. $\dfrac{\dfrac{1}{y^2}+\dfrac{2}{3}}{\dfrac{1}{y}-\dfrac{5}{6}}$ $\dfrac{6+4y^2}{6y-5y^2}$

21. $\dfrac{1+\dfrac{1}{y-2}}{y+\dfrac{1}{y-2}}$ $\dfrac{1}{y-1}$

22. $\dfrac{x-\dfrac{1}{2x+1}}{1-\dfrac{x}{2x+1}}$ $2x-1$

23. $\dfrac{\dfrac{4y-8}{16}}{\dfrac{6y-12}{4}}$ $\dfrac{1}{6}$

24. $\dfrac{\dfrac{7y+21}{3}}{\dfrac{3y+9}{8}}$ $\dfrac{56}{9}$

🔒 **25.** $\dfrac{\dfrac{x}{y}+1}{\dfrac{x}{y}-1}$ $\dfrac{x+y}{x-y}$

26. $\dfrac{\dfrac{3}{5y}+8}{\dfrac{3}{5y}-8}$ $\dfrac{3+40y}{3-40y}$

27. $\dfrac{1}{2+\dfrac{1}{3}}$ $\dfrac{3}{7}$

28. $\dfrac{3}{1-\dfrac{4}{3}}$ -9

29. $\dfrac{\dfrac{ax+ab}{x^2-b^2}}{\dfrac{x+b}{x-b}}$ $\dfrac{a}{x+b}$

30. $\dfrac{\dfrac{m+2}{m-2}}{\dfrac{2m+4}{m^2-4}}$ $\dfrac{m+2}{2}$

31. $\dfrac{\dfrac{-3+y}{4}}{\dfrac{8+y}{28}}$ $\dfrac{7(y-3)}{8+y}$

32. $\dfrac{\dfrac{-x+2}{18}}{\dfrac{8}{9}}$ $\dfrac{-x+2}{16}$

33. $\dfrac{3+\dfrac{12}{x}}{1-\dfrac{16}{x^2}}$ $\dfrac{3x}{x-4}$

34. $\dfrac{2+\dfrac{6}{x}}{1-\dfrac{9}{x^2}}$ $\dfrac{2x}{x-3}$

35. $\dfrac{\dfrac{8}{x+4}+2}{\dfrac{12}{x+4}-2}$ $-\dfrac{x+8}{x-2}$

36. $\dfrac{\dfrac{25}{x+5}+5}{\dfrac{3}{x+5}-5}$ $-\dfrac{5x+50}{5x+22}$

37. $\dfrac{\dfrac{s}{r}+\dfrac{r}{s}}{\dfrac{s}{r}-\dfrac{r}{s}}$ $\dfrac{s^2+r^2}{s^2-r^2}$

38. $\dfrac{\dfrac{2}{x}+\dfrac{x}{2}}{\dfrac{2}{x}-\dfrac{x}{2}}$ $\dfrac{4+x^2}{4-x^2}$

39. Explain how to simplify a complex fraction using method 1.

40. Explain how to simplify a complex fraction using method 2.

REVIEW AND PREVIEW   **39–40.** answers may vary

Simplify.

41. $\sqrt{81}$ 9   **42.** $\sqrt{16}$ 4   **43.** $\sqrt{1}$ 1   **44.** $\sqrt{0}$ 0

45. $\sqrt{\dfrac{1}{25}}$ $\dfrac{1}{5}$ **46.** $\sqrt{\dfrac{1}{49}}$ $\dfrac{1}{7}$ **47.** $\sqrt{\dfrac{4}{9}}$ $\dfrac{2}{3}$ **48.** $\sqrt{\dfrac{121}{100}}$ $\dfrac{11}{10}$

Concept Extensions

To find the average of two numbers, we find their sum and divide by 2. For example, the average of 65 and 81 is found by simplifying $\dfrac{65+81}{2}$. This simplifies to $\dfrac{146}{2}=73$.

49. Find the average of $\dfrac{1}{3}$ and $\dfrac{3}{4}$. $\dfrac{13}{24}$

50. Write the average of $\dfrac{3}{n}$ and $\dfrac{5}{n^2}$ as a simplified rational expression.

Solve.

51. In electronics, when two resistors R_1 (read R sub 1) and R_2 (read R sub 2) are connected in parallel, the total resistance is given by the complex fraction $\left(\dfrac{1}{\dfrac{1}{R_1}+\dfrac{1}{R_2}}\right)$ Simplify this expression.

50. $\dfrac{3n+5}{2n^2}$

51. $\dfrac{R_1 R_2}{R_2+R_1}$

Resistance R_1 R_2

52. Astronomers occasionally need to know the day of the week a particular date fell on. The complex fraction

$$\dfrac{J+\dfrac{3}{2}}{7},$$

where J is the *Julian day number*, is used to make this calculation. Simplify this expression. $\dfrac{2J+3}{14}$

53. If the distance formula $d=r\cdot t$ is solved for t, then $t=\dfrac{d}{r}$. Use this formula to find t if distance d is $\dfrac{20x}{3}$ miles and rate r is $\dfrac{5x}{9}$ miles per hour. Write t in simplified form.   12 hr

△ **54.** If the formula for area of a rectangle, $A=l\cdot w$, is solved for w, then $w=\dfrac{A}{l}$. Use this formula to find w if area A is $\dfrac{4x-2}{3}$ square meters and length l is $\dfrac{6x-3}{5}$ meters. Write w in simplified form. $\dfrac{10}{9}$ m

Simplify. First write each expression without exponents. Then simplify the complex fraction. The first step has been completed for Exercise 55.

55. $\dfrac{x^{-1}+2^{-1}}{x^{-2}-4^{-1}}=\dfrac{\dfrac{1}{x}+\dfrac{1}{2}}{\dfrac{1}{x^2}-\dfrac{1}{4}}$ $\dfrac{2x}{2-x}$ **56.** $\dfrac{3^{-1}-x^{-1}}{9^{-1}-x^{-2}}$ $\dfrac{3x}{x+3}$

57. $\dfrac{\dfrac{x + y^{-1}}{x}}{y} \quad \dfrac{xy + 1}{x}$

58. $\dfrac{\dfrac{x - xy^{-1}}{1 + x}}{y} \quad \dfrac{xy - x}{1 + x}$

59. $\dfrac{y^{-2}}{1 - y^{-2}} \quad \dfrac{1}{y^2 - 1}$

60. $\dfrac{4 + x^{-1}}{3 + x^{-1}} \quad \dfrac{4x + 1}{3x + 1}$

CHAPTER 7 PROJECT

Comparing Dosage Formulas

In this project, you will have the opportunity to investigate two well-known formulas for predicting the correct doses of medication for children. This project may be completed by working in groups or individually.

Young's Rule and Cowling's Rule are dose formulas for prescribing medicines to children. Unlike formulas for, say area or distance, these dose formulas describe only an approximate relationship. The formulas relate a child's age A in years and an adult dose D of medication to the proper child's dose C. The formulas are most accurate when applied to children between the ages of 2 and 13.

$$\text{Young's Rule:} \quad C = \frac{DA}{A + 12}$$

$$\text{Cowling's Rule:} \quad C = \frac{D(A + 1)}{24}$$

1. Let the adult dose $D = 1000$ mg. Complete the Young's Rule and Cowling's Rule columns of the following table comparing the doses predicted by both formulas for ages 2 through 13.

Age A	Young's Rule	Cowling's Rule	Difference
2			
3			
4			
5			
6			
7			
8			
9			
10			
11			
12			
13			

2. Use the data from the table in Question 1 to form sets of ordered pairs of the form (age, child's dose) for each formula. Graph the ordered pairs for each formula on the same graph. Describe the shapes of the graphed data.

3. Use your table, graph, or both, to decide whether either formula will consistently predict a larger dose than the other. If so, which one? If not, is there an age at which the doses predicted by one becomes greater than the doses predicted by the other? If so, estimate that age.

4. Use your graph to estimate for what age the difference in the two predicted doses is greatest.

5. Return to the table in Question 1 and complete the last column, titled "Difference," by finding the absolute value of the difference between the Young's dose and the Cowling's dose for each age. Use this column in the table to verify your graphical estimate found in Question 4.

6. Does Cowling's Rule ever predict exactly the adult dose? If so, at what age? Explain. Does Young's Rule ever predict exactly the adult dose? If so, at what age? Explain.

7. Many doctors prefer to use formulas that relate doses to factors other than a child's age. Why is age not necessarily the most important factor when predicting a child's dose? What other factors might be used?

7 CHAPTER VOCABULARY CHECK

Fill in each blank with one of the words or phrases listed below.

rational expression	complex fraction	ratio	proportion
cross products	direct variation	inverse variation	

1. A <u>ratio</u> is the quotient of two numbers.

2. $\dfrac{x}{2} = \dfrac{7}{16}$ is an example of a <u>proportion</u>.

3. If $\dfrac{a}{b} = \dfrac{c}{d}$, then ad and bc are called <u>cross products</u>.

4. A <u>rational expression</u> is an expression that can be written in the form $\dfrac{P}{Q}$, where P and Q are polynomials and Q is not 0.

5. In a <u>complex fraction</u>, the numerator or denominator or both may contain fractions.

6. The equation $y = \dfrac{k}{x}$ is an example of <u>inverse variation</u>.

7. The equation $y = kx$ is an example of <u>direct variation</u>.

CHAPTER 7 HIGHLIGHTS

Definitions and Concepts	Examples

Section 7.1 Simplifying Rational Expressions

A **rational expression** is an expression that can be written in the form $\frac{P}{Q}$, where P and Q are polynomials and Q does not equal 0.

Rational Expressions
$$\frac{7y^3}{4}, \quad \frac{x^2 + 6x + 1}{x - 3}, \quad \frac{-5}{s^3 + 8}$$

To find values for which a rational expression is undefined, find values for which the denominator is 0.

Find any values for which the expression $\dfrac{5y}{y^2 - 4y + 3}$ is undefined.

$$
\begin{aligned}
y^2 - 4y + 3 &= 0 && \text{Set denominator equal to 0.} \\
(y - 3)(y - 1) &= 0 && \text{Factor.} \\
y - 3 = 0 \text{ or } y - 1 &= 0 && \text{Set each factor equal to 0.} \\
y = 3 \text{ or } y &= 1 && \text{Solve.}
\end{aligned}
$$

The expression is undefined when y is 3 and when y is 1.

Fundamental Principle of Rational Expressions
If P and Q are polynomials, and Q and R are not 0, then
$$\frac{PR}{QR} = \frac{P}{Q}$$

By the fundamental principle,
$$\frac{(x - 3)(x + 1)}{x(x + 1)} = \frac{x - 3}{x}$$

as long as $x \neq 0$ and $x \neq -1$.

To simplify a rational expression,

Step 1. Factor the numerator and denominator.

Step 2. Apply the fundamental principle to divide out common factors.

Simplify: $\dfrac{4x + 20}{x^2 - 25}$
$$\frac{4x + 20}{x^2 - 25} = \frac{4(x + 5)}{(x + 5)(x - 5)} = \frac{4}{x - 5}$$

Definitions and Concepts	**Examples**

Section 7.2 Multiplying and Dividing Rational Expressions

To multiply rational expressions,

Step 1. Factor numerators and denominators.

Step 2. Multiply numerators and multiply denominators.

Step 3. Write the product in simplest form.

$$\frac{P}{Q} \cdot \frac{R}{S} = \frac{PR}{QS}$$

Multiply: $\dfrac{4x + 4}{2x - 3} \cdot \dfrac{2x^2 + x - 6}{x^2 - 1}$

$$\frac{4x + 4}{2x - 3} \cdot \frac{2x^2 + x - 6}{x^2 - 1} = \frac{4(x + 1)}{2x - 3} \cdot \frac{(2x - 3)(x + 2)}{(x + 1)(x - 1)}$$

$$= \frac{4(x + 1)(2x - 3)(x + 2)}{(2x - 3)(x + 1)(x - 1)}$$

$$= \frac{4(x + 2)}{x - 1}$$

To divide by a rational expression, multiply by the reciprocal.

$$\frac{P}{Q} \div \frac{R}{S} = \frac{P}{Q} \cdot \frac{S}{R} = \frac{PS}{QR}$$

Divide: $\dfrac{15x + 5}{3x^2 - 14x - 5} \div \dfrac{15}{3x - 12}$

$$\frac{15x + 5}{3x^2 - 14x - 5} \div \frac{15}{3x - 12} = \frac{5(3x + 1)}{(3x + 1)(x - 5)} \cdot \frac{3(x - 4)}{3 \cdot 5}$$

$$= \frac{x - 4}{x - 5}$$

Section 7.3 Adding and Subtracting Rational Expressions with Common Denominators and Least Common Denominator

To add or subtract rational expressions with the same denominator, add or subtract numerators, and place the sum or difference over a common denominator.

$$\frac{P}{R} + \frac{Q}{R} = \frac{P + Q}{R}$$

$$\frac{P}{R} - \frac{Q}{R} = \frac{P - Q}{R}$$

Perform indicated operations.

$$\frac{5}{x + 1} + \frac{x}{x + 1} = \frac{5 + x}{x + 1}$$

$$\frac{2y + 7}{y^2 - 9} - \frac{y + 4}{y^2 - 9} = \frac{(2y + 7) - (y + 4)}{y^2 - 9}$$

$$= \frac{2y + 7 - y - 4}{y^2 - 9}$$

$$= \frac{y + 3}{(y + 3)(y - 3)}$$

$$= \frac{1}{y - 3}$$

To find the least common denominator (LCD),

Step 1. Factor the denominators.

Step 2. The LCD is the product of all unique factors, each raised to a power equal to the greatest number of times that it appears in any one factored denominator.

Find the LCD for

$$\frac{7x}{x^2 + 10x + 25} \text{ and } \frac{11}{3x^2 + 15x}$$

$$x^2 + 10x + 25 = (x + 5)(x + 5)$$

$$3x^2 + 15x = 3x(x + 5)$$

LCD is $3x(x + 5)(x + 5)$ or $3x(x + 5)^2$

Definitions and Concepts	**Examples**

Section 7.4 Adding and Subtracting Rational Expressions with Unlike Denominators

To add or subtract rational expressions with unlike denominators,

Step 1. Find the LCD.

Step 2. Rewrite each rational expression as an equivalent expression whose denominator is the LCD.

Step 3. Add or subtract numerators and place the sum or difference over the common denominator.

Step 4. Write the result in simplest form.

Perform the indicated operation.

$$\frac{9x + 3}{x^2 - 9} - \frac{5}{x - 3}$$

$$= \frac{9x + 3}{(x + 3)(x - 3)} - \frac{5}{x - 3}$$

LCD is $(x + 3)(x - 3)$.

$$= \frac{9x + 3}{(x + 3)(x - 3)} - \frac{5(x + 3)}{(x - 3)(x + 3)}$$

$$= \frac{9x + 3 - 5(x + 3)}{(x + 3)(x - 3)}$$

$$= \frac{9x + 3 - 5x - 15}{(x + 3)(x - 3)}$$

$$= \frac{4x - 12}{(x + 3)(x - 3)}$$

$$= \frac{4(x - 3)}{(x + 3)(x - 3)} = \frac{4}{x + 3}$$

Section 7.5 Solving Equations Containing Rational Expressions

To solve an equation containing rational expressions,

Step 1. Multiply both sides of the equation by the LCD of all rational expressions in the equation.

Step 2. Remove any grouping symbols and solve the resulting equation.

Step 3. Check the solution in the original equation.

Solve: $\dfrac{5x}{x + 2} + 3 = \dfrac{4x - 6}{x + 2}$

$$(x + 2)\left(\frac{5x}{x + 2} + 3\right) = (x + 2)\left(\frac{4x - 6}{x + 2}\right)$$

$$(x + 2)\left(\frac{5x}{x + 2}\right) + (x + 2)(3) = (x + 2)\left(\frac{4x - 6}{x + 2}\right)$$

$$5x + 3x + 6 = 4x - 6$$

$$4x = -12$$

$$x = -3$$

The solution checks and the solution is -3.

Section 7.6 Proportion and Problem Solving with Rational Equations

A **ratio** is the quotient of two numbers or two quantities. A **proportion** is a mathematical statement that two ratios are equal.

Cross products:

$$\text{If } \frac{a}{b} = \frac{c}{d}, \text{ then } ad = bc.$$

Proportions

$$\frac{2}{3} = \frac{8}{12} \qquad \frac{x}{7} = \frac{15}{35}$$

Cross Products

$$\frac{2}{3} \diagdown\!\!\!\!\diagup \frac{8}{12} \quad \begin{array}{l} \rightarrow \quad 3 \cdot 8 \text{ or } 24 \\ \rightarrow \quad 2 \cdot 12 \text{ or } 24 \end{array}$$

Definitions and Concepts	Examples

Section 7.6 Proportion and Problem Solving with Rational Equations

Solve: $\dfrac{3}{4} = \dfrac{x}{x-1}$

$\dfrac{3}{4} \diagdown \dfrac{x}{x-1}$

$3(x-1) = 4x$ *Set cross products equal.*

$3x - 3 = 4x$

$-3 = x$

Problem-Solving Steps

1. UNDERSTAND. Read and reread the problem.

A small plane and a car leave Kansas City, Missouri, and head for Minneapolis, Minnesota, a distance of 450 miles. The speed of the plane is 3 times the speed of the car, and the plane arrives 6 hours ahead of the car. Find the speed of the car.

Let x = the speed of the car.
Then $3x$ = the speed of the plane.

	Distance	=	Rate	·	Time
Car	450		x		$\dfrac{450}{x} \left(\dfrac{\text{distance}}{\text{rate}}\right)$
Plane	450		$3x$		$\dfrac{450}{3x} \left(\dfrac{\text{distance}}{\text{rate}}\right)$

In words:

plane's time	+	6 hours	=	car's time
↓		↓		↓

2. TRANSLATE.

Translate: $\dfrac{450}{3x}$ $+$ 6 $=$ $\dfrac{450}{x}$

$\dfrac{450}{3x} + 6 = \dfrac{450}{x}$

3. SOLVE.

$3x\left(\dfrac{450}{3x}\right) + 3x(6) = 3x\left(\dfrac{450}{x}\right)$

$450 + 18x = 1350$

$18x = 900$

$x = 50$

4. INTERPRET.

Check the solution by replacing x with 50 in the original equation. *State* the conclusion: The speed of the car is 50 miles per hour.

Section 7.7 Direct and Inverse Variation and Problem Solving

y **varies directly as** x, or y is **directly proportional to** x, if there is a nonzero constant k such that
$$y = kx$$

y **varies inversely as** x, or y is **inversely proportional to** x, if there is a nonzero constant k such that
$$y = \dfrac{k}{x}$$

The circumference of a circle C varies directly as its radius r.
$$C = \underbrace{2\pi}_{k}\, r$$

Pressure P varies inversely with volume V.

$$P = \dfrac{k}{V}$$

Definitions and Concepts	**Examples**

Section 7.8 Simplifying Complex Fractions

Simplify.

Method 1: To Simplify a Complex Fraction

Step 1. Add or subtract fractions in the numerator and the denominator of the complex fraction.

Step 2. Perform the indicated division.

Step 3. Write the result in simplest form.

$$\frac{\dfrac{1}{x} + 2}{\dfrac{1}{x} - \dfrac{1}{y}} = \frac{\dfrac{1}{x} + \dfrac{2x}{x}}{\dfrac{y}{xy} - \dfrac{x}{xy}}$$

$$= \frac{\dfrac{1 + 2x}{x}}{\dfrac{y - x}{xy}}$$

$$= \frac{1 + 2x}{x} \cdot \frac{xy}{y - x}$$

$$= \frac{y(1 + 2x)}{y - x}$$

Method 2: To Simplify a Complex Fraction

Step 1. Find the LCD of all fractions in the complex fraction.

Step 2. Multiply the numerator and the denominator of the complex fraction by the LCD.

Step 3. Perform indicated operations and write in simplest form.

$$\frac{\dfrac{1}{x} + 2}{\dfrac{1}{x} - \dfrac{1}{y}} = \frac{xy\left(\dfrac{1}{x} + 2\right)}{xy\left(\dfrac{1}{x} - \dfrac{1}{y}\right)}$$

$$= \frac{xy\left(\dfrac{1}{x}\right) + xy(2)}{xy\left(\dfrac{1}{x}\right) - xy\left(\dfrac{1}{y}\right)}$$

$$= \frac{y + 2xy}{y - x} \quad \text{or} \quad \frac{y(1 + 2x)}{y - x}$$

2. $x = \frac{5}{2}, x = -\frac{3}{2}$ **8.** $\dfrac{5(x - 5)}{x - 3}$ **12.** $\dfrac{x + a}{x - c}$

7 CHAPTER REVIEW

(7.1) Find any real number for which each rational expression is undefined.

1. $\dfrac{x + 5}{x^2 - 4}$ $x = 2, x = -2$

2. $\dfrac{5x + 9}{4x^2 - 4x - 15}$

Find the value of each rational expression when $x = 5$, $y = 7$, and $z = -2$.

3. $\dfrac{z^2 - z}{z + xy}$ $\dfrac{2}{11}$

4. $\dfrac{x^2 + xy - z^2}{x + y + z}$ $\dfrac{28}{5}$

Simplify each rational expression.

5. $\dfrac{x + 2}{x^2 - 3x - 10}$ $\dfrac{1}{x - 5}$

6. $\dfrac{x + 4}{x^2 + 5x + 4}$ $\dfrac{1}{x + 1}$

7. $\dfrac{x^3 - 4x}{x^2 + 3x + 2}$ $\dfrac{x(x - 2)}{x + 1}$

8. $\dfrac{5x^2 - 125}{x^2 + 2x - 15}$

9. $\dfrac{x^2 - x - 6}{x^2 - 3x - 10}$ $\dfrac{x - 3}{x - 5}$

10. $\dfrac{x^2 - 2x}{x^2 + 2x - 8}$ $\dfrac{x}{x + 4}$

11. $\dfrac{x^2 + 6x + 5}{2x^2 + 11x + 5}$ $\dfrac{x + 1}{2x + 1}$

12. $\dfrac{x^2 + xa + xb + ab}{x^2 - xc + bx - bc}$

13. $\dfrac{x^2 + 5x - 2x - 10}{x^2 - 3x - 2x + 6}$ $\dfrac{x + 5}{x - 3}$

14. $\dfrac{x^2 - 9}{9 - x^2}$ -1

15. $\dfrac{4 - x}{x^3 - 64}$ $-\dfrac{1}{x^2 + 4x + 16}$

(7.2) Perform the indicated operations and simplify.

16. $\dfrac{15x^3y^2}{z} \cdot \dfrac{z}{5xy^3}$ $\dfrac{3x^2}{y}$

17. $\dfrac{-y^3}{8} \cdot \dfrac{9x^2}{y^3}$ $-\dfrac{9x^2}{8}$

18. $\dfrac{x^2-9}{x^2-4} \cdot \dfrac{x-2}{x+3}$ $\dfrac{x-3}{x+2}$

19. $\dfrac{2x+5}{x-6} \cdot \dfrac{2x}{-x+6}$

20. $\dfrac{x^2-5x-24}{x^2-x-12} \div \dfrac{x^2-10x+16}{x^2+x-6}$ $\dfrac{x+3}{x-4}$

21. $\dfrac{4x+4y}{xy^2} \div \dfrac{3x+3y}{x^2y}$ $\dfrac{4x}{3y}$

22. $\dfrac{x^2+x-42}{x-3} \cdot \dfrac{(x-3)^2}{x+7}$

  22. $(x-6)(x-3)$

23. $\dfrac{2a+2b}{3} \cdot \dfrac{a-b}{a^2-b^2}$ $\dfrac{2}{3}$

24. $\dfrac{x^2-9x+14}{x^2-5x+6} \cdot \dfrac{x+2}{x^2-5x-14}$

25. $(x-3) \cdot \dfrac{x}{x^2+3x-18}$

  25. $\dfrac{x}{x+6}$

26. $\dfrac{2x^2-9x+9}{8x-12} \div \dfrac{x^2-3x}{2x}$ $\dfrac{1}{2}$

27. $\dfrac{x^2-y^2}{x^2+xy} \div \dfrac{3x^2-2xy-y^2}{3x^2+6x}$ $\dfrac{3(x+2)}{3x+y}$

28. $\dfrac{x^2-y^2}{8x^2-16xy+8y^2} \div \dfrac{x+y}{4x-y}$ $\dfrac{4x-y}{8(x-y)}$

29. $\dfrac{x-y}{4} \div \dfrac{y^2-2y-xy+2x}{16x+24}$

30. $\dfrac{y-3}{4x+3} \div \dfrac{9-y^2}{4x^2-x-3}$

  30. $-\dfrac{x-1}{y+3}$

(7.3) Perform the indicated operations and simplify.

31. $\dfrac{5x-4}{3x-1} + \dfrac{6}{3x-1}$ $\dfrac{5x+2}{3x-1}$

32. $\dfrac{4x-5}{3x^2} - \dfrac{2x+5}{3x^2}$

33. $\dfrac{9x+7}{6x^2} - \dfrac{3x+4}{6x^2}$ $\dfrac{2x+1}{2x^2}$

Find the LCD of each pair of rational expressions.

34. $\dfrac{x+4}{2x}, \dfrac{3}{7x}$ $14x$

35. $\dfrac{x-2}{x^2-5x-24}, \dfrac{3}{x^2+11x+24}$

  $(x-8)(x+8)(x+3)$

Rewrite the following rational expressions as equivalent expressions whose denominator is the given polynomial.

36. $\dfrac{x+2}{x^2+11x+18}, (x+2)(x-5)(x+9)$

37. $\dfrac{3x-5}{x^2+4x+4}, (x+2)^2(x+3)$ $\dfrac{3x^2+4x-15}{(x+2)^2(x+3)}$

(7.4) Perform the indicated operations and simplify.

38. $\dfrac{4}{5x^2} - \dfrac{6}{y}$ $\dfrac{4y-30x^2}{5x^2y}$

39. $\dfrac{2}{x-3} - \dfrac{4}{x-1}$

40. $\dfrac{x+7}{x+3} - \dfrac{x-3}{x+7}$

41. $\dfrac{4}{x+3} - 2$ $\dfrac{-2x-2}{x+3}$

42. $\dfrac{3}{x^2+2x-8} + \dfrac{2}{x^2-3x+2}$

43. $\dfrac{2x-5}{6x+9} - \dfrac{4}{2x^2+3x}$

44. $\dfrac{x-1}{x^2-2x+1} - \dfrac{x+1}{x-1}$

45. $\dfrac{x-1}{x^2+4x+4} + \dfrac{x-1}{x+2}$

Find the perimeter and the area of each figure.

△ **46.**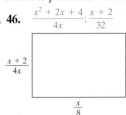

△ **47.**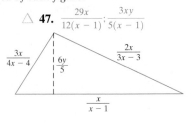

(7.5) Solve each equation for the variable or perform the indicated operation.   **51.** $-4, 3$ **53.** no solution

48. $\dfrac{x+4}{9} = \dfrac{5}{9}$ 1

49. $\dfrac{n}{10} = 9 - \dfrac{n}{5}$ 30

50. $\dfrac{5y-3}{7} = \dfrac{15y-2}{28}$ 2

51. $\dfrac{2}{x+1} - \dfrac{1}{x-2} = -\dfrac{1}{2}$

52. $\dfrac{1}{a+3} + \dfrac{1}{a-3} = -\dfrac{5}{a^2-9}$ $-\dfrac{5}{2}$

53. $\dfrac{y}{2y+2} + \dfrac{2y-16}{4y+4} = \dfrac{y-3}{y+1}$

54. $\dfrac{4}{x+3} + \dfrac{8}{x^2-9} = 0$ 1

55. $\dfrac{2}{x-3} - \dfrac{4}{x+3} = \dfrac{8}{x^2-9}$ 5

56. $\dfrac{x-3}{x+1} - \dfrac{x-6}{x+5} = 0$ $\dfrac{9}{7}$

57. $x + 5 = \dfrac{6}{x}$ $-6, 1$

Solve the equation for the indicated variable.   **59.** $y = \dfrac{560-8x}{7}$

58. $\dfrac{4A}{5b} = x^2$, for b $b = \dfrac{4A}{5x^2}$

59. $\dfrac{x}{7} + \dfrac{y}{8} = 10$, for y

(7.6) Solve each proportion.

60. $\dfrac{x}{2} = \dfrac{12}{4}$ $x = 6$

61. $\dfrac{20}{1} = \dfrac{x}{25}$ $x = 500$

62. $\dfrac{2}{x-1} = \dfrac{3}{x+3}$ $x = 9$

63. $\dfrac{4}{y-3} = \dfrac{2}{y-3}$
  no solution

Solve.

64. A machine can process 300 parts in 20 minutes. Find how many parts can be processed in 45 minutes.   675 parts

65. As his consulting fee, Mr. Visconti charges $90.00 per day. Find how much he charges for 3 hours of consulting. Assume an 8-hour work day. $33.75

66. One fundraiser can address 100 letters in 35 minutes. Find how many he can address in 55 minutes. $157\frac{1}{7}$ or 157 letters

67. Five times the reciprocal of a number equals the sum of $\frac{3}{2}$ times the reciprocal of the number and $\frac{7}{6}$. What is the number?   3

68. The reciprocal of a number equals the reciprocal of the difference of 4 and the number. Find the number.   2

69. A car travels 90 miles in the same time that a car traveling 10 miles per hour slower travels 60 miles. Find the speed of each car.   30 mph; 20 mph

70. The speed of a bayou near Lafayette, Louisiana, is 4 miles per hour. A paddleboat travels 48 miles upstream in the same amount of time it takes to travel 72 miles downstream. Find the speed of the boat in still water.   20 mph

71. When Mark and Maria manicure Mr. Stergeon's lawn, it takes them 5 hours. If Mark works alone, it takes 7 hours. Find how long it takes Maria alone. $17\frac{1}{2}$ hr

72. It takes pipe A 20 days to fill a fish pond. Pipe B takes 15 days. Find how long it takes both pipes together to fill the pond. $8\frac{4}{7}$ days

Given that the pairs of triangles are similar, find each missing length x.

△ **73.** $x = 20$

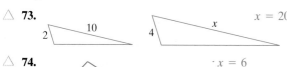

△ **74.** $x = 6$

△ **75.** $x = 15$

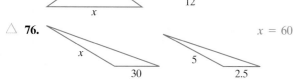

△ **76.** $x = 60$

(7.7) Solve.

77. y varies directly as x. If $y = 40$ when $x = 4$, find y when x is 11. $y = 110$

78. y varies inversely as x. If $y = 4$ when $x = 6$, find y when x is 48. $y = \frac{1}{2}$

79. y varies inversely as x^3. If $y = 12.5$ when $x = 2$, find y when x is 3. $y = \frac{100}{27}$

80. y varies directly as x^2. If $y = 175$ when $x = 5$, find y when $x = 10$. $y = 700$

81. The cost of manufacturing a certain medicine varies inversely as the amount of medicine manufactured increases. If 3000 milliliters can be manufactured for $6600, find the cost to manufacture 5000 milliliters. $3960

82. The distance a spring stretches varies directly with the weight attached to the spring. If a 150-pound weight stretches the spring 8 inches, find the distance that a 90-pound weight stretches the spring. $4\frac{4}{5}$ inches

(7.8) Simplify each complex fraction.

83. $\dfrac{\dfrac{5x}{27}}{\dfrac{10xy}{21}} - \dfrac{7}{18y}$

84. $\dfrac{\dfrac{8x}{x^2 - 9}}{\dfrac{4}{x + 3}} \cdot \dfrac{2x}{x - 3}$

85. $\dfrac{\dfrac{3}{5} + \dfrac{2}{7}}{\dfrac{1}{5} + \dfrac{5}{6}}$ $\dfrac{6}{7}$

86. $\dfrac{\dfrac{2}{a} + \dfrac{1}{2a}}{a + \dfrac{a}{2}}$ $\dfrac{5}{3a^2}$

87. $\dfrac{3 - \dfrac{1}{y}}{2 - \dfrac{1}{y}}$ $\dfrac{3y - 1}{2y - 1}$

88. $\dfrac{2 + \dfrac{1}{x^2}}{\dfrac{1}{x} + \dfrac{2}{x^2}}$ $\dfrac{2x^2 + 1}{x + 2}$

89. $\dfrac{\dfrac{1}{a} + \dfrac{1}{b}}{\dfrac{1}{ab}}$ $b + a$

90. $\dfrac{\dfrac{6}{x + 2} + 4}{\dfrac{8}{x + 2} - 4} \cdot \dfrac{7 + 2x}{2x}$

19. $-\dfrac{2x(2x + 5)}{(x - 6)^2}$ **24.** $\dfrac{1}{x - 3}$

29. $-\dfrac{2(2x + 3)}{y - 2}$ **32.** $\dfrac{2x - 10}{3x^2}$

36. $\dfrac{x^2 - 3x - 10}{(x + 2)(x - 5)(x + 9)}$

39. $\dfrac{-2x + 10}{(x - 3)(x - 1)}$

40. $\dfrac{14x + 58}{(x + 3)(x + 7)}$

42. $\dfrac{5x + 5}{(x + 4)(x - 2)(x - 1)}$

43. $\dfrac{x - 4}{3x}$

44. $-\dfrac{x}{x - 1}$ **45.** $\dfrac{x^2 + 2x - 3}{(x + 2)^2}$

STUDY SKILLS REMINDER

Are You Preparing for a Test on Chapter 7?

Below I have listed *a common trouble* area for topics covered in Chapter 7. After studying for your test, but before taking your test, read this.

Do you know the differences between how to perform operations such as $\frac{4}{x} + \frac{2}{3}$ or $\frac{4}{x} \div \frac{2}{x}$ and how to solve an equation such as $\frac{4}{x} + \frac{2}{3} = 1$?

$$\frac{\mathbf{4}}{\mathbf{x}} \uparrow + \frac{\mathbf{2}}{\mathbf{3}} = \frac{4 \cdot 3}{x \cdot 3} + \frac{2 \cdot x}{3 \cdot x} = \frac{12}{3x} + \frac{2x}{3x} = \frac{12 + 2x}{3x} \text{ or } \frac{2(6 + x)}{3x}, \text{ the sum.}$$

Addition—write each expression as an equivalent expression with the same LCD denominator.

$$\frac{\mathbf{4}}{\mathbf{x}} \uparrow \div \frac{\mathbf{2}}{\mathbf{x}} = \frac{4}{x} \cdot \frac{x}{2} = \frac{4 \cdot x}{x \cdot 2} = \frac{4}{2} = 2, \text{ the quotient.}$$

Division—multiply the first rational expression by the reciprocal of the second.

Solve for x: $\dfrac{4}{x} + \dfrac{2}{3} = 1$ Equation to be solved.

$3x\left(\dfrac{4}{x} + \dfrac{2}{3}\right) = 3x \cdot 1$ Multiply both sides of the equation by the LCD **3x**.

$3x\left(\dfrac{4}{x}\right) + 3x\left(\dfrac{2}{3}\right) = 3x \cdot 1$ Use the distributive property.

$12 + 2x = 3x$ Multiply and simplify.

$12 = x$ Subtract **2x** from both sides.

The solution is 12. Check the solution in the original equation.

Remember: This is simply a sampling of selected topics given to check your understanding. For a review of Chapter 7 in your text, see the material at the end of this chapter.

4. $\dfrac{1}{x+6}$ **6.** $\dfrac{2m(m+2)}{m-2}$ **10.** $\dfrac{y-2}{4}$ **12.** $\dfrac{3a-4}{(a-3)(a+2)}$ **18.** no solution   **19.** $-2, 5$

CHAPTER 7 TEST

Remember to use your Chapter Test Prep Video CD to help you study and view solutions to the test questions you need help with.

1. Find any real numbers for which the following expression is undefined.

$$\frac{x+5}{x^2+4x+3} \qquad x = -1, x = -3$$

2. For a certain computer desk, the average cost C (in dollars) per desk manufactured is

$$C = \frac{100x + 3000}{x}$$

where x is the number of desks manufactured.

  a. Find the average cost per desk when manufacturing 200 computer desks. $115

  b. Find the average cost per desk when manufacturing 1000 computer desks. $103

Simplify each rational expression.

3. $\dfrac{3x-6}{5x-10}$ $\dfrac{3}{5}$

4. $\dfrac{x+6}{x^2+12x+36}$

5. $\dfrac{x+3}{x^3+27} \cdot \dfrac{1}{x^2-3x+9}$

6. $\dfrac{2m^3-2m^2-12m}{m^2-5m+6}$

7. $\dfrac{ay+3a+2y+6}{ay+3a+5y+15} \qquad \dfrac{a+2}{a+5}$

8. $\dfrac{y-x}{x^2-y^2} - \dfrac{1}{x+y}$

Perform the indicated operation and simplify if possible.

9. $\dfrac{3}{x-1} \cdot (5x-5) \quad 15$

10. $\dfrac{y^2-5y+6}{2y+4} \cdot \dfrac{y+2}{2y-6}$

11. $\dfrac{5}{2x+5} - \dfrac{6}{2x+5} \quad -\dfrac{1}{2x+5}$

12. $\dfrac{5a}{a^2-a-6} - \dfrac{2}{a-3}$

13. $\dfrac{6}{x^2-1} + \dfrac{3}{x+1} \quad \dfrac{3}{x-1}$

14. $\dfrac{x^2-9}{x^2-3x} \div \dfrac{xy+5x+3y+15}{2x+10} \quad \dfrac{2(x+5)}{x(y+5)}$

15. $\dfrac{x+2}{x^2+11x+18} + \dfrac{5}{x^2-3x-10} \quad \dfrac{x^2+2x+35}{(x+9)(x+2)(x-5)}$

Solve each equation.

16. $\dfrac{4}{y} - \dfrac{5}{3} = \dfrac{-1}{5} \quad \dfrac{30}{11}$

17. $\dfrac{5}{y+1} = \dfrac{4}{y+2} \quad -6$

18. $\dfrac{a}{a-3} = \dfrac{3}{a-3} - \dfrac{3}{2}$

19. $x - \dfrac{14}{x-1} = 4 - \dfrac{2x}{x-1}$

Simplify each complex fraction.

20. $\dfrac{\dfrac{5x^2}{yz^2}}{\dfrac{10x}{z^3}} \quad \dfrac{xz}{2y}$

21. $\dfrac{5 - \dfrac{1}{y^2}}{\dfrac{1}{y} + \dfrac{2}{y^2}} \quad \dfrac{5y^2-1}{y+2}$

22. y varies directly as x. If $y = 10$ when $x = 15$, find y when x is 42.   28

23. y varies inversely as x^2. If $y = 8$ when $x = 5$, find y when x is 15. $\frac{8}{9}$

24. In a sample of 85 fluorescent bulbs, 3 were found to be defective. At this rate, how many defective bulbs should be found in 510 bulbs?   18 bulbs

25. One number plus five times its reciprocal is equal to six. Find the number.   5 or 1

26. A pleasure boat traveling down the Red River takes the same time to go 14 miles upstream as it takes to go 16 miles downstream. If the current of the river is 2 miles per hour, find the speed of the boat in still water.   30 mph

27. An inlet pipe can fill a tank in 12 hours. A second pipe can fill the tank in 15 hours. If both pipes are used, find how long it takes to fill the tank. $6\frac{2}{3}$ hr

△ **28.** Given that the two triangles are similar, find x. $x = 12$

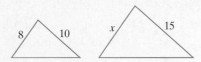

7 CHAPTER CUMULATIVE REVIEW

1. Write each sentence as an equation. Let x represent the unknown number.   **1.c.** $4x + 17 = 21$ (Sec. 1.4, Ex. 9)

  a. The quotient of 15 and a number is 4. $\frac{15}{x} = 4$

  b. Three subtracted from 12 is a number. $12 - 3 = x$

  c. Four times a number, added to 17, is 21.

2. Write each sentence as an equation. Let x represent the unknown number.   **2. a.** $12 - x = -45$

  a. The difference of 12 and a number is -45.

  b. The product of 12 and a number is -45. $12x = -45$

  c. A number less 10 is twice the number. $x - 10 = 2x$ (Sec. 1.4)

3. Rajiv Puri invested part of his \$20,000 inheritance in a mutual funds account that pays 7% simple interest yearly and the rest in a certificate of deposit that pays 9% simple interest yearly. At the end of one year, Rajiv's investments earned \$1550. Find the amount he invested at each rate.

4. The number of non-business bankruptcies has increased over the years. In 2002, the number of non-business bankruptcies was 80,000 less than twice the number in 1994. If the total of non-business bankruptcies for these two years is 2,290,000 find the number of non-business bankruptcies for each year. (*Source:* American Bankruptcy Institute)

5. Graph $x - 3y = 6$ by finding and plotting intercepts.
See graphing answer section. (Sec. 3.3, Ex. 2)

6. Find the slope of the line whose equation is $7x + 2y = 9$.

7. Use the product rule to simplify each expressing.

  a. $4^2 \cdot 4^5$ 4^7 **b.** $x^2 \cdot x^5$ x^7 **c.** $y^3 \cdot y$ y^4

  d. $y^3 \cdot y^2 \cdot y^7$ y^{12} **e.** $(-5)^7 \cdot (-5)^8$ **f.** $a^2 \cdot b^2$ a^2b^2

8. Simplify.

  a. $\dfrac{x^9}{x^7}$ **b.** $\dfrac{x^{19}y^5}{xy}$ **c.** $(x^5y^2)^3$ **d.** $(-3a^2b)(5a^3b)$

9. Subtract $(5z - 7)$ from the sum of $(8z + 11)$ and $(9z - 2)$.

10. Subtract $(9x^2 - 6x + 2)$ from $(x + 1)$. $-9x^2 + 7x - 1$
(Sec. 5.2)

11. Multiply: $(3a + b)^3$

12. Multiply: $(2x + 1)(5x^2 - x + 2)$

13. Use a special product to square each binomial.

  a. $(t + 2)^2$ $t^2 + 4t + 4$ **b.** $(p - q)^2$ $p^2 - 2pq + q^2$

  c. $(2x + 5)^2$ **d.** $(x^2 - 7y)^2$

14. Multiply.   **14.a.** $x^2 + 18x + 81$ **b.** $4x^2 - 1$

  a. $(x + 9)^2$ **b.** $(2x + 1)(2x - 1)$

  c. $8x(x^2 + 1)(x^2 - 1)$ $8x^5 - 8x$ (Sec. 5.4)

15. Simplify each expression. Write results using positive exponents only.

  a. $\dfrac{1}{x^{-3}}$ x^3 **b.** $\dfrac{1}{3^{-4}}$ 81

  c. $\dfrac{p^{-4}}{q^{-9}}$ $\dfrac{q^9}{p^4}$ **d.** $\dfrac{5^{-3}}{2^{-5}}$ $\dfrac{32}{125}$ (Sec. 5.5, Ex. 2)

16. Simplify. Write results with positive exponents.

  a. 5^{-3} $\dfrac{1}{125}$ **b.** $\dfrac{9}{x^{-7}}$ $9x^7$ **c.** $\dfrac{11^{-1}}{7^{-2}}$ $\dfrac{49}{11}$ (Sec. 5.5)

17. Divide: $\dfrac{4x^2 + 7 + 8x^3}{2x + 3}$

18. Divide $(4x^3 - 9x + 2)$ by $(x - 4)$. $4x^2 + 16x + 55 + \dfrac{222}{x - 4}$
(Sec. 5.6)

3. amount at 7%: \$12,500; amount at 9%: \$7500 (Sec. 2.7, Ex. 5)   **4.** 1994: 790,000; 2002: 1,500,000 (Sec. 2.7)

6. $-\dfrac{7}{2}$ (Sec. 3.3)   **7.e.** $(-5)^{15}$ (Sec. 5.1, Ex. 3)   **8.a.** x^2 **8. b.** $x^{18}y^4$ **8.c.** $x^{15}y^6$ **8.d.** $-15a^5b^2$ (Sec. 5.1)   **9.** $12z + 16$ (Sec. 5.2, Ex. 15)

11. $27a^3 + 27a^2b + 9ab^2 + b^3$ (Sec. 5.3, Ex. 8)   **12.** $10x^3 + 3x^2 + 3x + 2$ (Sec. 5.3)   **13.c.** $4x^2 + 20x + 25$

13.d. $x^4 - 14x^2y + 49y^2$ (Sec. 5.4, Ex 5)   **17.** $4x^2 - 4x + 6 + \dfrac{-11}{2x + 3}$ (Sec. 5.6, Ex. 6)

19. Find the GCF of each list of numbers.

    a. 28 and 40 4   **b.** 55 and 21 1

    c. 15, 18, and 66 3 (Sec. 6.1, Ex. 1)

20. Find the GCF of $9x^2, 6x^3$, and $21x^5$. $3x^2$ (Sec. 6.1)

Factor.

21. $-9a^5 + 18a^2 - 3a$ $-3a(3a^4 - 6a + 1)$ (Sec. 6.1, Ex. 5)

22. $7x^6 - 7x^5 + 7x^4$ $7x^4(x^2 - x + 1)$ (Sec. 6.1)

23. $3m^2 - 24m - 60$ $3(m + 2)(m - 10)$ (Sec. 6.2, Ex. 7)

24. $-2a^2 + 10a + 12$ $-2(a + 1)(a - 6)$ (Sec. 6.2)

25. $3x^3 + 11x + 6$ $(3x + 2)(x + 3)$ (Sec. 6.3, Ex. 1)

26. $10m^2 - 7m + 1$ $(5m - 1)(2m - 1)$ (Sec. 6.3)

27. $x^2 + 12x + 36$ $(x + 6)^2$ (Sec. 6.3, Ex. 7)

28. $4x^2 + 12x + 9$ $(2x + 3)^2$ (Sec. 6.3)

29. $x^2 + 4$ prime polynomial (Sec. 6.4, Ex. 5)

30. $x^2 - 4$ $(x + 2)(x - 2)$ (Sec. 6.4)

31. $x^3 + 8$ $(x + 2)(x^2 - 2x + 4)$ (Sec. 6.4, Ex. 6)

32. $27y^3 - 1$ $(3y - 1)(9y^2 + 3y + 1)$ (Sec. 6.4)

33. $2x^3 + 3x^2 - 2x - 3$

34. $3x^3 + 5x^2 - 12x - 20$ $(3x + 5)(x + 2)(x - 2)$ (Sec. 6.4)

35. $12m^2 - 3n^2$ $3(2m + n)(2m - n)$ (Int. Rev., Ex. 3)

36. $x^5 - x$ $x(x^2 + 1)(x + 1)(x - 1)$ (Sec. 6.4)

37. Solve: $x(2x - 7) = 4$ $-\frac{1}{2}, 4$ (Sec. 6.5, Ex. 3)

38. Solve: $3x^2 + 5x = 2$ $-2, \frac{1}{3}$ (Sec. 6.5)

39. Find the x-intercepts of the graph of $y = x^2 - 5x + 4$.

40. Find the x-intercepts of the graph of $y = x^2 - x - 6$

41. The height of a triangular sail is 2 meters less than twice the length of the base. If the sail has an area of 30 square meters, find the length of its base and the height.

42. The height of a parallelogram is 5 feet more than three times its base. If the area of the parallelogram is 182 square feet, find the length of its base and height.

43. Simplify: $\dfrac{5x - 5}{x^3 - x^2}$ $\dfrac{5}{x^2}$ (Sec. 7.1, Ex. 5)

44. Simplify: $\dfrac{2x^2 - 50}{4x^4 - 20x^3}$ $\dfrac{x + 5}{2x^3}$ (Sec. 7.1)

45. Divide: $\dfrac{6x + 2}{x^2 - 1} \div \dfrac{3x^2 + x}{x - 1}$ $\dfrac{2}{x(x + 1)}$ (Sec. 7.2, Ex. 6)

46. Multiply: $\dfrac{6x^2 - 18x}{3x^2 - 2x} \cdot \dfrac{15x - 10}{x^2 - 9}$ $\dfrac{30}{x + 3}$ (Sec. 7.2)

47. Simplify: $\dfrac{\dfrac{x + 1}{y}}{\dfrac{x}{y} + 2}$ $\dfrac{x + 1}{x + 2y}$ (Sec. 7.8, Ex. 5)

48. Simplify: $\dfrac{\dfrac{m}{3} + \dfrac{n}{6}}{\dfrac{m + n}{12}}$ $\dfrac{4m + 2n}{m + n}$ or $\dfrac{2(2m + n)}{m + n}$ (Sec. 7.8)

33. $(2x + 3)(x + 1)(x - 1)$ (Sec. 6.5, Ex. 2)   **39.** $(1, 0), (4, 0)$ (Sec. 6.5, Ex. 8)   **40.** $(-2, 0)(3, 0)$ (Sec. 6.5)

41. base: 6 m; height: 10 m (Sec. 6.6, Ex. 3)   **42.** base: 7ft; height: 26ft (Sec. 6.6)

Roots and Radicals

Having spent several chapters studying equations, we return now to algebraic expressions. We expand on your skills of operating on expressions—adding, subtracting, multiplying, dividing, and raising to powers—to include finding roots. Just as subtraction is defined by addition and division by multiplication, finding roots is defined by raising to powers. This chapter also includes working with equations that contain roots and solving problems that can be modeled by such equations.

A highway maintenance supervisor is an essential member of the vast network of people who keep American drivers moving. The on-site responsibilities of the supervisor include assignment and explanation of duties, allocation of the use of machinery and equipment, minimization of traffic delays, and safety of their employees and the public.

In addition to this, highway maintenance supervisors make decisions that require analytical thought as well as technical expertise. They should be task-oriented and organized, and capable of communicating well with laborers and professionals alike. Highway maintenance supervisors should be able to understand basic engineering plans and to work comfortably with mathematical formulas and concepts.

In the Spotlight on Decision Making feature on page 488, you will have the opportunity to make a decision about ramp speed limit as a highway maintenance supervisor.

Link: http://da.state.ks.us/ps/specs/specs/3082n3.htm
Source of text: http://da.state.ks.us/ps/specs/specs/3082n3.htm

8.1 INTRODUCTION TO RADICALS

Objectives

1. Find square roots of perfect squares.
2. Approximate irrational square roots.
3. Simplify square roots containing variables.
4. Find higher roots.

1 In this section, we define finding the *root* of a number by its reverse operation, raising a number to a power. We begin with squares and square roots.

$$\text{The square of 5 is } 5^2 = 25.$$
$$\text{The square of } -5 \text{ is } (-5)^2 = 25.$$
$$\text{The square of } \frac{1}{2} \text{ is } \left(\frac{1}{2}\right)^2 = \frac{1}{4}.$$

The reverse operation of squaring a number is finding the *square root* of a number. For example,

$$\text{A square root of 25 is 5, because } 5^2 = 25.$$
$$\text{A square root of 25 is also } -5, \text{ because } (-5)^2 = 25.$$
$$\text{A square root of } \frac{1}{4} \text{ is } \frac{1}{2}, \text{ because } \left(\frac{1}{2}\right)^2 = \frac{1}{4}.$$

In general, a number b is a square root of a number a if $b^2 = a$.

Notice that both 5 and -5 are square roots of 25. The symbol $\sqrt{}$ is used to denote the **positive** or **principal square root** of a number. For example,

$$\sqrt{25} = 5 \text{ since } 5^2 = 25 \text{ and 5 is positive.}$$

The symbol $-\sqrt{}$ is used to denote the **negative square root**. For example,

$$-\sqrt{25} = -5$$

The symbol $\sqrt{}$ is called a **radical** or **radical sign**. The expression within or under a radical sign is called the **radicand**. An expression containing a radical is called a **radical expression**.

$$\sqrt{a} \quad \text{radical sign} \atop \text{radicand}$$

Square Root

The positive or principal square root of a positive number a is written as $\sqrt{a}$. The negative square root of a is written as $-\sqrt{a}$.

$$\sqrt{a} = b \qquad \text{only if} \qquad b^2 = a \text{ and } b > 0$$

Also, the square root of 0, written as $\sqrt{0}$, is 0.

EXAMPLE 1

Find each square root.

a. $\sqrt{36}$ **b.** $\sqrt{64}$ **c.** $-\sqrt{16}$ **d.** $\sqrt{\dfrac{9}{100}}$ **e.** $\sqrt{0}$

Solution

a. $\sqrt{36} = 6$, because $6^2 = 36$ and 6 is positive.

b. $\sqrt{64} = 8$, because $8^2 = 64$ and 8 is positive.

c. $-\sqrt{16} = -4$. The negative sign in front of the radical indicates the negative square root of 16.

d. $\sqrt{\dfrac{9}{100}} = \dfrac{3}{10}$ because $\left(\dfrac{3}{10}\right)^2 = \dfrac{9}{100}$ and $\dfrac{3}{10}$ is positive.

e. $\sqrt{0} = 0$ because $0^2 = 0$.

Is the square root of a negative number a real number? For example, is $\sqrt{-4}$ a real number? To answer this question, we ask ourselves, is there a real number whose square is -4? Since there is no real number whose square is -4, we say that $\sqrt{-4}$ is not a real number. In general,

> A square root of a negative number is not a real number.

We will discuss numbers such as $\sqrt{-4}$ in Chapter 9.

2 Recall that numbers such as $1, 4, 9, 25,$ and $\frac{4}{25}$ are called **perfect squares**, since $1^2 = 1, 2^2 = 4, 3^2 = 9, 5^2 = 25,$ and $\left(\frac{2}{5}\right)^2 = \frac{4}{25}$. Square roots of perfect square radicands simplify to rational numbers. What happens when we try to simplify a root such as $\sqrt{3}$? Since 3 is not a perfect square, $\sqrt{3}$ is not a rational number. It cannot be written as a quotient of integers. It is called an **irrational number** and we can find a decimal **approximation** of it. To find decimal approximations, use a calculator or an appendix. (For calculator help, see the box at the end of this section.)

EXAMPLE 2

Use a calculator or an appendix to approximate $\sqrt{3}$ to three decimal places.

Solution

We may use an appendix or a calculator to approximate $\sqrt{3}$. To use a calculator, find the square root key $\boxed{\sqrt{}}$.

$\sqrt{3} \approx 1.732050808$

To three decimal places, $\sqrt{3} \approx 1.732$.

3 Radicals can also contain variables. To simplify radicals containing variables, special care must be taken. To see how we simplify $\sqrt{x^2}$, let's look at a few examples in this form.

If $x = 3$, we have $\sqrt{3^2} = \sqrt{9} = 3$, or x.

If x is 5, we have $\sqrt{5^2} = \sqrt{25} = 5$, or x.

From these two examples, you may think that $\sqrt{x^2}$ simplifies to x. Let's now look at an example where x is a negative number. If $x = -3$, we have $\sqrt{(-3)^2} = \sqrt{9} = 3$, not -3, our original x. To make sure that $\sqrt{x^2}$ simplifies to a nonnegative number, we have the following.

> For any real number a,
> $$\sqrt{a^2} = |a|.$$

Thus,

$$\sqrt{x^2} = |x|,$$
$$\sqrt{(-8)^2} = |-8| = 8$$
$$\sqrt{(7y)^2} = |7y|, \quad \text{and so on}.$$

To avoid this, for the rest of the chapter we assume that **if a variable appears in the radicand of a radical expression, it represents positive numbers only**. Then

$$\sqrt{x^2} = |x| = x \text{ since } x \text{ is a positive number}.$$

$$\sqrt{y^2} = y \qquad \text{Because } (y)^2 = y^2$$
$$\sqrt{x^8} = x^4 \qquad \text{Because } (x^4)^2 = x^8$$
$$\sqrt{9x^2} = 3x \qquad \text{Because } (3x)^2 = 9x^2$$

EXAMPLE 3

Simplify each expression. Assume that all variables represent positive numbers.

a. $\sqrt{x^2}$ **b.** $\sqrt{x^6}$ **c.** $\sqrt{16x^{16}}$ **d.** $\sqrt{\dfrac{x^4}{25}}$

Solution
a. $\sqrt{x^2} = x$ because x times itself equals x^2.
b. $\sqrt{x^6} = x^3$ because $(x^3)^2 = x^6$.
c. $\sqrt{16x^{16}} = 4x^8$ because $(4x^8)^2 = 16x^{16}$.
d. $\sqrt{\dfrac{x^4}{25}} = \dfrac{x^2}{5}$ because $\left(\dfrac{x^2}{5}\right)^2 = \dfrac{x^4}{25}$.

CLASSROOM EXAMPLE

Simplify each expression. Assume that all variables represent positive numbers.

a. $\sqrt{x^8}$ **b.** $\sqrt{x^{20}}$
c. $\sqrt{4x^6}$ **d.** $\sqrt{64y^{12}}$
e. $\sqrt{\dfrac{x^{10}}{25}}$

answer:

a. x^4 **b.** x^{10}
c. $2x^3$ **d.** $8y^6$
e. $\dfrac{x^5}{5}$

4   We can find roots other than square roots. For example, since $2^3 = 8$, we call 2 the **cube root** of 8. In symbols, we write

$$\sqrt[3]{8} = 2 \qquad \text{The number 3 is called the \textbf{index}.}$$

Also,

$$\sqrt[3]{27} = 3 \qquad \text{Since } 3^3 = 27$$
$$\sqrt[3]{-64} = -4 \qquad \text{Since } (-4)^3 = -64$$

Notice that unlike the square root of a negative number, the cube root of a negative number is a real number. This is so because while we cannot find a real number whose **square** is negative, we **can** find a real number whose **cube** is negative. In fact, the cube of a negative number is a negative number. Therefore, the cube root of a negative number is a negative number.

EXAMPLE 4

Find each cube root.

a. $\sqrt[3]{1}$ **b.** $\sqrt[3]{-27}$ **c.** $\sqrt[3]{\dfrac{1}{125}}$

Solution   **a.** $\sqrt[3]{1} = 1$ because $1^3 = 1$.

b. $\sqrt[3]{-27} = -3$ because $(-3)^3 = -27$.

c. $\sqrt[3]{\dfrac{1}{125}} = \dfrac{1}{5}$ because $\left(\dfrac{1}{5}\right)^3 = \dfrac{1}{125}$.

Just as we can raise a real number to powers other than 2 or 3, we can find roots other than square roots and cube roots. In fact, we can take the nth root of a number where n is any natural number. An ***nth root*** of a number a is a number whose nth power is a. The natural number n is called the **index**.

In symbols, the nth root of a is written as $\sqrt[n]{a}$. The index 2 is usually omitted for square roots.

> **Helpful Hint**
>
> If the index is even, such as $\sqrt{}$, $\sqrt[4]{}$, $\sqrt[6]{}$, and so on, the radicand must be nonnegative for the root to be a real number. For example,
>
> $$\sqrt[4]{16} = 2 \text{ but } \sqrt[4]{-16} \text{ is not a real number}$$
> $$\sqrt[6]{64} = 2 \text{ but } \sqrt[6]{-64} \text{ is not a real number}$$

✔ **CONCEPT CHECK**

Which of the following is a real number?

a. $\sqrt{-64}$

b. $\sqrt[4]{-64}$

c. $\sqrt[5]{-64}$

d. $\sqrt[6]{-64}$

EXAMPLE 5

Find each root.

a. $\sqrt[4]{16}$ **b.** $\sqrt[5]{-32}$ **c.** $-\sqrt[3]{8}$ **d.** $\sqrt[4]{-81}$

Solution   **a.** $\sqrt[4]{16} = 2$ because $2^4 = 16$ and 2 is positive.

b. $\sqrt[5]{-32} = -2$ because $(-2)^5 = -32$.

c. $-\sqrt[3]{8} = -2$ since $\sqrt[3]{8} = 2$.

d. $\sqrt[4]{-81}$ is not a real number since the index 4 is even and the radicand −81 is negative.

Calculator Explorations

To simplify or approximate square roots using a calculator, locate the key marked $\boxed{\sqrt{}}$. To simplify $\sqrt{25}$ using a scientific calculator, press $\boxed{25}$ $\boxed{\sqrt{}}$. The display should read $\boxed{5}$. To simplify $\sqrt{25}$ using a graphing calculator, press $\boxed{\sqrt{}}$ $\boxed{25}$ $\boxed{\text{ENTER}}$.

To approximate $\sqrt{30}$, press $\boxed{30}$ $\boxed{\sqrt{}}$ (or $\boxed{\sqrt{}}$ $\boxed{30}$ $\boxed{\text{ENTER}}$). The display should read $\boxed{5.4772256}$. This is an approximation for $\sqrt{30}$. A three-decimal-place approximation is

$$\sqrt{30} \approx 5.477$$

Is this answer reasonable? Since 30 is between perfect squares 25 and 36, $\sqrt{30}$ is between $\sqrt{25} = 5$ and $\sqrt{36} = 6$. The calculator result is then reasonable since 5.4772256 is between 5 and 6.

Use a calculator to approximate each expression to three decimal places. Decide whether each result is reasonable.

1. $\sqrt{7}$ 2.646; yes   **2.** $\sqrt{14}$ 3.742; yes   **3.** $\sqrt{11}$ 3.317; yes
4. $\sqrt{200}$ 14.142; yes   **5.** $\sqrt{82}$ 9.055; yes   **6.** $\sqrt{46}$ 6.782; yes

Many scientific calculators have a key, such as $\boxed{\sqrt[x]{y}}$, that can be used to approximate roots other than square roots. To approximate these roots using a graphing calculator, look under the $\boxed{\text{MATH}}$ menu or consult your manual.

Use a calculator to approximate each expression to three decimal places. Decide whether each result is reasonable.

7. $\sqrt[3]{40}$ 3.420; yes   **8.** $\sqrt[3]{71}$ 4.141; yes   **9.** $\sqrt[4]{20}$ 2.115; yes
10. $\sqrt[4]{15}$ 1.968; yes   **11.** $\sqrt[5]{18}$ 1.783; yes   **12.** $\sqrt[6]{2}$ 1.122; yes

Spotlight on

DECISION ✈ MAKING

Suppose you are a highway maintenance supervisor. You and your crew are heading out to post signs at a new cloverleaf exit ramp on the highway. One of the signs needed is a suggested ramp speed limit. Once you are at the ramp, you realize that you forgot to check with the highway engineers about which sign to post. You know that the formula $S = \sqrt{2.5r}$ can be used to estimate the maximum safe speed S, in miles per hour, at which a car can travel on a curved road with *radius of curvature, r*, in feet. Using a tape measure, your crew measures the radius of curvature as 400 feet. Which sign should you post? Explain your reasoning.

RAMP **25** M. P. H.   RAMP **30** M. P. H.   RAMP **35** M. P. H.   RAMP **40** M. P. H.

MENTAL MATH

Answer each exercise true or false.

1. $\sqrt{-16}$ simplifies to a real number.   false

2. $\sqrt{64} = 8$ while $\sqrt[3]{64} = 4$.   true

3. The number 9 has two square roots.   true

4. $\sqrt{0} = 0$ and $\sqrt{1} = 1$.   true

5. If x is a positive number, $\sqrt{x^{10}} = x^5$.   true

6. If x is a positive number, $\sqrt{x^{16}} = x^4$.   false

EXERCISE SET 8.1

STUDY GUIDE/SSM   CD/ VIDEO   PH MATH TUTOR CENTER   MathXL®Tutorials ON CD   MathXL®   MyMathLab®

Find each square root if it is a real number. See Example 1.

1. $\sqrt{16}$ 4

2. $\sqrt{4}$ 2

3. $\sqrt{81}$ 9

4. $\sqrt{49}$ 7

5. $\sqrt{\dfrac{1}{25}}$ $\dfrac{1}{5}$

6. $\sqrt{\dfrac{1}{64}}$ $\dfrac{1}{8}$

7. $-\sqrt{100}$ -10

8. $-\sqrt{36}$ -6

9. $\sqrt{-4}$ not a real number

10. $\sqrt{-25}$ not a real number

11. $-\sqrt{121}$ -11

12. $-\sqrt{49}$ -7

13. $\sqrt{\dfrac{9}{25}}$ $\dfrac{3}{5}$

14. $\sqrt{\dfrac{4}{81}}$ $\dfrac{2}{9}$

15. $\sqrt{144}$ 12

16. $\sqrt{169}$ 13

17. $\sqrt{\dfrac{49}{36}}$ $\dfrac{7}{6}$

18. $\sqrt{\dfrac{100}{121}}$ $\dfrac{10}{11}$

19. $-\sqrt{1}$ -1

20. $-\sqrt{225}$ -15

Approximate each square root to three decimal places. See Example 2.

21. $\sqrt{37}$ 6.083

22. $\sqrt{27}$ 5.196

23. $\sqrt{136}$ 11.662

24. $\sqrt{8}$ 2.828

25. A standard baseball diamond is a square with 90-foot sides connecting the bases. The distance from home plate to second base is $90 \cdot \sqrt{2}$ feet. Approximate $\sqrt{2}$ to two decimal places and use your result to approximate the distance $90 \cdot \sqrt{2}$ feet. $\sqrt{2} \approx 1.41$; 126.90 ft

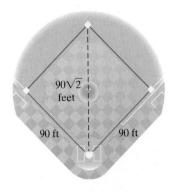

26. The roof of the warehouse shown needs to be shingled. The total area of the roof is exactly $240 \cdot \sqrt{41}$ square feet. Approximate this area to the nearest whole number.   1537 sq. ft

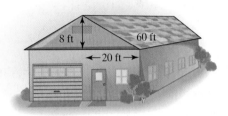

Find each root. Assume that all variables represent positive numbers. See Example 3.

27. $\sqrt{z^2}$ z

28. $\sqrt{y^{10}}$ y^5

29. $\sqrt{x^4}$ x^2

30. $\sqrt{x^6}$ x^3

31. $\sqrt{9x^8}$ $3x^4$

32. $\sqrt{36x^{12}}$ $6x^6$

33. $\sqrt{81x^2}$ $9x$

34. $\sqrt{100z^4}$ $10z^2$

35. $\sqrt{\dfrac{x^6}{36}}$ $\dfrac{x^3}{6}$

36. $\sqrt{\dfrac{y^8}{49}}$ $\dfrac{y^4}{7}$

37. $\sqrt{\dfrac{25y^2}{9}}$ $\dfrac{5y}{3}$

38. $\sqrt{\dfrac{4x^2}{81}}$ $\dfrac{2x}{9}$

Find each cube root. See Example 4.

39. $\sqrt[3]{125}$ 5

40. $\sqrt[3]{64}$ 4

41. $\sqrt[3]{-64}$ -4

42. $\sqrt[3]{-27}$ -3

43. $-\sqrt[3]{125}$ -5

44. $-\sqrt[3]{27}$ -3

45. $\sqrt[3]{\dfrac{1}{8}}$ $\dfrac{1}{2}$

46. $\sqrt[3]{\dfrac{1}{64}}$ $\dfrac{1}{4}$

47. $\sqrt[3]{-125}$ -5

48. $-\sqrt[3]{64}$ -4

49. $\sqrt[3]{-1000}$ -10

50. $\sqrt[3]{-8}$ -2

51. Explain why the square root of a negative number is not a real number.   answers may vary

52. Explain why the cube root of a negative number is a real number.   answers may vary

Find each root. See Example 5.   **55.–56.** not a real number

53. $\sqrt[5]{32}$ 2

54. $\sqrt[4]{81}$ 3

55. $\sqrt[4]{-16}$

56. $\sqrt{-9}$

57. $-\sqrt[4]{625}$ -5

58. $-\sqrt[5]{32}$ -2

59. $\sqrt[6]{1}$ 1

60. $\sqrt[5]{1}$ 1

61. $\sqrt[5]{-32}$ -2

62. $-\sqrt[4]{256}$ -4

63. $\sqrt[4]{256}$ 4

64. $\sqrt[6]{64}$ 2

REVIEW AND PREVIEW

Write each integer as a product of two integers such that one of the factors is a perfect square. For example, in $18 = 9 \cdot 2, 9$ *is a perfect square.*

65. 50 $25 \cdot 2$

66. 8 $4 \cdot 2$

67. 32 $16 \cdot 2$

68. 75 $25 \cdot 3$

69. 28 $4 \cdot 7$

70. 44 $4 \cdot 11$

71. 27 $9 \cdot 3$

72. 90 $9 \cdot 10$

Concept Extensions

The length of a side of a square is given by the expression $\sqrt{A}$ *units where A square units is the square's area. Use this expression for Exercises 73 through 76. Be sure to attach the appropriate units.*

△ **73.** The area of a square is 49 square miles. Find the length of a side of the square.   7 mi

$\sqrt{A}$

△ **74.** The area of a square is $\dfrac{1}{81}$ square meters. Find the length of a side of the square. $\dfrac{1}{9}$ m

△ **75.** The base of a Sharp mini disc player is in the shape of a square with area 9.61 square inches. Find the length of a side. (*Source: Guinness World Records,* 2001)   3.1 in.

△ **76.** A parking lot is in the shape of a square with area 2500 square yards. Find the length of a side.   50 yd

77. Simplify $\sqrt{\sqrt{81}}$.   3

78. Simplify $\sqrt[3]{\sqrt[3]{1}}$.   1

Recall from this section that $\sqrt{a^2} = |a|$ *for any real number a. Simplify the following given that x represents any real number.*

79. $\sqrt{x^2}$ $|x|$

80. $\sqrt{4x^2}$ $|2x|$

81. $\sqrt{(x+2)^2}$ $|x+2|$

82. $\sqrt{x^2 + 6x + 9}$ Hint: First factor $x^2 + 6x + 9$. $|x+3|$

△ **83.** If the amount of gold discovered by humankind could be assembled in one place, it would make a cube with a volume of 195,112 cubic feet. Each side of the cube would be $\sqrt[3]{195,112}$ feet long. How long would one side of the cube be? (*Source: Reader's Digest*)   58 ft

84. Graph $y = \sqrt{x}$. (*Hint:* Complete the table below, plot the ordered pair solutions, and draw a smooth curve through the points. Remember that since the radicand cannot be negative, this particular graph begins at the point with coordinates $(0,0)$.)

x	y	
0	0	
1	1	
3	1.7	(approximate)
4	2	
9	3	See graphing answer section.

85. Graph $y = \sqrt[3]{x}$ (Complete the table below, plot the ordered pair solutions, and draw a smooth curve through the points.)

x	y	
-8	-2	
-2	-1.3	(approximate)
-1	-1	
0	0	
1	1	
2	1.3	(approximate)
8	2	See graphing answer section.

Simplify.

86. $\sqrt[3]{x^{12}}$ x^4

87. $\sqrt[3]{x^{21}}$ x^7

88. $\sqrt[4]{x^{12}}$ x^3

89. $\sqrt[4]{x^{20}}$ x^5

Use a grapher and graph each function. Observe the graph from left to right and give the ordered pair that corresponds to the "beginning" of the graph. Then tell why the graph starts at that point.
answers may vary

90. $y = \sqrt{x-2}$ $(2,0)$

91. $y = \sqrt{x+3}$ $(-3,0)$

92. $y = \sqrt{x+4}$ $(-4,0)$

93. $y = \sqrt{x-5}$ $(5,0)$

8.2   *SIMPLIFYING RADICALS*

Objectives

1 Use the product rule to simplify square roots.

2 Use the quotient rule to simplify square roots.

3 Simplify radicals containing variables.

4 Simplify higher roots.

1 A square root is simplified when the radicand contains no perfect square factors (other than 1). For example, $\sqrt{20}$ is not simplified because $\sqrt{20} = \sqrt{4 \cdot 5}$ and 4 is a perfect square.

To begin simplifying square roots, we notice the following pattern.

$$\sqrt{9 \cdot 16} = \sqrt{144} = 12$$
$$\sqrt{9} \cdot \sqrt{16} = 3 \cdot 4 = 12$$

Since both expressions simplify to 12, we can write

$$\sqrt{9 \cdot 16} = \sqrt{9} \cdot \sqrt{16}$$

This suggests the following product rule for square roots.

Product Rule for Square Roots

If $\sqrt{a}$ and $\sqrt{b}$ are real numbers, then

$$\sqrt{a \cdot b} = \sqrt{a} \cdot \sqrt{b}$$

In other words, the square root of a product is equal to the product of the square roots.

To simplify $\sqrt{20}$, for example, we factor 20 so that one of its factors is a perfect square factor.

$$\sqrt{20} = \sqrt{4 \cdot 5} \qquad \text{Factor 20.}$$
$$= \sqrt{4} \cdot \sqrt{5} \qquad \text{Use the product rule.}$$
$$= 2\sqrt{5} \qquad \text{Write } \sqrt{4} \text{ as 2.}$$

The notation $2\sqrt{5}$ means $2 \cdot \sqrt{5}$. Since the radicand 5 has no perfect square factor other than 1 then $2\sqrt{5}$ is in simplest form.

Helpful Hint

A radical expression in simplest form does *not mean* a decimal approximation. The simplest form of a radical expression is an exact form and may still contain a radical.

$$\underbrace{\sqrt{20} = 2\sqrt{5}}_{\text{exact}} \qquad \underbrace{\sqrt{20} \approx 4.47}_{\text{decimal approximation}}$$

EXAMPLE 1

Simplify.

a. $\sqrt{54}$ **b.** $\sqrt{12}$ **c.** $\sqrt{200}$ **d.** $\sqrt{35}$

Solution

a. Try to factor 54 so that at least one of the factors is a perfect square. Since 9 is a perfect square and $54 = 9 \cdot 6$,

$$
\begin{aligned}
\sqrt{54} &= \sqrt{9 \cdot 6} && \text{Factor 54.} \\
&= \sqrt{9} \cdot \sqrt{6} && \text{Apply the product rule.} \\
&= 3\sqrt{6} && \text{Write } \sqrt{9} \text{ as 3.}
\end{aligned}
$$

b.
$$
\begin{aligned}
\sqrt{12} &= \sqrt{4 \cdot 3} && \text{Factor 12.} \\
&= \sqrt{4} \cdot \sqrt{3} && \text{Apply the product rule.} \\
&= 2\sqrt{3} && \text{Write } \sqrt{4} \text{ as 2.}
\end{aligned}
$$

c. The largest perfect square factor of 200 is 100.

$$
\begin{aligned}
\sqrt{200} &= \sqrt{100 \cdot 2} && \text{Factor 200.} \\
&= \sqrt{100} \cdot \sqrt{2} && \text{Apply the product rule.} \\
&= 10\sqrt{2} && \text{Write } \sqrt{100} \text{ as 10.}
\end{aligned}
$$

d. The radicand 35 contains no perfect square factors other than 1. Thus $\sqrt{35}$ is in simplest form.

In Example 1, part **(c)**, what happens if we don't use the largest perfect square factor of 200? Although using the largest perfect square factor saves time, the result is the same no matter what perfect square factor is used. For example, it is also true that $200 = 4 \cdot 50$. Then

$$
\begin{aligned}
\sqrt{200} &= \sqrt{4} \cdot \sqrt{50} \\
&= 2 \cdot \sqrt{50}
\end{aligned}
$$

Since $\sqrt{50}$ is not in simplest form, we continue.

$$
\begin{aligned}
\sqrt{200} &= 2 \cdot \sqrt{50} \\
&= 2 \cdot \sqrt{25} \cdot \sqrt{2} \\
&= 2 \cdot 5 \cdot \sqrt{2} \\
&= 10\sqrt{2}
\end{aligned}
$$

2 Next, let's examine the square root of a quotient.

$$
\sqrt{\frac{16}{4}} = \sqrt{4} = 2
$$

Also,

$$
\frac{\sqrt{16}}{\sqrt{4}} = \frac{4}{2} = 2
$$

Since both expressions equal 2, we can write

$$
\sqrt{\frac{16}{4}} = \frac{\sqrt{16}}{\sqrt{4}}
$$

This suggests the following quotient rule.

Quotient Rule for Square Roots

If $\sqrt{a}$ and $\sqrt{b}$ are real numbers and $b \neq 0$, then

$$\sqrt{\frac{a}{b}} = \frac{\sqrt{a}}{\sqrt{b}}$$

In other words, the square root of a quotient is equal to the quotient of the square roots.

EXAMPLE 2

Simplify.

a. $\sqrt{\dfrac{25}{36}}$ **b.** $\sqrt{\dfrac{3}{64}}$ **c.** $\sqrt{\dfrac{40}{81}}$

Solution   Use the quotient rule.

a. $\sqrt{\dfrac{25}{36}} = \dfrac{\sqrt{25}}{\sqrt{36}} = \dfrac{5}{6}$ **b.** $\sqrt{\dfrac{3}{64}} = \dfrac{\sqrt{3}}{\sqrt{64}} = \dfrac{\sqrt{3}}{8}$

c. $\sqrt{\dfrac{40}{81}} = \dfrac{\sqrt{40}}{\sqrt{81}}$ Use the quotient rule.

$= \dfrac{\sqrt{4} \cdot \sqrt{10}}{9}$ Apply the product rule and write $\sqrt{81}$ as **9**.

$= \dfrac{2\sqrt{10}}{9}$ Write $\sqrt{4}$ as **2**.

3   Recall that $\sqrt{x^6} = x^3$ because $(x^3)^2 = x^6$. If an odd exponent occurs, we write the exponential expression so that one factor is the greatest even power contained in the expression. Then we use the product rule to simplify.

EXAMPLE 3

Simplify. Assume that all variables represent positive numbers.

a. $\sqrt{x^5}$ **b.** $\sqrt{8y^2}$ **c.** $\sqrt{\dfrac{45}{x^6}}$

Solution   **a.** $\sqrt{x^5} = \sqrt{x^4 \cdot x} = \sqrt{x^4} \cdot \sqrt{x} = x^2\sqrt{x}$

b. $\sqrt{8y^2} = \sqrt{4 \cdot 2 \cdot y^2} = \sqrt{4y^2 \cdot 2} = \sqrt{4y^2} \cdot \sqrt{2} = 2y\sqrt{2}$

c. $\sqrt{\dfrac{45}{x^6}} = \dfrac{\sqrt{45}}{\sqrt{x^6}} = \dfrac{\sqrt{9 \cdot 5}}{x^3} = \dfrac{\sqrt{9} \cdot \sqrt{5}}{x^3} = \dfrac{3\sqrt{5}}{x^3}$

4   The product and quotient rules also apply to roots other than square roots. In general, we have the following product and quotient rules for radicals.

Product Rule for Radicals

If $\sqrt[n]{a}$ and $\sqrt[n]{b}$ are real numbers, then

$$\sqrt[n]{a \cdot b} = \sqrt[n]{a} \cdot \sqrt[n]{b}$$

Quotient Rule for Radicals

If $\sqrt[n]{a}$ and $\sqrt[n]{b}$ are real numbers and $b \neq 0$, then

$$\sqrt[n]{\frac{a}{b}} = \frac{\sqrt[n]{a}}{\sqrt[n]{b}}$$

To simplify cube roots, look for perfect cube factors of the radicand. For example, 8 is a perfect cube, since $2^3 = 8$.

To simplify $\sqrt[3]{48}$, factor 48 as $8 \cdot 6$.

$$\sqrt[3]{48} = \sqrt[3]{8 \cdot 6} \qquad \text{Factor 48.}$$
$$= \sqrt[3]{8} \cdot \sqrt[3]{6} \qquad \text{Apply the product rule.}$$
$$= 2\sqrt[3]{6} \qquad \text{Write } \sqrt[3]{8} \text{ as 2.}$$

$2\sqrt[3]{6}$ is in simplest form since the radicand 6 contains no perfect cube factors other than 1.

EXAMPLE 4

Simplify.

a. $\sqrt[3]{54}$ **b.** $\sqrt[3]{18}$ **c.** $\sqrt[3]{\frac{7}{8}}$ **d.** $\sqrt[3]{\frac{40}{27}}$

Solution   **a.** $\sqrt[3]{54} = \sqrt[3]{27 \cdot 2} = \sqrt[3]{27} \cdot \sqrt[3]{2} = 3\sqrt[3]{2}$

b. The number 18 contains no perfect cube factors, so $\sqrt[3]{18}$ cannot be simplified further.

c. $\sqrt[3]{\frac{7}{8}} = \frac{\sqrt[3]{7}}{\sqrt[3]{8}} = \frac{\sqrt[3]{7}}{2}$

d. $\sqrt[3]{\frac{40}{27}} = \frac{\sqrt[3]{40}}{\sqrt[3]{27}} = \frac{\sqrt[3]{8 \cdot 5}}{3} = \frac{\sqrt[3]{8} \cdot \sqrt[3]{5}}{3} = \frac{2\sqrt[3]{5}}{3}$

To simplify fourth roots, look for perfect fourth powers of the radicand. For example, 16 is a perfect fourth power since $2^4 = 16$.

To simplify $\sqrt[4]{32}$, factor 32 as $16 \cdot 2$.

$$\sqrt[4]{32} = \sqrt[4]{16 \cdot 2} \qquad \text{Factor 32.}$$
$$= \sqrt[4]{16} \cdot \sqrt[4]{2} \qquad \text{Apply the product rule.}$$
$$= 2\sqrt[4]{2} \qquad \text{Write } \sqrt[4]{16} \text{ as 2.}$$

> ## EXAMPLE 5

Simplify.

a. $\sqrt[4]{243}$ **b.** $\sqrt[4]{\dfrac{3}{16}}$ **c.** $\sqrt[5]{64}$

Solution

a. $\sqrt[4]{243} = \sqrt[4]{81 \cdot 3} = \sqrt[4]{81} \cdot \sqrt[4]{3} = 3\sqrt[4]{3}$

b. $\sqrt[4]{\dfrac{3}{16}} = \dfrac{\sqrt[4]{3}}{\sqrt[4]{16}} = \dfrac{\sqrt[4]{3}}{2}$

c. $\sqrt[5]{64} = \sqrt[5]{32 \cdot 2} = \sqrt[5]{32} \cdot \sqrt[5]{2} = 2\sqrt[5]{2}$

CLASSROOM EXAMPLE

Simplify.

a. $\sqrt[4]{80}$ **b.** $\sqrt[4]{\dfrac{5}{81}}$ **c.** $\sqrt[5]{96}$

answer: **a.** $2\sqrt[4]{5}$ **b.** $\dfrac{\sqrt[4]{5}}{3}$ **c.** $2\sqrt[5]{3}$

MENTAL MATH

Simplify each expression. Assume that all variables represent nonnegative real numbers.

1. $\sqrt{4 \cdot 9}$ 6

2. $\sqrt{9 \cdot 36}$ 18

3. $\sqrt{x^2}$ x

4. $\sqrt{y^4}$ y^2

5. $\sqrt{0}$ 0

6. $\sqrt{1}$ 1

7. $\sqrt{25x^4}$ $5x^2$

8. $\sqrt{49x^2}$ $7x$

EXERCISE SET 8.2

STUDY GUIDE/SSM   CD/VIDEO   PH MATH TUTOR CENTER   MathXL®Tutorials ON CD   MathXL®   MyMathLab®

Use the product rule to simplify each radical. See Example 1.

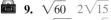 **1.** $\sqrt{20}$ $2\sqrt{5}$

2. $\sqrt{44}$ $2\sqrt{11}$

3. $\sqrt{18}$ $3\sqrt{2}$

4. $\sqrt{45}$ $3\sqrt{5}$

5. $\sqrt{50}$ $5\sqrt{2}$

6. $\sqrt{28}$ $2\sqrt{7}$

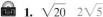 **7.** $\sqrt{33}$ $\sqrt{33}$

8. $\sqrt{98}$ $7\sqrt{2}$

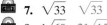 **9.** $\sqrt{60}$ $2\sqrt{15}$

10. $\sqrt{90}$ $3\sqrt{10}$

11. $\sqrt{180}$ $6\sqrt{5}$

12. $\sqrt{150}$ $5\sqrt{6}$

13. $\sqrt{52}$ $2\sqrt{13}$

14. $\sqrt{75}$ $5\sqrt{3}$

Use the quotient rule and the product rule to simplify each radical. See Example 2.

15. $\sqrt{\dfrac{8}{25}}$ $\dfrac{2\sqrt{2}}{5}$

16. $\sqrt{\dfrac{63}{16}}$ $\dfrac{3\sqrt{7}}{4}$

 17. $\sqrt{\dfrac{27}{121}}$ $\dfrac{3\sqrt{3}}{11}$

18. $\sqrt{\dfrac{24}{169}}$ $\dfrac{2\sqrt{6}}{13}$

19. $\sqrt{\dfrac{9}{4}}$ $\dfrac{3}{2}$

20. $\sqrt{\dfrac{100}{49}}$ $\dfrac{10}{7}$

21. $\sqrt{\dfrac{125}{9}}$ $\dfrac{5\sqrt{5}}{3}$

22. $\sqrt{\dfrac{27}{100}}$ $\dfrac{3\sqrt{3}}{10}$

23. $\sqrt{\dfrac{11}{36}}$ $\dfrac{\sqrt{11}}{6}$

24. $\sqrt{\dfrac{30}{49}}$ $\dfrac{\sqrt{30}}{7}$

25. $-\sqrt{\dfrac{27}{144}}$ $-\dfrac{\sqrt{3}}{4}$

26. $-\sqrt{\dfrac{84}{121}}$ $-\dfrac{2\sqrt{21}}{11}$

Simplify each radical. Assume that all variables represent positive numbers. See Example 3.

27. $\sqrt{x^7}$ $x^3\sqrt{x}$

28. $\sqrt{y^3}$ $y\sqrt{y}$

 29. $\sqrt{x^{13}}$ $x^6\sqrt{x}$

30. $\sqrt{y^{17}}$ $y^8\sqrt{y}$

31. $\sqrt{75x^2}$ $5x\sqrt{3}$

32. $\sqrt{72y^2}$ $6y\sqrt{2}$

33. $\sqrt{96x^4}$ $4x^2\sqrt{6}$

34. $\sqrt{40y^{10}}$ $2y^5\sqrt{10}$

 35. $\sqrt{\dfrac{12}{y^2}}$ $\dfrac{2\sqrt{3}}{y}$

36. $\sqrt{\dfrac{63}{x^4}}$ $\dfrac{3\sqrt{7}}{x^2}$

37. $\sqrt{\dfrac{9x}{y^2}}$ $\dfrac{3\sqrt{x}}{y}$

38. $\sqrt{\dfrac{6y^2}{x^4}}$ $\dfrac{y\sqrt{6}}{x^2}$

39. $\sqrt{\dfrac{88}{x^4}}$ $\dfrac{2\sqrt{22}}{x^2}$

40. $\sqrt{\dfrac{x^{11}}{81}}$ $\dfrac{x^5\sqrt{x}}{9}$

Simplify each radical. See Example 4.

41. $\sqrt[3]{24}$ $2\sqrt[3]{3}$

42. $\sqrt[3]{81}$ $3\sqrt[3]{3}$

 43. $\sqrt[3]{250}$ $5\sqrt[3]{2}$

44. $\sqrt[3]{40}$ $2\sqrt[3]{5}$

 45. $\sqrt[3]{\dfrac{5}{64}}$ $\dfrac{\sqrt[3]{5}}{4}$

46. $\sqrt[3]{\dfrac{32}{125}}$ $\dfrac{2\sqrt[3]{4}}{5}$

47. $\sqrt[3]{\dfrac{7}{8}}$ $\dfrac{\sqrt[3]{7}}{2}$

48. $\sqrt[3]{\dfrac{10}{27}}$ $\dfrac{\sqrt[3]{10}}{3}$

49. $\sqrt[3]{\dfrac{15}{64}}$ $\dfrac{\sqrt[3]{15}}{4}$

50. $\sqrt[3]{\dfrac{4}{27}}$ $\dfrac{\sqrt[3]{4}}{3}$

51. $\sqrt[3]{80}$ $2\sqrt[3]{10}$

52. $\sqrt[3]{108}$ $3\sqrt[3]{4}$

Simplify. See Example 5.

53. $\sqrt[4]{48}$ $2\sqrt[4]{3}$

54. $\sqrt[4]{405}$ $3\sqrt[4]{5}$

55. $\sqrt[4]{\dfrac{8}{81}}$ $\dfrac{\sqrt[4]{8}}{3}$

56. $\sqrt[4]{\dfrac{25}{256}}$ $\dfrac{\sqrt[4]{25}}{4}$

57. $\sqrt[5]{96}$ $2\sqrt[5]{3}$

58. $\sqrt[5]{128}$ $2\sqrt[5]{4}$

59. $\sqrt[5]{\dfrac{5}{32}}$ $\dfrac{\sqrt[5]{5}}{2}$

60. $\sqrt[5]{\dfrac{16}{243}}$ $\dfrac{\sqrt[5]{16}}{3}$

Simplify.

△ **61.** If a cube is to have a volume of 80 cubic inches, then each side must be $\sqrt[3]{80}$ inches long. Simplify the radical representing the side length. $2\sqrt[3]{10}$

△ **62.** Jeannie Boswell is swimming across a 40-foot-wide river, trying to head straight across to the opposite shore. However, the current is strong enough to move her downstream 100 feet by the time she reaches land. (See the figure.) Because of the current, the actual distance she swam is $\sqrt{11{,}600}$ feet. Simplify this radical. $20\sqrt{29}$

63. By using replacement values for a and b, show that $\sqrt{a^2 + b^2}$ does not equal $a + b$.   answers may vary

64. By using replacement values for a and b, show that $\sqrt{a + b}$ does not equal $\sqrt{a} + \sqrt{b}$.   answers may vary

REVIEW AND PREVIEW

Perform the following operations. See Sections 5.2 and 5.3.

65. $6x + 8x$ $14x$

66. $(6x)(8x)$ $48x^2$

67. $(2x + 3)(x - 5)$

68. $(2x + 3) + (x - 5)$

69. $9y^2 - 9y^2$ 0

70. $(9y^2)(-8y^2)$ $-72y^4$

67. $2x^2 - 7x - 15$

68. $3x - 2$

The following pairs of triangles are similar. Find the unknown lengths. See Section 7.6.

△ **71.**   8 cm

△ **72.** $\dfrac{21}{5}$ in.

Concept Extensions

Simplify each radical. Assume that all variables represent positive numbers.

73. $\sqrt{x^6 y^3}$ $x^3 y\sqrt{y}$

74. $\sqrt{98x^5 y^4}$ $7x^2 y^2\sqrt{2x}$

75. $\sqrt{x^2 + 4x + 4}$ (Hint: Factor the trinomial first.) $x + 2$

76. $\sqrt[3]{-8x^6}$ $-2x^2$

77. $\sqrt[3]{\dfrac{2}{x^9}}$ $\dfrac{\sqrt[3]{2}}{x^3}$

78. $\sqrt[3]{\dfrac{48}{x^{12}}}$ $\dfrac{2\sqrt[3]{6}}{x^4}$

The length of a side of a cube is given by the expression $\sqrt{\dfrac{A}{6}}$ units where A square units is the cube's surface area. Use this expression for Exercises 79 through 82. Be sure to attach the appropriate units.

△ **79.** The surface area of a cube is 120 square inches. Find the exact length of a side of the cube. $2\sqrt{5}$ in.

△ **80.** The surface area of a cube is 594 square feet. Find the exact length of a side of the cube. $3\sqrt{11}$ ft

$\sqrt{A/6}$

TEACHING TIP

For Exercises 63 and 64, point out to students that if $a = 0$ and $b = 0$, the two expressions are equal but to be equivalent expressions (interchangeable expressions), they must be true for all values of a and b. If you can find 1 pair of values for which the expressions are not equal, the expressions are not equivalent.

△ **81.** The Borg space ship on *Star Trek: The Next Generation* is in the shape of a cube. Suppose a model of this ship has a surface area of 150 square inches. Find the length of a side of the ship.   5 in.

△ **82.** A shipping crate in the shape of a cube is to be constructed. If the crate is to have a surface area of 486 square feet, find the length of a side of the crate.   9 ft

The cost C in dollars per day to operate a small delivery service is given by $C = 100\sqrt[3]{n} + 700$, where n is the number of deliveries per day.

83. Find the cost if the number of deliveries is 1000. $1700

84. Approximate the cost if the number of deliveries is 500. $1493.70

85. $\frac{26}{15}$ sq. m $\approx$ 1.7 sq. m

The Mosteller formula for calculating body surface area is $B = \sqrt{\dfrac{hw}{3600}}$, where B is an individual's body surface area in square meters, h is the individual's height in centimeters, and w is the individual's weight in kilograms. Use this formula in Exercises 85 and 86. Round answers to the nearest tenth.

85. Find the body surface area of a person who is 169 cm tall and weighs 64 kilograms.

86. Approximate the body surface area of a person who is 183 cm tall and weighs 85 kilograms.   2.1 sq. m

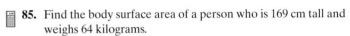

8.3   *ADDING AND SUBTRACTING RADICALS*

Objectives

1   Add or subtract like radicals.

2   Simplify radical expressions, and then add or subtract any like radicals.

1   To combine like terms, we use the distributive property.

$$5x + 3x = (5 + 3)x = 8x$$

The distributive property can also be applied to expressions containing radicals. For example,

$$5\sqrt{2} + 3\sqrt{2} = (5 + 3)\sqrt{2} = 8\sqrt{2}$$

Also,

$$9\sqrt{5} - 6\sqrt{5} = (9 - 6)\sqrt{5} = 3\sqrt{5}$$

Radical terms $5\sqrt{2}$ and $3\sqrt{2}$ are **like radicals**, as are $9\sqrt{5}$ and $6\sqrt{5}$.

Like Radicals

Like radicals are radical expressions that have the same index and the same radicand.

From the examples above, we can see that **only like radicals can be combined** in this way. For example, the expression $2\sqrt{3} + 3\sqrt{2}$ cannot be further simplified since the radicals are not like radicals. Also, the expression $4\sqrt{7} + 4\sqrt[3]{7}$ cannot be further simplified because the radicals are not like radicals since the indices are different.

EXAMPLE 1

Simplify by combining like radical terms.

a. $4\sqrt{5} + 3\sqrt{5}$ **b.** $\sqrt{10} - 6\sqrt{10}$ **c.** $2\sqrt[3]{7} - 5\sqrt[3]{7} - 3\sqrt[3]{7}$ **d.** $2\sqrt{6} + 2\sqrt[3]{6}$

Solution

a. $4\sqrt{5} + 3\sqrt{5} = (4 + 3)\sqrt{5} = 7\sqrt{5}$

b. $\sqrt{10} - 6\sqrt{10} = 1\sqrt{10} - 6\sqrt{10} = (1 - 6)\sqrt{10} = -5\sqrt{10}$

c. $2\sqrt[3]{7} - 5\sqrt[3]{7} - 3\sqrt[3]{7} = (2 - 5 - 3)\sqrt[3]{7} = -6\sqrt[3]{7}$

d. $2\sqrt{6} + 2\sqrt[3]{6}$ cannot be simplified further since the indices are not the same. ⬭

✔ **CONCEPT CHECK**

Which is true?

a. $2 + 3\sqrt{5} = 5\sqrt{5}$

b. $2\sqrt{3} + 2\sqrt{7} = 2\sqrt{10}$

c. $\sqrt{3} + \sqrt{5} = \sqrt{8}$

d. None of the above is true. In each case, the left-hand side cannot be simplified further.

2 At first glance, it appears that the expression $\sqrt{50} + \sqrt{8}$ cannot be simplified further because the radicands are different. However, the product rule can be used to simplify each radical, and then further simplification might be possible.

EXAMPLE 2

Add or subtract by first simplifying each radical.

a. $\sqrt{50} + \sqrt{8}$ **b.** $7\sqrt{12} - \sqrt{75}$ **c.** $\sqrt{25} - \sqrt{27} - 2\sqrt{18} - \sqrt{16}$

Solution **a.** First simplify each radical.

$$\sqrt{50} + \sqrt{8} = \sqrt{25 \cdot 2} + \sqrt{4 \cdot 2} \qquad \text{Factor radicands.}$$
$$= \sqrt{25} \cdot \sqrt{2} + \sqrt{4} \cdot \sqrt{2} \qquad \text{Apply the product rule.}$$
$$= 5\sqrt{2} + 2\sqrt{2} \qquad \text{Simplify } \sqrt{25} \text{ and } \sqrt{4}.$$
$$= 7\sqrt{2} \qquad \text{Add like radicals.}$$

b.
$$7\sqrt{12} - \sqrt{75} = 7\sqrt{4 \cdot 3} - \sqrt{25 \cdot 3} \qquad \text{Factor radicands.}$$
$$= 7\sqrt{4} \cdot \sqrt{3} - \sqrt{25} \cdot \sqrt{3} \qquad \text{Apply the product rule.}$$
$$= 7 \cdot 2\sqrt{3} - 5\sqrt{3} \qquad \text{Simplify } \sqrt{4} \text{ and } \sqrt{25}.$$
$$= 14\sqrt{3} - 5\sqrt{3} \qquad \text{Multiply.}$$
$$= 9\sqrt{3} \qquad \text{Subtract like radicals.}$$

c. $\sqrt{25} - \sqrt{27} - 2\sqrt{18} - \sqrt{16}$

$= 5 - \sqrt{9 \cdot 3} - 2\sqrt{9 \cdot 2} - 4$ Factor radicands.

$= 5 - \sqrt{9} \cdot \sqrt{3} - 2\sqrt{9} \cdot \sqrt{2} - 4$ Apply the product rule.

$= 5 - 3\sqrt{3} - 2 \cdot 3\sqrt{2} - 4$ Simplify.

$= 1 - 3\sqrt{3} - 6\sqrt{2}$ Write $5 - 4$ as 1 and $2 \cdot 3$ as 6.

If radical expressions contain variables, we proceed in a similar way. Simplify radicals using the product and quotient rules. Then add or subtract any like radicals.

EXAMPLE 3

Simplify $2\sqrt{x^2} - \sqrt{25x} + \sqrt{x}$. Assume variables represent positive numbers.

Solution $2\sqrt{x^2} - \sqrt{25x} + \sqrt{x}$

$= 2x - \sqrt{25} \cdot \sqrt{x} + \sqrt{x}$ Write $\sqrt{x^2}$ as x and apply the product rule.

$= 2x - 5\sqrt{x} + 1\sqrt{x}$ Simplify.

$= 2x - 4\sqrt{x}$ Add like radicals.

CLASSROOM EXAMPLE
Simplify $\sqrt{9x^4} - \sqrt{36x^3} + \sqrt{x^3}$.
answer: $3x^2 - 5x\sqrt{x}$

EXAMPLE 4

CLASSROOM EXAMPLE
Simplify $5\sqrt[3]{4} - \sqrt[3]{32}$.
answer: $3\sqrt[3]{4}$

Add or subtract by first simplifying each radical.

$2\sqrt[3]{27} - \sqrt[3]{54}$

Solution $2\sqrt[3]{27} - \sqrt[3]{54} = 2 \cdot 3 - \sqrt[3]{27 \cdot 2}$ Simplify $\sqrt[3]{27}$ and factor 54.

$= 6 - \sqrt[3]{27} \cdot \sqrt[3]{2}$ Apply the product rule.

$= 6 - 3\sqrt[3]{2}$ Simplify $\sqrt[3]{27}$.

> **Helpful Hint**
> These two terms may not be combined. They are unlike terms.

MENTAL MATH

Simplify each expression by combining like radicals.

1. $3\sqrt{2} + 5\sqrt{2}$ $8\sqrt{2}$ **2.** $2\sqrt{3} + 7\sqrt{3}$ $9\sqrt{3}$ **3.** $5\sqrt{x} + 2\sqrt{x}$ $7\sqrt{x}$

4. $8\sqrt{x} + 3\sqrt{x}$ $11\sqrt{x}$ **5.** $5\sqrt{7} - 2\sqrt{7}$ $3\sqrt{7}$ **6.** $8\sqrt{6} - 5\sqrt{6}$ $3\sqrt{6}$

EXERCISE SET 8.3

STUDY GUIDE/SSM   CD/ VIDEO   PH MATH TUTOR CENTER   MathXL®Tutorials ON CD   MathXL®   MyMathLab®

Simplify each expression by combining like radicals where possible. See Example 1.

1. $4\sqrt{3} - 8\sqrt{3}$ $-4\sqrt{3}$

2. $\sqrt{5} - 9\sqrt{5}$ $-8\sqrt{5}$

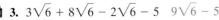 **3.** $3\sqrt{6} + 8\sqrt{6} - 2\sqrt{6} - 5$ $9\sqrt{6} - 5$

4. $12\sqrt{2} - 3\sqrt{2} + 8\sqrt{2} + 10$ $17\sqrt{2} + 10$

5. $6\sqrt{5} - 5\sqrt{5} + \sqrt{2}$ $\sqrt{5} + \sqrt{2}$

6. $4\sqrt{3} + \sqrt{5} - 3\sqrt{3}$ $\sqrt{3} + \sqrt{5}$

7. $2\sqrt[3]{3} + 5\sqrt[3]{3} - \sqrt{3}$ $7\sqrt[3]{3} - \sqrt{3}$

8. $8\sqrt[3]{4} + 2\sqrt[3]{4} + 4$ $10\sqrt[3]{4} + 4$

9. $2\sqrt[3]{2} - 7\sqrt[3]{2} - 6$ $-5\sqrt[3]{2} - 6$

10. $5\sqrt[3]{9} + 2 - 11\sqrt[3]{9}$ $2 - 6\sqrt[3]{9}$

11. Find the perimeter of the rectangular picture frame. $8\sqrt{5}$ in.

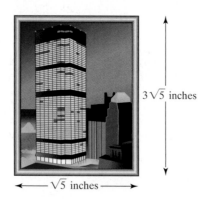

$3\sqrt{5}$ inches

$\sqrt{5}$ inches

12. Find the perimeter of the plot of land. $80\sqrt{6}$ ft

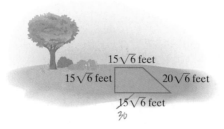

$15\sqrt{6}$ feet

$15\sqrt{6}$ feet $20\sqrt{6}$ feet

$15\sqrt{6}$ feet

13. In your own words, describe like radicals.   answers may vary

14. In the expression $\sqrt{5} + 2 - 3\sqrt{5}$, explain why 2 and -3 cannot be combined.   answers may vary

MIXED PRACTICE   **21.** $x + \sqrt{x}$ **22.** $-5x + 2\sqrt{x}$ **23.** 0

Add or subtract by first simplifying each radical and then combining any like radical terms. Assume that all variables represent positive real numbers. See Examples 2 and 3.

15. $\sqrt{12} + \sqrt{27}$ $5\sqrt{3}$ **16.** $\sqrt{50} + \sqrt{18}$ $8\sqrt{2}$

17. $\sqrt{45} + 3\sqrt{20}$ $9\sqrt{5}$ **18.** $\sqrt{28} + \sqrt{63}$ $5\sqrt{7}$

19. $2\sqrt{54} - \sqrt{20} + \sqrt{45} - \sqrt{24}$ $4\sqrt{6} + \sqrt{5}$

20. $2\sqrt{8} - \sqrt{128} + \sqrt{48} + \sqrt{18}$ $-\sqrt{2} + 4\sqrt{3}$

21. $4x - 3\sqrt{x^2} + \sqrt{x}$ **22.** $x - 6\sqrt{x^2} + 2\sqrt{x}$

23. $\sqrt{25x} + \sqrt{36x} - 11\sqrt{x}$ **24.** $3\sqrt{x^3} - x\sqrt{4x}$ $x\sqrt{x}$

25. $\sqrt{16x} - \sqrt{x^3}$ **26.** $\sqrt{8x^3} - \sqrt{x^2}$

27. $12\sqrt{5} - \sqrt{5} - 4\sqrt{5}$ **28.** $7\sqrt{3} + 2\sqrt{3} - 13\sqrt{3}$

29. $\sqrt{5} + \sqrt[3]{5}$ $\sqrt{5} + \sqrt[3]{5}$ **30.** $\sqrt{5} + \sqrt{5}$ $2\sqrt{5}$

31. $4 + 8\sqrt{2} - 9$ **32.** $6 - 2\sqrt{3} - \sqrt{3}$

25. $4\sqrt{x} - x\sqrt{x}$ **26.** $2x\sqrt{2x} - x$ **27.** $7\sqrt{5}$ **28.** $-4\sqrt{3}$ **31.** $-5 + 8\sqrt{2}$ **32.** $6 - 3\sqrt{3}$

33. $8 - \sqrt{2} - 5\sqrt{2}$ $8 - 6\sqrt{2}$ **34.** $\sqrt{75} + \sqrt{48}$ $9\sqrt{3}$

35. $5\sqrt{32} - \sqrt{72}$ $14\sqrt{2}$ **36.** $2\sqrt{80} - \sqrt{45}$ $5\sqrt{5}$

37. $\sqrt{8} + \sqrt{9} + \sqrt{18} + \sqrt{81}$ $5\sqrt{2} + 12$

38. $\sqrt{6} + \sqrt{16} + \sqrt{24} + \sqrt{25}$ $3\sqrt{6} + 9$

39. $\sqrt{\dfrac{5}{9}} + \sqrt{\dfrac{5}{81}}$ $\dfrac{4\sqrt{5}}{9}$ **40.** $\sqrt{\dfrac{3}{64}} + \sqrt{\dfrac{3}{16}}$ $\dfrac{3\sqrt{3}}{8}$

41. $\sqrt{\dfrac{3}{4}} - \sqrt{\dfrac{3}{64}}$ $\dfrac{3\sqrt{3}}{8}$ **42.** $\sqrt{\dfrac{7}{25}} - \sqrt{\dfrac{7}{100}}$ $\dfrac{\sqrt{7}}{10}$

43. $2\sqrt{45} - 2\sqrt{20}$ $2\sqrt{5}$ **44.** $5\sqrt{18} + 2\sqrt{32}$ $23\sqrt{2}$

45. $\sqrt{35} - \sqrt{140}$ $-\sqrt{35}$ **46.** $\sqrt{6} - \sqrt{600}$ $-9\sqrt{6}$

47. $5\sqrt{2x} + \sqrt{98x}$ $12\sqrt{2x}$ **48.** $3\sqrt{9x} + 2\sqrt{x}$ $11\sqrt{x}$

49. $5\sqrt{x} + 4\sqrt{4x} - 13\sqrt{x}$ 0   **50.** $\sqrt{9x} + \sqrt{81x} - 11\sqrt{x}$ $\sqrt{x}$

51. $\sqrt{3x^3} + 3x\sqrt{x}$ **52.** $x\sqrt{4x} + \sqrt{9x^3}$ $5x\sqrt{x}$

     $x\sqrt{3x} + 3x\sqrt{x}$

Add or subtract by first simplifying each radical and then combining any like radical terms. Assume that all variables represent positive real numbers. See Example 4.

53. $\sqrt[3]{81} + \sqrt[3]{24}$ $5\sqrt[3]{3}$ **54.** $\sqrt[3]{32} - \sqrt[3]{4}$ $\sqrt[3]{4}$

55. $4\sqrt[3]{9} - \sqrt[3]{243}$ $\sqrt[3]{9}$ **56.** $7\sqrt[3]{6} - \sqrt[3]{48}$ $5\sqrt[3]{6}$

57. $2\sqrt[3]{8} + 2\sqrt[3]{16}$ $4 + 4\sqrt[3]{2}$ **58.** $3\sqrt[3]{27} + 3\sqrt[3]{81}$

59. $\sqrt[3]{8} + \sqrt[3]{54} - 5$ **60.** $\sqrt[3]{64} + \sqrt[3]{14} - 9$

61. $\sqrt{32x^2} + \sqrt[3]{32} + \sqrt{4x^2}$ $4x\sqrt{2} + 2\sqrt[3]{4} + 2x$

62. $\sqrt{18x^2} + \sqrt[3]{24} + \sqrt{2x^2}$ $4x\sqrt{2} + 2\sqrt[3]{3}$

63. $\sqrt{40x} + \sqrt[3]{40} - 2\sqrt{10x} - \sqrt[3]{5}$ $\sqrt[3]{5}$

64. $\sqrt{72x^2} + \sqrt[3]{54} - x\sqrt{50} - 3\sqrt[3]{2}$ $x\sqrt{2}$

58. $9 + 9\sqrt[3]{3}$ **59.** $-3 + 3\sqrt[3]{2}$ **60.** $-5 + \sqrt[3]{14}$

REVIEW AND PREVIEW

Square each binomial. See Section 5.4.

65. $(x + 6)^2$ $x^2 + 12x + 36$

66. $(3x + 2)^2$ $9x^2 + 12x + 4$

67. $(2x - 1)^2$ $4x^2 - 4x + 1$

68. $(x - 5)^2$ $x^2 - 10x + 25$

Solve each system of linear equations. See Section 4.2.

69. $\begin{cases} x = 2y \\ x + 5y = 14 \end{cases}$ $(4, 2)$ **70.** $\begin{cases} y = -5x \\ x + y = 16 \end{cases}$ $(-4, 20)$

Concept Extensions

71. An 8-foot-long water trough is to be made of wood. Each of the two triangular end pieces has an area of $\dfrac{3\sqrt{27}}{4}$ square feet. The two side panels are both rectangular. In simplest radical form, find the total area of the wood needed. $\left(48 + \dfrac{9\sqrt{3}}{2}\right)$ sq. ft

72. Eight wooden braces are to be attached along the diagonals of the vertical sides of a storage bin. Each of four of these diagonals has a length of $\sqrt{52}$ feet, while each of the other four has a length of $\sqrt{80}$ feet. In simplest radical form, find the total length of the wood needed for these braces. $\left(8\sqrt{13} + 16\sqrt{5}\right)$ ft

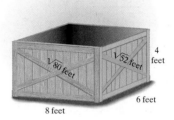

Simplify.

73. $\sqrt{\dfrac{x^3}{16}} - x\sqrt{\dfrac{9x}{25}} + \dfrac{\sqrt{81x^3}}{2}$ $\dfrac{83x\sqrt{x}}{20}$

74. $7\sqrt{x^{11}y^7} - x^2y\sqrt{25x^7y^5} + \sqrt{8x^8y^2}$ $2x^5y^3\sqrt{xy} + 2x^4y\sqrt{2}$

8.4 MULTIPLYING AND DIVIDING RADICALS

Objectives

1 Multiply radicals.

2 Divide radicals.

3 Rationalize denominators.

4 Rationalize using conjugates.

1 In Section 8.2 we used the product and quotient rules for radicals to help us simplify radicals. In this section, we use these rules to simplify products and quotients of radicals.

> ### Product Rule for Radicals
> If $\sqrt[n]{a}$ and $\sqrt[n]{b}$ are real numbers, then
> $$\sqrt[n]{a} \cdot \sqrt[n]{b} = \sqrt[n]{a \cdot b}$$

This property says that the product of the nth roots of two numbers is the nth root of the product of the two numbers. For example,

$$\sqrt{3} \cdot \sqrt{2} = \sqrt{3 \cdot 2} = \sqrt{6}$$

Also,

$$\sqrt[3]{5} \cdot \sqrt[3]{7} = \sqrt[3]{5 \cdot 7} = \sqrt[3]{35}$$

EXAMPLE 1

Multiply. Then simplify if possible.

a. $\sqrt{7} \cdot \sqrt{3}$ **b.** $\sqrt{3} \cdot \sqrt{15}$ **c.** $2\sqrt{6} \cdot 5\sqrt{2}$ **d.** $\left(3\sqrt{2}\right)^2$

Solution

a. $\sqrt{7} \cdot \sqrt{3} = \sqrt{7 \cdot 3} = \sqrt{21}$

b. $\sqrt{3} \cdot \sqrt{15} = \sqrt{45}$. Next, simplify $\sqrt{45}$.

$\sqrt{45} = \sqrt{9 \cdot 5} = \sqrt{9} \cdot \sqrt{5} = 3\sqrt{5}$

c. $2\sqrt{6} \cdot 5\sqrt{2} = 2 \cdot 5\sqrt{6 \cdot 2} = 10\sqrt{12}$. Next, simplify $\sqrt{12}$.

$10\sqrt{12} = 10\sqrt{4 \cdot 3} = 10\sqrt{4} \cdot \sqrt{3} = 10 \cdot 2 \cdot \sqrt{3} = 20\sqrt{3}$

d. $\left(3\sqrt{2}\right)^2 = 3^2 \cdot \left(\sqrt{2}\right)^2 = 9 \cdot 2 = 18$

CLASSROOM EXAMPLE

Multiply. Then simplify if possible.

a. $\sqrt{5} \cdot \sqrt{2}$ **b.** $\sqrt{6} \cdot \sqrt{3}$

c. $7\sqrt{3} \cdot 4\sqrt{15}$ **d.** $\left(5\sqrt{3}\right)^2$

answer:

a. $\sqrt{10}$ **b.** $3\sqrt{2}$ **c.** $84\sqrt{5}$ **d.** 75

EXAMPLE 2

Multiply $\sqrt[3]{4} \cdot \sqrt[3]{18}$. Then simplify if possible.

Solution

$\sqrt[3]{4} \cdot \sqrt[3]{18} = \sqrt[3]{4 \cdot 18} = \sqrt[3]{4 \cdot 2 \cdot 9} = \sqrt[3]{8 \cdot 9} = \sqrt[3]{8} \cdot \sqrt[3]{9} = 2\sqrt[3]{9}$

When multiplying radical expressions containing more than one term, use the same techniques we use to multiply other algebraic expressions with more than one term.

CLASSROOM EXAMPLE

Multiply. $\sqrt[3]{6} \cdot \sqrt[3]{18}$

answer: $3\sqrt[3]{4}$

EXAMPLE 3

Multiply. Then simplify if possible.

a. $\sqrt{5}\left(\sqrt{5} - \sqrt{2}\right)$

b. $\left(\sqrt{x} + \sqrt{2}\right)\left(\sqrt{3} - \sqrt{2}\right)$

Solution

a. Using the distributive property, we have

$$\sqrt{5}\left(\sqrt{5} - \sqrt{2}\right) = \sqrt{5} \cdot \sqrt{5} - \sqrt{5} \cdot \sqrt{2}$$
$$= 5 - \sqrt{10}$$

b. Use the FOIL method of multiplication.

$$\overset{F\qquad O\qquad I\qquad L}{\left(\sqrt{x} + \sqrt{2}\right)\left(\sqrt{3} - \sqrt{2}\right) = \sqrt{x} \cdot \sqrt{3} - \sqrt{x} \cdot \sqrt{2} + \sqrt{2} \cdot \sqrt{3} - \sqrt{2} \cdot \sqrt{2}}$$
$$= \sqrt{3x} - \sqrt{2x} + \sqrt{6} - \sqrt{4} \quad \text{Apply the product rule.}$$
$$= \sqrt{3x} - \sqrt{2x} + \sqrt{6} - 2 \quad \text{Simplify.}$$

CLASSROOM EXAMPLE

Multiply.

a. $\sqrt{7}\left(\sqrt{7} - \sqrt{3}\right)$

b. $\left(\sqrt{x} + \sqrt{5}\right)\left(\sqrt{x} - \sqrt{3}\right)$

answer: **a.** $7 - \sqrt{21}$

b. $x - \sqrt{3x} + \sqrt{5x} - \sqrt{15}$

Special products can be used to multiply expressions containing radicals.

Within Example 3, we found that

$$\sqrt{5} \cdot \sqrt{5} = 5 \quad \text{and} \quad \sqrt{2} \cdot \sqrt{2} = 2$$

This is true in general.

> If a is a positive number,
> $$\sqrt{a} \cdot \sqrt{a} = a$$

✔ **CONCEPT CHECK**

Identify the true statement(s).

a. $\sqrt{7} \cdot \sqrt{7} = 7$ **b.** $\sqrt{2} \cdot \sqrt{3} = 6$ **c.** $\sqrt{131} \cdot \sqrt{131} = 131$
d. $\sqrt{5x} \cdot \sqrt{5x} = 5x$ (Here x is a positive number.)

EXAMPLE 4

Multiply. Then simplify if possible.

a. $\left(\sqrt{5} - 7\right)\left(\sqrt{5} + 7\right)$ **b.** $\left(\sqrt{7x} + 2\right)^2$

Solution

a. Recall from Chapter 5 that $(a - b)(a + b) = a^2 - b^2$. Then

$$\left(\sqrt{5} - 7\right)\left(\sqrt{5} + 7\right) = \left(\sqrt{5}\right)^2 - 7^2$$
$$= 5 - 49$$
$$= -44$$

b. Recall that $(a + b)^2 = a^2 + 2ab + b^2$. Then

$$\left(\sqrt{7x} + 2\right)^2 = \left(\sqrt{7x}\right)^2 + 2\left(\sqrt{7x}\right)(2) + (2)^2$$
$$= 7x + 4\sqrt{7x} + 4$$

2   To simplify quotients of radical expressions, we use the quotient rule.

> ### Quotient Rule for Radicals
> If $\sqrt[n]{a}$ and $\sqrt[n]{b}$ are real numbers and $b \neq 0$, then
> $$\frac{\sqrt[n]{a}}{\sqrt[n]{b}} = \sqrt[n]{\frac{a}{b}}, \text{ providing } b \neq 0$$

EXAMPLE 5

Divide. Then simplify if possible.

a. $\dfrac{\sqrt{14}}{\sqrt{2}}$ **b.** $\dfrac{\sqrt{100}}{\sqrt{5}}$ **c.** $\dfrac{\sqrt{12x^3}}{\sqrt{3x}}$

Solution   Use the quotient rule and then simplify the resulting radicand.

a. $\dfrac{\sqrt{14}}{\sqrt{2}} = \sqrt{\dfrac{14}{2}} = \sqrt{7}$

b. $\dfrac{\sqrt{100}}{\sqrt{5}} = \sqrt{\dfrac{100}{5}} = \sqrt{20} = \sqrt{4 \cdot 5} = \sqrt{4} \cdot \sqrt{5} = 2\sqrt{5}$

c. $\dfrac{\sqrt{12x^3}}{\sqrt{3x}} = \sqrt{\dfrac{12x^3}{3x}} = \sqrt{4x^2} = 2x$

EXAMPLE 6

Divide $\dfrac{\sqrt[3]{32}}{\sqrt[3]{4}}$. Then simplify if possible.

Solution $\dfrac{\sqrt[3]{32}}{\sqrt[3]{4}} = \sqrt[3]{\dfrac{32}{4}} = \sqrt[3]{8} = 2$

CLASSROOM EXAMPLE

Divide. $\dfrac{\sqrt[3]{250}}{\sqrt[3]{2}}$

answer: 5

3 It is sometimes easier to work with radical expressions if the denominator does not contain a radical. To eliminate the radical in the denominator of a radical expression, we use the fact that we can multiply the numerator and the denominator of a fraction by the same nonzero number. This is equivalent to multiplying the fraction by 1. To eliminate the radical in the denominator of $\dfrac{\sqrt{5}}{\sqrt{2}}$, multiply the numerator and the denominator by $\sqrt{2}$. Then

$$\frac{\sqrt{5}}{\sqrt{2}} = \frac{\sqrt{5}\cdot\sqrt{2}}{\sqrt{2}\cdot\sqrt{2}} = \frac{\sqrt{10}}{2}$$

This process is called **rationalizing** the denominator.

EXAMPLE 7

Rationalize each denominator.

    a. $\dfrac{2}{\sqrt{7}}$ **b.** $\dfrac{\sqrt{5}}{\sqrt{12}}$ **c.** $\sqrt{\dfrac{1}{18x}}$

Solution **a.** To eliminate the radical in the denominator of $\dfrac{2}{\sqrt{7}}$, multiply the numerator and the denominator by $\sqrt{7}$.

$$\frac{2}{\sqrt{7}} = \frac{2\cdot\sqrt{7}}{\sqrt{7}\cdot\sqrt{7}} = \frac{2\sqrt{7}}{7}$$

CLASSROOM EXAMPLE

Rationalize each denominator.

  a. $\dfrac{5}{\sqrt{3}}$ **b.** $\dfrac{\sqrt{7}}{\sqrt{20}}$ **c.** $\sqrt{\dfrac{2}{45x}}$

answer:

  a. $\dfrac{5\sqrt{3}}{3}$ **b.** $\dfrac{\sqrt{35}}{10}$ **c.** $\dfrac{\sqrt{10x}}{15x}$

b. We can multiply the numerator and denominator by $\sqrt{12}$, but see what happens if we simplify first.

$$\frac{\sqrt{5}}{\sqrt{12}} = \frac{\sqrt{5}}{\sqrt{4\cdot 3}} = \frac{\sqrt{5}}{2\sqrt{3}}$$

To rationalize the denominator now, multiply the numerator and the denominator by $\sqrt{3}$.

$$\frac{\sqrt{5}}{2\sqrt{3}} = \frac{\sqrt{5}\cdot\sqrt{3}}{2\sqrt{3}\cdot\sqrt{3}} = \frac{\sqrt{15}}{2\cdot 3} = \frac{\sqrt{15}}{6}$$

c. $\sqrt{\dfrac{1}{18x}} = \dfrac{\sqrt{1}}{\sqrt{18x}} = \dfrac{1}{\sqrt{9}\cdot\sqrt{2x}} = \dfrac{1}{3\sqrt{2x}}$

To rationalize the denominator, multiply the numerator and denominator by $\sqrt{2x}$.

$$\frac{1}{3\sqrt{2x}} = \frac{1\cdot\sqrt{2x}}{3\sqrt{2x}\cdot\sqrt{2x}} = \frac{\sqrt{2x}}{3\cdot 2x} = \frac{\sqrt{2x}}{6x}$$

As a general rule, simplify a radical expression first and then rationalize the denominator.

EXAMPLE 8

Rationalize each denominator.

a. $\dfrac{5}{\sqrt[3]{4}}$ **b.** $\dfrac{\sqrt[3]{7}}{\sqrt[3]{3}}$

Solution

a. Since the denominator contains a cube root, we multiply the numerator and the denominator by a factor that gives the **cube root of a perfect cube** in the denominator. Recall that $\sqrt[3]{8} = 2$ and that the denominator $\sqrt[3]{4}$ multiplied by $\sqrt[3]{2}$ is $\sqrt[3]{4 \cdot 2}$ or $\sqrt[3]{8}$.

$$\frac{5}{\sqrt[3]{4}} = \frac{5 \cdot \sqrt[3]{2}}{\sqrt[3]{4} \cdot \sqrt[3]{2}} = \frac{5\sqrt[3]{2}}{\sqrt[3]{8}} = \frac{5\sqrt[3]{2}}{2}$$

b. Recall that $\sqrt[3]{27} = 3$. Multiply the denominator $\sqrt[3]{3}$ by $\sqrt[3]{9}$ and the result is $\sqrt[3]{3 \cdot 9}$ or $\sqrt[3]{27}$.

$$\frac{\sqrt[3]{7}}{\sqrt[3]{3}} = \frac{\sqrt[3]{7} \cdot \sqrt[3]{9}}{\sqrt[3]{3} \cdot \sqrt[3]{9}} = \frac{\sqrt[3]{63}}{\sqrt[3]{27}} = \frac{\sqrt[3]{63}}{3}$$

4   To rationalize a denominator that is a sum, such as the denominator in

$$\frac{2}{4 + \sqrt{3}}$$

we multiply the numerator and the denominator by $4 - \sqrt{3}$. The expressions $4 + \sqrt{3}$ and $4 - \sqrt{3}$ are called **conjugates** of each other. When a radical expression such as $4 + \sqrt{3}$ is multiplied by its conjugate $4 - \sqrt{3}$, the product simplifies to an expression that contains no radicals.

$$(a + b)(a - b) = a^2 - b^2$$
$$\left(4 + \sqrt{3}\right)\left(4 - \sqrt{3}\right) = 4^2 - \left(\sqrt{3}\right)^2 = 16 - 3 = 13$$

Then

$$\frac{2}{4 + \sqrt{3}} = \frac{2\left(4 - \sqrt{3}\right)}{\left(4 + \sqrt{3}\right)\left(4 - \sqrt{3}\right)} = \frac{2\left(4 - \sqrt{3}\right)}{13}$$

EXAMPLE 9

Rationalize each denominator and simplify.

a. $\dfrac{2}{1 + \sqrt{3}}$ **b.** $\dfrac{\sqrt{5} + 4}{\sqrt{5} - 1}$

Solution **a.** Multiply the numerator and the denominator of this fraction by the conjugate of $1 + \sqrt{3}$, that is, by $1 - \sqrt{3}$.

$$\frac{2}{1 + \sqrt{3}} = \frac{2(1 - \sqrt{3})}{(1 + \sqrt{3})(1 - \sqrt{3})}$$

$$= \frac{2(1 - \sqrt{3})}{1^2 - (\sqrt{3})^2}$$

$$= \frac{2(1 - \sqrt{3})}{1 - 3}$$

$$= \frac{2(1 - \sqrt{3})}{-2}$$

$$= -\frac{2(1 - \sqrt{3})}{2} \qquad \frac{a}{-b} = -\frac{a}{b}$$

$$= -1(1 - \sqrt{3}) \qquad \text{Simplify.}$$

$$= -1 + \sqrt{3}$$

b. $\dfrac{\sqrt{5} + 4}{\sqrt{5} - 1} = \dfrac{(\sqrt{5} + 4)(\sqrt{5} + 1)}{(\sqrt{5} - 1)(\sqrt{5} + 1)}$ Multiply the numerator and denominator by $\sqrt{5} + 1$, the conjugate of $\sqrt{5} - 1$.

$$= \frac{5 + \sqrt{5} + 4\sqrt{5} + 4}{5 - 1} \qquad \text{Multiply.}$$

EXAMPLE 10

Simplify $\dfrac{12 - \sqrt{18}}{9}$.

Solution First simplify $\sqrt{18}$.

$$\frac{12 - \sqrt{18}}{9} = \frac{12 - \sqrt{9 \cdot 2}}{9} = \frac{12 - 3\sqrt{2}}{9}$$

Next, factor out a common factor of 3 from the terms in the numerator and the denominator and simplify.

$$\frac{12 - 3\sqrt{2}}{9} = \frac{3(4 - \sqrt{2})}{3 \cdot 3} = \frac{4 - \sqrt{2}}{3}$$

MENTAL MATH

Find each product. Assume that variables represent nonnegative real numbers.

1. $\sqrt{2} \cdot \sqrt{3}$ $\sqrt{6}$

2. $\sqrt{5} \cdot \sqrt{7}$ $\sqrt{35}$

3. $\sqrt{1} \cdot \sqrt{6}$ $\sqrt{6}$

4. $\sqrt{7} \cdot \sqrt{x}$ $\sqrt{7x}$

5. $\sqrt{10} \cdot \sqrt{y}$ $\sqrt{10y}$

6. $\sqrt{x} \cdot \sqrt{y}$ $\sqrt{xy}$

EXERCISE SET 8.4

STUDY GUIDE/SSM   CD/VIDEO   PH MATH TUTOR CENTER   MathXL®Tutorials ON CD   MathXL®   MyMathLab®

Multiply and simplify. See Examples 1, 3, and 4.

1. $\sqrt{8} \cdot \sqrt{2}$ 4

2. $\sqrt{3} \cdot \sqrt{12}$ 6

3. $\sqrt{10} \cdot \sqrt{5}$ $5\sqrt{2}$

4. $3\sqrt{2} \cdot 5\sqrt{14}$ $30\sqrt{7}$

5. $\sqrt{10}(\sqrt{2} + \sqrt{5})$

6. $\sqrt{6}(\sqrt{3} + \sqrt{2})$

7. $(3\sqrt{5} - \sqrt{10})(\sqrt{5} - 4\sqrt{3})$ $15 - 12\sqrt{15} - 5\sqrt{2} + 4\sqrt{30}$

8. $(2\sqrt{3} - 6)(\sqrt{3} - 4\sqrt{2})$ $6 - 8\sqrt{6} - 6\sqrt{3} + 24\sqrt{2}$

9. $(\sqrt{x} + 6)(\sqrt{x} - 6)$

10. $(2\sqrt{5} + 1)(2\sqrt{5} - 1)$

11. $(\sqrt{3} + 8)^2$ $67 + 16\sqrt{3}$

12. $(\sqrt{x} - 7)^2$

13. Find the area of a rectangle whose length is $13\sqrt{2}$ meters and width is $5\sqrt{6}$ meters. $130\sqrt{3}$ sq. m

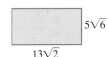

$5\sqrt{6}$

$13\sqrt{2}$

14. Find the volume of a cube whose length is $\sqrt{3}$ feet, width is $\sqrt{2}$ feet, and height is $\sqrt{2}$ feet. $2\sqrt{3}$ cu. ft

$\sqrt{2}$ feet

$\sqrt{3}$ feet

$\sqrt{2}$ feet

5. $2\sqrt{5} + 5\sqrt{2}$ **6.** $3\sqrt{2} + 2\sqrt{3}$

9. $x - 36$ **10.** 19   **12.** $x + 49 - 14\sqrt{x}$

Divide and simplify. See Example 5.

15. $\dfrac{\sqrt{32}}{\sqrt{2}}$ 4

16. $\dfrac{\sqrt{40}}{\sqrt{10}}$ 2

17. $\dfrac{\sqrt{90}}{\sqrt{5}}$ $3\sqrt{2}$

18. $\dfrac{\sqrt{96}}{\sqrt{8}}$ $2\sqrt{3}$

19. $\dfrac{\sqrt{75y^5}}{\sqrt{3y}}$ $5y^2$

20. $\dfrac{\sqrt{24x^7}}{\sqrt{6x}}$ $2x^3$

Rationalize each denominator and simplify. See Example 7.

21. $\sqrt{\dfrac{3}{5}}$ $\dfrac{\sqrt{15}}{5}$

22. $\sqrt{\dfrac{2}{3}}$ $\dfrac{\sqrt{6}}{3}$

23. $\dfrac{1}{\sqrt{6y}}$ $\dfrac{\sqrt{6y}}{6y}$

24. $\dfrac{1}{\sqrt{10z}}$ $\dfrac{\sqrt{10z}}{10z}$

25. $\sqrt{\dfrac{5}{18}}$ $\dfrac{\sqrt{10}}{6}$

26. $\sqrt{\dfrac{7}{12}}$ $\dfrac{\sqrt{21}}{6}$

Rationalize each denominator and simplify. See Example 9.

27. $\dfrac{3}{\sqrt{2} + 1}$ $3\sqrt{2} - 3$

28. $\dfrac{6}{\sqrt{5} + 2}$ $6\sqrt{5} - 12$

29. $\dfrac{2}{\sqrt{10} - 3}$ $2\sqrt{10} + 6$

30. $\dfrac{4}{2 - \sqrt{3}}$ $8 + 4\sqrt{3}$

31. $\dfrac{\sqrt{5} + 1}{\sqrt{6} - \sqrt{5}}$ $\sqrt{30} + 5 + \sqrt{6} + \sqrt{5}$

32. $\dfrac{\sqrt{3} + 1}{\sqrt{3} - \sqrt{2}}$ $3 + \sqrt{6} + \sqrt{3} + \sqrt{2}$

Simplify the following. See Example 10.

33. $\dfrac{6 + 2\sqrt{3}}{2}$ $3 + \sqrt{3}$

34. $\dfrac{9 + 6\sqrt{2}}{3}$ $3 + 2\sqrt{2}$

35. $\dfrac{18 - 12\sqrt{5}}{6}$ $3 - 2\sqrt{5}$

36. $\dfrac{8 - 20\sqrt{3}}{4}$ $2 - 5\sqrt{3}$

37. $\dfrac{15\sqrt{3} + 5}{5}$ $3\sqrt{3} + 1$

38. $\dfrac{8 + 16\sqrt{2}}{8}$ $1 + 2\sqrt{2}$

MIXED PRACTICE

Multiply or divide as indicated and simplify.

39. $2\sqrt{3} \cdot 4\sqrt{15}$ $24\sqrt{5}$

40. $3\sqrt{14} \cdot 4\sqrt{2}$ $24\sqrt{7}$

41. $(2\sqrt{5})^2$ 20

42. $(3\sqrt{10})^2$ 90

43. $(6\sqrt{x})^2$ $36x$

44. $(8\sqrt{y})^2$ $64y$

45. $\sqrt{6}(\sqrt{5} + \sqrt{7})$

46. $\sqrt{10}(\sqrt{3} - \sqrt{7})$

47. $4\sqrt{5x}(\sqrt{x} - 3\sqrt{5})$

48. $3\sqrt{7y}(\sqrt{y} - 2\sqrt{7})$

49. $(\sqrt{3} + \sqrt{5})(\sqrt{2} - \sqrt{5})$ $\sqrt{6} - \sqrt{15} + \sqrt{10} - 5$

50. $(\sqrt{6} + \sqrt{3})(\sqrt{6} - \sqrt{3})$ 3

51. $(\sqrt{7} - 2\sqrt{3})(\sqrt{7} + 2\sqrt{3})$ -5

52. $(\sqrt{2} - 4\sqrt{5})(\sqrt{2} + 4\sqrt{5})$ -78

53. $(\sqrt{x} - 3)(\sqrt{x} + 3)$ $x - 9$

54. $(2\sqrt{y} + 5)(2\sqrt{y} - 5)$ $4y - 25$

55. $(\sqrt{6} + 3)^2$ $15 + 6\sqrt{6}$

56. $(2 + \sqrt{7})^2$ $11 + 4\sqrt{7}$

57. $(3\sqrt{x} - 5)^2$

58. $(2\sqrt{x} - 7)^2$

45. $\sqrt{30} + \sqrt{42}$

46. $\sqrt{30} - \sqrt{70}$

47. $4x\sqrt{5} - 60\sqrt{x}$

48. $3y\sqrt{7} - 42\sqrt{y}$

57. $9x - 30\sqrt{x} + 25$

58. $4x - 28\sqrt{x} + 49$

59. $\dfrac{\sqrt{150}}{\sqrt{2}}$ $5\sqrt{3}$

60. $\dfrac{\sqrt{120}}{\sqrt{3}}$ $2\sqrt{10}$

61. $\dfrac{\sqrt{72y^5}}{\sqrt{3y^3}}$ $2y\sqrt{6}$

62. $\dfrac{\sqrt{54x^3}}{\sqrt{2x}}$ $3x\sqrt{3}$

63. $\dfrac{\sqrt{24x^3y^4}}{\sqrt{2xy}}$ $2xy\sqrt{3y}$

64. $\dfrac{\sqrt{96x^5y^3}}{\sqrt{3x^2y}}$ $4xy\sqrt{2x}$

Rationalize each denominator and simplify.

65. $\sqrt{\dfrac{2}{15}}$ $\quad \dfrac{\sqrt{30}}{15}$

66. $\sqrt{\dfrac{11}{14}}$ $\quad \dfrac{\sqrt{154}}{14}$

67. $\sqrt{\dfrac{3}{20}}$ $\quad \dfrac{\sqrt{15}}{10}$

68. $\sqrt{\dfrac{3}{50}}$ $\quad \dfrac{\sqrt{6}}{10}$

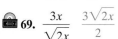 **69.** $\dfrac{3x}{\sqrt{2x}}$ $\quad \dfrac{3\sqrt{2x}}{2}$

70. $\dfrac{5y}{\sqrt{3y}}$ $\quad \dfrac{5\sqrt{3y}}{3}$

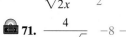 **71.** $\dfrac{4}{2 - \sqrt{5}}$ $\quad -8 - 4\sqrt{5}$

72. $\dfrac{2}{1 - \sqrt{2}}$ $\quad -2 - 2\sqrt{2}$

73. $\dfrac{5}{3 + \sqrt{10}}$ $\quad 5\sqrt{10} - 15$

74. $\dfrac{5}{\sqrt{6} + 2}$ $\quad \dfrac{5\sqrt{6}}{2} - 5$

75. $\dfrac{2\sqrt{3}}{\sqrt{15} + 2}$ $\quad \dfrac{6\sqrt{5} - 4\sqrt{3}}{11}$

76. $\dfrac{3\sqrt{2}}{\sqrt{10} + 2}$ $\quad \sqrt{5} - \sqrt{2}$

77. $\dfrac{\sqrt{3} + 1}{\sqrt{2} - 1}$

78. $\dfrac{\sqrt{2} - 2}{2 - \sqrt{3}}$

77. $\sqrt{6} + \sqrt{3} + \sqrt{2} + 1$

78. $2\sqrt{2} + \sqrt{6} - 2\sqrt{3} - 4$

Multiply or divide as indicated. See Examples 2 and 6.

79. $\sqrt[3]{12} \cdot \sqrt[3]{4}$ $\quad 2\sqrt[3]{6}$

80. $\sqrt[3]{9} \cdot \sqrt[3]{6}$ $\quad 3\sqrt[3]{2}$

81. $2\sqrt[3]{5} \cdot 6\sqrt[3]{2}$ $\quad 12\sqrt[3]{10}$

82. $8\sqrt[3]{4} \cdot 7\sqrt[3]{7}$ $\quad 56\sqrt[3]{28}$

83. $\sqrt[3]{15} \cdot \sqrt[3]{25}$ $\quad 5\sqrt[3]{3}$

84. $\sqrt[3]{4} \cdot \sqrt[3]{4}$ $\quad 2\sqrt[3]{2}$

85. $\dfrac{\sqrt[3]{54}}{\sqrt[3]{2}}$ $\quad 3$

86. $\dfrac{\sqrt[3]{80}}{\sqrt[3]{10}}$ $\quad 2$

87. $\dfrac{\sqrt[3]{120}}{\sqrt[3]{5}}$ $\quad 2\sqrt[3]{3}$

88. $\dfrac{\sqrt[3]{270}}{\sqrt[3]{5}}$ $\quad 3\sqrt[3]{2}$

Rationalize each denominator. See Example 8.

89. $\sqrt[3]{\dfrac{5}{4}}$ $\quad \dfrac{\sqrt[3]{10}}{2}$

90. $\sqrt[3]{\dfrac{7}{9}}$ $\quad \dfrac{\sqrt[3]{21}}{3}$

91. $\dfrac{6}{\sqrt[3]{2}}$ $\quad 3\sqrt[3]{4}$

92. $\dfrac{3}{\sqrt[3]{5}}$ $\quad \dfrac{3\sqrt[3]{25}}{5}$

93. $\sqrt[3]{\dfrac{1}{9}}$ $\quad \dfrac{\sqrt[3]{3}}{3}$

94. $\sqrt[3]{\dfrac{8}{11}}$ $\quad \dfrac{2\sqrt[3]{121}}{11}$

95. $\sqrt[3]{\dfrac{2}{9}}$ $\quad \dfrac{\sqrt[3]{6}}{3}$

96. $\sqrt[3]{\dfrac{3}{4}}$ $\quad \dfrac{\sqrt[3]{6}}{2}$

REVIEW AND PREVIEW

Simplify the following expressions. See Section 7.1.

97. $\dfrac{3x + 12}{3}$ $\quad x + 4$

98. $\dfrac{12x + 8}{4}$ $\quad 3x + 2$

99. $\dfrac{6x^2 - 3x}{3x}$ $\quad 2x - 1$

100. $\dfrac{8y^2 - 2y}{2y}$ $\quad 4y - 1$

Solve each equation. See Sections 2.4 and 5.3.

101. $x + 5 = 7^2$ $\quad 44$

102. $2y - 1 = 3^2$ $\quad 5$

103. $4z^2 + 6z - 12 = (2z)^2$ $\quad 2$

104. $9x^2 + 5x + 4 = (3x + 1)^2$ $\quad 3$

Concept Extensions

△ **105.** If a circle has area A, then the formula for the radius r of the circle is

$$r = \sqrt{\dfrac{A}{\pi}} \qquad \dfrac{\sqrt{A\pi}}{\pi}$$

Simplify this expression by rationalizing the denominator.

△ **106.** If a round ball has volume V, then the formula for the radius r of the ball is

$$r = \sqrt[3]{\dfrac{3V}{4\pi}} \qquad \dfrac{\sqrt[3]{6V\pi^2}}{2\pi}$$

Simplify this expression by rationalizing the denominator.

107. When rationalizing the denominator of $\dfrac{\sqrt{2}}{\sqrt{3}}$, explain why both the numerator and the denominator must be multiplied by $\sqrt{3}$.   answers may vary

108. In your own words, explain why $\sqrt{6} + \sqrt{2}$ cannot be simplified further, but $\sqrt{6} \cdot \sqrt{2}$ can be.   answers may vary

109. When rationalizing the denominator of $\dfrac{\sqrt[3]{2}}{\sqrt[3]{3}}$, explain why both the numerator and the denominator must be multiplied by $\sqrt[3]{9}$.   answers may vary

It is often more convenient to work with a radical expression whose numerator is rationalized. Rationalize the numerator of each expression by multiplying numerator and denominator by the conjugate of the numerator.

110. $\dfrac{\sqrt{3} + 1}{\sqrt{2} - 1}$

$\dfrac{2}{\sqrt{6} - \sqrt{2} - \sqrt{3} + 1}$

111. $\dfrac{\sqrt{2} - 2}{2 - \sqrt{3}}$

$\dfrac{2}{-2\sqrt{2} - 4 + \sqrt{6} + 2\sqrt{3}}$

SIMPLIFYING RADICALS

INTEGRATED REVIEW

Simplify. Assume that all variables represent positive numbers.

1. $\sqrt{36}$ 6
2. $\sqrt{48}$ $4\sqrt{3}$
3. $\sqrt{x^4}$ x^2
4. $\sqrt{y^7}$ $y^3\sqrt{y}$

5. $\sqrt{16x^2}$ $4x$
6. $\sqrt{18x^{11}}$ $3x^5\sqrt{2x}$
7. $\sqrt[3]{8}$ 2
8. $\sqrt[4]{81}$ 3

9. $\sqrt[3]{-27}$ -3
10. $\sqrt{-4}$ not a real number
11. $\sqrt{\frac{11}{9}}$ $\frac{\sqrt{11}}{3}$
12. $\sqrt[3]{\frac{7}{64}}$ $\frac{\sqrt[3]{7}}{4}$

13. $-\sqrt{16}$ -4
14. $-\sqrt{25}$ -5
15. $\sqrt{\frac{9}{49}}$ $\frac{3}{7}$
16. $\sqrt{\frac{1}{64}}$ $\frac{1}{8}$

17. $\sqrt{a^8b^2}$ a^4b
18. $\sqrt{x^{10}y^{20}}$ x^5y^{10}
19. $\sqrt{25m^6}$ $5m^3$
20. $\sqrt{9n^{16}}$ $3n^8$

Add or subtract as indicated.

21. $5\sqrt{7} + \sqrt{7}$ $6\sqrt{7}$
22. $\sqrt{50} - \sqrt{8}$ $3\sqrt{2}$

23. $5\sqrt{2} - 5\sqrt{3}$ cannot be simplified
24. $2\sqrt{x} + \sqrt{25x} - \sqrt{36x} + 3x$ $\sqrt{x} + 3x$

Multiply and simplify if possible.   **29.** $\sqrt{33} + \sqrt{3}$

25. $\sqrt{2}\cdot\sqrt{15}$ $\sqrt{30}$
26. $\sqrt{3}\cdot\sqrt{3}$ 3
27. $(2\sqrt{7})^2$ 28

28. $(3\sqrt{5})^2$ 45
29. $\sqrt{3}(\sqrt{11} + 1)$
30. $\sqrt{6}(\sqrt{3} - 2)$ $3\sqrt{2} - 2\sqrt{6}$

31. $\sqrt{8y}\cdot\sqrt{2y}$ $4y$
32. $\sqrt{15x^2}\cdot\sqrt{3x^2}$ $3x^2\sqrt{5}$
33. $(\sqrt{x} - 5)(\sqrt{x} + 2)$ $x - 3\sqrt{x} - 10$

34. $(3 + \sqrt{2})^2$ $11 + 6\sqrt{2}$

Divide and simplify if possible.

35. $\frac{\sqrt{8}}{\sqrt{2}}$ 2
36. $\frac{\sqrt{45}}{\sqrt{15}}$ $\sqrt{3}$
37. $\frac{\sqrt{24x^5}}{\sqrt{2x}}$ $2x^2\sqrt{3}$
38. $\frac{\sqrt{75a^4b^5}}{\sqrt{5ab}}$ $ab^2\sqrt{15a}$

Rationalize each denominator.

39. $\sqrt{\frac{1}{6}}$ $\frac{\sqrt{6}}{6}$
40. $\frac{x}{\sqrt{20}}$ $\frac{x\sqrt{5}}{10}$
41. $\frac{4}{\sqrt{6}+1}$ $\frac{4\sqrt{6}-4}{5}$
42. $\frac{\sqrt{2}+1}{\sqrt{x}-5}$ $\frac{\sqrt{2x}+5\sqrt{2}+\sqrt{x}+5}{x-25}$

8.5   *SOLVING EQUATIONS CONTAINING RADICALS*

Objectives

1   Solve radical equations by using the squaring property of equality once.

2   Solve radical equations by using the squaring property of equality twice.

1   In this section, we solve **radical equations** such as

$$\sqrt{x + 3} = 5 \quad \text{and} \quad \sqrt{2x + 1} = \sqrt{3x}$$

Radical equations contain variables in the radicand. To solve these equations, we rely on the following squaring property.

> ### The Squaring Property of Equality
> If $a = b$, then $a^2 = b^2$

Unfortunately, this squaring property does not guarantee that all solutions of the new equation are solutions of the original equation. For example, if we square both sides of the equation

$$x = 2$$

we have

$$x^2 = 4$$

This new equation has two solutions, 2 and -2, while the original equation $x = 2$ has only one solution. Thus, raising both sides of the original equation to the second power resulted in an equation that has an **extraneous solution** that isn't a solution of the original equation. For this reason, we must **always check proposed solutions of radical equations in the original equation**. If a proposed solution does not work, we call that value an **extraneous solution**.

EXAMPLE 1

Solve. $\sqrt{x + 3} = 5$

Solution To solve this radical equation, we use the squaring property of equality and square both sides of the equation.

$$
\begin{aligned}
\sqrt{x + 3} &= 5 \\
\left(\sqrt{x + 3}\right)^2 &= 5^2 \quad &&\text{Square both sides.} \\
x + 3 &= 25 \quad &&\text{Simplify.} \\
x &= 22 \quad &&\text{Subtract 3 from both sides.}
\end{aligned}
$$

Check We replace x with 22 in the original equation.

> **Helpful Hint**
> Don't forget to check the proposed solutions of radical equations in the original equation.

$$
\begin{aligned}
\sqrt{x + 3} &= 5 \quad &&\text{Original equation} \\
\sqrt{22 + 3} &\overset{?}{=} 5 \quad &&\text{Let } x = 22. \\
\sqrt{25} &\overset{?}{=} 5 \\
5 &= 5 \quad &&\text{True}
\end{aligned}
$$

Since a true statement results, 22 is the solution.

EXAMPLE 2

Solve. $\sqrt{x} + 6 = 4$

Solution First we set the radical by itself on one side of the equation. Then we square both sides.

$$
\begin{aligned}
\sqrt{x} + 6 &= 4 \\
\sqrt{x} &= -2 \quad &&\text{Subtract 6 from both sides to get the radical by itself.}
\end{aligned}
$$

Recall that $\sqrt{x}$ is the principal or nonnegative square root of x so that $\sqrt{x}$ cannot equal -2 and thus this equation has no solution. We arrive at the same conclusion if we continue by applying the squaring property.

$$\sqrt{x} = -2$$
$$\left(\sqrt{x}\right)^2 = (-2)^2 \quad \text{Square both sides.}$$
$$x = 4 \quad \text{Simplify.}$$

Check   We replace x with 4 in the original equation.

$$\sqrt{x} + 6 = 4 \quad \text{Original equation}$$
$$\sqrt{4} + 6 \stackrel{?}{=} 4 \quad \text{Let } x = 4.$$
$$2 + 6 = 4 \quad \text{False}$$

Since 4 *does not* satisfy the original equation, this equation has no solution.

Example 2 makes it very clear that we *must* check proposed solutions in the original equation to determine if they are truly solutions. Remember, if a proposed solution is not an actual solution, we say that the value is an **extraneous solution**.

The following steps can be used to solve radical equations containing square roots.

Solving a Radical Equation Containing Square Roots

Step 1: Arrange terms so that one radical is by itself on one side of the equation. That is, isolate a radical.

Step 2: Square both sides of the equation.

Step 3: Simplify both sides of the equation.

Step 4: If the equation still contains a radical term, repeat steps 1 through 3.

Step 5: Solve the equation.

Step 6: Check all solutions in the original equation for extraneous solutions.

EXAMPLE 3

Solve $\sqrt{x} = \sqrt{5x - 2}$.

Solution   Each of the radicals is already isolated, since each is by itself on one side of the equation. So we begin solving by squaring both sides.

$$\sqrt{x} = \sqrt{5x - 2} \quad \text{Original equation}$$
$$\left(\sqrt{x}\right)^2 = \left(\sqrt{5x - 2}\right)^2 \quad \text{Square both sides.}$$
$$x = 5x - 2 \quad \text{Simplify.}$$
$$-4x = -2 \quad \text{Subtract } 5x \text{ from both sides.}$$
$$x = \frac{-2}{-4} = \frac{1}{2} \quad \text{Divide both sides by } -4 \text{ and simplify.}$$

CLASSROOM EXAMPLE

Solve $\sqrt{6x - 1} = \sqrt{x}$.

answer: $\dfrac{1}{5}$

Check We replace x with $\dfrac{1}{2}$ in the original equation.

$$\sqrt{x} = \sqrt{5x - 2} \quad \textit{Original equation}$$

$$\sqrt{\dfrac{1}{2}} \overset{?}{=} \sqrt{5 \cdot \dfrac{1}{2} - 2} \quad \textit{Let } x = \dfrac{1}{2}.$$

$$\sqrt{\dfrac{1}{2}} \overset{?}{=} \sqrt{\dfrac{5}{2} - 2} \quad \textit{Multiply.}$$

$$\sqrt{\dfrac{1}{2}} \overset{?}{=} \sqrt{\dfrac{5}{2} - \dfrac{4}{2}} \quad \textit{Write 2 as } \dfrac{4}{2}.$$

$$\sqrt{\dfrac{1}{2}} = \sqrt{\dfrac{1}{2}} \quad \textit{True}$$

This statement is true, so the solution is $\dfrac{1}{2}$.

EXAMPLE 4

Solve. $\sqrt{4y^2 + 5y - 15} = 2y$

Solution The radical is already isolated, so we start by squaring both sides.

$$\sqrt{4y^2 + 5y - 15} = 2y$$
$$\left(\sqrt{4y^2 + 5y - 15}\right)^2 = (2y)^2 \quad \textit{Square both sides.}$$
$$4y^2 + 5y - 15 = 4y^2 \quad \textit{Simplify.}$$
$$5y - 15 = 0 \quad \textit{Subtract } 4y^2 \textit{ from both sides.}$$
$$5y = 15 \quad \textit{Add 15 to both sides.}$$
$$y = 3 \quad \textit{Divide both sides by 5.}$$

Check We replace y with 3 in the original equation.

$$\sqrt{4y^2 + 5y - 15} = 2y \quad \textit{Original equation}$$
$$\sqrt{4 \cdot 3^2 + 5 \cdot 3 - 15} \overset{?}{=} 2 \cdot 3 \quad \textit{Let } y = 3.$$
$$\sqrt{4 \cdot 9 + 15 - 15} \overset{?}{=} 6 \quad \textit{Simplify.}$$
$$\sqrt{36} \overset{?}{=} 6$$
$$6 = 6 \quad \textit{True}$$

This statement is true, so the solution is 3.

EXAMPLE 5

Solve. $\sqrt{x + 3} - x = -3$

Solution First we isolate the radical by adding x to both sides. Then we square both sides.

$$\sqrt{x + 3} - x = -3$$
$$\sqrt{x + 3} = x - 3 \quad \textit{Add } x \textit{ to both sides.}$$
$$\left(\sqrt{x + 3}\right)^2 = (x - 3)^2 \quad \textit{Square both sides.}$$
$$x + 3 = \underbrace{x^2 - 6x + 9}$$

> **Helpful Hint**
> Don't forget that $(x - 3)^2 = (x - 3)(x - 3) = x^2 - 6x + 9$

To solve the resulting quadratic equation, we write the equation in standard form by subtracting x and 3 from both sides.

$$3 = x^2 - 7x + 9 \qquad \text{Subtract } x \text{ from both sides.}$$
$$0 = x^2 - 7x + 6 \qquad \text{Subtract 3 from both sides.}$$
$$0 = (x - 6)(x - 1) \qquad \text{Factor.}$$
$$0 = x - 6 \quad \text{or} \quad 0 = x - 1 \qquad \text{Set each factor equal to zero.}$$
$$6 = x \qquad\qquad 1 = x \qquad \text{Solve for } x.$$

Check   We replace x with 6 and then x with 1 in the original equation.

Let $x = 6$.
$$\sqrt{x+3} - x = -3$$
$$\sqrt{6+3} - 6 \stackrel{?}{=} -3$$
$$\sqrt{9} - 6 \stackrel{?}{=} -3$$
$$3 - 6 \stackrel{?}{=} -3$$
$$-3 = -3 \quad \text{True}$$

Let $x = 1$.
$$\sqrt{x+3} - x = -3$$
$$\sqrt{1+3} - 1 \stackrel{?}{=} -3$$
$$\sqrt{4} - 1 \stackrel{?}{=} -3$$
$$2 - 1 \stackrel{?}{=} -3$$
$$1 = -3 \quad \text{False}$$

Since replacing x with 1 resulted in a false statement, 1 is an extraneous solution. The only solution is 6.

2   If a radical equation contains two radicals, we may need to use the squaring property twice.

EXAMPLE 6

Solve. $\sqrt{x - 4} = \sqrt{x} - 2$

Solution

$$\sqrt{x - 4} = \sqrt{x} - 2$$
$$\left(\sqrt{x - 4}\right)^2 = \left(\sqrt{x} - 2\right)^2 \qquad \text{Square both sides.}$$
$$x - 4 = \underline{x - 4\sqrt{x} + 4}$$
$$-8 = -4\sqrt{x}$$
$$2 = \sqrt{x} \qquad \text{Divide both sides by } -4.$$
$$4 = x \qquad \text{Square both sides again.}$$

Helpful Hint

$$\left(\sqrt{x} - 2\right)^2 = \left(\sqrt{x} - 2\right)\left(\sqrt{x} - 2\right)$$
$$= \sqrt{x} \cdot \sqrt{x} - 2\sqrt{x} - 2\sqrt{x} + 4$$
$$= x - 4\sqrt{x} + 4$$

Check the proposed solution in the original equation. The solution is 4.

EXERCISE SET 8.5

STUDY GUIDE/SSM · CD/ VIDEO · PH MATH TUTOR CENTER · MathXL®Tutorials ON CD · MathXL® · MyMathLab®

MIXED PRACTICE

Solve each equation. See Examples 1 through 5.

1. $\sqrt{x} = 9$ 81
2. $\sqrt{x} = 4$ 16
3. $\sqrt{x + 5} = 2$ -1
4. $\sqrt{x + 12} = 3$ -3
5. $\sqrt{2x + 6} = 4$ 5
6. $\sqrt{3x + 7} = 5$ 6
7. $\sqrt{x - 2} = 5$ 49
8. $4\sqrt{x} - 7 = 5$ 9
9. $3\sqrt{x} + 5 = 2$ no solution
10. $3\sqrt{x} + 5 = 8$ 1
11. $\sqrt{x + 6} + 1 = 3$ -2
12. $\sqrt{x + 5} + 2 = 5$ 4
13. $\sqrt{2x + 1} + 3 = 5$ $\frac{3}{2}$
14. $\sqrt{3x - 1} + 4 = 1$
15. $\sqrt{x + 3} = 7$ 16
16. $\sqrt{x + 5} = 10$ 25
17. $\sqrt{x + 6} + 5 = 3$
18. $\sqrt{2x - 1} + 7 = 1$
19. $\sqrt{4x - 3} = \sqrt{x + 3}$ 2
20. $\sqrt{5x - 4} = \sqrt{x + 8}$ 3
21. $\sqrt{x} = \sqrt{3x - 8}$ 4
22. $\sqrt{x} = \sqrt{4x - 3}$ 1
23. $\sqrt{4x} = \sqrt{2x + 6}$ 3
24. $\sqrt{5x + 6} = \sqrt{8x}$ 2
25. $\sqrt{9x^2 + 2x - 4} = 3x$ 2
26. $\sqrt{4x^2 + 3x - 9} = 2x$ 3
27. $\sqrt{16x^2 - 3x + 6} = 4x$ 2
28. $\sqrt{9x^2 - 2x + 8} = 3x$ 4
29. $\sqrt{16x^2 + 2x + 2} = 4x$
30. $\sqrt{4x^2 + 3x - 2} = 2x$ $\frac{2}{3}$
31. $\sqrt{2x^2 + 6x + 9} = 3$
32. $\sqrt{3x^2 + 6x + 4} = 2$
33. $\sqrt{x + 7} = x + 5$ -3
34. $\sqrt{x + 5} = x - 1$ 4
35. $\sqrt{x} = x - 6$ 9
36. $\sqrt{x} = x + 6$
37. $\sqrt{2x + 1} = x - 7$ 12
38. $\sqrt{2x + 5} = x - 5$ 10
39. $x = \sqrt{2x - 2} + 1$ 3, 1
40. $\sqrt{1 - 8x} + 2 = x$
41. $\sqrt{1 - 8x} - x = 4$ -1
42. $\sqrt{3x + 7} - x = 3$
43. $\sqrt{2x + 5} - 1 = x$ 2
44. $x = \sqrt{4x - 7} + 1$ 2, 4

14. no solution   **17.** no solution   **18.** no solution

Solve each equation. See Example 6.

45. $\sqrt{x - 7} = \sqrt{x} - 1$ 16
46. $\sqrt{x + 2} = \sqrt{x + 24}$ 25
47. $\sqrt{x + 3} = \sqrt{x + 15}$ 1
48. $\sqrt{x - 8} = \sqrt{x} - 2$ 9
49. $\sqrt{x + 8} = \sqrt{x} + 2$ 1
50. $\sqrt{x + 1} = \sqrt{x + 15}$ 49

REVIEW AND PREVIEW

Translate each sentence into an equation and then solve. See Section 2.5.

51. If 8 is subtracted from the product of 3 and x, the result is 19. Find x. $3x - 8 = 19$; $x = 9$

52. If 3 more than x is subtracted from twice x, the result is 11. Find x. $2x - (x + 3) = 11$; $x = 14$

△ **53.** The length of a rectangle is twice the width. The perimeter is 24 inches. Find the length. $2(2x + x) = 24$; length: 8 in.

△ **54.** The length of a rectangle is 2 inches longer than the width. The perimeter is 24 inches. Find the length.
$2x + 2(x + 2) = 24$; length: 7 in.

29. no solution   **31.** 0, -3 **32.** 0, -2 **36.** no solution

Concept Extensions

Solve.

55. A number is 6 more than its principal square root. Find the number.   9

56. A number is 4 more than the principal square root of twice the number. Find the number.   8

△ **57.** The formula $b = \sqrt{\dfrac{V}{2}}$ can be used to determine the length b of a side of the base of a square-based pyramid with height 6 units and volume V cubic units.

a. Find the length of the side of the base that produces a pyramid with each volume. (Round to the nearest tenth of a unit.)

V	20	200	2000
b	3.2	10	31.6

b. Notice in the table that volume V has been increased by a factor of 10 each time. Does the corresponding length b of a side increase by a factor of 10 each time also?   no

c. Solve the formula for V. $V = 2b^2$

△ **58.** The formula $r = \sqrt{\dfrac{V}{2\pi}}$ can be used to determine the radius r of a cylinder with height 2 units and volume V cubic units.

a. Find the radius needed to manufacture a cylinder with each volume. (Round to the nearest tenth of a unit.)

V	10	100	1000
r	1.3	4.0	12.6

2 units

b. Notice in the table that volume V has been increased by a factor of 10 each time. Does the corresponding radius increase by a factor of 10 each time also?   no

c. Solve the formula for V. $V = 2\pi r^2$

40. no solution   **42.** $-1, -2$

59. Explain why proposed solutions of radical equations must be checked in the original equation.   answers may vary

 Graphing calculators can be used to solve equations. To solve $\sqrt{x-2} = x-5$, for example, graph $y_1 = \sqrt{x-2}$ and $y_2 = x-5$ on the same set of axes. Use the TRACE and ZOOM features or an INTERSECT feature to find the point of intersec- *tion of the graphs. The x-value of the point is the solution of the equation. Use a graphing calculator to solve the equations below. Approximate solutions to the nearest hundredth.*

60. $\sqrt{x-2} = x-5$ 7.30   **61.** $\sqrt{x+1} = 2x-3$ 2.43

62. $-\sqrt{x+4} = 5x-6$ 0.76   **63.** $-\sqrt{x+5} = -7x+1$ 0.48

STUDY SKILLS REMINDER

Are You Prepared for a Test on Chapter 8?

Below I have listed some *common trouble areas* for topics covered in Chapter 8. After studying for your test—but before taking your test—read these.

▶ Do you understand the difference between $\sqrt{3} \cdot \sqrt{2}$ and $\sqrt{3} + \sqrt{2}$?

$$\sqrt{3} \cdot \sqrt{2} = \sqrt{3 \cdot 2} = \sqrt{6}$$

$\sqrt{3} + \sqrt{2}$ cannot be simplified further. The terms are unlike terms.

▶ Do you understand the difference between rationalizing the denominator of $\dfrac{\sqrt{3}}{\sqrt{7}}$ and rationalizing the denominator of $\dfrac{\sqrt{3}}{\sqrt{7}+1}$?

$$\frac{\sqrt{3}}{\sqrt{7}} = \frac{\sqrt{3} \cdot \sqrt{7}}{\sqrt{7} \cdot \sqrt{7}} = \frac{\sqrt{21}}{7}$$

$$\frac{\sqrt{3}}{\sqrt{7}+1} = \frac{\sqrt{3}(\sqrt{7}-1)}{(\sqrt{7}+1)(\sqrt{7}-1)}$$

$$= \frac{\sqrt{3}(\sqrt{7}-1)}{7-\sqrt{7}+\sqrt{7}-1} = \frac{\sqrt{3}(\sqrt{7}-1)}{6}$$

▶ To solve an equation containing a radical, don't forget to first isolate the radical

$$\sqrt{x} - 10 = -4$$
$$\sqrt{x} = 6 \qquad \text{Isolate the radical.}$$
$$(\sqrt{x})^2 = 6^2 \qquad \text{Square both sides.}$$
$$x = 36 \qquad \text{Simplify.}$$

Make sure you check the proposed solution in the original equation.

Remember: This is simply a listing of a few common trouble areas. For a review of Chapter 8, see the Highlights and Chapter Review at the end of the Chapter.

8.6 RADICAL EQUATIONS AND PROBLEM SOLVING

TEACHING TIP
After Example 1, demonstrate the Pythagorean theorem by drawing squares on each side of the triangle, finding the area of each square, and noting that the sum of the area of the squares of each leg equals the area of the square of the hypotenuse.

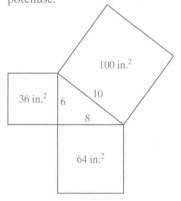

Objectives

1. Use the Pythagorean formula to solve problems.
2. Use the distance formula.
3. Solve problems using formulas containing radicals.

1 Applications of radicals can be found in geometry, finance, science, and other areas of technology. Our first application involves the Pythagorean theorem, giving a formula that relates the lengths of the three sides of a right triangle. We first studied the Pythagorean theorem in Chapter 6 and we review it here.

The Pythagorean Theorem

If a and b are lengths of the legs of a right triangle and c is the length of the hypotenuse, then $a^2 + b^2 = c^2$.

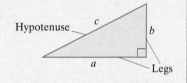

That is, the square of the length of the hypotenuse is equal to the sum of the squares of the lengths of the legs.

⚠ EXAMPLE 1

Find the length of the hypotenuse of a right triangle whose legs are 6 inches and 8 inches long.

Solution Because this is a right triangle, we use the Pythagorean theorem. We let $a = 6$ inches and $b = 8$ inches. Length c must be the length of the hypotenuse.

$$a^2 + b^2 = c^2 \quad \text{Use the Pythagorean theorem.}$$
$$6^2 + 8^2 = c^2 \quad \text{Substitute the lengths of the legs.}$$
$$36 + 64 = c^2 \quad \text{Simplify.}$$
$$100 = c^2$$

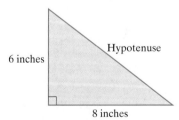

CLASSROOM EXAMPLE
Find the length of the hypotenuse of the right triangle shown.

answer: 5 cm

Since c represents a length, we know that c is positive and is the principal square root of 100.

$$100 = c^2$$
$$\sqrt{100} = c \quad \text{Use the definition of principal square root.}$$
$$10 = c \quad \text{Simplify.}$$

The hypotenuse has a length of 10 inches.

EXAMPLE 2

Find the length of the leg of the right triangle shown. Give the exact length and a two-decimal-place approximation.

Solution We let $a = 2$ meters and b be the unknown length of the other leg. The hypotenuse is $c = 5$ meters.

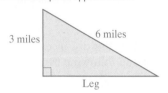
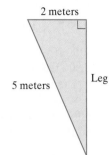

2 meters

5 meters

Leg

$a^2 + b^2 = c^2$ Use the Pythagorean theorem.

$2^2 + b^2 = 5^2$ Let $a = 2$ and $c = 5$.

$4 + b^2 = 25$

$b^2 = 21$

$b = \sqrt{21} \approx 4.58$ meters

The length of the leg is exactly $\sqrt{21}$ meters or approximately 4.58 meters.

EXAMPLE 3

FINDING A DISTANCE

A surveyor must determine the distance across a lake at points P and Q as shown in the figure. To do this, she finds a third point R perpendicular to line PQ. If the length of $\overline{PR}$ is 320 feet and the length of $\overline{QR}$ is 240 feet, what is the distance across the lake? Approximate this distance to the nearest whole foot.

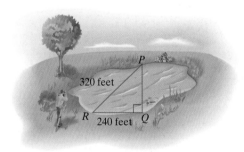

P

320 feet

R   240 feet   Q

Solution   1. UNDERSTAND.   Read and reread the problem. We will set up the problem using the Pythagorean theorem. By creating a line perpendicular to line PQ, the surveyor deliberately constructed a right triangle. The hypotenuse, $\overline{PR}$, has a length of 320 feet, so we let $c = 320$ in the Pythagorean theorem. The side $\overline{QR}$ is one of the legs, so we let $a = 240$ and $b = $ the unknown length.

2. TRANSLATE.

P

$c = 320$

b

R

$a = 240$ Q

$a^2 + b^2 = c^2$ Use the Pythagorean theorem.

$240^2 + b^2 = 320^2$ Let $a = 240$ and $c = 320$.

3. SOLVE.

$$57,600 + b^2 = 102,400$$
$$b^2 = 44,800 \qquad \text{Subtract } \textbf{57,600} \text{ from both sides.}$$
$$b = \sqrt{44,800} \qquad \text{Use the definition of principal square root.}$$

4. INTERPRET.

Check: See that $240^2 + \left(\sqrt{44,800}\right)^2 = 320^2$.

State: The distance across the lake is **exactly** $\sqrt{44,800}$ feet. The surveyor can now use a calculator to find that $\sqrt{44,800}$ feet is **approximately** 211.6601 feet, so the distance across the lake is roughly 212 feet.

2 A second important application of radicals is in finding the distance between two points in the plane. By using the Pythagorean theorem, the following formula can be derived.

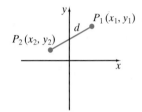

Distance Formula

The distance d between two points with coordinates (x_1, y_1) and (x_2, y_2) is given by

$$d = \sqrt{(x_2 - x_1)^2 + (y_2 - y_1)^2}$$

EXAMPLE 4

Find the distance between $(-1, 9)$ and $(-3, -5)$.

Solution Use the distance formula with $(x_1, y_1) = (-1, 9)$ and $(x_2, y_2) = (-3, -5)$.

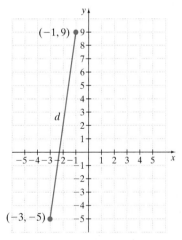

$$
\begin{aligned}
d &= \sqrt{(x_2 - x_1)^2 + (y_2 - y_1)^2} && \text{The distance formula.} \\
&= \sqrt{[-3 - (-1)]^2 + (-5 - 9)^2} && \text{Substitute known values.} \\
&= \sqrt{(-2)^2 + (-14)^2} && \text{Simplify.} \\
&= \sqrt{4 + 196} \\
&= \sqrt{200} = 10\sqrt{2} && \text{Simplify the radical.}
\end{aligned}
$$

The distance is **exactly** $10\sqrt{2}$ units or **approximately** 14.1 units.

3   The Pythagorean theorem is an extremely important result in mathematics and should be memorized. But there are other applications involving formulas containing radicals that are not quite as well known, such as the velocity formula used in the next example.

EXAMPLE 5

DETERMINING VELOCITY

CLASSROOM EXAMPLE

Use the formula in Example 5 to find the velocity of an object after falling 10 feet.
answer: $8\sqrt{10} \approx 25.3$ ft/sec

A formula used to determine the velocity v, in feet per second, of an object (neglecting air resistance) after it has fallen a certain height is $v = \sqrt{2gh}$, where g is the acceleration due to gravity, and h is the height the object has fallen. On Earth, the acceleration g due to gravity is approximately 32 feet per second per second. Find the velocity of a person after falling 5 feet.

Solution   We are told that $g = 32$ feet per second per second. To find the velocity v when $h = 5$ feet, we use the velocity formula.

$$v = \sqrt{2gh} \qquad \text{Use the velocity formula.}$$
$$= \sqrt{2 \cdot 32 \cdot 5} \quad \text{Substitute known values.}$$
$$= \sqrt{320}$$
$$= 8\sqrt{5} \qquad \text{Simplify the radicand.}$$

The velocity of the person after falling 5 feet is **exactly** $8\sqrt{5}$ feet per second, or **approximately** 17.9 feet per second.

Spotlight on
DECISION
∿ MAKING

Suppose you are a carpenter. You are installing a 10-foot by 14-foot wooden deck attached to a client's house. Before sinking the deck posts into the ground, you lay out the dimensions and deck placement with stakes and string. It is very important that the string layout is "square"—that is, that the edges of the deck layout meet at right angles. If not, then the deck posts could be sunk in the wrong positions and portions of the deck will not line up properly.

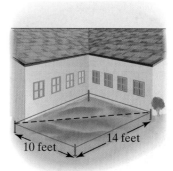

From the Pythagorean theorem, you know that in a right triangle, $a^2 + b^2 = c^2$. It's also true that in a triangle, if $a^2 + b^2 = c^2$, you know that the triangle is a right triangle. This can be used to check that the two edges meet at right angles.

What should the diagonal of the string layout measure if everything is square?

You measure one diagonal of the string layout as 17 feet, $2\frac{15}{32}$ inches. Should any adjustments be made to the string layout? Explain.

EXERCISE SET 8.6

     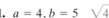

STUDY GUIDE/SSM CD/VIDEO PH MATH TUTOR CENTER MathXL®Tutorials ON CD MathXL® MyMathLab®

Use the Pythagorean theorem to find the unknown side of each right triangle. See Examples 1 and 2.

 1. $\sqrt{13}$ △ **2.** $\sqrt{34}$

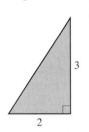 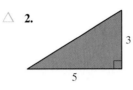

△ **3.** $3\sqrt{3}$

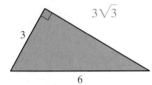

TEACHING TIP
A Group Activity for this section is available in the Instructor's Resource Manual.

△ **4.** $4\sqrt{3}$

△ **5.**

△ **6.**

△ **7.** $\sqrt{22}$ △ **8.** $\sqrt{31}$

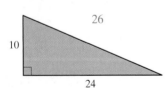 **9.** △ **10.**

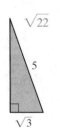 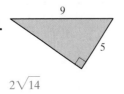

Find the length of the unknown side of each right triangle with sides a, b, and c, where c is the hypotenuse. See Examples 1 and 2.

△ **11.** $a = 4, b = 5$ $\sqrt{41}$ △ **12.** $a = 2, b = 7$ $\sqrt{53}$

△ **13.** $b = 2, c = 6$ $4\sqrt{2}$ △ **14.** $b = 1, c = 5$ $2\sqrt{6}$

△ **15.** $a = \sqrt{10}, c = 10$ $3\sqrt{10}$ △ **16.** $a = \sqrt{7}, c = \sqrt{35}$ $2\sqrt{7}$

Solve. See Examples 1 through 5.

△ **17.** Evan and Noah Saacks want to determine the distance at certain points across a pond on their property. They are able to measure the distances shown on the following diagram. Find how wide the pond is to the nearest tenth of a foot. 51.2 ft

△ **18.** Use the formula from Example 5 and find the velocity of an object after falling 20 feet. $16\sqrt{5} \approx 35.8$ ft per sec

19. A wire is used to anchor a 20-foot-high pole. One end of the wire is attached to the top of the pole. The other end is fastened to a stake five feet away from the bottom of the pole. Find the length of the wire, to the nearest tenth of a foot. 20.6 ft

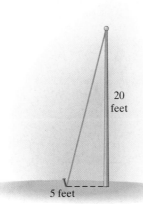

20 feet

5 feet

△ **20.** Jim Spivey needs to connect two underground pipelines, which are offset by 3 feet, as pictured in the diagram. Neglecting the joints needed to join the pipes, find the length of the shortest possible connecting pipe rounded to the nearest hundredth of a foot.   4.24 ft

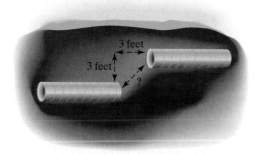

3 feet

3 feet

?

△ **21.** Robert Weisman needs to attach a diagonal brace to a rectangular frame in order to make it structurally sound. If the framework is 6 feet by 10 feet, find how long the brace needs to be to the nearest tenth of a foot.   11.7 ft

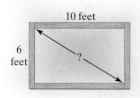

10 feet

6 feet

?

△ **22.** Elizabeth Kaster is flying a kite. She let out 80 feet of string and attached the string to a stake in the ground. The kite is now directly above her brother Mike, who is 32 feet away

from Elizabeth. Find the height of the kite to the nearest foot.   73 ft

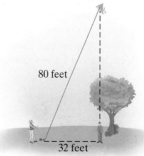

80 feet

32 feet

Use the distance formula to find the distance between the points given. See Example 4.

23. $(3,6),(5,11)$ $\sqrt{29}$
24. $(2,3),(9,7)$ $\sqrt{65}$
25. $(-3,1),(5,-2)$ $\sqrt{73}$
26. $(-2,6),(3,-2)$ $\sqrt{89}$
27. $(3,-2),(1,-8)$ $2\sqrt{10}$
28. $(-5,8),(-2,2)$ $3\sqrt{5}$
29. $\left(\frac{1}{2},2\right),(2,-1)$ $\frac{3\sqrt{5}}{2}$
30. $\left(\frac{1}{3},1\right),(1,-1)$ $\frac{2\sqrt{10}}{3}$
31. $(3,-2),(5,7)$ $\sqrt{85}$
32. $(-2,-3),(-1,4)$ $\sqrt{50}$

Solve each problem. See Example 5.

△ **33.** For a square-based pyramid, the formula $b=\sqrt{\frac{3V}{h}}$ describes the relationship between the length b of one side of the base, the volume V, and the height h. Find the volume if each side of the base is 6 feet long, and the pyramid is 2 feet high.   24 cu. ft

h

b

34. The formula $t=\frac{\sqrt{d}}{4}$ relates the distance d, in feet, that an object falls in t seconds, assuming that air resistance does not slow down the object. Find how long, to the nearest hundredth of a second, it takes an object to reach the ground from the top of the Sears Tower in Chicago, a distance of 1450 feet. (*Source: World Almanac and Book of Facts*)   9.52 sec

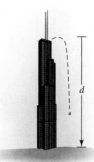

d

35. Police use the formula $s = \sqrt{30fd}$ to estimate the speed s of a car in miles per hour. In this formula, d represents the distance the car skidded in feet and f represents the coefficient of friction. The value of f depends on the type of road surface, and for wet concrete f is 0.35. Find how fast a car was moving if it skidded 280 feet on wet concrete, to the nearest mile per hour.   54 mph

36. The coefficient of friction of a certain dry road is 0.95. Use the formula in Exercise 35 to find how far a car will skid on this dry road if it is traveling at a rate of 60 mph. Round the length to the nearest foot.   126 ft

37. The formula $v = \sqrt{2.5r}$ can be used to estimate the maximum safe velocity, v, in miles per hour, at which a car can travel if it is driven along a curved road with a **radius of curvature**, r, in feet. To the nearest whole number, find the maximum safe speed if a cloverleaf exit on an expressway has a radius of curvature of 300 feet.   27 mph

38. Use the formula from Exercise 37 to find the radius of curvature if the safe velocity is 30 mph.   360 ft

39. The maximum distance d in kilometers that you can see from a height of h meters is given by $d = 3.5\sqrt{h}$. Find how far you can see from the top of the Texas Commerce Tower in Houston, a height of 305.4 meters. Round to the nearest tenth of a kilometer. (*Source: World Almanac and Book of Facts*)

61.2 km

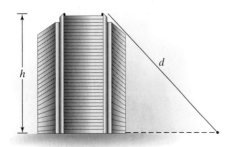

40. Use the formula from Exercise 39 to determine how high above the ground you need to be to see 40 kilometers. Round to the nearest tenth of a meter.   130.6 m

REVIEW AND PREVIEW

Simplify using rules for exponents. See Section 5.1.

41. 2^5 32

42. $(-3)^3$ -27

43. $\left(-\dfrac{1}{5}\right)^2$ $\dfrac{1}{25}$

44. $\left(\dfrac{2}{7}\right)^3$ $\dfrac{8}{343}$

45. $x^2 \cdot x^3$ x^5

46. $x^4 \cdot x^2$ x^6

47. $y^3 \cdot y$ y^4

48. $x \cdot x^7$ x^8

Concept Extensions

For each triangle, find the length of x.

△ **49.**

$2\sqrt{10} - 4$

△ **50.**

$10\sqrt{2} - 12$

Solve.

△ **51.** Mike and Sandra Hallahan leave the seashore at the same time. Mike drives northward at a rate of 30 miles per hour, while Sandra drives west at 60 mph. Find how far apart they are after 3 hours to the nearest mile.   201 mi

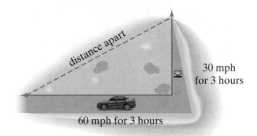

△ **52.** Railroad tracks are invariably made up of relatively short sections of rail connected by expansion joints. To see why this construction is necessary, consider a single rail 100 feet long (or 1200 inches). On an extremely hot day, suppose it expands 1 inch in the hot sun to a new length of 1201 inches. Theoretically, the track would bow upward as pictured.

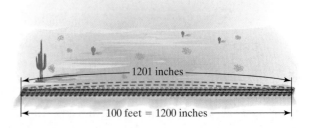

Let us approximate the bulge in the railroad this way. Calculate the height h of the bulge.   24.5 in.

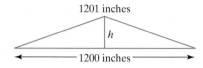

1201 inches

h

1200 inches

53. Based on the results of Exercise 52, explain why railroads use short sections of rail connected by expansion joints. answers may vary

---

8.7   RATIONAL EXPONENTS

Objectives

1. Evaluate exponential expressions of the form $a^{1/n}$.
2. Evaluate exponential expressions of the form $a^{m/n}$.
3. Evaluate exponential expressions of the form $a^{-m/n}$.
4. Use rules for exponents to simplify expressions containing fractional exponents.

1   Radical notation is widely used, as we've seen. In this section, we study an alternate notation, one that proves to be more efficient and compact. This alternate notation makes use of expressions containing an exponent that is a rational number but not necessarily an integer, for example,

$$3^{1/2}, \ 2^{-3/4}, \ \text{and} \ y^{5/6}$$

In giving meaning to rational exponents, keep in mind that we want the rules for operating with them to be the same as the rules for operating with integer exponents. For this to be true,

$$\left(3^{1/2}\right)^2 = 3^{1/2 \cdot 2} = 3^1 = 3$$

Also, we know that

$$\left(\sqrt{3}\right)^2 = 3$$

Since the square of both $3^{1/2}$ and $\sqrt{3}$ is 3, it would be reasonable to say that

$$3^{1/2} \ \text{means} \ \sqrt{3}$$

In general, we have the following.

> **Definition of $a^{1/n}$**
>
> If n is a positive integer and $\sqrt[n]{a}$ is a real number, then
> $$a^{1/n} = \sqrt[n]{a}$$

Notice that the denominator of the rational exponent is the same as the index of the corresponding radical.

 EXAMPLE 1

Write in radical notation. Then simplify.

a. $25^{1/2}$ **b.** $8^{1/3}$ **c.** $-16^{1/4}$ **d.** $(-27)^{1/3}$ **e.** $\left(\dfrac{1}{9}\right)^{1/2}$

Solution   **a.** $25^{1/2} = \sqrt{25} = 5$

b. $8^{1/3} = \sqrt[3]{8} = 2$

c. In $-16^{1/4}$, the base of the exponent is 16. Thus **the negative sign is not affected by the exponent**; so $-16^{1/4} = -\sqrt[4]{16} = -2$.

d. The parentheses show that -27 is the base. $(-27)^{1/3} = \sqrt[3]{-27} = -3$.

e. $\left(\dfrac{1}{9}\right)^{1/2} = \sqrt{\dfrac{1}{9}} = \dfrac{1}{3}$

2   In Example 1, each rational exponent has a numerator of 1. What happens if the numerator is some other positive integer? Consider $8^{2/3}$. Since $\frac{2}{3}$ is the same $\frac{1}{3} \cdot 2$, we reason that

$$8^{2/3} = 8^{(1/3)2} = (8^{1/3})^2 = (\sqrt[3]{8})^2 = 2^2 = 4$$

The denominator 3 of the rational exponent is the same as the index of the radical. The numerator 2 of the fractional exponent indicates that the radical base is to be squared.

> ## Definition of $a^{m/n}$
>
> If m and n are integers with $n > 0$ and if a is a positive number, then
> $$a^{m/n} = (a^{1/n})^m = \left(\sqrt[n]{a}\right)^m$$
>
> Also,
> $$a^{m/n} = (a^m)^{1/n} = \sqrt[n]{a^m}$$

EXAMPLE 2

Simplify each expression.

a. $4^{3/2}$ **b.** $27^{2/3}$ **c.** $-16^{3/4}$

Solution   **a.** $4^{3/2} = (4^{1/2})^3 = \left(\sqrt{4}\right)^3 = 2^3 = 8$ **b.** $27^{2/3} = (27^{1/3})^2 = \left(\sqrt[3]{27}\right)^2 = 3^2 = 9$

c. The negative sign is **not** affected by the exponent since the base of the exponent is 16. Thus, $-16^{3/4} = -(16^{1/4})^3 = -\left(\sqrt[4]{16}\right)^3 = -2^3 = -8$.

> ## Helpful Hint
>
> Recall that
> $$-3^2 = -(3 \cdot 3) = -9$$
> and
> $$(-3)^2 = (-3)(-3) = 9$$
> In other words, without parentheses the exponent 2 applies to the base 3, **not** -3. The same is true of rational exponents. For example,
> $$-16^{1/2} = -\sqrt{16} = -4$$
> and
> $$(-27)^{1/3} = \sqrt[3]{-27} = -3$$

3 If the exponent is a negative rational number, use the following definition.

Definition of $a^{-m/n}$

If $a^{-m/n}$ is a nonzero real number, then

$$a^{-m/n} = \frac{1}{a^{m/n}}$$

EXAMPLE 3

Write each expression with a positive exponent and then simplify.

a. $36^{-1/2}$ **b.** $16^{-3/4}$ **c.** $-9^{1/2}$ **d.** $32^{-4/5}$

Solution

a. $36^{-1/2} = \dfrac{1}{36^{1/2}} = \dfrac{1}{\sqrt{36}} = \dfrac{1}{6}$

b. $16^{-3/4} = \dfrac{1}{16^{3/4}} = \dfrac{1}{\left(\sqrt[4]{16}\right)^3} = \dfrac{1}{2^3} = \dfrac{1}{8}$

c. $-9^{1/2} = -\sqrt{9} = -3$

d. $32^{-4/5} = \dfrac{1}{32^{4/5}} = \dfrac{1}{\left(\sqrt[5]{32}\right)^4} = \dfrac{1}{2^4} = \dfrac{1}{16}$

4 It can be shown that the properties of integer exponents hold for rational exponents. By using these properties and definitions, we can now simplify products and quotients of expressions containing rational exponents.

EXAMPLE 4

Simplify each expression. Write results with positive exponents only. Assume that all variables represent positive numbers.

a. $3^{1/2} \cdot 3^{3/2}$ **b.** $\dfrac{5^{1/3}}{5^{2/3}}$ **c.** $\left(x^{1/4}\right)^{12}$ **d.** $\dfrac{x^{1/5}}{x^{-4/5}}$ **e.** $\left(\dfrac{y^{3/5}}{z^{1/4}}\right)^2$

Solution

a. $3^{1/2} \cdot 3^{3/2} = 3^{(1/2)+(3/2)} = 3^{4/2} = 3^2 = 9$

b. $\dfrac{5^{1/3}}{5^{2/3}} = 5^{(1/3)-(2/3)} = 5^{-1/3} = \dfrac{1}{5^{1/3}}$

c. $\left(x^{1/4}\right)^{12} = x^{(1/4)12} = x^3$

d. $\dfrac{x^{1/5}}{x^{-4/5}} = x^{(1/5)-(-4/5)} = x^{5/5} = x^1 \text{ or } x$

e. $\left(\dfrac{y^{3/5}}{z^{1/4}}\right)^2 = \dfrac{y^{(3/5)2}}{z^{(1/4)2}} = \dfrac{y^{6/5}}{z^{1/2}}$

EXERCISE SET 8.7

STUDY GUIDE/SSM · CD/VIDEO · PH MATH TUTOR CENTER · MathXL®Tutorials ON CD · MathXL® · MyMathLab®

Simplify each expression. See Examples 1 and 2.

1. $8^{1/3}$ 2

2. $16^{1/4}$ 2

3. $9^{1/2}$ 3

4. $16^{1/2}$ 4

5. $16^{3/4}$ 8

6. $27^{4/3}$ 81

 7. $32^{2/5}$ 4

8. $-64^{5/6}$ -32

Simplify each expression. See Example 3.

9. $-16^{-1/4}$ $-\dfrac{1}{2}$

10. $-8^{-1/3}$ $-\dfrac{1}{2}$

11. $16^{-3/2}$ $\dfrac{1}{64}$

12. $27^{-4/3}$ $\dfrac{1}{81}$

13. $81^{-3/2}$ $\dfrac{1}{729}$

14. $32^{-2/5}$ $\dfrac{1}{4}$

15. $\left(\dfrac{4}{25}\right)^{-1/2}$ $\dfrac{5}{2}$

16. $\left(\dfrac{8}{27}\right)^{-1/3}$ $\dfrac{3}{2}$

17. Explain the meaning of the numbers 2, 3, and 4 in the exponential expression $4^{3/2}$. answers may vary

18. Explain why $-4^{1/2}$ is a real number but $(-4)^{1/2}$ is not. answers may vary

Simplify each expression. Write each answer with positive exponents. Assume that all variables represent positive numbers. See Example 4.

19. $2^{1/3} \cdot 2^{2/3}$ 2

20. $4^{2/5} \cdot 4^{3/5}$ 4

21. $\dfrac{4^{3/4}}{4^{1/4}}$ 2

22. $\dfrac{9^{7/2}}{9^{3/2}}$ 81

23. $\dfrac{x^{1/6}}{x^{5/6}}$ $\dfrac{1}{x^{2/3}}$

24. $\dfrac{x^{1/4}}{x^{3/4}}$ $\dfrac{1}{x^{1/2}}$

25. $\left(x^{1/2}\right)^6$ x^3

26. $\left(x^{1/3}\right)^6$ x^2

27. Explain how simplifying $x^{1/2} \cdot x^{1/3}$ is similar to simplifying $x^2 \cdot x^3$. answers may vary

28. Explain how simplifying $\left(x^{1/2}\right)^{1/3}$ is similar to simplifying $\left(x^2\right)^3$. answers may vary

MIXED PRACTICE

Simplify each expression.

29. $81^{1/2}$ 9

30. $(-32)^{6/5}$ 64

31. $(-8)^{1/3}$ -2

32. $36^{1/2}$ 6

33. $-81^{1/4}$ -3

34. $-64^{1/3}$ -4

35. $\left(\dfrac{1}{81}\right)^{1/2}$ $\dfrac{1}{9}$

36. $\left(\dfrac{9}{16}\right)^{1/2}$ $\dfrac{3}{4}$

37. $\left(\dfrac{27}{64}\right)^{1/3}$ $\dfrac{3}{4}$

38. $\left(\dfrac{16}{81}\right)^{1/4}$ $\dfrac{2}{3}$

39. $9^{3/2}$ 27

40. $16^{3/2}$ 64

41. $64^{3/2}$ 512

42. $64^{2/3}$ 16

43. $-8^{2/3}$ -4

44. $8^{2/3}$ 4

45. $4^{5/2}$ 32

46. $9^{4/2}$ 81

47. $\left(\dfrac{4}{9}\right)^{3/2}$ $\dfrac{8}{27}$

48. $\left(\dfrac{8}{27}\right)^{2/3}$ $\dfrac{4}{9}$

49. $\left(\dfrac{1}{81}\right)^{3/4}$ $\dfrac{1}{27}$

50. $\left(\dfrac{1}{32}\right)^{3/5}$ $\dfrac{1}{8}$

51. $4^{-1/2}$ $\dfrac{1}{2}$

52. $9^{-1/2}$ $\dfrac{1}{3}$

53. $125^{-1/3}$ $\dfrac{1}{5}$

54. $216^{-1/3}$ $\dfrac{1}{6}$

55. $625^{-3/4}$ $\dfrac{1}{125}$

56. $256^{-5/8}$ $\dfrac{1}{32}$

Simplify each expression. Write each answer with positive exponents. Assume that all variables represent positive numbers.

57. $3^{4/3} \cdot 3^{2/3}$ 9

58. $2^{5/4} \cdot 2^{3/4}$ 4

59. $\dfrac{6^{2/3}}{6^{1/3}}$ $6^{1/3}$

60. $\dfrac{3^{3/5}}{3^{1/5}}$ $3^{2/5}$

61. $\left(x^{2/3}\right)^9$ x^6

62. $\left(x^6\right)^{3/4}$ $x^{9/2}$

63. $\dfrac{6^{1/3}}{6^{-5/3}}$ 36

64. $\dfrac{2^{-3/4}}{2^{5/4}}$ $\dfrac{1}{4}$

65. $\dfrac{3^{-3/5}}{3^{2/5}}$ $\dfrac{1}{3}$

66. $\dfrac{5^{1/4}}{5^{-3/4}}$ 5

67. $\left(\dfrac{x^{1/3}}{y^{3/4}}\right)^2$ $\dfrac{x^{2/3}}{y^{3/2}}$

68. $\left(\dfrac{x^{1/2}}{y^{2/3}}\right)^6$ $\dfrac{x^3}{y^4}$

69. $\left(\dfrac{x^{2/5}}{y^{3/4}}\right)^8$ $\dfrac{x^{16/5}}{y^6}$

70. $\left(\dfrac{x^{3/4}}{y^{1/6}}\right)^3$ $\dfrac{x^{9/4}}{y^{1/2}}$

REVIEW AND PREVIEW

Solve each system of linear inequalities by graphing on a single coordinate system. See Section 4.5. See graphing answer section.

71. $\begin{cases} x + y < 6 \\ y \geq 2x \end{cases}$

72. $\begin{cases} 2x - y \geq 3 \\ x < 5 \end{cases}$

Solve each quadratic equation. See Section 6.5.

73. $x^2 - 4 = 3x$ $-1, 4$

74. $x^2 + 2x = 8$ $-4, 2$

75. $2x^2 - 5x - 3 = 0$ $-\frac{1}{2}, 3$

76. $3x^2 + x - 2 = 0$ $\frac{2}{3}, -1$

Concept Extensions

77. If a population grows at a rate of 8% annually, the formula $P = P_O(1.08)^N$ can be used to estimate the total population P after N years have passed, assuming the original population is P_O. Find the population after $1\frac{1}{2}$ years if the original population of 10,000 people is growing at a rate of 8% annually.   11,224 people

78. Money grows in a certain savings account at a rate of 4% compounded annually. The amount of money A in the account at time t is given by the formula

$$A = P(1.04)^t$$

where P is the original amount deposited in the account. Find the amount of money in the account after $3\frac{3}{4}$ years if $200 was initially deposited. $231.69

Use a calculator and approximate each to three decimal places.

79. $5^{3/4}$ 3.344

80. $20^{1/8}$ 1.454

81. $18^{3/5}$ 5.665

82. $42^{3/10}$ 3.069

CHAPTER 8 PROJECT

Investigating the Dimensions of Cylinders

The volume V (in cubic units) of a cylinder is given by the formula $V = \pi r^2 h$, where r is the radius of the cylinder and h is its height. In this project, you will investigate the radii of several cylinders by completing the table below.

For this project, you will need several empty cans of different sizes, a 2-cup (16-fluid-ounce) transparent measuring cup with metric markings (in milliliters), a metric ruler, and water. This project may be completed by working in groups or individually.

△ **1.** For each can, measure its volume by filling it with water and pouring the water into the measuring cup. Find the volume of the water in milliliters (ml). Record the volumes of the cans in the table. (Remember that 1 ml = 1 cm³.)

△ **2.** Use a ruler to measure the height of each can in centimeters (cm). Record the heights in the table.

△ **3.** Solve the formula $V = \pi r^2 h$ for the radius r.

△ **4.** Use your formula from Question 3 to calculate an estimate of each can's radius based on the volume and height measurements recorded in the table. Record these calculated radii in the table.

△ **5.** Try to measure the radius of each can and record these measured radii in the table. (Remember that radius = $\frac{1}{2}$ diameter.)

△ **6.** How close are the values of the calculated radius and the measured radius of each can? What factors could account for the differences?

Can	Volume (ml)	Height (cm)	Calculated Radius (cm)	Measured Radius (cm)
A				
B				
C				
D				

8 CHAPTER VOCABULARY CHECK

Fill in each blank with one of the words or phrases listed below.

index	rationalizing the denominator	principal square root	radicand
conjugate	radical	like radicals	

1. The expressions $5\sqrt{x}$ and $7\sqrt{x}$ are examples of <u>like radicals</u>.

2. In the expression $\sqrt[3]{45}$ the number 3 is the <u>index</u>, the number 45 is the <u>radicand</u>, and $\sqrt{}$ is called the <u>radical</u> sign.

3. The <u>conjugate</u> of $(a + b)$ is $(a - b)$.

4. The <u>principal square root</u> of 25 is 5.

5. The process of eliminating the radical in the denominator of a radical expression is called <u>rationalizing the denominator</u>.

CHAPTER 8 HIGHLIGHTS

Definitions and Concepts	Examples
Section 8.1 Introduction to Radicals	
The **positive** or **principal square root** of a positive number a is written as $\sqrt{a}$. The **negative square root** of a is written as $-\sqrt{a}$. $\sqrt{a} = b$ only if $b^2 = a$ and $b > 0$.	$\sqrt{25} = 5 \qquad \sqrt{100} = 10$ $-\sqrt{9} = -3 \qquad \sqrt{\dfrac{4}{49}} = \dfrac{2}{7}$
A square root of a negative number is not a real number.	$\sqrt{-4}$ is not a real number.
The ***n*th root** of a number a is written as $\sqrt[n]{a}$ and $\sqrt[n]{a} = b$ only if $b^n = a$.	$\sqrt[3]{64} = 4 \qquad \sqrt[3]{-8} = -2$ $\sqrt[4]{81} = 3$
The natural number n is called the **index**, the symbol $\sqrt{}$ is called a **radical**, and the expression within the radical is called the **radicand**.	$\sqrt[5]{-32} = -2$
(Note: If the index is even, the radicand must be nonnegative for the root to be a real number.)	
Section 8.2 Simplifying Radicals	
Product rule for radicals If $\sqrt[n]{a}$ and $\sqrt[n]{b}$ are real numbers, then $\sqrt[n]{a} \cdot \sqrt[n]{b} = \sqrt[n]{a \cdot b}$. A square root is in **simplified form** if the radicand contains no perfect square factors other than 1. To simplify a square root, factor the radicand so that one of its factors is a perfect square factor.	$\sqrt{2} \cdot \sqrt{3} = \sqrt{6}$ $\sqrt[3]{7} \cdot \sqrt[3]{2} = \sqrt[3]{14}$ $\sqrt{45} = \sqrt{9 \cdot 5}$ $\phantom{\sqrt{45}} = \sqrt{9} \cdot \sqrt{5}$ $\phantom{\sqrt{45}} = 3\sqrt{5} \qquad$ in simplest form.

(continued)

Definitions and Concepts	**Examples**

Section 8.2 Simplifying Radicals

To simplify cube roots, factor the radicand so that one of its factors is a perfect cube.

$$\sqrt[3]{48} = \sqrt[3]{8 \cdot 6}$$
$$= \sqrt[3]{8} \cdot \sqrt[3]{6}$$
$$= 2\sqrt[3]{6}$$

Quotient rule for radicals

If $\sqrt[n]{a}$ and $\sqrt[n]{b}$ are real numbers and $b \neq 0$, then

$$\sqrt[n]{\frac{a}{b}} = \frac{\sqrt[n]{a}}{\sqrt[n]{b}}$$

$$\sqrt{\frac{18}{x^6}} = \frac{\sqrt{9 \cdot 2}}{\sqrt{x^6}} = \frac{\sqrt{9} \cdot \sqrt{2}}{x^3} = \frac{3\sqrt{2}}{x^3}$$

$$\sqrt[3]{\frac{18}{x^6}} = \frac{\sqrt[3]{18}}{\sqrt[3]{x^6}} = \frac{\sqrt[3]{18}}{x^2}$$

Section 8.3 Adding and Subtracting Radicals

Like radicals are radical expressions that have the same index and the same radicand.

To combine like radicals, use the distributive property.

Like Radicals

$$5\sqrt{2}, -7\sqrt{2}, \sqrt{2}$$
$$-\sqrt[3]{11}, 3\sqrt[3]{11}$$
$$2\sqrt{7} - 13\sqrt{7} = (2 - 13)\sqrt{7} = -11\sqrt{7}$$
$$\sqrt[3]{24} + \sqrt[3]{8} + \sqrt[3]{81}$$
$$= \sqrt[3]{8 \cdot 3} + 2 + \sqrt[3]{27 \cdot 3}$$
$$= \sqrt[3]{8} \cdot \sqrt[3]{3} + 2 + \sqrt[3]{27} \cdot \sqrt[3]{3}$$
$$= 2\sqrt[3]{3} + 2 + 3\sqrt[3]{3}$$
$$= (2 + 3)\sqrt[3]{3} + 2$$
$$= 5\sqrt[3]{3} + 2$$

Section 8.4 Multiplying and Dividing Radicals

The product and quotient rules for radicals may be used to simplify products and quotients of radicals.

Perform indicated operations and simplify.

$$\left(2\sqrt{5}\right)^2 = 2^2 \cdot \left(\sqrt{5}\right)^2 = 4 \cdot 5 = 20$$

Multiply.

$$\left(\sqrt{3x} + 1\right)\left(\sqrt{5} - \sqrt{3}\right)$$
$$= \sqrt{15x} - \sqrt{9x} + \sqrt{5} - \sqrt{3}$$
$$= \sqrt{15x} - 3\sqrt{x} + \sqrt{5} - \sqrt{3}$$

$$\frac{\sqrt[3]{56x^4}}{\sqrt[3]{7x}} = \sqrt[3]{\frac{56x^4}{7x}} = \sqrt[3]{8x^3} = 2x$$

The process of eliminating the radical in the denominator of a radical expression is called **rationalizing the denominator**.

Rationalize the denominator.

$$\frac{5}{\sqrt{11}} = \frac{5 \cdot \sqrt{11}}{\sqrt{11} \cdot \sqrt{11}} = \frac{5\sqrt{11}}{11}$$

(continued)

Definitions and Concepts	Examples

Section 8.4 Multiplying and Dividing Radicals

The **conjugate** of $a + b$ is $a - b$.

To rationalize a denominator that is a sum or difference of radicals, multiply the numerator and the denominator by the conjugate of the denominator.

The conjugate of $2 + \sqrt{3}$ is $2 - \sqrt{3}$.

Rationalize the denominator.

$$\frac{5}{6 - \sqrt{5}} = \frac{5(6 + \sqrt{5})}{(6 - \sqrt{5})(6 + \sqrt{5})}$$

$$= \frac{5(6 + \sqrt{5})}{36 + 6\sqrt{5} - 6\sqrt{5} - 5}$$

$$= \frac{5(6 + \sqrt{5})}{31}$$

Section 8.5 Solving Equations Containing Radicals

To solve a radical equation containing square roots

Step 1: Get one radical by itself on one side of the equation.

Step 2: Square both sides of the equation.

Step 3: Simplify both sides of the equation.

Step 4: If the equation still contains a radical term, repeat steps 1 through 3.

Step 5: Solve the equation.

Step 6: Check solutions in the original equation.

Solve $\sqrt{2x - 1} - x = -2$.

$$\sqrt{2x - 1} = x - 2$$
$$\left(\sqrt{2x - 1}\right)^2 = (x - 2)^2 \quad \text{Square both sides.}$$
$$2x - 1 = x^2 - 4x + 4$$
$$0 = x^2 - 6x + 5$$
$$0 = (x - 1)(x - 5) \quad \text{Factor.}$$
$$x - 1 = 0 \quad \text{or} \quad x - 5 = 0$$
$$x = 1 \quad \text{or} \quad x = 5 \quad \text{Solve.}$$

Check both proposed solutions in the original equation. 5 checks but 1 does not. The only solution is 5.

Section 8.6 Radical Equations and Problem Solving

Problem-solving steps

1. UNDERSTAND.   Read and reread the problem.

A gutter is mounted on the eaves of a house 15 feet above the ground. A garden is adjacent to the house so that the closest a ladder can be placed to the house is 6 feet. How long a ladder is needed for installing the gutter? Let x = the length of the ladder.

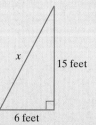

2. TRANSLATE.

Here, we use the Pythagorean theorem. The unknown length x is the hypotenuse.

In words:

$$(\text{leg})^2 \quad + \quad (\text{leg})^2 \quad = \quad (\text{hypotenuse})^2$$

(continued)

Definitions and Concepts	**Examples**

Section 8.6 Radical Equations and Problem Solving

3. Solve.

Translate:
$$6^2 + 15^2 = x^2$$
$$36 + 225 = x^2$$
$$261 = x^2$$
$$\sqrt{261} = x \quad \text{or} \quad x = 3\sqrt{29}$$

4. Interpret.

Check and state. The ladder needs to be $3\sqrt{29}$ feet or approximately 16.2 feet long.

Section 8.7 Rational Exponents

If n is a positive integer and $\sqrt[n]{a}$ is a real number, then $a^{1/n} = \sqrt[n]{a}$.

If m and n are integers with $n > 0$ and a is positive, then
$$a^{m/n} = (a^{1/n})^m = (\sqrt[n]{a})^m$$

Also
$$a^{m/n} = (a^m)^{1/n} = \sqrt[n]{a^m}$$

If $a^{m/n}$ is a nonzero real number, then
$$a^{-m/n} = \frac{1}{a^{m/n}}$$

Properties for integer exponents hold for rational exponents also.
$$a^m \cdot a^n = a^{m+n}$$
$$(a^m)^n = a^{mn}$$
$$\frac{a^m}{a^n} = a^{m-n}$$

$$9^{1/2} = \sqrt{9} = 3$$
$$(-8)^{1/3} = \sqrt[3]{-8} = -2$$

$$25^{3/2} = (25^{1/2})^3 = (\sqrt{25})^3 = 5^3 \text{ or } 125$$

$$81^{-3/4} = \frac{1}{81^{3/4}} = \frac{1}{(\sqrt[4]{81})^3} = \frac{1}{3^3} = \frac{1}{27}$$

$$x^{1/2} \cdot x^{1/4} = x^{(1/2)+(1/4)} = x^{(2/4)+(1/4)} = x^{3/4}$$
$$(x^{2/3})^{1/5} = x^{(2/3)\cdot(1/5)} = x^{2/15}$$
$$\frac{x^{5/6}}{x^{1/6}} = x^{(5/6)-(1/6)} = x^{4/6} = x^{2/3}$$

8 CHAPTER REVIEW

(8.1) *Find the root. Indicate if the expression is not a real number.*

1. $\sqrt{81}$ 9

2. $-\sqrt{49}$ -7

3. $\sqrt[3]{27}$ 3

4. $\sqrt[4]{16}$ 2

5. $-\sqrt{\dfrac{9}{64}}$ $-\dfrac{3}{8}$

6. $\sqrt{\dfrac{36}{81}}$ $\dfrac{2}{3}$

7. $\sqrt[4]{-\dfrac{16}{81}}$ not a real number

8. $\sqrt[3]{-\dfrac{27}{64}}$ $-\dfrac{3}{4}$

Determine whether each of the following is rational or irrational. If rational, find the exact value. If irrational, use a calculator to find an approximation accurate to three decimal places.

9. $\sqrt{76}$ irrational, 8.718

10. $\sqrt{576}$ rational, 24

Find the following roots. Assume that variables represent positive numbers only.

11. $\sqrt{x^{12}}$ x^6

12. $\sqrt{x^8}$ x^4

13. $\sqrt{9x^6}$ $3x^3$

14. $\sqrt{25x^4}$ $5x^2$

15. $\sqrt{\dfrac{16}{y^{10}}}$ $\dfrac{4}{y^5}$

16. $\sqrt{\dfrac{y^{12}}{49}}$ $\dfrac{y^6}{7}$

(8.2) *Simplify each expression using the product rule. Assume that variables represent nonnegative real numbers.*

17. $\sqrt{54}$ $3\sqrt{6}$

18. $\sqrt{88}$ $2\sqrt{22}$

19. $\sqrt{150x^3}$ $5x\sqrt{6x}$

20. $\sqrt{92y^5}$ $2y^2\sqrt{23y}$

21. $\sqrt[3]{54}$ $3\sqrt[3]{2}$

22. $\sqrt[3]{88}$ $2\sqrt[3]{11}$

23. $\sqrt[4]{48}$ $2\sqrt[4]{3}$

24. $\sqrt[4]{162}$ $3\sqrt[4]{2}$

Simplify each expression using the quotient rule. Assume that variables represent positive real numbers.

25. $\sqrt{\dfrac{18}{25}}$ $\dfrac{3\sqrt{2}}{5}$

26. $\sqrt{\dfrac{75}{64}}$ $\dfrac{5\sqrt{3}}{8}$

27. $\sqrt{\dfrac{45y^2}{4x^4}}$ $\dfrac{3y\sqrt{5}}{2x^2}$

28. $\sqrt{\dfrac{20x^5}{9x^2}}$ $\dfrac{2x\sqrt{5x}}{3}$

29. $\sqrt[4]{\dfrac{9}{16}}$ $\dfrac{\sqrt[4]{9}}{2}$

30. $\sqrt[3]{\dfrac{40}{27}}$ $\dfrac{2\sqrt[3]{5}}{3}$

31. $\sqrt[3]{\dfrac{3}{8}}$ $\dfrac{\sqrt[3]{3}}{2}$

32. $\sqrt[4]{\dfrac{5}{81}}$ $\dfrac{\sqrt[4]{5}}{3}$

(8.3) *Add or subtract by combining like radicals.*

33. $3\sqrt[3]{2} + 2\sqrt[3]{3} - 4\sqrt[3]{2}$ $2\sqrt[3]{3} - \sqrt[3]{2}$

34. $5\sqrt{2} + 2\sqrt[3]{2} - 8\sqrt{2}$ $-3\sqrt{2} + 2\sqrt[3]{2}$

35. $\sqrt{6} + 2\sqrt[3]{6} - 4\sqrt[3]{6} + 5\sqrt{6}$ $6\sqrt{6} - 2\sqrt[3]{6}$

36. $3\sqrt{5} - \sqrt[3]{5} - 2\sqrt{5} + 3\sqrt[3]{5}$ $\sqrt{5} + 2\sqrt[3]{5}$

Add or subtract by simplifying each radical and then combining like terms. Assume that variables represent nonnegative real numbers.

37. $\sqrt{28x} + \sqrt{63x} + \sqrt[3]{56}$ $5\sqrt{7x} + 2\sqrt[3]{7}$

38. $\sqrt{75y} + \sqrt{48y} - \sqrt[4]{16}$ $9\sqrt{3y} - 2$

39. $\sqrt{\dfrac{5}{9}} - \sqrt{\dfrac{5}{36}}$ $\dfrac{\sqrt{5}}{6}$

40. $\sqrt{\dfrac{11}{25}} + \sqrt{\dfrac{11}{16}}$ $\dfrac{9\sqrt{11}}{20}$

41. $2\sqrt[3]{125} - 5\sqrt[3]{8}$ 0

42. $3\sqrt[3]{16} - 2\sqrt[3]{2}$ $4\sqrt[3]{2}$

(8.4) *Find the product and simplify if possible.*

43. $3\sqrt{10} \cdot 2\sqrt{5}$ $30\sqrt{2}$

44. $2\sqrt[3]{4} \cdot 5\sqrt[3]{6}$ $20\sqrt[3]{3}$

45. $\sqrt{3}(2\sqrt{6} - 3\sqrt{12})$

46. $4\sqrt{5}(2\sqrt{10} - 5\sqrt{5})$

47. $(\sqrt{3} + 2)(\sqrt{6} - 5)$

48. $(2\sqrt{5} + 1)(4\sqrt{5} - 3)$

Find the quotient and simplify if possible. Assume that variables represent positive real numbers.

49. $\dfrac{\sqrt{96}}{\sqrt{3}}$ $4\sqrt{2}$

50. $\dfrac{\sqrt{160}}{\sqrt{8}}$ $2\sqrt{5}$

51. $\dfrac{\sqrt{15x^6}}{\sqrt{12x^3}}$ $\dfrac{x\sqrt{5x}}{2}$

52. $\dfrac{\sqrt{50y^8}}{\sqrt{72y^3}}$ $\dfrac{5y^2\sqrt{y}}{6}$

Rationalize each denominator and simplify.

53. $\sqrt{\dfrac{5}{6}}$ $\dfrac{\sqrt{30}}{6}$

54. $\sqrt{\dfrac{7}{10}}$ $\dfrac{\sqrt{70}}{10}$

55. $\sqrt{\dfrac{3}{2x}}$ $\dfrac{\sqrt{6x}}{2x}$

56. $\sqrt{\dfrac{6}{5y}}$ $\dfrac{\sqrt{30y}}{5y}$

57. $\sqrt{\dfrac{7}{20y^2}}$ $\dfrac{\sqrt{35}}{10y}$

58. $\sqrt{\dfrac{5z}{12x^2}}$ $\dfrac{\sqrt{15z}}{6x}$

59. $\sqrt[3]{\dfrac{7}{9}}$ $\dfrac{\sqrt[3]{21}}{3}$

60. $\sqrt[3]{\dfrac{3}{4}}$ $\dfrac{\sqrt[3]{6}}{2}$

61. $\sqrt[3]{\dfrac{3}{2}}$ $\dfrac{\sqrt[3]{12}}{2}$

62. $\sqrt[3]{\dfrac{5}{4}}$ $\dfrac{\sqrt[3]{10}}{2}$

63. $\dfrac{3}{\sqrt{5} - 2}$ $3\sqrt{5} + 6$

64. $\dfrac{8}{\sqrt{10} - 3}$ $8\sqrt{10} + 24$

65. $\dfrac{8}{\sqrt{6} + 2}$ $4\sqrt{6} - 8$

66. $\dfrac{12}{\sqrt{15} - 3}$ $2\sqrt{15} + 6$

67. $\dfrac{\sqrt{2}}{4 + \sqrt{2}}$ $\dfrac{2\sqrt{2} - 1}{7}$

68. $\dfrac{\sqrt{3}}{5 + \sqrt{3}}$ $\dfrac{5\sqrt{3} - 3}{22}$

69. $\dfrac{2\sqrt{3}}{\sqrt{3} - 5}$ $-\dfrac{3 + 5\sqrt{3}}{11}$

70. $\dfrac{7\sqrt{2}}{\sqrt{2} - 4}$ $-1 - 2\sqrt{2}$

(8.5) *Solve the following radical equations.*

71. $\sqrt{2x} = 6$ 18

72. $\sqrt{x + 3} = 4$ 13

73. $\sqrt{x + 3} = 8$ 25

74. $\sqrt{x + 8} = 3$ no solution

75. $\sqrt{2x + 1} = x - 7$ 12

76. $\sqrt{3x + 1} = x - 1$ 5

77. $\sqrt{x + 3} + x = 9$ 6

78. $\sqrt{2x} + x = 4$ 2

(8.6) *Use the Pythagorean theorem to find the length of the unknown side.*

△ **79.** $2\sqrt{14}$

△ **80.** $\sqrt{117}$

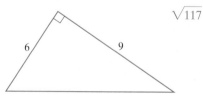

45. $6\sqrt{2} - 18$ **46.** $40\sqrt{2} - 100$ **47.** $3\sqrt{2} - 5\sqrt{3} + 2\sqrt{6} - 10$ **48.** $37 - 2\sqrt{5}$

Solve.

△ **81.** Romeo is standing 20 feet away from the wall below Juliet's balcony during a school play. Juliet is on the balcony, 12 feet above the ground. Find how far apart Romeo and Juliet are. $4\sqrt{34}$ ft

△ **82.** The diagonal of a rectangle is 10 inches long. If the width of the rectangle is 5 inches, find the length of the rectangle. $5\sqrt{3}$ in.

Use the distance formula to find the distance between the points.

83. $(6, -2)$ and $(-3, 5)$ $\sqrt{130}$ **84.** $(2, 8)$ and $(-6, 10)$ $2\sqrt{17}$

Use the formula $r = \sqrt{\dfrac{S}{4\pi}}$, where r = the radius of a sphere and S = the surface area of the sphere, for Exercises 85 and 86.

△ **85.** Find the radius of a sphere to the nearest tenth of an inch if the area is 72 square inches.   2.4 in.

△ **86.** Find the exact surface area of a sphere if its radius is 6 inches. (Do not approximate π.) 144π sq. in.

(8.7) *Write each of the following with fractional exponents and simplify if possible. Assume that variables represent nonnegative real numbers.*

87. $\sqrt{a^5}$ $a^{5/2}$ **88.** $\sqrt[5]{a^3}$ $a^{3/5}$

89. $\sqrt[6]{x^{15}}$ $x^{5/2}$ **90.** $\sqrt[4]{x^{12}}$ x^3

Simplify each of the following expressions.

91. $16^{1/2}$ 4   **92.** $36^{1/2}$ 6

93. $(-8)^{1/3}$ -2 **94.** $(-32)^{1/5}$ -2

95. $-64^{3/2}$ -512 **96.** $-8^{2/3}$ -4

97. $\left(\dfrac{16}{81}\right)^{3/4}$ $\dfrac{8}{27}$ **98.** $\left(\dfrac{9}{25}\right)^{3/2}$ $\dfrac{27}{125}$

99. $25^{-1/2}$ $\dfrac{1}{5}$ **100.** $64^{-2/3}$ $\dfrac{1}{16}$

Simplify each expression using positive exponents only. Assume that variables represent positive real numbers.

101. $8^{1/3} \cdot 8^{4/3}$ 32   **102.** $4^{3/2} \cdot 4^{1/2}$ 16

103. $\dfrac{3^{1/6}}{3^{5/6}}$ $\dfrac{1}{3^{2/3}}$ **104.** $\dfrac{2^{1/4}}{2^{-3/5}}$ $2^{17/20}$

105. $(x^{-1/3})^6$ $\dfrac{1}{x^2}$ **106.** $\left(\dfrac{x^{1/2}}{y^{1/3}}\right)^2$ $\dfrac{x}{y^{2/3}}$

STUDY SKILLS REMINDER

Are You Prepared for Your Final Exam?

To prepare for your final exam, try the following study techniques.

▶ Review the material that you will be responsible for on your exam. Also check your notebook for any lecture notes that you high-lighted.

▶ Review any formulas that you may need to memorize.

▶ Check to see if your instructor or math department will be conducting a final exam review.

▶ Check with your instructor to see whether there are final exams from previous semesters/quarters that are available to students for study.

▶ Use your previously taken tests as a practice final exam. To do so, rewrite the test questions in mixed order on blank sheets of paper. This will help you prepare for exam conditions.

▶ If you are unsure of a few topics, see your instructor or visit a learning lab for further assistance. Also, viewing the video segment of a troublesome section will help.

▶ If you need further exercises to work, try the chapter tests and cumulative reviews at the end of appropriate chapters.

Good luck!

CHAPTER 8 TEST   Remember to use your Chapter Test Prep Video CD to help you study and view solutions to the test questions you need help on.

Simplify the following. Indicate if the expression is not a real number.

1. $\sqrt{16}$ 4

2. $\sqrt[3]{125}$ 5

3. $16^{3/4}$ 8

4. $\left(\dfrac{9}{16}\right)^{1/2}$ $\dfrac{3}{4}$

5. $\sqrt[4]{-81}$ not a real number

6. $27^{-2/3}$ $\dfrac{1}{9}$

Simplify each radical expression. Assume that variables represent positive numbers only.   **15.** $3\sqrt[3]{2} - 2x\sqrt{2}$

7. $\sqrt{54}$ $3\sqrt{6}$

8. $\sqrt{92}$ $2\sqrt{23}$

9. $\sqrt{3x^6}$ $x^3\sqrt{3}$

10. $\sqrt{8x^4y^7}$ $2x^2y^3\sqrt{2y}$

11. $\sqrt{9x^9}$ $3x^4\sqrt{x}$

12. $\sqrt[3]{8}$ 2

13. $\sqrt[3]{40}$ $2\sqrt[3]{5}$

14. $\sqrt{12} - 2\sqrt{75}$ $-8\sqrt{3}$

15. $\sqrt{2x^2} + \sqrt[3]{54} - x\sqrt{18}$

16. $\sqrt{\dfrac{5}{16}}$ $\dfrac{\sqrt{5}}{4}$

17. $\sqrt[3]{\dfrac{2}{27}}$ $\dfrac{\sqrt[3]{2}}{3}$

18. $3\sqrt{8x}$ $6\sqrt{2x}$

Multiply.

19. $\sqrt{5}\left(\sqrt{5} + 2\sqrt{7}\right)$ $5 + 2\sqrt{35}$

20. $\left(2\sqrt{x} + 3\right)\left(2\sqrt{x} - 3\right)$ $4x - 9$

Rationalize the denominator.

21. $\sqrt{\dfrac{2}{3}}$ $\dfrac{\sqrt{6}}{3}$

22. $\sqrt[3]{\dfrac{5}{9}}$ $\dfrac{\sqrt[3]{15}}{3}$

23. $\sqrt{\dfrac{5}{12x^2}}$ $\dfrac{\sqrt{15}}{6x}$

24. $\dfrac{2\sqrt{3}}{\sqrt{3} - 3}$ $-1 - \sqrt{3}$

Solve each of the following radical equations.

25. $\sqrt{x} + 8 = 11$ 9

26. $\sqrt{3x - 6} = \sqrt{x + 4}$ 5

27. $\sqrt{2x - 2} = x - 5$ 9

△ **28.** Find the length of the unknown leg of a right triangle if the other leg is 8 inches long and the hypotenuse is 12 inches long. $4\sqrt{5}$ in.

29. Find the distance between $(-3, 6)$ and $(-2, 8)$. $\sqrt{5}$

Simplify each expression using positive exponents only.

30. $16^{-3/4} \cdot 16^{-1/4}$ $\dfrac{1}{16}$

31. $\left(\dfrac{x^{2/3}}{y^{2/5}}\right)^5$ $\dfrac{x^{10/3}}{y^2}$

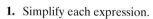 **CHAPTER CUMULATIVE REVIEW**

1. Simplify each expression.   **b.** 2 (Sec. 1.7, Ex. 9)

a. $\dfrac{(-12)(-3) + 3}{-7 - (-2)}$ $-\dfrac{39}{5}$

b. $\dfrac{2(-3)^2 - 20}{-5 + 4}$

2. Simplify each expression.

a. $\dfrac{4(-3) - (-6)}{-8 + 4}$ $\dfrac{3}{2}$

b. $\dfrac{3 + (-3)(-2)^3}{-1 - (-4)}$ 9 (Sec. 1.7)

3. Solve. $2x + 3x - 5 + 7 = 10x + 3 - 6x - 4$

4. Solve. $6y - 11 + 4 + 2y = 8 + 15y - 8y$ 15 (Sec. 2.2)

5. Complete the table for the equation $y = 3x$.

x	y	x	y
-1		-1	-3
	0	0	0
	-9	-3	-9

(Sec. 3.1. Ex. 5)

6. Complete the table for the equation $2x + y = 6$.

x	y	x	y
0		0	6
	-2	4	-2
3		3	0

(Sec. 3.1)

7. Find an equation of the line with y-intercept $(0, -3)$ and slope of $\frac{1}{4}$. $y = \frac{1}{4}x - 3$ (Sec. 3.5, Ex. 5)

8. Find an equation of a line with y-intercept $(0, 4)$ and slope of -2. $y = -2x + 4$ (Sec. 3.5)

9. Find an equation of the line parallel to the line $y = 5$ and passing through $(-2, -3)$. $y = -3$ (Sec. 3.6, Ex. 4)

10. Find an equation of the line perpendicular to $y = 2x + 4$ and passing through $(1, 5)$. $y = -\dfrac{1}{2}x + \dfrac{11}{2}$ (Sec. 3.6)

3. -3 (Sec. 2.2, Ex. 5)

11. Which of the following linear equations are functions?

   a. $y = x$

   b. $y = 2x + 1$

   c. $y = 5$

   d. $x = -1$ a, b, c, (Sec. 3.7, Ex. 5)

12. Which of the following linear equations are functions?

   a. $2x + 3 = y$ **19.** 5% saline solution: 2.5 L; 25% saline

   b. $x + 4 = 0$ solution: 7.5 L; (Sec. 4.4, Ex. 4)

   c. $\frac{1}{2}y = 2x$

   d. $y = 0$ a, c, d (Sec. 3.7)

13. Which of the following ordered pairs is a solution of the given system?

$$\begin{cases} 2x - 3y = 6 \\ x = 2y \end{cases}$$

   a. $(12, 6)$ solution

   b. $(0, -2)$ not a solution (Sec. 4.1, Ex. 1)

14. Which of the following ordered pairs is a solution of the given system?

$$\begin{cases} 2x + y = 4 \\ x + y = 2 \end{cases}$$

   a. $(1, 1)$ not a solution

   b. $(2, 0)$ solution (Sec. 4.1)

15. Solve the system:

$$\begin{cases} 2x + y = 10 \\ x = y + 2 \end{cases} \quad (4, 2) \text{ (Sec. 4.2, Ex. 1)}$$

16. Solve the system:

$$\begin{cases} 3y = x + 10 \\ 2x + 5y = 24 \end{cases} \quad (2, 4) \text{ (Sec. 4.2)}$$

17. Solve the system:

$$\begin{cases} -x - \dfrac{y}{2} = \dfrac{5}{2} \\ -\dfrac{x}{2} + \dfrac{y}{4} = 0 \end{cases} \left(-\dfrac{5}{4}, -\dfrac{5}{2}\right) \text{ (Sec. 4.3, Ex. 6.)}$$

18. Solve the system:

$$\begin{cases} \dfrac{x}{2} + y = \dfrac{5}{6} \\ 2x - y = \dfrac{5}{6} \end{cases} \left(\dfrac{2}{3}, \dfrac{1}{2}\right) \text{ (Sec. 4.3)}$$

19. Eric Daly, a chemistry teaching assistant, needs 10 liters of a 20% saline solution (salt water) for his 2 P.M. laboratory class. Unfortunately, the only mixtures on hand are a 5% saline solution and a 25% saline solution. How much of each solution should he mix to produce the 20% solution?

33. $\dfrac{x(3x - 1)}{(x + 1)^2(x - 1)}$ (Sec. 7.4, Ex. 7)   **34.** $\dfrac{3x^2 - 2x + 2}{(x + 2)(x + 3)(x - 1)}$ (Sec. 7.4)

20. Two streetcars are 11 miles apart and traveling toward each other on parallel tracks. They meet in 12 minutes. Find the speed of each streetcar if one travels 15 miles per hour faster than the other.   20 mph, 35 mph (Sec. 4.4)

21. Graph the solution of the system:

$$\begin{cases} 3x \geq y \\ x + 2y \leq 8 \end{cases} \text{ See graphing answer section. (Sec. 4.6, Ex. 1)}$$

22. Graph the solution of the system:

$$\begin{cases} x + y \leq 1 \\ 2x - y \geq 2 \end{cases} \text{ See graphing answer section. (Sec. 4.6)}$$

23. Combine like terms to simplify. $-9x^2 + 3xy - 5y^2 + 7xy$
$10xy - 5y^2 - 9x^2$ (Sec. 5.2, Ex. 7)

24. Combine like terms to simplify.
$4a^2 + 3a - 2a^2 + 7a - 5$ $2a^2 + 10a - 5$ (Sec. 5.2)

25. Factor: $x^2 + 5yx + 6y^2$ $(x + 2y)(x + 3y)$ (Sec. 6.2, Ex. 6)

26. Factor: $3x^2 + 15x + 18$ $3(x + 2)(x + 3)$ (Sec. 6.2)

27. Simplify: $\dfrac{4 - x^2}{3x^2 - 5x - 2}$ $\dfrac{-2 - x}{3x + 1}$ (Sec. 7.1, Ex. 8)

28. Simplify: $\dfrac{2x^2 + 7x + 3}{x^2 - 9}$ $\dfrac{2x + 1}{x - 3}$ (Sec. 7.1)

29. Divide: $\dfrac{3x^3y^7}{40} \div \dfrac{4x^3}{y^2}$ $\dfrac{3y^9}{160}$ (Sec. 7.2, Ex. 4)

30. Divide: $\dfrac{12x^2y^3}{5} \div \dfrac{3y^3}{x}$ $\dfrac{4x^3}{5}$ (Sec. 7.2)

31. Subtract: $\dfrac{2y}{2y - 7} - \dfrac{7}{2y - 7}$ 1 (Sec. 7.3, Ex. 2)

32. Subtract: $\dfrac{-4x^2}{x + 1} - \dfrac{4x}{x + 1}$ $-4x$ (Sec. 7.3)

33. Add: $\dfrac{2x}{x^2 + 2x + 1} + \dfrac{x}{x^2 - 1}$

34. Add: $\dfrac{3x}{x^2 + 5x + 6} + \dfrac{1}{x^2 + 2x - 3}$.

35. Solve: $\dfrac{x}{2} + \dfrac{8}{3} = \dfrac{1}{6}$ -5 (Sec. 7.5, Ex. 1)

36. Solve: $\dfrac{1}{21} + \dfrac{x}{7} = \dfrac{5}{3}$ $\dfrac{34}{3}$ (Sec. 7.5)

37. If the following two triangles are similar, find the missing length x.   15 yd (Sec. 7.6, Ex. 3)

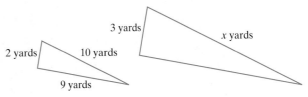

38. If the following two triangles are similar, find the missing length. $\dfrac{25}{2}$ (Sec. 7.6)

 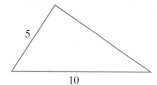

39. Simplify: $\dfrac{\dfrac{1}{z} - \dfrac{1}{2}}{\dfrac{1}{3} - \dfrac{z}{6}}$ $\dfrac{3}{z}$ (Sec. 7.8, Ex. 3)

40. Simplify. $\dfrac{x + 3}{\dfrac{1}{x} + \dfrac{1}{3}}$ $3x$ (Sec. 7.8)

41. Simplify.

    a. $\sqrt{54}$ $3\sqrt{6}$

    b. $\sqrt{12}$ $2\sqrt{3}$

    c. $\sqrt{200}$ $10\sqrt{2}$

    d. $\sqrt{35}$ $\sqrt{35}$ (Sec. 8.2, Ex. 1)

42. Simplify.

    a. $\sqrt{40}$ $2\sqrt{10}$

    b. $\sqrt{500}$ $10\sqrt{5}$

    c. $\sqrt{63}$ $3\sqrt{7}$

    d. $\sqrt{169}$ 13 (Sec. 8.2)

43. Multiply. Then simplify, if possible.

    a. $\left(\sqrt{5} - 7\right)\left(\sqrt{5} + 7\right)$ -44

    b. $\left(\sqrt{7x} + 2\right)^2$ $7x + 4\sqrt{7x} + 4$ (Sec. 8.4, Ex. 4)

44. Multiply. Then simplify, if possible.

    a. $\left(\sqrt{6} + 2\right)^2$ $10 + 4\sqrt{6}$

    b. $\left(\sqrt{x} + 5\right)\left(\sqrt{x} - 5\right)$ $x - 25$ (Sec. 8.4)

45. Solve. $\sqrt{x} + 6 = 4$ no solution (Sec. 8.5, Ex. 2)

46. Solve. $\sqrt{x + 4} = \sqrt{3x - 1}$ $\dfrac{5}{2}$ (Sec. 8.5)

47. Find the length of the hypotenuse of a right triangle whose legs are 6 inches and 8 inches long. 10 in. (Sec. 8.6, Ex. 1)

48. Find the length of the unknown leg of a right triangle whose other leg is 9 and whose hypotenuse is 13. $2\sqrt{22}$ (Sec. 8.6)

49. Simplify each expression.

    a. $4^{3/2}$ 8

    b. $27^{2/3}$ 9

    c. $-16^{3/4}$ -8 (Sec. 8.7, Ex. 2)

50. Simplify each expression.

    a. $9^{5/2}$ 243

    b. $-81^{1/4}$ -3

    c. $(-64)^{2/3}$ 16 (Sec. 8.7)

Solving Quadratic Equations

An important part of the study of algebra is learning to use methods for solving equations. In Chapter 2, we presented techniques for solving linear equations in one variable. In Chapter 6, we solved quadratic equations in one variable by factoring the quadratic expressions. We now present other methods for solving quadratic equations in one variable.

OUTDOOR OPPORTUNITIES

Professional landscapers, landscape technicians, and landscape architects work together to design, install, and maintain attractive outdoor planting schemes in parks, around buildings, and in homeowners' yards. There are over 70,000 professional lawn care and landscape companies in the United States, employing more than 623,000 people. In recent years, the demand for lawn care and landscape services has skyrocketed.

Landscapers find math and geometry useful in situations such as estimating and pricing landscaping jobs, mixing fertilizers or pesticides, and designing or laying out areas to be landscaped.

In the Spotlight on Decision Making feature on page 570, you will have the opportunity as a landscaper to make a decision about a flower bed design.

9.1   SOLVING QUADRATIC EQUATIONS BY THE SQUARE ROOT METHOD

Objectives

1. Use the square root property to solve quadratic equations.
2. Solve problems modeled by quadratic equations.

1   Recall that a quadratic equation is an equation that can be written in the form

$$ax^2 + bx + c = 0$$

where a, b, and c are real numbers and $a \neq 0$.

To solve quadratic equations by factoring, use the **zero factor theorem**: If the product of two numbers is zero, then at least one of the two numbers is zero. For example, to solve $x^2 - 4 = 0$, we first factor the left side of the equation and then set each factor equal to 0.

$$x^2 - 4 = 0$$
$$(x + 2)(x - 2) = 0 \quad \text{Factor.}$$

$$x + 2 = 0 \quad \text{or} \quad x - 2 = 0 \quad \text{Apply the zero factor theorem.}$$
$$x = -2 \quad \text{or} \quad x = 2 \quad \text{Solve each equation.}$$

The solutions are -2 and 2.

Now let's solve $x^2 - 4 = 0$ another way. First, add 4 to both sides of the equation.

$$x^2 - 4 = 0$$
$$x^2 = 4 \quad \text{Add 4 both sides.}$$

Now we see that the value for x must be a number whose square is 4. Therefore $x = \sqrt{4} = 2$ or $x = -\sqrt{4} = -2$. This reasoning is an example of the square root property.

Square Root Property

If $x^2 = a$ for $a \geq 0$, then

$$x = \sqrt{a} \quad \text{or} \quad x = -\sqrt{a}$$

EXAMPLE 1

Use the square root property to solve $x^2 - 9 = 0$.

Solution   First we solve for x^2 by adding 9 to both sides.

$$x^2 - 9 = 0$$
$$x^2 = 9 \quad \text{Add 9 to both sides.}$$

Next we use the square root property.

$$x = \sqrt{9} \quad \text{or} \quad x = -\sqrt{9}$$
$$x = 3 \qquad\qquad x = -3$$

Check:

$$x^2 - 9 = 0 \quad \textit{Original equation} \qquad\qquad x^2 - 9 = 0 \quad \textit{Original equation}$$

$$3^2 - 9 \overset{?}{=} 0 \quad \textit{Let } x = 3. \qquad\qquad (-3)^2 - 9 \overset{?}{=} 0 \quad \textit{Let } x = -3.$$

$$0 = 0 \quad \textit{True} \qquad\qquad\qquad 0 = 0 \quad \textit{True}$$

The solutions are 3 and -3.

EXAMPLE 2

Use the square root property to solve $2x^2 = 7$.

Solution First we solve for x^2 by dividing both sides by 2. Then we use the square root property.

$$2x^2 = 7$$

$$x^2 = \frac{7}{2} \qquad \textit{Divide both sides by } \textbf{2}.$$

$$x = \sqrt{\frac{7}{2}} \quad \text{or} \quad x = -\sqrt{\frac{7}{2}} \qquad \textit{Use the square root property.}$$

If the denominators are rationalized, we have

$$x = \frac{\sqrt{7} \cdot \sqrt{2}}{\sqrt{2} \cdot \sqrt{2}} \quad \text{or} \quad x = -\frac{\sqrt{7} \cdot \sqrt{2}}{\sqrt{2} \cdot \sqrt{2}} \qquad \textit{Rationalize the denominator.}$$

$$x = \frac{\sqrt{14}}{2} \qquad\qquad x = -\frac{\sqrt{14}}{2} \qquad \textit{Simplify.}$$

Remember to check both solutions in the original equation. The solutions are $\dfrac{\sqrt{14}}{2}$ and $-\dfrac{\sqrt{14}}{2}$.

EXAMPLE 3

Use the square root property to solve $(x - 3)^2 = 16$.

Solution Instead of x^2, here we have $(x - 3)^2$. But the square root property can still be used.

$$(x - 3)^2 = 16$$

$$x - 3 = \sqrt{16} \quad \text{or} \quad x - 3 = -\sqrt{16} \qquad \textit{Use the square root property.}$$

$$x - 3 = 4 \qquad\qquad x - 3 = -4 \qquad \textit{Write } \sqrt{\textbf{16}} \textit{ as } \textbf{4} \textit{ and } -\sqrt{\textbf{16}} \textit{ as } -\textbf{4}.$$

$$x = 7 \qquad\qquad x = -1 \qquad \textit{Solve.}$$

Check

$$(x - 3)^2 = 16 \quad \textit{Original equation} \qquad\qquad (x - 3)^2 = 16 \quad \textit{Original equation}$$

$$(7 - 3)^2 \overset{?}{=} 16 \quad \textit{Let } x = 7. \qquad\qquad (-1 - 3)^2 \overset{?}{=} 16 \quad \textit{Let } x = -1.$$

$$4^2 \overset{?}{=} 16 \quad \textit{Simplify.} \qquad\qquad (-4)^2 \overset{?}{=} 16 \quad \textit{Simplify.}$$

$$16 = 16 \quad \textit{True} \qquad\qquad\qquad 16 = 16 \quad \textit{True}$$

Both 7 and -1 are solutions.

EXAMPLE 4

Use the square root property to solve $(x + 1)^2 = 8$.

Solution $(x + 1)^2 = 8$

$$x + 1 = \sqrt{8} \quad \text{or} \quad x + 1 = -\sqrt{8} \qquad \text{Use the square root property.}$$

$$x + 1 = 2\sqrt{2} \qquad x + 1 = -2\sqrt{2} \qquad \text{Simplify the radical.}$$

$$x = -1 + 2\sqrt{2} \qquad x = -1 - 2\sqrt{2} \qquad \text{Solve for } x.$$

Check both solutions in the original equation. The solutions are $-1 + 2\sqrt{2}$ and $-1 - 2\sqrt{2}$. This can be written compactly as $-1 \pm 2\sqrt{2}$. The notation $\pm$ is read as "plus or minus."

> **Helpful Hint**
>
> read "plus or minus"
>
> The notation $-1 \pm \sqrt{5}$, for example, is just a shorthand notation for both $-1 + \sqrt{5}$ and $-1 - \sqrt{5}$.

EXAMPLE 5

Use the square root property to solve $(x - 1)^2 = -2$.

Solution This equation has no real solution because the square root of -2 is not a real number.

EXAMPLE 6

Use the square root property to solve $(5x - 2)^2 = 10$.

Solution $(5x - 2)^2 = 10$

$$5x - 2 = \sqrt{10} \quad \text{or} \quad 5x - 2 = -\sqrt{10} \qquad \text{Use the square root property.}$$

$$5x = 2 + \sqrt{10} \qquad 5x = 2 - \sqrt{10} \qquad \text{Add 2 to both sides.}$$

$$x = \frac{2 + \sqrt{10}}{5} \qquad x = \frac{2 - \sqrt{10}}{5} \qquad \text{Divide both sides by 5.}$$

Check both solutions in the original equation. The solutions are $\dfrac{2 + \sqrt{10}}{5}$ and $\dfrac{2 - \sqrt{10}}{5}$, which can be written as $\dfrac{2 \pm \sqrt{10}}{5}$.

> **Helpful Hint**
>
> For some applications and graphing purposes, decimal approximations of exact solutions to quadratic equations may be desired.
>
> | **Exact Solutions from Example 6** | | **Decimal Approximations** |
> |---|---|---|
> | $\dfrac{2 + \sqrt{10}}{5}$ | $\approx$ | 1.032 |
> | $\dfrac{2 - \sqrt{10}}{5}$ | $\approx$ | -0.232 |

2   Many real-world applications are modeled by quadratic equations.

EXAMPLE 7

FINDING THE LENGTH OF TIME OF A DIVE

The record for the highest dive into a lake was made by Harry Froboess of Switzerland. In 1936 he dove 394 feet from the airship Hindenburg into Lake Constance. To the nearest tenth of a second, how long did his dive take? (*Source: The Guiness Book of Records*)

Solution   1.   UNDERSTAND. To approximate the time of the dive, we use the formula $h = 16t^2$ * where t is time in seconds and h is the distance in feet, traveled by a free-falling body or object. For example, to find the distance traveled in 1 second, or 3 seconds, we let $t = 1$ and then $t = 3$.

$$\text{If } t = 1, h = 16(1)^2 = 16 \cdot 1 = 16 \text{ feet}$$
$$\text{If } t = 3, h = 16(3)^2 = 16 \cdot 9 = 144 \text{ feet}$$

Since a body travels 144 feet in 3 seconds, we now know the dive of 394 feet lasted longer than 3 seconds.

2.   TRANSLATE. Use the formula $h = 16t^2$, let the distance $h = 394$, and we have the equation $394 = 16t^2$.

3.   SOLVE.   To solve $394 = 16t^2$ for t, we will use the square root property.

$$394 = 16t^2$$

$$\frac{394}{16} = t^2 \qquad\qquad \text{Divide both sides by } \mathbf{16.}$$

$$24.625 = t^2 \qquad\qquad \text{Simplify.}$$

$$\sqrt{24.625} = t \quad \text{or} \quad -\sqrt{24.625} = t \qquad \text{Use the square root property.}$$

$$5.0 \approx t \quad \text{or} \qquad\qquad -5.0 \approx t \qquad \text{Approximate.}$$

4.   INTERPRET.

Check: We reject the solution -5.0 since the length of the dive is not a negative number.

State: The dive lasted approximately 5 seconds.

*The formula $h = 16t^2$ does not take into account air resistance.

CLASSROOM EXAMPLE

Use the formula $h = 16t^2$ to find how long it takes a free-falling body to fall 650 feet.

answer: 6.4 sec

EXERCISE SET 9.1

STUDY GUIDE/SSM   CD/ VIDEO   PH MATH TUTOR CENTER   MathXL®Tutorials ON CD   MathXL®   MyMathLab®

Use the square root property to solve each quadratic equation. See Examples 1, 2, and 5.

1. $x^2 = 64$ ± 8

2. $x^2 = 121$ ± 11

3. $x^2 = 21$ $\pm\sqrt{21}$

4. $x^2 = 22$ $\pm\sqrt{22}$

5. $x^2 = \dfrac{1}{25}$ $\pm\dfrac{1}{5}$

6. $x^2 = \dfrac{1}{16}$ $\pm\dfrac{1}{4}$

7. $x^2 = -4$ no real solution

8. $x^2 = -25$ no real solution

9. $3x^2 = 13$ $\pm\dfrac{\sqrt{39}}{3}$

10. $5x^2 = 2$ $\pm\dfrac{\sqrt{10}}{5}$

11. $7x^2 = 4$ $\pm\dfrac{2\sqrt{7}}{7}$

12. $2x^2 = 9$ $\pm\dfrac{3\sqrt{2}}{2}$

13. $x^2 - 2 = 0$ $\pm\sqrt{2}$

14. $x^2 - 15 = 0$ $\pm\sqrt{15}$

15. $2x^2 - 10 = 0$ $\pm\sqrt{5}$

16. $7x^2 - 21 = 0$ $\pm\sqrt{3}$

17. Explain why the equation $x^2 = -9$ has no real solution. answers may vary

18. Explain why the equation $x^2 = 9$ has two solutions. answers may vary

Use the square root property to solve each quadratic equation. See Examples 3 through 6. **29.** no real solution   **30.** no real solution

19. $(x - 5)^2 = 49$ $12, -2$

20. $(x + 2)^2 = 25$ $-7, 3$

21. $(x + 2)^2 = 7$ $-2 \pm \sqrt{7}$

22. $(x - 7)^2 = 2$ $7 \pm \sqrt{2}$

23. $\left(m - \dfrac{1}{2}\right)^2 = \dfrac{1}{4}$ $1, 0$

24. $\left(m + \dfrac{1}{3}\right)^2 = \dfrac{1}{9}$ $0, -\dfrac{2}{3}$

25. $(p + 2)^2 = 10$ $-2 \pm \sqrt{10}$

26. $(p - 7)^2 = 13$ $7 \pm \sqrt{13}$

27. $(3y + 2)^2 = 100$ $\dfrac{8}{3}, -4$

28. $(4y - 3)^2 = 81$ $3, -\dfrac{3}{2}$

29. $(z - 4)^2 = -9$

30. $(z + 7)^2 = -20$

31. $(2x - 11)^2 = 50$

32. $(3x - 17)^2 = 28$

33. $(3x - 7)^2 = 32$ $\dfrac{7 \pm 4\sqrt{2}}{3}$

34. $(5x - 11)^2 = 54$

35. $(2p - 5)^2 = 121$ $8, -3$

36. $(3p - 1)^2 = 4$ $1, -\dfrac{1}{3}$

Solve. See Example 7.

37. Neglecting air resistance, the distance d in feet that an object falls in t seconds is given by the equation $d = 16t^2$. If a sandblaster drops his goggles from a bridge 400 feet from the water below, find how long it takes for the goggles to hit the water.   5 sec

31. $\dfrac{11 \pm 5\sqrt{2}}{2}$

32. $\dfrac{17 \pm 2\sqrt{7}}{3}$

34. $\dfrac{11 \pm 3\sqrt{6}}{5}$

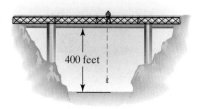

400 feet

38. The area of a circle is found by the equation $A = \pi r^2$. If the area A of a certain circle is 36π square inches, find its radius r. $r = 6$ in.

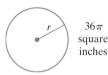

r

36π square inches

Solve. See Example 7. For Exercises 39 through 42, use the formula from $h = 16t^2$ and round answers to the nearest tenth of a second.

39. The highest regularly performed dives are made by professional divers from La Quebrada. This cliff in Acapulco has a height of 87.6 feet. Determine the time of a dive. (*Source: The Guinness Book of Records*)   2.3 sec

40. In 1988, Eddie Turner saved Frank Fanan, who became unconscious after an injury while jumping out of an airplane. Fanan fell 11,136 feet before Turner pulled his ripcord. Determine the time of Fanan's unconscious free-fall.   26.4 sec

41. In New Mexico, Joseph Kittinger fell 16 miles before opening his parachute on August 16, 1960. How long did Kittinger free-fall before opening his parachute? (*Hint:* First convert 16 miles to feet. Use 1 mile = 5280 feet.) (*Source: Guinness World Record*).   72.7 sec

42. In the Ukraine, Elvira Fomitcheva fell 9 miles 1056 feet before opening her parachute on October 26, 1977. How long did she free-fall before opening her parachute? (Use the hint from Exercise 41.) (*Source: Guinness World Record*).   55.1 sec

The formula for area of a square is $A = s^2$ where s is the length of a side. Use this formula for Exercises 43 through 46. For each exercise, give an exact answer and a two decimal place approximation.

43. If the area of a square is 20 square inches, find the length of a side. $2\sqrt{5}$ in. ≈ 4.47 in.

44. If the area of a square is 32 square meters, find the length of a side. $4\sqrt{2}$ m ≈ 5.66 m

45. The Washington Monument has a square base whose area is approximately 3039 square feet. Find the length of a side. (*Source: The World Almanac*, 2003) $\sqrt{3039}$ ft ≈ 55.13 ft

△ **46.** A giant strawberry shortcake was made by the Greater Plant City Chamber of Commerce in Plant City, Florida. Its base had an area of 827 square feet. If the base is in the shape of a square, find the length of a side. (*Source: Guinness World Records*, 2001) $\sqrt{827}$ ft $\approx$ 28.76 ft

REVIEW AND PREVIEW

Factor each perfect square trinomial. See Section 6.3.

47. $x^2 + 6x + 9$ $(x + 3)^2$ **48.** $y^2 + 10y + 25$ $(y + 5)^2$

49. $x^2 - 4x + 4$ $(x - 2)^2$ **50.** $x^2 - 20x + 100$ $(x - 10)^2$

Concept Extensions   **53.** $-7 \pm \sqrt{31}$ **54.** $5 \pm \sqrt{11}$

Solve each quadratic equation by first factoring the perfect square trinomial on the left side. Then apply the square root property.

51. $x^2 + 4x + 4 = 16$ $2, -6$ **52.** $z^2 - 6z + 9 = 25$ $-2, 8$

53. $x^2 + 14x + 49 = 31$ **54.** $y^2 - 10y + 25 = 11$

△ **55.** A 13-inch TV is advertised in the local paper. If 13 inches is the measure of the diagonal of the picture tube, use the Pythagorean theorem and the diagram below to find the measures of the sides of the picture tube. $x = 10.4$ in. $\frac{3}{4}x, = 7.8$ in.

56. The number of cattle y (in thousands) on farms in Kansas from 2000 through 2002 is given by the equation $y = -100x^2 + 6700$. In this equation, $x = 0$ represents the year 2001. Assume that this trend continues and find the year in which there are 5100 thousand cattle on Kansas farms. (*Hint:* Replace y with 5100 in the equation and solve for x.) (*Source:* Based on data from the U.S. Department of Agriculture)   2005

58. $-1.02, 3.76$

Solve each quadratic equation by using the square root property. Use a calculator and round each solution to the nearest hundredth.

57. $x^2 = 1.78$ ± 1.33 **58.** $(x - 1.37)^2 = 5.71$

59. The number y of Target stores open for business from 1998 through 2000 is given by the equation $y = 2(x + 14.75)^2 + 415.875$, where $x = 0$ represents the year 1998. Assume that this trend continues and find the year in which there will be 1451 Target stores open for business. (*Hint:* Replace y with 1451 in the equation and solve for x.) (*Source:* Based on data from Target Corporation)   2006

60. World cotton production y (in thousand metric tons) from 1998 through 2000 is given by the equation $y = -200(x - 1.75)^2 + 19,112.5$, where $x = 0$ represents the year 1998. Assume that this trend continues and find the year in which there will be 17,000 thousand metric tons. (*Hint:* Replace y with 17,000 in the equation and solve for x.) (*Source:* Based on data from the U.S. Department of Agriculture, Foreign Agricultural Service)   2003

9.2   *SOLVING QUADRATIC EQUATIONS BY COMPLETING THE SQUARE*

Objectives

1   Find perfect square trinomials.

2   Solve quadratic equations by completing the square.

1   In the last section, we used the square root property to solve equations such as

$$(x + 1)^2 = 8 \quad \text{and} \quad (5x - 2)^2 = 3$$

Notice that one side of each equation is a quantity squared and that the other side is a constant. To solve

$$x^2 + 2x = 4$$

notice that if we add 1 to both sides of the equation, the left side is a perfect square trinomial that can be factored.

$$x^2 + 2x + 1 = 4 + 1 \qquad \text{Add } \mathbf{1} \text{ to both sides.}$$
$$(x + 1)^2 = 5 \qquad \text{Factor.}$$

Now we can solve this equation as we did in the previous section by using the square root property.

$$x + 1 = \sqrt{5} \quad \text{or} \quad x + 1 = -\sqrt{5} \qquad \text{Use the square root property.}$$
$$x = -1 + \sqrt{5} \qquad x = -1 - \sqrt{5} \qquad \text{Solve.}$$

The solutions are $-1 \pm \sqrt{5}$.

Adding a number to $x^2 + 2x$ to form a perfect square trinomial is called **completing the square** on $x^2 + 2x$.

In general, we have the following.

Completing The Square

To complete the square on $x^2 + bx$, add $\left(\dfrac{b}{2}\right)^2$. To find $\left(\dfrac{b}{2}\right)^2$, **find half the coefficient of** x, **then square the result.**

EXAMPLE 1

Complete the square for each expression and then factor the resulting perfect square trinomial.

a. $x^2 + 10x$ **b.** $m^2 - 6m$ **c.** $x^2 + x$

Solution   **a.** The coefficient of the x-term is 10. Half of 10 is 5, and $5^2 = 25$. Add 25.

$$x^2 + 10x + 25 = (x + 5)^2$$

b. Half the coefficient of m is -3, and $(-3)^2$ is 9. Add 9.

$$m^2 - 6m + 9 = (m - 3)^2$$

c. Half the coefficient of x is $\dfrac{1}{2}$ and $\left(\dfrac{1}{2}\right)^2 = \dfrac{1}{4}$. Add $\dfrac{1}{4}$.

$$x^2 + x + \frac{1}{4} = \left(x + \frac{1}{2}\right)^2$$

2   By completing the square, a quadratic equation can be solved using the square root property.

EXAMPLE 2

Solve $x^2 + 6x + 3 = 0$ by completing the square.

Solution   First we get the variable terms alone by subtracting 3 from both sides of the equation.

$$x^2 + 6x + 3 = 0$$
$$x^2 + 6x = -3 \qquad \text{Subtract 3 from both sides.}$$

Next we find half the coefficient of the x-term, then square it. Add this result to **both sides** of the equation. This will make the left side a perfect square trinomial. The coefficient of x is 6, and half of 6 is 3. So we add 3^2 or 9 to both sides.

$$x^2 + 6x + 9 = -3 + 9 \qquad \text{Complete the square.}$$
$$(x + 3)^2 = 6 \qquad \text{Factor the trinomial } x^2 + 6x + 9.$$
$$x + 3 = \sqrt{6} \quad \text{or} \quad x + 3 = -\sqrt{6} \qquad \text{Use the square root property.}$$
$$x = -3 + \sqrt{6} \qquad x = -3 - \sqrt{6} \quad \text{Subtract 3 from both sides.}$$

Check by substituting $-3 + \sqrt{6}$ and $-3 - \sqrt{6}$ in the original equation. The solutions are $-3 \pm \sqrt{6}$.

CLASSROOM EXAMPLE

Solve $x^2 + 8x + 1 = 0$.
answer: $-4 \pm \sqrt{15}$

> **Helpful Hint**
>
> Remember, when completing the square, add the number that completes the square to **both sides of the equation**. In Example 2, we added 9 to both sides to complete the square.

EXAMPLE 3

Solve $x^2 - 10x = -14$ by completing the square.

Solution   The variable terms are already alone on one side of the equation. The coefficient of x is -10. Half of -10 is -5, and $(-5)^2 = 25$. So we add 25 to both sides.

> **Helpful Hint**
>
> Add 25 to *both* sides of the equation.

$$x^2 - 10x = -14$$
$$x^2 - 10x + 25 = -14 + 25$$

$$(x - 5)^2 = 11 \qquad \text{Factor the trinomial and simplify } -14 + 25.$$
$$x - 5 = \sqrt{11} \quad \text{or} \quad x - 5 = -\sqrt{11} \qquad \text{Use the square root property.}$$
$$x = 5 + \sqrt{11} \qquad x = 5 - \sqrt{11} \quad \text{Add 5 to both sides.}$$

The solutions are $5 \pm \sqrt{11}$.

CLASSROOM EXAMPLE

Solve $x^2 + 14x = -32$.
answer: $7 \pm \sqrt{17}$

The method of completing the square can be used to solve *any* quadratic equation whether the coefficient of the squared variable is 1 or not. When the coefficient of the squared variable is not 1, we first divide both sides of the equation by the coefficient of the squared variable so that the coefficient is 1. Then we complete the square.

> ### EXAMPLE 4

Solve $4x^2 - 8x - 5 = 0$ by completing the square.

Solution

$$4x^2 - 8x - 5 = 0$$

$$x^2 - 2x - \frac{5}{4} = 0 \qquad \text{Divide both sides by 4.}$$

$$x^2 - 2x = \frac{5}{4} \qquad \text{Get the variable terms alone on one side of the equation.}$$

The coefficient of x is -2. Half of -2 is -1, and $(-1)^2 = 1$. So we add 1 to both sides.

$$x^2 - 2x + 1 = \frac{5}{4} + 1$$

$$(x - 1)^2 = \frac{9}{4} \qquad \text{Factor } x^2 - 2x + 1 \text{ and simplify } \frac{5}{4} + 1.$$

$$x - 1 = \sqrt{\frac{9}{4}} \quad \text{or} \quad x - 1 = -\sqrt{\frac{9}{4}} \qquad \text{Use the square root property.}$$

$$x = 1 + \frac{3}{2} \qquad x = 1 - \frac{3}{2} \qquad \text{Add 1 to both sides and simplify the radical.}$$

$$x = \frac{5}{2} \qquad x = -\frac{1}{2} \qquad \text{Simplify.}$$

Both $\frac{5}{2}$ and $-\frac{1}{2}$ are solutions.

CLASSROOM EXAMPLE
Solve $4x^2 - 16x - 9 = 0$.
answer: $-\frac{1}{2}$, and $\frac{9}{2}$

The following steps may be used to solve a quadratic equation in x by completing the square.

> ### Solving a Quadratic Equation in *x* by Completing the Square
>
> **Step 1:** If the coefficient of x^2 is 1, go to Step 2. If not, divide both sides of the equation by the coefficient of x^2.
>
> **Step 2:** Get all terms with variables on one side of the equation and constants on the other side.
>
> **Step 3:** Find half the coefficient of x and then square the result. Add this number to both sides of the equation.
>
> **Step 4:** Factor the resulting perfect square trinomial.
>
> **Step 5:** Use the square root property to solve the equation.

> ### EXAMPLE 5

Solve $2x^2 + 6x = -7$ by completing the square.

Solution The coefficient of x^2 is not 1. We divide both sides by 2, the coefficient of x^2.

$$2x^2 + 6x = -7$$

$$x^2 + 3x = -\frac{7}{2} \qquad \text{Divide both sides by 2.}$$

$$x^2 + 3x + \frac{9}{4} = -\frac{7}{2} + \frac{9}{4} \qquad \text{Add } \left(\frac{3}{2}\right)^2 \text{ or } \frac{9}{4} \text{ to both sides.}$$

$$\left(x + \frac{3}{2}\right)^2 = -\frac{5}{4} \qquad \text{Factor the left side and simplify the right.}$$

CLASSROOM EXAMPLE
Solve $2x^2 + 10x = -13$.
answer: no real solution

There is no real solution to this equation since the square root of a negative number is not a real number.

EXAMPLE 6

Solve $2x^2 = 10x + 1$ by completing the square.

Solution   First we divide both sides of the equation by 2, the coefficient of x^2.

$$2x^2 = 10x + 1$$
$$x^2 = 5x + \frac{1}{2} \qquad \text{Divide both sides by 2.}$$

Next we get the variable terms alone by subtracting $5x$ from both sides.

$$x^2 - 5x = \frac{1}{2}$$

$$x^2 - 5x + \frac{25}{4} = \frac{1}{2} + \frac{25}{4} \qquad \text{Add } \left(-\frac{5}{2}\right)^2 \text{ or } \frac{25}{4} \text{ to both sides.}$$

$$\left(x - \frac{5}{2}\right)^2 = \frac{27}{4} \qquad \text{Factor the left side and simplify the right side.}$$

$$x - \frac{5}{2} = \sqrt{\frac{27}{4}} \quad \text{or} \quad x - \frac{5}{2} = -\sqrt{\frac{27}{4}} \qquad \text{Use the square root property.}$$

$$x - \frac{5}{2} = \frac{3\sqrt{3}}{2} \qquad\qquad x - \frac{5}{2} = -\frac{3\sqrt{3}}{2} \qquad \text{Simplify.}$$

$$x = \frac{5}{2} + \frac{3\sqrt{3}}{2} \qquad\qquad x = \frac{5}{2} - \frac{3\sqrt{3}}{2}$$

The solutions are $\dfrac{5 \pm 3\sqrt{3}}{2}$.

MENTAL MATH

Determine the number to add to make each expression a perfect square trinomial. See Example 1.

1. $p^2 + 8p$ 16
4. $x^2 + 18x$ 81

2. $p^2 + 6p$ 9
5. $y^2 + 14y$ 49

3. $x^2 + 20x$ 100
6. $y^2 + 2y$ 1

EXERCISE SET 9.2

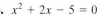
STUDY
GUIDE/SSM   CD/
VIDEO   PH MATH
TUTOR CENTER   MathXL®Tutorials
ON CD   MathXL®   MyMathLab®

13. $-1 \pm \sqrt{6}$ **14.** $-3 \pm 3\sqrt{2}$
15. $-3 \pm \sqrt{34}$ **19.** $1 \pm \sqrt{2}$ **21.** $-4, -1$

Complete the square for each expression and then factor the resulting perfect square trinomial. See Example 1.

 1. $x^2 + 4x$ $(x + 2)^2$
3. $k^2 - 12k$ $(k - 6)^2$
 5. $x^2 - 3x$ $\left(x - \frac{3}{2}\right)^2$
7. $m^2 - m$ $\left(m - \frac{1}{2}\right)^2$

2. $x^2 + 6x$ $(x + 3)^2$
4. $k^2 - 16k$ $(k - 8)^2$
6. $x^2 - 5x$ $\left(x - \frac{5}{2}\right)^2$
8. $y^2 + y$ $\left(y + \frac{1}{2}\right)^2$

Solve each quadratic equation by completing the square. See Examples 2 and 3.

9. $x^2 - 6x = 0$ $0, 6$
 11. $x^2 + 8x = -12$ $-6, -2$ **12.** $x^2 - 10x = -24$ $4, 6$

10. $y^2 + 4y = 0$ $-4, 0$

13. $x^2 + 2x - 5 = 0$
15. $x^2 + 6x - 25 = 0$
17. $z^2 + 5z = 7$ $\dfrac{-5 \pm \sqrt{53}}{2}$
 19. $x^2 - 2x - 1 = 0$
21. $y^2 + 5y + 4 = 0$
23. $x(x + 3) = 18$ $-6, 3$

14. $z^2 + 6z - 9 = 0$
16. $x^2 - 6x + 7 = 0$ $3 \pm \sqrt{2}$
18. $x^2 - 7x = 5$ $\dfrac{7 \pm \sqrt{69}}{2}$
20. $x^2 - 4x + 2 = 0$ $2 \pm \sqrt{2}$
22. $y^2 - 5y + 6 = 0$ $2, 3$
24. $x(x - 3) = 18$ $-3, 6$

Solve each quadratic equation by completing the square. See Examples 4 through 6.

25. $4x^2 - 24x = 13$ $-\frac{1}{2}, \frac{13}{2}$ **26.** $2x^2 + 8x = 10$ $-5, 1$

27.–28. no real solution

27. $5x^2 + 10x + 6 = 0$

28. $3x^2 - 12x + 14 = 0$

29. $2x^2 = 6x + 5$ $\quad \dfrac{3 \pm \sqrt{19}}{2}$

30. $4x^2 = -20x + 3$

31. $3x^2 - 6x = 24$ $\quad -2, 4$

32. $2x^2 + 18x = -40$ $\quad -5, -4$

33. $2y^2 + 8y + 5 = 0$

34. $3z^2 + 6z + 4 = 0$

35. $2y^2 - 3y + 1 = 0$ $\quad \frac{1}{2}, 1$

36. $2y^2 - y - 1 = 0$ $\quad -\frac{1}{2}, 1$

37. $3y^2 - 2y - 4 = 0$

38. $4y^2 - 2y - 3 = 0$

39. In your own words, describe a perfect square trinomial. answers may vary

40. Describe how to find the number to add to $x^2 - 7x$ to make a perfect square trinomial. answers may vary

30. $\dfrac{-5 \pm 2\sqrt{7}}{2}$

33. $\dfrac{-4 \pm \sqrt{6}}{2}$

34. no real solution

37. $\dfrac{1 \pm \sqrt{13}}{3}$

38. $\dfrac{1 \pm \sqrt{13}}{4}$

REVIEW AND PREVIEW

Simplify each expression. See Section 8.3.

41. $\dfrac{3}{4} - \sqrt{\dfrac{25}{16}}$ $\quad -\frac{1}{2}$

42. $\dfrac{3}{5} + \sqrt{\dfrac{16}{25}}$ $\quad \frac{7}{5}$

43. $\dfrac{1}{2} - \sqrt{\dfrac{9}{4}}$ $\quad -1$

44. $\dfrac{9}{10} - \sqrt{\dfrac{49}{100}}$ $\quad \frac{1}{5}$

Simplify each expression. See Section 8.4.

45. $\dfrac{6 + 4\sqrt{5}}{2}$ $\quad 3 + 2\sqrt{5}$

46. $\dfrac{10 - 20\sqrt{3}}{2}$ $\quad 5 - 10\sqrt{3}$

47. $\dfrac{3 - 9\sqrt{2}}{6}$ $\quad \frac{1 - 3\sqrt{2}}{2}$

48. $\dfrac{12 - 8\sqrt{7}}{16}$ $\quad \frac{3 - 2\sqrt{7}}{4}$

Concept Extensions

49. Find a value of k that will make $x^2 + kx + 16$ a perfect square trinomial. $k = 8$ or $k = -8$

50. Find a value of k that will make $x^2 + kx + 25$ a perfect square trinomial. $k = 10$ or $k = -10$

51. Retail sales y (in millions of dollars) for bookstores in the United States from 1998 through 2000 are given by the

equation $y = 268x^2 + 720x + 13{,}390$. In this equation, x is the number of years after 1998. Assume that this trend continues and predict the year after 1998 in which the retail sales for U.S. bookstores will be $47,390 million. (*Source: Based on data from the U.S. Bureau of the Census, Monthly Retail Surveys Branch*) 2008

52. The average price of gold y (in dollars per ounce) from 1996 through 2000 is given by the equation $y = 10x^2 - 67x + 389$. In this equation, x is the number of years after 1996. Assume that this trend continues and find the year after 2000 in which the price of gold will be $1025 per ounce. (*Source: Based on data from Platinum Guild International (USA) Inc.*) 2008

Recall that a graphing calculator may be used to solve an equation. For example, to solve $x^2 + 8x = -12$ (Exercise 11), graph

$$y_1 = x^2 + 8x \quad \text{(left side of equation) and}$$
$$y_2 = -12 \quad \text{(right side of equation)}$$

The x-coordinate of the point of intersection of the graphs is the solution. Use a graphing calculator and solve each equation. Round solutions to the nearest hundredth.

53. Exercise 11 $\quad -6, -2$ **54.** Exercise 12 $\quad 4, 6$

55. Exercise 29 $\quad -0.68, 3.68$ **56.** Exercise 18 $\quad -0.65, 7.65$

9.3 SOLVING QUADRATIC EQUATIONS BY THE QUADRATIC FORMULA

Objectives

1 Use the quadratic formula to solve quadratic equations.

2 Determine the number of solutions of a quadratic equation by using the discriminant.

1 We can use the technique of completing the square to develop a formula to find solutions of any quadratic equation. We develop and use the **quadratic formula** in this section.

Recall that a quadratic equation in **standard form** is

$$ax^2 + bx + c = 0, \quad a \neq 0$$

To develop the quadratic formula, let's complete the square for this quadratic equation in standard form.

First we divide both sides of the equation by the coefficient of x^2 and then get the variable terms alone on one side of the equation.

$$x^2 + \frac{b}{a}x + \frac{c}{a} = 0 \qquad \text{Divide by } a; \text{ recall that } a \text{ cannot be } 0.$$

$$x^2 + \frac{b}{a}x = -\frac{c}{a} \qquad \text{Get the variable terms alone on one side of the equation.}$$

The coefficient of x is $\dfrac{b}{a}$. Half of $\dfrac{b}{a}$ is $\dfrac{b}{2a}$ and $\left(\dfrac{b}{2a}\right)^2 = \dfrac{b^2}{4a^2}$. So we add $\dfrac{b^2}{4a^2}$ to both sides of the equation.

$$x^2 + \frac{b}{a}x + \frac{b^2}{4a^2} = -\frac{c}{a} + \frac{b^2}{4a^2} \qquad \text{Add } \frac{b^2}{4a^2} \text{ to both sides.}$$

$$\left(x + \frac{b}{2a}\right)^2 = -\frac{c}{a} + \frac{b^2}{4a^2} \qquad \text{Factor the left side.}$$

$$\left(x + \frac{b}{2a}\right)^2 = -\frac{4ac}{4a^2} + \frac{b^2}{4a^2} \qquad \begin{array}{l}\text{Multiply } -\dfrac{c}{a} \text{ by } \dfrac{4a}{4a} \text{ so that both terms on} \\ \text{the have a right side common denominator.}\end{array}$$

$$\left(x + \frac{b}{2a}\right)^2 = \frac{b^2 - 4ac}{4a^2} \qquad \text{Simplify the right side.}$$

Now we use the square root property.

$$x + \frac{b}{2a} = \sqrt{\frac{b^2 - 4ac}{4a^2}} \quad \text{or} \quad x + \frac{b}{2a} = -\sqrt{\frac{b^2 - 4ac}{4a^2}} \qquad \begin{array}{l}\text{Use the square root} \\ \text{property.}\end{array}$$

$$x + \frac{b}{2a} = \frac{\sqrt{b^2 - 4ac}}{2a} \qquad x + \frac{b}{2a} = -\frac{\sqrt{b^2 - 4ac}}{2a} \qquad \text{Simplify the radical.}$$

$$x = -\frac{b}{2a} + \frac{\sqrt{b^2 - 4ac}}{2a} \qquad x = -\frac{b}{2a} - \frac{\sqrt{b^2 - 4ac}}{2a} \qquad \begin{array}{l}\text{Subtract } \dfrac{b}{2a} \text{ from both} \\ \text{sides.}\end{array}$$

$$x = \frac{-b + \sqrt{b^2 - 4ac}}{2a} \qquad x = \frac{-b - \sqrt{b^2 - 4ac}}{2a} \qquad \text{Simplify.}$$

The solutions are $\dfrac{-b \pm \sqrt{b^2 - 4ac}}{2a}$. This final equation is called the **quadratic formula** and gives the solutions of any quadratic equation.

Quadratic Formula

If a, b, and c are real numbers and $a \neq 0$, a quadratic equation written in the form $ax^2 + bx + c = 0$ has solutions

$$x = \frac{-b \pm \sqrt{b^2 - 4ac}}{2a}$$

▶ Helpful Hint

Don't forget that to correctly identify a, b, and c in the quadratic formula, you should write the equation in standard form.

Quadratic Equations in Standard Form

$$5x^2 - 6x + 2 = 0 \qquad a = 5, b = -6, c = 2$$
$$4y^2 - 9 = 0 \qquad a = 4, b = 0, c = -9$$
$$x^2 + x = 0 \qquad a = 1, b = 1, c = 0$$
$$\sqrt{2}x^2 + \sqrt{5}x + \sqrt{3} = 0 \qquad a = \sqrt{2}, b = \sqrt{5}, c = \sqrt{3}$$

EXAMPLE 1

Solve $3x^2 + x - 3 = 0$ using the quadratic formula.

Solution This equation is in standard form with $a = 3, b = 1,$ and $c = -3$. By the quadratic formula, we have

$$x = \frac{-b \pm \sqrt{b^2 - 4ac}}{2a}$$

$$x = \frac{-1 \pm \sqrt{1^2 - 4 \cdot 3 \cdot (-3)}}{2 \cdot 3} \qquad \text{Let } a = 3, b = 1, \text{ and } c = -3.$$

$$= \frac{-1 \pm \sqrt{1 + 36}}{6} \qquad \text{Simplify.}$$

$$= \frac{-1 \pm \sqrt{37}}{6}$$

Check both solutions in the original equation. The solutions are $\dfrac{-1 + \sqrt{37}}{6}$ and $\dfrac{-1 - \sqrt{37}}{6}$.

CLASSROOM EXAMPLE
Solve $2x^2 - x - 5 = 0$.
answer: $\dfrac{1 \pm \sqrt{41}}{4}$

EXAMPLE 2

Solve $2x^2 - 9x = 5$ using the quadratic formula.

Solution First we write the equation in standard form by subtracting 5 from both sides.

$$2x^2 - 9x = 5$$
$$2x^2 - 9x - 5 = 0$$

Next we note that $a = 2, b = -9,$ and $c = -5$. We substitute these values into the quadratic formula.

$$x = \frac{-b \pm \sqrt{b^2 - 4ac}}{2a}$$

$$x = \frac{-(-9) \pm \sqrt{(-9)^2 - 4 \cdot 2 \cdot (-5)}}{2 \cdot 2} \qquad \text{Substitute in the formula.}$$

$$= \frac{9 \pm \sqrt{81 + 40}}{4} \qquad \text{Simplify.}$$

$$= \frac{9 \pm \sqrt{121}}{4} = \frac{9 \pm 11}{4}$$

Then,

$$x = \frac{9 - 11}{4} = -\frac{1}{2} \qquad \text{or} \qquad x = \frac{9 + 11}{4} = 5$$

Check $-\dfrac{1}{2}$ and 5 in the original equation. Both $-\dfrac{1}{2}$ and 5 are solutions.

CLASSROOM EXAMPLE
Solve $3x^2 + 8x = 3$.
answer: -3 and $\dfrac{1}{3}$

> **Helpful Hint**
>
> Notice that the fraction bar is under the entire numerator of
> $-b \pm \sqrt{b^2 - 4ac}$.

The following steps may be useful when solving a quadratic equation by the quadratic formula.

Solving a Quadratic Equation by the Quadratic Formula

Step 1: Write the quadratic equation in standard form: $ax^2 + bx + c = 0$.

Step 2: If necessary, clear the equation of fractions to simplify calculations.

Step 3: Identify a, b, and c.

Step 4: Replace a, b, and c in the quadratic formula with the identified values, and simplify.

✔ **CONCEPT CHECK**

For the quadratic equation $2x^2 - 5 = 7x$, if $a = 2$ and $c = -5$ in the quadratic formula, the value of b is which of the following?

a. $\dfrac{7}{2}$

b. 7

c. -5

d. -7

EXAMPLE 3

Solve $7x^2 = 1$ using the quadratic formula.

Solution First we write the equation in standard form by subtracting 1 from both sides.

$$7x^2 = 1$$
$$7x^2 - 1 = 0$$

Next we replace a, b, and c with the identified values: $a = 7$, $b = 0$ and $c = -1$.

$$x = \frac{0 \pm \sqrt{0^2 - 4 \cdot 7 \cdot (-1)}}{2 \cdot 7} \qquad \text{Substitute in the formula.}$$

$$= \frac{\pm\sqrt{28}}{14} \qquad \text{Simplify.}$$

$$= \frac{\pm 2\sqrt{7}}{14}$$

$$= \pm\frac{\sqrt{7}}{7}$$

The solutions are $\dfrac{\sqrt{7}}{7}$ and $-\dfrac{\sqrt{7}}{7}$.

> Notice that the equation in Example 3, $7x^2 = 1$, could have been easily solved by dividing both sides by 7 and then using the square root property. We solved the equation by the quadratic formula to show that this formula can be used to solve any quadratic equation.

CLASSROOM EXAMPLE

Solve $5x^2 = 2$.

answer: $\pm\dfrac{\sqrt{10}}{5}$

EXAMPLE 4

Solve $x^2 = -x - 1$ using the quadratic formula.

Solution First we write the equation in standard form.

$$x^2 + x + 1 = 0$$

Concept Check Answer:
d

Next we replace a, b, and c in the quadratic formula with $a = 1$, $b = 1$, and $c = 1$.

$$x = \frac{-1 \pm \sqrt{1^2 - 4 \cdot 1 \cdot 1}}{2 \cdot 1} \quad \text{Substitute in the formula.}$$

$$= \frac{-1 \pm \sqrt{-3}}{2} \quad \text{Simplify.}$$

There is no real number solution because $\sqrt{-3}$ is not a real number.

EXAMPLE 5

Solve $\frac{1}{2}x^2 - x = 2$ by using the quadratic formula.

Solution We write the equation in standard form and then clear the equation of fractions by multiplying both sides by the LCD, 2.

$$\frac{1}{2}x^2 - x = 2$$

$$\frac{1}{2}x^2 - x - 2 = 0 \quad \text{Write in standard form.}$$

$$x^2 - 2x - 4 = 0 \quad \text{Multiply both sides by 2.}$$

Here, $a = 1$, $b = -2$, and $c = -4$, so we substitute these values into the quadratic formula.

$$x = \frac{-(-2) \pm \sqrt{(-2)^2 - 4 \cdot 1 \cdot (-4)}}{2 \cdot 1}$$

$$= \frac{2 \pm \sqrt{20}}{2} = \frac{2 \pm 2\sqrt{5}}{2} \quad \text{Simplify.}$$

$$= \frac{2(1 \pm \sqrt{5})}{2} = 1 \pm \sqrt{5} \quad \text{Factor and simplify.}$$

The solutions are $1 - \sqrt{5}$ and $1 + \sqrt{5}$.

> **Helpful Hint**
>
> When simplifying expressions such as
>
> $$\frac{3 \pm 6\sqrt{2}}{6}$$
>
> first factor out a common factor from the terms of the numerator and then simplify.
>
> $$\frac{3 \pm 6\sqrt{2}}{6} = \frac{3(1 \pm 2\sqrt{2})}{2 \cdot 3} = \frac{1 \pm 2\sqrt{2}}{2}$$

2   In the quadratic formula, $x = \dfrac{-b \pm \sqrt{b^2 - 4ac}}{2a}$, the radicand $b^2 - 4ac$ is called the **discriminant** because, by knowing its value, we can **discriminate** among the possible number and type of solutions of a quadratic equation. Possible values of the discriminant and their meanings are summarized next.

Discriminant

The following table corresponds the discriminant $b^2 - 4ac$ of a quadratic equation of the form $ax^2 + bx + c = 0$ with the number of solutions of the equation.

$b^2 - 4ac$	NUMBER OF SOLUTIONS
Positive	Two distinct real solutions
Zero	One real solution
Negative	No real solution*

*In this case, the quadratic equation will have two complex (but not real) solutions. See Section 9.4 for a discussion of complex numbers.

EXAMPLE 6

Use the discriminant to determine the number of solutions of $3x^2 + x - 3 = 0$.

Solution In $3x^2 + x - 3 = 0$, $a = 3$, $b = 1$, and $c = -3$. Then

$$b^2 - 4ac = (1)^2 - 4(3)(-3) = 1 + 36 = 37$$

Since the discriminant is 37, a positive number, this equation has two distinct real solutions.

We solved this equation in Example 1 of this section, and the solutions are $\dfrac{-1 + \sqrt{37}}{6}$ and $\dfrac{-1 - \sqrt{37}}{6}$, two distinct real solutions.

CLASSROOM EXAMPLE

Use the discriminant to determine the number of solutions of
$$5x^2 + 2x - 3 = 0$$
answer: two distinct real solutions

EXAMPLE 7

Use the discriminant to determine the number of solutions of each quadratic equation.

a. $x^2 - 6x + 9 = 0$ **b.** $5x^2 + 4 = 0$

Solution **a.** In $x^2 - 6x + 9 = 0$, $a = 1$, $b = -6$, and $c = 9$.

$$b^2 - 4ac = (-6)^2 - 4(1)(9) = 36 - 36 = 0$$

Since the discriminant is 0, this equation has one real solution.

b. In $5x^2 + 4 = 0$, $a = 5$, $b = 0$, and $c = 4$.

$$b^2 - 4ac = 0^2 - 4(5)(4) = 0 - 80 = -80$$

Since the discriminant is -80, a negative number, this equation has no real solution.

CLASSROOM EXAMPLE

Use the discriminant to determine the number of solutions.

a. $x^2 + 2x + 2 = 0$ **b.** $x^2 + 2x + 1 = 0$

answer:

a. no real solution   **b.** one real solution

Spotlight on DECISION *MAKING*

Suppose you are an engineering technician in a manufacturing plant. The engineering department has been asked to upgrade any production lines that produce less than 3000 items per hour. Production records show the following:

Production Lines A and B

Lines A and B together: produce 3000 items in 32 minutes

Line A alone: takes 11.2 minutes longer than Line B to produce 3000 items

Decide whether either Line A or Line B should be upgraded. Explain your reasoning.

MENTAL MATH

Identify the value of a, b, and c in each quadratic equation.

1. $2x^2 + 5x + 3 = 0$ $a = 2, b = 5, c = 3$

2. $5x^2 - 7x + 1 = 0$ $a = 5, b = -7, c = 1$

3. $10x^2 - 13x - 2 = 0$ $a = 10, b = -13, c = -2$

4. $x^2 + 3x - 7 = 0$ $a = 1, b = 3, c = -7$

5. $x^2 - 6 = 0$ $a = 1, b = 0, c = -6$

6. $9x^2 - 4 = 0$ $a = 9, b = 0, c = -4$

EXERCISE SET 9.3

STUDY GUIDE/SSM   CD/VIDEO   PH MATH TUTOR CENTER   MathXL®Tutorials ON CD   MathXL®   MyMathLab®

Simplify the following.

1. $\dfrac{-1 \pm \sqrt{1^2 - 4(1)(-2)}}{2(1)}$ $-2, 1$

2. $\dfrac{-(-5) \pm \sqrt{(-5)^2 - 4(2)(3)}}{2(2)}$ $1, \dfrac{3}{2}$

3. $\dfrac{-5 \pm \sqrt{5^2 - 4(1)(2)}}{2(1)}$ $\dfrac{-5 \pm \sqrt{17}}{2}$

4. $\dfrac{-7 \pm \sqrt{7^2 - 4(2)(1)}}{2(2)}$ $\dfrac{-7 \pm \sqrt{41}}{4}$

5. $\dfrac{-(-4) \pm \sqrt{(-4)^2 - 4(2)(1)}}{2(2)}$ $\dfrac{2 \pm \sqrt{2}}{2}$

6. $\dfrac{-6 \pm \sqrt{6^2 - 4(3)(1)}}{2(3)}$ $-1 \pm \dfrac{\sqrt{6}}{3}$

9. $\dfrac{-7 \pm \sqrt{37}}{6}$

10. $\dfrac{-3 \pm \sqrt{37}}{14}$

25. $\dfrac{5 \pm \sqrt{33}}{2}$

29. $\dfrac{-9 \pm \sqrt{129}}{12}$

30. $\dfrac{9 \pm \sqrt{177}}{6}$

MIXED PRACTICE

Use the quadratic formula to solve each quadratic equation. See Examples 1 through 4.   **13.–14.** no real solution

7. $x^2 - 3x + 2 = 0$ $1, 2$

8. $x^2 - 5x - 6 = 0$ $-1, 6$

9. $3k^2 + 7k + 1 = 0$

10. $7k^2 + 3k - 1 = 0$

11. $49x^2 - 4 = 0$ $\pm\dfrac{2}{7}$

12. $25x^2 - 15 = 0$ $\pm\dfrac{\sqrt{15}}{5}$

13. $5z^2 - 4z + 3 = 0$

14. $3z^2 + 2z + 1 = 0$

15. $y^2 = 7y + 30$ $-3, 10$

16. $y^2 = 5y + 36$ $-4, 9$

17. $2x^2 = 10$ $\pm\sqrt{5}$

18. $5x^2 = 15$ $\pm\sqrt{3}$

19. $m^2 - 12 = m$ $-3, 4$

20. $m^2 - 14 = 5m$ $-2, 7$

21. $3 - x^2 = 4x$ $-2 \pm \sqrt{7}$

22. $10 - x^2 = 2x$ $-1 \pm \sqrt{11}$

23. $2a^2 - 7a + 3 = 0$ $\dfrac{1}{2}, 3$

24. $3a^2 - 7a + 2 = 0$ $\dfrac{1}{3}, 2$

25. $x^2 - 5x - 2 = 0$

26. $x^2 - 2x - 5 = 0$ $1 \pm \sqrt{6}$

27. $3x^2 - x - 14 = 0$ $-2, \dfrac{7}{3}$

28. $5x^2 - 13x - 6 = 0$ $-\dfrac{2}{5}, 3$

29. $6x^2 + 9x = 2$

30. $3x^2 - 9x = 8$

31. $7p^2 + 2 = 8p$ $\dfrac{4 \pm \sqrt{2}}{7}$

32. $11p^2 + 2 = 10p$ $\dfrac{5 \pm \sqrt{3}}{11}$

33. $a^2 - 6a + 2 = 0$

34. $a^2 - 10a + 19 = 0$

35. $2x^2 - 6x + 3 = 0$

36. $5x^2 - 8x + 2 = 0$

37. $3x^2 = 1 - 2x$ $-1, \dfrac{1}{3}$

38. $5y^2 = 4 - y$ $-1, \dfrac{4}{5}$

39. $20y^2 = 3 - 11y$ $-\dfrac{3}{4}, \dfrac{1}{5}$

40. $2z^2 = z + 3$ $-1, \dfrac{3}{2}$

41. $x^2 + x + 1 = 0$

42. $k^2 + 2k + 5 = 0$

43. $4y^2 = 6y + 1$

44. $6z^2 + 3z + 2 = 0$

Use the quadratic formula to solve each quadratic equation. See Example 5.

45. $3p^2 - \dfrac{2}{3}p + 1 = 0$

46. $\dfrac{5}{2}p^2 - p + \dfrac{1}{2} = 0$

47. $\dfrac{m^2}{2} = m + \dfrac{1}{2}$ $1 \pm \sqrt{2}$

48. $\dfrac{m^2}{2} = 3m - 1$ $3 \pm \sqrt{7}$

49. $4p^2 + \dfrac{3}{2} = -5p$ $-\dfrac{3}{4}, -\dfrac{1}{2}$

50. $4p^2 + \dfrac{3}{2} = 5p$ $\dfrac{1}{2}, \dfrac{3}{4}$

51. $5x^2 = \dfrac{7}{2}x + 1$

52. $2x^2 = \dfrac{5}{2}x + \dfrac{7}{2}$ $\dfrac{5 \pm \sqrt{137}}{8}$

53. $28x^2 + 5x + \dfrac{11}{4} = 0$

54. $\dfrac{2}{3}x^2 - 2x - \dfrac{2}{3} = 0$

55. $5z^2 - 2z = \dfrac{1}{5}$

56. $9z^2 + 12z = -1$

57. $x^2 + 3\sqrt{2}x - 5 = 0$

58. $y^2 - 2\sqrt{5}y - 1 = 0$

Use the discriminant to determine the number of solutions of each quadratic equation. See Example 7.

59. $x^2 + 3x - 1 = 0$

60. $x^2 - 5x - 3 = 0$

61. $3x^2 + x + 5 = 0$

62. $2x^2 + x + 4 = 0$

63. $4x^2 + 4x = -1$

64. $7x^2 - x = 0$

33. $3 \pm \sqrt{7}$ **34.** $5 \pm \sqrt{6}$ **35.** $\dfrac{3 \pm \sqrt{3}}{2}$ **36.** $\dfrac{4 \pm \sqrt{6}}{5}$ **41.–42.** no real solution   **43.** $\dfrac{3 \pm \sqrt{13}}{4}$ **44.–46.** no real solution   **51.** $\dfrac{7 \pm \sqrt{129}}{20}$

53. no real solution   **54.** $\dfrac{3 \pm \sqrt{13}}{2}$ **55.** $\dfrac{1 \pm \sqrt{2}}{5}$ **56.** $\dfrac{-2 \pm \sqrt{3}}{3}$ **57.** $\dfrac{-3\sqrt{2} \pm \sqrt{38}}{2}$ **58.** $\sqrt{5} \pm \sqrt{6}$ **59.–60.** 2 real solutions

61.–62. no real solution   **63.** 1 real solution   **64.** 2 real solutions

65. $9x^2 + 2x = 0$

66. $x^2 + 10x = -25$

67. $5x^2 + 1 = 0$

68. $4x^2 + 9 = 12x$

69. $x^2 + 36 = -12x$

70. $10x^2 + 2 = 0$

71. For the quadratic equation $2x^2 - 5 = 9x$, if $a = 2$ and $c = -5$ in the quadratic formula, the value of b is

   a. $\dfrac{9}{2}$ **b.** 9 **c.** -5 **d.** -9 d

72. Explain how the quadratic formula is derived and why it is useful.   answers may vary

REVIEW AND PREVIEW

Simplify each radical. See Section 8.2.

73. $\sqrt{48}$ $4\sqrt{3}$ **74.** $\sqrt{104}$ $2\sqrt{26}$

75. $\sqrt{50}$ $5\sqrt{2}$ **76.** $\sqrt{80}$ $4\sqrt{5}$

Solve the following. See Section 2.6.

77. The height of a triangle is 4 times the length of the base. The area of the triangle is 18 square feet. Find the height and base of the triangle.   base: 3 ft; height: 12 ft

65. 2 real solutions
66. 1 real solution
67. no real solution
68.–69. 1 real solution
70. no real solution

78. The length of a rectangle is 6 inches more than its width. The area of the rectangle is 391 square inches. Find the dimensions of the rectangle.   width: 17 in.; length: 23 in.

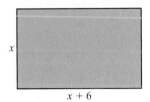

Concept Extensions

79. The largest chocolate bar was a Cadbury's Dairy Milk bar that weighed 1.1 tons. The bar had a base area of 35 square feet and its length was five feet longer than its width. Find the dimensions of the bar. Round each dimension to the nearest tenth. (*Source: Guinness World Records*, 2000)   3.9 ft by 8.9 ft

80. The area of a rectangular conference room table is 95 square feet. If its length is six feet longer than its width, find the dimensions of the table. Round each dimension to the nearest tenth.   7.2 ft by 13.2 ft

Use the quadratic formula and a calculator to solve each equation. Round solutions to the nearest tenth.

81. $x^2 + x = 15$ $-4.4, 3.4$ **82.** $y^2 - y = 11$ $3.9, -2.9$

83. $1.2x^2 - 5.2x - 3.9 = 0$ **84.** $7.3z^2 + 5.4z - 1.1 = 0$
83. $-0.7, 5.0$ **84.** $-0.9, 0.2$

A rocket is launched from the top of an 80-foot cliff with an initial velocity of 120 feet per second. The height, h, of the rocket after t seconds is given by the equation $h = -16t^2 + 120t + 80$. Use this for Exercises 85 and 86.

85. How long after the rocket is launched will it be 30 feet from the ground? Round to the nearest tenth of a second.   7.9 sec

86. How long after the rocket is launched will it strike the ground? Round to the nearest tenth of a second. (*Hint:* The rocket will strike the ground when its height $h = 0$.)   8.1 sec

87. The net sales y (in millions of dollars) of Goodyear Tire and Rubber Company from 2000 through 2002 is given by the equation $y = -14x^2 - 257x + 14,417$, where $x = 0$ represents 2000. Assume that this trend continues and predict the year in which Goodyear's net sales will be $12,371. (*Source:* Based on data from the Goodyear Tire and Rubber Company)   2006

88. The average annual salary y (in dollars) for NFL players for the years 1998 through 2000 is given by the equation $y = 57,000x^2 - 14,000x + 1,000,000$, where $x = 0$ represents 1998. Assume that this trend continues and predict the year in which the average NFL salary will be $3,695,000. (*Source:* Based on data from the NFL Players Association)   2005

89. The number of Sears stores y operating in North America from 1998 through 2002 is given by the equation $y = -x^2 + 11x + 846$. In this equation, x is the number of years after 1998. Assume that this trend continues and predict the year after 1998 in which the number of Sears stores will be 870. (*Source:* Based on data from Sears Roebuck & Co) 2001 or 2006

INTEGRATED REVIEW SUMMARY ON SOLVING QUADRATIC EQUATIONS

An important skill in mathematics is learning when to use one technique in favor of another. We now practice this by deciding which method to use when solving quadratic equations. Although both the quadratic formula and completing the square can be used to solve any quadratic equation, the quadratic formula is usually less tedious and thus preferred. The following steps may be used to solve a quadratic equation.

To Solve a Quadratic Equation

Step 1: If the equation is in the form $ax^2 = c$ or $(ax + b)^2 = c$, use the square root property and solve. If not, go to Step 2.

Step 2: Write the equation in standard form: $ax^2 + bx + c = 0$.

Step 3: Try to solve the equation by the factoring method. If not possible, go to Step 4.

Step 4: Solve the equation by the quadratic formula.

Study the examples below to help you review these steps.

EXAMPLE 1

Solve $m^2 - 2m - 7 = 0$.

Solution The equation is in standard form, but the quadratic expression $m^2 - 2m - 7$ is not factorable, so use the quadratic formula with $a = 1$, $b = -2$, and $c = -7$.

$$m^2 - 2m - 7 = 0$$

$$m = \frac{-(-2) \pm \sqrt{(-2)^2 - 4 \cdot 1 \cdot (-7)}}{2 \cdot 1} = \frac{2 \pm \sqrt{32}}{2}$$

$$m = \frac{2 \pm 4\sqrt{2}}{2} = \frac{2(1 \pm 2\sqrt{2})}{2} = 1 \pm 2\sqrt{2}$$

The solutions are $1 - 2\sqrt{2}$ and $1 + 2\sqrt{2}$.

EXAMPLE 2

Solve $(3x + 1)^2 = 20$.

Solution   This equation is in a form that makes the square root property easy to apply.

$$(3x + 1)^2 = 20$$
$$3x + 1 = \pm\sqrt{20} \qquad \text{Apply the square root property.}$$
$$3x + 1 = \pm 2\sqrt{5} \qquad \text{Simplify } \sqrt{20}.$$
$$3x = -1 \pm 2\sqrt{5}$$
$$x = \frac{-1 \pm 2\sqrt{5}}{3}$$

The solutions are $\dfrac{-1 - 2\sqrt{5}}{3}$ and $\dfrac{-1 + 2\sqrt{5}}{3}$.

EXAMPLE 3

Solve $x^2 - \dfrac{11}{2}x = -\dfrac{5}{2}$.

Solution   The fractions make factoring more difficult and also complicate the calculations for using the quadratic formula. Clear the equation of fractions by multiplying both sides of the equation by the LCD 2.

$$x^2 - \frac{11}{2}x = -\frac{5}{2}$$

$$x^2 - \frac{11}{2}x + \frac{5}{2} = 0 \qquad \text{Write in standard form.}$$

$$2x^2 - 11x + 5 = 0 \qquad \text{Multiply both sides by 2.}$$

$$(2x - 1)(x - 5) = 0 \qquad \text{Factor.}$$

$$2x - 1 = 0 \quad \text{or} \quad x - 5 = 0 \qquad \text{Apply the zero factor theorem.}$$

$$2x = 1 \qquad\qquad x = 5$$

$$x = \frac{1}{2} \qquad\qquad x = 5$$

The solutions are $\dfrac{1}{2}$ and 5.

Choose and use a method to solve each equation.

1. $5x^2 - 11x + 2 = 0$
$\dfrac{1}{5}, 2$

2. $5x^2 + 13x - 6 = 0$
$-3, \dfrac{2}{5}$

3. $x^2 - 1 = 2x$
$1 \pm \sqrt{2}$

4. $x^2 + 7 = 6x$
$3 \pm \sqrt{2}$

5. $a^2 = 20$
$\pm 2\sqrt{5}$

6. $a^2 = 72$
$\pm 6\sqrt{2}$

7. $x^2 - x + 4 = 0$
no real solution

8. $x^2 - 2x + 7 = 0$
no real solution

9. $3x^2 - 12x + 12 = 0$
2

10. $5x^2 - 30x + 45 = 0$
3

11. $9 - 6p + p^2 = 0$
3

12. $49 - 28p + 4p^2 = 0$
$\dfrac{7}{2}$

13. $4y^2 - 16 = 0$
$y = \pm 2$

14. $3y^2 - 27 = 0$
$y = \pm 3$

15. $x^4 - 3x^3 + 2x^2 = 0$
$0, 1, 2$

16. $x^3 + 7x^2 + 120x = 0$
$0, -3, -4$

🔒 **17.** $(2z + 5)^2 = 25$
$-5, 0$

18. $(3z - 4)^2 = 16$
$0, \dfrac{8}{3}$

19. $30x = 25x^2 + 2$
$\dfrac{3 \pm \sqrt{7}}{5}$

20. $12x = 4x^2 + 4$
$\dfrac{3 \pm \sqrt{5}}{2}$

21. $\dfrac{2}{3}m^2 - \dfrac{1}{3}m - 1 = 0$
$\dfrac{3}{2}, -1$

22. $\dfrac{5}{8}m^2 + m - \dfrac{1}{2} = 0$
$\dfrac{2}{5}, -2$

🔒 **23.** $x^2 - \dfrac{1}{2}x - \dfrac{1}{5} = 0$
$\dfrac{5 \pm \sqrt{105}}{20}$

24. $x^2 + \dfrac{1}{2}x - \dfrac{1}{8} = 0$
$\dfrac{-1 \pm \sqrt{3}}{4}$

25. $4x^2 - 27x + 35 = 0$
$5, \dfrac{7}{4}$

26. $9x^2 - 16x + 7 = 0$
$\dfrac{7}{9}, 1$

27. $(7 - 5x)^2 = 18$
$\dfrac{7 \pm 3\sqrt{2}}{5}$

28. $(5 - 4x)^2 = 75$
$\dfrac{5 \pm 5\sqrt{3}}{4}$

29. $3z^2 - 7z = 12$
$\dfrac{7 \pm \sqrt{193}}{6}$

30. $6z^2 + 7z = 6$
$\dfrac{-7 \pm \sqrt{193}}{12}$

🔒 **31.** $x = x^2 - 110$
$11, -10$

32. $x = 56 - x^2$
$-8, 7$

33. $\dfrac{3}{4}x^2 - \dfrac{5}{2}x - 2 = 0$
$-\dfrac{2}{3}, 4$

34. $x^2 - \dfrac{6}{5}x - \dfrac{8}{5} = 0$
$2, -\dfrac{4}{5}$

35. $x^2 - 0.6x + 0.05 = 0$
$0.5, 0.1$

36. $x^2 - 0.1x - 0.06 = 0$
$0.3, -0.2$

37. $10x^2 - 11x + 2 = 0$
$\dfrac{11 \pm \sqrt{41}}{20}$

38. $20x^2 - 11x + 1 = 0$
$\dfrac{11 \pm \sqrt{41}}{40}$

39. $\dfrac{1}{2}z^2 - 2z + \dfrac{3}{4} = 0$
$\dfrac{4 \pm \sqrt{10}}{2}$

40. $\dfrac{1}{5}z^2 - \dfrac{1}{2}z - 2 = 0$ $\dfrac{5 \pm \sqrt{185}}{4}$

41. Explain how you will decide what method to use when solving quadratic equations. answers may vary

9.4   COMPLEX SOLUTIONS OF QUADRATIC EQUATIONS

Objectives

1. Write complex numbers using i notation.
2. Add and subtract complex numbers.
3. Multiply complex numbers.
4. Divide complex numbers.
5. Solve quadratic equations that have complex solutions.

In Chapter 8, we learned that $\sqrt{-4}$, for example, is not a real number because there is no real number whose square is -4. However, our real number system can be extended to include numbers like $\sqrt{-4}$. This extended number system is called the **complex number** system. The complex number system includes the **imaginary unit i**, which is defined next.

> ### Imaginary Unit i
>
> The imaginary unit, written i, is the number whose square is -1. That is,
>
> $$i^2 = -1 \quad \text{and} \quad i = \sqrt{-1}$$

1 We use i to write numbers like $\sqrt{-6}$ as the product of a real number and i. Since $i = \sqrt{-1}$, we have

$$\sqrt{-6} = \sqrt{-1 \cdot 6} = \sqrt{-1} \cdot \sqrt{6} = i\sqrt{6}$$

EXAMPLE 1

Write each radical as the product of a real number and i.

   a. $\sqrt{-4}$ **b.** $\sqrt{-11}$ **c.** $\sqrt{-20}$

Solution Write each negative radicand as a product of a positive number and -1. Then write $\sqrt{-1}$ as i.

   a. $\sqrt{-4} = \sqrt{-1 \cdot 4} = \sqrt{-1} \cdot \sqrt{4} = i \cdot 2 = 2i$

   b. $\sqrt{-11} = \sqrt{-1 \cdot 11} = \sqrt{-1} \cdot \sqrt{11} = i\sqrt{11}$

   c. $\sqrt{-20} = \sqrt{-1 \cdot 20} = \sqrt{-1} \cdot \sqrt{20} = i \cdot 2\sqrt{5} = 2i\sqrt{5}$

    The numbers $2i$, $i\sqrt{11}$, and $2i\sqrt{5}$ are called **pure imaginary numbers**. Both real numbers and pure imaginary numbers are complex numbers.

CLASSROOM EXAMPLE

Write each radical as the product of a real number and i.

a. $\sqrt{-81}$ **b.** $\sqrt{-13}$ **c.** $\sqrt{-80}$

answer:

a. $9i$ **b.** $i\sqrt{13}$ **c.** $4i\sqrt{5}$

> ### Complex Numbers and Pure Imaginary Numbers
>
> A complex number is a number that can be written in the form
>
> $$a + bi$$
>
> where a and b are real numbers. A complex number that can be written in the form
>
> $$0 + bi$$
>
> $b \neq 0$, is also called a pure imaginary number.

A complex number written in the form $a + bi$ is in **standard form**. We call a the real part and bi the imaginary part of the complex number $a + bi$.

EXAMPLE 2

Identify each number as a complex number by writing it in standard form $a + bi$.

a. 7   **b.** 0   **c.** $\sqrt{20}$ **d.** $\sqrt{-27}$ **e.** $2 + \sqrt{-4}$

Solution

a. 7 is a complex number since $7 = 7 + 0i$.

b. 0 is a complex number since $0 = 0 + 0i$.

c. $\sqrt{20}$ is a complex number since $\sqrt{20} = 2\sqrt{5} = 2\sqrt{5} + 0i$.

d. $\sqrt{-27}$ is a complex number since $\sqrt{-27} = i \cdot 3\sqrt{3} = 0 + 3i\sqrt{3}$.

e. $2 + \sqrt{-4}$ is a complex number since $2 + \sqrt{-4} = 2 + 2i$.

2 We now present arithmetic operations—addition, subtraction, multiplication, and division—for the complex number system. Complex numbers are added and subtracted in the same way as we add and subtract polynomials.

EXAMPLE 3

Simplify the sum or difference. Write the result in standard form.

a. $(2 + 3i) + (-6 - i)$ **b.** $-i + (3 + 7i)$ **c.** $(5 - i) - 4$

Solution Add the real parts and then add the imaginary parts.

a. $(2 + 3i) + (-6 - i) = [2 + (-6)] + (3i - i) = -4 + 2i$

b. $-i + (3 + 7i) = 3 + (-i + 7i) = 3 + 6i$

c. $(5 - i) - 4 = (5 - 4) - i = 1 - i$

EXAMPLE 4

Subtract $(11 - i)$ from $(1 + i)$.

Solution $(1 + i) - (11 - i) = 1 + i - 11 + i = (1 - 11) + (i + i) = -10 + 2i$

3 Use the distributive property and the FOIL method to multiply complex numbers.

EXAMPLE 5

Find the following products and write in standard form.

a. $5i(2 - i)$ **b.** $(7 - 3i)(4 + 2i)$ **c.** $(2 + 3i)(2 - 3i)$

Solution   **a.** By the distributive property, we have

$$5i(2 - i) = 5i \cdot 2 - 5i \cdot i \qquad \text{Apply the distributive property.}$$

$$= 10i - 5i^2$$

$$= 10i - 5(-1) \qquad \text{Write } i^2 \text{ as } -1.$$

$$= 10i + 5$$

$$= 5 + 10i \qquad \text{Write in standard form.}$$

$$\overset{\text{F}\qquad\text{O}\qquad\text{I}\qquad\text{L}}{}$$

b. $(7 - 3i)(4 + 2i) = 28 + 14i - 12i - 6i^2$

$$= 28 + 2i - 6(-1) \qquad \text{Write } i^2 \text{ as } -1.$$

$$= 28 + 2i + 6$$

$$= 34 + 2i$$

c. $(2 + 3i)(2 - 3i) = 4 - 6i + 6i - 9i^2$

$$= 4 - 9(-1) \qquad \text{Write } i^2 \text{ as } -1.$$

$$= 13$$

The product in part (c) is the real number 13. Notice that one factor is the sum of 2 and $3i$, and the other factor is the difference of 2 and $3i$. When complex number factors are related as these two are, their product is a real number. In general,

$$(a + bi)(a - bi) = a^2 + b^2$$

$$\text{sum} \quad \text{difference} \quad \text{real number}$$

The complex numbers $a + bi$ and $a - bi$ are called **complex conjugates** of each other. For example, $2 - 3i$ is the conjugate of $2 + 3i$, and $2 + 3i$ is the conjugate of $2 - 3i$. Also,

The conjugate of $3 - 10i$ is $3 + 10i$.
The conjugate of 5 is 5. (Note that $5 = 5 + 0i$ and its conjugate is $5 - 0i = 5$.)
The conjugate of $4i$ is $-4i$. ($0 - 4i$ is the conjugate of $0 + 4i$.)

4   The fact that the product of a complex number and its conjugate is a real number provides a method for dividing by a complex number and for simplifying fractions whose denominators are complex numbers.

EXAMPLE 6

Write $\dfrac{4 + i}{3 - 4i}$ in standard form.

Solution   To write this quotient as a complex number in the standard form $a + bi$, we need to find an equivalent fraction whose denominator is a real number. By multiplying both numerator and denominator by the denominator's conjugate, we obtain a new fraction that is an equivalent fraction with a real number denominator.

$$\frac{4 + i}{3 - 4i} = \frac{(4 + i)}{(3 - 4i)} \cdot \frac{(3 + 4i)}{(3 + 4i)}$$ Multiply numerator and denominator by $3 + 4i$.

$$= \frac{12 + 16i + 3i + 4i^2}{9 - 16i^2}$$

$$= \frac{12 + 19i + 4(-1)}{9 - 16(-1)}$$ Recall that $i^2 = -1$.

$$= \frac{12 + 19i - 4}{9 + 16} = \frac{8 + 19i}{25}$$

$$= \frac{8}{25} + \frac{19}{25}i$$ Write in standard form.

Note that our last step was to write $\dfrac{4 + i}{3 - 4i}$ in standard form $a + bi$, where a and b are real numbers.

5 Some quadratic equations have complex solutions.

EXAMPLE 7

Solve $(x + 2)^2 = -25$.

Solution Begin by applying the square root property.

$$(x + 2)^2 = -25$$

$$x + 2 = \pm\sqrt{-25}$$ Apply the square root property.

$$x + 2 = \pm 5i$$ Write $\sqrt{-25}$ as $5i$.

$$x = -2 \pm 5i$$

The solutions are $-2 + 5i$ and $-2 - 5i$.

EXAMPLE 8

Solve $m^2 = 4m - 5$.

Solution Write the equation in standard form and use the quadratic formula to solve.

$$m^2 = 4m - 5$$

$$m^2 - 4m + 5 = 0$$ Write the equation in standard form.

Apply the quadratic formula with $a = 1$, $b = -4$, and $c = 5$.

$$m = \frac{4 \pm \sqrt{16 - 4 \cdot 1 \cdot 5}}{2 \cdot 1}$$

$$= \frac{4 \pm \sqrt{-4}}{2}$$

$$= \frac{4 \pm 2i}{2}$$ Write $\sqrt{-4}$ as $2i$.

$$= \frac{2(2 \pm i)}{2} = 2 \pm i$$

The solutions are $2 - i$ and $2 + i$.

EXAMPLE 9

Solve $x^2 + x = -1$.

Solution

$$x^2 + x = -1$$

$$x^2 + x + 1 = 0 \qquad \text{Write in standard form.}$$

$$x = \frac{-1 \pm \sqrt{1 - 4 \cdot 1 \cdot 1}}{2 \cdot 1} \qquad \text{Apply the quadratic formula with } a = 1, b = 1, \text{ and } c = 1.$$

$$= \frac{-1 \pm \sqrt{-3}}{2}$$

$$= \frac{-1 \pm i\sqrt{3}}{2}$$

The solutions are $\dfrac{-1 - i\sqrt{3}}{2}$ and $\dfrac{-1 + i\sqrt{3}}{2}$.

CLASSROOM EXAMPLE

Solve $x^2 + x = -2$.

answer: $\dfrac{-1 \pm i\sqrt{7}}{2}$

9. $-3 + 9i$ 10. $-2 - i$
11. $-12 + 6i$ 12. $7 + 9i$
13. $1 - 3i$ 14. -9 15. $9 - 4i$
16. $-7 - 2i$ 19. $26 - 2i$
20. $26 + 2i$ 29. $-1 \pm 3i$

EXERCISE SET 9.4

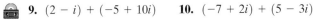

STUDY GUIDE/SSM   CD/VIDEO   PH MATH TUTOR CENTER   MathXL®Tutorials ON CD   MathXL®   MyMathLab®

Write each expression in i notation. See Example 1.

 1. $\sqrt{-9}$ $3i$
2. $\sqrt{-64}$ $8i$
3. $\sqrt{-100}$ $10i$
4. $\sqrt{-16}$ $4i$
5. $\sqrt{-50}$ $5i\sqrt{2}$
6. $\sqrt{-98}$ $7i\sqrt{2}$
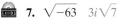 7. $\sqrt{-63}$ $3i\sqrt{7}$
8. $\sqrt{-44}$ $2i\sqrt{11}$

Add or subtract as indicated. See Examples 2 through 4.

9. $(2 - i) + (-5 + 10i)$
10. $(-7 + 2i) + (5 - 3i)$
11. $(-11 + 3i) - (1 - 3i)$
12. $(1 + i) - (-6 - 8i)$
13. $(3 - 4i) - (2 - i)$
14. $(-6 + i) - (3 + i)$
15. $(16 + 2i) + (-7 - 6i)$
16. $(-3 - i) + (-4 - i)$

Multiply. See Example 5.

17. $4i(3 - 2i)$ $8 + 12i$
18. $-2i(5 + 4i)$ $8 - 10i$
19. $(6 - 2i)(4 + i)$
20. $(6 + 2i)(4 - i)$
21. $(3 + 8i)(3 - 8i)$ 73
22. $(-9 + 2i)(-9 - 2i)$ 85

23. Earlier in this text, we learned that $\sqrt{-4}$ is not a real number. Explain what that means and explain what type of number $\sqrt{-4}$ is.   answers may vary

24. Describe how to find the conjugate of a complex number.   answers may vary

Divide. Write each answer in standard form. See Example 6.

25. $\dfrac{8 - 12i}{4}$ $2 - 3i$
26. $\dfrac{14 + 28i}{-7}$ $-2 - 4i$

27. $\dfrac{7 - i}{4 - 3i}$ $\dfrac{31}{25} + \dfrac{17}{25}i$
28. $\dfrac{4 - 3i}{7 - i}$ $\dfrac{31}{50} - \dfrac{17}{50}i$

Solve the following quadratic equations for complex solutions. See Example 7.

29. $(x + 1)^2 = -9$
30. $(y - 2)^2 = -25$ $2 \pm 5i$
31. $(2z - 3)^2 = -12$ $\dfrac{3 \pm 2i\sqrt{3}}{2}$
32. $(3p + 5)^2 = -18$ $\dfrac{-5 \pm 3i\sqrt{2}}{3}$

Solve the following quadratic equations for complex solutions. See Examples 8 and 9.

33. $y^2 + 6y + 13 = 0$
34. $y^2 - 2y + 5 = 0$ $1 \pm 2i$
35. $4x^2 + 7x + 4 = 0$
36. $8x^2 - 7x + 2 = 0$ $\dfrac{7 \pm i\sqrt{15}}{16}$
37. $2m^2 - 4m + 5 = 0$
38. $5m^2 - 6m + 7 = 0$ $\dfrac{3 \pm i\sqrt{26}}{5}$

33. $-3 \pm 2i$ 35. $\dfrac{-7 \pm i\sqrt{15}}{8}$ 37. $\dfrac{2 \pm i\sqrt{6}}{2}$

MIXED PRACTICE

Perform the indicated operations. Write results in standard form.

39. $3 + (12 - 7i)$ $15 - 7i$
40. $(-14 + 5i) + 3i$ $-14 + 8i$
41. $-9i(5i - 7)$ $45 + 63i$
42. $10i(4i - 1)$ $-40 - 10i$
43. $(2 - i) - (3 - 4i)$
44. $(3 + i) - (-6 + i)$ 9
45. $\dfrac{15 + 10i}{5i}$ $2 - 3i$
46. $\dfrac{-18 + 12i}{-6i}$ $-2 - 3i$

43. $-1 + 3i$

47. Subtract $(2 + 3i)$ from $(-5 + i)$. $-7 - 2i$
48. Subtract $(-8 - i)$ from $(7 - 4i)$. $15 - 3i$
49. $(4 - 3i)(4 + 3i)$ 25
50. $(12 - 5i)(12 + 5i)$ 169

51. $\dfrac{4 - i}{1 + 2i}$ $\dfrac{2}{5} - \dfrac{9}{5}i$ **52.** $\dfrac{9 - 2i}{-3 + i}$ $-\dfrac{29}{10} - \dfrac{3}{10}i$

53. $(5 + 2i)^2$ $21 + 20i$ **54.** $(9 - 7i)^2$ $32 - 126i$

Solve the following quadratic equations for complex solutions.

55. $(y - 4)^2 = -64$ **56.** $(x + 7)^2 = -1$

57. $4x^2 = -100$ $\pm 5i$ **58.** $7x^2 = -28$ $\pm 2i$

59. $z^2 + 6z + 10 = 0$ **60.** $z^2 + 4z + 13 = 0$

61. $2a^2 - 5a + 9 = 0$ **62.** $4a^2 + 3a + 2 = 0$

63. $(2x + 8)^2 = -20$ **64.** $(6z - 4)^2 = -24$

65. $3m^2 + 108 = 0$ $\pm 6i$ **66.** $5m^2 + 80 = 0$ $\pm 4i$

67. $x^2 + 14x + 50 = 0$ **68.** $x^2 + 8x + 25 = 0$

REVIEW AND PREVIEW

Graph the following linear equations in two variables. See Section 3.2. **69–72.** See graphing answer section.

69. $y = -3$ **70.** $x = 4$

71. $y = 3x - 2$ **72.** $y = 2x + 3$

The line graph at the top of the next column shows the percent of U.S. households with computers. Use this graph for Exercises 73 through 75. See Sections 3.5 and 7.2.

55. $4 \pm 8i$ **56.** $-7 \pm i$ **59.** $-3 \pm i$ **60.** $-2 \pm 3i$

61. $\dfrac{5 \pm i\sqrt{47}}{4}$ **62.** $\dfrac{-3 \pm i\sqrt{23}}{8}$ **63.** $-4 \pm i\sqrt{5}$

64. $\dfrac{2 \pm i\sqrt{6}}{3}$ **67.** $-7 \pm i$ **68.** $-4 \pm 3i$

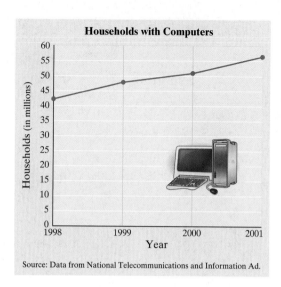

Households with Computers

Source: Data from National Telecommunications and Information Ad.

73. Estimate the percent of households with computers in 2000. 51%

74. The percent growth of computers in households has been almost linear since 1998. Approximate this growth with a linear equation. To do so, find an equation of the line through the ordered pairs $(0, 42.1)$ and $(3, 56.5)$ where x is the number of years since 1998 and y is the percent of households that have computers. Write the equation in slope-intercept form. $y = 4.8x + 42.1$

75. Use the equation found in Exercise 74 to predict the percent of households with computers in 2008. 90.1%

Concept Extensions

Answer the following true or false.

76. Every real number is a complex number. true
77. Every complex number is a real number. false
78. If a complex number such as $2 + 3i$ is a solution of a quadratic equation, then its conjugate $2 - 3i$ is also a solution. true
79. Some pure imaginary numbers are real numbers. false

9.5 GRAPHING QUADRATIC EQUATIONS

Objectives

1 Graph quadratic equations of the form $y = ax^2 + bx + c$.

2 Find the intercepts of a parabola.

3 Determine the vertex of a parabola.

1 Recall from Section 3.2 that the graph of a linear equation in two variables $Ax + By = C$ is a straight line. Also recall from Section 6.5 that the graph of a quadratic equation in two variables $y = ax^2 + bx + c$ is a parabola. In this section, we further investigate the graph of a quadratic equation.

To graph the quadratic equation $y = x^2$, select a few values for x and find the corresponding y-values. Make a table of values to keep track. Then plot the points corresponding to these solutions.

$y = x^2$

x	y
-3	9
-2	4
-1	1
0	0
1	1
2	4
3	9

If $x = -3$, then $y = (-3)^2$, or 9.

If $x = -2$, then $y = (-2)^2$, or 4.

If $x = -1$, then $y = (-1)^2$, or 1.

If $x = 0$, then $y = 0^2$, or 0.

If $x = 1$, then $y = 1^2$, or 1.

If $x = 2$, then $y = 2^2$, or 4.

If $x = 3$, then $y = 3^2$, or 9.

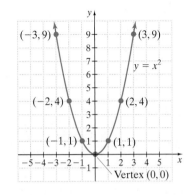

Clearly, these points are not on one straight line. As we saw in Chapter 6, the graph of $y = x^2$ is a smooth curve through the plotted points. This curve is called a **parabola**. The lowest point on a parabola opening upward is called the **vertex**. The vertex is $(0, 0)$ for the parabola $y = x^2$. If we fold the graph paper along the y-axis, the two pieces of the parabola match perfectly. For this reason, we say the graph is **symmetric about the y-axis**, and we call the y-axis the **axis of symmetry**.

Notice that the parabola that corresponds to the equation $y = x^2$ opens upward. This happens when the coefficient of x^2 is positive. In the equation $y = x^2$, the coefficient of x^2 is 1. Example 1 shows the graph of a quadratic equation whose coefficient of x^2 is negative.

EXAMPLE 1

Graph $y = -2x^2$.

Solution   Select x-values and calculate the corresponding y-values. Plot the ordered pairs found. Then draw a smooth curve through those points. When the coefficient of x^2 is negative, the corresponding parabola opens downward. When a parabola opens downward, the vertex is the highest point of the parabola. The vertex of this parabola is $(0, 0)$ and the axis of symmetry is again the y-axis.

CLASSROOM EXAMPLE

Graph $y = -3x^2$.

answer:

$y = -2x^2$

x	y
0	0
1	-2
2	-8
3	-18
-1	-2
-2	-8
-3	-18

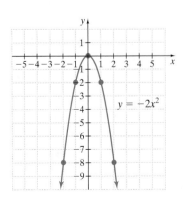

2   Just as for linear equations, we can use x- and y-intercepts to help graph quadratic equations. Recall from Chapter 3 that an x-intercept is the point where the graph intersects the x-axis. A y-intercept is the point where the graph intersects the y-axis.

> **Helpful Hint**
>
> Recall that:
>
> To find x-intercepts, let $y = 0$ and solve for x.
> To find y-intercepts, let $x = 0$ and solve for y.

EXAMPLE 2

Graph $y = x^2 - 4$.

Solution   First, find intercepts. To find the y-intercept, let $x = 0$. Then

$$y = 0^2 - 4 = -4$$

To find x-intercepts, we let $y = 0$.

$$0 = x^2 - 4$$
$$0 = (x - 2)(x + 2)$$
$$x - 2 = 0 \quad \text{or} \quad x + 2 = 0$$
$$x = 2 \qquad\qquad x = -2$$

CLASSROOM EXAMPLE

Graph $y = x^2 - 9$.

answer:

Thus far, we have the y-intercept $(0, -4)$ and the x-intercepts $(2, 0)$ and $(-2, 0)$. Now we can select additional x-values, find the corresponding y-values, plot the points, and draw a smooth curve through the points.

$$y = x^2 - 4$$

x	y
0	−4
1	−3
2	0
3	5
−1	−3
−2	0
−3	5

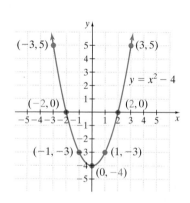

Notice that the vertex of this parabola is $(0, -4)$.

> **Helpful Hint**
>
> For the graph of $y = ax^2 + bx + c$,
>
> If a is positive, the parabola opens upward.
> If a is negative, the parabola opens downward.

✔ **CONCEPT CHECK**

For which of the following graphs of $y = ax^2 + bx + c$ would the value of a be negative?

a.

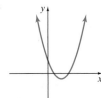

b.

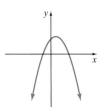

3   Thus far, we have accidentally stumbled upon the vertex of each parabola that we have graphed. It would be helpful if we could first find the vertex of a parabola, next determine whether the parabola opens upward or downward, and finally calculate additional points such as x- and y-intercepts as needed. In fact, there is a formula that may be used to find the vertex of a parabola.

One way to develop this formula is to notice that the x-value of the vertex of the parabolas that we are considering lies halfway between its x-intercepts. We can use this fact to find a formula for the vertex.

Recall that the x-intercepts of a parabola may be found by solving $0 = ax^2 + bx + c$. These solutions, by the quadratic formula, are

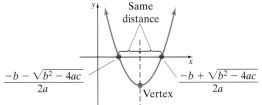

$$x = \frac{-b - \sqrt{b^2 - 4ac}}{2a}, \quad x = \frac{-b + \sqrt{b^2 - 4ac}}{2a}$$

The x-coordinate of the vertex of a parabola is halfway between its x-intercepts, so the x-value of the vertex may be found by computing the average, or $\frac{1}{2}$ of the sum of the intercepts.

$$x = \frac{1}{2}\left(\frac{-b - \sqrt{b^2 - 4ac}}{2a} + \frac{-b + \sqrt{b^2 - 4ac}}{2a}\right)$$

$$= \frac{1}{2}\left(\frac{-b - \sqrt{b^2 - 4ac} - b + \sqrt{b^2 - 4ac}}{2a}\right)$$

$$= \frac{1}{2}\left(\frac{-2b}{2a}\right)$$

$$= \frac{-b}{2a}$$

Vertex Formula

The vertex of the parabola $y = ax^2 + bx + c$ has x-coordinate

$$\frac{-b}{2a}$$

The corresponding y-coordinate of the vertex is found by substituting the x–coordinate into the equation and evaluating y.

Concept Check Answer:
b

EXAMPLE 3

Graph $y = x^2 - 6x + 8$.

Solution   In the equation $y = x^2 - 6x + 8$, $a = 1$ and $b = -6$. The x-coordinate of the vertex is

$$x = \frac{-b}{2a} = \frac{-(-6)}{2 \cdot 1} = 3 \qquad \text{Use the vertex formula, } x = \frac{-b}{2a}.$$

To find the corresponding y-coordinate, we let $x = 3$ in the original equation.

$$y = x^2 - 6x + 8 = 3^2 - 6 \cdot 3 + 8 = -1$$

The vertex is $(3, -1)$ and the parabola opens upward since a is positive. We now find and plot the intercepts.

To find the x-intercepts, we let $y = 0$.

$$0 = x^2 - 6x + 8$$

We factor the expression $x^2 - 6x + 8$ to find $(x - 4)(x - 2) = 0$. The x-intercepts are $(4, 0)$ and $(2, 0)$.

If we let $x = 0$ in the original equation, then $y = 8$ and the y-intercept is $(0, 8)$. Now we plot the vertex $(3, -1)$ and the intercepts $(4, 0)$, $(2, 0)$, and $(0, 8)$. Then we can sketch the parabola. These and two additional points are shown in the table.

$y = x^2 - 6x + 8$

x	y
3	-1
4	0
2	0
0	8
1	3
5	3
6	8

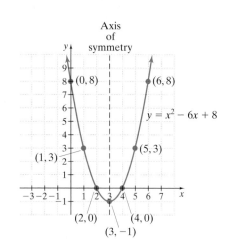

EXAMPLE 4

Graph $y = x^2 + 2x - 5$.

Solution   In the equation $y = x^2 + 2x - 5$, $a = 1$ and $b = 2$. Using the vertex formula, we find that the x-coordinate of the vertex is

$$x = \frac{-b}{2a} = \frac{-2}{2 \cdot 1} = -1$$

The y-coordinate of the vertex is

$$y = (-1)^2 + 2(-1) - 5 = -6$$

Thus the vertex is $(-1, -6)$.

To find the x-intercepts, we let $y = 0$.

$$0 = x^2 + 2x - 5$$

This cannot be solved by factoring, so we use the quadratic formula.

$$x = \frac{-2 \pm \sqrt{2^2 - 4(1)(-5)}}{2 \cdot 1}$$ Let $a = 1, b = 2,$ and $c = -5$.

$$x = \frac{-2 \pm \sqrt{24}}{2}$$

$$x = \frac{-2 \pm 2\sqrt{6}}{2}$$ Simplify the radical.

$$x = \frac{2(-1 \pm \sqrt{6})}{2} = -1 \pm \sqrt{6}$$

The x-intercepts are $(-1 + \sqrt{6}, 0)$ and $(-1 - \sqrt{6}, 0)$. We use a calculator to approximate these so that we can easily graph these intercepts.

$$-1 + \sqrt{6} \approx 1.4 \quad \text{and} \quad -1 - \sqrt{6} \approx -3.4$$

To find the y-intercept, we let $x = 0$ in the original equation and find that $y = -5$. Thus the y-intercept is $(0, -5)$.

$y = x^2 + 2x - 5$

x	y
-1	-6
$-1 + \sqrt{6}$	0
$-1 - \sqrt{6}$	0
0	-5
-2	-5

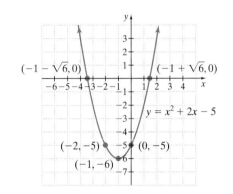

Helpful Hint

Notice that the number of x-intercepts of the graph of the parabola $y = ax^2 + bx + c$ is the same as the number of real solutions of $0 = ax^2 + bx + c$.

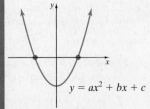 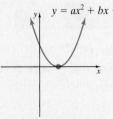 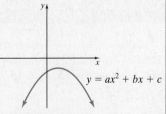

Two x-intercepts
Two real solutions of
$0 = ax^2 + bx + c$

One x-intercept
One real solution of
$0 = ax^2 + bx + c$

No x-intercepts
No real solutions of
$0 = ax^2 + bx + c$

Graphing Calculator Explorations

Recall that a graphing calculator may be used to solve quadratic equations. The x-intercepts of the graph of $y = ax^2 + bx + c$ are solutions of $0 = ax^2 + bx + c$. To solve $x^2 - 7x - 3 = 0$, for example, graph $y_1 = x^2 - 7x - 3$. The x-intercepts of the graph are the solutions of the equation.

Use a graphing calculator to solve each quadratic equation. Round solutions to two decimal places.

1. $x^2 - 7x - 3 = 0$ $x = -0.41, 7.41$
2. $2x^2 - 11x - 1 = 0$ $x = -0.09, 5.59$
3. $-1.7x^2 + 5.6x - 3.7 = 0$ $x = 0.91, 2.38$
4. $-5.8x^2 + 2.3x - 3.9 = 0$ no real solution
5. $5.8x^2 - 2.6x - 1.9 = 0$ $x = -0.39, 0.84$
6. $7.5x^2 - 3.7x - 1.1 = 0$ $x = -0.21, 0.70$

Spotlight on DECISION ⋆ MAKING

Suppose you are a landscape designer and you have some daffodil bulbs that you would like to plant in a homeowner's yard. It is recommended that daffodil bulbs be planted after the ground temperature has fallen below 50°F. Based on the following graph of normal ground temperatures in the area, when should you plant the bulbs? Explain.

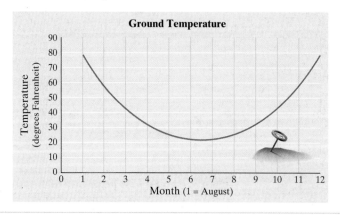

Ground Temperature

EXERCISE SET 9.5

| STUDY GUIDE/SSM | CD/ VIDEO | PH MATH TUTOR CENTER | MathXL®Tutorials ON CD | MathXL® | MyMathLab® |

Graph each quadratic equation by finding and plotting ordered pair solutions. See Example 1.1–12. See graphing answer section.

1. $y = 2x^2$
2. $y = 3x^2$
3. $y = -x^2$
4. $y = -4x^2$
5. $y = \frac{1}{3}x^2$
6. $y = -\frac{1}{2}x^2$

Sketch the graph of each equation. Identify the vertex and the intercepts. See Examples 2 through 4.

7. $y = x^2 - 1$
8. $y = x^2 - 16$
9. $y = x^2 + 4$
10. $y = x^2 + 9$
11. $y = x^2 + 6x$
12. $y = x^2 - 4x$

13. $y = x^2 + 2x - 8$ **14.** $y = x^2 - 2x - 3$

15. $y = -x^2 + x + 2$ **16.** $y = -x^2 - 2x - 1$

17. $y = x^2 + 5x + 4$ **18.** $y = x^2 + 7x + 10$

19. $y = -x^2 + 4x - 3$ **20.** $y = -x^2 + 6x - 8$

21. $y = x^2 + 2x - 2$ **22.** $y = x^2 - 4x - 3$

23. $y = x^2 - 3x + 1$ **24.** $y = x^2 - 2x - 5$

13–24. See graphing answer section.

REVIEW AND PREVIEW

Simplify the following complex fractions. See Section 7.8.

25. $\dfrac{\frac{1}{7} \quad \frac{5}{14}}{\frac{2}{5}}$ **26.** $\dfrac{\frac{3}{8} \quad \frac{21}{8}}{\frac{1}{7}}$

27. $\dfrac{\frac{1}{x} \quad \frac{x}{2}}{\frac{2}{x^2}}$ **28.** $\dfrac{\frac{x}{5} \quad \frac{x^2}{10}}{\frac{2}{x}}$

29. $\dfrac{\frac{2x}{1 - \frac{1}{x}} \quad \frac{2x^2}{x - 1}}{}$ **30.** $\dfrac{\frac{x}{x - \frac{1}{x}} \quad \frac{x^2}{x^2 - 1}}{}$

31. $\dfrac{\frac{a - b}{2b}}{\frac{b - a}{8b^2}}$ $-4b$ **32.** $\dfrac{\frac{2a^2}{a - 3}}{\frac{a}{3 - a}}$ $-2a$

Concept Extensions

The graph of a quadratic equation that takes the form $y = ax^2 + bx + c$ is the graph of a function. Write the domain and the range of each of the functions graphed.

33.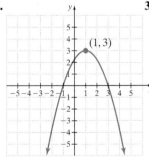

34.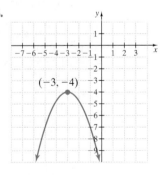

35.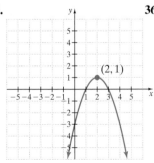

36.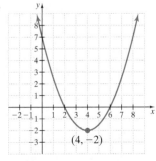

37. The height h of a fireball launched from a Roman candle with an initial velocity of 128 feet per second is given by the equation

$$h = -16t^2 + 128t$$

where t is time in seconds after launch.
Use the graph of this function to answer the questions.

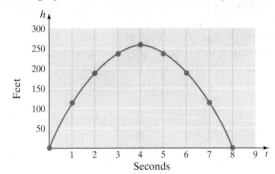

a. Estimate the maximum height of the fireball.   256 ft

b. Estimate the time when the fireball is at its maximum height. $t = 4$ sec

c. Estimate the time when the fireball returns to the ground. $t = 8$ sec

Match the values given with the correct graph of each quadratic equation of the form $y = a(x - h)^2 + k$.

38. $a > 0, h > 0, k > 0$ C   **39.** $a < 0, h > 0, k > 0$ B

40. $a > 0, h > 0, k < 0$ A   **41.** $a < 0, h > 0, k < 0$ D

A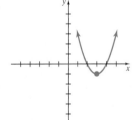

B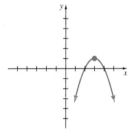

C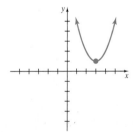

D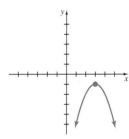

33. domain: $(-\infty, \infty)$; range: $(-\infty, 3]$

34. domain: $(-\infty, \infty)$; range: $(-\infty, -4]$

35. domain: $(-\infty, \infty)$; range: $(-\infty, 1]$

36. domain: $(-\infty, \infty)$; range: $[-2, \infty)$

CHAPTER 9 PROJECT

Modeling a Physical Situation

When water comes out of a water fountain, it initially heads upward, but then gravity causes the water to fall. The curve formed by the stream of water can be modeled using a quadratic equation (parabola).

In this project, you will have the opportunity to model the parabolic path of water as it leaves a drinking fountain. This project may be completed by working in groups or individually.

1. Using the figure above, collect data for the *x*-intercepts of the parabolic path. Let points *A* and *B* in the figure be on the *x*-axis and let the coordinates of point *A* be $(0, 0)$. Use a ruler to measure the distance between points *A* and *B* **on the figure** to the nearest even one-tenth centimeter, and use this information to determine the coordinates of point *B*. Record this data in the data table. (*Hint:* If the distance from *A* to *B* measures 8 one-tenth centimeters, then the coordinates of point *B* are $(8, 0)$.)

DATA TABLE

	x	*y*
Point A		
Point B		
Point V		

2. Next, collect data for the vertex *V* of the parabolic path. What is the relationship between the *x*-coordinate of the vertex and the *x*-intercepts found in Question 1? What is the line of symmetry? To locate point *V* in the figure, find the midpoint of the line segment joining points *A* and *B* and mark point *V* on the path of water directly above the midpoint. To approximate the *y*-coordinate of the vertex, use a ruler to measure its distance from the *x*-axis to the nearest one-tenth centimeter. Record this data in the data table.

3. Plot the points from the data table on a rectangular coordinate system. Sketch the parabola through your points *A*, *B*, and *V*.

4. Which of the following models best fits the data you collected? Explain your reasoning.

   a. $y = 16x + 18$

   b. $y = -13x^2 + 20x$

   c. $y = 0.13x^2 - 2.6x$

   d. $y = -0.13x^2 + 2.6x$

5. (Optional) Enter your data into a graphing calculator and use the quadratic curve-fitting feature to find a model for your data. How does the model compare with your selection from Question 4?

CHAPTER VOCABULARY CHECK

Fill in each blank with one of the words listed below.

square root	complex	imaginary	*i*
completing the square	quadratic	conjugate	vertex

1. If $x^2 = a$, then $x = \sqrt{a}$ or $x = -\sqrt{a}$. This property is called the <u>square root</u> property.

2. A number that can be written in the form $a + bi$ is called a(n) <u>complex</u> number.

3. The formula $\dfrac{-b}{2a}$ where $y = ax^2 + bx + c$ is called the <u>vertex</u> formula.

4. A complex number that can be written in the form $0 + bi$ is also called a(n) <u>imaginary</u> number.

5. The <u>conjugate</u> of $2 + 3i$ is $2 - 3i$.

6. $\sqrt{-1} = \underline{\quad i \quad}$

7. The process of solving a quadratic equation by writing it in the form $(x + a)^2 = c$ is called <u>completing the square</u>.

8. The formula $x = \dfrac{-b \pm \sqrt{b^2 - 4ac}}{2a}$ is called the <u>quadratic</u> formula.

STUDY SKILLS REMINDER

Are You Prepared for Your Final Exam?

To prepare for your final exam, try the following study techniques.

▶ Review the material that you will be responsible for on your exam. Also check your notebook for any lecture notes that you high-lighted.

▶ Review any formulas that you may need to memorize.

▶ Check to see if your instructor or math department will be conducting a final exam review.

▶ Check with your instructor to see whether there are final exams from previous semesters/quarters that are available for students to study.

▶ Use your previously taken tests as a practice final exam. To do so, rewrite the test questions in mixed order on blank sheets of paper. This will help you prepare for exam conditions.

▶ If you are unsure of a few topics, see your instructor or visit a learning lab for further assistance. Also, viewing the video segment of a troublesome section will help.

▶ If you need further exercises to work, try the chapter tests at the end of appropriate chapters.

Good luck! I hope you have enjoyed this textbook and your beginning algebra course.

CHAPTER 9   HIGHLIGHTS

Definitions and Concepts	**Examples**

Section 9.1 Solving Quadratic Equations by the Square Root Method

Square Root Property

If $x^2 = a$ for $a \geq 0$, then $x = \pm\sqrt{a}$

Solve the equation.

$$(x - 1)^2 = 15$$
$$x - 1 = \pm\sqrt{15}$$
$$x = 1 \pm \sqrt{15}$$

Section 9.2 Solving Quadratic Equations by Completing the Square

To Solve a Quadratic Equation by Completing the Square

Step 1: If the coefficient of x^2 is not 1, divide both sides of the equation by the coefficient.

Step 2: Isolate all terms with variables on one side.

Step 3: Complete the square by adding the square of half of the coefficient of x to both sides.

Step 4: Factor the perfect square trinomial.

Step 5: Apply the square root property to solve.

Solve $2x^2 + 12x - 10 = 0$ by completing the square.

$$\frac{2x^2}{2} + \frac{12x}{2} - \frac{10}{2} = \frac{0}{2} \qquad \text{Divide by 2.}$$
$$x^2 + 6x - 5 = 0 \qquad \text{Simplify.}$$
$$x^2 + 6x = 5 \qquad \text{Add 5.}$$

The coefficient of x is 6. Half of 6 is 3 and $3^2 = 9$. Add 9 to both sides.

$$x^2 + 6x + 9 = 5 + 9$$
$$(x + 3)^2 = 14 \qquad \text{Factor.}$$
$$x + 3 = \pm\sqrt{14}$$
$$x = -3 \pm \sqrt{14}$$

Section 9.3 Solving Quadratic Equations by the Quadratic Formula

Quadratic Formula

If a, b, and c are real numbers and $a \neq 0$, the quadratic equation $ax^2 + bx + c = 0$ has solutions

$$x = \frac{-b \pm \sqrt{b^2 - 4ac}}{2a}$$

To Solve a Quadratic Equation by the Quadratic Formula

Step 1: Write the equation in standard form: $ax^2 + bx + c = 0$.

Step 2: If necessary, clear the equation of fractions.

Step 3: Identify a, b, and c.

Step 4: Replace a, b, and c in the quadratic formula by known values, and simplify.

Identify a, b, and c in the quadratic equation

$$4x^2 - 6x = 5$$

First, subtract 5 from both sides.

$$4x^2 - 6x - 5 = 0$$
$$a = 4, b = -6, \text{ and } c = -5$$

Solve $3x^2 - 2x - 2 = 0$.

In this equation, $a = 3$, $b = -2$, and $c = -2$.

$$x = \frac{-(-2) \pm \sqrt{(-2)^2 - 4(3)(-2)}}{2 \cdot 3}$$
$$= \frac{2 \pm \sqrt{4 - (-24)}}{6}$$
$$= \frac{2 \pm \sqrt{28}}{6} = \frac{2 \pm \sqrt{4 \cdot 7}}{6} = \frac{2 \pm 2\sqrt{7}}{6}$$
$$= \frac{2(1 \pm \sqrt{7})}{2 \cdot 3} = \frac{1 \pm \sqrt{7}}{3}$$

(continued)

Definitions and Concepts	**Examples**

Section 9.3 Summary on Solving Quadratic Equations

To Solve a Quadratic Equation

Step 1: If the equation is in the form $(ax + b)^2 = c$, use the square root property and solve. If not, go to step 2.

Step 2: Write the equation in standard form: $ax^2 + bx + c = 0$.

Step 3: Try to solve by factoring. If not, go to step 4.

Step 4: Solve by the quadratic formula.

Solve $(3x - 1)^2 = 10$.

$$3x - 1 = \pm\sqrt{10} \quad \text{Square root property.}$$
$$3x = 1 \pm \sqrt{10} \quad \text{Add 1.}$$
$$x = \frac{1 \pm \sqrt{10}}{3} \quad \text{Divide by 3.}$$

Solve $x(2x + 9) = 5$.

$$2x^2 + 9x - 5 = 0$$
$$(2x - 1)(x + 5) = 0$$
$$2x - 1 = 0 \quad \text{or} \quad x + 5 = 0$$
$$2x = 1 \qquad\qquad x = -5$$
$$x = \frac{1}{2}$$

Section 9.4 Complex Solutions of Quadratic Equations

The **imaginary unit**, written i, is the number whose square is -1. That is,

$$i^2 = -1 \quad \text{and} \quad i = \sqrt{-1}$$

A **complex number** is a number that can be written in the form

$$a + bi$$

where a and b are real numbers. A complex number that can be written in the form $0 + bi$, $b \neq 0$, is also called an **imaginary number**.

Complex numbers are added and subtracted in the same way as polynomials are added and subtracted.

Use the distributive property to multiply complex numbers.

The complex numbers $a + bi$ and $a - bi$ are called **complex conjugates**.

To write a quotient of complex numbers in standard form $a + bi$, multiply numerator and denominator by the denominator's conjugate.

Write $\sqrt{-10}$ as the product of a real number and i.
$$\sqrt{-10} = \sqrt{-1 \cdot 10} = \sqrt{-1} \cdot \sqrt{10} = i\sqrt{10}$$
Identify each number as a complex number by writing it in **standard form** $a + bi$.
$$7 = 7 + 0i$$
$$\sqrt{-5} = 0 + i\sqrt{5} \quad \text{Also an imaginary number}$$
$$1 - \sqrt{-9} = 1 - 3i$$
Simplify the sum or difference.
$$(2 + 3i) - (1 - 6i) = 2 + 3i - 1 + 6i$$
$$= 1 + 9i$$
$$2i + (5 - 3i) = 2i + 5 - 3i$$
$$= 5 - i$$

Multiply: $(4 - i)(-2 + 3i)$
$$= -8 + 12i + 2i - 3i^2$$
$$= -8 + 14i + 3$$
$$= -5 + 14i$$

The conjugate of $5 - 6i$ is $5 + 6i$.

Write $\dfrac{3 - 2i}{2 + i}$ in standard form.
$$\frac{3 - 2i}{2 + i} \cdot \frac{(2 - i)}{(2 - i)} = \frac{6 - 7i + 2i^2}{4 - i^2}$$
$$= \frac{6 - 7i + 2(-1)}{4 - (-1)}$$
$$= \frac{4 - 7i}{5} \quad \text{or} \quad \frac{4}{5} - \frac{7}{5}i$$

(continued)

Definitions and Concepts	**Examples**

Section 9.4 Complex Solutions of Quadratic Equations

Some quadratic equations have complex solutions.	Solve $x^2 - 3x = -3$ $$x^2 - 3x + 3 = 0$$ $$a = 1, b = -3, c = 3$$ $$x = \frac{-(-3) \pm \sqrt{(-3)^2 - 4(1)(3)}}{2(1)}$$ $$= \frac{3 \pm \sqrt{-3}}{2} = \frac{3 \pm i\sqrt{3}}{2}$$

Section 9.5 Graphing Quadratic Equations

The graph of a quadratic equation $y = ax^2 + bx + c, a \neq 0$, is called a **parabola**. The lowest point on a parabola opening upward or the highest point on a parabola opening downward is called the **vertex**. The vertical line through the vertex is the **axis of symmetry**.

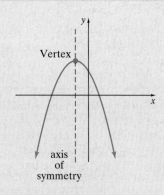

The vertex of the parabola $y = ax^2 + bx + c$ has x-value $\frac{-b}{2a}$.

Graph $y = 2x^2 - 6x + 4$.

The x-value of the vertex is

$$x = \frac{-b}{2a} = \frac{-(-6)}{2(2)} = \frac{6}{4} = \frac{3}{2}$$

The y-value is

$$y = 2\left(\frac{3}{2}\right)^2 - 6\left(\frac{3}{2}\right) + 4 = -\frac{1}{2}$$

The vertex is $\left(\frac{3}{2}, -\frac{1}{2}\right)$.

The y-intercept is

$$y = 2 \cdot 0^2 - 6 \cdot 0 + 4 = 4$$

The x-intercepts are found by solving

$$0 = 2x^2 - 6x + 4$$
$$0 = 2(x^2 - 3x + 2)$$
$$0 = 2(x - 2)(x - 1)$$
$$x - 2 = 0 \quad \text{or} \quad x - 1 = 0$$
$$x = 2 \quad \text{or} \quad x = 1$$

(continued)

Definitions and Concepts	**Examples**

Section 9.5 Graphing Quadratic Equations

Find more ordered pair solutions as needed.

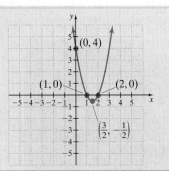

---

17. $a^2 + 4a + 4 = (a + 2)^2$ **18.** $a^2 - 12a + 36 = (a - 6)^2$ **19.** $m^2 - 3m + \frac{9}{4} = \left(m - \frac{3}{2}\right)^2$ **20.** $m^2 + 5m + \frac{25}{4} = \left(m + \frac{5}{2}\right)^2$

21. $3 \pm \sqrt{2}$ **22.** $-3 \pm \sqrt{2}$ **24.** $\dfrac{-3 \pm \sqrt{13}}{2}$ **25.** $5 \pm 3\sqrt{2}$ **26.** $-2 \pm \sqrt{11}$ **28.** $\dfrac{-3 \pm \sqrt{13}}{2}$ **33.–34.** no real solution

35. 2 real solutions **36.** no real solution **37.** 1 real solution **38.** 2 real solutions

CHAPTER REVIEW

39. no real solution **40.** 1 real solution **41.** $\dfrac{-1 \pm \sqrt{21}}{10}$ **42.** $\dfrac{-7 \pm \sqrt{65}}{8}$ **53.** $\dfrac{3 \pm \sqrt{5}}{20}$

(9.1) *Solve each quadratic equation by factoring or using the square root property.*

1. $(x - 4)(5x + 3) = 0$ $-\frac{3}{5}, 4$
2. $(x + 7)(3x + 4) = 0$ $-7, -\frac{4}{3}$
3. $3m^2 - 5m = 2$ $-\frac{1}{3}, 2$
4. $7m^2 + 2m = 5$ $\frac{5}{7}, -1$
5. $k^2 = 50$ $\pm 5\sqrt{2}$
6. $k^2 = 45$ $\pm 3\sqrt{5}$
7. $(x - 5)(x - 1) = 12$ $-1, 7$
8. $(x - 3)(x + 2) = 6$ $-3, 4$
9. $(x - 11)^2 = 49$ $4, 18$
10. $(x + 3)^2 = 100$ $-13, 7$
11. $6x^3 - 54x = 0$ $0, \pm 3$
12. $2x^2 - 8 = 0$ ± 2
13. $(4p + 2)^2 = 100$ $-3, 2$
14. $(3p + 6)^2 = 81$ $-5, 1$

Solve. For Exercises 15 and 16, use the formula $h = 16t^2$.

15. If Kara Washington dives from a height of 100 feet, how long before she hits the water? 2.5 sec

16. How long does a 5-mile free-fall take? Round your result to the nearest tenth of a second. 40.6 sec

(9.2) *Complete the square for the following expressions and then factor the resulting perfect square trinomial.*

17. $a^2 + 4a$
18. $a^2 - 12a$
19. $m^2 - 3m$
20. $m^2 + 5m$

Solve each quadratic equation by completing the square.

21. $x^2 - 6x + 7 = 0$
22. $x^2 + 6x + 7 = 0$
23. $2y^2 + y - 1 = 0$ $-1, \frac{1}{2}$
24. $y^2 + 3y - 1 = 0$

(9.3) *Solve each quadratic equation by using the quadratic formula.*

25. $x^2 - 10x + 7 = 0$
26. $x^2 + 4x - 7 = 0$
27. $2x^2 + x - 1 = 0$ $-1, \frac{1}{2}$
28. $x^2 + 3x - 1 = 0$
29. $9x^2 + 30x + 25 = 0$ $-\frac{5}{3}$
30. $16x^2 - 72x + 81 = 0$ $\frac{9}{4}$
31. $15x^2 + 2 = 11x$ $\frac{2}{5}, \frac{1}{3}$
32. $15x^2 + 2 = 13x$ $\frac{2}{3}, \frac{1}{5}$
33. $2x^2 + x + 5 = 0$
34. $7x^2 - 3x + 1 = 0$

Use the discriminant to determine the number of solutions of each quadratic equation.

35. $x^2 - 7x - 1 = 0$
36. $x^2 + x + 5 = 0$
37. $9x^2 + 1 = 6x$
38. $x^2 + 6x = 5$
39. $5x^2 + 4 = 0$
40. $x^2 + 25 = 10x$

(Summary) *Solve the following equations by using the most appropriate method.*

41. $5z^2 + z - 1 = 0$
42. $4z^2 + 7z - 1 = 0$
43. $4x^4 = x^2$ $0, \pm\frac{1}{2}$
44. $9x^3 = x$ $0, \pm\frac{1}{3}$
45. $2x^2 - 15x + 7 = 0$ $\frac{1}{2}, 7$
46. $x^2 - 6x - 7 = 0$ $-1, 7$
47. $(3x - 1)^2 = 0$ $\frac{1}{3}$
48. $(2x - 3)^2 = 0$ $\frac{3}{2}$
49. $x^2 = 6x - 9$ 3
50. $x^2 = 10x - 25$ 5
51. $\left(\frac{1}{2}x - 3\right)^2 = 64$ $-10, 22$
52. $\left(\frac{1}{3}x + 1\right)^2 = 49$ $-24, 18$
53. $x^2 - 0.3x + 0.01 = 0$
54. $x^2 + 0.6x - 0.16 = 0$ $-\frac{4}{5}, \frac{1}{5}$
55. $\frac{1}{10}x^2 + x - \frac{1}{2} = 0$

$-5 \pm \sqrt{30}$
56. $\frac{1}{12}x^2 - \frac{1}{2}x + \frac{1}{3} = 0$

$3 \pm \sqrt{5}$

57. The average price of silver (in cents per ounce) from 1999 to 2001 is given by the equation $y = -8x^2 - 13x + 552$. In this equation, x is the number of years since 1999. Assume that this trend continues and find the year after 1999 in which the price of silver will be 186 cents per ounce. (*Source:* U.S. Bureau of Mines)   2005

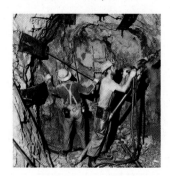

58. The average price of platinum (in dollars per ounce) from 1999 to 2001 is given by the equation $y = -93x^2 + 263x + 379$. In this equation, x is the number of years since 1999. Assume that this trend continues and find the year after 1999 in which the price of platinum is 331 dollars per ounce. (*Source:* U.S. Bureau of Mines)   2002

(9.4) *Perform the indicated operations. Write the resulting complex number in standard form.* **65.** $21 - 10i$ **66.** $19 - 25i$

59. $\sqrt{-144}$ $12i$

60. $\sqrt{-36}$ $6i$

61. $\sqrt{-108}$ $6i\sqrt{3}$

62. $\sqrt{-500}$ $10i\sqrt{5}$

63. $2i(3 - 5i)$ $10 + 6i$

64. $i(-7 - i)$ $1 - 7i$

65. $(7 - i) + (14 - 9i)$

66. $(10 - 4i) + (9 - 21i)$

67. $3 - (11 + 2i)$ $-8 - 2i$

68. $(-4 - 3i) + 5i$ $-4 + 2i$

69. $(2 - 3i)(3 - 2i)$ $-13i$

70. $(2 + 5i)(5 - i)$ $15 + 23i$

71. $(3 - 4i)(3 + 4i)$ 25

72. $(7 - 2i)(7 - 2i)$ $45 - 28i$

73. $\dfrac{2 - 6i}{4i}$ $-\dfrac{3}{2} - \dfrac{1}{2}i$

74. $\dfrac{5 - i}{2i}$ $-\dfrac{1}{2} - \dfrac{5}{2}i$

75. $\dfrac{4 - i}{1 + 2i}$ $\dfrac{2}{5} - \dfrac{9}{5}i$

76. $\dfrac{1 + 3i}{2 - 7i}$ $-\dfrac{19}{53} + \dfrac{13}{53}i$

Solve each quadratic equation.

77. $3x^2 = -48$ $\pm 4i$

78. $5x^2 = -125$ $\pm 5i$

79. $x^2 - 4x + 13 = 0$

80. $x^2 + 4x + 11 = 0$

79. $2 \pm 3i$ **80.** $-2 \pm i\sqrt{7}$

(9.5) *Identify the vertex, axis of symmetry, and whether the parabola opens upward or downward for the given quadratic equation.* **81.–88.** See graphing answer section.

81. $y = -3x^2$

82. $y = -\dfrac{1}{2}x^2$

83. $y = (x - 3)^2$

84. $y = (x - 5)^2$

85. $y = 3x^2 - 7$

86. $y = -2x^2 + 25$

87. $y = -5(x - 72)^2 + 14$

88. $y = 2(x - 35)^2 - 21$

Graph the following quadratic equations. Label the vertex and the intercepts with their coordinates.

89.–96. See graphing answer section.

89. $y = -x^2$

90. $y = 4x^2$

91. $y = \dfrac{1}{2}x^2$

92. $y = \dfrac{1}{4}x^2$

93. $y = x^2 + 5x + 6$

94. $y = x^2 - 4x - 8$

95. $y = 2x^2 - 11x - 6$

96. $y = 3x^2 - x - 2$

Quadratic equations $y = ax^2 + bx + c$ are graphed below. Determine the number of real solutions for the related equation $0 = ax^2 + bx + c$ from each graph. List the solutions.

97.

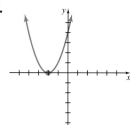

1 solution; $x = -2$

81. vertex: $(0, 0)$; axis of symmetry: $x = 0$; opens downward
82. vertex: $(0, 0)$; axis of symmetry: $x = 0$; opens downward

98.

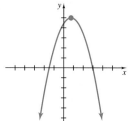

2 solutions; $x = -1.5$, $x = 3$

83. vertex: $(3, 0)$; axis of symmetry: $x = 3$; opens upward
84. vertex: $(5, 0)$; axis of symmetry: $x = 5$; opens upward
85. vertex: $(0, -7)$; axis of symmetry: $x = 0$; opens upward

99.

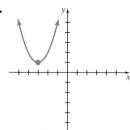

no real solution

86. vertex: $(0, 25)$; axis of symmetry: $x = 0$; opens downward
87. vertex: $(72, 14)$; axis of symmetry: $x = 72$; opens downward

100.

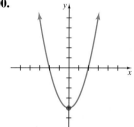

2 solutions; $x = \pm 2$

88. vertex: $(35, -21)$; axis of symmetry: $x = 35$; opens upward

CHAPTER 9 TEST

Remember to use your Chapter Test Prep Video CD to help you study and view solutions to the test questions you need help with.

Solve using the square root property.

1. $5k^2 = 80$ ± 4 **2.** $(3m - 5)^2 = 8$ $\dfrac{5 \pm 2\sqrt{2}}{3}$

Solve by completing the square.

3. $x^2 - 26x + 160 = 0$ **4.** $3x^2 + 12x - 4 = 0$

Solve using the quadratic formula.

5. $x^2 - 3x - 10 = 0$ **6.** $p^2 - \dfrac{5}{3}p - \dfrac{1}{3} = 0$

Solve by the most appropriate method.

7. $(3x - 5)(x + 2) = -6$ **8.** $(3x - 1)^2 = 16$ $-1, \dfrac{5}{3}$

9. $3x^2 - 7x - 2 = 0$ **10.** $x^2 - 4x + 5 = 0$ $2 \pm i$

11. $3x^2 - 7x + 2 = 0$ $\dfrac{1}{3}, 2$ **12.** $2x^2 - 6x + 1 = 0$

13. $9x^3 = x$ $0, \pm\dfrac{1}{3}$

3. $10, 16$ **4.** $\dfrac{-6 \pm 4\sqrt{3}}{3}$ **5.** $-2, 5$ **6.** $\dfrac{5 \pm \sqrt{37}}{6}$ **7.** $-\dfrac{4}{3}, 1$ **9.** $\dfrac{7 \pm \sqrt{73}}{6}$ **12.** $\dfrac{3 \pm \sqrt{7}}{2}$ **16.** $8 + i$

20.–21. See graphing answer section.

Perform the indicated operations. Write the resulting complex number in standard form.

14. $\sqrt{-25}$ $5i$ **15.** $\sqrt{-200}$ $10i\sqrt{2}$

16. $(3 + 2i) + (5 - i)$ **17.** $(3 + 2i) - (3 - 2i)$ $4i$

18. $(3 + 2i)(3 - 2i)$ 13 **19.** $\dfrac{3 - i}{1 + 2i}$ $\dfrac{1}{5} - \dfrac{7}{5}i$

Graph the quadratic equations. Label the vertex and the intercept points with their coordinates.

20. $y = -3x^2$ **21.** $y = x^2 - 7x + 10$

Solve.

22. The highest dive from a diving board by a woman was made by Lucy Wardle of the United States. She dove from a height of 120.75 feet at Ocean Park, Hong Kong, in 1985. To the nearest tenth of a second, how long did the dive take? Use the formula $h = 16t^2$.   2.7 sec

CHAPTER CUMULATIVE REVIEW

1. Find the value of each expression when $x = 2$ and $y = -5$.

   a. $\dfrac{x - y}{12 + x}$ $\dfrac{1}{2}$ **b.** $x^2 - y$ 9 (Sec. 1.6, Ex. 6)

2. Find the value of each expression when $x = -4$ and $y = 7$.

   a. $\dfrac{x - y}{7 - x}$ -1 **b.** $x^2 + 2y$ 30 (Sec. 1.6)

3. Simplify each expression by combining like terms.

   a. $2x + 3x + 5 + 2$ $5x + 7$

   b. $-5a - 3 + a + 2$ $-4a - 1$

   c. $4y - 3y^2$ $4y - 3y^2$

   d. $2.3x + 5x - 6$ $7.3x - 6$

   e. $-\dfrac{1}{2}b + b$ $\dfrac{1}{2}b$ (Sec. 2.1, Ex. 4)

4. Simplify each expression by combining like terms.

   a. $4x - 3 + 7 - 5x$ $-x + 4$

   b. $-6y + 3y - 8 + 8y$ $5y - 8$

   c. $2 + 8.1a + a - 6$ $9.1a - 4$

   d. $2x^2 - 2x$ $2x^2 - 2x$ (Sec. 2.1)

5. Identify the *x*- and *y*-intercepts.

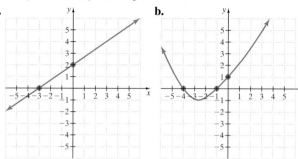

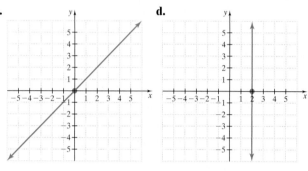

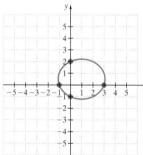

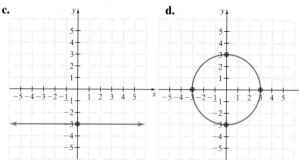

c. *x*-int: none; *y*-int: $(0,-3)$

d. *x*-int: $(-3,0)$, $(3,0)$; *y*-int: $(0,-3)$, $(0,3)$

a. *x*-int: $(-3, 0)$; *y*-int: $(0, 2)$

b. *x*-int: $(-4, 0)$, $(-1, 0)$; *y*-int: $(0, 1)$

c. *x*-int and *y*-int: $(0,0)$

d. *x*-int: $(2,0)$; *y*-int: none

e. *x*-int: $(-1, 0)$, $(3, 0)$; *y*-int: $(0, -1)$, $(0, 2)$ (Sec. 3.3, Ex. 1)

7. Determine whether the graphs of $y = -\dfrac{1}{5}x + 1$ and $2x + 10y = 30$ are parallel lines, perpendicular lines, or neither.   parallel (Sec. 3.5, Ex. 4)

8. Determine whether the graphs of $y = 3x + 7$ and $x + 3y = -15$ are parallel lines, perpendicular lines, or neither.   perpendicular (Sec. 3.5)

9. Solve the following system of equations by graphing.
$$\begin{cases} 2x + y = 7 \\ 2y = -4x \end{cases}$$ no solution (Sec. 4.1, Ex. 3)

10. Solve the following system by graphing.
$$\begin{cases} y = x + 2 \\ 2x + y = 5 \end{cases}$$ $(1, 3)$ (Sec. 4.1)

11. Solve the system.
$$\begin{cases} 7x - 3y = -14 \\ -3x + y = 6 \end{cases}$$ $(-2, 0)$ (Sec. 4.2, Ex. 3)

12. Solve the system.
$$\begin{cases} 5x + y = 3 \\ y = -5x \end{cases}$$ no solution (Sec. 4.2)

13. Solve the system.
$$\begin{cases} 3x - 2y = 2 \\ -9x + 6y = -6 \end{cases}$$ infinite number of solutions (Sec. 4.3, Ex. 4)

14. Solve the system.
$$\begin{cases} -2x + y = 7 \\ 6x - 3y = -21 \end{cases}$$ infinite number of solutions (Sec. 4.3)

15. As part of an exercise program, Albert and Louis started walking each morning. They live 15 miles away from each other and decided to meet one day by walking toward one another. After 2 hours they meet. If Louis walks one mile per hour faster than Albert, find both walking speeds.   Albert: 3.25 mph; Louis: 4.25 mph (Sec. 4.4, Ex. 3)

16. A coin purse contains dimes and quarters only. There are 15 coins totalling $2.85. How many dimes and how many quarters are in the purse?   6 dimes, 9 quarters (Sec. 4.4)

6. Identify the *x*- and *y*-intercepts.

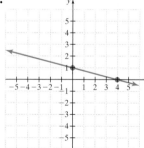

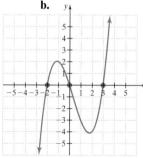

a. *x*-int: $(4,0)$; *y*-int: $(0,1)$ (Sec. 3.3)

b. *x*-int: $(-2,0)$, $(0,0)$, $(3,0)$; *y*-int: $(0,0)$

17. Graph the solution of the system

$$\begin{cases} -3x + 4y < 12 \\ \quad\quad x \ge 2 \end{cases}$$ See graphing answer section.
(Sec. 4.6, Ex. 3)

18. Graph the solution of the system.

$$\begin{cases} 2x - y \le 6 \\ \quad\quad y \ge 2 \end{cases}$$ See graphing answer section. (Sec. 4.6)

19. Simplify the following.

a. $\left(\dfrac{-5x^2}{y^3}\right)^2$ $\dfrac{25x^4}{y^6}$

b. $\dfrac{(x^3)^4 x}{x^7}$ x^6

c. $\dfrac{(2x)^5}{x^3}$ $32x^2$

d. $\dfrac{(a^2b)^3}{a^3b^2}$ a^3b (Sec. 5.1, Ex. 10)

20. Simplify.

a. $\left(\dfrac{-6x}{y^3}\right)^3$ $-\dfrac{216x^3}{y^9}$

b. $\dfrac{a^2b^7}{(2b^2)^5}$ $\dfrac{a^2}{32b^3}$

c. $\dfrac{(3y)^2}{y^2}$ 9

d. $\dfrac{(x^2y^4)^2}{xy^3}$ x^3y^5 (Sec. 5.1)

21. Solve $(5x - 1)(2x^2 + 15x + 18) = 0.$ $-6, -\frac{3}{2}, \frac{1}{5}$ (Sec. 6.5, Ex. 6)

22. Solve $(x + 1)(2x^2 - 3x - 5) = 0.$ $-1, \dfrac{5}{2}$ (Sec. 6.5)

23. Solve $\dfrac{45}{x} = \dfrac{5}{7}.$ 63 (Sec. 7.6, Ex. 1)

24. Solve $\dfrac{2x + 7}{3} = \dfrac{x - 6}{2}.$ -32 (Sec. 7.6)

25. Find each root.

a. $\sqrt[4]{16}$ 2

b. $\sqrt[5]{-32}$ -2

c. $-\sqrt[3]{8}$ -2

d. $\sqrt[4]{-81}$ not a real number (Sec. 8.1, Ex. 5)

26. Find each root.

a. $\sqrt[3]{27}$ 3

b. $\sqrt[4]{256}$ -4

c. $\sqrt[3]{-125}$ -5

d. $\sqrt[5]{1}$ 1 (Sec. 8.1)

27. Simplify the following expressions.

a. $\sqrt{\dfrac{25}{36}}$ $\dfrac{5}{6}$

b. $\sqrt{\dfrac{3}{64}}$ $\dfrac{\sqrt{3}}{8}$

c. $\sqrt{\dfrac{40}{81}}$ $\dfrac{2\sqrt{10}}{9}$ (Sec. 8.2, Ex. 2)

28. Simplify the following expressions.

a. $\sqrt{\dfrac{4}{25}}$ $\dfrac{2}{5}$

b. $\sqrt{\dfrac{16}{121}}$ $\dfrac{4}{11}$

c. $\sqrt{\dfrac{2}{49}}$ $\dfrac{\sqrt{2}}{7}$ (Sec. 8.2, Ex. 2)

29. Add or subtract by first simplifying each radical.

a. $\sqrt{50} + \sqrt{8}$ $7\sqrt{2}$

b. $7\sqrt{12} - \sqrt{75}$ $9\sqrt{3}$

c. $\sqrt{25} - \sqrt{27} - 2\sqrt{18} - \sqrt{16}$
$1 - 3\sqrt{3} - 6\sqrt{2}$ (Sec. 8.3, Ex. 2)

30. Add or subtract by first simplifying each radical.

a. $\sqrt{80} + \sqrt{20}$ $6\sqrt{5}$

b. $2\sqrt{98} - 2\sqrt{18}$ $8\sqrt{2}$

c. $\sqrt{32} + \sqrt{121} - \sqrt{12}$ $11 - 2\sqrt{3} + 4\sqrt{2}$ (Sec. 8.3)

31. Multiply. Then simplify if possible.

a. $\sqrt{7} \cdot \sqrt{3}$ $\sqrt{21}$

b. $\sqrt{3} \cdot \sqrt{15}$ $3\sqrt{5}$

c. $2\sqrt{6} \cdot 5\sqrt{2}$ $20\sqrt{3}$

d. $(3\sqrt{2})^2$ 18 (Sec. 8.4, Ex. 1)

32. Multiply. Then simplify of possible.

a. $\sqrt{2} \cdot \sqrt{5}$ $\sqrt{10}$

b. $\sqrt{56} \cdot \sqrt{7}$ $14\sqrt{2}$

c. $(4\sqrt{3})^2$ 48

d. $3\sqrt{8} \cdot 7\sqrt{2}$ 84 (Sec. 8.4)

33. Solve $\sqrt{x} = \sqrt{5x - 2}.$ $\frac{1}{2}$ (Sec. 8.5, Ex. 3)

34. Solve $\sqrt{x - 4} + 7 = 2.$ no solution (Sec. 8.5)

35. A surveyor must determine the distance across a lake at points P and Q. To do this, she finds a third point R perpendicular to line PQ. If the length of $\overline{PR}$ is 320 feet and the length of $\overline{QR}$ is 240 feet, what is the distance across the lake? Approximate this distance to the nearest whole foot.
$\sqrt{44,800} \approx 212$ ft (Sec. 8.6, Ex. 3)

36. Find the distance between $(-7, 4)$ and $(2, 5)$.
$\sqrt{82}$ (Sec. 8.6)

37. Write in radical notation. Then simplify.

   a. $25^{1/2}$ $\sqrt{25} = 5$

   b. $8^{1/3}$ $\sqrt[3]{8} = 2$

   c. $-16^{1/4}$ $-\sqrt[4]{16} = -2$

   d. $(-27)^{1/3}$ $\sqrt[3]{-27} = -3$

   e. $\left(\dfrac{1}{9}\right)^{1/2}$ $\sqrt{\dfrac{1}{9}} = \dfrac{1}{3}$ (Sec. 8.7, Ex. 1)

38. Write in radical notation, then simplify.

   a. $-49^{1/2}$ $-\sqrt{49} = -7$

   b. $256^{1/4}$ $\sqrt[4]{256} = 4$

   c. $(-64)^{1/3}$ $\sqrt[3]{-64} = -4$

   d. $\left(\dfrac{25}{36}\right)^{1/2}$ $\sqrt{\dfrac{25}{36}} = \dfrac{5}{6}$

   e. $(32)^{1/5}$ $\sqrt[5]{32} = 2$ (Sec. 8.7)

39. Use the square root property to solve $2x^2 = 7$.
$\dfrac{\sqrt{14}}{2}, -\dfrac{\sqrt{14}}{2}$ (Sec. 9.1, Ex. 2)

40. Use the square root property to solve
$3(x - 4)^2 = 9$. $4 \pm \sqrt{3}$ (Sec. 9.1)

41. Solve $x^2 - 10x = -14$ by completing the square.
$5 \pm \sqrt{11}$ (Sec. 9.2, Ex. 3)

42. Solve $x^2 + 4x = 8$ by completing the square.
$-2 \pm 2\sqrt{3}$ (Sec. 9.2)

43. Solve $2x^2 - 9x = 5$ using the quadratic formula.
$-\dfrac{1}{2}, 5$ (Sec. 9.3, Ex. 2)

44. Solve $2x^2 + 5x = 7$ using the quadratic formula.
$-\dfrac{7}{2}, 1$ (Sec. 9.3)

45. Write each radical as the product of a real number and i.

   a. $\sqrt{-4}$ $2i$

   b. $\sqrt{-11}$ $i\sqrt{11}$

   c. $\sqrt{-20}$ $2i\sqrt{5}$ (Sec. 9.4, Ex. 1)

46. Write each radical as the product of a real number and i.

   a. $\sqrt{-7}$ $i\sqrt{7}$

   b. $\sqrt{-16}$ $4i$

   c. $\sqrt{-27}$ $3i\sqrt{3}$ (Sec. 9.4)

47. Graph $y = x^2 - 4$.
See graphing answer section. (Sec. 9.5, Ex. 2)

48. Graph $y = x^2 + 2x + 3$.
See graphing answer section. (Sec. 9.5)

APPENDIX A

Operations on Decimals

To **add** or **subtract** decimals, write the numbers vertically with decimal points lined up. Add or subtract as with whole numbers and place the decimal point in the answer directly below the decimal points in the problem.

EXAMPLE 1

Add $5.87 + 23.279 + 0.003$.

Solution

$$
\begin{array}{r}
5.87 \\
23.279 \\
+\,0.003 \\
\hline
29.152
\end{array}
$$

EXAMPLE 2

Subtract $32.15 - 11.237$.

Solution

$$
\begin{array}{r}
3\ \overset{1}{\cancel{2}}\ .\ \overset{11}{\cancel{1}}\ \overset{4}{\cancel{5}}\ \overset{10}{\cancel{0}} \\
-\ 1\ \ 1\ .\ 2\ \ 3\ \ 7 \\
\hline
2\ \ 0\ .\ 9\ \ 1\ \ 3
\end{array}
$$

To **multiply** decimals, multiply the numbers as if they were whole numbers. The decimal point in the product is placed so that the number of decimal places in the product is the same as the sum of the number of decimal places in the factors.

EXAMPLE 3

Multiply 0.072×3.5.

Solution

$$
\begin{array}{r}
0.072 \quad \text{3 decimal places}\\
\times\ \ \ 3.5 \quad \text{1 decimal place}\\
\hline
360\\
216\ \ \\
\hline
0.2520 \quad \text{4 decimal places}
\end{array}
$$

To **divide** decimals, move the decimal point in the divisor to the right of the last digit. Move the decimal point in the dividend the same number of places that the decimal point in the divisor was moved. The decimal point in the quotient lies directly above the decimal point in the dividend.

583

EXAMPLE 4

Divide $9.46 \div 0.04$.

Solution

$$
\begin{array}{r}
236.5 \\
04.\overline{)946.0} \\
-8 \\
\hline
14 \\
-12 \\
\hline
26 \\
-24 \\
\hline
20 \\
-20 \\
\hline
\end{array}
$$

APPENDIX A EXERCISE SET

Perform the indicated operations.

1. $9.076 + 8.004$ 17.08

2. $\begin{array}{r} 6.3 \\ \times\ 0.05 \end{array}$ 0.315

3. $\begin{array}{r} 27.004 \\ -14.2 \end{array}$ 12.804

4. $\begin{array}{r} 0.0036 \\ 7.12 \\ 32.502 \\ +\ 0.05 \end{array}$ 39.6756

5. $\begin{array}{r} 107.92 \\ +3.04 \end{array}$ 110.96

6. $7.2 \div 4$ 1.8

7. $10 - 7.6$ 2.4

8. $40 \div 0.25$ 160

9. $126.32 - 97.89$ 28.43

10. $\begin{array}{r} 3.62 \\ 7.11 \\ 12.36 \\ 4.15 \\ +2.29 \end{array}$ 29.53

11. $\begin{array}{r} 3.25 \\ \times\ 70 \end{array}$ 227.5

12. $\begin{array}{r} 26.014 \\ -\ 7.8 \end{array}$ 18.214

13. $8.1 \div 3$ 2.7

14. $\begin{array}{r} 1.2366 \\ 0.005 \\ 15.17 \\ +\ 0.97 \end{array}$ 17.3816

15. $55.405 - 6.1711$ 49.2339

16. $8.09 + 0.22$ 8.31

17. $60 \div 0.75$ 80

18. $20 - 12.29$ 7.71

19. $7.612 \div 100$ 0.07612

20. $\begin{array}{r} 8.72 \\ 1.12 \\ 14.86 \\ 3.98 \\ +\ 1.99 \end{array}$ 30.67

21. $12.312 \div 2.7$ 4.56

22. $0.443 \div 100$ 0.00443

23. $\begin{array}{r} 569.2 \\ 71.25 \\ +\ 8.01 \end{array}$ 648.46

24. $3.706 - 2.91$ 0.796

25. $768 - 0.17$ 767.83

26. $63 \div 0.28$ 225

27. $12 + 0.062$ 12.062

28. $0.42 + 18$ 18.42

29. $76 - 14.52$ 61.48

30. $1.1092 \div 0.47$ 2.36

31. $3.311 \div 0.43$ 7.7

32. $7.61 + 0.0004$ 7.6104

33. $\begin{array}{r} 762.12 \\ 89.7 \\ +\ 11.55 \end{array}$ 863.37

34. $444 \div 0.6$ 740

35. $23.4 - 0.821$ 22.579

36. $3.7 + 5.6$ 9.3

37. $476.12 - 112.97$ 363.15

38. $19.872 \div 0.54$ 36.8

39. $0.007 + 7$ 7.007

40. $\begin{array}{r} 51.77 \\ +\ 3.6 \end{array}$ 55.37

Review of Angles, Lines, and Special Triangles

The word **geometry** is formed from the Greek words, **geo**, meaning earth, and **metron**, meaning measure. Geometry literally means to measure the earth.

   This section contains a review of some basic geometric ideas. It will be assumed that fundamental ideas of geometry such as point, line, ray, and angle are known. In this appendix, the notation $\angle 1$ is read "angle 1" and the notation $m\angle 1$ is read "the measure of angle 1."

   We first review types of angles.

Angles

A **right angle** is an angle whose measure is 90°. A right angle can be indicated by a square drawn at the vertex of the angle, as shown below.

An angle whose measure is more than 0° but less than 90° is called an **acute angle**.

An angle whose measure is greater than 90° but less than 180° is called an **obtuse angle**.

An angle whose measure is 180° is called a **straight angle**.

Two angles are said to be **complementary** if the sum of their measures is 90°. Each angle is called the **complement** of the other.

Two angles are said to be **supplementary** if the sum of their measures is 180°. Each angle is called the **supplement** of the other.

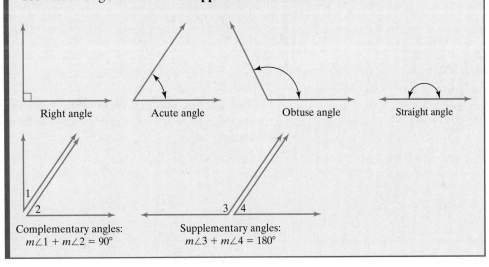

Right angle   Acute angle   Obtuse angle   Straight angle

Complementary angles:
$m\angle 1 + m\angle 2 = 90°$

Supplementary angles:
$m\angle 3 + m\angle 4 = 180°$

EXAMPLE 1

If an angle measures 28°, find its complement.

Solution Two angles are complementary if the sum of their measures is 90°. The complement of a 28° angle is an angle whose measure is $90° − 28° = 62°$. To check, notice that $28° + 62° = 90°$.

Plane is an undefined term that we will describe. A plane can be thought of as a flat surface with infinite length and width, but no thickness. A plane is two dimensional. The arrows in the following diagram indicate that a plane extends indefinitely and has no boundaries.

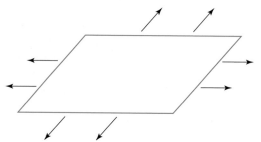

Figures that lie on a plane are called **plane figures**. (See the description of common plane figures in Appendix C.) Lines that lie in the same plane are called **coplanar**.

Lines

Two lines are **parallel** if they lie in the same plane but never meet.

Intersecting lines meet or cross in one point.

Two lines that form right angles when they intersect are said to be **perpendicular**.

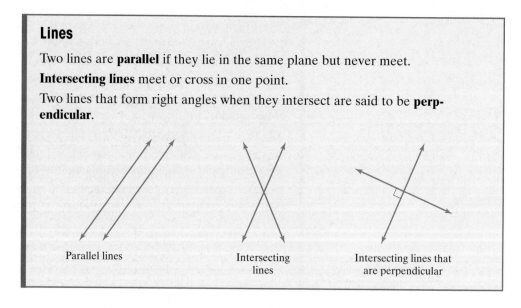

Parallel lines   Intersecting lines   Intersecting lines that are perpendicular

---

Two intersecting lines form **vertical angles**. Angles 1 and 3 are vertical angles. Also angles 2 and 4 are vertical angles. It can be shown that **vertical angles have equal measures**.

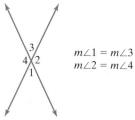

$$m\angle 1 = m\angle 3$$
$$m\angle 2 = m\angle 4$$

Adjacent angles have the same vertex and share a side. Angles 1 and 2 are adjacent angles. Other pairs of adjacent angles are angles 2 and 4, angles 3 and 4, and angles 3 and 1.

A **transversal** is a line that intersects two or more lines in the same plane. Line l is a transversal that intersects lines m and n. The eight angles formed are numbered and certain pairs of these angles are given special names.

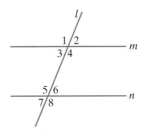

Corresponding angles: $\angle 1$ and $\angle 5, \angle 3$ and $\angle 7, \angle 2$ and $\angle 6$, and $\angle 4$ and $\angle 8$.

Exterior angles: $\angle 1, \angle 2, \angle 7$, and $\angle 8$.

Interior angles: $\angle 3, \angle 4, \angle 5$, and $\angle 6$.

Alternate interior angles: $\angle 3$ and $\angle 6, \angle 4$ and $\angle 5$.

These angles and parallel lines are related in the following manner.

Parallel Lines Cut by a Transversal

1. If two parallel lines are cut by a transversal, then

   a. corresponding angles are equal and
   b. alternate interior angles are equal.

2. If corresponding angles formed by two lines and a transversal are equal, then the lines are parallel.
3. If alternate interior angles formed by two lines and a transversal are equal, then the lines are parallel.

EXAMPLE 2

Given that lines m and n are parallel and that the measure of angle 1 is 100°, find the measures of angles 2, 3, and 4.

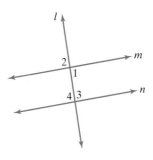

Solution $m\angle 2 = 100°$, since angles 1 and 2 are vertical angles.

$m\angle 4 = 100°$, since angles 1 and 4 are alternate interior angles.

$m\angle 3 = 180° - 100° = 80°$, since angles 4 and 3 are supplementary angles.

A **polygon** is the union of three or more coplanar line segments that intersect each other only at each end point, with each end point shared by exactly two segments.

A **triangle** is a polygon with three sides. The sum of the measures of the three angles of a triangle is 180°. In the following figure, $m\angle 1 + m\angle 2 + m\angle 3 = 180°$.

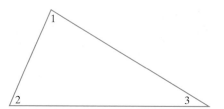

EXAMPLE 3

Find the measure of the third angle of the triangle shown.

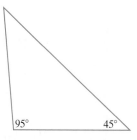

Solution   The sum of the measures of the angles of a triangle is 180°. Since one angle measures 45° and the other angle measures 95°, the third angle measures $180° - 45° - 95° = 40°$.

Two triangles are **congruent** if they have the same size and the same shape. In congruent triangles, the measures of corresponding angles are equal and the lengths of corresponding sides are equal. The following triangles are congruent.

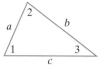 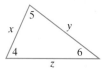

Corresponding angles are equal: $m\angle 1 = m\angle 4$, $m\angle 2 = m\angle 5$, and $m\angle 3 = m\angle 6$. Also, lengths of corresponding sides are equal: $a = x$, $b = y$, and $c = z$.

Any one of the following may be used to determine whether two triangles are congruent.

Congruent Triangles

1. If the measures of two angles of a triangle equal the measures of two angles of another triangle and the lengths of the sides between each pair of angles are equal, the triangles are congruent.

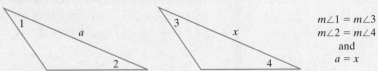

$$m\angle 1 = m\angle 3$$
$$m\angle 2 = m\angle 4$$
and
$$a = x$$

2. If the lengths of the three sides of a triangle equal the lengths of corresponding sides of another triangle, the triangles are congruent.

$$a = x$$
$$b = y$$
and
$$c = z$$

3. If the lengths of two sides of a triangle equal the lengths of corresponding sides of another triangle, and the measures of the angles between each pair of sides are equal, the triangles are congruent.

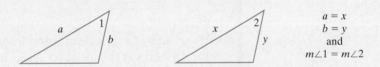

$$a = x$$
$$b = y$$
and
$$m\angle 1 = m\angle 2$$

Two triangles are **similar** if they have the same shape. In similar triangles, the measures of corresponding angles are equal and corresponding sides are in proportion. The following triangles are similar. (All similar triangles drawn in this appendix will be oriented the same.)

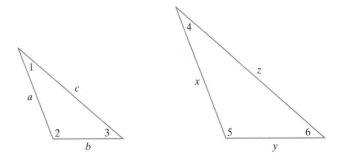

Corresponding angles are equal: $m\angle 1 = m\angle 4$, $m\angle 2 = m\angle 5$, and $m\angle 3 = m\angle 6$.

Also, corresponding sides are proportional: $\dfrac{a}{x} = \dfrac{b}{y} = \dfrac{c}{z}$.

Any one of the following may be used to determine whether two triangles are similar.

Similar Triangles

1. If the measures of two angles of a triangle equal the measures of two angles of another triangle, the triangles are similar.

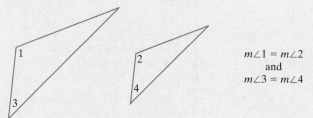

$$m\angle 1 = m\angle 2$$
and
$$m\angle 3 = m\angle 4$$

2. If three sides of one triangle are proportional to three sides of another triangle, the triangles are similar.

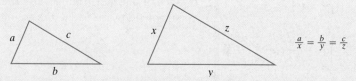

$$\frac{a}{x} = \frac{b}{y} = \frac{c}{z}$$

3. If two sides of a triangle are proportional to two sides of another triangle and the measures of the included angles are equal, the triangles are similar.

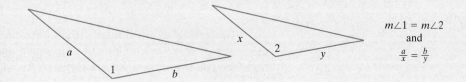

$$m\angle 1 = m\angle 2$$
and
$$\frac{a}{x} = \frac{b}{y}$$

EXAMPLE 4

Given that the following triangles are similar, find the missing length x.

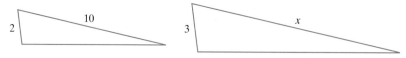

Solution   Since the triangles are similar, corresponding sides are in proportion. Thus, $\frac{2}{3} = \frac{10}{x}$. To solve this equation for x, we multiply both sides by the LCD, $3x$.

$$3x\left(\frac{2}{3}\right) = 3x\left(\frac{10}{x}\right)$$
$$2x = 30$$
$$x = 15$$

The missing length is 15 units.

A **right triangle** contains a right angle. The side opposite the right angle is called the **hypotenuse**, and the other two sides are called the **legs**. The **Pythagorean theorem** gives a formula that relates the lengths of the three sides of a right triangle.

The Pythagorean Theorem

If a and b are the lengths of the legs of a right triangle, and c is the length of the hypotenuse, then $a^2 + b^2 = c^2$.

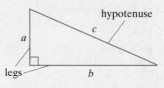

EXAMPLE 5

Find the length of the hypotenuse of a right triangle whose legs have lengths of 3 centimeters and 4 centimeters.

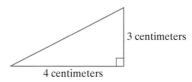

3 centimeters

4 centimeters

Solution   Because we have a right triangle, we use the Pythagorean theorem. The legs are 3 centimeters and 4 centimeters, so let $a = 3$ and $b = 4$ in the formula.

$$a^2 + b^2 = c^2$$
$$3^2 + 4^2 = c^2$$
$$9 + 16 = c^2$$
$$25 = c^2$$

Since c represents a length, we assume that c is positive. Thus, if c^2 is 25, c must be 5. The hypotenuse has a length of 5 centimeters.

APPENDIX B EXERCISE SET

Find the complement of each angle. See Example 1.

1. $19°$ $71°$

2. $65°$ $25°$

3. $70.8°$ $19.2°$

4. $45\frac{2}{3}°$ $44\frac{1}{3}°$

5. $11\frac{1}{4}°$ $78\frac{3}{4}°$

6. $19.6°$ $70.4°$

Find the supplement of each angle.

7. $150°$ $30°$

8. $90°$ $90°$

9. $30.2°$ $149.8°$

10. $81.9°$ $98.1°$

11. $79\frac{1}{2}°$ $100\frac{1}{2}°$

12. $165\frac{8}{9}°$ $14\frac{1}{9}°$

13. If lines m and n are parallel, find the measures of angles 1 through 7. See Example 2. $m\angle 1 = m\angle 5 = m\angle 7 = 110°$
$m\angle 2 = m\angle 3 = m\angle 4 = m\angle 6 = 70°$

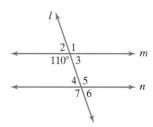

14. If lines *m* and *n* are parallel, find the measures of angles 1 through 5. See Example 2.

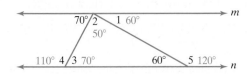

In each of the following, the measures of two angles of a triangle are given. Find the measure of the third angle. See Example 3.

15. 11°, 79°   90°

16. 8°, 102°   70°

17. 25°, 65°   90°

18. 44°, 19°   117°

19. 30°, 60°   90°

20. 67°, 23°   90°

In each of the following, the measure of one angle of a right triangle is given. Find the measures of the other two angles.

21. 45°   45°, 90°

22. 60°   30°, 90°

23. 17°   73°, 90°

24. 30°   60°, 90°

25. $39\frac{3}{4}°$ $50\frac{1}{4}°$, 90°

26. 72.6°   17.4°, 90°

Given that each of the following pairs of triangles is similar, find the missing lengths. See Example 4.

27.   *x* = 6

28.   *x* = 6

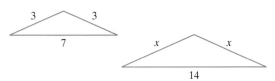

29.   *x* = 4.5

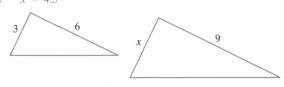

30.   *x* = 36

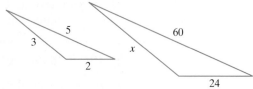

Use the Pythagorean theorem to find the missing lengths in the right triangles. See Example 5.

31.   10

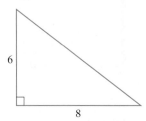

32.   13

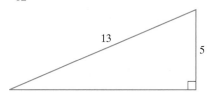

33.   12

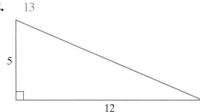

34.   16

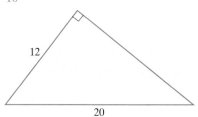

Review of Geometric Figures

Plane figures have length and width but no thickness or depth

Name	Description	Figure
Polygon	Union of three or more coplanar line segments that intersect with each other only at each end point, with each end point shared by two segments.	
Triangle	Polygon with three sides (sum of measures of three angles is 180°).	
Scalene Triangle	Triangle with no sides of equal length.	
Isosceles Triangle	Triangle with two sides of equal length.	
Equilateral Triangle	Triangle with all sides of equal length.	
Right Triangle	Triangle that contains a right angle.	leg, hypotenuse, leg
Quadrilateral	Polygon with four sides (sum of measures of four angles is 360°).	

Plane figures have length and width but no thickness or depth

Name	Description	Figure
Trapezoid	Quadrilateral with exactly one pair of opposite sides parallel.	
Isosceles Trapezoid	Trapezoid with legs of equal length.	
Parallelogram	Quadrilateral with both pairs of opposite sides parallel and equal in length.	
Rhombus	Parallelogram with all sides of equal length.	
Rectangle	Parallelogram with four right angles.	
Square	Rectangle with all sides of equal length.	
Circle	All points in a plane the same distance from a fixed point called the **center**.	

Solid figures have length, width, and depth

Name	Description	Figure
Rectangular Solid	A solid with six sides, all of which are rectangles.	
Cube	A rectangular solid whose six sides are squares.	
Sphere	All points the same distance from a fixed point, called the center.	
Right Circular Cylinder	A cylinder consisting of two circular bases that are perpendicular to its altitude.	
Right Circular Cone	A cone with a circular base that is perpendicular to its altitude.	

Mean, Median, and Mode

It is sometimes desirable to be able to describe a set of data, or a set of numbers, by a single "middle" number. Three such **measures of central tendency** are the mean, the median, and the mode.

The most common measure of central tendency is the mean (sometimes called the arithmetic mean or the average). The **mean** of a set of data items, denoted by $\bar{x}$, is the sum of the items divided by the number of items.

EXAMPLE 1

Seven students in a psychology class conducted an experiment on mazes. Each student was given a pencil and asked to successfully complete the same maze. The timed results are below.

Student	Ann	Thanh	Carlos	Jesse	Melinda	Ramzi	Dayni
Time (Seconds)	13.2	11.8	10.7	16.2	15.9	13.8	18.5

a. Who completed the maze in the shortest time? Who completed the maze in the longest time?

b. Find the mean.

c. How many students took longer than the mean time? How many students took shorter than the mean time?

Solution

a. Carlos completed the maze in 10.7 seconds, the shortest time. Dayni completed the maze in 18.5 seconds, the longest time.

b. To find the mean, $\bar{x}$, find the sum of the data items and divide by 7, the number of items.

$$\bar{x} = \frac{13.2 + 11.8 + 10.7 + 16.2 + 15.9 + 13.8 + 18.5}{7} = \frac{100.1}{7} = 14.3$$

c. Three students, Jesse, Melinda, and Dayni, had times longer than the mean time. Four students, Ann, Thanh, Carlos, and Ramzi, had times shorter than the mean time.

Two other measures of central tendency are the median and the mode.

The **median** of an ordered set of numbers is the middle number. If the number of items is even, the median is the mean of the two middle numbers. The **mode** of a set of numbers is the number that occurs most often. It is possible for a data set to have no mode or more than one mode.

EXAMPLE 2

Find the median and the mode of the following list of numbers. These numbers were high temperatures for fourteen consecutive days in a city in Montana.

$$76, 80, 85, 86, 89, 87, 82, 77, 76, 79, 82, 89, 89, 92$$

Solution First, write the numbers in order.

$$76, 76, 77, 79, 80, 82, 82, 85, 86, 87, 89, 89, 89, 92$$

two middle numbers

mode

Since there are an even number of items, the median is the mean of the two middle numbers.

$$\text{median} = \frac{82 + 85}{2} = 83.5$$

The mode is 89, since 89 occurs most often.

APPENDIX D EXERCISE SET

For each of the following data sets, find the mean, the median, and the mode. If necessary, round the mean to one decimal place.

1. 21, 28, 16, 42, 38
2. 42, 35, 36, 40, 50
3. 7.6, 8.2, 8.2, 9.6, 5.7, 9.1
4. 4.9, 7.1, 6.8, 6.8, 5.3, 4.9
5. 0.2, 0.3, 0.5, 0.6, 0.6, 0.9, 0.2, 0.7, 1.1
6. 0.6, 0.6, 0.8, 0.4, 0.5, 0.3, 0.7, 0.8, 0.1
7. 231, 543, 601, 293, 588, 109, 334, 268
8. 451, 356, 478, 776, 892, 500, 467, 780

1. mean: 29, median: 28, no mode
2. mean: 40.6, median: 40, no mode
3. mean: 8.1, median: 8.2, mode: 8.2
4. mean: 6.0, median: 6.05, mode: 6.8 and 4.9
5. mean: 0.6, median: 0.6, mode: 0.2 and 0.6
6. mean: 0.5, median: 0.6, mode: 0.6 and 0.8
7. mean: 370.9, median: 313.5, no mode
8. mean: 587.5, median: 489, no mode

The eight tallest buildings in the United States are listed below. Use this table for Exercises 9–12.

Building	Height (feet)
Sears Tower, Chicago, IL	1454
Empire State, New York, NY	1250
Amoco, Chicago, IL	1136
John Hancock Center, Chicago, IL	1127
First Interstate World Center, Los Angeles, CA	1107
Chrysler, New York, NY	1046
NationsBank Tower, Atlanta, GA	1023
Texas Commerce Tower, Houston, TX	1002

9. Find the mean height for the five tallest buildings.   1214.8 ft

10. Find the median height for the five tallest buildings.   1136 ft

11. Find the median height for the eight tallest buildings.   1117 ft

12. Find the mean height for the eight tallest buildings.   1143.1 ft

During an experiment, the following times (in seconds) were recorded: 7.8, 6.9, 7.5, 4.7, 6.9, 7.0.

13. Find the mean. Round to the nearest tenth.   6.8

14. Find the median.   6.95   **15.** Find the mode.   6.9

In a mathematics class, the following test scores were recorded for a student: 86, 95, 91, 74, 77, 85.   18. no mode

16. Find the mean. Round to the nearest hundredth.   84.67

17. Find the median.   85.5   **18.** Find the mode.

The following pulse rates were recorded for a group of fifteen students: 78, 80, 66, 68, 71, 64, 82, 71, 70, 65, 70, 75, 77, 86, 72.

19. Find the mean.   73   **20.** Find the median.   71

21. Find the mode.   70 and 71

22. How many rates were higher than the mean?   6

23. How many rates were lower than the mean?   9

24. Have each student in your algebra class take his/her pulse rate. Record the data and find the mean, the median, and the mode.   answers may vary

Find the missing numbers in each list of numbers. (These numbers are not necessarily in numerical order)

25. <u>21</u>, <u>21</u>, 16, 18, <u>24</u>

The mode is 21.

The mean is 20.

26. <u>35</u>, <u>35</u>, <u>37</u>, <u>43</u>, 40

The mode is 35.

The median is 37.

The mean is 38.

Table of Squares and Square Roots

n	n^2	$\sqrt{n}$	n	n^2	$\sqrt{n}$
1	1	1.000	51	2,601	7.141
2	4	1.414	52	2,704	7.211
3	9	1.732	53	2,809	7.280
4	16	2.000	54	2,916	7.348
5	25	2.236	55	3,025	7.416
6	36	2.449	56	3,136	7.483
7	49	2.646	57	3,249	7.550
8	64	2.828	58	3,364	7.616
9	81	3.000	59	3,481	7.681
10	100	3.162	60	3,600	7.746
11	121	3.317	61	3,721	7.810
12	144	3.464	62	3,844	7.874
13	169	3.606	63	3,969	7.937
14	196	3.742	64	4,096	8.000
15	225	3.873	65	4,225	8.062
16	256	4.000	66	4,356	8.124
17	289	4.123	67	4,489	8.185
18	324	4.243	68	4,624	8.246
19	361	4.359	69	4,761	8.307
20	400	4.472	70	4,900	8.367
21	441	4.583	71	5,041	8.426
22	484	4.690	72	5,184	8.485
23	529	4.796	73	5,329	8.544
24	576	4.899	74	5,476	8.602
25	625	5.000	75	5,625	8.660
26	676	5.099	76	5,776	8.718
27	729	5.196	77	5,929	8.775
28	784	5.292	78	6,084	8.832
29	841	5.385	79	6,241	8.888
30	900	5.477	80	6,400	8.944

n	n^2	$\sqrt{n}$	n	n^2	$\sqrt{n}$
31	961	5.568	81	6,561	9.000
32	1,024	5.657	82	6,724	9.055
33	1,089	5.745	83	6,889	9.110
34	1,156	5.831	84	7,056	9.165
35	1,225	5.916	85	7,225	9.220
36	1,296	6.000	86	7,396	9.274
37	1,369	6.083	87	7,569	9.327
38	1,444	6.164	88	7,744	9.381
39	1,521	6.245	89	7,921	9.434
40	1,600	6.325	90	8,100	9.487
41	1,681	6.403	91	8,281	9.539
42	1,764	6.481	92	8,464	9.592
43	1,849	6.557	93	8,649	9.644
44	1,936	6.633	94	8,836	9.695
45	2,025	6.708	95	9,025	9.747
46	2,116	6.782	96	9,216	9.798
47	2,209	6.856	97	9,409	9.849
48	2,304	6.928	98	9,604	9.899
49	2,401	7.000	99	9,801	9.950
50	2,500	7.071	100	10,000	10.000

Review of Volume and Surface Area

A **convex solid** is a set of points, S, not all in one plane, such that for any two points A and B in S, all points between A and B are also in S. In this appendix, we will find the volume and surface area of special types of solids called polyhedrons. A solid formed by the intersection of a finite number of planes is called a **polyhedron**. The box below is an example of a polyhedron.

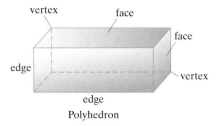

Polyhedron

Each of the plane regions of the polyhedron is called a **face** of the polyhedron. If the intersection of two faces is a line segment, this line segment is an **edge** of the polyhedron. The intersections of the edges are the **vertices** of the polyhedron.

Volume is a measure of the space of a solid. The volume of a box or can, for example, is the amount of space inside. Volume can be used to describe the amount of juice in a pitcher or the amount of concrete needed to pour a foundation for a house.

The volume of a solid is the number of **cubic units** in the solid. A cubic centimeter and a cubic inch are illustrated.

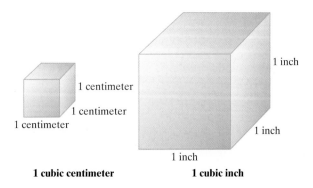

1 cubic centimeter   **1 cubic inch**

The **surface area** of a polyhedron is the sum of the areas of the faces of the polyhedron. For example, each face of the cube to the left above has an area of 1 square centimeter. Since there are 6 faces of the cube, the sum of the areas of the faces is 6 square centimeters. Surface area can be used to describe the amount of material needed to cover or form a solid. Surface area is measured in square units.

Formulas for finding the volumes, *V*, and surface areas, *SA*, of some common solids are given next.

Volume and Surface Area Formulas of Common Solids

Solid	*Formulas*

RECTANGULAR SOLID

$V = lwh$

$SA = 2lh + 2wh + 2lw$

where h = height, w = width, l = length

CUBE

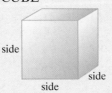

$V = s^3$

$SA = 6s^2$

where s = side

SPHERE

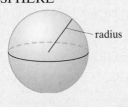

$V = \dfrac{4}{3}\pi r^3$

$SA = 4\pi r^2$

where r = radius

CIRCULAR CYLINDER

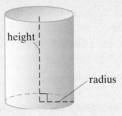

$V = \pi r^2 h$

$SA = 2\pi rh + 2\pi r^2$

where h = height, r = radius

CONE

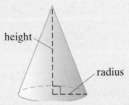

$V = \dfrac{1}{3}\pi r^2 h$

$SA = \pi r \sqrt{r^2 + h^2} + \pi r^2$

where h = height, r = radius

SQUARE-BASED PYRAMID

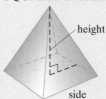

$V = \dfrac{1}{3}s^2 h$

$SA = B + \dfrac{1}{2}pl$

where B = area of base; p = perimeter of base, h = height, s = side, l = slant height

> **Helpful Hint**
> Volume is measured in cubic units. Surface area is measured in square units.

EXAMPLE 1

Find the volume and surface area of a rectangular box that is 12 inches long, 6 inches wide, and 3 inches high.

3 in.

FRAGILE

6 in.

12 in.

Solution   Let $h = 3$ in., $l = 12$ in., and $w = 6$ in.
$V = lwh$
$V = 12$ inches $\cdot 6$ inches $\cdot 3$ inches $= 216$ cubic inches

The volume of the rectangular box is 216 cubic inches.

$$SA = 2lh + 2wh + 2lw$$
$$= 2(12 \text{ in.})(3 \text{ in.}) + 2(6 \text{ in.})(3 \text{ in.}) + 2(12 \text{ in.})(6 \text{ in.})$$
$$= 72 \text{ sq. in.} + 36 \text{ sq. in.} + 144 \text{ sq. in.}$$
$$= 252 \text{ sq. in.}$$

The surface area of rectangular box is 252 square inches.

EXAMPLE 2

2 in.

Find the volume and surface area of a ball of radius 2 inches. Give the exact volume and surface area and then use the approximation $\frac{22}{7}$ for π.

Solution

$$V = \frac{4}{3}\pi r^3 \qquad \text{Formula for volume of a sphere.}$$

$$V = \frac{4}{3} \cdot \pi (2 \text{ in.})^3 \qquad \text{Let } r = 2 \text{ inches.}$$

$$= \frac{32}{3}\pi \text{ cu. in.} \qquad \text{Simplify.}$$

$$\approx \frac{32}{3} \cdot \frac{22}{7} \text{cu. in.} \qquad \text{Approximate } \pi \text{ with } \frac{22}{7}.$$

$$= \frac{704}{21} \text{ or } 33\frac{11}{21} \text{cu. in.}$$

The volume of the sphere is exactly $\frac{32}{3}\pi$ cubic inches or approximately $33\frac{11}{21}$ cubic inches.

$$SA = 4\pi r^2 \qquad \text{Formula for surface area.}$$
$$SA = 4 \cdot \pi (2 \text{ in.})^2 \qquad \text{Let } r = 2 \text{ inches.}$$
$$= 16\pi \text{ sq. in.} \qquad \text{Simplify.}$$
$$\approx 16 \cdot \frac{22}{7} \text{ sq. in.} \qquad \text{Approximate } \pi \text{ with } \frac{22}{7}.$$
$$= \frac{352}{7} \text{ or } 50\frac{2}{7} \text{ sq. in.}$$

The surface area of the sphere is exactly 16π square inches or approximately $50\frac{2}{7}$ square inches.

APPENDIX F EXERCISE SET

Find the volume and surface area of each solid. See Examples 1 and 2. For formulas that contain π, give an exact answer and then approximate using $\frac{22}{7}$ for π.

 1.

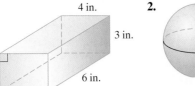
4 in.
3 in.
6 in.

2.

3 mi

3.

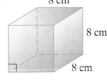
8 cm
8 cm
8 cm

4.

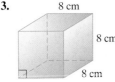
8 cm
4 cm
4 cm

5. (For surface area, use 3.14 for π and round to two decimal places.)

3 yd
2 yd

6.

10 ft
6 ft

7.

10 in.

8. Find the volume only.

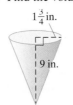
$1\frac{3}{4}$ in.
9 in.

9.

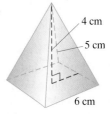
4 cm
5 cm
6 cm

10.

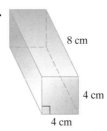
1 ft

Solve.

11. Find the volume of a cube with edges of $1\frac{1}{3}$ inches.

$1\frac{1}{3}$ in.
$2\frac{10}{27}$ cu. in.

12. A water storage tank is in the shape of a cone with the pointed end down. If the radius is 14 ft and the depth of the

1. $V = 72$ cu. in.; $SA = 108$ sq. in.
2. $V = 36\pi$ cu. mi $\approx 113\frac{1}{7}$ cu. mi; $SA = 36\pi$ sq. mi $\approx 113\frac{1}{7}$ sq. mi
3. $V = 512$ cu. cm; $SA = 384$ sq. cm   4. $V = 128$ cu. cm; $SA = 160$ sq. cm
5. $V = 4\pi$ cu. yd $\approx 12\frac{4}{7}$ cu. yd; $SA = \left(2\sqrt{13}\pi + 4\pi\right)$ sq. yd ≈ 35.20 sq. yd
6. $V = 150\pi$ cu. ft $\approx 471\frac{3}{7}$ cu. ft; $SA = 110\pi$ sq. ft $\approx 345\frac{5}{7}$ sq. ft   7. $V = \frac{500}{3}\pi$ cu. in. $\approx 523\frac{17}{21}$ cu. in.; $SA = 100\pi$ sq. in. $\approx 314\frac{2}{7}$ sq. in.
8. $V = \frac{147}{16}\pi$ cu. in. $\approx 28\frac{7}{8}$ cu. in.   9. $V = 48$ cu. cm; $SA = 96$ sq. cm   10. $V = 18$ cu. ft; $SA = 42$ sq. ft

tank is 15 ft, approximate the volume of the tank in cubic feet. Use $\frac{22}{7}$ for π.   3080 cu. ft

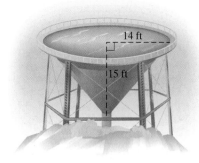

13. Find the surface area of a rectangular box 2 ft by 1.4 ft by 3 ft.   26 sq. ft

14. Find the surface area of a box in the shape of a cube that is 5 ft on each side.   150 sq. ft

15. Find the volume of a pyramid with a square base 5 in. on a side and a height of $1\frac{3}{10}$ in. $10\frac{5}{6}$ cu. in.

16. Approximate to the nearest hundredth the volume of a sphere with a radius of 2 cm. Use 3.14 for π.   33.49 cu. cm

17. A paperweight is in the shape of a square-based pyramid 20 cm tall. If an edge of the base is 12 cm, find the volume of the paperweight.   960 cu. cm

18. A bird bath is made in the shape of a hemisphere (half-sphere). If its radius is 10 in., approximate the volume. Use $\frac{22}{7}$ for π. $2095\frac{5}{21}$ cu. in.

19. Find the exact surface area of a sphere with a radius of 7 in. 196π sq. in.

20. A tank is in the shape of a cylinder 8 ft tall and 3 ft in radius. Find the exact surface area of the tank. 66π sq. ft

21. Find the volume of a rectangular block of ice 2 ft by $2\frac{1}{2}$ ft by $1\frac{1}{2}$ ft. $7\frac{1}{2}$ cu. ft

22. Find the capacity (volume in cubic feet) of a rectangular ice chest with inside measurements of 3 ft by $1\frac{1}{2}$ ft by $1\frac{3}{4}$ ft. $7\frac{7}{8}$ cu. ft

23. An ice cream cone with a 4-cm diameter and 3-cm depth is filled exactly level with the top of the cone. Approximate how much ice cream (in cubic centimeters) is in the cone. Use $\frac{22}{7}$ for π. $12\frac{4}{7}$ cu. cm

24. A child's toy is in the shape of a square-based pyramid 10 in. tall. If an edge of the base is 7 in., find the volume of the toy. $163\frac{1}{3}$ cu. in.

ANSWERS TO SELECTED EXERCISES

CHAPTER 1   REVIEW OF REAL NUMBERS

Exercise Set 1.2   1. $>$ **3.** $=$ **5.** $<$ **7.** $<$ **9.** $32 < 212$ **11.** $44,300 > 34,611$ **13.** true   **15.** false   **17.** false   **19.** true
21. $30 \le 45$ **23.** $8 < 12$ **25.** $5 \ge 4$ **27.** $15 \ne -2$ **29.** $535; -8$ **31.** $-34,841$ **33.** $350; -126$ **35.** 1993   **37.** 1993, 1994   **39.** $827 \ge 818$
41. whole, integers, rational, real   **43.** integers, rational, real   **45.** natural, whole, integers, rational, real   **47.** rational, real   **49.** irrational, real
51. false   **53.** true   **55.** true   **57.** true   **59.** false   **61.** $>$ **63.** $>$ **65.** $<$ **67.** $<$ **69.** $>$ **71.** $=$ **73.** $<$ **75.** $<$
77. $-0.04 > -26.7$ **79.** Sun   **81.** Sun   **83.** $20 \le 25$ **85.** $6 > 0$ **87.** $-12 < -10$ **89.** answers may vary

Mental Math   1. $\dfrac{3}{8}$ **3.** $\dfrac{5}{7}$ **5.** numerator; denominator

Exercise Set 1.3   1. $3 \cdot 11$ **3.** $2 \cdot 7 \cdot 7$ **5.** $2 \cdot 2 \cdot 5$ **7.** $3 \cdot 5 \cdot 5$ **9.** $3 \cdot 3 \cdot 5$ **11.** $\dfrac{1}{2}$ **13.** $\dfrac{2}{3}$ **15.** $\dfrac{3}{7}$ **17.** $\dfrac{3}{5}$ **19.** $\dfrac{3}{8}$ **21.** $\dfrac{1}{2}$ **23.** $\dfrac{6}{7}$ **25.** 15   **27.** $\dfrac{1}{6}$
29. $\dfrac{25}{27}$ **31.** $\dfrac{11}{20}$ sq. mi   **33.** $\dfrac{3}{5}$ **35.** 1   **37.** $\dfrac{1}{3}$ **39.** $\dfrac{9}{35}$ **41.** $\dfrac{21}{30}$ **43.** $\dfrac{4}{18}$ **45.** $\dfrac{16}{20}$ **47.** $\dfrac{23}{21}$ **49.** $1\dfrac{2}{3}$ **51.** $\dfrac{5}{66}$ **53.** $\dfrac{7}{5}$ **55.** $\dfrac{1}{5}$ **57.** $\dfrac{3}{8}$ **59.** $\dfrac{1}{9}$ **61.** $\dfrac{5}{7}$
63. $\dfrac{65}{21}$ **65.** $\dfrac{2}{5}$ **67.** $\dfrac{9}{7}$ **69.** $\dfrac{3}{4}$ **71.** $\dfrac{17}{3}$ **73.** $\dfrac{7}{26}$ **75.** 1   **77.** $\dfrac{1}{5}$ **79.** $5\dfrac{1}{6}$ **81.** $\dfrac{17}{18}$ **83.** $55\dfrac{1}{4}$ ft   **85.** $6\dfrac{3}{20}$ m   **87.** answers may vary   **89.** $3\dfrac{3}{8}$ mi
91. $\dfrac{7}{50}$ **93.** $\dfrac{1}{4}$ **95.** $\dfrac{111}{613}$ **97.** $\dfrac{7}{36}$ sq. ft

Calculator Explorations   1. 625   **3.** 59,049   **5.** 30   **7.** 9857   **9.** 2376

Mental Math   1. multiply   **3.** subtract

Exercise Set 1.4   1. 243   **3.** 27   **5.** 1   **7.** 5   **9.** $\dfrac{1}{125}$ **11.** $\dfrac{16}{81}$ **13.** 49   **15.** 16   **17.** 1.44   **19.** 17   **21.** 20   **23.** 10   **25.** 21   **27.** 45   **29.** 0   **31.** $\dfrac{2}{7}$
33. 30   **35.** 2   **37.** $\dfrac{7}{18}$ **39.** $\dfrac{27}{10}$ **41.** $\dfrac{7}{5}$ **43.** no   **45. a.** 64   **b.** 43   **c.** 19   **d.** 22   **47.** 9   **49.** 1   **51.** 1   **53.** 11   **55.** 45   **57.** 27   **59.** 132   **61.** $\dfrac{37}{18}$
63. $16, 64, 144, 256$ **65.** yes   **67.** no   **69.** no   **71.** yes   **73.** no   **75.** $x + 15$ **77.** $x - 5$ **79.** $3x + 22$ **81.** $1 + 2 = 9 \div 3$ **83.** $3 \ne 4 \div 2$
85. $5 + x = 20$ **87.** $13 - 3x = 13$ **89.** $\dfrac{12}{x} = \dfrac{1}{2}$ **91.** answers may vary   **93.** $(20 - 4) \cdot 4 \div 2$ **95.** 28 m   **97.** 12,000 sq. ft   **99.** 6.5%   **101.** $19.96

Mental Math   1. negative   **3.** 0   **5.** negative

Exercise Set 1.5   1. 9   **3.** -14 **5.** 1   **7.** -12 **9.** -5 **11.** -12 **13.** -4 **15.** 7   **17.** -2 **19.** 0   **21.** -19 **23.** 31   **25.** -47 **27.** -2.1 **29.** -8
31. 38   **33.** -13.1 **35.** $\dfrac{2}{8} = \dfrac{1}{4}$ **37.** $-\dfrac{3}{16}$ **39.** $-\dfrac{13}{10}$ **41.** -8 **43.** -59 **45.** -9 **47.** 5   **49.** 11   **51.** -18 **53.** 19   **55.** -0.7 **57.** $-6°$
59. -654 ft   **61.** $-\$2127$ million   **63.** -2 **65.** -6 **67.** 2   **69.** 0   **71.** -6 **73.** answers may vary.   **75.** -2 **77.** 0   **79.** $-\dfrac{2}{3}$
81. answers may vary   **83.** yes   **85.** no   **87.** July   **89.** October   **91.** 4.7°F   **93.** negative   **95.** positive

Exercise Set 1.6   1. -10 **3.** -5 **5.** 19   **7.** $\dfrac{1}{6}$ **9.** 2   **11.** -11 **13.** 11   **15.** 5   **17.** 37   **19.** -6.4 **21.** -71 **23.** 0   **25.** 4.1   **27.** $\dfrac{2}{11}$ **29.** $-\dfrac{11}{12}$
31. 8.92   **33.** 13   **35.** -5 **37.** -1 **39.** -23 **41.** answers may vary   **43.** -26 **45.** -24 **47.** 3   **49.** -45 **51.** -4 **53.** 13   **55.** 6   **57.** 9   **59.** -9
61. -7 **63.** $\dfrac{7}{5}$ **65.** 21   **67.** $\dfrac{1}{4}$ **69.** 100°   **71.** -23 yd or 23 yd loss   **73.** -569 or 569 B.C.   **75.** -308 ft   **77.** 19,852 ft   **79.** 130°   **81.** 30°   **83.** no
85. no   **87.** yes   **89.** $-4.4°; 2.6°; 12°; 23.5°; 15.3°; 3.9°; -0.3°; -6.3°; -18.2°; -15.7°; -10.3°$ **91.** October   **93.** true   **95.** true
97. negative, -2.6466

Calculator Explorations   1. 38   **3.** -441 **5.** $163.\overline{3}$ **7.** 54,499   **9.** 15,625

Mental Math   1. positive   **3.** negative   **5.** positive

Exercise Set 1.7   1. -24 **3.** -2 **5.** 50   **7.** -12 **9.** 0   **11.** -18 **13.** $\dfrac{3}{10}$ **15.** $\dfrac{2}{3}$ **17.** -7 **19.** 0.14   **21.** -800 **23.** -28 **25.** 25   **27.** $-\dfrac{8}{27}$
29. -121 **31.** $-\dfrac{1}{4}$ **33.** -30 **35.** 23   **37.** -7 **39.** true   **41.** false   **43.** 16   **45.** -1 **47.** 25   **49.** -49 **51.** $\dfrac{1}{9}$ **53.** $\dfrac{3}{2}$ **55.** $-\dfrac{1}{14}$ **57.** $-\dfrac{11}{3}$
59. $\dfrac{1}{0.2}$ **61.** -6.3 **63.** -9 **65.** 4   **67.** -4 **69.** 0   **71.** -5 **73.** undefined   **75.** 3   **77.** -15 **79.** $-\dfrac{18}{7}$ **81.** $\dfrac{20}{27}$ **83.** -1 **85.** $-\dfrac{9}{2}$ **87.** -4
89. 16   **91.** -3 **93.** $-\dfrac{16}{7}$ **95.** 2   **97.** $\dfrac{6}{5}$ **99.** -5 **101.** $\dfrac{3}{2}$ **103.** -21 **105.** 41   **107.** -134 **109.** 3   **111.** 0   **113.** $-\$5088$ million   **115.** yes
117. no   **119.** yes   **121.** answers may vary   **123.** $1, -1$ **125.** positive   **127.** not possible   **129.** negative   **131.** $-2 + \dfrac{-15}{3}; -7$ **133.** $2[-5 + (-3)]; -16$

Integrated Review   1. positive   **2.** positive   **3.** negative   **4.** negative   **5.** positive   **6.** negative   **7.** negative   **8.** positive   **9.** -35 **10.** 30
11. 5   **12.** -5 **13.** 10   **14.** -18 **15.** -2 **16.** -2 **17.** $\frac{3}{8}$ **18.** $-\frac{11}{42}$ **19.** -60 **20.** 1.9   **21.** -42 **22.** -7 **23.** 2   **24.** -39 **25.** 64   **26.** -81
27. -27 **28.** 16   **29.** -1 **30.** 1   **31.** -32 **32.** -32 **33.** 48   **34.** -30 **35.** -6 **36.** 9   **37.** -30 **38.** -44 **39.** 4   **40.** -3 **41.** 2   **42.** 16
43. 0   **44.** 19   **45.** $-\frac{32}{15}$ **46.** $\frac{54}{5}$

Exercise Set 1.8   1. $16 + x$ **3.** $y \cdot (-4)$ **5.** yx **7.** $13 + 2x$ **9.** $x \cdot (yz)$ **11.** $(2 + a) + b$ **13.** $(4a) \cdot b$ **15.** $a + (b + c)$ **17.** $17 + b$
19. $24y$ **21.** y **23.** $26 + a$ **25.** $-72x$ **27.** s **29.** answers may vary   **31.** $4x + 4y$ **33.** $9x - 54$ **35.** $6x + 10$ **37.** $28x - 21$
39. $18 + 3x$ **41.** $-2y + 2z$ **43.** $-21y - 35$ **45.** $5x + 20m + 10$ **47.** $-4 + 8m - 4n$ **49.** $-5x - 2$ **51.** $-r + 3 + 7p$ **53.** $3x + 4$
55. $-x + 3y$ **57.** $6r + 8$ **59.** $-36x - 70$ **61.** $-16x - 25$ **63.** $4(1 + y)$ **65.** $11(x + y)$ **67.** $-1(5 + x)$ **69.** $30(a + b)$
71. commutative property of multiplication   **73.** associative property of addition   **75.** distributive property   **77.** associative property of multiplication
79. identity element of addition   **81.** distributive property   **83.** commutative and associative properties of multiplication   **85.** $-8; \frac{1}{8}$ **87.** $-x; \frac{1}{x}$
89. $-2x; \frac{1}{2x}$ **91.** no   **93.** yes   **95.** answers may vary

Exercise Set 1.9   1. approx. 7.8 million   **3.** 2002   **5.** red; 23 shades   **7.** 9 shades   **9.** France   **11.** France, United States, Spain   **13.** 39 million
15. Africa/Middle East; 11 million users   **17.** 187 million users   **19.** approx. 59 beats per min   **21.** approx. 26 beats per min   **23.** 74,800
25. 2000; 76,600   **27.** 2003   **29.** answers may vary   **31.** 20   **33.** 1985   **35.** 1997   **37.** 18 million   **39.** 63 million   **41.** 1900   **43.** 27 million
45. answers may vary   **47.** 2001, 2002, 2003   **49.** 46 million   **51.** 30° north, 90° west   **53.** 40° north, 104° west

Chapter 1 Review   1. $<$ **3.** $>$ **5.** $<$ **7.** $=$ **9.** $>$ **11.** $4 \geq -3$ **13.** $0.03 < 0.3$ **15. a.** 1, 3   **b.** 0, 1, 3   **c.** $-6, 0, 1, 3$ **d.** $-6, 0, 1, 1\frac{1}{2}, 3, 9.62$
e. π **f.** $-6, 0, 1, 1\frac{1}{2}, 3, \pi, 9.62$ **17.** Friday   **19.** $2 \cdot 2 \cdot 3 \cdot 3$ **21.** $\frac{12}{25}$ **23.** $\frac{13}{10}$ **25.** $9\frac{3}{8}$ **27.** 15   **29.** $\frac{7}{12}$ **31.** $A = \frac{34}{121}$ sq. in.; $P = 2\frac{4}{11}$ in.   **33.** $2\frac{15}{16}$ lb
35. $11\frac{5}{16}$ lb   **37.** Odera   **39.** $3\frac{7}{8}$ lb   **41.** 16   **43.** $\frac{4}{49}$ **45.** 70   **47.** 37   **49.** $\frac{18}{7}$ **51.** $20 - 12 = 2 \cdot 4$ **53.** 18   **55.** 5   **57.** 63°   **59.** no   **61.** $-\frac{2}{3}$
63. 7   **65.** -17 **67.** -5 **69.** 3.9   **71.** -14 **73.** 5   **75.** -19 **77.** 15   **79.** $-\frac{1}{6}$ **81.** -48 **83.** 3   **85.** undefined   **87.** undefined   **89.** -12
91. 9   **93.** -5 **95.** commutative property of addition   **97.** distributive property   **99.** associative property of addition   **101.** distributive property
103. multiplicative inverse   **105.** commutative property of addition   **107.** 19 million   **109.** number of subscribers is increasing
111. sleeping, 15 calories   **113.** 10 more calories

Chapter 1 Test   1. $|-7| > 5$ **2.** $9 + 5 \geq 4$ **3.** -5 **4.** -11 **5.** -3 **6.** -39 **7.** 12   **8.** -2 **9.** undefined   **10.** -8 **11.** $-\frac{1}{3}$ **12.** $4\frac{5}{8}$ **13.** $-\frac{5}{2}$ or $-2\frac{1}{2}$
14. -32 **15.** -48 **16.** 3   **17.** 0   **18.** $>$ **19.** $>$ **20.** $<$ **21.** $=$ **22.** $2221 < 10,993$ **23. a.** 1, 7   **b.** 0, 1, 7   **c.** $-5, -1, 0, 1, 7$
d. $-5, -1, 0, \frac{1}{4}, 1, 7, 11.6$ **e.** $\sqrt{7}, 3\pi$ **f.** $-5, -1, 0, \frac{1}{4}, 1, 7, 11.6, \sqrt{7}, 3\pi$ **24.** 40   **25.** 12   **26.** 22   **27.** -1 **28.** associative property of addition
29. commutative property of multiplication   **30.** distributive property   **31.** multiplicative inverse   **32.** 9   **33.** -3 **34.** second down   **35.** yes
36. 17°   **37.** $650 million   **38.** $420   **39.** 2000; 27 billion pounds   **40.** 1940; 7 billion pounds   **41.** 1970, 1980, 1990, 2000   **42.** 1970
43. Indiana; 25.2 million tons   **44.** Texas; 5 million tons   **45.** 16 million tons   **46.** 3 million tons

CHAPTER 2   EQUATIONS, INEQUALITIES, AND PROBLEM SOLVING

Mental Math   1. -7 **3.** 1   **5.** 17   **7.** $\frac{1}{8}$ **9.** $-\frac{2}{3}$ **11.** unlike   **13.** like   **15.** unlike

Exercise Set 2.1   1. $15y$ **3.** $-15n$ **5.** $-1t$ or $-t$ **7.** $13w$ **9.** $-7b - 9$ **11.** $-m - 6$ **13.** $5y - 20$ **15.** $7d - 11$ **17.** $-3x + 2y - 1$ **19.** $2x + 14$
21. answers may vary   **23.** $(6x + 7) + (4x - 10) = 10x - 3$ **25.** $(3x - 8) - (7x + 1) = -4x - 9$ **27.** $5x^2$ **29.** $4x - 3$
31. $8x - 53$ **33.** $7.2x - 5.2$ **35.** $k - 6$ **37.** $0.9m + 1$ **39.** $-12y + 16$ **41.** $x + 5$ **43.** $1.3x + 3.5$ **45.** $x + 2$ **47.** $-15x + 18$
49. $2k + 10$ **51.** $-3x + 5$ **53.** $2x - 4$ **55.** $\frac{3}{4}x + 12$ **57.** $5x + (-2) + 7x = -2 + 12x$ **59.** $(m - 9) - (5m - 6) = -4m - 3$
61. $8(x + 6) = 8x + 48$ **63.** $2x - (x + 10) = x - 10$ **65.** $7\left(\frac{x}{6}\right) = \frac{7x}{6}$ **67.** $2 + 3x + (-9) + 4x = 7x - 7$ **69.** $(18x - 2)$ ft
71. 2   **73.** -23 **75.** -25 **77.** balanced   **79.** balanced   **81.** $(15x + 23)$ in.   **83.** $5b^2c^3 + b^3c^2$ **85.** $5x^2 + 9x$ **87.** $-7x^2y$

Mental Math   1. 2   **3.** 12   **5.** 17

Exercise Set 2.2   1. 3   **3.** -2 **5.** 3   **7.** 0.5   **9.** $\frac{5}{12}$ **11.** -0.7 **13.** 3   **15.** answers may vary   **17.** -4 **19.** -3 **21.** -10 **23.** 11   **25.** $\frac{2}{3}$
27. -9 **29.** -17.9 **31.** $-\frac{1}{2}$ **33.** 11   **35.** -30 **37.** -7 **39.** 2   **41.** $-\frac{3}{4}$ **43.** 21   **45.** 25   **47.** 0   **49.** 1.83   **51.** $20 - p$ **53.** $(10 - x)$ ft
55. $(180 - x)°$ **57.** $n + 284$ **59.** $m + 178.5$ **61.** $7x$ sq.mi   **63.** $\frac{8}{5}$ **65.** $\frac{1}{2}$ **67.** -9 **69.** x **71.** y **73.** x **75.** $(173 - 3x)°$ **77.** 250 ml
79. answers may vary   **81.** solution   **83.** not a solution

Mental Math **1.** 9 **3.** 2 **5.** −5

Exercise Set 2.3 **1.** −4 **3.** 0 **5.** 12 **7.** −12 **9.** 3 **11.** 2 **13.** 0 **15.** 6.3 **17.** −20 **19.** 0 **21.** −5 **23.** $-\frac{3}{2}$ **25.** 1 **27.** $-\frac{1}{4}$ **29.** −30 **31.** 6
33. −5.5 **35.** $\frac{14}{3}$ **37.** 10 **39.** 0 **41.** −9 **43.** $\frac{11}{2}$ **45.** −21 **47.** −30 **49.** $\frac{9}{10}$ **51.** 2 **53.** −2 **55.** answers may vary **57.** answers may vary
59. $2x + 2$ **61.** $2x + 2$ **63.** $7x - 12$ **65.** 1 **67.** > **69.** = **71.** < **73.** $\frac{700}{3}$ mg **75.** −2.95 **77.** 0.02

Calculator Explorations **1.** solution **3.** not a solution **5.** solution

Exercise Set 2.4 **1.** 1 **3.** $\frac{9}{2}$ **5.** $\frac{3}{2}$ **7.** 0 **9.** 2 **11.** −5 **13.** 10 **15.** 18 **17.** 1 **19.** 50 **21.** 0.2 **23.** all real numbers **25.** no solution
27. no solution **29.** answers may vary **31.** answers may vary **33.** 4 **35.** −4 **37.** 3 **39.** −2 **41.** 4 **43.** $\frac{7}{3}$ **45.** no solution **47.** $\frac{9}{5}$ **49.** $\frac{4}{19}$
51. 1 **53.** no solution **55.** $\frac{7}{2}$ **57.** −17 **59.** $\frac{19}{6}$ **61.** all real numbers **63.** 3 **65.** 13 **67.** $2x + 7 = x + 6; -1$ **69.** $3x - 6 = 2x + 8; 14$
71. $\frac{1}{3}x = \frac{5}{6}; \frac{5}{2}$ **73.** $\frac{x}{4} + \frac{1}{2} = \frac{3}{4}; 1$ **75.** $10 - 5x = 3x; \frac{5}{4}$ **77.** $x = 4$ cm, $2x = 8$ cm **79.** −1 **81.** $\frac{1}{5}$ **83.** $(6x - 8)$ m **85.** Fairview
87. 155 **89.** 15.3 **91.** −0.2 **93.** $-\frac{7}{8}$ **95.** no solution

Integrated Review **1.** 6 **2.** −17 **3.** 12 **4.** −26 **5.** −3 **6.** −1 **7.** $\frac{27}{2}$ **8.** $\frac{25}{2}$ **9.** 8 **10.** −64 **11.** 2 **12.** −3 **13.** no solution
14. no solution **15.** −2 **16.** −2 **17.** $-\frac{5}{6}$ **18.** $\frac{1}{6}$ **19.** 1 **20.** 6 **21.** 4 **22.** 1 **23.** $\frac{9}{5}$ **24.** $-\frac{6}{5}$ **25.** all real numbers **26.** all real numbers
27. 0 **28.** −1.6 **29.** $\frac{4}{19}$ **30.** $-\frac{5}{19}$ **31.** $\frac{7}{2}$ **32.** $-\frac{1}{4}$

Exercise Set 2.5 **1.** −25 **3.** $-\frac{3}{4}$ **5.** 234, 235 **7.** Belgium: 32; France: 33; Spain: 34 **9.** 1st piece: 5 in.; 2nd piece: 10 in.; 3rd piece: 25 in.
11. governor of Nebraska: $65,000; governor of Washington: $130,000 **13.** 172 mi **15.** 25 mi **17.** 1st angle: 37.5°; 2nd angle: 37.5°; 3rd angle: 105°
19. A: 60°; B: 120°; C: 120°; D: 60° **21.** 5 ft, 12 ft **23.** 1997: 15.1 million prescriptions; 2001: 20.6 million prescriptions **25.** 45°, 135°
27. 58°, 60°, 62° **29.** 280 mi **31.** Tampa Bay: 48; Oakland: 21 **33.** China: 59 medals; Australia: 58 medals; Germany: 57 medals
35. California: 55; Texas: 34 **37.** Neptune: 8 moons; Uranus: 21 moons; Saturn: 18 moons **39.** −16 **41.** Brown: 66,362; Randall: 53,074 **43.** Illinois
45. Texas: $29.4 million; Florida: $27.2 million **47.** answers may vary **49.** −10 **51.** −9 **53.** −15 **55.** yes: 30°, 60°, 90° **57.** answers may vary
59. answers may vary **61.** c

Exercise Set 2.6 **1.** $h = 3$ **3.** $h = 3$ **5.** $h = 20$ **7.** $c = 12$ **9.** $r = 2.5$ **11.** $T = 3$ **13.** $h = 15$ **15.** $h = \frac{f}{5g}$ **17.** $W = \frac{V}{LH}$ **19.** $y = 7 - 3x$
21. $R = \frac{A - P}{PT}$ **23.** $A = \frac{3V}{h}$ **25.** $a = P - b - c$ **27.** $h = \frac{S - 2\pi r^2}{2\pi r}$ **29.** 131 ft **31.** −10°C **33.** 6.25 hr **35.** length: 78 ft; width: 52 ft
37. 18 ft, 36 ft, 48 ft **39.** 137.5 mi **41.** 96 piranhas **43.** 2 bags **45.** one 16-in. pizza **47.** 4.65 min **49.** 13 in. **51.** 2.25 hr **53.** 12,090 ft **55.** 50°C
57. 515,509.5 cu. in. **59.** 449 cu. in. **61.** 332.6°F **63.** $\frac{9}{x + 5}$ **65.** $3(x + 4)$ **67.** $3(x - 12)$ **69.** −109.3°F **71.** 500 sec or $8\frac{1}{3}$ min **73.** 608.33 ft
75. 565.5 cu. in. **77.** $25\frac{5}{9}$°C **79.** It multiplies the area by 4.

Exercise Set 2.7 **1.** $64 decrease; $192 sale price **3.** $104 **5.** 73% **7.** $666\frac{2}{3}$ mi **9.** 160 mi **11.** 2 gal **13.** $6\frac{2}{3}$ lb **15.** no
17. $11,500 @ 8%; $13,500 @ 9% **19.** $7000 @ 11% profit; $3000 @ 4% loss **21.** $3900 **23.** $30,000 @ 8%; $24,000 @ 10% **25.** $2\frac{2}{9}$ hr **27.** 400 oz
29. 75% increase **31.** $4500 **33.** 230 million **35.** 2.2 mph; 3.3 mph **37.** 1.45% increase **39.** $1\frac{5}{7}$ hr **41.** −4 **43.** $\frac{9}{16}$ **45.** −4 **47.** >
49. = **51.** 25 skateboards **53.** 800 books **55.** answers may vary **57.** 7.7% **59.** 19.3% **61.** answers may vary

Mental Math **1.** $x > 2$ **3.** $x \geq 8$

Exercise Set 2.8 **1.** $x \geq 2$ **3.** $x < -5$ **5.** $(-\infty, -1]$ **7.** $\left(-\infty, \frac{1}{2}\right)$
9. $[5, \infty)$ **11.** $x < -3$, **13.** $x \geq -5$, **15.** $x \geq -2, [2, \infty)$
17. $x > -3, (-3, \infty)$ **19.** $x \leq 1, (-\infty, 1]$ **21.** $x > -5, (-5, \infty)$
23. $x \leq -2, (-\infty, -2]$ **25.** $x \leq -8, (-\infty, -8]$ **27.** $x > 4, (4, \infty)$
29. $x \geq 20, (20, \infty)$ **31.** $x > 16, (16, \infty)$ **33.** $x > -3, (-3, \infty)$
35. $x \leq -\frac{2}{3}, \left(-\infty, -\frac{2}{3}\right]$ **37.** $x > \frac{8}{3}, \left(\frac{8}{3}, \infty\right)$ **39.** $x > -13, (-13, \infty)$

41. $x < 0, (-\infty, 0)$ **43.** $x \le 0, (-\infty, 0]$ **45.** $x > 3, (3, \infty)$

47. $x \le 0, (-\infty, 0]$ **49.** answers may vary **51.** $(-1, 3)$ **53.** $[2, 3]$

55. $-1 < x < 2, (-1, 2)$ **57.** $4 \le x \le 5, [4, 5]$ **59.** $1 < x \le 5, (1, 5]$

61. $1 < x < 4, (1, 4)$ **63.** $0 < x \le \frac{14}{3}, \left(0, \frac{14}{3}\right]$ **65.** answers may vary **67.** $x > -10$

69. 86 people **71.** $x \le 35$ **73.** at least 10% **75.** at least 193 **77.** $x < 200$ recommended; $200 \le x \le 240$ borderline; $x > 240$ high

79. $-3 < x < 3$ **81.** $-38.2°F$ to $113°F$ **83.** 8 **85.** 1 **87.** $\frac{16}{49}$ **89.** \$3395 **91.** 1986–1990 **93.** $0.924 \le d \le 0.987$ **95.** $(1, \infty)$

97. $\left(-\infty, \frac{5}{8}\right)$

Chapter 2 Review **1.** $6x$ **3.** $4x - 2$ **5.** $3n - 18$ **7.** $-6x + 7$ **9.** $3x - 7$ **11.** 4 **13.** -6 **15.** -9 **17.** $5; 5$ **19.** $10 - x$ **21.** $(175 - x)°$
23. 4 **25.** -1 **27.** -1 **29.** 6 **31.** 2 **33.** no solution **35.** $\frac{3}{4}$ **37.** 20 **39.** 0 **41.** $\frac{20}{7}$ **43.** $\frac{23}{7}$ **45.** 102 **47.** 3 **49.** 1052 ft **51.** $307; 955$
53. $w = 9$ **55.** $m = \frac{y - b}{x}$ **57.** $x = \frac{2y - 7}{5}$ **59.** $\pi = \frac{C}{D}$ **61.** 15 m **63.** 1 hr and 20 min **65.** 80 nickels **67.** 48 mi **69.** $(0, \infty)$
71. $[0.5, 1.5)$ **73.** $(-\infty, -4)$ **75.** $(-\infty, 4]$ **77.** $\left(-\frac{1}{2}, \frac{3}{4}\right)$
79. $\left(-\infty, \frac{19}{3}\right]$ **93.** score must be less than 83

Chapter 2 Test **1.** $y - 10$ **2.** $5.9x + 1.2$ **3.** $-2x + 10$ **4.** $10y + 1$ **5.** -5 **6.** 8 **7.** $\frac{7}{10}$ **8.** 0 **9.** 27 **10.** 3 **11.** 0.25 **12.** $\frac{25}{7}$ **13.** no solution
14. 21 **15.** 7 gal **16.** \$8500 @ 10%; \$17,000 @ 12% **17.** $2\frac{1}{2}$ hr **18.** $x = 6$ **19.** $h = \frac{V}{\pi r^2}$ **20.** $y = \frac{3x - 10}{4}$ **21.** $(-\infty, -2]$
22. $(-\infty, 4)$ **23.** $\left(-1, \frac{7}{3}\right)$ **24.** $\left(\frac{2}{5}, \infty\right)$

Chapter 2 Cumulative Review **1.** **a.** $11, 112$ **b.** $0, 11, 112$ **c.** $-3, -2, 0, 11, 112$ **d.** $-3, -2, 0, \frac{1}{4}, 11, 112$ **e.** $\sqrt{2}$ **f.** all numbers in the given set;
Sec. 1.2, Ex. 5 **3.** **a.** 4 **b.** 5 **c.** 0; Sec. 1.2, Ex. 7 **5.** **a.** $2 \cdot 2 \cdot 2 \cdot 5$ **b.** $3 \cdot 3 \cdot 7$; Sec. 1.3, Ex. 1 **7.** $\frac{8}{20}$; Sec. 1.3, Ex. 6

9. 36; Sec. 1.4, Ex. 4 **11.** 2 is a solution; Sec. 1.4, Ex. 7 **13.** -3; Sec. 1.5, Ex. 2 **15.** 2; Sec. 1.5, Ex. 4 **17.** **a.** 10 **b.** $\frac{1}{2}$ **c.** $2x$ **d.** -6; Sec. 1.5, Ex. 10
19. **a.** 9.9 **b.** $-\frac{4}{5}$ **c.** $\frac{2}{15}$; Sec. 1.6, Ex. 2 **21.** **a.** $52°$ **b.** $118°$; Sec. 1.6, Ex. 8 **23.** **a.** -0.06 **b.** $-\frac{7}{15}$; Sec. 1.7, Ex. 3 **25.** **a.** 6 **b.** -12
c. $-\frac{8}{15}$; Sec. 1.7, Ex. 7 **27.** **a.** $5 + x$ **b.** $x \cdot 3$; Sec. 1.8, Ex. 1 **29.** **a.** $8(2 + x)$ **b.** $7(s + t)$; Sec. 1.8, Ex. 5 **31.** $-2x - 1$; Sec. 2.1, Ex. 7
33. $\frac{5}{4}$; Sec. 2.2, Ex. 3 **35.** 19; Sec. 2.2, Ex. 6 **37.** 140; Sec. 2.3, Ex. 3 **39.** 2; Sec. 2.4, Ex. 1 **41.** 10; Sec. 2.5, Ex. 1 **43.** $\frac{V}{wh} = l$; Sec. 2.6, Ex. 5
45. $(-\infty, -10]$; Sec. 2.8, Ex. 2

CHAPTER 3   GRAPHING

Mental Math **1.** answers may vary; Ex. $(5, 5), (7, 3)$ **3.** answers may vary; Ex. $(3, 5), (3, 0)$

Exercise Set 3.1

1. quadrant I **3.** no quadrant, x-axis **5.** quadrant IV **7.** no quadrant, x-axis **9.** no quadrant, origin **11.** no quadrant, y-axis

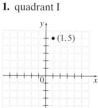

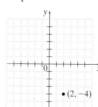
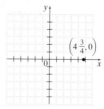
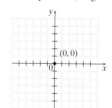
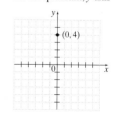

13. $A(0, 0); B\left(3\frac{1}{2}, 0\right); C(3, 2); D(-1, 3); E(-2, -2); F(0, -1); G(2, -1)$ **15.** 26 units

17. a. $(1994, 1.11), (1995, 1.15), (1996, 1.23), (1997, 1.23), (1998, 1.06), (1999, 1.17), (2000, 1.51), (2001, 1.46), (2002, 1.29)$

b.

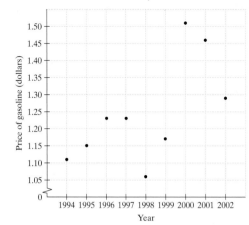

19. a. $(1998, 690), (1999, 733), (2000, 818), (2001, 789), (2002, 827)$

b. 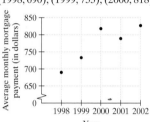   **c.** Average monthly mortgage payment increases each year.

21. a. $(2313, 2), (2085, 1), (2711, 21), (2869, 39), (2920, 42), (4038, 99), (1783, 0), (2493, 9)$

b. 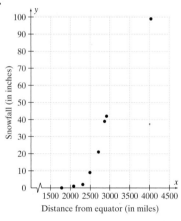   **c.** The farther from the equator, the more snowfall.

23. yes; no; yes   **25.** no; yes; yes   **27.** no; yes; yes   **29.** yes; no   **31.** no; no   **33.** yes; yes   **35.** $(-4, -2), (4, 0)$ **37.** $(0, 9), (3, 0)$

39. $(11, -7)$; answers may vary; Ex. $(2, -7)$ **41.** $(5, 1), \left(1, -\dfrac{1}{7}\right)$

43. $(0, 2), (6, 0), (3, 1)$ **45.** $(0, -12), (5, -2), (-3, -18)$ **47.** $\left(0, \dfrac{5}{7}\right), \left(\dfrac{5}{2}, 0\right), (-1, 1)$ **49.** $(3, 0), (3, -0.5), \left(3, \dfrac{1}{4}\right)$

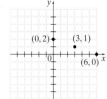
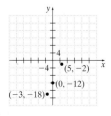
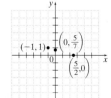
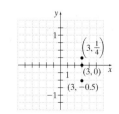

51. $(0,0), (-5,1), (10,-2)$ **53.** Answers may vary.   **55. a.** $13{,}000; 21{,}000; 29{,}000$ **b.** 45 desks

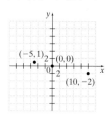

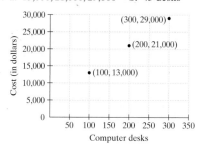

57. a. $20.98, 13.08, 5.18$ **b.** 1997   **59.** In 1995, there were 670 Target stores.   **61.** year 6: 66 stores; year 7: 60 stores; year 8: 55 stores   **63.** $a = b$

65. $y = 5 - x$ **67.** $y = -\dfrac{1}{2}x + \dfrac{5}{4}$ **69.** $y = -2x$ **71.** $y = \dfrac{1}{3}x - 2$ **73.** quadrant IV   **75.** quadrants II or III   **77. a.** $(-2,6)$ **b.** 28 units   **c.** 45 sq. units

Calculator Explorations   **1.**

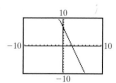

3.

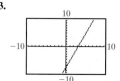

5.

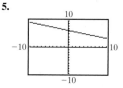

Exercise Set 3.2   **1.** yes   **3.** yes   **5.** no   **7.** yes

9.

11.

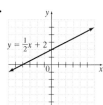

13.

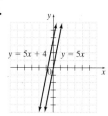

15.

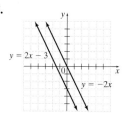

17.

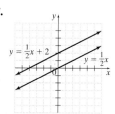

19.

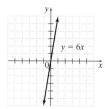

21.

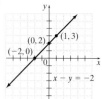

23.

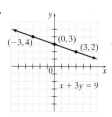

25.

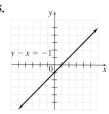

27.

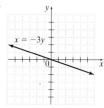

29.

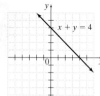

31.

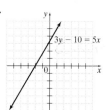

33.

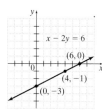

35.

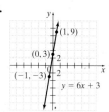

37.

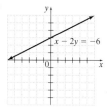

39. c   **41.** d   **43.** $2790 million dollars   **45.**

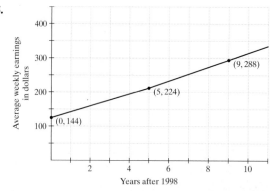

47. $(4, -1)$ **49.** -5 **51.** $-\dfrac{1}{10}$ **53.** $(0, 3), (-3, 0)$ **55.** $(0, 0), (0, 0)$

57. $y = x + 5$ **59.** $2x + 3y = 6$ **61.** answers may vary **63.** answers may vary **65.** $0, 1, 1, 4, 4$

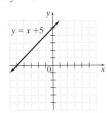

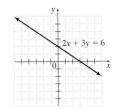

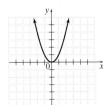

Calculator Explorations **1.** 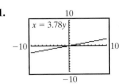 **3.** 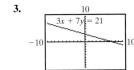 **5.**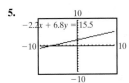

Mental Math **1.** false **3.** true **5.** false

Exercise Set 3.3 **1.** $(-1, 0); (0, 1)$ **3.** $(-2, 0)$ **5.** $(-1, 0); (1, 0); (0, 1); (0, -2)$ **7.** infinite **9.** 0

11. 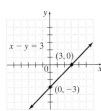 **13.** 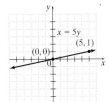 **15.** 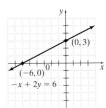 **17.** 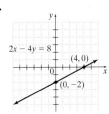 **19.**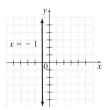

21. 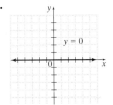 **23.** 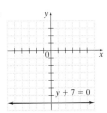 **25.** 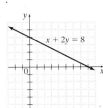 **27.** 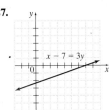 **29.**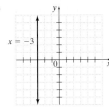

31. 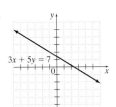 **33.** 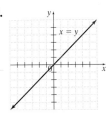 **35.** 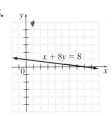 **37.** 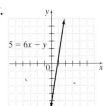 **39.**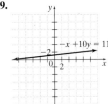

41. 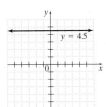 **43.** 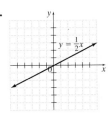 **45.** 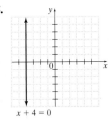 **47.** 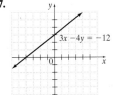 **49.**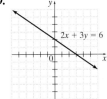

51. C **53.** E **55.** B **57.** $\dfrac{3}{2}$ **59.** 6 **61.** $-\dfrac{6}{5}$ **63. a.** $(0, 569.9)$

b. In 1999, the average price of a digital camera was \$569.90. **65. a.** $(7.1, 0)$ **b.** 7.1 years after 1995, no music cassettes will be shipped.
c. answers may vary

67. a. $(0, 200)$; answers may vary   **b.** $(400, 0)$; answers may vary

c. 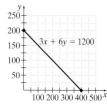   **d.** 300 chairs   **69.** $y = -4$ **71.** answers may vary   **73.** answers may vary

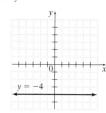

Calculator Explorations   1. 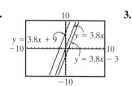   **3.**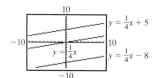

Mental Math   1. upward   **3.** horizontal

Exercise Set 3.4   1. $m = -\dfrac{4}{3}$ **3.** undefined slope   **5.** $m = \dfrac{5}{2}$ **7.** $\dfrac{8}{7}$ **9.** -1 **11.** $-\dfrac{1}{4}$ **13.** undefined   **15.** $-\dfrac{2}{3}$ **17.** undefined   **19.** 0   **21.** line 1

23. line 2   **25.** D   **27.** B   **29.** E   **31.** undefined slope   **33.** $m = 0$ **35.** undefined slope   **37.** $m = 0$ **39. a.** 1   **b.** -1 **41. a.** $\dfrac{9}{11}$ **b.** $-\dfrac{11}{9}$

43. parallel   **45.** perpendicular   **47.** parallel   **49.** neither   **51.** perpendicular   **53.** $-\dfrac{1}{5}$ **55.** $\dfrac{1}{3}$ **57.** 1   **59.** $\dfrac{3}{5}$ **61.** 12.5%   **63.** 40%   **65.** 0.02

67. Every 1 year, there will be 5 million more cell phone users.   **69.** It costs \$0.36 per 1 mile to own and operate a compact car.   **71.** $y = -x + 10$

73. $y = -\dfrac{1}{2}x - 6$ **75.** $y = 5x - 17$ **77.** 28.3   **79.** 1992; 27.6   **81.** from 1992 to 1993   **83.** $x = 6$ **85. a.** $(1994, 782), (2001, 1132)$ **b.** 50

c. For the years 1994 through 2001, the price per acre of U.S. farmland rose \$50 every year.   **87.** The slope through $(-3, 0)$ and $(1, 1)$ is $\dfrac{1}{4}$. The slope
through $(-3, 0)$ and $(-4, 4)$ is -4. The product of the slopes is -1, so the sides are perpendicular.   **89.** -0.25 **91.** 0.875
93. The line becomes steeper.

Integrated Review   1. $m = 2$ **2.** $m = 0$ **3.** $m = -\dfrac{2}{3}$ **4.** undefined

5. 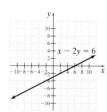   **6.** 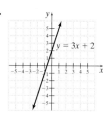   **7.** 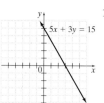   **8.**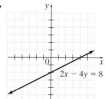

9. 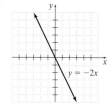   **10.** 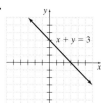   **11.** 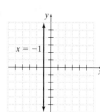   **12.**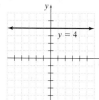

13. parallel   **14.** neither   **15. a.** $(0, 1407)$ **b.** In 2000, there were 1407 million admissions to movie theaters in the U.S.   **c.** 110
d. For the years 2000 through 2002, the number of movie theater admissions increased at a rate of 110 million per year.

Calculator Explorations

1. 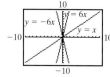   **3.** 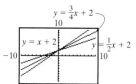   **5.**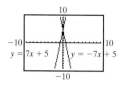

Mental Math **1.** $m = 2; (0, -1)$ **3.** $m = 1; \left(0, \dfrac{1}{3}\right)$ **5.** $m = \dfrac{5}{7}; (0, -4)$

Exercise Set 3.5 **1.** $m = -2; (0, 4)$ **3.** $m = -\dfrac{1}{9}; \left(0, \dfrac{1}{9}\right)$ **5.** $m = \dfrac{4}{3}; (0, -4)$ **7.** $m = -1; (0, 0)$ **9.** $m = 0; (0, -3)$ **11.** $m = \dfrac{1}{5}; (0, 4)$ **13.** B

15. D **17.** neither **19.** neither **21.** perpendicular **23.** parallel **25.** answers may vary **27.** $y = -x + 1$ **29.** $y = 2x + \dfrac{3}{4}$ **31.** $y = \dfrac{2}{7}x$

33. 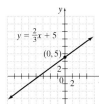 **35.** 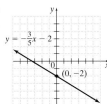 **37.** 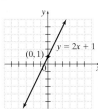 **39.**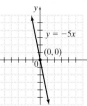

41. 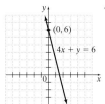 **43.** 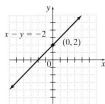 **45.** 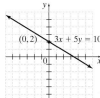 **47.**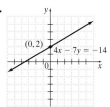

49. $y = 2x - 14$ **51.** $y = -6x - 11$ **53.** D **55.** B **57.** E **59. a.** $(0, 21), (22, 45)$ **b.** $y = \dfrac{12}{11}x + 21$ **61.** $y = -2x + 5$

63. a. The temperature $100°$ Celsius is equivalent to $212°$ Fahrenheit. **b.** $68°F$ **c.** $27°C$ **d.** $F = \dfrac{9}{5}C + 32$ **65.** answers may vary

Mental Math **1.** $m = 3$; answers may vary; Ex. $(4, 8)$ **3.** $m = -2$; answers may vary; Ex. $(10, -3)$ **5.** $m = \dfrac{2}{5}$; answers may vary; Ex. $(-1, 0)$

Exercise Set 3.6 **1.** $6x - y = 10$ **3.** $8x + y = -13$ **5.** $x - 2y = 17$ **7.** $2x - y = 4$ **9.** $8x - y = -11$ **11.** $4x - 3y = -1$ **13.** $x = 0$

15. $y = 3$ **17.** $x = -\dfrac{7}{3}$ **19.** $y = 2$ **21.** $y = 5$ **23.** $x = 6$ **25.** $3x + 6y = 10$ **27.** $x - y = -16$ **29.** $x + y = 17$ **31.** $y = 7$

33. $4x + 7y = -18$ **35.** $x + 8y = 0$ **37.** $3x - y = 0$ **39.** $x - y = 0$ **41.** $5x + y = 7$ **43.** $11x + y = -6$ **45.** $x = -\dfrac{3}{4}$ **47.** $y = -3$

49. $7x - y = 4$ **51. a.** $y = 640x + 4760$ **b.** 11,160 vehicles **53. a.** $s = 32t$ or $y = 32x$ **b.** 128 ft/sec **55. a.** $y = 0.93x + 70.3$

b. 86.11 persons per sq. mi **57. a.** $(0, 191), (5, 260)$ **b.** $y = 13.8x + 191$ **c.** \$246.2 million **59.** $31x - 5y = -5$ **61. a.** $S = -1000p + 13,000$

b. 9500 Fun Noodles **63.** -1 **65.** 5 **67.** no **69.** yes **71.** answers may vary **73. a.** $3x - y = -5$ **b.** $x + 3y = 5$ **75. a.** $3x + 2y = -1$

b. $2x - 3y = 21$

Exercise Set 3.7 **1.** domain: $\{-7, 0, 2, 10\}$; range: $\{-7, 0, 4, 10\}$ **3.** domain: $\{0, 1, 5\}$; range: $\{-2\}$ **5.** yes **7.** no **9.** no **11.** yes **13.** yes **15.** no

17. no **19.** yes **21.** yes **23.** yes **25.** yes **27.** no **29.** no **31.** 5:20 A.M. **33.** answers may vary **35.** \$4.75 per hour **37.** 1996

39. answers may vary **41.** $-9, -5, 1$ **43.** $6, 2, 11$ **45.** $-8, 0, 27$ **47.** $2, 0, 3$ **49.** $h(-1) = -5, h(0) = 0, h(4) = 20; (-1, -5), (0, 0), (4, 20)$

51. $h(-1) = 5, h(0) = 3, h(4) = 35; (-1, 5), (0, 3), (4, 35)$ **53.** $h(-1) = 4, h(0) = 3, h(4) = -21; (-1, 4), (0, 3), (4, -21)$

55. $h(-1) = 6, h(0) = 6, h(4) = 6; (-1, 6), (0, 6), (4, 6)$ **57.** $(-\infty, \infty)$ **59.** all real numbers except -5 or $(-\infty, -5) \cup (-5, \infty)$ **61.** $(-\infty, \infty)$

63. domain: $(-\infty, \infty)$; range: $[-4, \infty)$ **65.** domain: $(-\infty, \infty)$; range: $(-\infty, \infty)$ **67.** domain: $(-\infty, \infty)$; range: $\{2\}$ **69.** $(-2, 1)$ **71.** $(-3, -1)$

73. a. 166.38 cm **b.** 148.25 cm **75.** answers may vary **77.** $f(x) = x + 7$ **79. a.** $-3s + 12$ **b.** $-3r + 12$ **81. a.** 132 **b.** $a^2 - 12$

Chapter 3 Review

1. 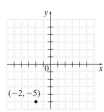 **3.** 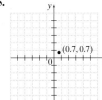 **5.** **7. a.** $(8.00, 1); (7.50, 10); (6.50, 25); (5.00, 50); (2.00, 100)$

b.

9. no; yes **11.** yes; yes **13.** $(7, 44)$

15. $(-3, 0)$; $(1, 3)$; $(9, 9)$ **17.** $(0, 0)$; $(10, 5)$; $(-10, -5)$ **19.** **21.** **23.**

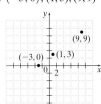 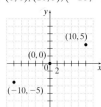 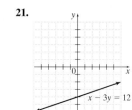 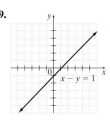 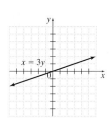

25. **27.** \$135 billion **29.** $(0, -3)$ **33.**

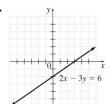 **31.** $(-1, 0)$; $(2, 0)$; $(3, 0)$; $(0, -2)$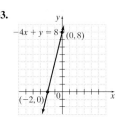

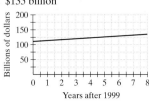

35. **37.** **39.** **41.** $m = \dfrac{1}{5}$ **43.** b **45.** a **47.** $\dfrac{3}{4}$ **49.** 4 **51.** undefined **53.** 0

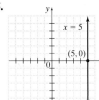 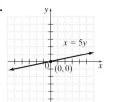 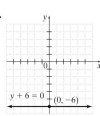 **55.** parallel **57.** neither

59. Every 1 year, monthly day care costs increase by \$17.75.

61. $m = -3$; $(0, 7)$ **63.** $m = 0$; $(0, 2)$ **65.** perpendicular

67. neither

69. $y = \dfrac{2}{3}x + 6$ **71.** **73.** **75.** C **77.** B **79.** $3x + y = -5$ **81.** $y = -3$ **83.** $6x + y = 11$

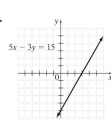 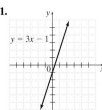 **85.** $x + y = 6$ **87.** $x = 5$ **89.** $x = 6$ **91. a.** $3x + y = 15$

b. $x - 3y = 5$ **93.** yes **95.** yes **97.** yes **99.** no **101. a.** 6 **b.** 10

c. 5 **103. a.** 45 **b.** -35 **c.** 0 **105.** all real numbers

107. domain: $[-3, 5]$ range: $[-4, 2]$

109. domain: $\{3\}$ range: $(-\infty, \infty)$

Chapter 3 Test **1. a.** $(1980, 38)$, $(1984, 47)$, $(1988, 51)$, $(1992, 54)$, $(1996, 59)$, $(2000, 55)$

b. **2.** **3.** **4.** **5.**

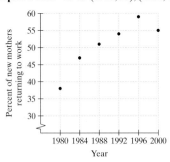 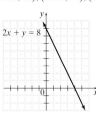 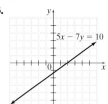 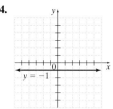 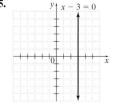

6. $\dfrac{2}{5}$ **7.** 0 **8.** -1 **9.** 3 **10.** undefined **11.** $m = \dfrac{7}{3}$; $\left(0, -\dfrac{2}{3}\right)$ **12.** neither **13.** $x + 4y = 10$ **14.** $7x + 6y = 0$ **15.** $8x + y = 11$

16. $x = -5$ **17.** $x - 8y = -96$ **18.** yes **19.** no **20. a.** 0 **b.** 0 **c.** 60 **21.** all real numbers except -1 or $(-\infty, -1) \cup (-1, \infty)$

22. domain: $(-\infty, \infty)$; range: $(-\infty, 4)$ **23.** domain: $(-\infty, \infty)$; range: $(-\infty, \infty)$ **24.** 9 P.M. **25.** 4 P.M. **26.** January 1st and December 1st

27. June 1st and end of July **28.** yes; it passes the vertical line test **29.** yes; every location has exactly 1 sunset time per day.

Chapter 3 Cumulative Review **1. a.** $<$ **b.** $>$ **c.** $>$; Sec. 1.2, Ex. 1 **3.** $\dfrac{2}{39}$; Sec. 1.3, Ex. 3 **5.** $\dfrac{8}{3}$; Sec. 1.4, Ex. 3 **7. a.** -19 **b.** 30 **c.** -0.5

d. $-\dfrac{4}{5}$ **e.** 6.7 **f.** $\dfrac{1}{40}$; Sec. 1.5, Ex. 6 **9. a.** -6; **b.** 6.3; Sec. 1.6, Ex. 4 **11. a.** -6 **b.** 0 **c.** $\dfrac{3}{4}$; Sec. 1.7, Ex. 10 **13. a.** $22 + x$ **b.** $-21x$; Sec. 1.8, Ex. 3

15. a. -3 **b.** 22 **c.** 1 **d.** -1 **e.** $\dfrac{1}{7}$; Sec. 2.1, Ex. 1 **17.** -1.6; Sec. 2.2, Ex. 2 **19.** $\dfrac{15}{4}$; Sec. 2.3, Ex. 5 **21.** $3x + 3$; Sec. 2.3, Ex. 8

23. 0; Sec. 2.4, Ex. 4 **25.** 208 Democratic representatives; 223 Republican representatives; Sec. 2.5, Ex. 4 **27.** 40 ft; Sec. 2.6, Ex. 2

29. $\dfrac{y - b}{m} = x$; Sec. 2.6, Ex. 6 **31.** 40% solution: 8 l; 70% solution: 4 l; Sec. 2.8, Ex. 4 **33.** ⟵———[———⟶ -1 ; Sec. 2.8, Ex. 1

35. ⟵——[———]——⟶ 1 4 $\{x \mid 1 \le x < 4\}$, ; Sec. 2.8, Ex. 10 **37. a.** solution **b.** not a solution **c.** solution; Sec. 3.1, Ex. 3 **39. a.** linear **b.** linear

c. not linear **d.** linear; Sec. 3.2, Ex. 1 **41.** 0; Sec. 3.4, Ex. 3 **43.** $\dfrac{3}{5}$; Sec. 3.4, Ex. 6

CHAPTER 4   SOLVING SYSTEMS OF LINEAR EQUATIONS AND INEQUALITIES

Calculator Explorations **1.** $(0.37, 0.23)$ **3.** $(0.03, -1.89)$

Mental Math **1.** 1 solution, $(-1, 3)$ **3.** infinite number of solutions **5.** no solution **7.** 1 solution, $(3, 2)$

Exercise Set 4.1 **1. a.** no **b.** yes **c.** no **3. a.** no **b.** yes **c.** no **5. a.** yes **b.** yes **c.** yes **7.** answers may vary

9. $(2, 3)$; consistent;   **11.** $(1, -2)$; consistent;   **13.** $(-2, 1)$; consistent;   **15.** $(4, 2)$; consistent;   **17.** no solution; inconsistent;
independent   independent   independent   independent   independent

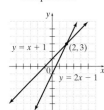 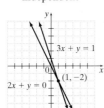 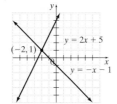 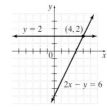 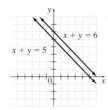

19. infinite number of solutions;   **21.** $(0, -1)$; consistent;   **23.** $(4, -3)$; consistent;   **25.** $(-5, -7)$; consistent;   **27.** $(5, 2)$; consistent;
consistent; dependent   independent   independent   independent   independent

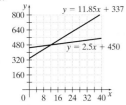 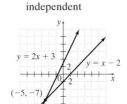 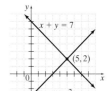 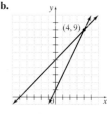

29. answers may vary **31.** intersecting, one solution **33.** parallel, no solution   **61. a.** $(4, 9)$ **b.**
35. identical lines, infinite number of solutions **37.** intersecting, one solution **39.** intersecting, one solution

41. identical lines, infinite number of solutions **43.** parallel, no solution **45.** 2 **47.** $-\dfrac{2}{5}$ **49.** 2
51. answers may vary **53.** $1984, 1988$ **55.** 1996 **57.** answers may vary **59.** answers may vary

Mental Math **1.** $(1, 4)$ **3.** infinite number of solutions **5.** $(0, 0)$

Exercise Set 4.2 **1.** $(2, 1)$ **3.** $(-3, 9)$ **5.** $(4, 2)$ **7.** $(10, 5)$ **9.** $(2, 7)$ **11.** $(-2, 4)$ **13.** $(-2, -1)$ **15.** no solution

17. infinite number of solutions **19.** $(3, -1)$ **21.** $(3, 5)$ **23.** $\left(\dfrac{2}{3}, -\dfrac{1}{3}\right)$ **25.** $(-1, -4)$ **27.** $(-6, 2)$ **29.** $(2, 1)$ **31.** no solution **33.** $\left(-\dfrac{1}{5}, \dfrac{43}{5}\right)$

35. answers may vary **37.** $(1, -3)$ **39.** $-6x - 4y = -12$ **41.** $-12x + 3y = 9$ **43.** $5n$ **45.** $-15b$ **47. a.** $(12, 480)$ **b.** answers may vary

c.   **49.** $(-2.6, 1.3)$ **51.** $(3.28, 2.11)$

Exercise Set 4.3   **1.** $(1,2)$ **3.** $(2,-3)$ **5.** $(6,0)$ **7.** no solution   **9.** $\left(2,-\dfrac{1}{2}\right)$ **11.** $(6,-2)$ **13.** infinite number of solutions   **15.** $(-2,-5)$

17. $(5,-2)$ **19.** $(-7,5)$ **21.** $\left(\dfrac{12}{11},-\dfrac{4}{11}\right)$ **23.** no solution   **25.** $\left(\dfrac{3}{2},3\right)$ **27.** $(1,6)$ **29.** infinite number of solutions   **31.** $(-2,0)$

33. answers may vary   **35.** $(2,5)$ **37.** $(-3,2)$ **39.** $(0,3)$ **41.** $(5,7)$ **43.** $\left(\dfrac{1}{3},1\right)$ **45.** infinite number of solutions   **47.** $(-8.9,10.6)$

49. $2x+6=x-3$ **51.** $20-3x=2$ **53.** $4(n+6)=2n$ **55. a.** 2004
b. skiers: 6.5 million; snow boarders: 6.98 million; rounding in part a caused a difference   **57. a.** $b=15$ **b.** any real number except 15
59. a. answers may vary   **b.** answers may vary

Integrated Review   **1.** $(2,5)$ **2.** $(4,2)$ **3.** $(5,-2)$ **4.** $(6,-14)$ **5.** $(-3,2)$ **6.** $(-4,3)$ **7.** $(0,3)$ **8.** $(-2,4)$ **9.** $(5,7)$ **10.** $(-3,-23)$

11. $\left(\dfrac{1}{3},1\right)$ **12.** $\left(-\dfrac{1}{4},2\right)$ **13.** no solution   **14.** infinite number of solutions   **15.** $(0.5,3.5)$ **16.** $(-0.75,1.25)$ **17.** infinite number of solutions

18. no solution   **19.** answers may vary   **20.** answers may vary

Exercise Set 4.4   **1.** c   **3.** b   **5.** a   **7.** $\begin{cases} x+y=15 \\ x-y=7 \end{cases}$ **9.** $\begin{cases} x+y=6500 \\ x=y+800 \end{cases}$ **11.** 33 and 50   **13.** 14 and -3

15. Jackson: 501 points: Smith 489 points   **17.** child's ticket: \$18; adult's ticket: \$29   **19.** quarters: 53; nickels: 27
21. IBM: \$84.92; GA Financial: \$25.80   **23.** still water: 6.5 mph; current: 2.5 mph   **25.** still air: 455 mph; wind: 65 mph

27. 12% solution: $7\dfrac{1}{2}$ oz; 4% solution: $4\dfrac{1}{2}$ oz   **29.** \$4.95 beans: 113 lb; \$2.65 beans; 87 lb   **31.** $60°,30°$ **33.** $20°,70°$

35. 20% solution: 10 l; 70% solution: 40 l   **37.** number sold at \$9.50: 23; number sold at \$7.50: 67   **39.** width: 30 in.; length: 42 in.   **41.** $4\dfrac{1}{2}$ hr

43. westbound: 80 mph; eastbound: 40 mph   **45.** $2\dfrac{1}{4}$ mph and $2\dfrac{3}{4}$ mph   **47.** 30%: 50 gal; 60%: 100 gal   **49.** $(3,\infty)$ **51.** $\left[\dfrac{1}{2},\infty\right)$
53. width: 9ft; length: 15 ft   **55. a.** $(5.3,38.2)$
b. In 5.3 years after 1995, cable viewers equaled network viewers. The number of viewers for each was 38.2 million.
c. increasing: cable; decreasing: networks

Mental Math   **1.** yes   **3.** yes   **5.** yes   **7.** no

Exercise Set 4.5   **1.** no; yes   **3.** no; no   **5.** no; yes

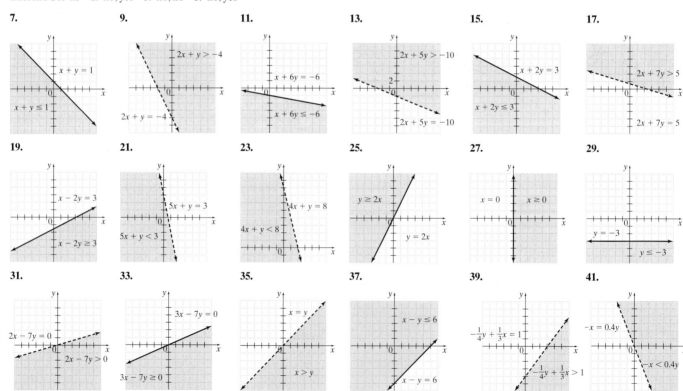

43. e   **45.** c   **47.** f   **49.** 8   **51.** -32 **53.** 48   **55.** 25   **57.** -2

59. $x + y \geq 13$

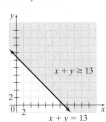
$x + y \geq 13$
$x + y = 13$

61. answers may vary

63.

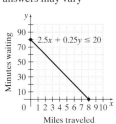
$2.5x + 0.25y \leq 20$
Minutes waiting
Miles traveled

65. answers may vary

67. a. $30x + 0.15y \leq 500$

b.

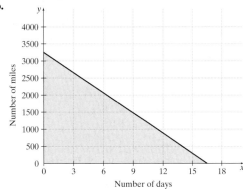
Number of miles
Number of days

Exercise Set 4.6

1.

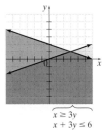
$y \geq x + 1$
$y \geq 3 - x$

3.

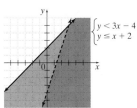
$\begin{cases} y < 3x - 4 \\ y \leq x + 2 \end{cases}$

5.

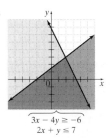
$y \leq -2x - 2$
$y \geq x + 4$

7.

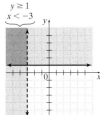
$y \geq -x + 2$
$y \leq 2x + 5$

9.

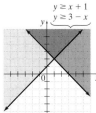
$x \geq 3y$
$x + 3y \leq 6$

11.

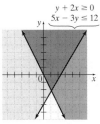
$y + 2x \geq 0$
$5x - 3y \leq 12$

13.

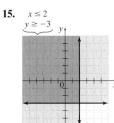
$3x - 4y \geq -6$
$2x + y \leq 7$

15.

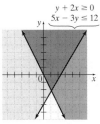
$x \leq 2$
$y \geq -3$

17.

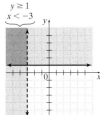
$y \geq 1$
$x < -3$

19.

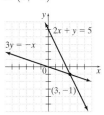
$2x + 3y < -8$
$x \geq -4$

21.

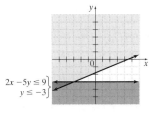
$\begin{cases} 2x - 5y \leq 9 \\ y \leq -3 \end{cases}$

23.

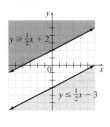
$y \geq \frac{1}{2}x + 2$
$y \leq \frac{1}{2}x - 3$

25. 16   **27.** $36x^2$ **29.** $100y^6$ **31.** C   **33.** D
35. answers may vary
37.

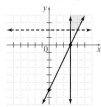

Chapter 4 Review   **1. a.** no   **b.** yes   **c.** no   **3. a.** no   **b.** no   **c.** yes

5. $(3, -1)$

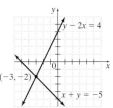
$2x + y = 5$
$3y = -x$
$(3, -1)$

7. $(-3, -2)$

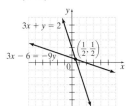
$y - 2x = 4$
$(-3, -2)$
$x + y = -5$

9. $\left(\frac{1}{2}, \frac{1}{2}\right)$

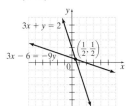
$3x + y = 2$
$3x - 6 = -9y$
$\left(\frac{1}{2}, \frac{1}{2}\right)$

11. intersecting, one solution **13.** identical, infinite number of solutions **15.** $(-1, 4)$ **17.** $(3, -2)$ **19.** no solution; inconsistent **21.** $(3, 1)$

23. $(8, -6)$ **25.** $(-6, 2)$ **27.** $(3, 7)$ **29.** infinite number of solutions; dependent **31.** $\left(2, -2\frac{1}{2}\right)$ **33.** $(-6, 15)$ **35.** $(-3, 1)$ **37.** -6 and 22

39. ship: 21.1 mph; current: 3.2 mph **41.** width: 1.15 ft; length: 1.85 ft **43.** one egg: \$0.40; one strip of bacon: \$0.65

45. 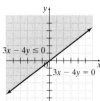 **47.** 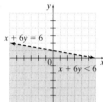 **49.** 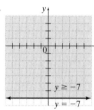 **51.**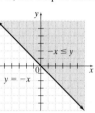

53. 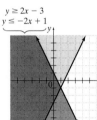 **55.** 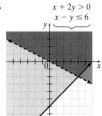 **57.** 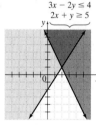 **59.**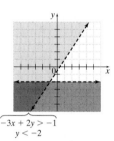

Chapter 4 Test **1.** false **2.** false **3.** true **4.** false **5.** no **6.** yes

7. $(-4, 2)$ 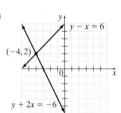 **8.** $(-4, 1)$ **9.** $\left(\frac{1}{2}, -2\right)$ **10.** $(4, -2)$ **11.** no solution

12. $(4, -5)$ **13.** $(7, 2)$ **14.** $(5, -2)$ **15.** 120 cc **16.** \$1225 at 5\%; \$2775 at 9\%

17. Texas: 226 thousand; Missouri: 110 thousand

18. 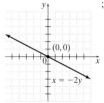 **19.** 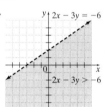 **20.** 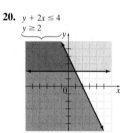 **21.**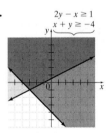

Chapter 4 Cumulative Review **1. a.** $<$ **b.** $=$ **c.** $>$; Sec. 1.2, Ex. 6 **3. a.** commutative property of multiplication **b.** associative property of addition **c.** identity element for addition **d.** commutative property of multiplication **e.** multiplicative inverse property **f.** additive inverse property **g.** commutative and associative properties of multiplication; Sec. 1.8, Ex. 6 **5.** $-2x - 1$; Sec. 2.1, Ex. 7 **7.** -7; Sec. 2.2, Ex. 4 **9.** 6; Sec. 2.3, Ex. 1

11. 12; Sec. 2.4, Ex. 3 **13.** 10; Sec. 2.5, Ex. 1 **15.** $x = \dfrac{y - b}{m}$; Sec. 2.6, Ex. 6 **17.** $[2, \infty)$; Sec. 2.8, Ex. 3

19. 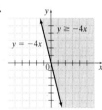 ; Sec. 3.3, Ex. 3 **21.** $-\dfrac{8}{3}$; Sec. 3.4, Ex. 1 **23.** $\dfrac{3}{4}$; Sec. 3.1, Ex. 1 **25.** slope: $\dfrac{3}{4}$; y-intercept: -1; Sec. 3.5, Ex. 3

27. $2x + y = 3$; Sec. 3.6, Ex. 1 **29.** $x = -1$; Sec. 3.6, Ex. 3 **31.** domain: $\{-1, 0, 3\}$; range: $\{-2, 0, 2, 3\}$; Sec. 3.7, Ex. 1

33. a. function **b.** not a function; Sec. 3.7, Ex. 2 **35.** one solution; Sec. 4.1, Ex. 6 **37.** $\left(6, \dfrac{1}{2}\right)$; Sec. 4.2, Ex. 2

39. $(6, 1)$; Sec. 4.3, Ex. 1 **41.** 29 and 8; Sec. 4.4, Ex. 1 **43.** 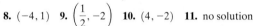 ; Sec. 4.5, Ex. 2

CHAPTER 5   EXPONENTS AND POLYNOMIALS

Mental Math   1. base: 3; exponent: 2   **3.** base: -3; exponent: 6   **5.** base: 4; exponent: 2   **7.** base: 5; exponent: 1; base: 3; exponent: 4   **9.** base: 5; exponent: 1; base: x; exponent: 2

Exercise Set 5.1   1. 49   **3.** -5 **5.** -16 **7.** 16   **9.** 0.00001   **11.** $\dfrac{1}{81}$ **13.** 224   **15.** -250 **17.** answers may vary   **19.** 4   **21.** 135   **23.** 150

25. $\dfrac{32}{5}$ **27.** 343 cu. m   **29.** volume   **31.** x^{10} **33.** $(-3)^{12}$ **35.** $15y^5$ **37.** $-24z^{20}$ **39.** p^7q^7 **41.** $\dfrac{m^9}{n^9}$ **43.** $x^{10}y^{15}$ **45.** $\dfrac{4x^2z^2}{y^{10}}$ **47.** x^2 **49.** 4

51. p^6q^5 **53.** $\dfrac{y^3}{2}$ **55.** 1   **57.** -2 **59.** 2   **61.** $-\dfrac{27a^6}{b^9}$ **63.** x^{39} **65.** $\dfrac{z^{14}}{625}$ **67.** $7776m^4n^3$ **69.** -25 **71.** $\dfrac{1}{64}$ **73.** $81x^2y^2$ **75.** 1   **77.** 40   **79.** b^6

81. a^9 **83.** $-16x^7$ **85.** $64a^3$ **87.** $36x^2y^2z^6$ **89.** $\dfrac{y^{15}}{8x^{12}}$ **91.** x **93.** $2x^2y$ **95.** $243x^7y^{24}$ **97.** answers may vary   **99.** $2y - 10$ **101.** $-x - 4$

103. $-x + 5$ **105.** $20x^5$ sq. ft   **107.** $25\pi y^2$ sq. cm   **109.** $27y^{12}$ cu. ft   **111.** x^{9a} **113.** x^{5a} **115.** $x^{5a}y^{5ab}z^{5ac}$ **117.** \$1045.85

Mental Math   1. $-14y$ **3.** $7y^3$ **5.** $7x$ **7.** $5m^2 + 2m$

Exercise Set 5.2   1. 1; binomial   **3.** 3; none of these   **5.** 6; trinomial   **7.** 2; binomial   **9.** 3   **11.** 2   **13.** answers may vary   **15.** answers may vary
17. a. 6   **b.** 5   **19. a.** -2 **b.** 4   **21. a.** -15 **b.** -16 **23.** -146 ft; the object has reached the ground.   **25.** $23x^2$ **27.** $12x^2 - y$

29. $7s$ **31.** $-1.1y^2 + 4.8$ **33.** $-\dfrac{7}{12}x^3 + \dfrac{7}{5}x^2 + 6$ **35.** $5a^2 - 9ab + 16b^2$ **37.** $12x + 12$ **39.** $-3x^2 + 10$ **41.** $-x^2 + 14$ **43.** $-2x + 9$

45. $2x^2 + 7x - 16$ **47.** $(x^2 + 7x + 4)$ ft   **49.** $(3y^2 + 4y + 11)$ m   **51.** $8t^2 - 4$ **53.** $-2z^2 - 16z + 6$ **55.** $2x^3 - 2x^2 + 7x + 2$
57. $62x^2 + 5$ **59.** $12x + 2$ **61.** $-11x$ **63.** $7x - 13$ **65.** $-y^2 - 3y - 1$ **67.** $2x^2 + 11x$ **69.** $-16x^4 + 8x + 9$ **71.** $7x^2 + 14x + 18$
73. $3x - 3$ **75.** $7x^2 - 2x + 2$ **77.** $4y^2 + 12y + 19$ **79.** $6x^2 - 5x + 21$ **81.** $3x^3 - x^2 + x + 6$ **83.** $6x^2$ **85.** $-12x^8$ **87.** $200x^3y^2$
89. $4x^2 + 7x + x^2 + 5x; 5x^2 + 12x$ **91.** $9x + 10 + 3x + 12 + 4x + 15 + 2x + 7; 18x + 44$ **93.** $2x^2 + 4xy$ **95.** $-2a - b + 1$ **97.** $3x^2 + 5$
99. $6x^2 - 2xy + 19y^2$ **101.** $8r^2s + 16rs - 8 + 7r^2s^2$ **103.** $-5.42x^2 + 7.75x - 19.61$ **105.** $3.7y^4 - 0.7y^3 + 2.2y - 4$ **107. a.** 184 ft   **b.** 600 ft
c. 595.84 ft   **d.** 362.56 ft   **109.** 1500   **111.** $0.05x^2 - 6.65x + 49.6$ **113.** 128.43 lb

Mental Math   1. $10xy$ **3.** x^7 **5.** $18x^3$ **7.** $-27x^5$ **9.** a^7 **11.** cannot simplify

Exercise Set 5.3   1. $4a^2 - 8a$ **3.** $7x^3 + 14x^2 - 7x$ **5.** $6x^4 - 3x^3$ **7.** $x^2 + 3x$ **9.** $a^2 + 5a - 14$ **11.** $4y^2 - 16y + 16$ **13.** $30x^2 - 79xy + 45y^2$
15. $4x^4 - 20x^2 + 25$ **17.** $x^2 + 5x + 6$ **19.** $x^3 - 5x^2 + 13x - 14$ **21.** $x^4 + 5x^3 - 3x^2 - 11x + 20$ **23.** $10a^3 - 27a^2 + 26a - 12$
25. $x^3 + 6x^2 + 12x + 8$ **27.** $8y^3 - 36y^2 + 54y - 27$ **29.** $2x^3 + 10x^2 + 11x - 3$ **31.** $x^5 - 4x^3 - 7x^2 - 45x + 63$ **33. a.** 25; 13   **b.** 324; 164
c. no; answers may vary   **35.** $2a^2 + 8a$ **37.** $6x^3 - 9x^2 + 12x$ **39.** $15x^2 + 37xy + 18y^2$ **41.** $x^3 + 7x^2 + 16x + 12$ **43.** $49x^2 + 56x + 16$
45. $-6a^4 + 4a^3 - 6a^2$ **47.** $x^3 + 10x^2 + 33x + 36$ **49.** $a^3 + 3a^2 + 3a + 1$ **51.** $x^2 + 2xy + y^2$ **53.** $x^2 - 13x + 42$ **55.** $3a^3 + 6a$
57. $-4y^3 - 12y^2 + 44y$ **59.** $25x^2 - 1$ **61.** $5x^3 - x^2 + 16x + 16$ **63.** $8x^3 - 60x^2 + 150x - 125$ **65.** $32x^3 + 48x^2 - 6x - 20$
67. $49x^2y^2 - 14xy + y^2$ **69.** $5y^4 - 16y^3 - 4y^2 - 7y - 6$ **71.** $6x^4 - 8x^3 - 7x^2 + 22x - 12$ **73.** $25x^2$ **75.** $9y^6$ **77.** \$7000   **79.** \$500
81. answers may vary   **83.** $(4x^2 - 25)$ sq. yd   **85.** $(6x^2 - 4x)$ sq. in.   **87.** $(x^2 + 6x + 5)$ sq. units   **89. a.** $a^2 - b^2$ **b.** $4x^2 - 9y^2$ **c.** $16x^2 - 49$
d. answers may vary

Mental Math   1. false   **3.** false

Exercise Set 5.4   1. $x^2 + 7x + 12$ **3.** $x^2 + 5x - 50$ **5.** $5x^2 + 4x - 12$ **7.** $4y^2 - 25y + 6$ **9.** $6x^2 + 13x - 5$ **11.** $x^2 - 4x + 4$
13. $4x^2 - 4x + 1$ **15.** $9a^2 - 30a + 25$ **17.** $25x^2 + 90x + 81$ **19.** answers may vary   **21.** $a^2 - 49$ **23.** $9x^2 - 1$ **25.** $9x^2 - \dfrac{1}{4}$ **27.** $81x^2 - y^2$
29. $4x^2 - 0.01$ **31.** $a^2 + 9a + 20$ **33.** $a^2 + 14a + 49$ **35.** $12a^2 - a - 1$ **37.** $x^2 - 4$ **39.** $9a^2 + 6a + 1$ **41.** $4x^3 - x^2y^4 + 4xy - y^5$
43. $x^3 - 3x^2 - 17x + 3$ **45.** $4a^2 - 12a + 9$ **47.** $25x^2 - 36z^2$ **49.** $x^{10} - 8x^5 + 15$ **51.** $x^2 - \dfrac{1}{9}$ **53.** $a^7 - 3a^3 + 11a^4 - 33$

55. $3x^2 - 12x + 12$ **57.** $6b^2 - b - 35$ **59.** $49p^2 - 64$ **61.** $\dfrac{1}{9}a^4 - 49$ **63.** $15x^4 - 5x^3 + 10x^2$ **65.** $4r^2 - 9s^2$ **67.** $9x^2 - 42xy + 49y^2$

69. $16x^2 - 25$ **71.** $64x^2 + 64x + 16$ **73.** $a^2 - \dfrac{1}{4}y^2$ **75.** $\dfrac{1}{25}x^2 - y^2$ **77.** $3a^3 + 2a^2 + 1$ **79.** $\dfrac{5b^5}{7}$ **81.** $-2a^{10}b^5$ **83.** $\dfrac{2y^8}{3}$ **85.** $\dfrac{1}{3}$ **87.** 1

89. $(4x^2 + 4x + 1)$ sq. ft   **91.** $\left(\dfrac{25}{2}a^2 - \dfrac{1}{2}b^2\right)$ sq. units   **93.** $(24x^2 - 32x + 8)$ sq. m   **95.** $x^2 + 10x + 25$ **97.** answers may vary
99. $a^2 + 2ac + c^2 - 25$ **101.** $x^2 - 4x + 4 - y^2$

Integrated Review   1. $35x^5$ **2.** $32y^9$ **3.** -16 **4.** 16   **5.** $2x^2 - 9x - 5$ **6.** $3x^2 + 13x - 10$ **7.** $3x - 4$ **8.** $4x + 3$ **9.** $7x^6y^2$ **10.** $\dfrac{10b^6}{7}$

11. $144m^{14}n^{12}$ **12.** $64y^{27}z^{30}$ **13.** $48y^2 - 27$ **14.** $98x^2 - 2$ **15.** $x^{63}y^{45}$ **16.** $27x^{27}$ **17.** $2x^2 - 2x - 6$ **18.** $6x^2 + 13x - 11$ **19.** $2.5y^2 - 6y - 0.2$
20. $8.4x^2 - 6.8x - 5.7$ **21.** $x^2 + 8xy + 16y^2$ **22.** $y^2 - 18yz + 81z^2$ **23.** $2x + 8y$ **24.** $2y - 18z$ **25.** $7x^2 - 10xy + 4y^2$ **26.** $-a^2 - 3ab + 6b^2$
27. $x^3 + 2x^2 - 16x + 3$ **28.** $x^3 - 2x^2 - 5x - 2$ **29.** $6x^5 + 20x^3 - 21x^2 - 70$ **30.** $20x^7 + 25x^3 - 4x^4 - 5$ **31.** $2x^3 - 19x^2 + 44x - 7$

32. $5x^3 + 9x^2 - 17x + 3$ **33.** cannot simplify   **34.** $25x^3y^3$ **35.** $125x^9$ **36.** $\dfrac{x^3}{y^3}$ **37.** $2x$ **38.** x^2

Calculator Explorations   1. 5.31 EE 3   **3.** 6.6 EE -9 **5.** 1.5×10^{13} **7.** 8.15×10^{19}

Mental Math   **1.** $\dfrac{5}{x^2}$ **3.** y^6 **5.** $4y^3$

Exercise Set 5.5   **1.** $\dfrac{1}{64}$ **3.** $\dfrac{1}{16}$ **5.** $\dfrac{7}{x^3}$ **7.** 32   **9.** -64 **11.** $\dfrac{5}{6}$ **13.** p^3 **15.** $\dfrac{q^4}{p^5}$ **17.** $\dfrac{1}{x^3}$ **19.** $\dfrac{4}{3}$ **21.** $-p^4$ **23.** -2 **25.** x^4 **27.** p^4 **29.** m^{11}

31. r^6 **33.** $\dfrac{1}{x^{15}y^9}$ **35.** $\dfrac{1}{x^4}$ **37.** $\dfrac{1}{a^2}$ **39.** $4k^3$ **41.** $3m$ **43.** $-\dfrac{4a^5}{b}$ **45.** $-\dfrac{6x^2}{y^3}$ **47.** $\dfrac{a^{30}}{b^{12}}$ **49.** $\dfrac{1}{x^{10}y^6}$ **51.** $\dfrac{z^2}{4}$ **53.** $\dfrac{1}{32x^5}$ **55.** $\dfrac{49a^4}{b^6}$ **57.** $a^{24}b^8$

59. x^9y^{19} **61.** $-\dfrac{y^8}{8x^2}$ **63.** $-\dfrac{6x}{7x^2}$ **65.** 7.8×10^4 **67.** 1.67×10^{-6} **69.** 6.35×10^{-3} **71.** 1.16×10^6 **73.** 2.0×10^7 **75.** 1.56×10^7

77. 1.36×10^4 **79.** 2.92×10^8 **81.** 0.0000000008673 **83.** 0.033 **85.** $20{,}320$ **87.** $6{,}250{,}000{,}000{,}000{,}000{,}000$ **89.** $9{,}460{,}000{,}000{,}000$

91. 0.000036 **93.** 0.0000000000000000028 **95.** 0.0000005 **97.** $200{,}000$ **99.** $\dfrac{5x^3}{3}$ **101.** $\dfrac{5z^3y^2}{7}$ **103.** $5y - 6 + \dfrac{5}{y}$ **105.** $\dfrac{27}{x^6z^3}$ cu. in.   **107.** -394.5

109. 1.3 sec   **111.** answers may vary   **113.** 2.7×10^9 gal   **115.** answers may vary   **117.** a^m **119.** y^{5a} **121.** $\dfrac{1}{z^{6a+4}}$

Mental Math   **1.** a^2 **3.** a^2 **5.** k^3 **7.** p^5 **9.** k^2

Exercise Set 5.6   **1.** $5p^2 + 6p$ **3.** $-\dfrac{3}{2x} + 3$ **5.** $-3x^2 + x - \dfrac{4}{x^3}$ **7.** $-1 + \dfrac{3}{2x} - \dfrac{7}{4x^4}$ **9.** $5x^3 - 3x + \dfrac{1}{x^2}$ **11.** $(3x^3 + x - 4)$ ft   **13.** $x + 1$

15. $2x + 3$ **17.** $2x + 1 + \dfrac{7}{x-4}$ **19.** $4x + 9$ **21.** $3a^2 - 3a + 1 + \dfrac{2}{3a+2}$ **23.** $2b^2 + b + 2 - \dfrac{12}{b+4}$ **25.** answers may vary   **27.** $(2x+5)$ m

29. $\dfrac{4}{x} + \dfrac{1}{x^2} + \dfrac{9}{5x^3}$ **31.** $5x - 2 + \dfrac{2}{x+6}$ **33.** $x^2 - \dfrac{12x}{5} - 1$ **35.** $6x - 1 - \dfrac{1}{x+3}$ **37.** $4x^2 + 1$ **39.** $2x^2 + 6x - 5 - \dfrac{2}{x-2}$ **41.** $6x - 1$

43. $-x^3 + 3x^2 - \dfrac{4}{x}$ **45.** $4x + 3 - \dfrac{2}{2x+1}$ **47.** $2x + 9$ **49.** $2x^2 + 3x - 4$ **51.** $x^2 + 3x + 9$ **53.** $x^2 - x + 1$ **55.** $-3x + 6 - \dfrac{11}{x+2}$

57. $2b - 1 - \dfrac{6}{2b-1}$ **59.** $2a^3 + 2a$ **61.** $2x^3 + 14x^2 - 10x$ **63.** $-3x^2y^3 - 21x^3y^2 - 24xy$ **65.** $9a^2b^3c + 36ab^2c - 72ab$
67. The Rolling Stones (1994)   **69.** \$110 million   **71.** $x^3 - x^2 + x$ **73.** $9x^{7a} - 6x^{5a} + 7x^{2a} - 1$

Chapter 5 Review   **1.** base: 3; exponent: 2   **3.** base: 5; exponent: 4   **5.** 36   **7.** -65 **9.** 1   **11.** 8   **13.** $-10x^5$ **15.** $\dfrac{b^4}{16}$ **17.** $\dfrac{x^6y^6}{4}$ **19.** $40a^{19}$

21. $\dfrac{3}{64}$ **23.** $\dfrac{1}{x}$ **25.** 5   **27.** 1   **29.** $6a^6b^9$ **31.** 7   **33.** 8   **35.** 5   **37.** 5   **39. a.** $3, -2, 1, -1, -6; 3, 2, 2, 2, 0$ **b.** 3   **41.** $22; 78; 154.02; 400$
43. $22x^2y^3 + 3xy + 6$ **45.** cannot be combined   **47.** $2s^5 + 3s^4 + 4s^3 + s^2 - 7s - 6$ **49.** $8r^2 + 11rs + 7s^2$ **51.** $4x - 13y$
53. $28{,}770{,}000$ trademark registrations   **55.** $-56xz^2$ **57.** $-12x^2a^2y^4$ **59.** $9x - 63$ **61.** $54a - 27$ **63.** $-32y^3 + 48y$
65. $-3a^3b - 3a^2b - 3ab^2$ **67.** $42b^4 - 28b^2 + 14b$ **69.** $6x^2 - 11x - 10$ **71.** $42a^2 + 11a - 3$ **73.** $x^6 + 2x^5 + x^2 + 3x + 2$
75. $x^6 + 8x^4 + 16x^2 - 16$ **77.** $8x^3 - 60x^2 + 150x - 125$ **79.** $15y^3 - 3y^2 + 6y$ **81.** $12a^5 + 28a^3 - 3a^2 - 7$ **83.** $x^6 - 10x^3y + 25y^2$
85. $16x^2 + 16x + 4$ **87.** $x^3 - 3x^2 + 4$ **89.** $25x^2 - 1$ **91.** $a^2 - 4b^2$ **93.** $16a^4 - 4b^2$ **95.** $-\dfrac{1}{49}$ **97.** $\dfrac{1}{16x^4}$ **99.** $\dfrac{9}{4}$ **101.** $\dfrac{1}{42}$ **103.** $-q^3r^3$

105. $\dfrac{r^5}{s^3}$ **107.** $\dfrac{12}{y}$ **109.** $\dfrac{y^6}{3}$ **111.** $\dfrac{x^6y^9}{64z^3}$ **113.** $\dfrac{y^8}{49}$ **115.** $\dfrac{x^3}{y^3}$ **117.** $\dfrac{a^{10}}{b^{10}}$ **119.** $-\dfrac{4}{27}$ **121.** x^{10+3b} **123.** a^{2m+5} **125.** 8.868×10^{-1} **127.** 8.68×10^5

129. 1.5×10^5 **131.** 0.00386 **133.** $893{,}600$ **135.** 0.0000000001 m   **137.** $400{,}000{,}000{,}000$ **139.** $-a^2 + 3b - 4$ **141.** $4x + \dfrac{7}{x+5}$

143. $3b^2 - 4b - \dfrac{1}{3b-2}$ **145.** $-x^2 - 16x - 117 - \dfrac{684}{x-6}$

Chapter 5 Test   **1.** 32   **2.** 81   **3.** -81 **4.** $\dfrac{1}{64}$ **5.** $-15x^{11}$ **6.** y^5 **7.** $\dfrac{1}{r^5}$ **8.** $\dfrac{y^{14}}{x^2}$ **9.** $\dfrac{1}{6xy^8}$ **10.** 5.63×10^5 **11.** 8.63×10^{-5} **12.** 0.0015
13. $62{,}300$ **14.** 0.036 **15. a.** $4, 7, 1, -2; 3, 3, 4, 0$ **b.** 4   **16.** $-2x^2 + 12xy + 11$ **17.** $16x^3 + 7x^2 - 3x - 13$ **18.** $-3x^3 + 5x^2 + 4x + 5$
19. $x^3 + 8x^2 + 3x - 5$ **20.** $3x^3 + 22x^2 + 41x + 14$ **21.** $6x^4 - 9x^3 + 21x^2$ **22.** $3x^2 + 16x - 35$ **23.** $16x^2 - 16x + 4$ **24.** $x^4 - 81b^2$
25. 1001 ft; 985 ft; 857 ft; 601 ft   **26.** $\dfrac{x}{2y} + 3 - \dfrac{7}{8y}$ **27.** $x + 2$ **28.** $9x^2 - 6x + 4 - \dfrac{16}{3x+2}$

Chapter 5 Cumulative Review   **1. a.** true   **b.** true   **c.** false   **d.** true; Sec. 1.2, Ex. 2   **3. a.** $\dfrac{64}{25}$ **b.** $\dfrac{1}{20}$ **c.** $\dfrac{5}{4}$; Sec. 1.3, Ex. 4   **5. a.** 9   **b.** 125   **c.** 16

d. 7   **e.** $\dfrac{9}{49}$; Sec. 1.4, Ex. 1   **7. a.** -10 **b.** -21 **c.** -12; Sec. 1.5, Ex. 3   **9.** -12; Sec. 1.6, Ex. 3   **11. a.** $\dfrac{1}{22}$ **b.** $\dfrac{16}{3}$ **c.** $-\dfrac{1}{10}$ **d.** $-\dfrac{13}{9}$; Sec. 1.7, Ex. 5
13. a. $(5+4)+6$ **b.** $-1 \cdot (2 \cdot 5)$; Sec. 1.8, Ex. 2   **15. a.** Alaska Village Electric   **b.** American Electric Power   **c.** American Electric Power: 5¢ per kilowatt-hour; Green Mountain Power: 11¢ per kilowatt-hour; Montana Power Co.: 6¢ per kilowatt-hour; Alaska Village Electric: 42¢ per kilowatt-hour   **d.** 37¢; Sec. 1.9, Ex. 1   **17. a.** $5x + 10$ **b.** $-2y - 0.6z + 2$ **c.** $-x - y + 2z - 6$; Sec. 2.1, Ex. 5   **19.** 17; Sec. 2.2, Ex. 1
21. 6; Sec. 2.3, Ex. 1   **23.** -10; Sec. 2.4, Ex. 8   **25.** 10; Sec. 2.5, Ex. 1   **27.** width: 4 ft; length: 10 ft; Sec. 2.6, Ex. 4   **29.** $\dfrac{5F - 160}{9} = C$; Sec. 2.6, Ex. 8
31. ◄———|———|———► ; Sec. 2.8, Ex. 9   2   4   **33. a.** $(0, 12)$ **b.** $(2, 6)$ **c.** $(-1, 15)$; Sec. 3.1, Ex. 4

35. 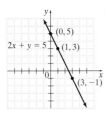 ; Sec. 3.2, Ex. 2   **37.** 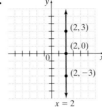 ; Sec. 3.3, Ex. 5   **39.** undefined slope; Sec. 3.4, Ex. 4

41. 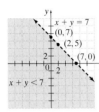 ;Sec. 4.5, Ex. 1   **43.** $-6x^7$; Sec. 5.1, Ex. 4   **45.** $-4x^2 + 6x + 2$; Sec. 5.2, Ex. 9   **47.** $4x^2 - 4xy + y^2$; Sec. 5.3, Ex. 6
49. $3m + 1$; Sec. 5.6, Ex. 1

CHAPTER 6   FACTORING POLYNOMIALS

Mental Math   1. $2 \cdot 7$ **3.** $2 \cdot 5$ **5.** 3   **7.** 3

Exercise Set 6.1   1. 4   **3.** 6   **5.** y^2 **7.** xy^2 **9.** 4   **11.** $4y^3$ **13.** $3x^3$ **15.** $9x^2y$ **17.** $15(2x - 1)$ **19.** $6cd(4d^2 - 3c)$ **21.** $-6a^3x(4a - 3)$
23. $4x(3x^2 + 4x - 2)$ **25.** $5xy(x^2 - 3x + 2)$ **27.** answers may vary   **29.** $(x + 2)(y + 3)$ **31.** $(y - 3)(x - 4)$ **33.** $(x + y)(2x - 1)$
35. $(x + 3)(5 + y)$ **37.** $(y - 4)(2 + x)$ **39.** $(y - 2)(3x + 8)$ **41.** $(y + 3)(y^2 + 1)$ **43.** $12x^3 - 2x; 2x(6x^2 - 1)$
45. $200x + 25\pi; 25(8x + \pi)$ **47.** $3(x - 2)$ **49.** $2x(16y - 9x)$ **51.** $4(x - 2y + 1)$ **53.** $(x + 2)(8 - y)$ **55.** $-8x^8y^5(5y + 2x)$
57. $-3(x - 4)$ **59.** $6x^3y^2(3y - 2 + x^2)$ **61.** $(x - 2)(y^2 + 1)$ **63.** $(y + 3)(5x + 6)$ **65.** $(x - 2y)(4x - 3)$ **67.** $42yz(3x^3 + 5y^3z^2)$
69. $(3 - x)(5 + y)$ **71.** $2(3x^2 - 1)(2y - 7)$ **73.** 2, 6   **75.** $-1, -8$ **77.** $-2, 5$ **79.** $-8, 3$ **81.** $x^2 + 7x + 10$ **83.** $a^2 - 15a + 56$
85. answers may vary   **87.** factored   **89.** not factored   **91.** $(n^3 - 6)$ units   **93.** $(x^n + 2)(x^n + 3)$ **95.** $(3x^n - 5)(x^n + 7)$
97. a. 44 million   **b.** 124 million   **c.** $4(2x^2 - 14x + 31)$

Mental Math   1. $+5$ **3.** -3 **5.** $+2$

Exercise Set 6.2   1. $(x + 6)(x + 1)$ **3.** $(x + 5)(x - 4)$ **5.** $(x - 5)(x - 3)$ **7.** $(x - 9)(x - 1)$ **9.** prime   **11.** $(x - 6)(x + 3)$ **13.** prime
15. $(x + 5y)(x + 3y)$ **17.** $(x - y)(x - y)$ **19.** $(x - 4y)(x + y)$ **21.** $2(z + 8)(z + 2)$ **23.** $2x(x - 5)(x - 4)$ **25.** $7(x + 3y)(x - y)$
27. product; sum   **29.** $(x + 12)(x + 3)$ **31.** $(x - 2)(x + 1)$ **33.** $(r - 12)(r - 4)$ **35.** $(x - 7)(x + 3)$ **37.** $(x + 5y)(x + 2y)$
39. prime   **41.** $2(t + 8)(t + 4)$ **43.** $x(x - 6)(x + 4)$ **45.** $(x - 9)(x - 7)$ **47.** $(x + 2y)(x - y)$ **49.** $3(x - 18)(x - 2)$
51. $(x - 24)(x + 6)$ **53.** $6x(x + 4)(x + 5)$ **55.** $2t^3(t - 4)(t - 3)$ **57.** $5xy(x - 8y)(x + 3y)$ **59.** $4y(x^2 + x - 3)$
61. $2b(a - 7b)(a - 3b)$ **63.** $2x^2 + 11x + 5$ **65.** $15y^2 - 17y + 4$ **67.** $9a^2 + 23a - 12$ **69.** 8; 16   **71.** 6; 26   **73.** 5; 8; 9   **75.** 3; 4
77. $2x^2 + 28x + 66; 2(x + 3)(x + 11)$ **79.** answers may vary   **81.** $2y(x + 5)(x + 10)$ **83.** $-12y^3(x^2 + 2x + 3)$ **85.** $(x + 1)(y - 5)(y + 3)$
87. $(x^n + 2)(x^n + 3)$

Mental Math   1. yes   **3.** no   **5.** yes

Exercise Set 6.3   1. $(2x + 3)(x + 5)$ **3.** $(2x + 1)(x - 5)$ **5.** $(2y + 3)(y - 2)$ **7.** $(4a - 3)^2$ **9.** $(9r - 8)(4r + 3)$ **11.** $(5x + 1)(2x + 3)$
13. $3(7x + 5)(x - 3)$ **15.** $2(2x - 3)(3x + 1)$ **17.** $x(4x + 3)(x - 3)$ **19.** $(x + 11)^2$ **21.** $(x - 8)^2$ **23.** $(4y - 5)^2$ **25.** $(xy - 5)^2$
27. answers may vary   **29.** $(2x + 11)(x - 9)$ **31.** $(2x - 7)(2x + 3)$ **33.** $(6x - 7)(5x - 3)$ **35.** $(4x - 9)(6x - 1)$ **37.** $(3x - 4y)^2$
39. $(x - 7y)^2$ **41.** $(2x + 5)(x + 1)$ **43.** prime   **45.** $(-2y + 5)(y + 2)$ **47.** $(4x + 3y)^2$ **49.** $2y(4x - 7)(x + 6)$ **51.** $(3x - 2)(x + 1)$
53. $(xy + 2)^2$ **55.** $(7y + 3x)^2$ **57.** $3(x^2 - 14x + 21)$ **59.** $(7a - 6)(6a - 1)$ **61.** $(6x - 7)(3x + 2)$ **63.** $(5p - 7q)^2$ **65.** $(5x + 3)(3x - 5)$
67. $(7t + 1)(t - 4)$ **69.** $a^2 + 2ab + b^2$ **71.** $x^2 - 4$ **73.** $a^3 + 27$ **75.** $y^3 - 125$ **77.** \$75,000 and above   **79.** answers may vary   **81.** 2; 14
83. 5; 13   **85.** 2   **87.** 4; 5   **89.** $-3xy^2(4x - 5)(x + 1)$ **91.** $-2pq(p - 3q)(15p + q)$ **93.** $(y - 1)^2(4x^2 + 10x + 25)$
95. $(3x^n + 2)(x^n + 5)$

Calculator Explorations

16	14	16
16	14	16
2.89	0.89	2.89
171.61	169.61	171.61
1	-1	1

Mental Math   1. 1^2 **3.** 9^2 **5.** 3^2 **7.** 1^3 **9.** 2^3

Exercise Set 6.4   1. $(x + 2)(x - 2)$ **3.** $(y + 7)(y - 7)$ **5.** $(5y + 3)(5y - 3)$ **7.** $(11 + 10x)(11 - 10x)$ **9.** $3(2x + 3)(2x - 3)$
11. $(13a + 7b)(13a - 7b)$ **13.** $(xy + 1)(xy - 1)$ **15.** $(x^2 + 3)(x^2 - 3)$ **17.** $(7a^2 + 4)(7a^2 - 4)$ **19.** $(x^2 + y^5)(x^2 - y^5)$ **21.** $(x + 6)$
23. $(a + 3)(a^2 - 3a + 9)$ **25.** $(2a + 1)(4a^2 - 2a + 1)$ **27.** $5(k + 2)(k^2 - 2k + 4)$ **29.** $(xy - 4)(x^2y^2 + 4xy + 16)$
31. $(x + 5)(x^2 - 5x + 25)$ **33.** $3x(2x - 3y)(4x^2 + 6xy + 9y^2)$ **35.** $(2x + y)$ **37.** $(x + 11)(x - 11)$ **39.** $(9 + p)(9 - p)$
41. $(2r + 1)(2r - 1)$ **43.** $(3x + 4)(3x - 4)$ **45.** prime   **47.** $(3 - t)(9 + 3t + t^2)$ **49.** $8(r - 2)(r^2 + 2r + 4)$ **51.** $(t - 7)(t^2 + 7t + 49)$
53. $(x + 13y)(x - 13y)$ **55.** $(xy + z)(xy - z)$ **57.** $(xy + 1)(x^2y^2 - xy + 1)$ **59.** $(s - 4t)(s^2 + 4st + 16t^2)$ **61.** $2(3r + 2)(3r - 2)$
63. $x(3y + 2)(3y - 2)$ **65.** $25y^2(y + 2)(y - 2)$ **67.** $xy(x + 2y)(x - 2y)$ **69.** $4s^3t^3(2s^3 + 25t^3)$ **71.** $xy^2(27xy - 1)$ **73.** $4x^3 + 2x^2 - 1 + \dfrac{3}{x}$
75. $2x + 1$ **77.** $3x + 4 - \dfrac{2}{x + 3}$ **79.** $(x + 2 + y)(x + 2 - y)$ **81.** $(a + 4)(a - 4)(b - 4)$ **83.** $(x + 3 + 2y)(x + 3 - 2y)$
85. $(x^n + 10)(x^n - 10)$ **87. a.** 777 ft   **b.** 441 ft   **c.** 7 sec   **d.** $(29 + 4t)(29 - 4t)$ **89. a.** 1300 ft   **b.** 660 ft   **c.** 10 sec   **d.** $4(19 + 2t)(19 - 2t)$
91. answer may vary

Integrated Review   1. $(a + b)^2$ **2.** $(a - b)^2$ **3.** $(a - 3)(a + 4)$ **4.** $(a - 5)(a - 2)$ **5.** $(a + 2)(a - 3)$ **6.** $(a + 1)^2$ **7.** $(x + 1)^2$
8. $(x + 2)(x - 1)$ **9.** $(x + 1)(x + 3)$ **10.** $(x + 3)(x - 2)$ **11.** $(x + 3)(x + 4)$ **12.** $(x + 4)(x - 3)$ **13.** $(x + 4)(x - 1)$
14. $(x - 5)(x - 2)$ **15.** $(x + 5)(x - 3)$ **16.** $(x + 6)(x + 5)$ **17.** $(x - 6)(x + 5)$ **18.** $(x + 8)(x + 3)$ **19.** $2(x + 7)(x - 7)$
20. $3(x + 5)(x - 5)$ **21.** $(x + 3)(x + y)$ **22.** $(y - 7)(3 + x)$ **23.** $(x + 8)(x - 2)$ **24.** $(x - 7)(x + 4)$ **25.** $4x(x + 7)(x - 2)$
26. $6x(x - 5)(x + 4)$ **27.** $2(3x + 4)(2x + 3)$ **28.** $(2a - b)(4a + 5b)$ **29.** $(2a + b)(2a - b)$ **30.** $(4 - 3x)(7 + 2x)$ **31.** $(5 - 2x)(4 + x)$
32. prime   **33.** prime   **34.** $(3y + 5)(2y - 3)$ **35.** $(4x - 5)(x + 1)$ **36.** $y(x + y)(x - y)$ **37.** $4(t^2 + 9)$ **38.** $(x + 1)(x + y)$
39. $(x + 1)(a + 2)$ **40.** $9x(2x^2 - 7x + 1)$ **41.** $4a(3a^2 - 6a + 1)$ **42.** $(x + 16)(x - 2)$ **43.** prime   **44.** $(4a - 7b)^2$ **45.** $(5p - 7q)^2$
46. $(7x + 3y)(x + 3y)$ **47.** $(5 - 2y)(25 + 10y + 4y^2)$ **48.** $(4x + 3)(16x^2 - 12x + 9)$ **49.** $-(x - 5)(x + 6)$ **50.** $-(x - 2)(x - 4)$
51. $(7 - x)(2 + x)$ **52.** $(3 + x)(1 - x)$ **53.** $3x^2y(x + 6)(x - 4)$ **54.** $2xy(x + 5y)(x - y)$ **55.** $5xy^2(x - 7y)(x - y)$
56. $4x^2y(x - 5)(x + 3)$ **57.** $3xy(4x^2 + 81)$ **58.** $2xy^2(3x^2 + 4)$ **59.** $(2 + x)(2 - x)$ **60.** $(3 + y)(3 - y)$ **61.** $(s + 4)(3r - 1)$
62. $(x - 2)(x^2 + 3)$ **63.** $(4x - 3)(x - 2y)$ **64.** $(2x - y)(2x + 7z)$ **65.** $6(x + 2y)(x + y)$ **66.** $2(x + 4y)(6x - y)$
67. $(x + 3)(y + 2)(y - 2)$ **68.** $(y + 3)(y - 3)(x^2 + 3)$ **69.** $(5 + x)(x + y)$ **70.** $(x - y)(7 + y)$ **71.** $(7t - 1)(2t - 1)$ **72.** prime
73. $(3x + 5)(x - 1)$ **74.** $(7x - 2)(x + 3)$ **75.** $(x + 12y)(x - 3y)$ **76.** $(3x - 2y)(x + 4y)$ **77.** $(1 - 10ab)(1 + 2ab)$
78. $(1 + 5ab)(1 - 12ab)$ **79.** $(3 + x)(3 - x)(1 + x)(1 - x)$ **80.** $(3 + x)(3 - x)(2 + x)(2 - x)$ **81.** $(x + 4)(x - 4)(x^2 + 2)$
82. $(x + 5)(x - 5)(x^2 + 3)$ **83.** $(x - 15)(x - 8)$ **84.** $(y + 16)(y + 6)$ **85.** $2x(3x - 2)(x - 4)$ **86.** $2y(3y + 5)(y - 3)$
87. $(3x - 5y)(9x^2 + 15xy + 25y^2)$ **88.** $(6y - z)(36y^2 + 6yz + z^2)$ **89.** $(xy + 2z)(x^2y^2 - 2xyz + 4z^2)$ **90.** $(3ab + 2)(9a^2b^2 - 6ab + 4)$
91. $2xy(1 + 6x)(1 - 6x)$ **92.** $2x(x + 3)(x - 3)$ **93.** $(x + 2)(x - 2)(x + 6)$ **94.** $(x - 2)(x + 6)(x - 6)$ **95.** $2a^2(3a + 5)$
96. $2n(2n - 3)$ **97.** $(a^2 + 2)(a + 2)$ **98.** $(a - b)(1 + x)$ **99.** $(x + 2)(x - 2)(x + 7)$ **100.** $(a + 3)(a - 3)(a + 5)$
101. $(x - y + z)(x - y - z)$ **102.** $(x + 2y + 3)(x + 2y - 3)$ **103.** $(9 + 5x + 1)(9 - 5x - 1)$ **104.** $(b + 4a + c)(b - 4a - c)$
105. answers may vary   **106.** yes; $9(x^2 + 9y^2)$ **107.** a, c

Calculator Explorations   1. $-0.9, 2.2$ **3.** no real solution   **5.** $-1.8, 2.8$

Mental Math   1. $3, 7$ **3.** $-8, -6$ **5.** $-1, 3$

Exercise Set 6.5   1. $2, -1$ **3.** $0, -6$ **5.** $-\dfrac{3}{2}, \dfrac{5}{4}$ **7.** $\dfrac{7}{2}, -\dfrac{2}{7}$ **9.** $(x - 6)(x + 1) = 0$ **11.** $9, 4$ **13.** $-4, 2$ **15.** $8, -4$ **17.** $\dfrac{7}{3}, -2$ **19.** $\dfrac{8}{3}, -9$
21. $x^2 - 12x + 35 = 0$ **23.** $0, 8, 4$ **25.** $\dfrac{3}{4}$ **27.** $0, \dfrac{1}{2}, -\dfrac{1}{2}$ **29.** $0, \dfrac{1}{2}, -\dfrac{3}{8}$ **31.** $0, -7$ **33.** $-5, 4$ **35.** $-5, 6$ **37.** $-\dfrac{4}{3}, 5$ **39.** $-\dfrac{3}{2}, -\dfrac{1}{2}, 3$
41. $-5, 3$ **43.** $0, 16$ **45.** $-\dfrac{9}{2}, \dfrac{8}{3}$ **47.** $\dfrac{5}{4}, \dfrac{11}{3}$ **49.** $-\dfrac{2}{3}, \dfrac{1}{6}$ **51.** $-\dfrac{5}{3}, \dfrac{1}{2}$ **53.** $\dfrac{17}{2}$ **55.** $2, -\dfrac{4}{5}$ **57.** $-4, 3$ **59.** $\dfrac{1}{2}, -\dfrac{1}{2}$ **61.** $-2, -11$ **63.** 3
65. -3 **67.** $\dfrac{3}{4}, -\dfrac{4}{3}$ **69.** $0, \dfrac{5}{2}, -1$ **71.** $-\dfrac{4}{3}, 1$ **73.** $-2, 5$ **75.** $-6, \dfrac{1}{2}$ **77.** e   **79.** b   **81.** c   **83.** $\dfrac{47}{45}$ **85.** $\dfrac{17}{60}$ **87.** $\dfrac{15}{8}$ **89.** $\dfrac{7}{10}$
91. a. $300; 304; 276; 216; 124; 0; -156$ **b.** 5 sec   **c.** 304 ft   **d.**

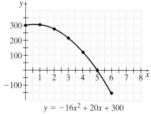

answers may vary

$$y = -16x^2 + 20x + 300$$

93. $0, \dfrac{1}{2}$ **95.** $0, -15$

Exercise Set 6.6   1. width $= x$; length $= x + 4$ **3.** x and $x + 2$ if x is an odd integer   **5.** base $= x$; height $= 4x + 1$ **7.** 11 units
9. 15 cm, 13 cm, 70 cm, 22 cm   **11.** base $= 16$ mi; height $= 6$ mi   **13.** 5 sec   **15.** length $= 5$ cm; width $= 6$ cm   **17.** 54 diagonals   **19.** 10 sides
21. -12 or 11   **23.** slow boat: 8 mph; fast boat: 15 mph   **25.** 13 and 7   **27.** 5 in.   **29.** 12 mm, 16 mm, 20 mm   **31.** 10 km   **33.** 36 ft   **35.** 6.25 sec
37. 20%   **39.** length: 15 mi; width: 8 mi   **41.** 105 units   **43.** 175 acres   **45.** 6.25 million   **47.** 1966   **49.** answers may vary   **51.** $\dfrac{3}{4}$ **53.** $\dfrac{5}{9}$
55. $\dfrac{9}{10}$ **57.** answers may vary   **59.** 3 yd

Chapter 6 Review   1. $2x - 5$ **3.** $4x(5x + 3)$ **5.** $-2x^2y(4x - 3y)$ **7.** $(x + 1)(5x - 1)$ **9.** $(2x - 1)(3x + 5)$ **11.** $(x + 4)(x + 2)$ **13.** prime
15. $(x + 4)(x - 2)$ **17.** $(x + 5y)(x + 3y)$ **19.** $2(3 - x)(12 + x)$ **21.** $(2x - 1)(x + 6)$ **23.** $(2x + 3)(2x - 1)$ **25.** $(6x - y)(x - 4y)$

27. $(2x + 3y)(x - 13y)$ **29.** $(6x + 5y)(3x - 4y)$ **31.** $(2x + 3)(2x - 3)$ **33.** prime **35.** $(2x + 3)(4x^2 - 6x + 9)$
37. $2(3 - xy)(9 + 3xy + x^2y^2)$ **39.** $(4x^2 + 1)(2x + 1)(2x - 1)$ **41.** $(2x - 3)(x + 4)$ **43.** $(x - 1)(x + 3)$ **45.** $2xy(2x - 3y)$
47. $(5x + 3)(25x^2 - 15x + 9)$ **49.** $(x + 7 + y)(x + 7 - y)$ **51.** $-6, 2$ **53.** $-\dfrac{1}{5}, -3$ **55.** $-4, 6$ **57.** $2, 8$. **59.** $-\dfrac{2}{7}, \dfrac{3}{8}$ **61.** $-\dfrac{2}{5}$ **63.** 3
65. $0, -\dfrac{7}{4}, 3$ **67.** 36 yd **69. a.** 17.5 sec and 10 sec; The rocket reaches a height of 2800 ft on its way up and on its way back down. **b.** 27.5 sec

Chapter 6 Test **1.** $(y - 12)(y + 4)$ **2.** prime **3.** $3x(3x + 1)(x + 4)$ **4.** $(3a - 7)(a + b)$ **5.** $(3x - 2)(x - 1)$ **6.** $(x + 12y)(x + 2y)$
7. $5(6 + x)(6 - x)$ **8.** $(6t + 5)(t - 1)$ **9.** $(y + 2)(y - 2)(x - 7)$ **10.** $x(1 + x^2)(1 + x)(1 - x)$ **11.** $-xy(y^2 + x^2)$
12. $(4x - 1)(16x^2 + 4x + 1)$ **13.** $8(y - 2)(y^2 + 2y + 4)$ **14.** $-7, 2$ **15.** $-7, 1$ **16.** $0, \dfrac{3}{2}, -\dfrac{4}{3}$ **17.** $0, 3, -3$ **18.** $-3, 5$ **19.** $0, \dfrac{5}{2}$ **20.** 17 ft
21. 8 and 9 **22.** 7 sec

Chapter 6 Cumulative Review **1. a.** $9 \le 11$ **b.** $8 > 1$ **c.** $3 \ne 4$; Sec. 1.2, Ex. 3 **3. a.** $\dfrac{6}{7}$ **b.** $\dfrac{11}{27}$ **c.** $\dfrac{22}{5}$; Sec. 1.3, Ex. 2 **5.** $\dfrac{14}{3}$; Sec. 1.4, Ex. 5
7. a. -12 **b.** -1; Sec. 1.5, Ex. 7 **9. a.** -24 **b.** -2 **c.** 50; Sec. 1.7, Ex. 1 **11. a.** $4x$ **b.** $11y^2$ **c.** $8x^2 - x$; Sec. 2.1, Ex. 3
13. -4; Sec. 2.2, Ex. 7 **15.** -11; Sec. 2.3, Ex. 2 **17.** $\dfrac{16}{3}$; Sec. 2.4, Ex. 2 **19.** shorter: 2 ft; longer: 8 ft; Sec. 2.5, Ex. 3
21.

;Sec. 3.3, Ex. 5 **23.** yes; Sec. 3.4, Ex. 5 **25. a.** 250 **b.** 1; Sec. 5.1, Ex. 2 **27. a.** 2 **b.** 5 **c.** 0; Sec. 5.2, Ex. 1
29. $9x^2 - 6x - 1$; Sec. 5.2, Ex. 12 **31.** $6x^2 - 11x - 10$; Sec. 5.3, Ex. 5 **33.** $9y^2 + 6y + 1$; Sec. 5.4, Ex. 4 **35. a.** $\dfrac{1}{9}$
b. $\dfrac{2}{x^3}$ **c.** $\dfrac{3}{4}$ **d.** $\dfrac{1}{16}$; Sec. 5.5, Ex. 1 **37. a.** 3.67×10^6 **b.** 3.0×10^{-6} **c.** 2.052×10^{10}
d. 8.5×10^{-4}; Sec. 5.5, Ex. 5 **39.** $x + 4$; Sec. 5.6, Ex. 4 **41. a.** x^3 **b.** y; Sec. 6.1, Ex. 2
43. $(x + 3)(x + 4)$; Sec. 6.2, Ex. 1 **45.** $(4x - 1)(2x - 5)$; Sec. 6.3, Ex. 2 **47.** $(5a + 3b)(5a - 3b)$; Sec. 6.4, Ex. 2b
49. $3, -1$; Sec. 6.6, Ex. 1

CHAPTER 7 RATIONAL EXPRESSIONS

Mental Math **1.** $x = 0$ **3.** $x = 0, x = 1$ **5.** no **7.** yes

Exercise Set 7.1 **1.** $\dfrac{7}{4}$ **3.** $\dfrac{13}{3}$ **5.** $-\dfrac{11}{2}$ **7.** $\dfrac{7}{4}$ **9.** $-\dfrac{8}{3}$ **11.** $y = 0$ **13.** $x = -2$ **15.** $x = 4$ **17.** $x = -2$ **19.** $x = -1, x = 0, x = 1$ **21.** none
23. answers may vary **25.** $\dfrac{2}{x^4}$ **27.** $\dfrac{5}{x + 1}$ **29.** -5 **31.** $\dfrac{1}{x - 9}$ **33.** $5x + 1$ **35.** $\dfrac{x + 4}{2x + 1}$ **37.** answers may vary **39.** -1 **41.** $-\dfrac{y}{2}$ **43.** $\dfrac{2 - x}{x + 2}$
45. $-\dfrac{3y^5}{x^4}$ **47.** $\dfrac{x - 2}{5}$ **49.** -6 **51.** $\dfrac{x + 2}{2}$ **53.** $\dfrac{11x}{6}$ **55.** $\dfrac{1}{x - 2}$ **57.** $x + 2$ **59.** $\dfrac{x + 1}{x - 1}$ **61.** $\dfrac{m - 3}{m + 3}$ **63.** $\dfrac{2(a - 3)}{a + 3}$ **65.** -1 **67.** $-x - 1$
69. $\dfrac{x + 5}{x - 5}$ **71.** $\dfrac{x + 2}{x + 4}$ **73.** $\dfrac{3}{11}$ **75.** $\dfrac{4}{3}$ **77.** $\dfrac{50}{99}$ **79.** $\dfrac{117}{40}$ **81.** $x + y$ **83.** $\dfrac{5 - y}{2}$ **85.** $x^2 - 2x + 4$ **87.** $-x^2 - x - 1$ **89. a.** \$37.5 million
b. \$85.7 million **c.** \$48.2 million **91.** 400 mg **93.** no; $B \approx 24$ **95.** a, c; answers may vary **97.** 58.3%
99.

101.

Mental Math **1.** $\dfrac{2x}{3y}$ **3.** $\dfrac{5y^2}{7x^2}$ **5.** $\dfrac{9}{5}$

Exercise Set 7.2 **1.** x^4 **3.** $-\dfrac{b^2}{6}$ **5.** $\dfrac{4x}{7(x + 2)}$ **7.** $\dfrac{x^2}{10}$ **9.** $\dfrac{1}{3}$ **11.** 1 **13.** $\dfrac{x + 5}{x}$ **15.** $\dfrac{2}{9x^2(x - 5)}$ sq. ft **17.** x^4 **19.** $\dfrac{12}{y^6}$ **21.** $x(x + 4)$
23. $\dfrac{3(x + 1)}{x^3(x - 1)}$ **25.** $m^2 - n^2$ **27.** $-\dfrac{x + 2}{x - 3}$ **29.** $-\dfrac{x + 2}{x - 3}$ **31.** answers may vary **33.** $\dfrac{1}{6b^4}$ **35.** $\dfrac{9}{7x^2y^7}$ **37.** $\dfrac{5}{6}$ **39.** $\dfrac{3x}{8}$ **41.** $\dfrac{3}{2}$ **43.** $\dfrac{3x + 4y}{2(x + 2y)}$
45. $-2(x + 3)$ **47.** $\dfrac{2(x + 2)}{x - 2}$ **49.** $\dfrac{(a + 5)(a + 3)}{(a + 2)(a + 1)}$ **51.** $-\dfrac{1}{x}$ **53.** $\dfrac{2(x + 3)}{x - 4}$ **55.** $-(x^2 + 2)$ **57.** -1 **59.** $4x^3(x - 3)$ **61.** 1440
63. 411,972 sq. yd **65.** 73 **67.** 3424.8 mph **69.** 1 **71.** $-\dfrac{10}{9}$ **73.** $-\dfrac{1}{5}$

75.

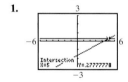

$x - 2y = 6$

77. $\dfrac{(a + b)^2}{a - b}$ **79.** $\dfrac{3x + 5}{x^2 + 4}$ **81.** $\dfrac{4}{x - 2}$ **83.** $\dfrac{a - b}{6(a^2 + ab + b^2)}$ **85.** $\dfrac{x}{2}$ **87.** $\dfrac{5a(2a + b)(3a - 2b)}{b^2(a - b)(a + 2b)}$

89. 1834.86 euros

Mental Math 1. 1 **3.** $\dfrac{7x}{9}$ **5.** $\dfrac{1}{9}$ **7.** $\dfrac{7 - 10y}{5}$

Exercise Set 7.3 1. $\dfrac{a + 9}{13}$ **3.** $\dfrac{y + 10}{3 + y}$ **5.** $\dfrac{3m}{n}$ **7.** $\dfrac{5x + 7}{x - 3}$ **9.** $\dfrac{1}{2}$ **11.** 4 **13.** $x + 5$ **15.** $x + 4$ **17.** 1 **19.** $\dfrac{12}{x^3}$ **21.** $-\dfrac{5}{x + 4}$ **23.** $\dfrac{x - 2}{x + y}$

25. 4 **27.** 3 **29.** $\dfrac{1}{a + 5}$ **31.** $\dfrac{1}{x - 6}$ **33.** $\dfrac{20}{x - 2}$m **35.** answers may vary **37.** 33 **39.** $4x^3$ **41.** $8x(x + 2)$ **43.** $6(x + 1)^2$ **45.** $8 - x$ or $x - 8$

47. $40x^3(x - 1)^2$ **49.** $(2x + 1)(2x - 1)$ **51.** $(2x - 1)(x + 4)(x + 3)$ **53.** answers may vary **55.** $\dfrac{6x}{4x^2}$ **57.** $\dfrac{24b^2}{12ab^2}$ **59.** $\dfrac{18}{2(x + 3)}$ **61.** $\dfrac{9ab + 2b}{5b(a + 2)}$

63. $\dfrac{x^2 + x}{(x + 4)(x + 2)(x + 1)}$ **65.** $\dfrac{18y - 2}{30x^2 - 60}$ **67.** $\dfrac{15x(x - 7)}{3x(2x + 1)(x - 7)(x - 5)}$ **69.** $\dfrac{29}{21}$ **71.** $-\dfrac{5}{12}$ **73.** 0, 3 **75.** $-5, -1$ **77.** c **79.** b

81. $-\dfrac{5}{x - 2}$ **83.** $\dfrac{7 + x}{x - 2}$ **85.** 95,304 Earth days **87.** answers may vary

Mental Math 1. D **3.** A

Exercise Set 7.4 1. $\dfrac{5}{x}$ **3.** $\dfrac{75a + 6b^2}{5b}$ **5.** $\dfrac{6x + 5}{2x^2}$ **7.** $\dfrac{21}{2(x + 1)}$ **9.** $\dfrac{17x + 30}{2(x - 2)(x + 2)}$ **11.** $\dfrac{35x - 6}{4x(x - 2)}$ **13.** $\dfrac{5 + 10y - y^3}{y^2(2y + 1)}$ **15.** answers may vary **17.** $-\dfrac{2}{x - 3}$

19. $-\dfrac{1}{x^2 - 1}$ **21.** $\dfrac{1}{x - 2}$ **23.** $\dfrac{5 + 2x}{x}$ **25.** $\dfrac{6x - 7}{x - 2}$ **27.** $-\dfrac{y + 4}{y + 3}$ **29.** $\left(\dfrac{90x - 40}{x}\right)^\circ$ **31.** 2 **33.** $3x^3 - 4$ **35.** $\dfrac{x + 2}{(x + 3)^2}$ **37.** $\dfrac{9b - 4}{5b(b - 1)}$

39. $\dfrac{2 + m}{m}$ **41.** $\dfrac{10}{1 - 2x}$ **43.** $\dfrac{15x - 1}{(x + 1)^2(x - 1)}$ **45.** $\dfrac{x^2 - 3x - 2}{(x - 1)^2(x + 1)}$ **47.** $\dfrac{a + 2}{2(a + 3)}$ **49.** $\dfrac{x - 10}{2(x - 2)}$ **51.** $\dfrac{-3 - 2y}{(y - 1)(y - 2)}$ **53.** $\dfrac{-5x + 23}{(x - 3)(x - 2)}$

55. $\dfrac{12x - 32}{(x + 2)(x - 2)(x - 3)}$ **57.** $\dfrac{6x + 9}{(3x - 2)(3x + 2)}$ **59.** $\dfrac{2x^2 - 2x - 46}{(x + 1)(x - 6)(x - 5)}$ **61.** $\dfrac{2}{3}$ **63.** $-\dfrac{1}{2}, 1$ **65.** 1 **67.** -1

69. $\dfrac{10(y - 2)}{y(y - 5)}$ ft; $\dfrac{6}{y(y^2 - 5y)}$ sq. ft **71.** 10 **73.** 2 **75.** $\dfrac{25a}{9(a - 2)}$ **77.** $\dfrac{x + 4}{(x - 2)(x - 1)}$ **79.** answers may vary

81. $\dfrac{4x^2 - 15x + 6}{(x - 2)^2(x + 2)(x - 3)}$ **83.** $\dfrac{2(x + 1)}{(x - 3)(x^2 + 3x + 9)}$ **85.** $\dfrac{11DA - DA^2 - 12D}{24(A + 12)}$

Calculator Explorations

1.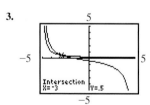

3.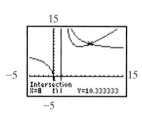

5.

Mental Math

1. $x = 10$ **3.** $z = 36$

Exercise Set 7.5 1. 30 **3.** 0 **5.** $-\dfrac{14}{19}$ **7.** $-5, 2$ **9.** 5 **11.** 3 **13.** 0 **15.** $\dfrac{1}{4}$ **17.** no solution **19.** $-1, 12$ **21.** 5 **23.** $-\dfrac{10}{9}$ **25.** $\dfrac{11}{14}$

27. no solution **29.** $-\dfrac{21}{11}$ **31.** 1 **33.** $\dfrac{9}{2}$ **35.** no solution **37.** $-\dfrac{9}{4}$ **39.** $-\dfrac{3}{2}, 4$ **41.** -2 **43.** $\dfrac{12}{5}$ **45.** $-2, 8$ **47.** $\dfrac{1}{5}$ **49.** $R = \dfrac{D}{T}$

51. $y = \dfrac{3x + 6}{5x}$ **53.** $b = -\dfrac{3a^2 + 2a - 4}{6}$ **55.** $B = \dfrac{2A}{H}$ **57.** $r = \dfrac{C}{2\pi}$ **59.** $a = \dfrac{bc}{c + b}$ **61.** $n = \dfrac{m^2 - 3p}{2}$ **63.** $(2, 0), (0, -2)$

65. $(-4, 0), (-2, 0), (3, 0), (0, 4)$ **67.** answers may vary **69.** $\dfrac{5x + 9}{9x}$ **71.** no solution **73.** $100°, 80°$ **75.** $22.5°, 67.5°$ **77.** $\dfrac{17}{4}$

Integrated Review 1. expression; $\dfrac{3 + 2x}{3x}$ **2.** expression; $\dfrac{18 + 5a}{6a}$ **3.** equation; 3 **4.** equation; 18 **5.** expression; $\dfrac{x + 1}{x(x - 1)}$

6. expression; $\dfrac{3(x + 1)}{x(x - 3)}$ **7.** equation; no solution **8.** equation; 1 **9.** expression; 10 **10.** expression; $\dfrac{z}{3(9z - 5)}$ **11.** expression; $\dfrac{5x + 7}{x - 3}$

12. expression; $\dfrac{7p + 5}{2p + 7}$ **13.** equation; 23   **14.** equation; 5   **15.** expression; $\dfrac{25a}{9(a - 2)}$ **16.** expression; $\dfrac{4x + 5}{(x + 1)(x - 1)}$

17. expression; $\dfrac{3x^2 + 5x + 3}{(3x - 1)^2}$ **18.** expression; $\dfrac{2x^2 - 3x - 1}{(2x - 5)^2}$ **19.** expression; $\dfrac{4x - 37}{5x}$ **20.** equation; $-\dfrac{7}{3}$ **21.** equation; $\dfrac{8}{5}$

22. expression; $\dfrac{29x - 23}{3x}$

Exercise Set 7.6   **1.** 4   **3.** $\dfrac{50}{9}$ **5.** -3 **7.** $\dfrac{14}{9}$ **9.** a   **11.** 123 lb   **13.** 165 cal   **15.** $y = 21.25$ **17.** $y = 5\dfrac{5}{7}$ ft   **19.** 2   **21.** -3 **23.** $2\dfrac{2}{9}$ hr

25. $1\dfrac{1}{2}$ min   **27.** trip to park rate: r; to park time: $\dfrac{12}{r}$; return trip rate: r; return time: $\dfrac{18}{r}$; $r = 6$ mph   **29.** 1st portion: 10 mph; cooldown: 8 mph

31. 360 sq. ft   **33.** 2   **35.** \$108.00   **37.** 63 mph   **39.** $37\dfrac{1}{2}$ ft   **41.** 337 yd/game   **43.** 5   **45.** 217 mph   **47.** 9 gal   **49.** 8 mph   **51.** 3 hr   **53.** $26\dfrac{2}{3}$ ft

55. 216 other nuts   **57.** 20 hr   **59.** $1\dfrac{1}{5}$ hr   **61.** $5\dfrac{1}{4}$ hr   **63.** first pump: 28 min; second pump: 84 min   **65.** $m = 3$; upward   **67.** $m = -\dfrac{9}{5}$; downward

69. $m = 0$; horizontal   **71.** 2035 megawatts   **73.** answers may vary   **75.** 84 yr   **77.** 30 yr   **79.** 3.75 min

Mental Math   **1.** direct   **3.** inverse   **5.** inverse   **7.** direct

Exercise Set 7.7   **1.** $y = \dfrac{1}{2}x$ **3.** $y = 6x$ **5.** $y = 3x$ **7.** $y = \dfrac{2}{3}x$ **9.** $y = \dfrac{7}{x}$ **11.** $y = \dfrac{0.5}{x}$ **13.** $y = kx$ **15.** $h = \dfrac{k}{t}$ **17.** $z = kx^2$ **19.** $y = \dfrac{k}{z^3}$

21. $x = \dfrac{k}{\sqrt{y}}$ **23.** $y = 40$ **25.** $y = 3$ **27.** $z = 54$ **29.** $a = \dfrac{4}{9}$ **31.** \$62.50   **33.** \$6   **35.** $5\dfrac{1}{3}$ in.   **37.** 179.1 lb   **39.** 1600 ft   **41.** $\dfrac{1}{2}$ **43.** $\dfrac{3}{7}$

45. multiplied by 3   **47.** it is doubled

Mental Math   **1.** $\dfrac{y}{5x}$ **3.** $\dfrac{3x}{5}$

Exercise Set 7.8   **1.** $\dfrac{2}{3}$ **3.** $\dfrac{2}{3}$ **5.** $\dfrac{1}{2}$ **7.** $-\dfrac{21}{5}$ **9.** $\dfrac{27}{16}$ **11.** $\dfrac{4}{3}$ **13.** $\dfrac{1}{21}$ **15.** $-\dfrac{4x}{15}$ **17.** $\dfrac{m - n}{m + n}$ **19.** $\dfrac{2x(x - 5)}{7x^2 + 10}$ **21.** $\dfrac{1}{y - 1}$ **23.** $\dfrac{1}{6}$ **25.** $\dfrac{x + y}{x - y}$

27. $\dfrac{3}{7}$ **29.** $\dfrac{a}{x + b}$ **31.** $\dfrac{7(y - 3)}{8 + y}$ **33.** $\dfrac{3x}{x - 4}$ **35.** $-\dfrac{x + 8}{x - 2}$ **37.** $\dfrac{s^2 + r^2}{s^2 - r^2}$ **39.** answers may vary   **41.** 9   **43.** 1   **45.** $\dfrac{1}{5}$ **47.** $\dfrac{2}{3}$ **49.** $\dfrac{13}{24}$

51. $\dfrac{R_1 R_2}{R_2 + R_1}$ **53.** 12 hr   **55.** $\dfrac{2x}{2 - x}$ **57.** $\dfrac{xy + 1}{x}$ **59.** $\dfrac{1}{y^2 - 1}$

Chapter 7 Review   **1.** $x = 2, x = -2$ **3.** $\dfrac{2}{11}$ **5.** $\dfrac{1}{x - 5}$ **7.** $\dfrac{x(x - 2)}{x + 1}$ **9.** $\dfrac{x - 3}{x - 5}$ **11.** $\dfrac{x + 1}{2x + 1}$ **13.** $\dfrac{x + 5}{x - 3}$ **15.** $-\dfrac{1}{x^2 + 4x + 16}$ **17.** $-\dfrac{9x^2}{8}$

19. $-\dfrac{2x(2x + 5)}{(x - 6)^2}$ **21.** $\dfrac{4x}{3y}$ **23.** $\dfrac{2}{3}$ **25.** $\dfrac{x}{x + 6}$ **27.** $\dfrac{3(x + 2)}{3x + y}$ **29.** $-\dfrac{2(2x + 3)}{y - 2}$ **31.** $\dfrac{5x + 2}{3x - 1}$ **33.** $\dfrac{2x + 1}{2x^2}$ **35.** $(x - 8)(x + 8)(x + 3)$

37. $\dfrac{3x^2 + 4x - 15}{(x + 2)^2(x + 3)}$ **39.** $\dfrac{-2x + 10}{(x - 3)(x - 1)}$ **41.** $\dfrac{-2x - 2}{x + 3}$ **43.** $\dfrac{x - 4}{3x}$ **45.** $\dfrac{x^2 + 2x - 3}{(x + 2)^2}$ **47.** $\dfrac{29x}{12(x - 1)}; \dfrac{3xy}{5(x - 1)}$ **49.** 30   **51.** $-4, 3$

53. no solution   **55.** 5   **57.** $-6, 1$ **59.** $y = \dfrac{560 - 8x}{7}$ **61.** $x = 500$ **63.** no solution   **65.** \$33.75   **67.** 3   **69.** 30 mph; 20 mph

71. $17\dfrac{1}{2}$ hr   **73.** $x = 15$ **75.** $x = 15$ **77.** $y = 110$ **79.** $y = \dfrac{100}{27}$ **81.** \$3960   **83.** $-\dfrac{7}{18y}$ **85.** $\dfrac{6}{7}$ **87.** $\dfrac{3y - 1}{2y - 1}$ **89.** $b + a$

Chapter 7 Test   **1.** $x = -1, x = -3$ **2. a.** \$115   **b.** \$103   **3.** $\dfrac{3}{5}$ **4.** $\dfrac{1}{x + 6}$ **5.** $\dfrac{1}{x^2 - 3x + 9}$ **6.** $\dfrac{2m(m + 2)}{m - 2}$ **7.** $\dfrac{a + 2}{a + 5}$ **8.** $-\dfrac{1}{x + y}$ **9.** 15

10. $\dfrac{y - 2}{4}$ **11.** $-\dfrac{1}{2x + 5}$ **12.** $\dfrac{3a - 4}{(a - 3)(a + 2)}$ **13.** $\dfrac{3}{x - 1}$ **14.** $\dfrac{2(x + 5)}{x(y + 5)}$ **15.** $\dfrac{x^2 + 2x + 35}{(x + 9)(x + 2)(x - 5)}$ **16.** $\dfrac{30}{11}$ **17.** -6 **18.** no solution

19. $-2, 5$ **20.** $\dfrac{xz}{2y}$ **21.** $\dfrac{5y^2 - 1}{y + 2}$ **22.** 28   **23.** $\dfrac{8}{9}$ **24.** 18 bulbs   **25.** 5 or 1   **26.** 30 mph   **27.** $6\dfrac{2}{3}$ hr   **28.** $x = 12$

Chapter 7 Cumulative Review   **1. a.** $\dfrac{15}{x} = 4$ **b.** $12 - 3 = x$ **c.** $4x + 17 = 21$; Sec. 1.4, Ex. 9   **3.** amount at 7%: \$12,500; amount at 9%: \$7500;

Sec. 2.7, Ex. 5   **5.** 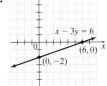 ; Sec. 3.3, Ex. 2

7. a. 4^7 **b.** x^{10} **c.** y^4 **d.** y^{12} **e.** $(-5)^{15}$;   **f.** a^2b^2 Sec. 5.1, Ex. 3   **9.** $12z + 16$; Sec. 5.2, Ex. 15   **11.** $27a^3 + 27a^2b + 9ab^2 + b^3$; Sec. 5.3, Ex. 8

13. a. $t^2 + 4t + 4$ **b.** $p^2 - 2qp + q^2$ **c.** $4x^2 + 20x + 25$ **d.** $x^4 - 14x^2y + 49y^2$; Sec. 5.4, Ex. 5   **15. a.** x^3 **b.** 81   **c.** $\dfrac{q^9}{p^4}$ **d.** $\dfrac{32}{125}$; Sec. 5.5, Ex. 2

17. $4x^2 - 4x + 6 + \dfrac{-11}{2x + 3}$; Sec. 5.6, Ex. 6   **19. a.** 4   **b.** 1   **c.** 3; Sec. 6.1, Ex. 1   **21.** $-3a(3a^4 - 6a + 1)$; Sec. 6.1, Ex. 5

23. $3(m + 2)(m - 10)$; Sec. 6.2, Ex. 7   **25.** $(3x + 2)(x + 3)$; Sec. 6.3, Ex. 1   **27.** $(x + 6)^2$; Sec. 6.3, Ex. 7   **29.** prime polynomial; Sec. 6.4, Ex. 5

31. $(x + 2)(x^2 - 2x + 4)$; Sec. 6.4, Ex. 6   **33.** $(2x + 3)(x + 1)(x - 1)$; Sec. 6.5, Ex. 2   **35.** $3(2m + n)(2m - n)$; Integrated Review, Ex. 3

37. $-\dfrac{1}{2}, 4$; Sec. 6.5, Ex. 3   **39.** $(1, 0), (4, 0)$; Sec. 5.6, Ex. 8   **41.** base: 6 m; height: 10 m; Sec. 6.6, Ex. 3   **43.** $\dfrac{5}{x^2}$; Sec. 7.1, Ex. 5

45. $\dfrac{2}{x(x + 1)}$; Sec. 7.2, Ex. 6   **47.** $\dfrac{x + 1}{x + 2y}$; Sec. 7.8, Ex. 5

CHAPTER 8   ROOTS AND RADICALS

Calculator Explorations   1. 2.646; yes   **3.** 3.317; yes   **5.** 9.055; yes   **7.** 3.420; yes   **9.** 2.115; yes   **11.** 1.783; yes

Mental Math   1. false   **3.** true   **5.** true

Exercise Set 8.1   1. 4   **3.** 9   **5.** $\dfrac{1}{5}$ **7.** -10 **9.** not a real number   **11.** -11 **13.** $\dfrac{3}{5}$ **15.** 12   **17.** $\dfrac{7}{6}$ **19.** -1 **21.** 6.083   **23.** 11.662

25. $\sqrt{2} \approx 1.41$; 126.90 ft   **27.** z **29.** x^2 **31.** $3x^4$ **33.** $9x$ **35.** $\dfrac{x^3}{6}$ **37.** $\dfrac{5y}{3}$ **39.** 5   **41.** -4 **43.** -5 **45.** $\dfrac{1}{2}$ **47.** -5 **49.** -10

51. answers may vary   **53.** 2   **55.** not a real number   **57.** -5 **59.** 1   **61.** -2 **63.** 4   **65.** $25 \cdot 2$ **67.** $16 \cdot 2$ **69.** $4 \cdot 7$ **71.** $9 \cdot 3$ **73.** 7 mi   **75.** 3.1 in.   **77.** 3   **79.** $|x|$ **81.** $|x + 2|$ **83.** 58 ft   **85.**

$-2, -1.3, -1, 0, 1, 1.3, 2$

$y = \sqrt[3]{x}$

87. x^7 **89.** x^5 **91.** $(-3, 0)$; answers may vary   **93.** $(5, 0)$; answers may vary

Mental Math   1. 6   **3.** x **5.** 0   **7.** $5x^2$

Exercise Set 8.2   1. $2\sqrt{5}$ **3.** $3\sqrt{2}$ **5.** $5\sqrt{2}$ **7.** $\sqrt{33}$ **9.** $2\sqrt{15}$ **11.** $6\sqrt{5}$ **13.** $2\sqrt{13}$ **15.** $\dfrac{2\sqrt{2}}{5}$ **17.** $\dfrac{3\sqrt{3}}{11}$ **19.** $\dfrac{3}{2}$ **21.** $\dfrac{5\sqrt{5}}{3}$ **23.** $\dfrac{\sqrt{11}}{6}$

25. $-\dfrac{\sqrt{3}}{4}$ **27.** $x^3\sqrt{x}$ **29.** $x^6\sqrt{x}$ **31.** $5x\sqrt{3}$ **33.** $4x^2\sqrt{6}$ **35.** $\dfrac{2\sqrt{3}}{y}$ **37.** $\dfrac{3\sqrt{x}}{y}$ **39.** $\dfrac{2\sqrt{22}}{x^2}$ **41.** $2\sqrt[3]{3}$ **43.** $5\sqrt[3]{2}$ **45.** $\dfrac{\sqrt[3]{5}}{4}$ **47.** $\dfrac{\sqrt[3]{7}}{2}$

49. $\dfrac{\sqrt[3]{15}}{4}$ **51.** $2\sqrt[3]{10}$ **53.** $2\sqrt[4]{3}$ **55.** $\dfrac{\sqrt[4]{8}}{3}$ **57.** $2\sqrt[5]{3}$ **59.** $\dfrac{\sqrt[5]{5}}{2}$ **61.** $2\sqrt[3]{10}$ **63.** answers may vary   **65.** $14x$ **67.** $2x^2 - 7x - 15$ **69.** 0

71. 8 cm   **73.** $x^3y\sqrt{y}$ **75.** $x + 2$ **77.** $\dfrac{\sqrt[3]{2}}{x^3}$ **79.** $2\sqrt{5}$ in.   **81.** 5 in.   **83.** \$1700   **85.** $\dfrac{26}{15}$ sq. m ≈ 1.7 sq. m

Mental Math   1. $8\sqrt{2}$ **3.** $7\sqrt{x}$ **5.** $3\sqrt{7}$

Exercise Set 8.3   1. $-4\sqrt{3}$ **3.** $9\sqrt{6} - 5$ **5.** $\sqrt{5} + \sqrt{2}$ **7.** $7\sqrt[3]{3} - \sqrt{3}$ **9.** $-5\sqrt[3]{2} - 6$ **11.** $8\sqrt{5}$ in.   **13.** answers may vary   **15.** $5\sqrt{3}$

17. $9\sqrt{5}$ **19.** $4\sqrt{6} + \sqrt{5}$ **21.** $x + \sqrt{x}$ **23.** 0   **25.** $4\sqrt{x} - x\sqrt{x}$ **27.** $7\sqrt{5}$ **29.** $\sqrt{5} + \sqrt[3]{5}$ **31.** $-5 + 8\sqrt{2}$ **33.** $8 - 6\sqrt{2}$ **35.** $14\sqrt{2}$

37. $5\sqrt{2} + 12$ **39.** $\dfrac{4\sqrt{5}}{9}$ **41.** $\dfrac{3\sqrt{3}}{8}$ **43.** $2\sqrt{5}$ **45.** $-\sqrt{35}$ **47.** $12\sqrt{2x}$ **49.** 0   **51.** $x\sqrt{3x} + 3x\sqrt{x}$ **53.** $5\sqrt[3]{3}$ **55.** $\sqrt[3]{9}$ **57.** $4 + 4\sqrt[3]{2}$

59. $-3 + 3\sqrt[3]{2}$ **61.** $4x\sqrt{2} + 2\sqrt[3]{4} + 2x$ **63.** $\sqrt[3]{5}$ **65.** $x^2 + 12x + 36$ **67.** $4x^2 - 4x + 1$ **69.** $(4, 2)$ **71.** $\left(48 + \dfrac{9\sqrt{3}}{2}\right)$ sq. ft   **73.** $\dfrac{83x\sqrt{x}}{20}$

Mental Math   1. $\sqrt{6}$ **3.** $\sqrt{6}$ **5.** $\sqrt{10y}$

Exercise Set 8.4   1. 4   **3.** $5\sqrt{2}$ **5.** $2\sqrt{5} + 5\sqrt{2}$ **7.** $15 - 12\sqrt{15} - 5\sqrt{2} + 4\sqrt{30}$ **9.** $x - 36$ **11.** $67 + 16\sqrt{3}$ **13.** $130\sqrt{3}$ sq. m   **15.** 4

17. $3\sqrt{2}$ **19.** $5y^2$ **21.** $\dfrac{\sqrt{15}}{5}$ **23.** $\dfrac{\sqrt{6y}}{6y}$ **25.** $\dfrac{\sqrt{10}}{6}$ **27.** $3\sqrt{2} - 3$ **29.** $2\sqrt{10} + 6$ **31.** $\sqrt{30} + 5 + \sqrt{6} + \sqrt{5}$ **33.** $3 + \sqrt{3}$ **35.** $3 - 2\sqrt{5}$

37. $3\sqrt{3} + 1$ **39.** $24\sqrt{5}$ **41.** 20   **43.** $36x$ **45.** $\sqrt{30} + \sqrt{42}$ **47.** $4x\sqrt{5} - 60\sqrt{x}$ **49.** $\sqrt{6} - \sqrt{15} + \sqrt{10} - 5$ **51.** -5 **53.** $x - 9$

55. $15 + 6\sqrt{6}$ **57.** $9x - 30\sqrt{x} + 25$ **59.** $5\sqrt{3}$ **61.** $2y\sqrt{6}$ **63.** $2xy\sqrt{3y}$ **65.** $\dfrac{\sqrt{30}}{15}$ **67.** $\dfrac{\sqrt{15}}{10}$ **69.** $\dfrac{3\sqrt{2x}}{2}$ **71.** $-8 - 4\sqrt{5}$ **73.** $5\sqrt{10} - 15$

75. $\dfrac{6\sqrt{5} - 4\sqrt{3}}{11}$ **77.** $\sqrt{6} + \sqrt{3} + \sqrt{2} + 1$ **79.** $2\sqrt[3]{6}$ **81.** $12\sqrt[3]{10}$ **83.** $5\sqrt[3]{3}$ **85.** 3   **87.** $2\sqrt[3]{3}$ **89.** $\dfrac{\sqrt[3]{10}}{2}$ **91.** $3\sqrt[3]{4}$ **93.** $\dfrac{\sqrt[3]{3}}{3}$ **95.** $\dfrac{\sqrt[3]{6}}{3}$

97. $x + 4$ **99.** $2x - 1$ **101.** 44   **103.** 2   **105.** $\dfrac{\sqrt{A\pi}}{\pi}$ **107.** answers may vary   **109.** answers may vary   **111.** $\dfrac{2}{-2\sqrt{2} - 4 + \sqrt{6} + 2\sqrt{3}}$

Integrated Review **1.** 6 **2.** $4\sqrt{3}$ **3.** x^2 **4.** $y^3\sqrt{y}$ **5.** $4x$ **6.** $3x^5\sqrt{2x}$ **7.** 2 **8.** 3 **9.** -3 **10.** not a real number **11.** $\dfrac{\sqrt{11}}{3}$ **12.** $\dfrac{\sqrt[3]{7}}{4}$ **13.** -4 **14.** -5 **15.** $\dfrac{3}{7}$ **16.** $\dfrac{1}{8}$ **17.** a^4b **18.** x^5y^{10} **19.** $5m^3$ **20.** $3n^8$ **21.** $6\sqrt{7}$ **22.** $3\sqrt{2}$ **23.** cannot be simplified **24.** $\sqrt{x}+3x$ **25.** $\sqrt{30}$ **26.** 3 **27.** 28 **28.** 45 **29.** $\sqrt{33}+\sqrt{3}$ **30.** $3\sqrt{2}-2\sqrt{6}$ **31.** $4y$ **32.** $3x^2\sqrt{5}$ **33.** $x-3\sqrt{x}-10$ **34.** $11+6\sqrt{2}$ **35.** 2 **36.** $\sqrt{3}$ **37.** $2x^2\sqrt{3}$ **38.** $ab^2\sqrt{15a}$ **39.** $\dfrac{\sqrt{6}}{6}$ **40.** $\dfrac{x\sqrt{5}}{10}$ **41.** $\dfrac{4\sqrt{6}-4}{5}$ **42.** $\dfrac{\sqrt{2x}+5\sqrt{2}+\sqrt{x}+5}{x-25}$

Exercise Set 8.5 **1.** 81 **3.** -1 **5.** 5 **7.** 49 **9.** no solution **11.** -2 **13.** $\dfrac{3}{2}$ **15.** 16 **17.** no solution **19.** 2 **21.** 4 **23.** 3 **25.** 2 **27.** 2 **29.** no solution **31.** $0,-3$ **33.** -3 **35.** 9 **37.** 12 **39.** $3,1$ **41.** -1 **43.** 2 **45.** 16 **47.** 1 **49.** 1 **51.** $3x-8=19$; $x=9$ **53.** $2(2x+x)=24$; length: 8 in. **55.** 9 **57. a.** $3.2, 10.0, 31.6$ **b.** no **c.** $V=2b^2$ **59.** answers may vary **61.** 2.43 **63.** 0.48

Exercise Set 8.6 **1.** $\sqrt{13}$ **3.** $3\sqrt{3}$ **5.** 25 **7.** $\sqrt{22}$ **9.** $3\sqrt{17}$ **11.** $\sqrt{41}$ **13.** $4\sqrt{2}$ **15.** $3\sqrt{10}$ **17.** 51.2 ft **19.** 20.6 ft **21.** 11.7 ft **23.** $\sqrt{29}$ **25.** $\sqrt{73}$ **27.** $2\sqrt{10}$ **29.** $\dfrac{3\sqrt{5}}{2}$ **31.** $\sqrt{85}$ **33.** 24 cu. ft **35.** 54 mph **37.** 27 mph **39.** 61.2 km **41.** 32 **43.** $\dfrac{1}{25}$ **45.** x^5 **47.** y^4 **49.** $2\sqrt{10}-4$ **51.** 201 mi **53.** answers may vary

Exercise Set 8.7 **1.** 2 **3.** 3 **5.** 8 **7.** 4 **9.** $-\dfrac{1}{2}$ **11.** $\dfrac{1}{64}$ **13.** $\dfrac{1}{729}$ **15.** $\dfrac{5}{2}$ **17.** answers may vary **19.** 2 **21.** 2 **23.** $\dfrac{1}{x^{2/3}}$ **25.** x^3 **27.** answers may vary **29.** 9 **31.** -2 **33.** -3 **35.** $\dfrac{1}{9}$ **37.** $\dfrac{3}{4}$ **39.** 27 **41.** 512 **43.** -4 **45.** 32 **47.** $\dfrac{8}{27}$ **49.** $\dfrac{1}{27}$ **51.** $\dfrac{1}{2}$ **53.** $\dfrac{1}{5}$ **55.** $\dfrac{1}{125}$ **57.** 9 **59.** $6^{1/3}$ **61.** x^6 **63.** 36 **65.** $\dfrac{1}{3}$ **67.** $\dfrac{x^{2/3}}{y^{3/2}}$ **69.** $\dfrac{x^{16/5}}{y^6}$ **71.**

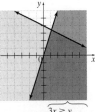

73. $-1,4$ **75.** $-\dfrac{1}{2},3$ **77.** 11,224 people **79.** 3.344 **81.** 5.665

Chapter 8 Review **1.** 9 **3.** 3 **5.** $-\dfrac{3}{8}$ **7.** not a real number **9.** irrational, 8.718 **11.** x^6 **13.** $3x^3$ **15.** $\dfrac{4}{y^5}$ **17.** $3\sqrt{6}$ **19.** $5x\sqrt{6x}$ **21.** $3\sqrt[3]{2}$ **23.** $2\sqrt[4]{3}$ **25.** $\dfrac{3\sqrt{2}}{5}$ **27.** $\dfrac{3y\sqrt{5}}{2x^2}$ **29.** $\dfrac{\sqrt[4]{9}}{2}$ **31.** $\dfrac{\sqrt[3]{3}}{2}$ **33.** $2\sqrt[3]{3}-\sqrt[3]{2}$ **35.** $6\sqrt{6}-2\sqrt[3]{6}$ **37.** $5\sqrt{7x}+2\sqrt[3]{7}$ **39.** $\dfrac{\sqrt{5}}{6}$ **41.** 0 **43.** $30\sqrt{2}$ **45.** $6\sqrt{2}-18$ **47.** $3\sqrt{2}-5\sqrt{3}+2\sqrt{6}-10$ **49.** $4\sqrt{2}$ **51.** $\dfrac{x\sqrt{5x}}{2}$ **53.** $\dfrac{\sqrt{30}}{6}$ **55.** $\dfrac{\sqrt{6x}}{2x}$ **57.** $\dfrac{\sqrt{35}}{10y}$ **59.** $\dfrac{\sqrt[3]{21}}{3}$ **61.** $\dfrac{\sqrt[3]{12}}{2}$ **63.** $3\sqrt{5}+6$ **65.** $4\sqrt{6}-8$ **67.** $\dfrac{2\sqrt{2}-1}{7}$ **69.** $-\dfrac{3+5\sqrt{3}}{11}$ **71.** 18 **73.** 25 **75.** 12 **77.** 6 **79.** $2\sqrt{14}$ **81.** $4\sqrt{34}$ ft **83.** $\sqrt{130}$ **85.** 2.4 in. **87.** $a^{5/2}$ **89.** $x^{5/2}$ **91.** 4 **93.** -2 **95.** -512 **97.** $\dfrac{8}{27}$ **99.** $\dfrac{1}{5}$ **101.** 32 **103.** $\dfrac{1}{3^{2/3}}$ **105.** $\dfrac{1}{x^2}$

Chapter 8 Test **1.** 4 **2.** 5 **3.** 8 **4.** $\dfrac{3}{4}$ **5.** not a real number **6.** $\dfrac{1}{9}$ **7.** $3\sqrt{6}$ **8.** $2\sqrt{23}$ **9.** $x^3\sqrt{3}$ **10.** $2x^2y^3\sqrt{2y}$ **11.** $3x^4\sqrt{x}$ **12.** 2 **13.** $2\sqrt[3]{5}$ **14.** $-8\sqrt{3}$ **15.** $3\sqrt[3]{2}-2x\sqrt{2}$ **16.** $\dfrac{\sqrt{5}}{4}$ **17.** $\dfrac{\sqrt[3]{2}}{3}$ **18.** $6\sqrt{2x}$ **19.** $5+2\sqrt{35}$ **20.** $4x-9$ **21.** $\dfrac{\sqrt{6}}{3}$ **22.** $\dfrac{\sqrt[3]{15}}{3}$ **23.** $\dfrac{\sqrt{15}}{6x}$ **24.** $-1-\sqrt{3}$ **25.** 9 **26.** 5 **27.** 9 **28.** $4\sqrt{5}$ in. **29.** $\sqrt{5}$ **30.** $\dfrac{1}{16}$ **31.** $\dfrac{x^{10/3}}{y^2}$

Chapter 8 Cumulative Review **1. a.** $-\dfrac{39}{5}$ **b.** 2; Sec. 1.7, Ex. 9 **3.** -3; Sec. 2.2, Ex. 5

5.

x	y
-1	-3
0	0
-3	-9; Sec. 3.1, Ex. 5

7. $y=\dfrac{1}{4}x-3$; (Sec. 3.5, Ex. 5) **9.** $y=-3$; Sec. 3.6, Ex. 4 **11.** a, b, c; Sec. 3.7, Ex. 5

13. a. solution **b.** not a solution; Sec. 4.1, Ex. 1 **15.** $(4,2)$; Sec. 4.2, Ex. 1 **17.** $\left(-\dfrac{5}{4},-\dfrac{5}{2}\right)$; Sec. 4.3, Ex. 6

19. 5% saline solution: 2.5 l; 25% saline solution: 7.5 l; Sec. 4.4, Ex. 4 **21.** ; Sec. 4.6, Ex. 1 **23.** $10xy-5y^2-9x^2$; Sec. 5.2, Ex. 7

25. $(x + 2y)(x + 3y)$; Sec. 6.2, Ex. 6 **27.** $\dfrac{-2 - x}{3x + 1}$; Sec. 7.1, Ex. 8 **29.** $\dfrac{3y^9}{160}$; Sec. 7.2, Ex. 4 **31.** 1; Sec. 7.3, Ex. 2 **33.** $\dfrac{x(3x - 1)}{(x + 1)^2(x - 1)}$; Sec. 7.4, Ex. 7

35. -5; Sec. 7.5, Ex. 1 **37.** 15 yd; Sec. 7.6, Ex. 3 **39.** $\dfrac{3}{z}$; Sec. 7.8, Ex. 3 **41. a.** $3\sqrt{6}$ **b.** $2\sqrt{3}$ **c.** $10\sqrt{2}$ **d.** $\sqrt{35}$; Sec. 8.2, Ex. 1

43. a. -44 **b.** $7x + 4\sqrt{7x} + 4$; Sec. 8.4, Ex. 4 **45.** no solution; Sec. 8.5, Ex. 2 **47.** 10 in.; Sec. 8.6, Ex. 1 **49. a.** 8 **b.** 9 **c.** -8; Sec. 8.7, Ex. 2

Integrated Review 1. $x = \dfrac{1}{5}, 2$ **2.** $x = -3, \dfrac{2}{5}$ **3.** $x = 1 \pm \sqrt{2}$ **4.** $x = 3 \pm \sqrt{2}$ **5.** $a = \pm 2\sqrt{5}$

CHAPTER 9 SOLVING QUADRATIC EQUATIONS

Exercise Set 9.1 1. ± 8 **3.** $\pm \sqrt{21}$ **5.** $\pm\dfrac{1}{5}$ **7.** no real solution **9.** $\pm\dfrac{\sqrt{39}}{3}$ **11.** $\pm\dfrac{2\sqrt{7}}{7}$ **13.** $\pm\sqrt{2}$ **15.** $\pm\sqrt{5}$ **17.** answers may vary

19. $12, -2$ **21.** $-2 \pm \sqrt{7}$ **23.** $1, 0$ **25.** $-2 \pm \sqrt{10}$ **27.** $\dfrac{8}{3}, -4$ **29.** no real solution **31.** $\dfrac{11 \pm 5\sqrt{2}}{2}$ **33.** $\dfrac{7 \pm 4\sqrt{2}}{3}$ **35.** $8, -3$ **37.** 5 sec

39. 2.3 sec **41.** 72.7 sec **43.** $2\sqrt{5}$ in. ≈ 4.47 in. **45.** $\sqrt{3039}$ ft ≈ 55.13 ft **47.** $(x + 3)^2$ **49.** $(x - 2)^2$ **51.** $2, -6$ **53.** $-7 \pm \sqrt{31}$

55. $x = 10.4$ in.; $\dfrac{3}{4} x = 1.8$ in. **57.** ± 1.33 **59.** 2006

Exercise Set 9.2 1. $x^2 + 4x + 4 = (x + 2)^2$ **3.** $k^2 - 12k + 36 = (k - 6)^2$ **5.** $x^2 - 3x + \dfrac{9}{4} = \left(x - \dfrac{3}{2}\right)^2$ **7.** $m^2 - m + \dfrac{1}{4} = \left(m - \dfrac{1}{2}\right)^2$

9. $0, 6$ **11.** $-6, -2$ **13.** $-1 \pm \sqrt{6}$ **15.** $-3 \pm \sqrt{34}$ **17.** $\dfrac{-5 \pm \sqrt{53}}{2}$ **19.** $1 \pm \sqrt{2}$ **21.** $-4. -1$ **23.** $-6, 3$ **25.** $-\dfrac{1}{2}, \dfrac{13}{2}$ **27.** no real solution

29. $\dfrac{3 \pm \sqrt{19}}{2}$ **31.** $-2, 4$ **33.** $\dfrac{-4 \pm \sqrt{6}}{2}$ **35.** $\dfrac{1}{2}, 1$ **37.** $\dfrac{1 \pm \sqrt{13}}{3}$ **39.** answers may vary **41.** $-\dfrac{1}{2}$ **43.** -1 **45.** $3 + 2\sqrt{5}$ **47.** $\dfrac{1 - 3\sqrt{2}}{2}$

49. $k = 8$ or $k = -8$ **51.** 2008 **53.** $-6, -2$ **55.** $-0.68, 3.68$

Mental Math 1. $a = 2, b = 5, c = 3$ **3.** $a = 10, b = 13, c = -2$ **5.** $a = 1, b = 0, c = -6$

Exercise Set 9.3 1. $-2, 1$ **3.** $\dfrac{-5 \pm \sqrt{17}}{2}$ **5.** $\dfrac{2 \pm \sqrt{2}}{2}$ **7.** $1, 2$ **9.** $\dfrac{-7 \pm \sqrt{37}}{6}$ **11.** $\pm\dfrac{2}{7}$ **13.** no real solution **15.** $-3, 10$ **17.** $\pm\sqrt{5}$ **19.** $-3, 4$

21. $-2 \pm \sqrt{7}$ **23.** $\dfrac{1}{2}, 3$ **25.** $\dfrac{5 \pm \sqrt{33}}{2}$ **27.** $-2, \dfrac{7}{3}$ **29.** $\dfrac{-9 \pm \sqrt{129}}{12}$ **31.** $\dfrac{4 \pm \sqrt{2}}{7}$ **33.** $3 \pm \sqrt{7}$ **35.** $\dfrac{3 \pm \sqrt{3}}{2}$ **37.** $-1, \dfrac{1}{3}$ **39.** $-\dfrac{3}{4}, \dfrac{1}{5}$

41. no real solution **43.** $\dfrac{3 \pm \sqrt{13}}{4}$ **45.** no real solution **47.** $1 \pm \sqrt{2}$ **49.** $-\dfrac{3}{4}, -\dfrac{1}{2}$ **51.** $\dfrac{7 \pm \sqrt{129}}{20}$ **53.** no real solution **55.** $\dfrac{1 \pm \sqrt{2}}{5}$

57. $\dfrac{-3\sqrt{2} \pm \sqrt{38}}{2}$ **59.** 2 real solutions **61.** no real solution **63.** 1 real solution **65.** 2 real solutions **67.** no real solution **69.** 1 real solution

71. d **73.** $4\sqrt{3}$ **75.** $5\sqrt{2}$ **77.** base : 3 ft; height : 12 ft **79.** 3.9 ft by 8.9 ft **81.** $-4.4, 3.4$ **83.** $-0.7, 5.0$ **85.** 7.9 sec **87.** 2006 **89.** 2001 or 2006

Integrated Review 1. $x = \dfrac{1}{5}, 2$ **2.** $x = 3, \dfrac{2}{5}$ **3.** $x = 1 \pm \sqrt{2}$ **4.** $x = 3 \pm \sqrt{2}$ **5.** $a = \pm 2\sqrt{5}$ **6.** $a = \pm 6\sqrt{2}$ **7.** no real solution

8. no real solution **9.** $x = 2$ **10.** $x = 3$ **11.** $p = 3$ **12.** $p = \dfrac{7}{2}$ **13.** $y = \pm 2$ **14.** $y = \pm 3$ **15.** $x = 0, 1, 2$ **16.** $x = 0, -3, -4$

17. $z = -5, 0$ **18.** $z = 0, \dfrac{8}{3}$ **19.** $x = \dfrac{3 \pm \sqrt{7}}{5}$ **20.** $x = \dfrac{3 \pm \sqrt{5}}{2}$ **21.** $m = \dfrac{3}{2}, -1$ **22.** $m = \dfrac{2}{5}, -2$ **23.** $x = \dfrac{5 \pm \sqrt{105}}{20}$ **24.** $x = \dfrac{-1 \pm \sqrt{3}}{4}$

25. $x = 5, \dfrac{7}{4}$ **26.** $x = \dfrac{7}{9}, 1$ **27.** $x = \dfrac{7 \pm 3\sqrt{2}}{5}$ **28.** $x = \dfrac{5 \pm 5\sqrt{3}}{4}$ **29.** $z = \dfrac{7 \pm \sqrt{193}}{6}$ **30.** $z = \dfrac{-7 \pm \sqrt{193}}{12}$ **31.** $x = 11, -10$ **32.** $x = -8, 7$

33. $x = -\dfrac{2}{3}, 4$ **34.** $x = 2, -\dfrac{4}{5}$ **35.** $x = 0.5, 0.1$ **36.** $x = 0.3, -0.2$ **37.** $x = \dfrac{11 \pm \sqrt{41}}{20}$ **38.** $x = \dfrac{11 \pm \sqrt{41}}{40}$ **39.** $z = \dfrac{4 \pm \sqrt{10}}{2}$

40. $z = \dfrac{5 \pm \sqrt{185}}{4}$ **41.** answers may vary

Exercise Set 9.4 1. $3i$ **3.** $10i$ **5.** $5i\sqrt{2}$ **7.** $3i\sqrt{7}$ **9.** $-3 + 9i$ **11.** $-12 + 6i$ **13.** $1 - 3i$ **15.** $9 - 4i$ **17.** $8 + 12i$ **19.** $26 - 2i$ **21.** 73

23. answers may vary **25.** $2 - 3i$ **27.** $\dfrac{31}{25} + \dfrac{17}{25}i$ **29.** $-1 \pm 3i$ **31.** $\dfrac{3 \pm 2i\sqrt{3}}{2}$ **33.** $-3 \pm 2i$ **35.** $\dfrac{-7 \pm i\sqrt{15}}{8}$ **37.** $\dfrac{2 \pm i\sqrt{6}}{2}$ **39.** $15 - 7i$

41. $45 + 63i$ **43.** $-1 + 3i$ **45.** $2 - 3i$ **47.** $-7 - 2i$ **49.** 25 **51.** $\dfrac{2}{5} - \dfrac{9}{5}i$ **53.** $21 + 20i$ **55.** $4 \pm 8i$ **57.** $\pm 5i$ **59.** $-3 \pm i$ **61.** $\dfrac{5 \pm i\sqrt{47}}{4}$

63. $-4 \pm i\sqrt{5}$ **65.** $\pm 6i$ **67.** $-7 \pm i$ **69.**

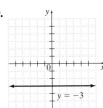

71.

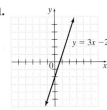

73. 51%   **75.** 90.1%   **77.** false   **79.** false

Calculator Explorations   **1.** $x = -0.41, 7.41$ **3.** $x = 0.91, 2.38$ **5.** $x = -0.39, 0.84$

Exercise Set 9.5   **1.**

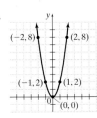

3.

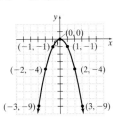

5.

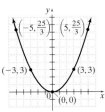

7.

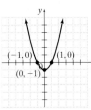

9.

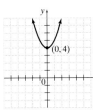

11.

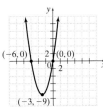

13.

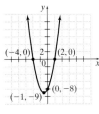

15.

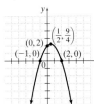

17.

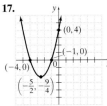

19.

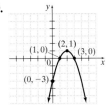

21.

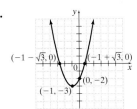

23.

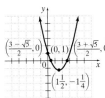

25. $\dfrac{5}{14}$ **27.** $\dfrac{x}{2}$ **29.** $\dfrac{2x^2}{x-1}$ **31.** $-4b$

33. domain: $(-\infty, \infty)$; range: $(-\infty, 3]$

35. domain: $(-\infty, \infty)$; range: $(-\infty, 1]$ **37. a.** 256 ft

b. $t = 4$ sec   **c.** $t = 8$ sec   **39.** B   **41.** D

Chapter 9 Review   **1.** $-\dfrac{3}{5}, 4$ **3.** $-\dfrac{1}{3}, 2$ **5.** $\pm 5\sqrt{2}$ **7.** $-1, 7$ **9.** $4, 18$ **11.** $0, \pm 3$ **13.** $-3, 2$ **15.** 2.5 sec   **17.** $a^2 + 4a + 4 = (a+2)^2$

19. $m^2 - 3m + \dfrac{9}{4} = \left(m - \dfrac{3}{2}\right)^2$ **21.** $3 \pm \sqrt{2}$ **23.** $-1, \dfrac{1}{2}$ **25.** $5 \pm 3\sqrt{2}$ **27.** $-1, \dfrac{1}{2}$ **29.** $-\dfrac{5}{3}$ **31.** $\dfrac{2}{5}, \dfrac{1}{3}$ **33.** no real solution

35. 2 real solutions   **37.** 1 real solution   **39.** no real solution   **41.** $\dfrac{-1 \pm \sqrt{21}}{10}$ **43.** $0, \pm\dfrac{1}{2}$ **45.** $\dfrac{1}{2}, 7$ **47.** $\dfrac{1}{3}$ **49.** 3   **51.** $-10, 22$ **53.** $\dfrac{3 \pm \sqrt{5}}{20}$

55. $-5 \pm \sqrt{30}$ **57.** 2005   **59.** $12i$ **61.** $6i\sqrt{3}$ **63.** $10 + 6i$ **65.** $21 - 10i$ **67.** $-8 - 2i$ **69.** $-13i$ **71.** 25   **73.** $-\dfrac{3}{2} - \dfrac{1}{2}i$ **75.** $\dfrac{2}{5} - \dfrac{9}{5}i$ **77.** $\pm 4i$

79. $2 \pm 3i$ **81.** vertex: $(0, 0)$; axis of symmetry: $x = 0$; opens downward   **83.** vertex: $(3, 0)$; axis of symmetry: $x = 3$; opens upward

85. vertex: $(0, -7)$; axis of symmetry: $x = 0$; opens upward   **87.** vertex: $(72, 14)$; axis of symmetry: $x = 72$; opens downward

89.

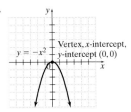

91.

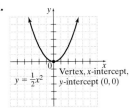

93.

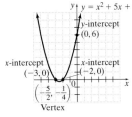

95.

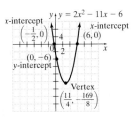

97. 1 solution; $x = -2$ **99.** no real solution

Chapter 9 Test **1.** ± 4 **2.** $\dfrac{5 \pm 2\sqrt{2}}{3}$ **3.** $10, 16$ **4.** $\dfrac{-6 \pm 4\sqrt{3}}{3}$ **5.** $-2, 5$ **6.** $\dfrac{5 \pm \sqrt{37}}{6}$ **7.** $-\dfrac{4}{3}, 1$ **8.** $-1, \dfrac{5}{3}$ **9.** $\dfrac{7 \pm \sqrt{73}}{6}$ **10.** $2 \pm i$ **11.** $\dfrac{1}{3}, 2$

12. $\dfrac{3 \pm \sqrt{7}}{2}$ **13.** $0, \pm\dfrac{1}{3}$ **14.** $5i$ **15.** $10i\sqrt{2}$ **16.** $8 + i$ **17.** $4i$ **18.** 13 **19.** $\dfrac{1}{5} - \dfrac{7}{5}i$ **20.**

21.

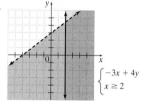

22. 2.7 sec

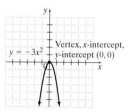

Chapter 9 Cumulative Review

1. a. $\dfrac{1}{2}$ **b.** 9; Sec. 1.6, Ex. 6 **3. a.** $5x + 7$ **b.** $-4a - 1$ **c.** $4y - 3y^2$ **d.** $7.3x - 6$; **e.** $\dfrac{1}{2}b$; Sec. 2.1, Ex. 4 **5. a.** x-int: $(-3, 0)$; y-int: $(0, 2)$

b. x-int: $(-4, 0), (-1, 0)$; y-int: $(0, 1)$ **c.** x-int and y-int: $(0, 0)$ **d.** x-int: $(2, 0)$; y-int: none **e.** x-int: $(-1, 0), (3, 0)$; y-int: $(0, -1), (0, 2)$; Sec. 3.3, Ex. 1
7. parallel; Sec. 3.5, Ex. 4 **9.** no solution; Sec. 4.1, Ex. 3 **11.** $(-2, 0)$; Sec. 4.2, Ex. 3 **13.** infinite number of solutions; Sec. 4.3, Ex. 4

15. Albert: 3.25 mph; Louis: 4.25 mph; Sec. 4.4, Ex. 3 **17.**   Sec. 4.6, Ex. 3

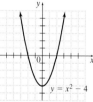

$$\begin{cases} -3x + 4y < 12 \\ x \ge 2 \end{cases}$$

19. a. $\dfrac{25x^4}{y^6}$ **b.** x^6 **c.** $32x^2$ **d.** a^3b; Sec. 5.1, Ex. 10 **21.** $-6, -\dfrac{3}{2}, \dfrac{1}{5}$; Sec. 6.5, Ex. 6 **23.** $\dfrac{31}{2}$; Sec. 7.7, Ex. 1

25. a. 2 **b.** -2 **c.** not a real number; Sec. 8.1, Ex. 5 **27. a.** $\dfrac{5}{6}$ **b.** $\dfrac{\sqrt{3}}{8}$ **c.** $\dfrac{2\sqrt{10}}{9}$; Sec. 8.2, Ex. 2 **29. a.** $7\sqrt{2}$ **b.** $9\sqrt{3}$

c. $1 - 3\sqrt{3} - 6\sqrt{2}$; Sec. 8.3, Ex. 2 **31. a.** $\sqrt{21}$ **b.** $3\sqrt{5}$ **c.** $20\sqrt{3}$ **d.** 18; Sec. 8.4, Ex. 1 **33.** $\dfrac{1}{2}$; Sec. 8.5, Ex. 3

35. $\sqrt{44{,}800} \approx 212$ ft; Sec. 8.6, Ex. 3 **37. a.** $\sqrt{25} = 5$ **b.** $\sqrt[3]{8} = 2$ **c.** $-\sqrt[4]{16} = -2$ **d.** $\sqrt[3]{-27} = -3$ **e.** $\sqrt{\dfrac{1}{9}} = \dfrac{1}{3}$; Sec. 8.7, Ex. 1

39. $\dfrac{\sqrt{14}}{2}, -\dfrac{\sqrt{14}}{2}$; Sec. 9.1, Ex. 2 **41.** $5 \pm \sqrt{11}$; Sec. 9.2, Ex. 3 **43.** $-\dfrac{1}{2}, 5$; Sec. 9.3, Ex. 2 **45. a.** $2i$ **b.** $i\sqrt{11}$

c. $2i\sqrt{5}$; Sec. 9.5, Ex. 1 **47.**   ; Sec. 9.6, Ex. 2

APPENDIX A   OPERATIONS ON DECIMALS

Appendix A Exercise Set **1.** 17.08 **3.** 12.804 **5.** 110.96 **7.** 2.4 **9.** 28.43 **11.** 227.5 **13.** 2.7 **15.** 49.2339 **17.** 80 **19.** 0.07612 **21.** 4.56
23. 648.46 **25.** 767.83 **27.** 12.062 **29.** 61.48 **31.** 7.7 **33.** 863.37 **35.** 22.579 **37.** 363.15 **39.** 7.007

APPENDIX B   REVIEW OF ANGLES, LINES, AND SPECIAL TRIANGLES

Appendix B Exercise Set

1. $71°$ **3.** $19.2°$ **5.** $78\frac{3}{4}°$ **7.** $30°$ **9.** $149.8°$ **11.** $100\frac{1}{2}°$ **13.** $m\angle 1 = m\angle 5 = m\angle 7 = 110°; m\angle 2 = m\angle 3 = m\angle 4 = m\angle 6 = 70°$ **15.** $90°$

17. $90°$ **19.** $90°$ **21.** $45°, 90°$ **23.** $73°, 90°$ **25.** $50\frac{1}{4}°, 90°$ **27.** $x = 6$ **29.** $x = 4.5$ **31.** 10 **33.** 12

APPENDIX D   MEAN, MEDIAN, AND MODE

Appendix D Exercise Set
1. mean: 29, median: 28, no mode **3.** mean: 8.1, median: 8.2, mode: 8.2 **5.** mean: 0.6, median: 0.6, mode: 0.2 and 0.6
7. mean: 370.9, median: 313.5, no mode **9.** 1136 ft **11.** 1117 ft **13.** 6.8 **15.** 6.9 **17.** 85.5 **19.** 73 **21.** 70 and 71 **23.** 9
25. 21, 21, 24

APPENDIX F   REVIEW OF VOLUME AND SURFACE AREA

Appendix F Exercise Set

1. $V = 72$ cu. in.; $SA = 108$ sq. in.   **3.** $V = 512$ cu. cm; $SA = 384$ sq. cm

5. $V = 4\pi$ cu. yd $\approx 12\frac{4}{7}$ cu. yd.; $SA = (2\sqrt{13}\pi + 4\pi)$ sq. yd ≈ 35.20 sq. yd

7. $V = \frac{500}{3}\pi$ cu. in. $\approx 523\frac{17}{21}$ cu. in.; $SA = 100\pi$ sq. in. $\approx 314\frac{2}{7}$ sq. in.

9. $V = 48$ cu. cm; $SA = 96$ sq. cm   **11.** $2\frac{10}{27}$ cu. in.   **13.** 26 sq. ft   **15.** $10\frac{5}{6}$ cu. in.

17. 960 cu. cm   **19.** 196π sq. in.   **21.** $7\frac{1}{2}$ cu. ft   **23.** $12\frac{4}{7}$ cu. cm

GRAPHING ANSWER SECTION

The Graphing Answer Section contains answers to exercises requiring graphical solutions, Spotlight on Decision Making exercises, and Chapter Projects.

CHAPTER 1   REVIEW OF REAL NUMBERS

1.2 Spotlight on Decision Making

The new machinery is working properly.

1.3 Spotlight on Decision Making

You did not set a new world's record; you missed the existing world record by $\frac{7}{32}$ pound.

1.4 Spotlight on Decision Making

The LAN should operate efficiently.

1.6 Spotlight on Decision Making

Yes; there has been an increase in the patient's gum tissue pocket depth for teeth 24, 25, 26, and 27.

Chapter Project

Student answers may vary based on survey results.

CHAPTER 2   EQUATIONS, INEQUALITIES, AND PROBLEM SOLVING

2.6 Spotlight on Decision Making

Answers may vary. The amount saved does not increase at the same rate the deductible increases, so it may not be worth it in the long run to have such a high deductible.

2.7 Spotlight on Decision Making

Yes, they are eligible to deduct their medical expenses since their total medical expenses are 8.6% of their adjusted gross income.

Chapter Project

Answers may vary. Students may prepare organized tables of information to complete each step.

Chapter 2 Review

68.

69.

70.

71.

72.

73.

74.

75.

76.

77.

78.

79.

Chapter 2 Test

21. 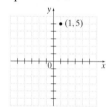 -2

22. 4

23. -1 $\frac{7}{3}$

25. $\frac{2}{5}$

Chapter 2 Cumulative Review

45. -10

46. 5

CHAPTER 3   GRAPHING

3.1 Spotlight on Decision Making

Answers may vary. Based on the graph, you should consider expanding your business to include taking orders over the Internet. You should consider factors such as whether your resources to supply your product can keep up with the increased demand due to Internet orders.

Exercise Set 3.1

1. quadrant I

$(1, 5)$

2. quadrant III

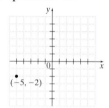
$(-5, -2)$

3. no quadrant, x-axis

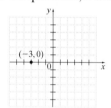
$(-3, 0)$

4. no quadrant, y-axis

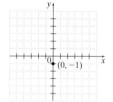
$(0, -1)$

5. quadrant IV

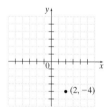
$(2, -4)$

6. quadrant II

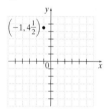
$\left(-1, 4\frac{1}{2}\right)$

7. no quadrant, x-axis

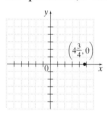
$\left(4\frac{3}{4}, 0\right)$

8. no quadrant, y-axis

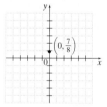
$\left(0, \frac{7}{8}\right)$

9. no quadrant, origin

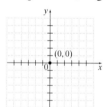
$(0, 0)$

10. no quadrant; x-axis

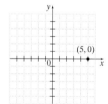
$(5, 0)$

11. no quadrant; y-axis

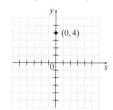
$(0, 4)$

12. quadrant III

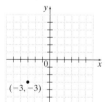
$(-3, -3)$

17. b.

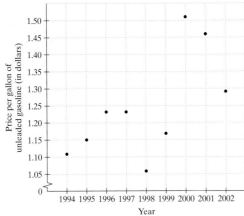

18. b.

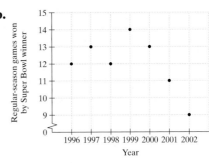

19. b.

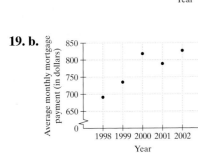

20. b.

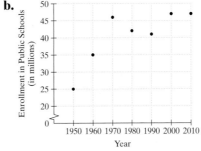

21. b.

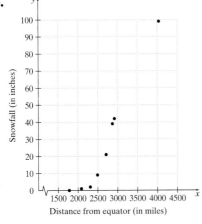

22. b.

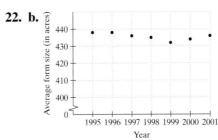

43.

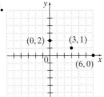

44.

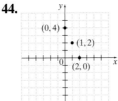

45.

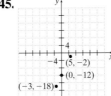

46.

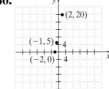

47.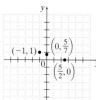

48.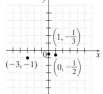

49.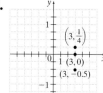

50.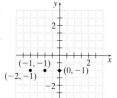

51.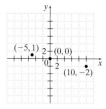

52.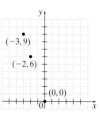

55.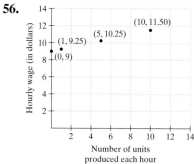

56.

Calculator Explorations

1.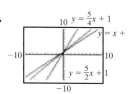

2.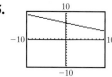

3.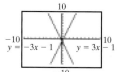

4.

5.

6.

Exercise Set 3.2

9.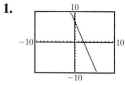

10.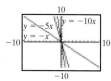

11.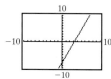

12.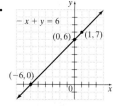

13.
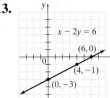

14.
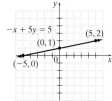

15.
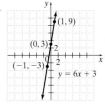

16.
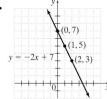

17.
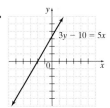

18.
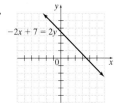

19.
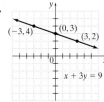

20.
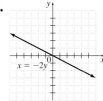

21.
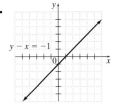

22.
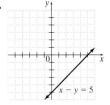

23.
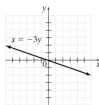

24.

25.
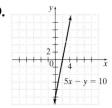

26.
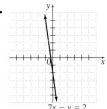

27.
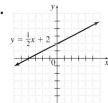

28.

29.
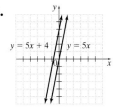

30.
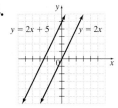

31.
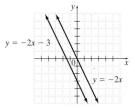

32.
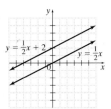

33.
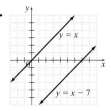

34.
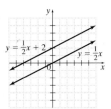

35.
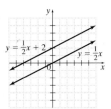

36.

37.
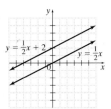

38.
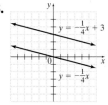

44.

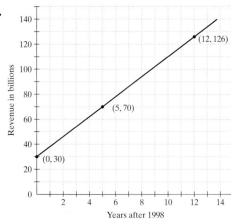

45.

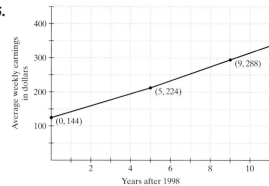

57.

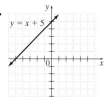

58.

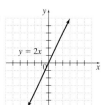

59.

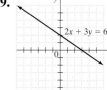

60.

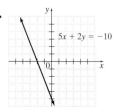

65.

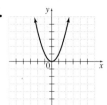

66.

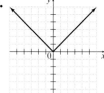

Calculator Explorations

1.

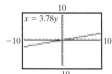

2.

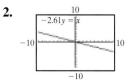

3.

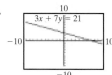

4.

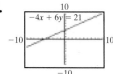

5.

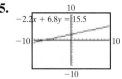

6.

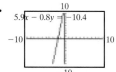

Exercise Set 3.3

11.

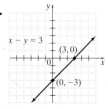

12.

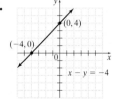

13.

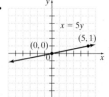

14.

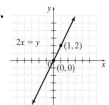

15.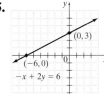

16.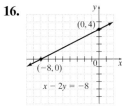

17.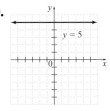

18.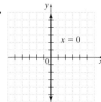

19.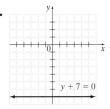

20.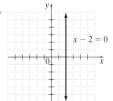

21.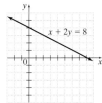

22.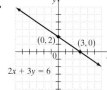

23.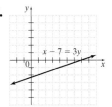

24.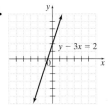

25.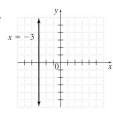

26.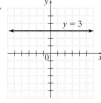

27.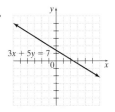

28.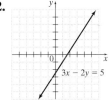

29.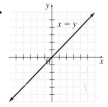

30.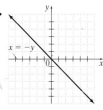

31.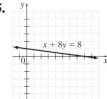

32.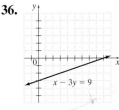

33.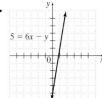

34.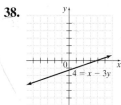

35.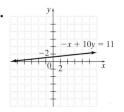

36.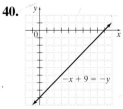

37.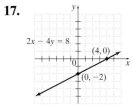

38.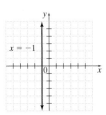

39.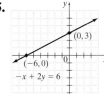

40.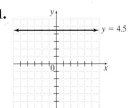

41.

42.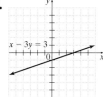

43.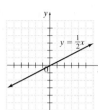
$y = \frac{1}{2}x$

44.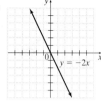
$y = -2x$

45.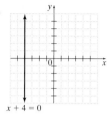
$x + 4 = 0$

46.
$y - 5.2 = 0$

47.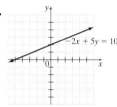
$3x - 4y = -12$

48.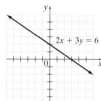
$-2x + 5y = 10$

49.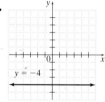
$2x + 3y = 6$

50.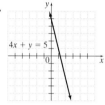
$4x + y = 5$

67. c.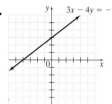
$3x + 6y = 1200$

68.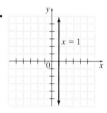
$x = 1$

69.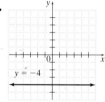
$y = -4$

3.4 Spotlight on Decision Making

Answers may vary. Check into First Choice Realty first, since their annual sales increased by $10 million each year, as opposed to Clarion's annual sales which increased by $5 million each year.

Calculator Explorations

1.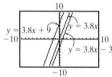
$y = 3.8x + 9$
$y = 3.8x$
$y = 3.8x - 3$

2.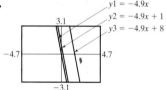
$y1 = -4.9x$
$y2 = -4.9x + 1$
$y3 = -4.9x + 8$

3.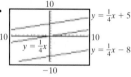
$y = \frac{1}{4}x + 5$
$y = \frac{1}{4}x$
$y = \frac{1}{4}x - 8$

4.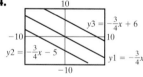
$y3 = -\frac{3}{4}x + 6$
$y2 = -\frac{3}{4}x - 5$
$y1 = -\frac{3}{4}x$

Integrated Review

5.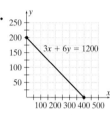
$y = -2x$

6.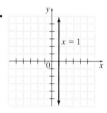
$x + y = 3$

7.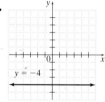
$x = -1$

8.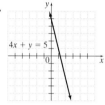
$y = 4$

9.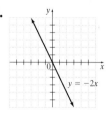
$x - 2y = 6$

10.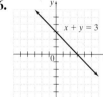
$y = 3x + 2$

11.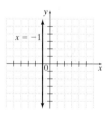
$5x + 3y = 15$

12.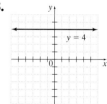
$2x - 4y = 8$

Calculator Explorations

1.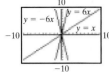

2.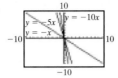

3.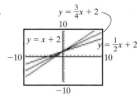

4.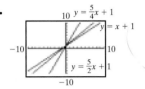

5.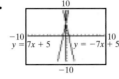

6.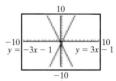

Exercise Set 3.5

33.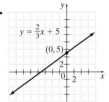

34.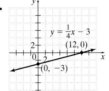

35.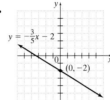

36.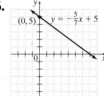

37.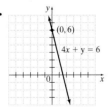

38.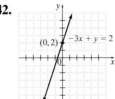

39.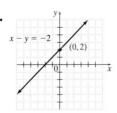

40.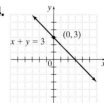

41.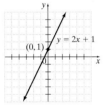

42.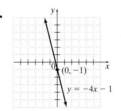

43.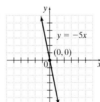

44.

45.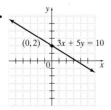

46.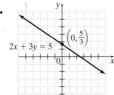

47.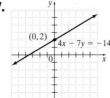

48.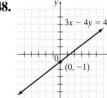

3.7 Spotlight on Decision Making

No; the average monthly temperature minus 10° (from the current ventilation system) is well within the comfortable working range. You may want to consider investing in fans for the days when the temperature is above average.

Chapter Project

1. Only the company Schering-Plough had a loss.

2.–3.

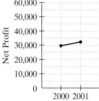
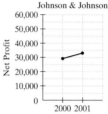
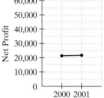
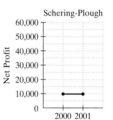

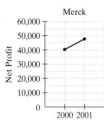
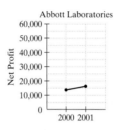
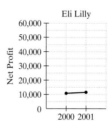
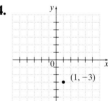

4. All of the lines except the one for Schering-Plough have positive slope. In this context, positive slope means net profit increased from 2000 to 2001 and negative slope means net profit decreased from 2000 to 2001.

5. Answers may vary. Merck had the steepest positive slope; their net profit increased the most in the one year period.

6. Answers may vary. Other factors to consider may include company's longevity, the number of products the company offers and what they are, etc.

Chapter 3 Review

1.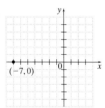

2.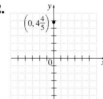

3.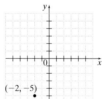

4.

5.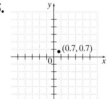

6.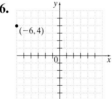

7. b.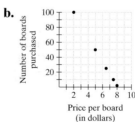

8. b.

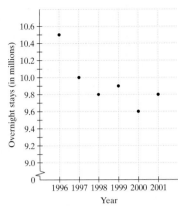

15.

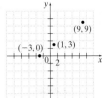

16.

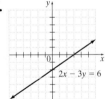

17.

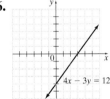

19.

20.

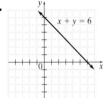

21.

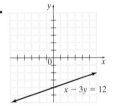

22.

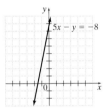

23.

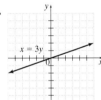

24.

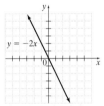

25.

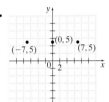

26.

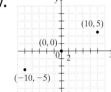

27.

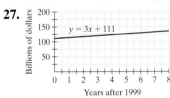

32.

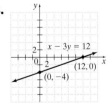

33.

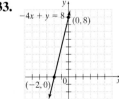

34.

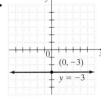

35.

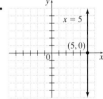

36.

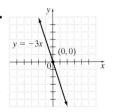

37.

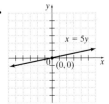

38.

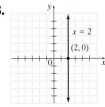

39.

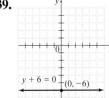

70.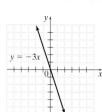

71.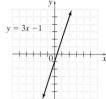

72.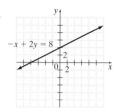

73.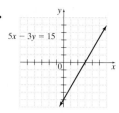

Chapter 3 Test

1. b.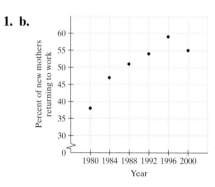

2.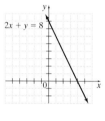

3.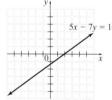

4.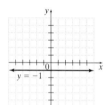

5.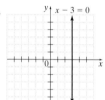

CHAPTER 4   SOLVING SYSTEMS OF LINEAR EQUATIONS AND INEQUALITIES

Exercise Set 4.1

9.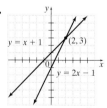

10.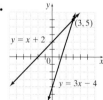

11.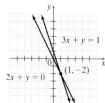

12.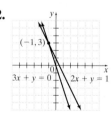

13.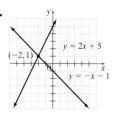

14.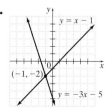

15.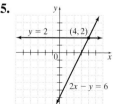

16.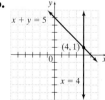

17.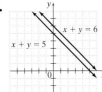

18.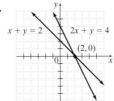

19.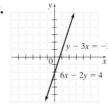

20.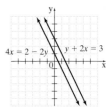

21.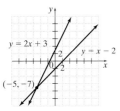

22.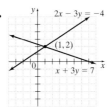

23.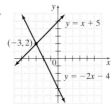

24.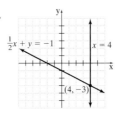

25.

26.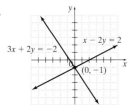

27.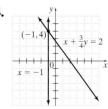

28.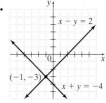

61. b.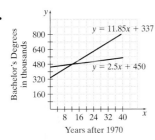

Exercise Set 4.2

47. c.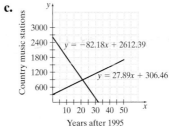

48. c.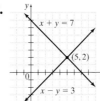

4.3 Spotlight on Decision Making

Answers may vary. You should consider what your average monthly sales are likely to be, as well as your confidence in attaining that average each month. It does make a difference if the position is at a car dealership or a shoestore, since the commissions will vary greatly.

Exercise Set 4.5

7.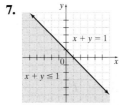

8.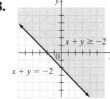

9.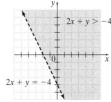

10.

$x + 3y = 3$

$x + 3y \leq 3$

11.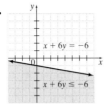

$x + 6y = -6$

$x + 6y \leq -6$

12.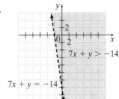

$7x + y > -14$

$7x + y = -14$

13.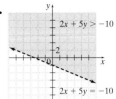

$2x + 5y > -10$

$2x + 5y = -10$

14.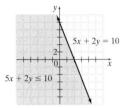

$5x + 2y = 10$

$5x + 2y \leq 10$

15.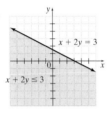

$x + 2y = 3$

$x + 2y \leq 3$

16.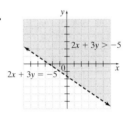

$2x + 3y > -5$

$2x + 3y = -5$

17.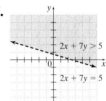

$2x + 7y > 5$

$2x + 7y = 5$

18.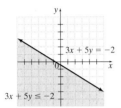

$3x + 5y = -2$

$3x + 5y \leq -2$

19.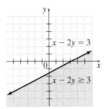

$x - 2y = 3$

$x - 2y \geq 3$

20.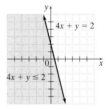

$4x + y = 2$

$4x + y \leq 2$

21.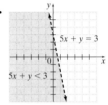

$5x + y = 3$

$5x + y < 3$

22.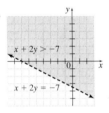

$x + 2y > -7$

$x + 2y = -7$

23.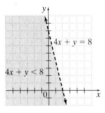

$4x + y = 8$

$4x + y < 8$

24.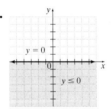

$9x + 2y \geq -9$

$9x + 2y = -9$

25.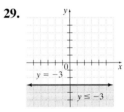

$y \geq 2x$

$y = 2x$

26.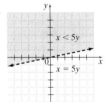

$x < 5y$

$x = 5y$

27.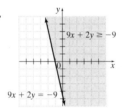

$x = 0$ $x \geq 0$

28.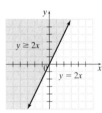

$y = 0$

$y \leq 0$

29.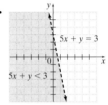

$y = -3$

$y \leq -3$

30.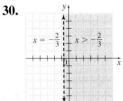

$x = -\frac{2}{3}$ $x > -\frac{2}{3}$

31.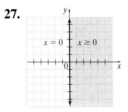

$2x - 7y = 0$

$2x - 7y > 0$

32.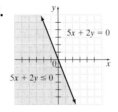

$5x + 2y = 0$

$5x + 2y \leq 0$

33.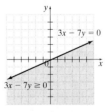

$3x - 7y = 0$

$3x - 7y \geq 0$

34.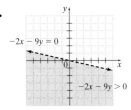

$-2x - 9y = 0$

$-2x - 9y > 0$

35.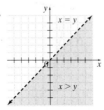

$x = y$

$x > y$

36.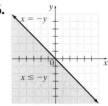

$x = -y$

$x \leq -y$

37.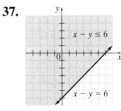

$x - y \leq 6$

$x - y = 6$

38.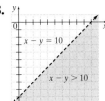

39.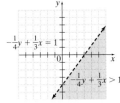

40.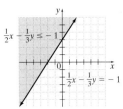

41.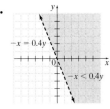

42.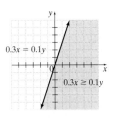

59.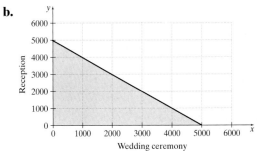

60.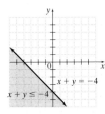

66. b.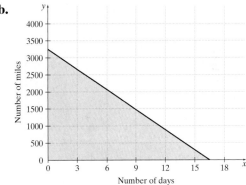

67. b.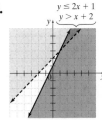

4.6 Spotlight on Decision Making

The attendee's blood pressure indicates moderate hypertension. Recommend that the attendee see his/her primary care physician within 1 month.

Exercise Set 4.6

1.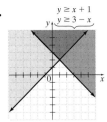

2.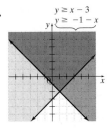

3.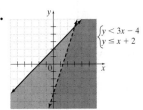

4.

5.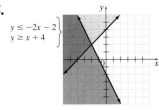

6.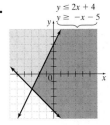

7.

8.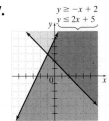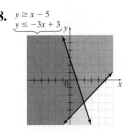

9.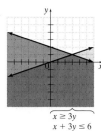

$x \geq 3y$
$x + 3y \leq 6$

10.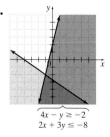

$-2x < y$
$x + 2y < 3$

11. $y + 2x \geq 0$
$5x - 3y \leq 12$

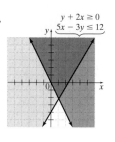

12. $y + 2x \leq 0$
$5x + 3y \geq -2$

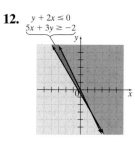

13.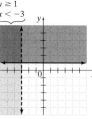

$3x - 4y \geq -6$
$2x + y \leq 7$

14.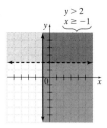

$4x - y \geq -2$
$2x + 3y \leq -8$

15. $x \leq 2$
$y \geq -3$

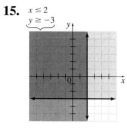

16. $x \geq -3$
$y \geq -2$

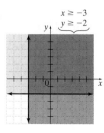

17. $y \geq 1$
$x < -3$

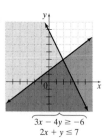

18. $y > 2$
$x \geq -1$

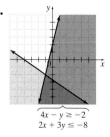

19.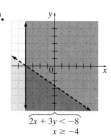

$2x + 3y < -8$
$x \geq -4$

20. $3x + 2y \leq 6$
$x < 2$

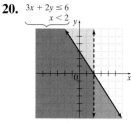

21.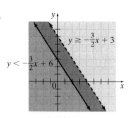

$2x - 5y \leq 9$
$y \leq -3$

22. $2x + 5y \leq -10$
$y \geq 1$

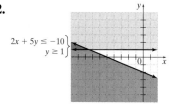

23.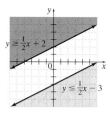

$y \geq \frac{1}{2}x + 2$
$y \leq \frac{1}{2}x - 3$

24.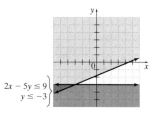

$y \geq -\frac{3}{2}x + 3$
$y < -\frac{3}{2}x + 6$

37.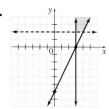

38.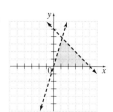

Chapter Project

1. $(46, 90)$ **2.** The speed of each ship

3. Ship A: $y = \frac{5}{2}x - 25$; Ship B: $y = -\frac{10}{3}x + \frac{730}{3}$

4. $(46, 90)$ **5.** $(46, 90)$

6. Ship A is about 86 miles from the collision point. Ship B is about 73 miles from the collision point. The ships will collide if Ship A travels at 86 mph and Ship B travels at 73 mph.

Chapter 4 Review

5.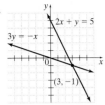

6.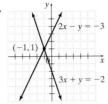

7.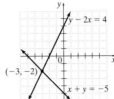

8.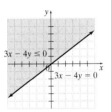

9.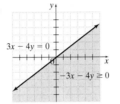

10.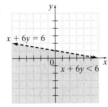

45.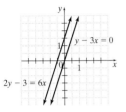

46.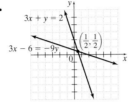

47.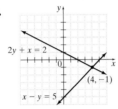

48.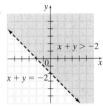

49.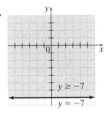

50.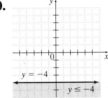

51.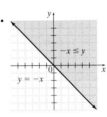

52.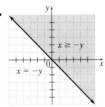

53. $y \geq 2x - 3$
$y \leq -2x + 1$

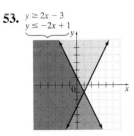

54.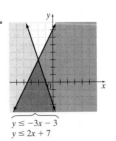

$y \leq -3x - 3$
$y \leq 2x + 7$

55. $x + 2y > 0$
$x - y \leq 6$

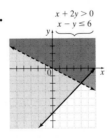

56.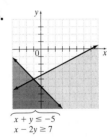

$x + y \leq -5$
$x - 2y \geq 7$

57. $3x - 2y \leq 4$
$2x + y \geq 5$

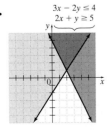

58.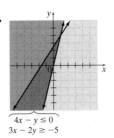

$4x - y \leq 0$
$3x - 2y \geq -5$

59.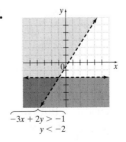

$-3x + 2y > -1$
$y < -2$

60. $-2x + 3y > -7$
$x \geq -2$

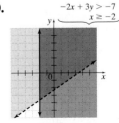

Chapter 4 Test

7.
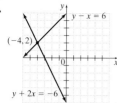

18.
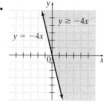

19.
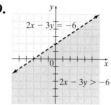

20. $y + 2x \leq 4$
    $y \geq 2$
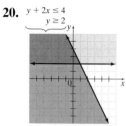

21. $2y - x \geq 1$
    $x + y \geq -4$
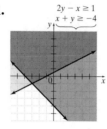

Chapter 4 Cumulative Review

19.
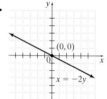

42.
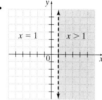

43.
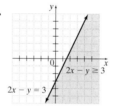

44.
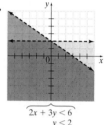

CHAPTER 5   EXPONENTS AND POLYNOMIALS

5.1 Spotlight on Decision Making

Enter your car in class C, since the displacement is approximately 2527.05 cc.

5.5 Spotlight on Decision Making

Answers may vary. Since each estimate shows that the size of the reserve on the client's property is 40% of the entire reserve, you may want to use the estimate of Expert A since (1.84×10^7) cu. ft is greater than (2.68×10^6) cu. ft.

Chapter Project

1.

Year	x	Consumer Spending per Person per Year on ALL U.S. Media	Consumer Spending per Person per Year on Subscription TV	Difference
1990	0	$364.27	$ 88.98	$275.29
1992	2	$400.41	$ 99.04	$301.37
1994	4	$449.75	$115.58	$334.17
1996	6	$512.29	$138.60	$373.69

While consumer spending is increasing in both areas from 1990 to 1996, the differences are also increasing. Consumers are spending more on all U.S. media at a faster rate than on subscription TV.

2. $0.84x^2 + 11.36x + 275.29$;

Year	x	Consumer Spending per Person per Year on Media Other than Subscription TV
1990	0	$275.29
1992	2	$301.37
1994	4	$334.17
1996	6	$373.69

3. The values are the same. Consumers are spending more money on media other than subscription TV each year from 1990 to 1996.

4. (a) $588.03
   (b) $168.10
   (c) $419.93

5. (a) $676.97
   (b) $204.08
   (c) $472.89

6.

Chapter 5 Cumulative Review

35.

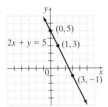

37.

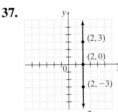

41.

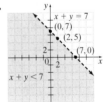

CHAPTER 6 FACTORING POLYNOMIALS

6.4 Spotlight on Decision Making

Answers may vary. Have students discuss their final choice, and which factors were most important to them.

Exercise Set 6.5

91. d.

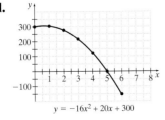

$$y = -16x^2 + 20x + 300$$

92. d.

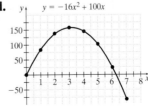

$$y = -16x^2 + 100x$$

6.6 Spotlight on Decision Making

The border should be 4 feet wide on all sides.

Chapter Project

1. 150 sq. ft   **2.** $(2x + 10)(2x + 15)$ sq. ft   **3.** $(4x^2 + 50x)$ sq. ft   **4.** 50 ft
5. cement: $5(4x^2 + 50x)$ dollars; brick: $[7.5(4x^2 + 50x) + 30]$ dollars; carpet: $[4.5(4x^2 + 50x) + 10.86(50)]$ dollars
6. 7.5 ft   **7.** 5.5 ft   **8.** 7 ft
9. Answers may vary. Cement will result in the widest patio, but may not be as visually pleasing as brick or outdoor carpeting.

Chapter 6 Cumulative Review

21.

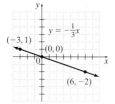

CHAPTER 7   RATIONAL EXPRESSIONS

7.1 Spotlight on Decision Making

The density of the metal is 8.04 g/ml. This is closest to the density of iron, so the metal is likely made of mostly iron with a small amount of something more dense than iron.

Exercise Set 7.1

98.

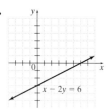

99.

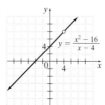

100.

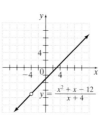

101.

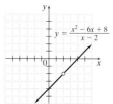

Exercise Set 7.2

75.

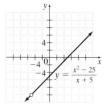

76.

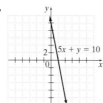

7.5 Calculator Explorations

1.

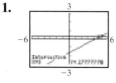

2.

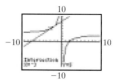

3.

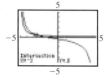

4.

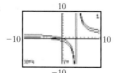

5.

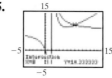

7.6 Spotlight on Decision Making - A

Answers may vary. Employee B has a better batting average, but you may want to consider each employee's abilities on the field.

7.6 Spotlight on Decision Making - B

Answers may vary, depending on the factors that are more important to you personally. Center B is less expensive, but it is farther from home and the number of adults per child is less than at Center A.

Chapter Project

1., 5.

Age A	Young's Rule	Cowling's Rule	Difference
2	142.86	125	17.86
3	200	166.67	33.33
4	250	208.33	41.67
5	294.12	250	44.12
6	333.33	291.67	41.67
7	368.42	333.33	35.09
8	400	375	25
9	428.57	416.67	11.9
10	454.55	458.33	−3.79
11	478.26	500	−21.74
12	500	541.67	−41.67
13	520	583.33	−63.33

2.

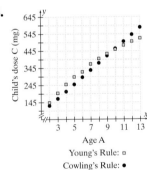

The data from Young's Rule is curved while the data from Cowling's Rule is linear (with positive slope).

3. Before age 9, Young's rule predicts a greater dose. Cowling's Rule predicts a greater dose for ages 10 and above.

4. For the data graphed, the difference is greatest at age 13.

6. Cowling's Rule predicts exactly the adult dose at age 23. There is no age at which Young's Rule predicts exactly the adult dose.

7. Answers may vary. Other factors such as a child's weight should be considered.

Chapter 7 Cumulative Review

5.

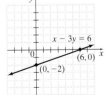

CHAPTER 8   ROOTS AND RADICALS

8.1 Spotlight on Decision Making

Using the formula, the maximum safe speed for the ramp is $\sqrt{1000} \approx 31.6$ mph.
Post the sign that says "30 M.P.H." since it is the closest to $\sqrt{1000}$ mph.

Exercise Set 8.1

80. 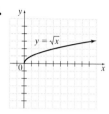   **81.**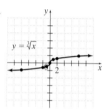

8.6 Spotlight on Decision Making

The diagonal should be $\sqrt{296}$ feet, or 17 feet, 2.4558 inches, which is within $\dfrac{1}{100}$ inch of the measured length. This is probably the closest length obtainable.

Exercise Set 8.7

71. 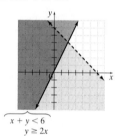   **72.**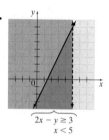

Chapter Project

1–2. Answers may vary. Students should measure the volumes and the heights as carefully as possible.

3. $r = \sqrt{\dfrac{V}{\pi h}}$

4. Answers may vary. Make sure students use a consistent value for π.
5. Answers may vary.
6. Calculated values should be very close to measured values, with the differences due to inaccurate measurements and rounding.

Chapter 8 Cumulative Review

21. 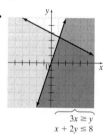   **22.**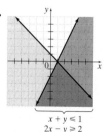

CHAPTER 9   SOLVING QUADRATIC EQUATIONS

9.3 Spotlight on Decision Making

It takes Line B about 58.9 minutes to produce 3000 items and Line A about 70.1 minutes. Line A should be upgraded since it produces fewer than 3000 items per hour.

Exercise Set 9.4

69.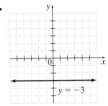

70.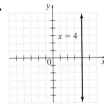

71.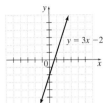

72.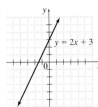

9.5 Spotlight on Decision Making

Plant between mid-September and mid-May.

Exercise Set 9.5

1.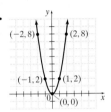

2.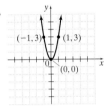

3.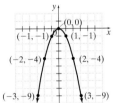

4.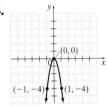

5.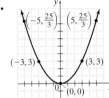

6.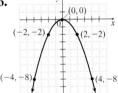

7.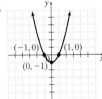

8.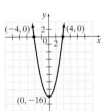

9.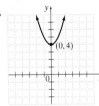

10.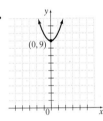

11.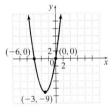

12.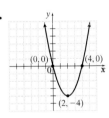

13.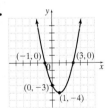

14.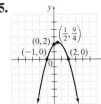

15.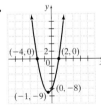

16.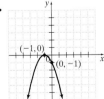

17.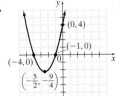

18.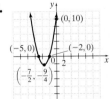

19.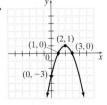

20.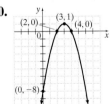

21.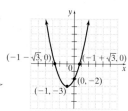

22.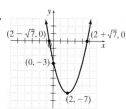

23.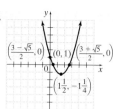

24.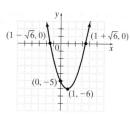

Chapter Project

1. Answers may vary. Coordinates of point B should be in the vicinity of $(20, 0)$.
2. The x-coordinate of the vertex is halfway between the x-intercepts. The axis of symmetry is the vertical line whose x-intercept is the same as the x-coordinate of the vertex. The coordinates of the vertex should be in the vicinity of $(10, 13)$.
3. Graphs will vary depending on ordered pairs.   **4.** (d)   **5.** Answers may vary.

Chapter 9 Review

89.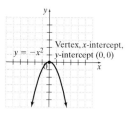

90.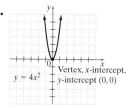

91.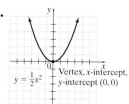

92.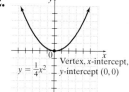

93.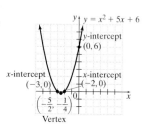

94.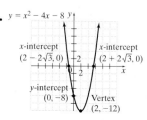

95.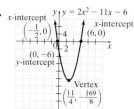

96.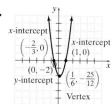

Chapter 9 Test

20.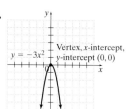

21.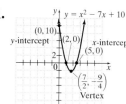

Chapter 9 Cumulative Review

17.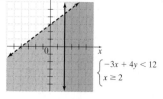

18.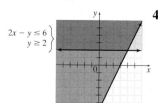

47.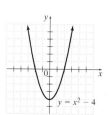

48.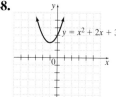

PHOTO CREDITS

Contents © Garry Black/Masterfile, © MTPA/Masterfile, © Royalty Free/CORBIS, © George Disario/CORBIS, © Tom Stewart/CORBIS, © Ariel Skelley/CORBIS, © Spencer Grant/Photo Edit, Dennis Brack/Aurora & Quanta Products, © Doug Scott/age fotostock.

Chapter 1 © Garry Black/Masterfile, (p. 2) Wagne Sepp Seitz/Woodfin Camp & Associates, (p. 4) © Rachel Epstein/Photo Edit, (p. 10) Tim Davis/Photo Researchers, Inc., (p. 15) Photo Researchers, Inc., (p. 27) Joseph McBride/Stone/Getty Images, (p. 39) © Freeman Patterson/Masterfile, (p. 39) © ZEFA/Masterfile, (p. 46) © Sharna Balfour/Gallo Images/Corbis, (p. 46) Bishop Airport

Chapter 2 © MTPA/Masterfile, (p. 93) Laurence R. Lowry/Stock Boston, (p. 93) Ed Pritchard/Stone/Getty Images, (p. 95) Hugh Sitton/Stone/Getty Images, (p. 95) Klaus G. Hinkelmann, (p. 102) © Spencer Grant/Photo Edit, (p. 113) © Rolf Bruderer/Masterfile, (p. 115) Bruce Hoertel/Getty Images, Inc., (p. 116) AP/Wide World Photos, (p. 116) © Rick Gomez/Masterfile, (p. 120) Nasa/Science Source/Photo Researchers, Inc., (p. 121) Bill Bachmann/Stock Boston, (p. 123) Sean Reid/Alaska Stock, (p. 132) © Norbert Wu/www.norbertwu.com, (p. 131) Colin Braley/Hulton Archive/Getty Images, (p. 132) John Elk III/Stock Boston, (p. 133) © Jonathan Nourok/Photo Edit, (p. 134) First Light/ImageState/International, (p. 139) AP/Wide World Photos, (p. 148) Jeff Greenberg/The Image Works, (p. 158) Jean-Claude LeJeune/Stock Boston

Chapter 3 © Royalty Free/CORBIS, (p. 170) David Young-Wolff/Stone/Getty Images, (p. 182) © Michael Newman/Photo Edit, (p. 184) © Amy C. Etra/Photo Edit, (p. 184) © Tony Freeman/Photo Edit, (p. 192) © Tony Freeman/Photo Edit, (p. 192) © Bill Aron/Photo Edit, (p. 199) © John Neubauer/Photo Edit, (p. 205) Miles Ertman/Masterfile, (p. 214) © Amy C. Etra/Photo Edit, (p. 214) © Cathy Melloan Resources/Photo Edit, (p. 218) Doug Menuez/Photodisc/Getty Images, (p. 220) © Spencer Grant/Photo Edit, (p. 220) © Mary Kate Denny/Photo Edit, (p. 221) © Ann States/Corbis/SABA Press Photos, Inc.

Chapter 4 © George Disario/CORBIS, (p. 257) © Peter Hvizdak/The Image Works, (p. 257) © Michael Newman/Photo Edit, (p. 263) © Michael Kevin Daly/CORBIS, (p. 264) © David Young-Wolff/Photo Edit, (p. 267) Photo of the Banquine act from the Cirque du Soleil show Quidam: © Cirque du Soleil Inc. Photo by Al Seib. Costumes by Dominique Lemieux, (p. 272) AP/Wide World Photos, (p. 272) AP/Wide World Photos, (p. 273) AP/Wide World Photos, (p. 293) AP/Wide World Photos

Chapter 5 © Tom Stewart/CORBIS, (p. 335) G. Brad Lewis/Innerspace Visions, (p. 335) Jim Pickerell/Stock Boston, (p. 349) Alvis Upitis/Image Bank/Getty Images, (p. 349) NASA, (p. 349) NASA

Chapter 6 © Ariel Skelley/CORBIS, (p. 361) Nick Koudis/Photodisc/Getty Images (p. 380) R. Ian Lloyd/Masterfile Corporation, (p. 381) John Freeman/Stone/Getty Images

Chapter 7 © Spencer Grant/Photo Edit, (p. 418) Chuck Keeler/Frozen Images/The Image Works, (p. 418) AP/Wide World Photos, (p. 423) Tebo Photography, (p. 425) © Bettmann/CORBIS, (p. 425) AP/Wide World Photos, (p. 451) © Charles O'Rear/CORBIS, (p. 458) © Spencer Grant/Photo Edit, (p. 465) © Susan Van Etten/Photo Edit, (p. 466) Brian Erler/Taxi/Getty Images, (p. 480) Keith Brofsky/Photodisc/Getty Images

Chapter 8 Dennis Brack/Aurora & Quanta Products, (p. 490) Cory Langley, (p. 497) Photofest, (p. 497) © Cindy Charles/Photo Edit, (p. 519) Andy Belcher/Image State/International

Chapter 9 © Doug Scott/age fotostock, (p. 541) Hulton Archive/Getty Images, (p. 542) AP/Wide World Photos, (p. 543) © Deborah Davis/Photo Edit, (p. 543) AP/Wide World Photos, (p. 548) © Tom Grill/age fotostock, (p. 556) © Dennis Nett/Syracuse Newspapers/The Image Works, (p. 572) © Karen Preuss/The Image Works, (p. 578) Ted Clutter/Photo Researchers, Inc.

Single PC Site License

1. **GRANT OF LICENSE and OWNERSHIP:** The enclosed computer programs and any data ("Software") are licensed, not sold, to you by Pearson Education, Inc. publishing as Pearson Prentice Hall ("We" or the "Company") in consideration of your adoption of the accompanying Company textbooks and/or other materials, and your agreement to these terms. You own only the disk(s) but we and/or our licensors own the Software itself. This license allows instructors and students enrolled in the course using the Company textbook that accompanies this Software (the "Course") to use and display the enclosed copy of the Software on an unlimited number of computers, for academic use only, so long as you comply with the terms of this Agreement. You may make one copy for back up only. We reserve any rights not granted to you.

2. **USE RESTRICTIONS:** You may <u>not</u> sell or license copies of the Software or the Documentation to others. You may <u>not</u> transfer, distribute or make available the Software or the Documentation. You may <u>not</u> reverse engineer, disassemble, decompile, modify, adapt, translate or create derivative works based on the Software or the Documentation. You may be held legally responsible for any copying or copyright infringement that is caused by your failure to abide by the terms of these restrictions.

3. **TERMINATION:** This license is effective until terminated. This license will terminate automatically without notice from the Company if you fail to comply with any provisions or limitations of this license. Upon termination, you shall destroy the Documentation and all copies of the Software. All provisions of this Agreement as to limitation and disclaimer of warranties, limitation of liability, remedies or damages, and our ownership rights shall survive termination.

4. **DISCLAIMER OF WARRANTY: THE COMPANY AND ITS LICENSORS MAKE <u>NO</u> WARRANTIES ABOUT THE SOFTWARE, WHICH IS PROVIDED "<u>AS-IS</u>." IF THE DISK IS DEFECTIVE IN MATERIALS OR WORKMANSHIP, YOUR ONLY REMEDY IS TO RETURN IT TO THE COMPANY WITHIN 30 DAYS FOR REPLACEMENT UNLESS THE COMPANY DETERMINES IN GOOD FAITH THAT THE DISK HAS BEEN MISUSED OR IMPROPERLY INSTALLED, REPAIRED, ALTERED OR DAMAGED. THE COMPANY DISCLAIMS ALL WARRANTIES, EXPRESS OR IMPLIED, INCLUDING WITHOUT LIMITATION, THE IMPLIED WARRANTIES OF MERCHANTABILITY AND FITNESS FOR A PARTICULAR PURPOSE. THE COMPANY DOES NOT WARRANT, GUARANTEE OR MAKE ANY REPRESENTATION REGARDING THE ACCURACY, RELIABILITY, CURRENTNESS, USE, OR RESULTS OF USE, OF THE SOFTWARE.**

5. **LIMITATION OF REMEDIES AND DAMAGES: IN NO EVENT, SHALL THE COMPANY OR ITS EMPLOYEES, AGENTS, LICENSORS OR CONTRACTORS BE LIABLE FOR ANY INCIDENTAL, INDIRECT, SPECIAL OR CONSEQUENTIAL DAMAGES ARISING OUT OF OR IN CONNECTION WITH THIS LICENSE OR THE SOFTWARE, INCLUDING, WITHOUT LIMITATION, LOSS OF USE, LOSS OF DATA, LOSS OF INCOME OR PROFIT, OR OTHER LOSSES SUSTAINED AS A RESULT OF INJURY TO ANY PERSON, OR LOSS OF OR DAMAGE TO PROPERTY, OR CLAIMS OF THIRD PARTIES, EVEN IF THE COMPANY OR AN AUTHORIZED REPRESENTATIVE OF THE COMPANY HAS BEEN ADVISED OF THE POSSIBILITY OF SUCH DAMAGES.** SOME JURISDICTIONS DO NOT ALLOW THE LIMITATION OF DAMAGES IN CERTAIN CIRCUMSTANCES, SO THE ABOVE LIMITATIONS MAY NOT ALWAYS APPLY.

6. **GENERAL:** THIS AGREEMENT SHALL BE CONSTRUED IN ACCORDANCE WITH THE LAWS OF THE UNITED STATES OF AMERICA AND THE STATE OF NEW YORK, APPLICABLE TO CONTRACTS MADE IN NEW YORK, AND SHALL BENEFIT THE COMPANY, ITS AFFILIATES AND ASSIGNEES. This Agreement is the complete and exclusive statement of the agreement between you and the Company and supersedes all proposals, prior agreements, oral or written, and any other communications between you and the company or any of its representatives relating to the subject matter. If you are a U.S. Government user, this Software is licensed with "restricted rights" as set forth in subparagraphs (a)-(d) of the Commercial Computer-Restricted Rights clause at FAR 52.227-19 or in subparagraphs (c)(1)(ii) of the Rights in Technical Data and Computer Software clause at DFARS 252.227-7013, and similar clauses, as applicable.

Should you have any questions concerning this agreement or if you wish to contact the Company for any reason, please contact in writing: Customer Service Pearson Prentice Hall, 200 Old Tappan Road, Old Tappan NJ 07675.

Minimum System Requirements

Windows

    Pentium II 300 MHz processor
    Windows 98 or later
    64 MB RAM
    800 x 600 resolution
    8x or faster CD-ROM drive
    QuickTime 6.0 or later

Macintosh

    Power PC G3 233 MHz or better
    Mac OS 9.x or 10.x
    64 MB RAM
    800 x 600 resolution
    8x or faster CD-ROM drive
    QuickTime 6.0 or later